CAUSAL INFERENCE

for Statistics, Social, and Biomedical Sciences
An Introduction

Most questions in social and biomedical sciences are causal in nature: what would happen to individuals, or to groups, if part of their environment were changed? In this groundbreaking text, two world-renowned experts present statistical methods for studying such questions.

This book starts with the notion of potential outcomes, each corresponding to the outcome that would be realized if a subject were exposed to a particular treatment or regime. In this approach, causal effects are comparisons of such potential outcomes. The fundamental problem of causal inference is that we can observe only one of the potential outcomes for a particular subject. The authors discuss how randomized experiments allow us to assess causal effects and then turn to observational studies. They lay out the assumptions needed for causal inference and describe the leading analysis methods, including matching, propensity-score methods, and instrumental variables. Many detailed applications are included, with special focus on practical aspects for the empirical researcher.

Guido W. Imbens is Applied Econometrics Professor and Professor of Economics at the Graduate School of Business at Stanford University. He has previously held tenured positions at the University of California at Los Angeles, the University of California at Berkeley, and Harvard University. He is a Fellow of the Econometric Society and the American Academy of Arts and Sciences. He holds an honorary doctorate from the University of St. Gallen, Switzerland. Imbens has done extensive research in econometrics and statistics, specializing in causal inference. He has published widely in leading economics and statistics journals, including the *American Economic Review*, *Econometrica*, the *Review of Economic Studies*, the *Journal of the American Statistical Association*, the *Annals of Statistics*, *Biometrika*, and the *Journal of the Royal Statistical Society, Series A*.

Donald B. Rubin is John L. Loeb Professor of Statistics at Harvard University, where he has been professor since 1983 and Department chair for 13 of those years. He has been elected to be a Fellow/Member/Honorary Member of the Woodrow Wilson Society, the Guggenheim Memorial Foundation, the Alexander von Humboldt Foundation, the American Statistical Association, the Institute of Mathematical Statistics, the International Statistical Institute, the American Association for the Advancement of Science, the American Academy of Arts and Sciences, the European Association of Methodology, the British Academy, and the U.S. National Academy of Sciences. As of 2014, he has authored/coauthored nearly 400 publications (including ten books), has five joint patents, and for many years has been one of the most highly cited authors in the world, with currently over 150,000 citations (Google Scholar). He is also the recipient of honorary doctorates from Otto Friedrich University, Bamberg, Germany; the University of Ljubljana, Slovenia; and Universidad Santa Tómas, Bogotá, Colombia, and has been named an honorary professor at four universities.

Advance Praise for *Causal Inference for Statistics, Social, and Biomedical Sciences*

"This thorough and comprehensive book uses the 'potential outcomes' approach to connect the breadth of theory of causal inference to the real-world analyses that are the foundation of evidence-based decision making in medicine, public policy, and many other fields. Imbens and Rubin provide unprecedented guidance for designing research on causal relationships, and for interpreting the results of that research appropriately."

– Dr. Mark McClellan, Director of the Health Care Innovation and Value Initiative, the Brookings Institution

"Clarity of thinking about causality is of central importance in financial decision making. Imbens and Rubin provide a rigorous foundation allowing practitioners to learn from the pioneers in the field."

– Dr. Stephen Blyth, Managing Director, Head of Public Markets, Harvard Management Company

"A masterful account of the potential outcomes approach to causal inference from observational studies that Rubin has been developing since he pioneered it 40 years ago."

– Adrian Raftery, Blumstein-Jordan Professor of Statistics and Sociology, University of Washington

"Correctly drawing causal inferences is critical in many important applications. Congratulations to Professors Imbens and Rubin, who have drawn on their decades of research in this area, along with the work of several others, to produce this impressive book covering concepts, theory, methods, and applications. I especially appreciate their clear exposition of conceptual issues, which are important to understand in the context of either a designed experiment or an observational study, and their use of real applications to motivate the methods described."

– Nathaniel Schenker, Former President of the American Statistical Association

CAUSAL INFERENCE

for Statistics, Social, and Biomedical Sciences
An Introduction

Guido W. Imbens

Stanford University

Donald B. Rubin

Harvard University

CAMBRIDGE
UNIVERSITY PRESS

CAMBRIDGE
UNIVERSITY PRESS

University Printing House, Cambridge CB2 8BS, United Kingdom

One Liberty Plaza, 20th Floor, New York, NY 10006, USA

477 Williamstown Road, Port Melbourne, VIC 3207, Australia

314–321, 3rd Floor, Plot 3, Splendor Forum, Jasola District Centre, New Delhi – 110025, India

103 Penang Road, #05–06/06, Visioncrest Commercial, Singapore 238467

Cambridge University Press is part of the University of Cambridge.

It furthers the University's mission by disseminating knowledge in the pursuit of education, learning, and research at the highest international levels of excellence.

www.cambridge.org
Information on this title: www.cambridge.org/9780521885881

First published 2015
12th printing 2022

Printed in the United Kingdom by TJ Books Limited, Padstow Cornwall

A catalog record for this publication is available from the British Library.

Library of Congress Cataloging in Publication Data
Imbens, Guido.
Causal inference for statistics, social, and biomedical sciences :
an introduction / Guido W. Imbens & Donald B. Rubin.
   pages   cm
1. Social sciences–Research. 2. Causation. 3. Inference. I. Rubin, Donald B. II. Title.
H62.I537 2014
519.5′4–dc23   2014020988

ISBN 978-0-521-88588-1 Hardback

To my parents, Annie Imbens-Fransen and Gerard Imbens,
for all their support and encouragement over the years; to
my children, Carleton, Annalise and Sylvia, who have
provided so much joy in recent years; and to my wife
and best friend, Susan Athey.

Guido W. Imbens

To my family, colleagues, and students, and Elizabeth Zell,
who has provided ten years of crucial support.

Donald B. Rubin

Contents

Preface

In many applications of statistics, a large proportion of the questions of interest are fundamentally questions of causality rather than simply questions of description or association. For example, a medical researcher may wish to find out whether a new drug is effective against a disease. An economist may be interested in uncovering the effects of a job-training program on an individual's employment prospects, or the effects of a new tax or regulation on economic activity. A sociologist may be concerned about the effects of divorce on children's subsequent education. In this text we discuss statistical methods for studying such questions.

The book arose out of a conversation we had in 1992 while we were both on the faculty at Harvard University. We found that although we were both interested in questions of causality, we had difficulty communicating our ideas because, coming from different disciplines, we were used to different terminology and conventions. However, the excitement about the ideas in these different areas motivated us to capitalize on these difficulties, which led to a long collaboration, including research projects, graduate and undergraduate teaching, and thesis advising. The book is a reflection of this collaboration.

The book is based directly on many semester and quarter-length courses we, initially jointly, and later separately, taught for a number of years, starting in 1995 at Harvard University, followed by the University of California at Los Angeles, the University of California at Berkeley, and Stanford University, to audiences of graduate and undergraduate students from statistics, economics, business, and other disciplines using applied statistics. In addition, we have taught shorter versions of such courses in Barcelona, Beijing, Berlin, Bern, Geneva, Maastricht, Mexico City, Miami, Montevideo, Santiago, Stockholm, Uppsala, Wuppertal, Zurich, and at the World Bank as well as other associations and agencies.

There are a number of key features of the approach taken in this book. First of all, the perspective we take is that all causal questions are tied to specific interventions or treatments. Second, causal questions are viewed as comparisons of potential outcomes, with each potential outcome corresponding to a level of the treatment. Each of these potential outcomes could have been observed had the treatment taken on the corresponding level. After the treatment has taken on a specific level, only the potential outcome corresponding to that level is realized and can actually be observed. Causal effects involve the comparison of the outcome actually observed with other potential outcomes that could have been observed had the treatment taken on a different level, but that are not, in fact, observed. Causal inference is therefore fundamentally a missing data problem and, as in all missing data problems, a key role is played by the mechanism that determines which data values are observed and which are missing. In causal inference, this mechanism is referred to as the assignment mechanism, the mechanism that determines levels of the treatment taken by the units studied.

The book is organized in seven parts. In the first part we set out the basic philosophy underlying our approach to causal inference and describe the potential outcomes framework. The next three parts of the book are distinguished by the assumptions maintained about the assignment mechanism. In Part II we assume that the assignment mechanism corresponds to a classical randomized experiment.

In Part III we assume that the assignment mechanism is "regular" in a well-defined sense, which generalizes randomized experiments. In this part of the book we discuss what we call the "design" phase of an observational study, which we view as extremely important for credible conclusions. In the next part, Part IV, we discuss data analysis for studies with regular assignment mechanisms. Here we consider matching and subclassification procedures, as well as model-based and weighting methods.

In Part V we relax this regularity assumption and discuss more general assignment mechanisms. First we assess the key assumption required for regularity, unconfoundedness. We also explore in this part of the text sensitivity analyses where we relax some of the key features of a regular assignment mechanism.

Next, in Part VI of the text, we consider settings where the assignment mechanism is regular, but compliance with the assignment is imperfect. As a result, the probability of receipt of treatment may depend on both observed and unobserved characteristics and outcomes of the units. To address these complications, we turn to instrumental variables methods. Part VII of the book concludes.

As with all books, ours has limitations. Foremost is our focus on binary treatments. Although many of the results can easily be extended to multi-valued treatments, we focus on the binary treatment case because many critical conceptual issues arise already in that setting. Second, throughout most of the book we make the "stability" assumption that treatments applied to one unit do not affect outcomes for other units and that there are no unrepresented versions of the treatments. There is a growing literature on interactions through networks and peer effects that builds on the notions of causality discussed in this book. Finally, although we designed the book to be theoretically tight and principled, we focus on practical rather than mathematical results, including detailed applications with real data sets, consistent with our target audience of researchers in applied fields.

ACKNOWLEDGMENTS

We are deeply indebted to many collaborators, colleagues, and students from whom we have learned much about the topics discussed in this book over the years. These include Alberto Abadie, Nikola Andric, Joshua Angrist, Susan Athey, Thomas Barrios, David Card, Matias Cattaneo, Gary Chamberlain, Raj Chetty, Rajeev Dehejia, Mike Dickstein,

Peng Ding, David Drukker, Dean Eckles, Valeria Espinosa, Avi Feller, Sergio Firpo, Andrew Gelman, Paul Gift, Paul Goldsmith-Pinkham, Bryan Graham, Roee Gutman, Gary Harvey, Jane Herr, Kei Hirano, Nate Holt, Joe Hotz, Wilbert van der Klaauw, Jacob Klerman, Alan Krueger, Tony Lancaster, Joseph Lee, Victoria Liublinska, Chuck Manski, Eric Maskin, Fabrizia Mealli, Kari Lock Morgan, Eduardo Morales, Julie Mortimer, Eleanor Murray, Whitney Newey, Cassandra Wolos Pattanayak, Jack Porter, Geert Ridder, Paul Rosenbaum, Bruce Sacerdote, Hal Stern, Elizabeth Stuart, Neal Thomas, Sadek Wahba, David Watson, Jeff Wooldridge, Alan Zaslavsky, and Elizabeth Zell.

We are grateful for stimulating discussions over the years with David Freedman, Sander Greenland, James Heckman, Nick Jewell, Mark van der Laan, Judea Pearl, Jamie Robins, Jasjeet Sekhon, Jeffrey Smith, and Edward Vytlacil.

We are also grateful to Joshua Angrist, David Card, Raj Chetty, Rajeev Dehejia, Esther Duflo, Brad Efron, Alan Krueger, Robert Lalonde, June Reinisch, Bruce Sacerdote, Sadek Wahba, and Scott Zeger for making their data available.

Finally, we would like to extend our gratitude to Dean Eckles and Karthik Rajkumar for comments on the first edition, and to Michael Pollman for a remarkably thorough check for errors and typos.

## PART I

Introduction

## CHAPTER 1

Causality: The Basic Framework

1.1 INTRODUCTION

In this introductory chapter we set out our basic framework for causal inference. We discuss three key notions underlying our approach. The first notion is that of *potential outcomes*, each corresponding to one of the levels of a *treatment* or *manipulation*, following the dictum "no causation without manipulation" (Rubin, 1975, p. 238). Each of these potential outcomes is *a priori* observable, in the sense that it could be observed if the unit were to receive the corresponding treatment level. But, *a posteriori*, that is, once a treatment is applied, at most one potential outcome can be observed. Second, we discuss the necessity, when drawing causal inferences, of observing *multiple units*, and the utility of the related *stability* assumption, which we use throughout most of this book to exploit the presence of multiple units. Finally, we discuss the central role of the *assignment mechanism*, which is crucial for inferring causal effects, and which serves as the organizing principle for this book.

1.2 POTENTIAL OUTCOMES

In everyday life, causal language is widely used in an informal way. One might say: "My headache went away because I took an aspirin," or "She got a good job last year because she went to college," or "She has long hair because she is a girl." Such comments are typically informed by observations on past exposures, for example, of headache outcomes after taking aspirin or not, or of characteristics of jobs of people with or without college educations, or the typical hair length of boys and girls. As such, these observations generally involve informal statistical analyses, drawing conclusions from associations between measurements of different quantities that vary from individual to individual, commonly called *variables* or *random variables* – language apparently first used by Yule (1897). Nevertheless, statistical theory has been relatively silent on questions of causality. Many, especially older, textbooks avoid any mention of the term other than in settings of randomized experiments. Some mention it mainly to stress that correlation or association is not the same as causation, and some even caution their readers to avoid using causal language in statistics. Nevertheless, for many users of statistical methods, causal statements are exactly what they seek.

The fundamental notion underlying our approach is that causality is tied to an *action* (or manipulation, treatment, or intervention), applied to a *unit*. A unit here can be a physical object, a firm, an individual person, or collection of objects or persons, such as a classroom or a market, at a particular point in time. For our purposes, the same physical object or person at a different time is a different unit. From this perspective, a causal statement presumes that, although a unit was (at a particular point in time) subject to, or exposed to, a particular action, treatment, or regime, the same unit could have been exposed to an alternative action, treatment, or regime (at the same point in time). For instance, when deciding to take an aspirin to relieve your headache, you could also have chosen not to take the aspirin, or you could have chosen to take an alternative medicine. In this framework, articulating with precision the nature and timing of the action sometimes requires a certain amount of imagination. For example, if we define race solely in terms of skin color, the action might be a pill that alters only skin color. Such a pill may not currently exist (but, then, neither did surgical procedures for heart transplants hundreds of years ago), but we can still imagine such an action.

This book primarily considers settings with two actions, although many of the extensions to multi-valued treatments are conceptually straightforward. Often one of these actions corresponds to a more active treatment (e.g., taking an aspirin) in contrast to a more passive action (e.g., not taking the aspirin). In such cases we sometimes refer to the first action as the *active treatment* as opposed to the *control treatment*, but these are merely labels and formally the two treatments are viewed symmetrically. In some cases, when it is clear from the context, we refer to the more active treatment simply as the "treatment" and the other treatment as the "control."

Given a unit and a set of actions, we associate each action-unit pair with a *potential outcome*. We refer to these outcomes as potential outcomes because only one will ultimately be realized and therefore possibly observed: the potential outcome corresponding to the action actually taken. *Ex post*, the other potential outcomes cannot be observed because the corresponding actions that would lead to them being realized were not taken. The causal effect of one action or treatment relative to another involves the comparison of these potential outcomes, one realized (and perhaps, though not necessarily, observed), and the others not realized and therefore not observable. Any treatment must occur temporally before the observation of any associated potential outcome is possible.

Although the preceding argument may appear obvious, its force is revealed by its ability to clarify otherwise murky concepts, as can be demonstrated by considering the three examples of informal "because" statements presented in the first paragraph of this section. In the first example, it is clear what the action is: I took an aspirin, but at the time that I took the aspirin, I could have followed the alternate course of not taking an aspirin. In that case, a different outcome might have resulted, and the "because" statement is causal in the perspective taken in this book as it reflects the comparison of those two potential outcomes. In the second example, it is less clear what the treatment and its alternative are: she went to college, and at the point in time when she decided to go to college, she could have decided not to go to college. In that case, she might have had a different job a year ago, and the implied causal statement compares the quality of the job she actually had then to the quality of the job she would have had a year ago, had she not gone to college. However, in this example, the alternative treatment is somewhat murky: had she not enrolled in college, would she have enrolled in the military, or would she have joined an artist's colony? As a result, the potential outcome under the alternative action, the job obtained a year ago without enrolling in college, is not as well defined as in the first example.

In the third example, the alternative action is not at all clear. The informal statement is "she has long hair because she is a girl." In some sense the implicit treatment is being a girl, and the implicit alternative is being a boy, but there is no action articulated that would have made her a boy and allowed us to observe the alternate potential outcome of hair length for this person as a boy. We could clarify the causal effect by defining such an action in terms of surgical procedures, or hormone treatments, all with various ages at which the action to be taken is specified, but clearly the causal effect is likely to depend on the particular alternative action and timing being specified. As stated, however, there is no clear action described that would have allowed us to observe the unit exposed to the alternative treatment. Hence, in our approach, this "because" statement is ill-defined as a causal statement.

It may seem restrictive to exclude from consideration such causal questions. However, the reason to do so in our framework is that without further explication of the intervention being considered, the causal question is not well defined. One can make many of these questions well posed in our framework by explicitly articulating the alternative intervention. For example, if the question concerns the causal effect of "race," then an ethnicity change on a *curriculum vitae* (or its perception, as in Bertrand and Mullainathan, 2004) defines one causal effect being contemplated, whereas if the question concerns a futuristic "at conception change of chromosomes determining skin color," there is a different causal effect being contemplated. With either manipulation, the explicit description of the intervention makes the question a plausible causal one in our framework.

A closely related way of interpreting the qualitative difference between the three "causal" statements is to consider, after application of the actual treatment, the counterfactual value of the potential outcome corresponding to the treatment not applied. In the first statement, the treatment applied is "aspirin taken," and the counterfactual potential outcome is the state of your headache under "aspirin not taken"; here it appears unambiguous to consider the counterfactual outcome. In the second example, the counterfactual outcome is her job a year ago had she decided not to go to college, which is not as well defined. In the last example, the counterfactual outcome – the person's hair length if she were a boy rather than a girl (note the lack of an action in this statement) – is not at all well defined, and therefore the causal statement is correspondingly poorly defined. In practice, the distinction between well and poorly defined causal statements is one of degree. The important point is, however, that causal statements become more clearly defined by more precisely articulating the intervention that would have made the alternative potential outcome the realized one.

1.3   DEFINITION OF CAUSAL EFFECTS

Let us consider the case of a single unit, I, at a particular point in time, contemplating whether or not to take an aspirin for my headache. That is, there are two treatment levels,

Table 1.1. *Example of Potential Outcomes and Causal Effect with One Unit*

Unit	Potential Outcomes		Causal Effect
	Y(Aspirin)	Y(No Aspirin)	
You	No Headache	Headache	Improvement due to Aspirin

taking an aspirin, and not taking an aspirin. If I take the aspirin, my headache may be gone, or it may remain, say, an hour later; we denote this outcome, which can be either "Headache" or "No Headache," by Y(Aspirin). (We could use a finer measure of the status of my headache an hour later, for example, rating my headache on a ten-point scale, but that does not alter the fundamental issues involved here.) Similarly, if I do not take the aspirin, my headache may remain an hour later, or it may not; we denote this potential outcome by Y(No Aspirin), which also can be either "Headache," or "No Headache." There are therefore two potential outcomes, Y(Aspirin) and Y(No Aspirin), one for each level of the treatment. The causal effect of the treatment involves the comparison of these two potential outcomes.

Because in this example each potential outcome can take on only two values, the unit-level causal effect – the comparison of these two outcomes for the same unit – involves one of four (two by two) possibilities:

1. Headache gone only with aspirin:
   Y(Aspirin) = No Headache, Y(No Aspirin) = Headache
2. No effect of aspirin, with a headache in both cases:
   Y(Aspirin) = Headache, Y(No Aspirin) = Headache
3. No effect of aspirin, with the headache gone in both cases:
   Y(Aspirin) = No Headache, Y(No Aspirin) = No Headache
4. Headache gone only without aspirin:
   Y(Aspirin) = Headache, Y(No Aspirin) = No Headache

Table 1.1 illustrates this situation assuming the values Y(Aspirin) = No Headache, Y(No Aspirin) = Headache. There is a zero causal effect of taking aspirin in the second and third possibilities. In the other two cases the aspirin has a causal effect, making the headache go away in one case and not allowing it to go away in the other.

There are two important aspects of this definition of a causal effect. First, the definition of the causal effect depends on the potential outcomes, but it does *not* depend on which outcome is actually observed. Specifically, whether I take an aspirin (and am therefore unable to observe the state of my headache with no aspirin) or do not take an aspirin (and am thus unable to observe the outcome with an aspirin) does not affect the definition of the causal effect. Second, the causal effect is the comparison of potential outcomes, for the same unit, at the same moment in time post-treatment. In particular, the causal effect is *not* defined in terms of comparisons of outcomes at different times, as in a before-and-after comparison of my headache before and after deciding to take or not to take the aspirin. "The fundamental problem of causal inference" (Holland, 1986, p. 947) is therefore the problem that at most one of the potential outcomes can be realized and thus observed. If the action you take is Aspirin, you observe Y(Aspirin) and

Table 1.2. *Example of Potential Outcomes, Causal Effect, Actual Treatment, and Observed Outcome with One Unit*

Unit	Not Observable		Causal Effect	Known	
	Potential Outcomes			Actual Treatment	Observed Outcome
	Y(Aspirin)	Y(No Aspirin)			
You	No Headache	Headache	Improvement due to Aspirin	Aspirin	No Headache

will never know the value of Y(No Aspirin) because you cannot go back in time. Similarly, if your action is No Aspirin, you observe Y(No Aspirin) but cannot know the value of Y(Aspirin). Likewise, for the college example, we know the outcome given college attendance because the woman actually went to college, but we will never know what job she would have had if she had not gone to college. In general, therefore, even though the unit-level causal effect (the comparison of the two potential outcomes) may be well defined, by definition we cannot learn its value from just the single realized potential outcome. Table 1.2 illustrates this concept for the aspirin example, assuming the action taken was that you took the aspirin.

For the *estimation* of causal effects, as opposed to the *definition* of causal effects, we will need to make different comparisons from the comparisons made for their definitions. For estimation and inference, we need to compare *observed* outcomes, that is, observed realizations of potential outcomes, and because there is only one realized potential outcome per unit, we will need to consider multiple units. For example, a before-and-after comparison of the same physical object involves distinct units in our framework, and also the comparison of two different physical objects at the same time involves distinct units. Such comparisons are critical for *estimating* causal effects, but they do not *define* causal effects in our approach. For estimation it will also be critical to know about, or make assumptions about, the reason why certain potential outcomes were realized and not others. That is, we will need to think about the *assignment mechanism*, which we introduce in Section 1.7. However, we do not need to think about the assignment mechanism for defining causal effects: we merely need to do the thought experiment of the manipulations leading to the definition of the potential outcomes.

1.4   CAUSAL EFFECTS IN COMMON USAGE

The definition of a causal effect given in the previous section may appear a bit formal, and the discussion a bit ponderous, but the presentation is simply intended to capture the way we use the concept in everyday life. Also, implicitly this definition of causal effect as the comparison of potential outcomes is frequently used in contemporary culture, for example, in the movies. Many of us have seen the movie *It's a Wonderful Life*, with Jimmy Stewart as George Bailey. In this movie George Bailey becomes very depressed and states that the world would have been a better place had he never been born. At the appropriate moment an angel appears and shows him what the world would have been like had he not been born. The actual world is the real, observed outcome, but the angel shows George the other potential outcome, had George not been born. Not only are there obvious consequences, like his own children not existing, but there are many other untoward events. For example, his younger brother, who was in actual life a World War II hero, in the counterfactual world drowns in a skating accident at age eight because George was not there to save him. In the counterfactual world a pharmacist fills in a wrong prescription and is convicted of manslaughter because George was not there to catch the error as he did in the actual world. The causal effect of George not being born is the comparison of the entire stream of events in the actual world with George in it, with the entire stream of events in the counterfactual world without George in it. In reality we would never be able to see both worlds, but in the movie George gets to observe both.

Another interesting comparison is to the "but-for" concept in legal settings. Suppose someone committed an action that is harmful, and a second person suffered damages. From a legal perspective, the damage that the second person is entitled to collect is the difference between the economic position of the plaintiff had the harmful event not occurred (the economic position "but-for" the harmful action) and the actual economic position of the plaintiff. Clearly, this is a comparison of the potential outcome that was not realized and the realized potential outcome, this difference being the causal effect of the harmful action.

1.5   LEARNING ABOUT CAUSAL EFFECTS: MULTIPLE UNITS

Although the *definition* of causal effects does not require more than one unit, *learning* about causal effects typically requires multiple units. Because with a single unit we can at most observe a single potential outcome, we must rely on multiple units to make causal inferences. More specifically, we must observe multiple units, some exposed to the active treatment, some exposed to the alternative (control) treatment.

One option is to observe the same physical object under different treatment levels at different points in time. This type of data set is a common source for personal, informal assessments of causal effects. For example, I might feel confident that an aspirin is going to relieve my headache within an hour, based on previous experiences, including episodes when my headache went away when I took an aspirin, and episodes when my headache did not go away when I did not take aspirin. In that situation, my views are shaped by comparisons of multiple units: myself at different times, taking and not taking aspirin. There is sometimes a tendency to view the same physical object at different times as the same unit. We view this as a fundamental mistake. The same physical unit, "myself at different times," is not the same unit in our approach to causality. Time matters for many reasons. For example, I may become more or less sensitive to aspirin, evenings may differ from mornings, or the initial intensity of my headache may affect the result. It is often reasonable to assume that time makes little difference for inanimate objects – we may feel confident, from past experience, that turning on a faucet will cause water to flow from that tap – but this assumption is typically less reasonable with human subjects, and it is never correct to confuse assumptions (e.g., about similarities between different units), with definitions (e.g., of a unit, or of a causal effect).

As an alternative to observing the same physical object repeatedly, one might observe different physical objects at approximately the same time. This situation is another common source for informal assessments of causal effects. For example, if both you and I have headaches, but only one of us takes an aspirin, we may attempt to infer the efficacy of taking aspirin by comparing our subsequent headaches. It is more obvious here that "you" and "I" at the same point in time are different units. Your headache status after taking an aspirin can obviously differ from what my headache status would have been had I taken an aspirin. I may be more or less sensitive to aspirin, or I may have started with a more or less severe headache. This type of comparison, often involving many different individuals, is widely used in informal assessments of causal effects, but it is also the basis for many formal studies of causal effects in the social and biomedical sciences. For example, many people view a college education as economically beneficial to future career outcomes based on comparisons of the careers of individuals with, and individuals without, college educations.

By itself, however, the presence of multiple units does not solve the problem of causal inference. Consider the aspirin example with two units, You and I, and two possible treatments for each unit, aspirin or no aspirin. For simplicity, assume that the two available aspirin tablets are equally effective. There are now a total of four treatment levels: you take an aspirin and I do not, I take an aspirin and you do not, we both take an aspirin, or neither of us does. There are therefore four potential outcomes for each of us. For "I" these four potential outcomes are the state of my headache (i) if neither of us takes an aspirin, (ii) if I take an aspirin and you do not, (iii) if you take an aspirin and I do not, and (iv) if both of us take an aspirin. "You," of course, have the corresponding set of four potential outcomes. We can still only observe at most one of these four potential outcomes for each unit, namely the one realized corresponding to whether you and I took, or did not take, an aspirin. Thus each level of the treatment now indicates both whether you take an aspirin and whether I do. In this situation, there are six different comparisons defining causal effects for each of us, depending on which two of the four potential outcomes for each unit are conceptually compared $\left(6 = \binom{4}{2}\right)$. For example, we can compare the status of my headache if we both take aspirin with the status of my headache if neither of us takes an aspirin, or we can compare the status of my headache if only you take an aspirin to the status of my headache if we both do.

Although we typically make the assumption that whether you take an aspirin does not affect my headache status, it is important to understand the force of such an assumption. One should not lose sight of the fact that it is an assumption, often a strong and controversial one, not a fact, and therefore may be false. Consider a setting where I take aspirin, and I will have a headache if you do not take an aspirin, whereas I will not have a headache if you do take an aspirin: we are in the same room, and unless you take an aspirin to ease your own headache, your incessant complaining will maintain my headache! Such interactions or spillover effects are an important feature of many educational programs, and often motivate changing the unit of analysis from individual children to schools or other groups of individuals.

1.6   THE STABLE UNIT TREATMENT VALUE ASSUMPTION

In many situations it may be reasonable to assume that treatments applied to one unit do not affect the outcome for another unit. For example, if we are in different locations and have no contact with each other, it would appear reasonable to assume that whether you take an aspirin has no effect on the status of my headache. (But, as the example in the previous section illustrates, this assumption need not hold if we are in the same location, and your behavior, itself affected by whether you take an aspirin, may affect the status of my headache, or if we communicate by extrasensory perception.) The stable unit treatment value assumption, or SUTVA (Rubin, 1980a) incorporates both this idea that units do not interfere with one another and the concept that for each unit there is only a single version of each treatment level (ruling out, in this case, that a particular individual could take aspirin tablets of varying efficacy):

Assumption 1.1 (SUTVA)
The potential outcomes for any unit do not vary with the treatments assigned to other units, and, for each unit, there are no different forms or versions of each treatment level, which lead to different potential outcomes.

These two elements of the stability assumption enable us to exploit the presence of multiple units for estimating causal effects.

SUTVA is the first of a number of assumptions discussed in this book that are referred to generally as *exclusion restrictions*: assumptions that rely on external, substantive, information to rule out the existence of a causal effect of a particular treatment relative to an alternative. For instance, in the aspirin example, in order to help make an assessment of the causal effect of aspirin on headaches, we could exclude the possibility that your taking or not taking aspirin has any effect on my headache. Similarly, we could exclude the possibility that the aspirin tablets available to me are of different strengths. Note, however, that these assumptions, and other restrictions discussed later, are not directly informed by observations – they are assumptions. That is, they rely on previously acquired knowledge of the subject matter for their justification. Causal inference is generally impossible without such assumptions, and thus it is critical to be explicit about their content and their justifications.

1.6.1 SUTVA: No Interference

Consider, first, the no-interference component of SUTVA – the assumption that the treatment applied to one unit does not affect the outcome for other units. Researchers have long been aware of the importance of this concept. For example, when studying the effect of different types of fertilizers in agricultural experiments on plot yields, traditionally researchers have taken care to separate plots using "guard rows," unfertilized strips of land between fertilized areas. By controlling the leaching of different fertilizers across experimental plots, these guard rows make SUTVA more credible; without them we might suspect that the fertilizer applied to one plot affected the yields in contiguous plots.

In our headache example, in order to address the no-interference assumption, one has to argue, on the basis of a prior knowledge of medicine and physiology, that someone else taking an aspirin in a different location cannot have an effect on my headache. You might think that we could learn about the magnitude of such interference from a separate experiment. Suppose people are paired, with each pair placed in a separate room. In each pair one randomly chosen individual is selected to be the "designated treated" individual and the other the "designated control" individual. Half the pairs are then randomly selected to be the "treatment pairs" and the other half selected to be "control pairs," with the "designated treated" individual in the treatment pairs given aspirin and the "designated treated" individual in the control pairs given a placebo. The outcome would then be the status of the headache of the "control" person in each pair. Although such an experiment could shed some light on the plausibility of our no-interference assumption, this experiment relies itself on a more distant version of SUTVA – that treatments assigned to one pair do not affect the results for other pairs. As this example reveals, in order to make any assessment of causal effects, the researcher has to rely on assumed existing knowledge of the current subject matter to assert that some treatments do not affect outcomes for some units.

There exist settings, moreover, in which the no-interference part of SUTVA is controversial. In large-scale job training programs, for example, the outcomes for one individual may well be affected by the number of people trained when that number is sufficiently large to create increased competition for certain jobs. In an extreme example, the effect on your future earnings of going to a graduate program in statistics would surely be very different if everybody your age also went to a graduate program in statistics. Economists refer to this concept as a *general equilibrium* effect, in contrast to a *partial equilibrium* effect, which is the effect on your earnings of a statistics graduate degree under the *ceteris paribus* assumption that "everything else" stayed equal. Another classic example of interference between units arises in settings with immunizations against infectious diseases. The causal effect of your immunization versus no immunization will surely depend on the immunization of others: if everybody else is already immunized with a perfect vaccine, and others can therefore neither get the disease nor transmit it, your immunization is superfluous. However, if no one else is immunized, your treatment (immunization with a perfect vaccine) would be effective relative to no immunization. In such cases, sometimes a more restrictive form of SUTVA can be considered by defining the unit to be the community within which individuals interact, for example, schools in educational settings, or specifically limiting the number of units assigned to a particular treatment.

1.6.2 SUTVA: No Hidden Variations of Treatments

The second component of SUTVA requires that an individual receiving a specific treatment level cannot receive different forms of that treatment. Consider again our assessment of the causal effect of aspirin on headaches. For the potential outcome with both of us taking aspirin, we obviously need more than one aspirin tablet. Suppose, however, that one of the tablets is old and no longer contains a fully effective dose, whereas the other is new and at full strength. In that case, each of us may have three treatments available: no aspirin, the ineffective tablet, and the effective tablet. There are thus two forms of the active treatment, both nominally labeled "aspirin": aspirin+ and aspirin−. Even with no interference we can now think of there being three potential outcomes for each of us, the no aspirin outcome Y_i(No Aspirin), the weak aspirin outcome Y_i(Aspirin−) and the strong aspirin outcome Y_i(Aspirin+), with i indexing "I" or "You." The second part of SUTVA either requires that the two aspirin outcomes are identical: Y_i(Aspirin+) $= Y_i$(Aspirin−), or that I can only get Aspirin+ and you can only get Aspirin− (or *vice versa*). Alternatively we can redefine the treatment as taking a randomly selected aspirin (either Aspirin− or Aspirin+). In that case SUTVA might be satisfied for the redefined stochastic treatment.

Another example of variation in the treatment that is ruled out by SUTVA occurs when differences in the method of administering the treatment matter. The effect of taking a drug for a particular individual may differ depending on whether the individual was assigned to receive it or chose to take it. For example, taking it after being given the choice may lead the individual to take actions that differ from those that would be taken if the individual had no choice in the taking of the drug.

Fundamentally, the second component of SUTVA is again an exclusion restriction. The requirement is that the label of the aspirin tablet, or the nature of the administration of the treatment, cannot alter the potential outcome for any unit. This assumption does *not* require that all forms of each level of the treatment are identical across all units, but only that unit i exposed to treatment level w specifies a well-defined potential outcome, $Y_i(w)$, for all i and w. One strategy to make SUTVA more plausible relies on redefining the represented treatment levels to comprise a larger set of treatments, for example, Aspirin−, Aspirin+, and no-aspirin instead of only Aspirin and no-aspirin. A second strategy involves coarsening the outcome; for example, SUTVA may be more plausible if the outcome is defined to be dead or alive rather than to be a detailed measurement of health status. The point is that SUTVA implies that the potential outcomes for each unit and each treatment are well-defined functions (possibly with stochastic images) of the unit index and the treatment.

1.6.3 Alternatives to SUTVA

To summarize the previous discussion, assessing the causal effect of a binary treatment requires observing more than a single unit, because we must have observations of potential outcomes under both treatments: those associated with the receipt of the treatment on some units and those associated with no receipt of it on some other units. However, with more than one unit, we face two immediate complications. First, there exists the possibility that the units interfere with one another, such that one unit's potential outcome when exposed to a specific treatment level, may also depend on the treatment received by another unit. Second, because in multi-unit settings, we must have available more than one copy of each treatment, we may face circumstances in which a unit's potential outcome when receiving the same nominal level of a treatment could vary with different versions of that treatment. These are serious complications, serious in the sense that unless we restrict them by assumptions, combined with careful study design to make these assumptions more realistic, any causal inference will have only limited credibility.

Throughout most of this book, we shall maintain SUTVA. In some cases, however, specific information may suggest that alternative assumptions are more appropriate. For example, in some early AIDS drug trial settings, many patients took some of their assigned drug and shared the remainder with other patients in hopes of avoiding placebos. Given this knowledge, it is clearly no longer appropriate to assert the no-interference element of SUTVA – that treatments assigned to one unit do not affect the outcomes for others. We can, however, use this specific information to model how treatments are received across patients in the study, making alternative – and in this case, more appropriate – assumptions that allow some inference. For example, SUTVA may be more appropriate using subgroups of people as units in such AIDS drug trials. Similarly, in educational settings, SUTVA may be more plausible with classrooms or schools as the units of analysis than with students as the units of analysis. In many economic examples, interactions between units are often modeled through assumptions on market structure, again avoiding the no-interference element of SUTVA. Consequently, SUTVA is only one candidate exclusion restriction for modeling the potentially complex interactions between units and the entire set of treatment levels in a particular experiment. In many settings, however, it appears that SUTVA is the leading choice.

1.7   THE ASSIGNMENT MECHANISM: AN INTRODUCTION

If we are willing to accept SUTVA, our complicated "You" and "I" aspirin example simplifies to the situation depicted in Table 1.3. Now You and I each face only two treatment levels (e.g., for "You" whether or not "You" take an aspirin), and the accompanying potential outcomes are a function of only our individual actions. This extends readily to many units. To accommodate this generalization, and also the discussion of other examples beyond that of taking or not taking aspirin, as introduced in Section 1.6, let us index the units in the population of size N by i, taking on values $1, \ldots, N$, and let the treatment indicator W_i take on the values 0 (the control treatment, e.g., no aspirin) and 1 (the active treatment, e.g., aspirin). We have one realized (and possibly observed) potential outcome for each unit. For unit i, now $i \in \{1, \ldots, N\}$, let Y_i^{obs} denote this realized (and possibly observed) outcome:

$$Y_i^{\text{obs}} = Y_i(W_i) = \begin{cases} Y_i(0) & \text{if } W_i = 0, \\ Y_i(1) & \text{if } W_i = 1. \end{cases}$$

For each unit we also have one missing potential outcome, for unit i denoted by Y_i^{mis}:

$$Y_i^{\text{mis}} = Y_i(1 - W_i) = \begin{cases} Y_i(1) & \text{if } W_i = 0, \\ Y_i(0) & \text{if } W_i = 1. \end{cases}$$

Many writers replace the potential outcomes and treatment indicator with simply the treatment indicator, W_i, and the observed outcome Y_i^{obs}. This "observed-value" notation confuses the objects of inference and the assignment mechanism and can lead to mistakes as we see in Section 1.9.

This information alone, still, does not allow us to infer the causal effect of taking an aspirin on headaches. Suppose, in the two-person headache example, that the person who chose not to take the aspirin did so because he had only a minor headache. Suppose then that an hour later both headaches have faded: the headache for the first person possibly faded because of the aspirin (it would still be there without the aspirin), and the headache of the second person faded simply because it was not a serious headache (it would be gone even without the aspirin). When comparing these two observed potential outcomes, we might conclude that the aspirin had no effect, whereas in fact it may have been the cause of easing the more serious headache. The key piece of information that

Table 1.3. *Example of Potential Outcomes and Causal Effects under SUTVA with Two Units*

Unit	Unknown			Known	
	Potential Outcomes		Causal Effect	Actual	Observed
	Y(Aspirin)	Y(No Aspirin)		Treatment W_i	Outcome Y_i^{obs}
You	No Headache	Headache	Improvement due to Aspirin	Aspirin	No Headache
I	No Headache	No Headache	None	No Aspirin	No Headache

Table 1.4. *Medical Example with Two Treatments, Four Units, and SUTVA: Surgery (S) and Drug Treatment (D)*

Unit	Potential Outcomes		Causal Effect
	$Y_i(0)$	$Y_i(1)$	$Y_i(1) - Y_i(0)$
Patient #1	1	7	6
Patient #2	6	5	−1
Patient #3	1	5	4
Patient #4	8	7	−1
Average	4	6	2

we lack is how each individual came to receive the treatment level actually received: in our language of causation, the *assignment mechanism*.

Because causal effects are defined by comparing potential outcomes (only one of which can ever be observed), they are well defined irrespective of the actions actually taken. But, because we observe at most half of all potential outcomes, and none of the unit-level causal effects, there is an inferential problem associated with assessing causal effects. In this sense, the problem of causal inference is, as pointed out in Rubin (1974), a *missing data problem*: given any treatment assigned to an individual unit, the potential outcome associated with any alternate treatment is missing. A key role is therefore played by the missing data mechanism, or, as we refer to it in the causal inference context, the assignment mechanism. How is it determined which units get which treatments or, equivalently, which potential outcomes are realized and which are not? This mechanism is, in fact, so crucial to the problem of causal inference that Parts II through VI of this book are organized by varying assumptions concerning this mechanism.

To illustrate the critical role of the assignment mechanism, consider the simple hypothetical example in Table 1.4. This example involves four units, in this case patients, and two possible medical procedures labeled 0 (Drug) and 1 (Surgery). Assuming SUTVA, Table 1.4 displays each patient's potential outcomes, in terms of years of post-treatment survival, under each treatment. From Table 1.4, it is clear that on average, Surgery is better than Drug by two years' life expectancy, that is, the average causal effect of Surgery versus Drug is two years for these four individuals.

Suppose now that the doctor, through expertise or magic, knows enough about these potential outcomes and so assigns each patient to the treatment that is more beneficial to that patient. In this scenario, Patients 1 and 3 will receive surgery, and Patients 2 and

Table 1.5. *Ideal Medical Practice: Patients Assigned to the Individually Optimal Treatment; Example from Table 1.4*

Unit i	Treatment W_i	Observed Outcome Y_i^{obs}
Patient #1	1	7
Patient #2	0	6
Patient #3	1	5
Patient #4	0	8

4 will receive the drug treatment. The observed treatments and outcomes will then be as displayed in Table 1.5, where the average observed outcome with surgery is one year less than the average observed outcome with the drug treatment. Thus, a casual observer might be led to believe that, on average, the drug treatment is superior to surgery. In fact, the opposite is true: as shown in Table 1.4, if the drug treatment were uniformly applied to a population like these four patients, the average survival would be four years, as can be seen from the "$Y(0)$" column in Table 1.4, as opposed to six years if all patients were treated with surgery, as can be seen from the "$Y(1)$" column in the same table. Based on this example, we can see that we cannot simply look at the observed values of potential outcomes under different treatments, that is, $\{Y_i^{obs}|i : \ \text{s.t. } W_i = 0\}$ and $\{Y_i^{obs}|i : \ \text{s.t. } W_i = 1\}$, and reach valid causal conclusions irrespective of the assignment mechanism. In order to draw valid causal inferences, we must consider why some units received one treatment rather than another. In Parts II through VI of this text, we will discuss in greater detail various assignment mechanisms and the accompanying analyses for drawing valid causal inferences.

1.8   ATTRIBUTES, PRE-TREATMENT VARIABLES, OR COVARIATES

Consider a study of causal effects involving many units, which we assume satisfies the stability assumption, SUTVA. At least half of all potential outcomes will be unobserved or missing, because only one potential outcome can be observed for each unit, namely the potential outcome corresponding to the realized level of the treatment or action. To estimate the causal effect for any particular unit, we will generally need to predict, or impute, the missing potential outcome. Comparing the imputed missing outcome to the realized and observed outcome for this unit allows us to estimate the unit-level causal effect. In general, creating such predictions is difficult. They involve assumptions about the assignment mechanism and about comparisons between different units, each exposed to only one of the treatments. Often the presence of unit-specific background attributes, also referred to as pre-treatment variables, or covariates, and denoted in this text by the K-component row vector X_i for unit i, can assist in making these predictions. For instance, in our headache example, such variables could include the intensity of the headache before making the decision to take aspirin or not. Similarly, in an evaluation of the effect of job training on future earnings, these attributes may include age, previous educational achievement, family, and socio-economic status, or pre–training earnings.

As these examples illustrate, sometimes a covariate (e.g., pre-training earnings) differs from the potential outcome (post-training earnings) solely in the timing of measurement, in which case the covariates can be highly predictive of the potential outcomes.

The key characteristic of these covariates is that they are *a priori* known to be unaffected by the treatment assignment. This knowledge often comes from the fact that they are permanent characteristics of units, or that they took on their values prior to the treatment being assigned, as reflected in the label "pre-treatment" variables.

The information available in these covariates can be used in three ways. First, covariates commonly serve to make estimates more precise by explaining some of the variation in outcomes. For instance, in the headache example, holding constant the intensity of the headache before receiving the treatment by studying units with the same initial headache intensity should give more precise estimates of the effect of aspirin, at least for units with that level of headache intensity. Second, for substantive reasons, the researcher may be interested in the typical (e.g., average) causal effect of the treatment on subgroups (as defined by a covariate) in the population of interest. For example, we may want to evaluate the effects of a job-training program separately for people with different education levels, or the effect of a medical drug separately for women and men. The final and most important role for covariates in our context, however, concerns their effect on the assignment mechanism. Young unemployed individuals may be more interested in training programs aimed at acquiring new skills, or high-risk groups may be more likely to take flu shots. As a result, those taking the active treatment may differ in the values of their background characteristics from those taking the control treatment. At the same time, these characteristics may be associated with the potential outcomes. As a result, assumptions about the assignment mechanism and its possible freedom from dependence on potential outcomes are typically more plausible within subpopulations that are homogeneous with respect to some covariates, that is, conditionally given the covariates, rather than unconditionally.

1.9   POTENTIAL OUTCOMES AND LORD'S PARADOX

To illustrate the clarity that comes with the potential outcomes interpretation of causality, we consider a problem from the literature that is known as Lord's paradox:

> A large university is interested in investigating the effects on the students of the diet provided in the university dining halls and any sex differences in these effects. Various types of data are gathered. In particular, the weight of each student at the time of his arrival in September and his weight the following June are recorded. (Lord, 1967, p. 304)

The results of the hypothetical study described in Lord's paper include the finding that for the males the average weight is identical at the end of the school year to what it was at the beginning; in fact, the whole distribution of weights is unchanged, although some males lost weight and some males gained weight – the gains and losses exactly balance. The same thing is true for the females. The only difference is that the females started and ended the year lighter on average than the males. On average, there is no weight gain or weight loss for either males or females. From Lord's quoted description of the problem, the object of interest, what we will generally call the *estimand*, is the difference between the causal effect of the university diet on males and the causal effect of the university diet on females. That is, the causal estimand is the difference between the causal effects for males and females, the "differential" causal effect.

The paradox is generated by considering the contradictory conclusions of two statisticians asked to comment on the data. Statistician 1 observes that there are no differences between the September and June weight distributions for either males or females. Thus, Statistician 1 concludes that

> as far as these data are concerned, there is no evidence of any interesting effect of diet (or of anything else) on student weight. In particular, there is no evidence of any differential effect on the two sexes, since neither group shows any systematic change. (Lord, 1967, p. 305)

Statistician 2 looks at the data in a more "sophisticated" way. Effectively, he examines males and females with the same initial weight in September, say a subgroup of "overweight" females (meaning simply above-average-weight females) and a subgroup of "underweight" males (analogously defined). He notices that these males tended to gain weight on average and these females tended to lose weight on average. He also notices that this result is true no matter what the value of initial weight he focuses on. (Actually, Lord's Statistician 2 used a technique known as covariance adjustment or regression adjustment described in Chapter 7.) His conclusion, therefore, is that after "controlling for" initial weight, the diet has a differential positive effect on males relative to females because for males and females with the same initial weight, on average the males gain more than the females.

Who's right? Statistician 1 or Statistician 2? Notice the focus of both statisticians on before-after or gain scores and recall that such gain scores are not causal effects because they do not compare potential outcomes at the same time post-treatment; rather, they compare changes over time. If both statisticians confined their comments to *describing* the data, both would be correct, but for causal inference, both are wrong because these data cannot support any conclusions about the causal effect of the diet without making some very strong, and arguably implausible, assumptions.

Back to the basics. The units are obviously the students, and the time of application of active treatment (the university diet) is clearly September and the time of the recording of the outcome Y is clearly June. Let us accept the stability assumption. Now, what are the potential outcomes, and what is the assignment mechanism? Notice that Lord's statement of the problem uses the already criticized notation with a treatment indicator and the observed variable, Y_i^{obs}, rather than the potential outcome notation being advocated. The potential outcomes are June weight under the university diet $Y_i(1)$ and under the "control" diet $Y_i(0)$. The covariates are sex of students, male versus female, and September weight. But the assignment mechanism has assigned everyone to the new treatment! There is no one, male or female, who is assigned to the control treatment. Hence, there is absolutely no purely empirical basis on which to compare the effects, either raw or differential, of the university diet with the control diet. By making the problem complicated with the introduction of the covariates "male/female" and "initial weight," Lord has created partial confusion. But the point here is that the "paradox" is immediately resolved through the explicit use of potential outcomes. Either answer could be correct for causal inference depending on what we are willing to assume about the (never-observed) potential outcome under the control diet and its relation to the (observed) potential outcome given the university diet.

1.10 CAUSAL ESTIMANDS

Let us now be a little more formal when describing causal estimands, the ultimate object of interest in our analyses. We start with a population of units, indexed by $i = 1, \ldots, N$, which is our focus. Each unit in this population can be exposed to one of a set of treatments. In the most general case, let $\mathbb{T}_i$ denote the set of treatments to which unit i can be exposed. In most cases, this set will be identical for all units. Exceptions include settings where the treatment is defined as the peer group for each individual. In the current text, the set $\mathbb{T}_i$ consists of the same two treatments for each unit (e.g., taking or not taking a drug),

$$\mathbb{T}_i = \mathbb{T} = \{0, 1\},$$

for all $i = 1, \ldots, N$. Generalizations of most of the discussion in this text to finite sets of treatments are conceptually straightforward.

For each unit i, and for each treatment in the common set of treatments, $\mathbb{T} = \{0, 1\}$, there are corresponding potential outcome, $Y_i(0)$ and $Y_i(1)$. Comparisons of $Y_i(1)$ and $Y_i(0)$ are *unit-level causal effects*. Often these are simple differences,

$$Y_i(1) - Y_i(0), \qquad \text{or ratios } Y_i(1)/Y_i(0),$$

but in general the comparisons can take different forms. There are many such unit-level causal effects, and we often wish to summarize them for the finite sample or for subpopulations. A leading example of what we in general refer to as a *causal estimand* is the average difference of the pair of potential outcomes, averaged over the entire population,

$$\tau_{\text{fs}} = \frac{1}{N} \sum_{i=1}^{N} \big(Y_i(1) - Y_i(0) \big),$$

where the subscript "fs" indicates that we average over the finite sample.

We can generalize this example in a number of ways. Here we discuss two of these generalizations, maintaining in each case the setting with $\mathbb{T} = \{0, 1\}$ for all units. First, we can average over subpopulations rather than over the full population. The subpopulation that we average over may be defined in terms of different sets of variables. First, it can be defined in terms of pre-treatment variables, or covariates, denoted by X_i. Recall these are variables measured on the units that, unlike outcomes, are *a priori* known to be unaffected by the treatment. For example, we may be interested in the average effect of a new drug only for females:

$$\tau_{\text{fs}}(f) = \frac{1}{N(f)} \sum_{i: X_i = f} \big(Y_i(1) - Y_i(0) \big).$$

Here $X_i \in \{f, m\}$ is an indicator for being female, and $N(f) = \sum_{i=1}^{N} \mathbf{1}_{X_i=f}$ is the number of females in the finite population, where $\mathbf{1}_A$ is the indicator function for the event A, equal to 1 if A is true and zero otherwise. Second, one can focus on the average effect of the treatment for those who were exposed to it:

$$\tau_{\mathrm{fs},t} = \frac{1}{N_{\mathrm{t}}} \sum_{i:W_i=1} \left(Y_i(1) - Y_i(0) \right),$$

where N_{t} is the number of units exposed to the active treatment. For example, we may be interested in the average effect of serving in the military on subsequent earnings in the civilian labor market for those who served in the military, or the average effect of exposure to asbestos on health for those exposed to it. In both examples, there is less interest in the average effect for units not exposed to the treatment. A third way of defining the relevant subpopulation is to do so partly in terms of potential outcomes. As an example, one may be interested in the average effect of a job-training program on hourly wages, averaged only over those individuals who would have been employed (with positive hourly wages) irrespective of the level of the treatment:

$$\tau_{\mathrm{fs,pos}} = \frac{1}{N_{\mathrm{pos}}} \sum_{i:Y_i(0)>0,Y_i(1)>0} \left(Y_i(1) - Y_i(0) \right),$$

where $N_{\mathrm{pos}} = \sum_{i=1}^{N} \mathbf{1}_{Y_i(0)>0,Y_i(1)>0}$. Because the conditioning variable (being employed irrespective of the treatment level) is a function of potential outcomes, the conditioning is (partly) on potential outcomes.

As a second generalization of the average treatment effect, we can focus on more general functions of potential outcomes. For example, we may be interested in the median (over the entire population or over a subpopulation) of $Y_i(1)$ versus the median of $Y_i(0)$. One may also be interested in the median of the difference $Y_i(1) - Y_i(0)$, which generally differs from the difference in medians.

In all cases with $\mathbb{T} = \{0, 1\}$, we can write the causal estimand as a row-exchangeable function of all potential outcomes for all units, all treatment assignments, and pre-treatment variables:

$$\tau = \tau(\mathbf{Y}(0), \mathbf{Y}(1), \mathbf{X}, \mathbf{W}).$$

In this expression $\mathbf{Y}(0)$ and $\mathbf{Y}(1)$ are the N-component column vectors of potential outcomes with ith elements equal to $Y_i(0)$ and $Y_i(1)$, $\mathbf{W}$ is the N-component column vector of treatment assignments, with ith element equal to W_i, and $\mathbf{X}$ is the $N \times K$ matrix of covariates with ith row equal to X_i. Not all such functions necessarily have a causal interpretation, but the converse is true: all the causal estimands we consider in this book can be written in this form, and all such estimands are comparisons of $Y_i(0)$ and $Y_i(1)$ for all units in a common set whose definition, as the previous examples illustrate, may depend on $\mathbf{Y}(0)$, $\mathbf{Y}(1)$, $\mathbf{X}$, and $\mathbf{W}$.

1.11   STRUCTURE OF THE BOOK

The remainder of Part I of this text includes a brief historical overview of the development of our framework for causal inference (Chapter 2) and some mathematical definitions that characterize assignment mechanisms (Chapter 3).

Parts II through V of this text cover different situations corresponding to different assumptions concerning the assignment mechanism. Part II deals with the inferentially simplest setting of randomized assignment, specifically what we call *classical randomized experiments*. In these settings, the assignment mechanism is under the control of the experimenter, and the probability of any assignment of treatments across the units in the experiment is entirely knowable before the experiment begins.

In Parts III and IV we discuss *regular assignment mechanisms*, where the assignment mechanism is not necessarily under the control of the experimenter, and the knowledge of the probabilities of assignment is incomplete in a very specific and limited way: within subpopulations of units defined by fixed values of the covariates, the assignment probabilities are known to be identical for all these units and known to be strictly between zero and one; the probabilities themselves need not be known. Moreover, in practice, we typically have few units with the same values for the covariates, so that the methods discussed in the chapters on classical randomized experiments are not directly applicable.

Finally, Parts V and VI concern *irregular assignment mechanisms*, which allow the assignment to depend on covariates *and* on potential outcomes, both observed and unobserved, or which allow the unit-level assignment probabilities to be equal to zero or one. Such assignment mechanisms present special challenges, and without further assumptions, only limited progress can be made. In this part of the text, we discuss several strategies for addressing these complications in specific settings. For example, we discuss investigating the sensitivity of the inferential results to violations of the critical "unconfoundedness" assumption on the assignment mechanism. We also discuss some specific cases where this unconfoundedness assumption is supplemented by, or replaced by, assumptions linking various potential outcomes. These assumptions are again exclusion restrictions, where specific treatments are assumed *a priori* not to have any, or limited, effects on outcomes. Because of the complications arising from these irregular assignment mechanisms, and the many forms such assignment mechanisms can take in practice, this area remains a fertile field for methodological research.

1.12   SAMPLES, POPULATIONS, AND SUPER-POPULATIONS

In much of the discussion in this text, the finite set of units for which we observe covariates, treatments, and realized outcomes is the set of units we are interested in, and we will refer to this as the *population*. It does not matter how this population was selected, or where it came from. All conclusions are conditional on this population, and we do not attempt to draw inferences for other populations. For part of the discussion, however, it is useful to view the set of units for which we observe values as drawn randomly from a larger population. In that case we typically take the population that the units were drawn from as infinite. When it is important to make this distinction, we will refer to the set of units for which we observe values as the *finite sample* (often using the subscript "fs"), and the infinite population that these were drawn from as the *super-population* (using subscript "sp") to distinguish between this case and the previous case where we observed values for all units in the population.

1.13 CONCLUSION

In this chapter we present the three basic concepts in our framework for causal inference. The first concept is that of potential outcomes, one for each unit for each level of the treatment. Causal estimands are defined in terms of these potential outcomes, possibly also involving the treatment assignments and pre-treatment variables. We discuss that, because at most only one of the potential outcomes can be observed, there is a need for observing multiple units to be able to conduct causal inference. In order to exploit the presence of multiple units, we use the stability assumption, SUTVA, which is the second basic concept in our framework. The third fundamental concept is that of the assignment mechanism, which determines which units receive which treatment. In Chapter 3 we provide a classification of assignment mechanisms that will serve as the organizing principle of the text.

NOTES

Note that the manipulation underlying our view of causality does not have to take place, merely that one has to be able to do the thought experiment in order for the causal effects to be well defined. Rubin (1978, p. 38) writes: "The fundamental problem facing inference for causal effects is that if treatment t is assigned to the ith experimental unit (i.e., $W_i = t$), only values in Y^t can be observed, Y^j for $j \neq t$ being unobservable (or missing)." Holland (1986, p. 947) puts it similarly when he describes the causal inference problem as arising from the fact that "It is impossible to *observe* the value of $Y_t(u)$ and $Y_c(u)$ on the same unit and, therefore, it is impossible to *observe* the effect of t on u" (emphasis in original). In Holland's notation, u denotes the unit, and $Y_t(u)$ and $Y_c(u)$ denote the two potential outcomes for unit u under the two levels of the treatment. See also Rubin (1977, 2004, 2012).

Following Holland (1986), we refer to the general potential outcomes approach taken in this book as the Rubin Causal Model, although it has precursors in the work by Neyman (1923). Their work explicitly uses potential outcomes ("potential yields" in Neyman, 1990, translation of the 1923 original, p. 467), although Neyman focused exclusively on what we call here completely randomized experiments. In Chapter 2 we discuss in more detail the historical background to the potential outcomes framework.

The Stable Unit Treatment Value Assumption (SUTVA) was formally introduced in Rubin (1980a). See also the discussions in Rubin (1986a, 1990b, 2010). It is implicit in the notation used by Neyman (1923, 1990) where the potential outcomes are indexed only by the treatment assigned to that unit. Cox (1958, p. 19) is explicit about the need for the no-interference part of SUTVA but does not address the part of SUTVA that requires a single version of each treatment for each unit. Fisher does not explicitly address the issue, but under the null hypothesis of no effect of the treatment whatsoever, SUTVA automatically holds.

For more statistical details of the resolution of Lord's paradox, see Lord (1967) and Holland and Rubin (1983), and for earlier related discussion, see, for example, Lindley and Novick (1981).

There is an extensive econometric literature concerned with causality and methods for inferring causal effects, often in settings with complex selection. For recent reviews, see Angrist and Krueger (2000), Leamer (1988), Heckman and Robb (1984), Heckman, Ichimura, Smith, and Todd, (1998), Heckman, Lalonde, and Smith (2000), and Angrist and Pischke (2008).

Recent textbooks discussing causal inference in various detail and from various points of view include Rosenbaum (1995, 2002, 2009), Shadish, Campbell, and Cook (2002), Van Der Laan and Robins (2003), Lee (2005), Caliendo (2006), Gelman and Hill (2006), Morgan and Winship (2007), Angrist and Pischke (2008), Guo and Fraser (2010), Morton and Williams (2010), Murnane and Willett (2011); and for collected papers, see Rubin (2006) and Freedman (2009). For a more philosophical perspective, see Beebee, Hitchcock, and Menzies (2009). The Rosenbaum books are closest to the current text in terms of the perspective on causality.

There are some approaches to causality that take conceptually different perspectives. In the analysis of time series, economists have found it useful to consider "Granger-Sims causality," which essentially views causality as a prediction property. Suppose we have two time series, one measuring the money supply ("money"), and one measuring gross domestic product (GDP). Money "causes" GDP in the Granger sense if, conditional on the past values of GDP, and possibly conditional on other variables, past values of money predict future values of GDP. Money does not "cause" GDP in the Sims sense if, when predicting money from past, present, and future values of GDP, the future values have no predictive power. See Granger (1969) and Sims (1972). For a recent analysis of the causal links between the money supply (or, more specifically, actions by the Federal Reserve Bank), and GDP, from a perspective that is, at least in spirit, closer to the potential outcome approach taken in this text, see Romer and Romer (2004). Angrist and Kuersteiner (2011) provide some discussion on the link with the potential outcome approach.

Dawid (2000) develops an interesting approach to causality that avoids potential outcomes, and which focuses primarily on a decision-oriented perspective. There has not been much experience with this approach in applications so far.

Pearl (1995, 2000, 2009) advocates a different approach to causality. Pearl combines aspects of structural equations models and path diagrams. In this approach, assumptions underlying causal statements are coded as missing links in the path diagrams. Mathematical methods are then used to infer, from these path diagrams, which causal effects can be inferred from the data, and which cannot. See Pearl (2000, 2009) for details and many examples. Pearl's work is interesting, and many researchers find his arguments that path diagrams are a natural and convenient way to express assumptions about causal structures appealing. In our own work, perhaps influenced by the type of examples arising in social and medical sciences, we have not found this approach to aid drawing of causal inferences, and we do not discuss it further in this text.

## CHAPTER 2

A Brief History of the Potential Outcomes Approach to Causal Inference

2.1 INTRODUCTION

The approach to causal inference outlined in the first chapter has important antecedents in the literature. In this chapter we review some of these antecedents to put the potential outcomes approach in perspective. The two most important early developments, in quick succession in the 1920s, are the introduction of potential outcomes in randomized experiments by Neyman (Neyman, 1923, translated and reprinted in Neyman, 1990), and the introduction of randomization as the "reasoned basis" for inference by Fisher (Fisher 1935, p. 14).

Once introduced, the basic idea that causal effects are the comparisons of potential outcomes may seem so obvious that one might expect it to be a long-established tenet of scientific thought. Yet, although the seeds of the idea can be traced back at least to the eighteenth century, the formal notation for potential outcomes was not introduced until 1923 by Neyman. Even then, however, the concept of potential outcomes was used exclusively in the context of randomized experiments, not in observational studies. The same statisticians, analyzing both experimental and observational data with the goal of inferring causal effects, would regularly use the notation of potential outcomes in experimental studies but switch to a notation purely in terms of realized and observed outcomes for observational studies. It is only more recently, starting in the early seventies with the work of Donald Rubin (1974), that the language and reasoning of potential outcomes was put front and center in observational study settings, and it took another quarter century before it found widespread acceptance as a natural way to define and assess causal effects, irrespective of the setting.

Moreover, before the twentieth century there appears to have been only limited awareness of the concept of the assignment mechanism. Although by the 1930s randomized experiments were firmly established in some areas of scientific investigation, notably in agricultural experiments, there was no formal statement for a general assignment mechanism and, moreover, not even formal arguments in favor of randomization until Fisher (1925).

2.2   POTENTIAL OUTCOMES AND THE ASSIGNMENT MECHANISM BEFORE NEYMAN

Before the twentieth century we can find seeds of the potential outcomes definition of causal effects among both experimenters and philosophers. For example, one can see some idea of potential outcomes, although as yet unlabeled as such, in discussions by the philosopher and economist Mill (1973, p. 327), who offers:

> If a person eats of a particular dish, and dies in consequence, that is, would not have died if he had not eaten of it, people would be apt to say that eating of that dish was the source of his death.

Applying the potential outcomes notation to this quotation, Mill appears to be considering the two potential outcomes, Y(eat dish) and Y(not eat dish) for the same person. In this case the observed outcome, Y(eat dish), is "death," and Mill appears to posit that if the alternative potential outcome, Y(not eat dish), is "not death," then one could infer that eating the dish was the source (cause) of the death.

Similarly, in the early twentieth century, the father of much of modern statistics, Fisher (1918, p. 214), argued:

> If we say, "This boy has grown tall because he has been well fed," . . . we are suggesting that he might quite probably have been worse fed, and that in this case he would have been shorter.

Here again we see a, somewhat implicit, reference to two potential outcomes, Y(well fed) = tall and Y(not well fed) = shorter, associated with a single unit, a boy.

Despite the insights we may perceive in these quotations, their authors may or may not have intended their words to mean as we choose to interpret them. For instance, in his argument, Mill goes on to require "constant conjunction" in order to assign causality – that is, for the dish to be the cause of death, this outcome must occur every time it is consumed, by this person, or perhaps by any person. Curiously, an early tobacco industry argument used a similar notion of causality: not everyone who smokes two or more packs of cigarettes a day gets lung cancer, therefore smoking does not cause lung cancer. Jerome Cornfield, the well-known American epidemiologist who studied smoking and lung cancer also struggled with this: "If cigarettes are carcinogenic, why don't all smokers get lung cancer?" (Cornfield, 1959, p. 242) without the benefits of the potential outcomes framework. See also Rubin (2012).

No matter how interpreted, however, we have found no early writer who formally pursued these intuitive insights about potential outcomes defining causal effects; in particular, until Neyman did so in 1923, no one developed a formal notation for the idea of potential outcomes. Nor did anyone discuss the importance of the assignment mechanism, which is necessary for the evaluation of causal effects. The first such formal mathematical use of the idea of potential outcomes was introduced by Jerzey Neyman (1923), and then only in the context of an urn model for assigning treatments to plots. The general formal definition of causal effects in terms of potential outcomes, as well as the formal definition of the assignment mechanism, was still another half century away.

2.3 NEYMAN'S (1923) POTENTIAL OUTCOME NOTATION IN RANDOMIZED EXPERIMENTS

Neyman (in the translated 1990 version) begins with a description of a field experiment with m plots on which v varieties might be applied. Neyman introduces what he calls "potential yield" U_{ik}, where i indexes the variety, $i = 1, \ldots, v$, and k indexes the plot, $k = 1, \ldots, m$. The potential yields are not equal to the actual or observed yield because i indexes all varieties and k indexes all plots, and each plot is exposed to only one variety. Throughout, the collection of potential outcomes, $\mathbf{U} = \{U_{ik} : i = 1, \ldots, v;\ k = 1, \ldots, m\}$ is considered *a priori* fixed but unknown. The "best estimate" (Neyman's term) of the yield of the ith variety in the field is the average potential outcomes for that variety over all m plots,

$$a_i = \frac{1}{m} \sum_{k=1}^{m} U_{ik}.$$

Neyman calls a_i the "best estimate" because of his concern with the definition of "true yield," something that he struggled with again in Neyman (1935). As we define potential outcomes, they are the "true" values under SUTVA, not estimates of them.

Neyman then goes on to describe an urn model for determining which variety each plot receives; this model is stochastically identical to the completely randomized experiment with $n = m/v$ plots exposed to each variety. He notes the lack of independence between assignments for different plots implied by this restricted sampling of treatments without replacement (i.e., if plot k receives variety i, then plot l is less likely to receive variety i), and he goes on to note that certain formulas for this situation that have been justified on the basis of independence (i.e., treating the U_{ik} as independent normal random variables given some parameters) need more careful consideration.

Now, still using Neyman's notation, let x_i be the sample average of the n plots actually exposed to the ith variety, as opposed to a_i, the average of the potential outcomes over all m plots. Neyman shows that the expectation of $x_i - x_j$, that is, the average value of $x_i - x_j$ over all assignments that are possible under his urn drawings, is $a_i - a_j$. Thus, the standard estimate of the effect of variety i versus variety j, the difference in observed means, $x_i - x_j$, is unbiased (over repeated randomizations on the m plots) for the causal estimand, $a_i - a_j$, the average effect of variety i versus variety j across all m plots.

Neyman's formalism made three contributions: (*i*) explicit notation for potential outcomes, (*ii*) implicit consideration of something like the stability assumption, and (*iii*) implicit consideration of a model for the assignment of treatments to units that corresponds to the completely randomized experiment. But as Speed (1990, p. 464) writes in his introduction to the translation of Neyman (1923): "Implicit is not explicit; randomization as a physical act, and later as a basis for analysis, was yet to be introduced by Fisher." Nevertheless, the explicit provision of mathematical notation for potential outcomes was a great advance, and after Fisher's introduction of randomized experiments in 1925, Neyman's notation quickly became standard for defining average causal effects in randomized experiments. See, for example, Pitman (1937), Welch (1937), McCarthy (1939), Anscombe (1948), Kempthorne (1952, 1955), Brillinger, Jones, and Tukey (1978), Hedges and Lehman (1970, sec. 9.4), and dozens of other places, often assuming additivity as in Cox (1956, 1958), and even in introductory texts (Freedman, Pisani, and Purves, 1978, pp. 456–458). Neyman himself, in hindsight, felt that the mathematical model was an advance:

> Neyman has always depreciated the statistical works which he produced in Bydogszcz [which is where Neyman (1923) was done], saying that if there is any merit in them, it is not in the few formulas giving various mathematical expectations but in the construction of a probabilistic model of agricultural trials which, at that time, was a novelty. (Reid, 1982, p. 45)

2.4   EARLIER HINTS FOR PHYSICAL RANDOMIZING

The notion of the central role of randomization, even if not actual randomized experiments, seems to have been "in the air" in the 1920s before it was explicitly introduced by Fisher. For example, "Student" (Gossett, 1923, pp. 281–282) writes: "If now the plots had been randomly placed . . . ," and Fisher and MacKenzie (1923, p. 473) write "Furthermore, if all the plots were undifferentiated, as if the numbers had been mixed up and written down in random order" (see Rubin, 1990, p. 477). Somewhat remarkably, however, an American psychologist and philosopher, Charles Sanders Peirce, appears to have proposed physical randomization decades earlier, although not as a basis for inference, as in Fisher (1925). Specifically, Peirce and Jastrow (1885, reprinted in Stigler, 1980, pp. 75–83) used physical randomization to create sequences of binary treatment conditions (heavier versus lighter weights) in a repeated-measures psychological experiment. The purpose of the randomization was to create sequences such that "any possible psychological guessing of what changes the operator [experimenter] was likely to select was avoided" (Stigler, pp. 79–80).[1] Peirce also appears to have anticipated, in the late nineteenth century, Neyman's concept of unbiased estimation when using simple random samples and appears to have even thought of randomization as a physical process to be implemented in practice (Peirce, 1931).[2] But we can find no suggestion for the physical randomizing of treatments to units as a basis for inference under Fisher (1925).

2.5   FISHER'S (1925) PROPOSAL TO RANDOMIZE TREATMENTS TO UNITS

An interesting aspect of Neyman's analysis was that, as just mentioned, although he developed his notation to treat data as if they arose from what was later called a completely randomly assigned experiment, he did not take the further step of proposing the necessity of physical randomization for credibly assessing causal effects. It was instead Ronald Fisher, in 1925, who first grasped this. Although the distinction may seem trivial in hindsight, Neyman did not see it as such:

---

[1] Thanks to Stephen Stigler for noting this, possibly first, use of randomization in formal experiments, in correspondence with the second author.

[2] Thanks to Keith O'Rourke and Stephen Stigler for pointing this out.

> On one occasion, when someone perceived him as anticipating the English statistician
> R. A. Fisher in the use of randomization, he objected strenuously:
> "I treated *theoretically* an unrestrictedly randomized agricultural experiment and the
> randomization was considered a prerequisite to probabilistic treatment of the results. This
> is not the same as the recognition that without randomization an experiment has little
> value irrespective of the subsequent treatment. The latter point is due to Fisher, and I
> consider it as one of the most valuable of Fisher's achievements" (Reid, 1982, p. 45)

Also,

> Owing to the work of R. A. Fisher, "Student" and their followers, it is hardly possible to
> add anything essential to the present knowledge concerning local experiments . . . . One of
> the most important achievements of the English School is their method of planning field
> experiments known as the method of Randomized Blocks and Latin Squares. (Neyman,
> 1935, p. 109)

Thus, independent of Neyman's work, Fisher (1925) proposed the physical randomization of units and furthermore developed a distinct method of inference for this special
class of assignment mechanisms, that is, randomized experiments. The random assignments can be made, for instance, by choosing balls from an urn, as described by Neyman
(1923). Fisher's "significance levels" (i.e., p-values), in the current text introduced and
discussed in Chapter 5, remain the accepted rigorous standard for the analysis of randomized clinical trials at the start of the twenty-first century and validate so-called
intent-to-treat analyses, as discussed in Chapters 5 and 23.

2.6   THE OBSERVED OUTCOME NOTATION IN OBSERVATIONAL
STUDIES FOR CAUSAL EFFECTS

Despite the almost immediate acceptance of randomized experiments, Fisher's p-values,
and Neyman's notation for potential outcomes in agricultural work and mathematical
statistics by 1930 within such experiments, these same elements were not used for
causal inference in observational studies. Among social scientists, who were using
almost exclusively observational data, the work on randomized experiments by Fisher,
Neyman, and others, received little or no attention, and researchers continued building
models for observed outcomes rather than thinking in terms of potential outcomes. Even
among statisticians involved in the analysis of both randomized and non-randomized
data for causal effects, the ideas and mathematical language used for causal inference
in the setting of randomized experiments were completely excluded from causal inference in the non-randomized settings. The approach in the latter continued to involve
building statistical models relating the observed value of the outcome variable to covariates and indicator variables for treatment levels, with the causal effects defined in
terms of the parameters of these models, a tradition that appears to originate with
Yule (1897).

This approach estimated associations, for example, correlations, between observed
variables, and then attempted, using various external arguments about temporal ordering of the variables, to infer causation, that is, to assess which of these associations
might be reflecting a causal mechanism. In particular, the pair of the potential outcomes

$(Y_i(1), Y_i(0))$, which in our approach is fundamental for defining causal effects, was replaced by the observed value of Y for unit i, introduced in Section 1.7.

$$Y_i^{\text{obs}} = Y_i(W_i) = W_i \cdot Y_i(1) + (1 - W_i) \cdot Y_i(0) = \begin{cases} Y_i(0) & \text{if } W_i = 0, \\ Y_i(1) & \text{if } W_i = 1. \end{cases}$$

The observed outcome Y_i^{obs} was then typically regressed, using ordinary least squares methods, as in Yule (1897), on covariates X_i and the indicator for treatment exposure, W_i. The regression coefficient of W_i in this regression was then interpreted as estimating the causal effect of $W_i = 1$ versus $W_i = 0$. Somewhat remarkably, under very specific conditions, this approach works as outlined in Chapter 7. But in broad generality it does not. This tradition dominated economics, sociology, psychology, education, and other social sciences, as well as the biomedical sciences, such as epidemiology, for most of a century.

In fact, for the half century following Neyman (1923), statisticians who wrote with great clarity and insight on randomized experiments using the potential outcomes notation did not use it when discussing non-randomized studies for causal effects. For example, contrast the discussion in Cochran and Cox (1956) on experiments with that in Cochran (1965) on observational studies, and the discussion in Cox (1958) on randomized experiments with that in Cox and McCullagh (1982) on Lord's paradox (which we discussed using the potential outcome framework in Chapter 1).

2.7   EARLY USES OF POTENTIAL OUTCOMES IN OBSERVATIONAL STUDIES IN SOCIAL SCIENCES

Although the potential outcome notation did not find widespread adoption in observational studies until recently, in some specific settings researchers used frameworks for causal inference that are similar. One of the most interesting examples is the use of potential outcomes in the analysis of demand and supply functions specifically, and the analysis of simultaneous equations models in economics in general. In the 1930s and 1940s, economists Tinbergen (1930) and Haavelmo (1944) formulated causal questions in such settings in terms that now appear very modern. Tinbergen writes:

> Let π be any imaginable price; and call total demand at this price $n(\pi)$, and total supply $a(\pi)$. Then the actual price p is determined by the equation $a(p) = n(p)$, so that the actual quantity demanded, or supplied, obeys the condition $u = a(p) = n(p)$, where u is this actual quantity. ... The problem of determining demand and supply curves ... may generally be put as follows: Given p and u as functions of time, what are the functions $n(\pi)$ and $a(\pi)$? (Tinbergen, 1930, translated in Hendry and Morgan, 1994, p. 233)

This quotation clearly describes the potential outcomes and the specific assignment mechanism corresponding to market clearing, closely following the treatment of such questions in economic theory. Note the clear distinction in notation between the price as an argument in the demand-and-supply function ("any imaginable price π") and the actual price p.

Similarly, Haavelmo (1934) writes:

> If the group of all consumers in society were repeatedly furnished with the total income, or purchasing power r per year, they would, on average or "normally" spend a total amount $\bar{u}$ for consumption per year, equal to $\bar{u} = \alpha r + \beta$. (Haavelmo, 1943, p. 3, reprinted in Hendry and Morgan, 1994, p. 456)

Although more ambiguous than the Tinbergen quote, this certainly suggests that Haavelmo viewed laws or structural equations in terms of potential outcomes that could have been observed by arranging an experiment.

There are two interesting aspects of the Haavelmo work and the link with potential outcomes. First, it appears that Haavelmo was directly influenced by Neyman (see Hendry and Morgan, 1994, p. 67) and in fact studied with him for a couple of months at Berkeley: "I then had the privilege of studying with the world famous statistician Jerzey Neyman for a couple of months in California. . . . When I met him for that second talk I had lost most of my illusions regarding my understanding of how to do econometrics" (Haavelmo, 1989). Second, the close connection between the Tinbergen and Haavelmo work and potential outcomes disappeared in later work. In the work by Koopmans and others associated with the Cowles Commission (e.g., the papers in Koopmans, 1950, and Hood and Koopmans, 1953), statistical models are formulated for observed outcomes in terms of observed explanatory variables. No distinction is made between variables that Cox describes as "treatments . . . potentially causal" and "intrinsic properties of the [units] under study" (Cox, 1992, p. 296) that are characteristics or attributes of the units. This observed outcome framework for analyzing causal questions dominated economics and other social sciences and continues to dominate the textbooks in econometrics, with few exceptions, until very recently.

2.8 POTENTIAL OUTCOMES AND THE ASSIGNMENT MECHANISM IN OBSERVATIONAL STUDIES: RUBIN (1974)

Rubin (1974, 1975, 1978) makes two key contributions. First, Rubin (1974) puts the potential outcomes center stage in the analysis of causal effects, irrespective of whether the study is an experimental one or an observational one. Second, he discusses the assignment mechanism in terms of the potential outcomes.

Rubin starts by *defining* the causal effect at the unit level in terms of the pair of potential outcomes:

> . . . define the causal effect of the E versus C treatment on Y for a particular trial (i.e., a particular unit . . .) as follows: Let $y(E)$ be the value of Y measured at t_2 on the unit, given that the unit received the experimental Treatment E initiated at t_1; Let $y(C)$ be the value of Y measured at t_2 on the unit given that the unit received the control Treatment C initiated at t_1. Then $y(E) - y(C)$ is the causal effect of the E versus C treatment on Y . . . for that particular unit. (Rubin, 1974, p. 639)

This definition fits perfectly with Neyman's framework for analyzing randomized experiments but shows that the definition has nothing to do with the assignment mechanism: it applies equally to observational studies as well as to randomized experiments.

Rubin (1975, 1978) then discusses the benefits of randomization in terms of eliminating systematic differences between treated and control units and formulates the assignment mechanism in general mathematical terms as possibly depending on the potential outcomes. Our formal consideration of the assignment mechanism begins in Chapter 3.

NOTES

When one of us (Rubin) was visiting the Department of Statistics at Berkeley in the mid-1970s, where Neyman was Professor Emeritus, he asked Neyman why no one ever used the potential outcomes notation from randomized experiments to define causal effects more generally. This meeting was fifteen years before the (re-)publication of Neyman (1923, 1990). Somewhat remarkably in hindsight, at this meeting, Neyman never mentioned that he invented the notation; his reply to the question as to why it was not used outside experiments was to the effect that defining causal effects in non-randomized settings was too speculative, and in such settings, statisticians should stick with statements concerning descriptions and associations (see Rubin, 2010, p. 42). This fits in with the Neyman quote given in Section 2.5: "without randomization, an experiment has little value irrespective of the subsequent treatment" (Reid, 1982, p. 45). The term "assignment mechanism," and its formal definition, including possible dependence on the potential outcomes, was introduced in Rubin (1975).

For discussions on the intention-to-treat principle, see Davies (1954), Fisher et al. (1990), Meier (1992), Cook and DeMets (2008), Wu and Hamada (2009), Altman (1991), Sheiner and Rubin (1995), and Lui (2011).

## CHAPTER 3

A Classification of Assignment Mechanisms

3.1 INTRODUCTION

As discussed in Chapter 1, the fundamental problem of causal inference is the presence of missing data – for each unit we can observe at most one of the potential outcomes. A key component in a causal analysis is, therefore, what we call the *assignment mechanism*: the process that determines which units receive which treatments, hence which potential outcomes are realized and thus can be observed, and, conversely, which potential outcomes are missing. In this chapter we introduce a taxonomy of assignment mechanisms that will serve as the organizing principle for this text. Formally, the assignment mechanism describes, as a function of all covariates and of all potential outcomes, the probability of any vector of assignments. We consider three basic restrictions on assignment mechanisms:

1. *Individualistic assignment*: This limits the dependence of a particular unit's assignment probability on the values of covariates and potential outcomes for other units.
2. *Probabilistic assignment*: This requires the assignment mechanism to imply a non-zero probability for each treatment value, for every unit.
3. *Unconfounded assignment*: This disallows dependence of the assignment mechanism on the potential outcomes.

Following Cochran (1965), we also make a distinction between experiments, where the assignment mechanism is both known and controlled by the researcher, and observational studies, where the assignment mechanism is not known to, or not under the control of, the researcher.

We consider three classes of assignment mechanisms, covered in Parts II, III, IV, V, and VI of this book. The first class, studied in Part II, corresponds to what we call *classical randomized experiments*. Here the assignment mechanism satisfies all three restrictions on the assignment process, and, moreover, the researcher knows and controls the functional form of the assignment mechanism. Such designs are well understood, and in such settings causal effects are often relatively straightforward to estimate, and, moreover, it is often possible to do finite sample inference.

We refer to the second class of assignment mechanisms, studied in Parts III and IV of this text, as *regular assignment mechanisms*. This class comprises assignment mechanisms that, like classical randomized experiments, are individualistic, probabilistic, and unconfounded, but, in contrast to classical randomized experiments, the assignment mechanism need not be under the control of, or known by, the researcher. When the assignment mechanism is not under the control of the researcher, the restrictions on the assignment mechanism that make it regular are now usually assumptions, and they are typically not satisfied by design, as they are in classical randomized experiments. In general, we will not be sure whether these assumptions hold in any specific application, and in later chapters we will discuss methods for assessing their plausibility, as well as investigating the sensitivity to violations of them.

In practice, the regular observational study is a setting of great importance. It has been studied extensively from a theoretical perspective and is widely used in empirical work. Many, but not all, of the methods applicable to randomized experiments can be used, but often modifications to the specific methods are critical to enhance the credibility of the results. The simple methods that suffice in the context of randomized experiments tend to be more controversial when applied with regular assignment mechanisms. The concerns these simple methods raise are particularly serious if the covariate distributions under the various treatment regimes are substantially different, or *unbalanced* in our terminology. In that case, it can be very important, for the purpose of making credible causal inferences, to have an initial, what we call *design* stage of the study. In this design stage, the data on covariate values and treatment assignment (but, importantly, not the final outcome data) are analyzed in order to assemble samples with improved balance in covariate distributions, somewhat in parallel with the design stage of randomized experiments. Often in this setting, the number of pre-treatment variables is substantial, typically because, conditional on a large number of pre-treatment variables, unconfoundedness is more plausible. Although this creates no conceptual problems, it makes the practical problem of drawing credible causal inferences more challenging.

In Part V of the book we discuss methods for assessing the plausibility of the unconfoundedness assumption, and sensitivity analyses for assessing the implications of violations of it. In Part VI we analyze a number of assignment mechanisms where the assignment itself is regular, but the treatment received is not equal to the treatment assigned for all units. Thus, although the treatment assigned *is* unconfounded, the treatment received *is not* unconfounded, because the probability of receiving the active versus control treatment depends on potential outcomes. Such settings have arisen in the econometric literature to account for settings where individuals choose the treatment regime, at least partly based on expected benefits associated with the two treatment regimes. Although, as a general matter, such optimizing behavior is not inconsistent with regular assignment mechanisms, in some cases it suggests assignment mechanisms associated with so-called *instrumental variable* methods.

The rest of this chapter is organized as follows. In the next section we introduce additional notation. In Section 3.3 we define the assignment mechanism, unit-level assignment probabilities, and the propensity score. In Section 3.4 we formally introduce the three general restrictions we consider imposing on assignment mechanisms. We then use those restrictions to define classical randomized experiments in Section 3.6. In Section 3.7 we define regular assignment mechanisms as a special class of observational studies. The next section, Section 3.8, discusses some non-regular assignment mechanisms. Section 3.9 concludes.

3.2 NOTATION

Continuing the potential outcomes discussion in Chapter 1, let us consider a population of N units, indexed by $i = 1, \ldots, N$. The ith unit in this population is characterized by a K-component row vector of covariates (also referred to as pre-treatment variables or attributes), X_i, with $\mathbf{X}$ the $N \times K$ matrix of covariates in the population with i^{th} row equal to X_i. In social science applications, the elements of X_i may include an individual's age, education, socio-economic status, labor market history, pre-test scores, sex, and marital status. In biomedical applications, the covariates may also include measures of an individual's medical history, and family background information. Most important is that covariates are known *a priori* to be unaffected by the assignment of treatment.

For each unit there is also a pair of potential outcomes, $Y_i(0)$ and $Y_i(1)$, denoting its outcome values under the two values of the treatment: $Y_i(0)$ denotes the outcome under the control treatment, and $Y_i(1)$ denotes the outcome under the active treatment. Notice that when using this notation, we tacitly accept the Stable Unit Treatment Value Assumption (SUTVA) that treatment assignments for other units do not affect the outcomes for unit i, and that each treatment defines a unique outcome for each unit. The latter requirement implies that there is only a single version of the active and control treatments for each unit. Let $\mathbf{Y}(0)$ and $\mathbf{Y}(1)$ denote the N-component vectors (or the N-vectors for short) of the potential outcomes. More generally, the potential outcomes could themselves be multi-component row vectors, in which case $\mathbf{Y}(0)$ and $\mathbf{Y}(1)$ would be matrices with the ith rows equal to $Y_i(0)$ and $Y_i(1)$, respectively. Here, we largely focus on the situation where the potential outcomes are scalars, although in most cases extensions to vector-valued outcomes are conceptually straightforward.

Next, the N-component column vector of treatment assignments is denoted by $\mathbf{W}$, with i^{th} element $W_i \in \{0, 1\}$, with $W_i = 0$ if unit i received the control treatment, and $W_i = 1$ if this unit received the active treatment. Let $N_c = \sum_{i=1}^{N} (1 - W_i)$ and $N_t = \sum_{i=1}^{N} W_i$ be the number of units assigned to the control and active treatment respectively, with $N_c + N_t = N$.

In Chapter 1 we defined the realized and possibly observed outcomes

$$Y_i^{\text{obs}} = Y_i(W_i) = \begin{cases} Y_i(0) & \text{if } W_i = 0, \\ Y_i(1) & \text{if } W_i = 1, \end{cases} \tag{3.1}$$

and the missing outcomes:

$$Y_i^{\text{mis}} = Y_i(1 - W_i) = \begin{cases} Y_i(1) & \text{if } W_i = 0, \\ Y_i(0) & \text{if } W_i = 1. \end{cases} \tag{3.2}$$

$\mathbf{Y}^{\text{obs}}$ and $\mathbf{Y}^{\text{mis}}$ are the corresponding N-vectors (or matrices in the case with multiple outcomes). We can invert these relations and characterize the potential outcomes in terms of the observed and missing outcomes:

$$Y_i(0) = \begin{cases} Y_i^{\text{mis}} & \text{if } W_i = 1, \\ Y_i^{\text{obs}} & \text{if } W_i = 0, \end{cases} \quad \text{and} \quad Y_i(1) = \begin{cases} Y_i^{\text{mis}} & \text{if } W_i = 0, \\ Y_i^{\text{obs}} & \text{if } W_i = 1. \end{cases} \tag{3.3}$$

This characterization illustrates that the causal inference problem is fundamentally a missing data problem: if we impute the missing outcomes, we "know" all the potential outcomes and thus the value of any causal estimand in the population of N units.

3.3 ASSIGNMENT PROBABILITIES

To introduce the taxonomy of assignment mechanisms used in this text requires some formal mathematical terms. First, we define the assignment mechanism to be the function that assigns probabilities to all 2^N possible values for the N-vector of assignments $\mathbf{W}$ (each unit can be assigned to treatment or control), given the N-vectors of potential outcomes $\mathbf{Y}(0)$ and $\mathbf{Y}(1)$, and given the $N \times K$ matrix of covariates $\mathbf{X}$:

Definition 3.1 (Assignment Mechanism)
Given a population of N units, the assignment mechanism is a row-exchangeable function $\Pr(\mathbf{W}|\mathbf{X}, \mathbf{Y}(0), \mathbf{Y}(1))$, *taking on values in* $[0, 1]$, *satisfying*

$$\sum_{\mathbf{W} \in \{0,1\}^N} \Pr(\mathbf{W}|\mathbf{X}, \mathbf{Y}(0), \mathbf{Y}(1)) = 1,$$

for all $\mathbf{X}$, $\mathbf{Y}(0)$, *and* $\mathbf{Y}(1)$.

The set $\mathbb{W} = \{0, 1\}^N$ is the set of all N-vectors with all elements equal to 0 or 1. By the assumption that the function $\Pr(\,\cdot\,)$ is row exchangeable, we mean that the order in which we list the N units within the vectors or matrices is irrelevant. Note that this probability $\Pr(\mathbf{W}|\mathbf{X}, \mathbf{Y}(0), \mathbf{Y}(1))$ is *not* the probability of a particular unit receiving the treatment. Instead, it is the probability that a particular value for the full assignment – first two units treated, third a control, fourth treated, etc. – will occur. The definition requires that the probabilities across the full set of 2^N possible assignment vectors $\mathbf{W}$ sum to one. Note also that some assignment vectors $\mathbf{W}$ may have zero probability. For example, if we were to design a study to evaluate a new drug, it is likely that we would want to rule out the possibility that all subjects received the control drug. We could do so by assigning zero probability to the vector of assignments $\mathbf{W}$ with $W_i = 0$ for all i, or perhaps even assign zero probability to all vectors of assignments other than those with $\sum_{i=1}^{N} W_i = N/2$, for even values of the population size N.

In addition to the probability of joint assignment for the entire population, we are often interested in the probability of an individual unit being assigned to the active treatment:

Definition 3.2 (Unit Assignment Probability)
The unit-level assignment probability for unit i is

$$p_i(\mathbf{X}, \mathbf{Y}(0), \mathbf{Y}(1)) = \sum_{\mathbf{W}:W_i=1} \Pr(\mathbf{W}|\mathbf{X}, \mathbf{Y}(0), \mathbf{Y}(1)).$$

Here we sum the probabilities across all possible assignment vectors $\mathbf{W}$ for which $W_i = 1$. Out of the set of 2^N different assignment vectors, half (that is 2^{N-1}) have the property that $W_i = 1$. The probability that unit i is assigned to the control treatment is $1 - p_i(\mathbf{X}, \mathbf{Y}(0), \mathbf{Y}(1))$. Note that according to this definition, the probability that unit i

receives the treatment can be a function of its own covariates X_i and potential outcomes $Y_i(0)$ and $Y_i(1)$, and it generally is also a function of the covariate values, and potential outcomes, and treatment assignments of the other units in the population.

We are also often interested in the average of the unit-level assignment probabilities for subpopulations with a common value of the covariates, for example, $X_i = x$. We label this function the *propensity score* at x. In the finite population case the definition of the propensity score follows.

Definition 3.3 (Finite Population Propensity Score)
The propensity score at x is the average unit assignment probability for units with $X_i = x$,

$$e(x) = \frac{1}{N(x)} \sum_{i:X_i=x} p_i(\mathbf{X}, \mathbf{Y}(0), \mathbf{Y}(1))$$

where $N(x) = \sum_{i=1}^{N} \mathbf{1}_{X_i=x}$ is the number of units with $X_i = x$. For values x with $N(x) = 0$, the propensity score is defined to be zero.

To illustrate these definitions more concretely, consider four examples, the first three with two units, and the last one with three units.

EXAMPLE 1 Suppose we have two units. Then there are four (2^2) possible values for $\mathbf{W}$,

$$\mathbf{W} \in \left\{ \begin{pmatrix} 0 \\ 0 \end{pmatrix}, \begin{pmatrix} 0 \\ 1 \end{pmatrix}, \begin{pmatrix} 1 \\ 0 \end{pmatrix}, \begin{pmatrix} 1 \\ 1 \end{pmatrix} \right\}.$$

We conduct a randomized experiment where all treatment assignments have equal probability. Then the assignment mechanism is equal to

$$\Pr(\mathbf{W}|\mathbf{X}, \mathbf{Y}(0), \mathbf{Y}(1)) = 1/4, \quad \text{for } \mathbf{W} \in \left\{ \begin{pmatrix} 0 \\ 0 \end{pmatrix}, \begin{pmatrix} 0 \\ 1 \end{pmatrix}, \begin{pmatrix} 1 \\ 0 \end{pmatrix}, \begin{pmatrix} 1 \\ 1 \end{pmatrix} \right\}. \tag{3.4}$$

In this case the unit assignment probability $p_i(\mathbf{X}, \mathbf{Y}(0), \mathbf{Y}(1))$ is equal to $1/2$ for both units $i = 1, 2$. In a randomized experiment with no covariates, the propensity score is equal to the unit assignment probabilities, here all equal to $1/2$.

EXAMPLE 2 We conduct a randomized experiment with two units where only those assignments with exactly one treated and one control unit are allowed. Then the assignment mechanism is

$$\Pr(\mathbf{W}|\mathbf{X}, \mathbf{Y}(0), \mathbf{Y}(1)) = \begin{cases} 1/2 & \text{if } \mathbf{W} \in \left\{ \begin{pmatrix} 0 \\ 1 \end{pmatrix}, \begin{pmatrix} 1 \\ 0 \end{pmatrix} \right\}, \\ 0 & \text{if } \mathbf{W} \in \left\{ \begin{pmatrix} 0 \\ 0 \end{pmatrix}, \begin{pmatrix} 1 \\ 1 \end{pmatrix} \right\}. \end{cases} \tag{3.5}$$

This does not change the unit-level assignment probabilities, which remains equal to $1/2$ for both units, and so does the propensity score.

EXAMPLE 3 A third, more complicated, assignment mechanism with two units is the following. The unit with more to gain from the active treatment (using a coin toss in the case of a tie) is assigned to the treatment group, and the other to the control group. This leads to

$$\Pr(\mathbf{W}|\mathbf{X}, \mathbf{Y}(0), \mathbf{Y}(1)) = \begin{cases} 1 & \text{if } Y_2(1) - Y_2(0) > Y_1(1) - Y_1(0) \text{ and } \mathbf{W} = \begin{pmatrix} 0 \\ 1 \end{pmatrix}, \\[2ex] 1 & \text{if } Y_2(1) - Y_2(0) < Y_1(1) - Y_1(0) \text{ and } \mathbf{W} = \begin{pmatrix} 1 \\ 0 \end{pmatrix}, \\[2ex] 1/2 & \text{if } Y_2(1) - Y_2(0) = Y_1(1) - Y_1(0) \text{ and } \mathbf{W} \in \left\{ \begin{pmatrix} 0 \\ 1 \end{pmatrix}, \begin{pmatrix} 1 \\ 0 \end{pmatrix} \right\}, \\[2ex] 0 & \text{if } \mathbf{W} \in \left\{ \begin{pmatrix} 0 \\ 0 \end{pmatrix}, \begin{pmatrix} 1 \\ 1 \end{pmatrix} \right\}, \\[2ex] 0 & \text{if } Y_2(1) - Y_2(0) < Y_1(1) - Y_1(0) \text{ and } \mathbf{W} = \begin{pmatrix} 0 \\ 1 \end{pmatrix}, \\[2ex] 0 & \text{if } Y_2(1) - Y_2(0) > Y_1(1) - Y_1(0) \text{ and } \mathbf{W} = \begin{pmatrix} 1 \\ 0 \end{pmatrix}. \end{cases}$$

$$(3.6)$$

In this example the unit-level treatment probabilities $p_i(\mathbf{X}, \mathbf{Y}(0), \mathbf{Y}(1))$ are equal to zero, one, or a half, depending whether the gain for unit i is smaller or larger than for the other unit, or equal. Given that there are no covariates, the propensity score remains a constant, equal to $1/2$ in this case. This is a type of assignment mechanism that we often rule out when attempting to infer causal effects.   □

EXAMPLE 4 A sequential randomized experiment allows for dependence of the assignment mechanism on the potential outcomes, thus violating some of the assumptions we consider later. In this example, there are three units, and thus eight possible values for $\mathbf{W}$:

$$\mathbf{W} \in \left\{ \begin{pmatrix} 0 \\ 0 \\ 0 \end{pmatrix}, \begin{pmatrix} 0 \\ 0 \\ 1 \end{pmatrix}, \begin{pmatrix} 0 \\ 1 \\ 0 \end{pmatrix}, \begin{pmatrix} 0 \\ 1 \\ 1 \end{pmatrix}, \begin{pmatrix} 1 \\ 0 \\ 0 \end{pmatrix}, \begin{pmatrix} 1 \\ 0 \\ 1 \end{pmatrix}, \begin{pmatrix} 1 \\ 1 \\ 0 \end{pmatrix}, \begin{pmatrix} 1 \\ 1 \\ 1 \end{pmatrix} \right\}.$$

Suppose there is a covariate X_i measuring the order in which the units entered the experiment, $X_i \in \{1, 2, 3\}$. Without loss of generality, let us assume that $X_i = i$. For the first unit, with $X_i = 1$, a fair coin toss determines the treatment. The second unit, with $X_i = 2$, is assigned to the alternative treatment. Let the observed outcomes for the first and second unit be Y_1^{obs} and Y_2^{obs}. The third unit, with $X_i = 3$, is assigned to the active or control treatment that appears better, based on a comparison of observed outcomes by treatment status for the first two units. If both treatments appear equally beneficial, the third unit is assigned to the active treatment. For example, if $W_1 = 0$, $W_2 = 1$, and $Y_1^{\text{obs}} > Y_2^{\text{obs}}$, then the third unit gets assigned to the control group; if $W_1 = 0$, $W_2 = 1$, and $Y_1^{\text{obs}} \leq Y_2^{\text{obs}}$, the third units gets assigned to the treatment group; and similarly given the alternative assignments for the first two units. Formally:

$$\Pr(\mathbf{W}|\mathbf{X}, \mathbf{Y}(0), \mathbf{Y}(1), \mathbf{X}) = \begin{cases} 1/2 & \text{if } Y_1(0) > Y_2(1), \text{ and } \mathbf{W} = \begin{pmatrix} 0 \\ 1 \\ 0 \end{pmatrix}, \\ \\ 1/2 & \text{if } Y_1(1) \geq Y_2(0), \text{ and } \mathbf{W} = \begin{pmatrix} 1 \\ 0 \\ 1 \end{pmatrix}, \\ \\ 1/2 & \text{if } Y_1(0) \leq Y_2(1), \text{ and } \mathbf{W} = \begin{pmatrix} 0 \\ 1 \\ 1 \end{pmatrix}, \\ \\ 1/2 & \text{if } Y_1(1) < Y_2(0), \text{ and } \mathbf{W} = \begin{pmatrix} 1 \\ 0 \\ 0 \end{pmatrix}. \end{cases} \qquad (3.7)$$

In this case the unit assignment probability is equal to $1/2$ for the first two units,

$$p_2(\mathbf{X}, \mathbf{Y}(0), \mathbf{Y}(1)) = p_2(\mathbf{X}, \mathbf{Y}(0), \mathbf{Y}(1)) = 1/2,$$

and, for unit 3, equal to

$$p_3(\mathbf{X}, \mathbf{Y}(0), \mathbf{Y}(1)) = \begin{cases} 0 & \text{if } Y_1(0) > Y_2(1) \text{ and } Y_1(1) < Y_2(0), \\ 1 & \text{if } Y_1(1) \geq Y_2(0) \text{ and } Y_1(0) \leq Y_2(1), \\ 1/2 & \text{otherwise.} \end{cases}$$

Because the covariates identify the unit, the propensity score is equal to the unit assignment probabilities. Thus, for $x = 1$ and $x = 2$ the propensity score is equal to $1/2$. If $x = 3$, the propensity score is equal to $p_3(\mathbf{X}, \mathbf{Y}(0), \mathbf{Y}(1))$. □

3.4 RESTRICTIONS ON THE ASSIGNMENT MECHANISM

Before classifying the various types of assignment mechanisms that are the basis of the organization of this text, we present three general properties that assignment mechanisms may satisfy. These properties restrict the dependence of the unit-level assignment probabilities on values of covariates and potential outcomes for other units, or restrict the range of values of the unit-level assignment probabilities, or restrict the dependence of the assignment mechanism on potential outcomes.

The first property we consider is *individualistic assignment*, which limits the dependence of the treatment assignment for unit i on the outcomes and assignments for other units:

Definition 3.4 (Individualistic Assignment)
An assignment mechanism $\Pr(\mathbf{W}|\mathbf{X}, \mathbf{Y}(0), \mathbf{Y}(1))$ *is individualistic if, for some function* $q(\,\cdot\,) \in [0, 1]$,

$$p_i(\mathbf{X}, \mathbf{Y}(0), \mathbf{Y}(1)) = q(X_i, Y_i(0), Y_i(1)), \quad \text{for all } i = 1, \dots, N,$$

and

$$\Pr(\mathbf{W}|\mathbf{X}, \mathbf{Y}(0), \mathbf{Y}(1)) = c \cdot \prod_{i=1}^{N} q(X_i, Y_i(0), Y_i(1))^{W_i} \, (1 - q(X_i, Y_i(0), Y_i(1)))^{1-W_i},$$

for $(\mathbf{W}, \mathbf{X}, \mathbf{Y}(0), \mathbf{Y}(1)) \in \mathbb{A}$, *for some set* $\mathbb{A}$, *and zero elsewhere (c is the constant that ensures that the probabilities sum to unity).*

Individualistic assignment is violated in sequential experiments such as Example 4. Given individualistic assignment, the propensity score simplifies to:

$$e(x) = \frac{1}{N(x)} \sum_{i:X_i=x} q(X_i, Y_i(0), Y_i(1)).$$

Next, we define *probabilistic assignment*, which requires every unit to have positive probability of being assigned to treatment level 0 and to treatment level 1:

Definition 3.5 (Probabilistic Assignment)
An assignment mechanism $\Pr(\mathbf{W}|\mathbf{X}, \mathbf{Y}(0), \mathbf{Y}(1))$ *is probabilistic if the probability of assignment to treatment for unit i is strictly between zero and one:*

$$0 < p_i(\mathbf{X}, \mathbf{Y}(0), \mathbf{Y}(1)) < 1, \quad \text{for each possible } \mathbf{X}, \mathbf{Y}(0), \mathbf{Y}(1),$$

for all $i = 1, \dots, N$.

Note that this merely requires that every unit has the possibility of being assigned to the active treatment and the possibility of being assigned to the control treatment.

The third property is a restriction on the dependence of the assignment mechanism on potential outcomes:

Definition 3.6 (Unconfounded Assignment)
An assignment mechanism is unconfounded if it does not depend on the potential outcomes:

$$\Pr(\mathbf{W}|\mathbf{X}, \mathbf{Y}(0), \mathbf{Y}(1)) = \Pr(\mathbf{W}|\mathbf{X}, \mathbf{Y}'(0), \mathbf{Y}'(1)),$$

for all $\mathbf{W}$, $\mathbf{X}$, $\mathbf{Y}(0)$, $\mathbf{Y}(1)$, $\mathbf{Y}'(0)$, *and* $\mathbf{Y}'(1)$.

If an assignment mechanism is unconfounded, we can drop the two potential outcomes as arguments and write the assignment mechanism as $\Pr(\mathbf{W}|\mathbf{X})$. The assignment mechanisms in Examples 1 and 2 are, but those in in Examples 3 and 4 are not, unconfounded.

The combination of unconfoundedness and individualistic assignment plays a very important role. In that case,

$$\text{Pr}(\mathbf{W}|\mathbf{X}, \mathbf{Y}(0), \mathbf{Y}(1)) = c \cdot \prod_{i=1}^{N} q(X_i)^{W_i} \cdot (1 - q(X_i))^{1-W_i}. \qquad (3.8)$$

so that

$$e(x) = q(x),$$

so that the assignment mechanism is the product of the propensity scores. Note that, under unconfoundedness, the propensity score is no longer just the average assignment probability for units with covariate value $X_i = x$; it can also be interpreted as the unit-level assignment probability for such units.

Given individualistic assignment, the combination of probabilistic and unconfounded assignment is referred to as *strongly ignorable treatment assignment* (Rosenbaum and Rubin, 1983a). More generally, *ignorable treatment assignment* refers to the weaker restriction where the assignment mechanism can be written in terms of $\mathbf{W}$, $\mathbf{X}$, and $\mathbf{Y}^{\text{obs}}$ only, without dependence on $\mathbf{Y}^{\text{mis}}$ (Rubin, 1978).

3.5   ASSIGNMENT MECHANISMS AND SUPER-POPULATIONS

In part of this text we view our sample of size N as a random sample from an infinite super-population. In that case we employ slightly different formulations of the restrictions on the assignment mechanism. Sampling from the super-population generates a joint sampling distribution on the quadruple of unit-level variables $(Y_i(0), Y_i(1), W_i, X_i)$, $i = 1, \ldots, N$. More explicitly, we assume the $(Y_i(0), Y_i(1), W_i, X_i)$ are independently and identically distributed draws from a distribution indexed by a global parameter. We write this in factored form as

$$f_{W|Y(0),Y(1),X}(W_i|Y_i(0), Y_i(1), X_i, \phi) f_{Y(0),Y(1)|X}(Y_i(0), Y_i(1)|X_i, \theta) f_X(X_i|\psi), \qquad (3.9)$$

where the parameters are in their respective parameter spaces, and the full parameter vector is (ϕ, θ, ψ), where each of these components is generally a function of the global parameter.

In this setting we define the propensity score as

Definition 3.7 (Super-Population Propensity Score)
The propensity score at x is the population average unit assignment probability for units with $X_i = x$,

$$e(x) = \mathbb{E}_{\text{SP}} \left[f_{W|Y(0),Y(1),X}(1|Y_i(0), Y_i(1), X_i, \phi) f_{Y(0),Y(1)|X}(Y_i(0), Y_i(1)|X_i, \theta) \,\middle|\, X_i = x \right],$$

for all x in the support of X_i; $e(x)$ is here a function of ϕ and θ, a dependence that we usually suppress in notation.

The "SP" subscript on the expectations operator indicates that the expectation is taken over the distribution generated by random sampling. In this case the expectation is taken over the potential outcomes $(Y_i(0), Y_i(1))$. By iterated expectations the propensity score in the super-population setting is also equal to $\Pr(W_i = 1|X_i = x, \phi, \theta)$ where the probability is taken both over the assignment mechanism and over the random sampling.

Note that with our definition of super-populations the assignment mechanism is automatically individualistic (of course, given (ϕ, θ)).

Definition 3.8 (Super-Population Probabilistic Assignment)

An assignment mechanism is super-population probabilistic if the probability of assignment to treatment for unit i is strictly between zero and one:

$$0 < f_{W|Y(0),Y(1),X}(1|Y_i(0), Y_i(1), X_i, \phi) < 1, \quad \text{for each possible } X_i, Y_i(0), Y_i(1).$$

Definition 3.9 (Super-Population Unconfounded Assignment)

An assignment mechanism is super-population unconfounded if it does not depend on the potential outcomes:

$$f_{W|Y(0),Y(1),X}(w|y_0, y_1, x, \phi) = f_{W|Y(0),Y(1),X}(w|y'_0, y'_1, x, \phi),$$

for all y_0, y_1, x, y'_0, y'_1, ϕ, and for $w = 0, 1$.

3.6   RANDOMIZED EXPERIMENTS

Part II of this text deals with the inferentially most straightforward class of assignment mechanisms, randomized assignment. Randomized experimental designs have traditionally been viewed as the most credible basis for causal inference, as reflected in the typical reliance of the U.S. Food and Drug Administration on such experiments in its approval process for pharmaceutical treatments.

Definition 3.10 (Randomized Experiment)

A randomized experiment is an assignment mechanism that

 (i) is probabilistic, and
(ii) has a known functional form that is controlled by the researcher.

In Part II of this text we will be concerned with a special case – what we call classical randomized experiments:

Definition 3.11 (Classical Randomized Experiment)

A classical randomized experiment is a randomized experiment with an assignment mechanism that is

 (i) individualistic, and
(ii) unconfounded.

The definition of a classical randomized experiment rules out sequential experiments as in Example 4. In sequential experiments, the assignment for units assigned in a later stage of the experiment generally depends on observed outcomes for units assigned earlier in the experiment.

A leading case of a classical randomized experiment is a *completely randomized experiment*, where, *a priori*, the number of treated units, N_t, is fixed (and thus the number of control units $N_c = N - N_t$ is fixed as well). In such a design, N_t units are randomly selected, from a population of N units, to receive the active treatment, with the remaining N_c assigned to the control group. In this case, each unit has unit assignment probability $q = N_t/N$, and the assignment mechanism equals

$$\Pr(\mathbf{W}|\mathbf{X}, \mathbf{Y}(0), \mathbf{Y}(1)) = \begin{cases} 1 \Big/ \dbinom{N}{N_t} & \text{if } \sum_{i=1}^{N} W_i = N_t, \\ 0 & \text{otherwise,} \end{cases}$$

where the number of distinct values of the assignment vector with N_t units out of N assigned to the active treatment is

$$\binom{N}{N_t} = \frac{N!}{N_t!(N - N_t)!}, \qquad \text{with } J! = J(J-1)\ldots 1.$$

Other prominent examples of classical randomized experiments include stratified randomized experiments and paired randomized experiments, discussed in Chapters 9 and 10.

3.7   OBSERVATIONAL STUDIES: REGULAR ASSIGNMENT MECHANISMS

In Parts III and IV of this text, we discuss cases where the exact assignment probabilities may be unknown to the researcher, but the researcher still has substantial information concerning the assignment mechanism. For instance, a leading case is where the researcher knows the set of variables that enters into the assignment mechanism but does not know the functional form of the dependence. Such information will generally come from subject-matter knowledge. For example, medical decisions in some situations are made solely using patients' medical records, but precisely how may be unknown. In general we refer to designs with unknown assignment mechanisms as *observational studies*:

Definition 3.12 (Observational Study)
An assignment mechanism corresponds to an observational study if the functional form of the assignment mechanism is unknown.

The special case of an assignment mechanism that is the focus of Part III of the book is a *regular assignment mechanism*:

Definition 3.13 (Regular Assignment Mechanism)
An assignment mechanism is regular if

  (i) the assignment mechanism is individualistic,
 (ii) the assignment mechanism is probabilistic, and
(iii) the assignment mechanism is unconfounded.

If, in addition, the functional form of a regular assignment mechanism is known, the assignment mechanism corresponds to a classical randomized experiment. If the functional form is not known, the assignment mechanism corresponds to an observational study with a regular assignment mechanism.

In Part III of this book we focus on the design stage of studies where the assumption of a regular assignment mechanism is viewed as plausible. In this design stage we focus on the data on treatment assignment and pre-treatment variables only, without seeing the outcome data. The concern at this stage is balance in the covariate distributions between treated and control groups. In completely and stratified randomized experiments, balance is guaranteed by design, but in observational studies this needs to be done by special analyses. We assess balance, and in cases where initially there is insufficient balance, we develop methods for improving balance.

In Part IV we discuss methods of analysis for causal inference with regular assignment mechanisms in some detail. Even if in many cases it may appear too strong to assume that an assignment mechanism is regular, we will argue that, in practice, it is a very important starting point for many studies. There are two main reasons for this. The first is that in many well-designed observational studies, researchers have attempted to record all the relevant covariates, that is, all the variables that may be associated with both outcomes and assignment to treatment. If they have been successful in this endeavor, or at least approximately so, a regular assignment mechanism may be a reasonable approximation to the true assignment mechanism. The second reason is that specific alternatives to regular assignment mechanisms are typically even less credible. Under a regular assignment mechanism, it will be sufficient to adjust appropriately for differences between treated and control units' covariate values to draw valid causal inferences. Any alternative method involves causal interpretations of comparisons of units with different treatments who also are observed to *differ* systematically in their values for covariates. It is relatively uncommon to find a convincing argument in support of such alternatives, although there are some notable exceptions, such as instrumental variables analyses discussed in Part VI of the book. More details of these arguments are presented in Chapter 12.

3.8   OBSERVATIONAL STUDIES: IRREGULAR ASSIGNMENT MECHANISMS

In Part VI of this book, we discuss another class of assignment mechanisms. We focus on settings where assignment to treatment may differ for some units from the receipt of treatment. We assume that assignment to treatment itself is unconfounded, but allow receipt of treatment to be confounded. This class of assignment mechanisms includes noncompliance in randomized experiments and sometimes utilizes *instrumental variables* analyses. Often in these designs, the receipt of treatment can be viewed as "latently regular" – that is, it would be regular given some additional covariates that are not fully observed. To conduct inference in such settings, it is often useful to invoke additional conditions, in particular *exclusion restrictions*, which rule out the presence of particular causal effects.

The remainder of this text provides more detailed discussion of methods of causal inference given each of these types of assignment mechanisms. In the next part of the book, Chapters 4–11, we start with classical randomized experiments.

3.9   CONCLUSION

This chapter presents the taxonomy of assignment mechanisms that serves as the organizing principle for this text. Using three restrictions on the assignment mechanism – individualistic assignment, probabilistic assignment, and unconfoundedness – we define regular assignment mechanisms and the special case of classical randomized experiments. In the next part of the book, we study classical randomized experiments, followed in Parts III and IV by the study of observational studies with regular assignment mechanisms. In Parts V and VI of the text we analyze some additional assignment mechanisms where receipt of treatment is confounded.

NOTES

Of the restrictions on assignment mechanisms we discuss in the current chapter, the first one, individualistic assignment, is often made implicitly, but the term is new. The notion of probabilistic assignment is often stated formally, although it is rarely given a formal label. The term unconfoundedness was coined by Rubin (1990a). It is sometimes referred to as the *conditional independence assumption* (Lechner, 2001; Angrist and Pischke, 2009). In the econometrics literature it is also closely related to the notion of *exogeneity* (Manski, Sandefur, McLanahan, and Powers, 1992), although formal definitions of exogeneity do not coincide with unconfoundedness (see Imbens, 2004, for some discussion). The combination of probabilistic assignment and unconfoundedness is referred to as *Strong Ignorability* or *Strongly Ignorable Treatment Assignment* by Rosenbaum and Rubin (1984). There is a close link between some of the assumptions used in the context of causal inference and the terminology in missing data problems. In the missing data literature, strong ignorability is closely linked with *Missing at Random* missingness mechanisms (Rubin, 1976c; Little and Rubin, 2002; Frumento, Mealli, Pacini, and Rubin, 2012).

Instrumental variables methods originate in the econometrics literature and go back to the 1920s and 1940s (P. Wright, 1928; S. Wright 1921, 1923; Tinbergen, 1928; Haavelmo, 1943). For a historical perspective, see Stock and Trebbi (2003) and Imbens (2014). For modern approaches see Imbens and Angrist (1994), and Angrist, Imbens, and Rubin (1996). For textbook discussions, see Wooldridge (2010) and Angrist and Pischke (2008).

Some methods for assignment mechanisms not covered in this edition of the book include *Principal Stratification, Regression Discontinuity Designs, Difference In Differences* methods, and case-control designs. The notion of *Principal Stratification* generalizes the binary-treatment version of instrumental variables. It was introduced by Frangakis and Rubin (2002). *Regression discontinuity designs* originate in the psychology literature (Thistlewaite and Campbell, 1960). See for a historical overview Cook (2008), and for recent surveys Imbens and Lemieux (2008) and Lee and Lemieux (2010). *Difference in Differences* (DID) methods are another set of methods intended for irregular designs. DID methods are widely used in the econometric literature. See Angrist and Pischke (2008) for a general discussion and references. Case-control designs, more accurately called case-noncase designs, are commonly used in epidemiology, especially when looking for exposures that lead to rare diseases (i.e., the cases).

Classical Randomized Experiments

## CHAPTER 4

A Taxonomy of Classical Randomized Experiments

4.1 INTRODUCTION

In this chapter we introduce four specific examples of classical randomized assignment mechanisms, and we relate these examples to the general taxonomy of assignment mechanisms described in the previous chapter. The four examples, Bernoulli trials, completely randomized experiments, stratified randomized experiments (randomized blocks), and paired randomized experiments, all satisfy the four criteria necessary for assignment mechanisms to be classified as classical randomized experiments. These criteria, as discussed in more detail in Chapter 3, require that the assignment mechanism (*i*) is *individualistic*, with the dependence on values of covariates and potential outcomes for other units limited; (*ii*) is *probabilistic* – each experimental unit has a positive probability of being assigned to the active treatment and a positive probability of being assigned to the control treatment; (*iii*) is *unconfounded* – that is, given covariates, does not depend on potential outcomes; and (*iv*) has a known functional form that is controlled by the researcher.

The key difference between the four types of classical randomized experiments we consider in this chapter is in the set of assignment vectors $\mathbf{W}$ (the N-dimensional vector with elements $W_i \in \{0, 1\}$) with positive probability. Let the set of all possible values be denoted by $\mathbb{W} = \{0, 1\}^N$, with cardinality 2^N, and let the subset of values for $\mathbf{W}$ with positive probability be denoted by $\mathbb{W}^+$. In the first example of randomized experiments, Bernoulli trials, each of the 2^N possible vectors $\mathbf{W}$ defining the treatment assignments of the full population of size N has positive probability. However, such trials put positive probability on assignments in which all units receive the same treatment, thereby compromising our ability to draw credible and precise inferences regarding the causal effect of one treatment versus another from the resulting data. The remaining three types of classical randomized experiments impose increasingly restrictive sets of conditions on the set $\mathbb{W}^+$ of values of $\mathbf{W}$ with positive probability. If imposed judiciously, these restrictions can lead to more precise inferences by reducing the possibility of unhelpful assignment vectors (i.e., assignment vectors that *a priori* are unlikely to lead to useful inferences regarding the causal effects of interest).

4.2 NOTATION

In this section we briefly review the definition of, and notation for, classical random-ized experiments, introduced in Chapter 3. The requirements for classical randomized experiments are that the assignment mechanism must be individualistic, probabilistic, and unconfounded and that the assignment mechanism is known to and controlled by the researcher. As a result of the first and third conditions, by Theorem 3.1, the assignment mechanism in a classical randomized experiment can be written as

$$\Pr(\mathbf{W}|\mathbf{X}, \mathbf{Y}(0), \mathbf{Y}(1)) = c \cdot \prod_{i=1}^{N} e(X_i)^{W_i} \cdot (1 - e(X_i))^{1-W_i},$$

for $\mathbf{W} \in \mathbb{W}^+$, and zero elsewhere. Here $\mathbb{W}^+ \subset \mathbb{W}$ is the subset of the set of possible values for $\mathbf{W}$ with positive probability, and $e(x)$ is the propensity score, which, by prob-abilistic assignment, is strictly between zero and one. The constant c ensures that the probabilities add to unity:

$$c = \left(\sum_{\mathbf{W} \in \mathbb{W}^+} \prod_{i=1}^{N} e(X_i)^{W_i} \cdot (1 - e(X_i))^{1-W_i} \right)^{-1}.$$

Because of the fourth condition, the propensity score $e(x)$ is a known function of the covariates. In this chapter we discuss four common classes of assignment mecha-nisms that fit into this framework: Bernoulli trials, completely randomized experiments, stratified randomized experiments, and pairwise randomized experiments.

4.3 BERNOULLI TRIALS

The simplest Bernoulli experiment tosses a fair coin for each unit: if the coin is heads, the unit is assigned the active treatment, and if it is tails, the unit is assigned the control treatment. Because the coin is fair, the unit-level probabilities and the propensity scores are all 0.5. Because the tosses are independent, the probability of any $\mathbf{W}$ for the N units in the study is the product of the individual probabilities; thus

$$\Pr(\mathbf{W}|\mathbf{X}, \mathbf{Y}(0), \mathbf{Y}(1)) = 0.5^N, \tag{4.1}$$

for all $\mathbf{W} \in \mathbb{W}^+$. Here $\mathbb{W}^+ = \{0, 1\}^N = \mathbb{W}$.

Slightly more generally, we allow the probability of assignment to the treatment – that is, the propensity score – to be different from $1/2$, say $q \in (0, 1)$. Then Equation (4.1) becomes

$$\Pr(\mathbf{W}|\mathbf{X}, \mathbf{Y}(0), \mathbf{Y}(1)) = q^{N_t} \cdot (1 - q)^{N_c}, \tag{4.2}$$

where $N_t = \sum_{i=1}^{N} W_i$, and $N_c = N - N_t = \sum_{i=1}^{N} (1 - W_i)$ are the number of treated and control units, respectively. Here, the probabilities of the different $\mathbf{W}$ vectors depend solely on the number of treated and control units, but still $\mathbb{W}^+ = \{0, 1\}^N$. Such an assignment mechanism, where say, $q \in (0.5, 1)$, may be attractive, for example, when trying to induce people with a serious disease to enroll in a placebo-controlled experiment of a promising new drug for that disease. When the probability of assignment to the treatment group is higher than the probability of assignment to the control group, it would be more attractive for individuals to enroll in this trial than in one where the placebo or control treatment is as likely to be assigned as the active treatment.

Our final generalization of Bernoulli trials allows the unit probabilities to vary with the unit's covariate values. This situation can occur, for example, when certain types of patients are thought to do better on one treatment than another, and the strength of this belief about the better treatment varies with characteristics of the person (e.g., age, sex, race). Here, each unit has a special coin tossed, with the probability that the coin comes up heads equal to the probability that the unit is treated: the unit's propensity score. Consequently,

$$\mathrm{Pr}(\mathbf{W}|\mathbf{X}, \mathbf{Y}(0), \mathbf{Y}(1)) = \prod_{i=1}^{N} \left[e(X_i)^{W_i} \cdot (1 - e(X_i))^{1-W_i} \right]. \tag{4.3}$$

Here again $\mathbb{W}^+ = \mathbb{W}$. Our formal definition of a Bernoulli trial requires that assignments to treatment are independent across all units in the population:

Definition 4.1 (Bernoulli Trial)
A Bernoulli trial is a classical randomized experiment with an assignment mechanism such that the assignments for all units are independent.

Theorem 4.1 (Assignment Mechanism for a Bernoulli Trial)
If the assignment mechanism is a Bernoulli trial, then

$$\mathrm{Pr}(\mathbf{W}|\mathbf{X}, \mathbf{Y}(0), \mathbf{Y}(1)) = \prod_{i=1}^{N} \left[e(X_i)^{W_i} \cdot (1 - e(X_i))^{1-W_i} \right],$$

where $e(x)$ is the propensity score, which must be strictly between zero and one for all i, implying $\mathbb{W}^+ = \{0, 1\}^N$.

Proof. If assignment to treatment is independent across all observations in the population, then the probability of observing a specific assignment vector $\mathbf{W}$, $\mathrm{Pr}(\mathbf{W}|\mathbf{X}, \mathbf{Y}(0), \mathbf{Y}(1))$, will simply equal the product of each unit's probability of assignment:

$$\mathrm{Pr}(\mathbf{W}|\mathbf{X}, \mathbf{Y}(0), \mathbf{Y}(1)) = \prod_{i=1}^{N} \left[p_i(\mathbf{X}, \mathbf{Y}(0), \mathbf{Y}(1))^{W_i} \cdot (1 - p_i(\mathbf{X}, \mathbf{Y}(0), \mathbf{Y}(1)))^{1-W_i} \right].$$

Combined with the fact that $p_i(\mathbf{X}, \mathbf{Y}(0), \mathbf{Y}(1)) = e(X_i)$ for all i, implied by the fact that a Bernoulli trial is a classical randomized experiment, it follows that the normalizing constant is $c = 1$ and that the general form of the assignment mechanism for this type of randomized experiment is

$$\Pr(\mathbf{W}|\mathbf{X}, \mathbf{Y}(0), \mathbf{Y}(1)) = \prod_{i=1}^{N} \left[e(X_i)^{W_i} \cdot (1 - e(X_i))^{1-W_i} \right],$$

as in Equation (4.3).   □

One common disadvantage of Bernoulli trials is that, because of the independence of the assignment across all units, there is always a positive probability (although small even in modest samples, and essentially zero in large samples) that all units will receive the same treatment. In that case, there will be no evidence in the data about the potential outcome values under the treatment that is not represented in the data. Even when there is a single unit being assigned one treatment and many assigned the other treatment, there will be limited evidence about the potential outcomes under the former treatment. Next, we therefore consider alternative classical randomized experiments that ensure that there are "enough" treated and control units under each assignment, beginning with the completely randomized experiment.

4.4   COMPLETELY RANDOMIZED EXPERIMENTS

In the second design we consider, the *completely randomized experiment*, a fixed number of subjects is assigned to receive the active treatment. The simplest completely randomized experiment takes an even number of units and divides them at random in two groups, with exactly one-half of the sample receiving the active treatment and the remaining units receiving the control treatment. This is accomplished, for example, by putting labels for the N units in an urn and drawing $N_t = N/2$ at random to be treated. The assignment mechanism is:

$$\Pr(\mathbf{W}|\mathbf{X}, \mathbf{Y}(0), \mathbf{Y}(1)) = \begin{cases} \binom{N}{N_t}^{-1} & \text{if } \sum_{i=1}^{N} W_i = N_t, \\ 0 & \text{otherwise,} \end{cases} \tag{4.4}$$

where

$$\binom{N}{N_t} = \frac{N!}{N_t!(N - N_t)!}.$$

The notation in (4.4) reveals that N_t does not have to equal $N/2$, but can be any positive integer less than N, fixed in advance. These designs are common in many applied settings, both because they assure that some units will be assigned each treatment, and because analyses using such designs are particularly straightforward in many circumstances. One reason for this simplicity is that the propensity scores are equal for all units, namely N_t/N.

Definition 4.2 (Completely Randomized Experiment)
A completely randomized experiment is a classical randomized experiment with an assignment mechanism satisfying

$$\mathbb{W}^+ = \left\{ \mathbf{W} \in \mathbb{W} \,\middle|\, \sum_{i=1}^{N} W_i = N_t \right\},$$

for some preset $N_t \in \{1, 2, \ldots, N-1\}$.

In other words, given a population of size N, we fix the number of units assigned to the treatment, N_t such that $1 \le N_t \le N - 1$. Out of the population of N, we draw N_t units at random to receive the treatment. Each unit therefore has probability $q = N_t/N$ of receiving the treatment. The number of possible assignment vectors, the cardinality of the set $\mathbb{W}^+$, is under this design $\binom{N}{N_t}$. All $\binom{N}{N_t}$ assignment vectors in $\mathbb{W}^+$ are equally likely; thus, the probability for any one is equal to $\binom{N}{N_t}^{-1}$, whence in completely randomized experiments, the assignment mechanism is given by Equation (4.4).

Although often very sensible, completely randomized experiments are not without drawbacks, especially when important covariates are available. Important covariates here means covariates *a priori* thought to be possibly highly associated with the potential outcomes. Consider, for example, a study with $N = 20$ units, ten men and ten women, where the potential treatment and control outcomes are *a priori* thought to vary substantially by sex. Then, although a completely randomized design with $N_t = 10$ would ensure that ten units get treated, there is the possibility that all ten of them are men (or women). In that case, average differences in the potential outcomes for active and control treatments could be due to sex differences rather than treatment effects. Related complications with relatively unhelpful (in the sense of being uninformative) experiments occur when only a single man is treated and nine men are in the control group, and so forth. The design studied in the next section addresses this issue in some circumstances.

4.5 STRATIFIED RANDOMIZED EXPERIMENTS

With the stratified randomized experiment, the population of units in the study is first partitioned into *blocks* or *strata* so that the units within each block are similar with respect to some (functions of) covariates thought to be predictive of potential outcomes. Then, within each block, we conduct a completely randomized experiment, with assignments independent across blocks.

The simplest randomized block experiment involves two blocks, say males and females, where independent completely randomized experiments are conducted for each group. There is no requirement that the numbers of males and females are the same. Thus, the assignment mechanism is the product of one expression like (4.4) for males, with $N(m)$ and $N_t(m)$ replacing N and N_t, and one expression like (4.4) for women, with $N(f)$ and $N_t(f)$ replacing N and N_t, with the experiment having a total of $N_t(m) + N_t(f)$ units assigned to the active treatment and has a total of $N(m) + N(f) - N_t(m) - N_t(f)$ units assigned to the control treatment.

In general, more strata can be used. Let $B_i \in \{1, \ldots, J\}$ indicate the block or stratum of the ith unit, with $B_i = B(X_i)$ a function of the pre-treatment variables X_i, with a total of J blocks or strata, and let $B_i(j)$ be the binary indicator for the event $B_i = j$. Then the assignment mechanism is the product of J versions of expression (4.4), each version having N and N_t indexed by the J distinct values of $B_i \in \{1, \ldots, J\}$. The unit-level probabilities are common for all units within a block but can vary across blocks. The main reason for generally preferring randomized blocks designs to completely randomized designs is that the former designs control balance in the covariates used to define blocks in treatment and control groups.

Formally, our definition of stratified randomized experiments is as follows:

Definition 4.3 (Stratified Randomized Experiment)
A stratified randomized experiment with J blocks is a classical randomized experiment with an assignment mechanism satisfying

$$\mathbb{W}^+ = \left\{ \mathbf{W} \in \mathbb{W} \,\middle|\, \sum_{i:B_i=j}^{N} W_i = N_t(j), \text{for } j = 1, 2, \ldots, J \right\},$$

and

$$\Pr(\mathbf{W}|\mathbf{X}, \mathbf{Y}(0), \mathbf{Y}(1)) = \begin{cases} \prod_{j=1}^{J} \begin{pmatrix} N(j) \\ N_t(j) \end{pmatrix}^{-1} & \text{if } \mathbf{W} \in \mathbb{W}^+, \\ \\ 0 & \text{otherwise,} \end{cases}$$

for some preset $N_t(j)$ such that $N(j) > N_t(j) > 0$, for $j = 1, \ldots, J$.

In this setting, the unit-level assignment probability or, equivalently in our situation with a classical randomized experiment, the propensity score, $e(X_i)$, is equal to $N_t(j)/N(j)$ for all units with $B_i = j$. As this representation makes explicit, this probability can vary with the stratum indicator. Often, however, the unit-level assignment probabilities are identical across the strata so that $e(x) = q$ for all x. In this case, the only difference between the stratified and completely randomized experiment is that in the former the relative sample size for treatment and control groups is constant across strata, whereas in the latter it may vary. If the covariates defining B_i correspond to substantive information about the units, in the sense that B_i is predictive of the potential outcomes, $(Y_i(0), Y_i(1))$, randomizing within the strata will lead to more precise inferences by eliminating the possibility that all or most units of a certain type, as defined by the blocks, are assigned to the same level of the treatment. Furthermore, even if there is no predictive power of the blocking indicator B_i, stratification does not reduce actual precision, though it reduces the number of allowable values of the assignment vector; see the notes to this chapter for some additional comments on this issue.

4.6   PAIRED RANDOMIZED EXPERIMENTS

The paired comparison, or randomized paired design, is an extreme version of the randomized block experiment in which there are exactly two units within each block, and a fair coin is tossed to decide which member of the pair gets the active treatment and which gets the control treatment. As an example, consider an educational experiment with a covariate, a pre-test score, and the students are ranked from high to low on their scores on this pre-test. The top two form the first pair, the next two form the next pair, and so forth. Within each pair, one of the two units is randomly assigned to the treatment, with the probability of assignment equal to $1/2$.

Definition 4.4 (Paired Randomized Experiment)
A paired randomized experiment is a stratified randomized experiment with $N(j) = 2$ and $N_t(j) = 1$ for $j = 1, \ldots, N/2$, so that

$$\mathbb{W}^+ = \left\{ \mathbf{W} \in \mathbb{W} \,\middle|\, \sum_{i:B_i=j}^{N} W_i = 1, \text{for } j = 1, 2, \ldots, N/2 \right\},$$

and

$$\Pr(\mathbf{W}|\mathbf{X}, \mathbf{Y}(0), \mathbf{Y}(1)) = \begin{cases} 2^{-N/2} & \text{if } \mathbf{W} \in \mathbb{W}^+, \\ 0 & \text{otherwise.} \end{cases}$$

In this design, each unit has probability $1/2$ of being assigned to the treatment group.

4.7   DISCUSSION

All four types of designs described in this chapter satisfy the four conditions for classical randomized experiments. In each case the assignment mechanism is individualistic, probabilistic, unconfounded, and known to the researcher. The way in which these four designs differ is in the set of values allowed for the vector of treatment indicators, $\mathbb{W}^+$. Reducing this set can be of great importance for the precision of estimated treatment effects. To illustrate this, consider the following example. Let N be even, and let the single pre-treatment variable X_i take on $N/2$ different values, with the number of units with $X_i = x$ equal to 2 for all $x \in \{1, \ldots, N/2\}$. Also assume identical unit-level assignment probabilities, that is, a constant propensity score, $e(x) = 1/2$ for all x. In Table 4.1 we report the number of values for the assignment vector that have positive probability under the various types of randomized experiments, for different sample sizes.

First, consider a Bernoulli trial. In this case, there are 2^N different values for the assignment vector. The first row in Table 4.1 shows that with $N = 4$ units, this corresponds to 16 assignment vectors. With $N = 16$, the number of possible treatment assignment combinations increases to more than 65,000.

Next consider a completely randomized experiment with $N_t = N/2$ units assigned to treatment and $N_c = N/2$ assigned to control. The number of allowed values for the assignment vector is now $\binom{N}{N/2}$, which is strictly less than the 2^N values allowed under the Bernoulli design. With $N = 4$ units, we now have only six possible assignment vectors; with a sample of $N = 16$, we have 12,870 possible assignment vectors, or roughly one-fifth the number possible with the Bernoulli trial.

Table 4.1. *Number of Possible Values for the Assignment Vector by Design and Sample Size*

Type of Experiment and Design	Number of Possible Assignments Cardinality of $\mathbb{W}^+$	Number of Units (N) in Sample			
		4	8	16	32
Bernoulli trial	2^N	16	256	65,536	4.2×10^9
Completely randomized experiment	$\binom{N}{N/2}$	6	70	12,870	0.6×10^9
Stratified randomized experiment	$\binom{N/2}{N/4}^2$	4	36	4,900	0.2×10^9
Paired randomized experiment	$2^{N/2}$	4	16	256	65,536

Third, consider a randomized block design, with two blocks, each consisting of $N/2$ units. Given our data set of N observations with the number of units with $X_i = x$ equal to 2 for all $x = 1, \ldots, N/2$, let the first block consist of all units with $X_i \leq N/4$, and the second block consist of the remainder. In terms of the notation introduced in Section 4.5,

$$B_i = \begin{cases} 1 & \text{if } X_i \leq N/4, \\ 2 & \text{if } X_i > N/4. \end{cases}$$

Suppose that within each block, the number of units assigned to the treatment group is equal to the number of units assigned to the control group, $N/4$. Now the number of values for the assignment vector within the first block is $\binom{N/2}{N/4}$, where this assignment vector $\mathbf{W}^{(1)}$ has $N/2$ components. In the second block the number of units is the same, $N/2$, so that the assignment vector for this block is also an $N/2$ component vector, $\mathbf{W}^{(2)}$, and the number of possible assignment vectors is again $\binom{N/2}{N/4}$. Therefore, the total number of values for the full assignment vector, $\mathbf{W} = (\mathbf{W}^{(1)}, \mathbf{W}^{(2)})$, possible under this design is the product of the within-block number of possibilities, $\binom{N/2}{N/4}^2$. Note that this is a strict subset of the set of possible values under the previous two designs. With $N = 4$ units, we now have only 4 possible assignment vectors; with a sample of 16, the number of possible assignment vectors is 4,900.

Fourth, consider the paired randomized experiment where units with the same value of X_i are paired, so $B_i = X_i$. Now there will be $2^{N/2}$ different possible values of the assignment vector with positive probability. This design is a randomized block experiment in which each stratum (block, or subclass) contains only two units. This assignment mechanism is also a paired randomized experiment. Note also that in a paired randomized experiment, using the same argument as above, any value of the assignment vector with positive probability under this design also has positive probability under the stratified randomized design. With only 4 units, the number of assignment vectors with positive probability under a paired randomized experiment is, in fact, identical to that with possible probability under a stratified randomized experiment. With only $N = 4$ units, in the stratified design there can be at most 2 strata, each with the 2 units of a pair, and within each, only one observation assigned to the treatment. With 16 units, however, under a paired randomized experiment there are 256 assignment vectors with positive probability, compared to the 4,900 with positive probability under a randomized block design with two blocks, or a total of 65,536 values for the assignment vector with positive probability under the Bernoulli design.

In this particular sequence of designs with fixed N, the number of distinct values of the assignment vector with positive probability, that is, the cardinality of the set $\mathbb{W}^+$, gradually decreases. The argument for choosing successively more restrictive designs is to eliminate "unhelpful" assignment vectors that are *a priori* unlikely to lead to precise causal inferences. Imposing the first restriction – from Bernoulli trials to completely randomized experiments – is obvious. An assignment vector with all, or almost all, units assigned to one of the treatment levels is typically not as informative as an assignment vector with more balance between the number of treated and control units. Hence, a completely randomized design will tend to be more informative than a Bernoulli trial. The further restrictions to stratified and paired randomized experiments have similar advantages, when the grouping into strata or pairs is based on covariates that are related to the potential outcomes. Formally, if the information used in defining the blocks or pairs is relevant for predicting the potential outcomes, $(Y_i(0), Y_i(1))$, then these designs can improve on completely randomized experiments in terms of the precision of the estimates obtained, often considerably so. In an extreme case, if the pre-treatment variable, X_i, upon which the stratification or pairing is based, perfectly predicts both potential outcomes, there will be no uncertainty remaining regarding the treatment effect across the N units or within the subgroups defined by the covariate. On the other hand, if the blocks or pairs are formed in a way unrelated to the potential outcomes (e.g., by randomly drawing units to assign block labels B_i), the eliminated assignment vectors are just as likely to be helpful as the retained ones, and in such cases, the precision of estimators for treatment effects in stratified or paired randomized experiments is usually no greater than that for the corresponding estimators under completely randomized experiments.

In the next chapters, we discuss analyzing results from the various types of classical randomized experiments in more detail and illustrate these analyses with real data. The methods for analyzing these randomized experiments are useful for two major reasons. First, they are valuable in their own right for analyzing randomized experiments. For many questions in the biomedical and social sciences, however, we must rely on data from observational studies. The second use of these methods, as templates for the analysis of data from observational studies, are therefore even more important for us. In Parts III through VI of this text, we extend these methods for analyzing specific types of classical randomized experiments to assessing data from observational studies and show that observational data can often be analyzed as if they fit the assumptions of one of the randomized experiments discussed here.

4.8   CONCLUSION

In this chapter we discuss four special cases of classical randomized experiments: Bernoulli trials, completely randomized experiments, stratified randomized experiments, and paired randomized experiments. In the next seven chapters we discuss and illustrate methods for estimation and inference in these settings. This is important for substantive reasons but also because understanding the analysis of such relatively simple cases is important for analyzing the more complex observational studies that are the subject of Parts III through VI of this text.

NOTES

There is a large classical literature on experimental design and the analyses of randomized experiments, including Cochran and Cox (1957), Cox (1958), Kempthorne (1952), and Box, Hunter, and Hunter (2005). Much of the design literature focuses on the optimal design of more complex studies with multiple treatments. Such questions are beyond the scope of the current text. Rosenbaum (2000) discusses the structure of the set of assignment vectors using results for finite distributive lattices. Morgan and Rubin (2012) discuss an additional class of designs for randomized experiments. The idea is to start with a completely randomized design. Then, given the assignments, balance of the covariates is assessed according to some well-defined criterion, articulated *prior* to the randomization. If the balance is deemed inadequate, the assignment is rejected and a new vector of assignments is drawn. This is repeated until an assignment vector is drawn that is deemed adequately balanced. Such designs can lead to more precise inferences than completely randomized designs, and they can be more attractive than stratification in settings with many covariates. A similar but different design is described by Morris (1979).

   For general discussions of the literature on analyses of randomized experiments, see Altman (1991), Wu and Hamada (2009), Cook and DeMets (2008), Davies (1954), Cox (1958), Cochran and Cox (1957), Kempthorne (1957), and Box, Hunter, and Hunter (2005).

   Imbens (2011) analyzes the gains from the stratification and shows that even in the absence of any dependence between the potential outcomes and the stratum indicators, stratification, in expectation, in settings with random draws from large strata, does not increase the actual sampling variance of simple estimators of the average treatment effect, thus showing that there is no cost in expected precision of estimation when using stratification even when the samples drawn from the strata are small. There are, however, fewer "degrees of freedom" to estimate that precision, and so the resulting inference is somewhat less precise, an issue studied first in Fisher (1935, pp. 248–250) from a fiducial-likelihood perspective. Specifically, Fisher suggests using the expected information, that is, the expected second derivative of the log-likelihood to adjust for this effect by multiplying the estimated sampling variances by $(K + 3)/(K + 1)$, where K is the number of degrees of freedom used to estimate each sampling variance. It is important here that the strata are large. If the strata are small in the population, it is possible that outcomes within strata are negatively correlated. Snedecor and Cochran (1967, p. 294) discuss examples where this may be relevant (e.g., rats' weights within a litter).

## CHAPTER 5

Fisher's Exact P-Values for Completely Randomized Experiments

5.1 INTRODUCTION

As discussed in Chapter 2, Fisher appears to have been the first to grasp fully the importance of physical randomization for credibly assessing causal effects (1925, 1936). A few years earlier, Neyman (1923) had introduced the language and the notation of potential outcomes, using this notation to define causal effects *as if* the assignments were determined by random draws from an urn, but he did not take the next logical step of appreciating the importance of actually randomizing. It was instead Fisher who made this leap.

Given data from a completely randomized experiment, Fisher was intent on assessing the *sharp null hypothesis* (or *exact null hypothesis*, Fisher, 1935) of no effect of the active versus control treatment, that is, the null hypothesis under which, for each unit in the experiment, both values of the potential outcomes are identical. In this setting, Fisher developed methods for calculating "p-values." We refer to them as *Fisher Exact P-values* (FEPs), although we use them more generally than Fisher originally proposed. Note that Fisher's null hypothesis of no effect of the treatment versus control whatsoever is distinct from the possibly more practical question of whether the *typical* (e.g., average) treatment effect across all units is zero. The latter is a weaker hypothesis, because the average treatment effect may be zero even when for some units the treatment effect is positive, as long as for some others the effect is negative. We discuss the testing of hypotheses on, and inference for, average treatment effects in Chapter 6. Under Fisher's null hypothesis, and under sharp null hypotheses more generally, for units with either potential outcome observed, the other potential outcome is known; and so, under such a sharp null hypothesis, both potential outcomes are "known" for each unit in the sample – being either directly observed or inferred through the sharp null hypothesis.

Consider any test statistic T: a function of the stochastic assignment vector, $\mathbf{W}$; the observed outcomes, $\mathbf{Y}^{\text{obs}}$; and any pre-treatment variables, $\mathbf{X}$. As we discuss in more detail shortly, the fact that the null hypothesis is sharp allows us to determine the distribution of T, generated by the complete randomization of units across treatments. The test statistic is stochastic solely through the stochastic nature of the assignment vector. We refer to the distribution of the statistic determined by the randomization as the *randomization distribution* of the test statistic T. Using this distribution, we can compare the actually observed value of the test statistic, T^{obs}, against the distribution of T under the null hypothesis. An observed value that is "very unlikely," given the null hypothesis and the induced distribution for the test statistic, will be taken as evidence against the null hypothesis in what is, essentially, a stochastic version of the mathematician's "proof by contradiction."

How unusual the observed value is under the null hypothesis will be measured by the probability that a value as extreme or more extreme (in practice, as large or larger) would have been observed – the significance level or p-value. Hence, the FEP approach entails two steps: (*i*) the choice of a sharp null hypothesis (in Fisher's original version, always the null hypothesis of no effect whatsoever, but easily generalized to any sharp null hypothesis, that is, a null hypothesis that allows us to infer all the missing potential outcomes from the observed potential outcomes), and (*ii*) the choice of test statistic. The scientific nature of the problem should govern these choices. In particular, although in Fisher's analysis the null hypothesis was always the one with no treatment effect whatsoever, in general the null hypothesis should follow from the substantive question of interest. The statistic should then be chosen to be sensitive to the difference between the null and some alternative hypothesis that the researcher wants to assess for its scientific interest. That is, the statistic should be chosen to have, what is now commonly referred to as, *statistical power* against a scientifically interesting alternative hypothesis.

An important characteristic of this approach is that it is truly nonparametric, in the sense that it does not rely on a model specified in terms of a set of unknown parameters. In particular, we do not model the distribution of the outcomes: the vectors of potential outcomes $\mathbf{Y}(0)$ and $\mathbf{Y}(1)$ are regarded as fixed but *a priori* unknown quantities. The only reason that the observed outcomes, $\mathbf{Y}^{\mathrm{obs}}$, and thus the statistic, T^{obs}, are random is that a stochastic assignment mechanism determines which of the two potential outcomes we observe for each unit. This assignment mechanism is, by definition, known for a classical randomized experiment. In addition, given the null hypothesis, all potential outcomes are known. Thus, we do not need modeling assumptions to calculate the randomization distribution of any test statistic; instead, the assignment mechanism completely determines the randomization distribution of the test statistic. The validity of any resulting p-value is therefore not dependent on assumptions concerning the distribution of the potential outcomes. This freedom from reliance on modeling assumptions does not mean, of course, that the values of the potential outcomes do not affect the properties of the test. These values will certainly affect the distribution of the p-value when the null hypothesis is false (i.e., the statistical power of the test). They will not, however, affect the validity of the test, which depends solely on the randomized assignment mechanism.

The remainder of this chapter begins with a brief description of the data that we will use to illustrate this approach. The data set is from a completely randomized evaluation of the effect of honey on nocturnal cough and resulting sleep quality for coughing children. Next, in Section 5.3, we start with a simple example using data from only six of the seventy-two children in the experiment. After that follows a detailed discussion of the two choices necessary for calculating FEPs: in Section 5.4 we discuss the choice of the null hypothesis, and in Section 5.5 we discuss the choice of the test statistic. In Section 5.6 we carry out a small simulation study to illustrate the properties of the method. Next, in Section 5.7 we discuss how the FEP approach can be extended to construct interval estimates. We then continue in Section 5.8 with a discussion of how to estimate,

Table 5.1. *Summary Statistics for Observed Honey Data*

Variable	Mean	(S.D.)	Mean Controls	Mean Treated
Cough frequency prior to treatment (`cfp`)	3.86	(0.92)	3.73	4.00
Cough frequency after treatment (`cfa`)	2.47	(1.61)	2.81	2.11
Cough severity prior to treatment (`csp`)	3.99	(1.03)	3.97	4.00
Cough severity after treatment (`csa`)	2.54	(1.74)	2.86	2.20

rather than calculate exactly, the p-value – the level of significance associated with a given observed value of the test statistic – when N is so large that such exact calculations are tedious at best and possibly infeasible. Next, in Section 5.9, we discuss how to use covariates to refine the choice of statistic. In Section 5.10, we expand the analysis to apply this approach to the full sample in which a random subset of the group of seventy-two children was given honey as a cough treatment. Section 5.11 concludes.

5.2   THE PAUL ET AL. HONEY EXPERIMENT DATA

The data used in this chapter are from a randomized experiment by Paul et al. (2007) on the evaluation of the effect of three treatments on nocturnal cough and sleep difficulties associated with childhood upper respiratory tract infections. The three treatments are (*i*) a single dose of buckwheat honey; (*ii*) a single dose of honey-flavored dextromethorphan, an over-the-counter drug; and (*iii*) no active treatment. The subjects were 105 children between two and eighteen years of age. Here we only use data on the $N = 72$ children receiving buckwheat honey ($N_t = 35$) or no active treatment ($N_c = 37$). The authors measure six different outcomes. We focus on two of them, cough frequency afterwards (`cfa`), and cough severity afterwards (`csa`), referring to measures of cough frequency and severity the night after being randomly assigned or not to the administration of the treatment. Both outcomes are measured on a scale from zero ("not at all frequent/severe") to six ("extremely frequent/severe"). We also use two covariates, measured on the night prior to the randomized assignment: cough frequency prior (`cfp`) and cough severity prior (`cfp`), both measured on the same scale as the outcomes.

Table 5.1 presents some summary statistics (means and standard deviations, and means by treatment status) for the four observed variables (`cfp`, `cfa`, `csp`, `csa`), for the 72 children receiving honey or no active treatment in this study. In Table 5.2 we also present cumulative frequencies for the two outcomes variables (`cfa` and `csa`) by treatment group for the seven levels of the outcome scale.

5.3   A SIMPLE EXAMPLE WITH SIX UNITS

Initially let us consider, for relative ease of exposition and data display, a subsample from the honey data set, with six children. Table 5.3 gives the observed data on cough frequency for these six children in the potential outcome form. A key part of the table is the pair of columns listing the potential outcomes, observed and missing. The first child (unit 1) was assigned to the (buckwheat honey) treatment group ($W_1 = 1$). Hence we

Table 5.2. *Cumulative Distribution Functions for Cough Frequency and Severity after Treatment Assignment for the Honey Study*

Value	cfa		csa	
	Controls	Treated	Controls	Treated
0	0.14	0.14	0.16	0.17
1	0.19	0.40	0.22	0.46
2	0.32	0.63	0.35	0.54
3	0.73	0.83	0.59	0.77
4	0.89	0.91	0.86	0.91
5	0.92	0.97	0.95	0.94
6	1.00	1.00	1.00	1.00

Table 5.3. *Cough Frequency for the First Six Units from the Honey Study*

Unit	Potential Outcomes				
	Cough Frequency (cfa)		Observed Variables		
	$Y_i(0)$	$Y_i(1)$	W_i	X_i (cfp)	Y_i^{obs} (cfa)
1	?	3	1	4	3
2	?	5	1	6	5
3	?	0	1	4	0
4	4	?	0	4	4
5	0	?	0	1	0
6	1	?	0	5	1

observe $Y_1^{obs} = Y_1(1)$ (equal to 3 for this child). We do not observe $Y_1(0)$, and in the table this missing potential outcome is represented by a question mark. The second child was also assigned to the treatment ($W_2 = 1$), and again we observe $Y_2^{obs} = Y_2(1)$ (equal to 5), and we do not observe $Y_2(0)$ (represented again by a question mark). Table 5.3 directly shows the fundamental problem of causal inference: many of the potential outcomes (in this particular case exactly half) are missing.

Using this subset of the honey data, we first calculate the p-value for the sharp null hypothesis that the treatment had absolutely no effect on coughing outcomes, that is:

$$H_0: \quad Y_i(0) = Y_i(1) \quad \text{for } i = 1, \dots, 6.$$

Under this null hypothesis, for each child, the missing potential outcomes, Y_i^{mis} are identical to the observed outcomes for the same child, Y_i^{obs}, or $Y_i^{mis} = Y_i^{obs}$ for all $i = 1, \dots, N$. Thus, we can fill in all six of the missing entries in Table 5.3 using the observed data; Table 5.4 lists the fully expanded data set under Fisher's sharp null hypothesis. This step is the first key insight of the FEP approach; under the sharp null hypothesis, all the missing values can be inferred from the observed ones.

Table 5.4. *Cough Frequency for the First Six Units from Honey Study with Missing Potential Outcomes in Parentheses Filled in under the Null Hypothesis of No Effect of the Treatment*

Unit	Potential Outcomes					
	Cough Frequency (cfa)		Observed Variables			
	$Y_i(0)$	$Y_i(1)$	Treatment	X_i	Y_i^{obs}	rank(Y_i^{obs})
1	(3)	3	1	4	3	4
2	(5)	5	1	6	5	6
3	(0)	0	1	4	0	1.5
4	4	(4)	0	4	4	5
5	0	(0)	0	1	0	1.5
6	1	(1)	0	5	1	3

We use the absolute value of the difference in average outcomes by treatment status as our test statistic:

$$T(\mathbf{W}, \mathbf{Y}^{\text{obs}}) = T^{\text{dif}} = \left| \overline{Y}_t^{\text{obs}} - \overline{Y}_c^{\text{obs}} \right|,$$

where $\overline{Y}_t^{\text{obs}} = \sum_{i:W_i=1} Y_i^{\text{obs}}/N_t$ and $\overline{Y}_c^{\text{obs}} = \sum_{i:W_i=0} Y_i^{\text{obs}}/N_c$ are the average of the observed outcomes in the treatment and control groups, respectively, and $N_c = \sum_{i=1}^{N}(1 - W_i)$ and $N_t = \sum_{i=1}^{N} W_i$ are the number of units in the control and treatment groups respectively. This test statistic is likely to be sensitive to deviations from the null hypothesis corresponding to a constant additive effect of the treatment. For the observed data in Table 5.3, the value of the test statistic is

$$T^{\text{obs}} = T(\mathbf{W}, \mathbf{Y}^{\text{obs}}) = |\overline{Y}_t^{\text{obs}} - \overline{Y}_c^{\text{obs}}|$$

$$= |(Y_1^{\text{obs}} + Y_2^{\text{obs}} + Y_3^{\text{obs}})/3 - (Y_4^{\text{obs}} + Y_5^{\text{obs}} + Y_6^{\text{obs}})/3| = |8/3 - 5/3| = 1.00.$$

Under the null hypothesis, we can calculate the value of this statistic under each vector of treatment assignments, $\mathbf{W}$. Suppose for example, that instead of the observed assignment vector $\mathbf{W}^{\text{obs}} = (1, 1, 1, 0, 0, 0)$, the assignment vector had been $\tilde{\mathbf{W}} = (0, 1, 1, 0, 0, 1)$. That would *not* have changed any of the values of the observed outcomes Y_i^{obs}, because under the null hypothesis, for each unit, $Y_i(0) = Y_i(1) = Y_i^{\text{obs}}$, but it *could* have changed the value of the test statistic because different units would have been assigned to the treatment and control groups. For example, under the assignment vector, $\tilde{\mathbf{W}} = (0, 1, 1, 0, 1, 0)$, the test statistic would have been $T(\tilde{\mathbf{W}}, \mathbf{Y}^{\text{obs}}) = |(Y_2^{\text{obs}} + Y_3^{\text{obs}} + Y_5^{\text{obs}})/3 - (Y_1^{\text{obs}} + Y_4^{\text{obs}} + Y_6^{\text{obs}})/3| = |6/3 - 7/3| = 0.33$, different from $T^{\text{obs}} = 1.00$. We can repeat this calculation for each possible assignment vector. Given that we have a population of six children with three assigned to treatment, there are $\binom{6}{3} = 20$ different possible assignment vectors. Table 5.5 lists all twenty possible assignment vectors for these six children. For the moment, focus on the first unit, $i = 1$. For all assignment vectors, Y_1^{obs} remains the same, but given our null hypothesis of no effect, Y_1^{obs} is associated with $Y_1(0)$ for those assignment vectors with $W_1 = 0$, and is associated

Table 5.5. *Randomization Distribution for Two Statistics for the Honey Data from Table 5.3*

						Statistic: Absolute Value of Difference in Average	
W_1	W_2	W_3	W_4	W_5	W_6	Levels (Y_i)	Ranks (R_i)
0	0	0	1	1	1	−1.00	−0.67
0	0	1	0	1	1	−3.67	−3.00
0	0	1	1	0	1	−1.00	−0.67
0	0	1	1	1	0	−1.67	−1.67
0	1	0	0	1	1	−0.33	0.00
0	1	0	1	0	1	2.33	2.33
0	1	0	1	1	0	1.67	1.33
0	1	1	0	0	1	−0.33	0.00
0	1	1	0	1	0	−1.00	−1.00
0	1	1	1	0	0	1.67	1.33
1	0	0	0	1	1	−1.67	−1.33
1	0	0	1	0	1	1.00	1.00
1	0	0	1	1	0	0.33	0.00
1	0	1	0	0	1	−1.67	−1.33
1	0	1	0	1	0	−2.33	−2.33
1	0	1	1	0	0	0.33	0.00
1	1	0	0	0	1	1.67	1.67
1	1	0	0	1	0	1.00	0.67
1	1	0	1	0	0	3.67	3.00
1	1	1	0	0	0	**1.00**	**0.67**

Note: Observed values in boldface (R_i is rank(Y_i)). Data based on cough frequency for first six units from honey study.

with $Y_1(1)$ for those assignment vectors with $W_1 = 1$; likewise for the other units. Thus the value of the corresponding statistics $T(\mathbf{W}, \mathbf{Y}^{\text{obs}})$ varies with $\mathbf{W}$.

For each vector of assignments, we calculate the corresponding value of the statistic. The last row of Table 5.5 lists the actual assignment vector, corresponding to the data in Table 5.4. In this case, $T^{\text{obs}} = 1.00$; in the sample of six children, the measure of the average cough frequency for the three children who had been given honey differs by one unit of measurement from the average for the three children who had not been given any active treatment for their coughing. The other rows list the value of the statistic under the alternative values of the assignment vector for the expanded data of Table 5.4. Under random assignment, each assignment vector has prior probability 1/20. Thus we can derive the prior probabilities for each of the twenty values of the test statistic under Fisher's null hypothesis.

Given the distribution of the test statistic, we can ask the following question: How unusual or extreme is the observed absolute average difference between children who had been given honey versus nothing (the number 1.00) assuming the null hypothesis is true? That is, how unusual is this observed difference, assuming that there is, in fact, absolutely no causal effect of giving honey on cough frequency? One way to implement this calculation is to ask how likely it is, according to the randomization distribution, to observe a value of the test statistic that is as large as the one actually observed, or even larger. This calculation clearly underestimates the likelihood of the observed result because it bundles it with all rarer events. Simply counting from Table 5.5 we see that there are sixteen assignment vectors with at least a difference in absolute value of 1.00 between children in the treated and control groups, out of a set of twenty possible assignment vectors. This corresponds to a p-value of $16/20 = 0.80$ for the given combination of the sharp null hypothesis and the test statistic. Under the null hypothesis of absolutely no effect of administering honey, the observed difference could, therefore, well be due to chance. If there were no effect of giving honey at all, we could have seen an effect as large as, or larger than, the one we actually observed for eighty out of every hundred times that we randomly assigned the honey. Note that, with three children out of six receiving the treatment, the most extreme p-value that we could have for this statistic for any values of the data is $2/20 = 0.10$; if $T = t$ is a possible value for the test statistic, then t will also be the value of the test statistic obtained by using the opposite assignment vector. Hence the sample of size six is generally too small to be able to assess, with any reasonable certainty, the existence of some effect of honey versus nothing – the sample size is not sufficient to have adequate statistical power to reach any firm conclusion.

In the next three sections we go over these three steps, specifying the null hypothesis, choosing the statistic, and measuring the extremeness, in more detail and generality.

5.4   THE CHOICE OF NULL HYPOTHESIS

The first choice that arises when calculating the FEP is the choice of null hypothesis. Fisher himself only focused on what is arguably the most obvious sharp null hypothesis, that of no effect whatsoever of the active treatment:

$$H_0 : Y_i(0) = Y_i(1), \quad \text{for } i = 1, \ldots, N. \tag{5.1}$$

We need not necessarily believe such a null hypothesis, but we may wish to see how strongly the data can speak against it. Note again that this sharp null hypothesis of no effect whatsoever is very different from the null hypothesis that the *average* effect of the treatment in the sample of N units is zero. This "average null" hypothesis is *not* a sharp null hypothesis, because it does not allow the researcher to infer values for all potential outcomes in the sample. The "average null" therefore does not fit into the framework that originates with Fisher, or its direct extensions. This does not imply that the average null hypothesis is less relevant than the hypothesis that the treatment effect is zero for all units. As we will see in Chapter 6, Neyman, whose approach focused on estimating the average effect of the treatment, was criticized, perhaps unfairly, by Fisher for his (Neyman's) questioning of the relative importance of the sharp null of absolutely no effect that was the focus of Fisher's analysis, compared to the null hypothesis of no average effect.

Although Fisher's approach cannot accommodate a null hypothesis of an average treatment effect of zero, it can accommodate sharp null hypotheses other than the null hypothesis of no effect whatsoever. Fisher did not actually take this step, but it is a natural one. An obvious alternative to the null hypothesis of no effect whatsoever, is the hypothesis that there is a constant additive treatment effect, $Y_i(1) = Y_i(0) + C$, possibly after some transformation of the outcomes, (e.g., by taking logarithms, so that the null hypothesis is that $Y_i(1)/Y_i(0) = C$ for all units) for some pre-specified value C. Once we depart from the world of no effect, however, we encounter several possible complications, among them, why the treatment effect should be additive in levels rather than in logarithms, or after some other transformation of the basic outcome.

The most general case that fits into the FEP framework is the null hypothesis that $Y_i(1) = Y_i(0) + C_i$ for some set of pre-specified treatment effects C_i for $i = 1, \ldots, N$. In practice, however, it is rare to have a meaningful and interesting null hypothesis precise enough to specify individual treatment effects for each unit, without these treatment effects being identical for all units (again, possibly after some transformation).

Although the FEP approach can allow for general sharp null hypotheses, we focus in the following discussion on the implementation of the case where the null hypothesis is that of no effect whatsoever, $Y_i(1) = Y_i(0)$ for all $i = 1, \ldots, N$, thereby implying that $Y_i^{\text{mis}} = Y_i^{\text{obs}}$. This limitation is without essential loss of generality.

5.5   THE CHOICE OF STATISTIC

The second decision in the FEP approach, the choice of test statistic, is typically more difficult than the choice of the null hypothesis. First let us formally define a statistic:

Definition 5.1 (Statistic)
A statistic T is a known, real-valued function $T(\mathbf{W}, \mathbf{Y}^{\text{obs}}, \mathbf{X})$ of: the vector of assignments, $\mathbf{W}$; the vector of observed outcomes, $\mathbf{Y}^{\text{obs}}$ (itself a function of $\mathbf{W}$ and the potential outcomes $\mathbf{Y}(0)$ and $\mathbf{Y}(1)$); and the matrix of pre-treatment variables, $\mathbf{X}$.

Any statistic that satisfies this definition can be used in the FEP approach in the sense that we can calculate its exact distribution under the null hypothesis. When such a statistic is scalar and used to find a p-value, we call it a "test statistic." However, not all statistics are sensible. We also want the test statistic to have the ability to distinguish between the null hypothesis and an interesting alternative hypothesis. Using the statistical term already introduced, we want the resulting test statistic to have *power* against alternatives, that is, to be likely to have a value, when the null hypothesis is false, that would be unusually large if the null hypothesis were true. Our desire for statistical power is complicated by the fact that there may be many alternative hypotheses of interest, and it is typically difficult, or even impossible, to specify a single test statistic that has substantial power against all interesting alternatives. We therefore look for statistics that lead to tests that have power against those alternative hypotheses that are viewed as the most interesting from a substantive point of view. Let us now introduce some test statistics and then return to the question of choosing among them.

The most popular choice of test statistic, although not necessarily the most highly recommended, is the one we also used in Section 5.3, the absolute value of the difference in average outcomes by treatment status:

$$T^{\text{dif}} = \left| \overline{Y}_{\text{t}}^{\text{obs}} - \overline{Y}_{\text{c}}^{\text{obs}} \right| = \left| \frac{\sum_{i:W_i=1} Y_i^{\text{obs}}}{N_{\text{t}}} - \frac{\sum_{i:W_i=0} Y_i^{\text{obs}}}{N_{\text{c}}} \right|. \tag{5.2}$$

This test statistic is relatively attractive if the most interesting alternative hypothesis corresponds to an additive treatment effect, and the frequency distributions of $Y_i(0)$ and $Y_i(1)$ have few outliers.

This particular test statistic, without the absolute value, also has an interpretation as an "unbiased" estimator for the average effect of the treatment under any alternative hypothesis, as we shall discuss in detail in the next chapter. However, this is somewhat coincidental and largely irrelevant here. In general, the test statistic need not have a direct interpretation in terms of estimating causal effects. Such an interpretation may be an attractive property, but it is not essential, and in this FEP approach, focusing only on such statistics can at times divert attention from generally more powerful test statistics.

Before discussing alternative statistics, we should add one note of caution. Although there are many choices for the statistic, the validity of the FEP approach and its p-value hinges on using one statistic and its p-value only. If one calculates multiple statistics and their corresponding p-values, the probability of observing at least one p-value less than a fixed value of p, say p^*, is larger than p^*. We return to this issue of multiple comparisons in Section 5.5.7.

5.5.1 Transformations

An obvious alternative to the simple difference in average outcomes by treatment status in (5.2) is to transform the outcomes before comparing average differences between treatment levels. This procedure would be an attractive option if a plausible alternative hypothesis corresponds to an additive treatment effect after such a transformation. For example, it may be interesting to consider a constant multiplicative effect of the treatment. In that case, the treatment effect would be an additive constant after taking logarithms, and so we might compare the average difference on a logarithmic scale by treatment status using the following test statistic:

$$T^{\text{log}} = \left| \frac{\sum_{i:W_i=1} \ln(Y_i^{\text{obs}})}{N_{\text{t}}} - \frac{\sum_{i:W_i=0} \ln(Y_i^{\text{obs}})}{N_{\text{c}}} \right|. \tag{5.3}$$

Such a transformation could also be sensible if the raw data have skewed distributions, which is typically the case for positive variables such as earnings or wealth, or levels of a pathogen, and treatment effects are more likely to be multiplicative than additive, although one needs to take care in case there are units with zero values. In such a case, the test statistic based on taking the average difference, after transforming to logarithms, would likely be more powerful than the test based on the simple average difference, as we illustrate later.

5.5.2 Quantiles

Motivated by the same concerns that led to test statistics based on logarithms, one may be led to test statistics based on trimmed means or other "robust" estimates of location, which are not sensitive to outliers. For example, one could use the absolute value of the difference in medians in the two samples,

$$T^{\text{median}} = \left| \text{med}_t(Y_i^{\text{obs}}) - \text{med}_c(Y_i^{\text{obs}}) \right|, \tag{5.4}$$

where $\text{med}_t(Y_i^{\text{obs}})$ and $\text{med}_c(Y_i^{\text{obs}})$ are the observed sample medians of the subsamples with $W_i = 0$, $\{Y_i^{\text{obs}} : W_i = 0\}$, and $W_i = 1$, $\{Y_i^{\text{obs}} : W_i = 1\}$, respectively. Other test statistics based on robust estimates of location include the average in each subsample after trimming (i.e., deleting) the lower and upper 5% or 25% of the two subsamples. Another way of generalizing the statistic based on the difference in medians is to use differences in other quantiles:

$$T^{\text{quant}} = \left| q_{\delta,t}(Y_i^{\text{obs}}) - q_{\delta,c}(Y_i^{\text{obs}}) \right|, \tag{5.5}$$

where $q_{\delta,t}(Y_i^{\text{obs}})$ and $q_{\delta,c}(Y_i^{\text{obs}})$, for $\delta \in (0,1)$, are the δ quantiles of the empirical distribution of Y_i^{obs} in the subsample with $W_i = 0$ and $W_i = 1$ respectively, so that, $\sum_{i:W_i=0} \mathbf{1}_{Y_i^{\text{obs}} \leq q_{\delta,c}(Y_i^{\text{obs}})}/N_c \geq \delta$, and $\sum_{i:W_i=0} \mathbf{1}_{Y_i^{\text{obs}} < q_{\delta,c}(Y_i^{\text{obs}})}/N_c < \delta$. Here $\mathbf{1}_E$ is the indicator function, equal to 1 if the event E is true and equal to 0 otherwise.

5.5.3 T-Statistics

Another choice for the test statistic is the conventional t-statistic for the test of the null hypothesis of equal means, with unequal variances in the two groups,

$$T^{\text{t-stat}} = \left| \frac{\overline{Y}_t^{\text{obs}} - \overline{Y}_c^{\text{obs}}}{\sqrt{s_c^2/N_c + s_t^2/N_t}} \right|, \tag{5.6}$$

where $s_c^2 = \sum_{i:W_i=0} (Y_i^{\text{obs}} - \overline{Y}_c^{\text{obs}})^2/(N_c - 1)$ and $s_t^2 = \sum_{i:W_i=1} (Y_i^{\text{obs}} - \overline{Y}_t^{\text{obs}})^2/(N_t - 1)$. Note that, in the approach of this chapter, we do not compare this test statistic to a student-t or normal distribution. Rather, we use the randomization distribution to obtain the exact distribution of the test statistic $T^{\text{t-stat}}$ under the null hypothesis given the potential outcomes. In many cases, the conventional normal or student-t approximation may be excellent in moderate to large samples, but in small samples, and with thick-tailed or skewed distributions for the potential outcomes, these approximations can be poor, and generally there is no need to rely on them in our era of fast computing, as we illustrate in Section 5.8.

5.5.4 Rank Statistics

An important class of test statistics involves transforming the outcomes to *ranks* before considering differences by treatment status. Such a transformation is particularly attractive when the raw outcomes have a distribution with a substantial number of outliers.

Assuming no ties, the rank of unit i, for $i = 1, \ldots, N$, is defined as the number of units, out of the sample of size N, with an observed outcome less than or equal to Y_i^{obs}. Without ties, the rank will take on all integer values from 1 to N, with a discrete uniform distribution, irrespective of the observed potential outcomes. This transformation leads to inferences that are insensitive to outliers, without requiring consideration of which continuous transformation would lead to a well-behaved distribution of potential outcomes. Formally the basic definition of rank in the absence of ties is

$$\tilde{R}_i = \tilde{R}_i(Y_1^{\text{obs}}, \ldots, Y_N^{\text{obs}}) = \sum_{j=1}^{N} \mathbf{1}_{Y_j^{\text{obs}} \leq Y_i^{\text{obs}}}.$$

We often subtract $(N+1)/2$ from each rank to obtain a normalized rank that has average value equal to zero in the sample:

$$\dot{R}_i = \tilde{R}_i(Y_1^{\text{obs}}, \ldots, Y_N^{\text{obs}}) - \frac{N+1}{2} = \sum_{j=1}^{N} \mathbf{1}_{Y_j^{\text{obs}} \leq Y_i^{\text{obs}}} - \frac{N+1}{2}.$$

When there are ties in outcomes within the sample, the definition is typically modified, for instance, by averaging all possible ranks across the tied observations. Suppose we have two units with outcomes both equal to y; if there are L units with outcomes smaller than y, the two possible ranks for these two units are $L+1$ and $L+2$. Hence we assign each of these units the average rank $(L+1)/2 + (L+2)/2 = L + 3/2$. More generally, if there are M observations with the same outcome value, and L observations with a strictly smaller value, the rank for the M observations with the same outcome value is $L + (1+M)/2$. Formally, after again subtracting the mean rank, we use the following definition for the normalized rank:

$$R_i = R_i(Y_1^{\text{obs}}, \ldots, Y_N^{\text{obs}}) = \sum_{j=1}^{N} \mathbf{1}_{Y_j^{\text{obs}} < Y_i^{\text{obs}}} + \frac{1}{2}\left(1 + \sum_{j=1}^{N} \mathbf{1}_{Y_j^{\text{obs}} = Y_i^{\text{obs}}}\right) - \frac{N+1}{2}.$$

Given the N ranks R_i, $i = 1, \ldots, N$, an obvious test statistic is the absolute value of the difference in average ranks for treated and control units:

$$T^{\text{rank}} = \left|\overline{R}_t - \overline{R}_c\right| = \left|\frac{\sum_{i:W_i=1} R_i}{N_t} - \frac{\sum_{i:W_i=0} R_i}{N_c}\right|, \qquad (5.7)$$

where $\overline{R}_t$ and $\overline{R}_c$ are the average rank in the treatment and control group respectively. In the absence of ties, the p-value for this test statistic is closely related to that based on the Wilcoxon rank sum test statistic, which is defined as $T^{\text{wilcoxon}} = \sum_{i=1}^{N} \tilde{R}_i$, because T^{rank} is a simple transformation of T^{wilcoxon}:

$$T^{\text{rank}} = \left|\frac{T^{\text{wilcoxon}} - N(N+1)/2}{N_t} - \frac{N(N-1)/2 - T^{\text{wilcoxon}}}{N_c}\right|.$$

Let us return to the first six units from the honey data in Table 5.3. The observed cough frequency for the first child is 3. There are three units with a smaller value for the outcome, so the rank for the first child's value of the outcome is 4. The second child has an observed outcome equal to 5, which is the largest observed value, so the rank for this child's value is 6. The cough frequency for the third child is zero, tied for the smallest value with one other child, so that the non-normalized rank is $(1 + 2)/2 = 1.5$. The ranks for all six units are reported in Table 5.4. We then calculate the test statistic as the average difference in rank between the three treated and the three control units, which leads to a test statistic of 0.67. To obtain the FEP for this test statistic, we count the number of times we get a test statistic equal to, or larger than, 0.67, across all randomized allocations. With all values reported in Table 5.5, this number is 16, so that the p-value is $16/20 = 0.80$.

Unlike the simple difference in means, or the difference in logarithms, the rank-based statistics do not have a direct interpretation as a meaningful treatment effect. Nevertheless, rank-based statistics can in practice lead to more powerful tests than statistics that have an interpretation as an estimated causal effect, due to their insensitivity to thick-tailed or skewed distributions. We will illustrate this feature when we look at an example with real data.

5.5.5 Model-Based Statistics

A rich class of possible test statistics with a form very different from a simple difference of averages outcomes, possibly after some transformation, is motivated by parametric models of the potential outcomes. Other uses of such models will be discussed in greater detail in Chapter 8. Here we briefly discuss their role in motivating statistics in the FEP approach.

Suppose we have two models, one for the distribution of the potential control outcomes $Y_i(0)$ and the other for the distribution of the potential treated outcomes $Y_i(1)$, governed by unknown parameters θ_c and θ_t respectively, where both θ_c and θ_t generally are vectors. For ease of exposition, let us assume that both models have a common functional form so that θ_c and θ_t have the same number of components. Let us estimate θ_c using the observed outcomes from the units assigned to the control group and denote the estimator by $\hat{\theta}_c$. We can use a variety of methods for estimation here, for example, method of moments, least squares, or maximum likelihood estimation. Similarly, let us estimate the parameter θ_t using outcomes from the units assigned to the treatment group, with estimator $\hat{\theta}_t$. Now, take any scalar function of the resulting estimates, say the difference in one of the components of the two vectors $\hat{\theta}_c$ and $\hat{\theta}_t$, or the sum of the squared differences between elements of the vectors $\hat{\theta}_c$ and $\hat{\theta}_t$. Because $\hat{\theta}_c$ and $\hat{\theta}_t$ are functions of the observed data $(\mathbf{W}, \mathbf{Y}^{obs}, \mathbf{X})$, they are statistics according to Definition 5.1. Hence any scalar function of the estimated parameters $\hat{\theta}_c$ and $\hat{\theta}_t$ is a test statistic that can be used to obtain a p-value for a sharp null hypothesis.

Although these test statistics are motivated by statistical models, the validity of an FEP based on any one of them does not rely on the validity of these models. In fact, these models are purely descriptive given that the potential outcomes are considered fixed quantities. The reason such models may be useful, however, is that they may provide good descriptive approximations to the sample distribution of the potential outcomes under some alternative hypothesis. If so, the models can suggest a test statistic that is relatively powerful against such alternatives.

Let us consider two examples. First, suppose the model for $Y_i(0)$ is normal with mean μ_c and variance σ_c^2. Similarly, suppose the model for $Y_i(1)$ is also normal but with a generally different mean μ_t and variance σ_t^2. Thus, $\theta_c = (\mu_c, \sigma_c^2)$, and $\theta_t = (\mu_t, \sigma_t^2)$. The natural estimates for μ_c and μ_t are the two subsample means by treatment status $\hat{\mu}_c = \overline{Y}_c^{\text{obs}}$ and $\hat{\mu}_t = \overline{Y}_t^{\text{obs}}$. Hence, if we use the statistic

$$T^{\text{model}} = \left| \hat{\mu}_t - \hat{\mu}_c \right| = \left| \overline{Y}_t^{\text{obs}} - \overline{Y}_c^{\text{obs}} \right| = T^{\text{dif}},$$

we return to the familiar territory of using the difference in averages by treatment status for the test statistic.

Second, suppose that the model for $Y_i(0)$ is a normal distribution with mean μ_c and variance σ_c^2, censored from above at C, and similarly that $Y_i(1)$ has a normal distribution with mean μ_t and variance σ_t^2, also censored from above at a known value C, so that again, $\theta_c = (\mu_c, \sigma_c^2)$, and $\theta_t = (\mu_t, \sigma_t^2)$. We can estimate the parameters μ_c, μ_t, σ_c^2, and σ_t^2 by maximum likelihood as $\hat{\mu}_{\text{ml},c}$, $\hat{\mu}_{\text{ml},t}$, $\hat{\sigma}_{\text{ml},c}^2$, and $\hat{\sigma}_{\text{ml},t}^2$ respectively, or by the method of moments. There are no analytic solutions for the maximum likelihood estimates in this case, but the FEP based on a test statistic using such estimates, for example, $T^{\text{model}} = \left| \hat{\mu}_{\text{ml},t} - \hat{\mu}_{\text{ml},c} \right|$, is still valid.

5.5.6 The Kolmogorov-Smirnov Statistic

The test statistics discussed so far focus on difference in particular features of the outcome distributions between treated and control units. Initially this was the difference in averages, and later we considered differences in averages after taking transformations of outcomes, including ranks. Focusing on a single, or even multiple, features of these distributions may lead the researcher to miss differences in other aspects. For example, suppose we focus on the difference in average outcomes by treatment status. If the true distribution for the potential outcomes given treatment is normal with mean zero and unit variance, and the true distribution for the potential outcome given no treatment is normal with the same mean, zero, but a different variance, say, two, focusing solely on the average difference will not generate extreme p-values very often, even in large samples, despite the null hypothesis not holding. Formally, the test based on the difference in averages will have little power against an alternative hypothesis with different variances. We may, therefore, be interested in test statistics that would be able to detect, given sufficiently large samples, any differences in distributions between treated and control units. An example of such a test statistic is the Kolmogorov-Smirnov statistic.

Let $\hat{F}_c(y)$ and $\hat{F}_t(y)$ be the empirical distribution functions based on units with treatment $W_i = 0$ and $W_i = 1$, respectively:

$$\hat{F}_c(y) = \frac{1}{N_c} \sum_{i:W_i=0} \mathbf{1}_{Y_i^{\text{obs}} \leq y}, \quad \text{and} \quad \hat{F}_t(y) = \frac{1}{N_t} \sum_{i:W_i=1} \mathbf{1}_{Y_i^{\text{obs}} \leq y},$$

for all $-\infty < y < \infty$. Then the Kolmogorov-Smirnov test statistic is

$$T^{\text{ks}} = \sup_y \left| \hat{F}_t(y) - \hat{F}_c(y) \right| = \max_{i=1,\ldots,N} \left| \hat{F}_t\left(Y_i^{\text{obs}}\right) - \hat{F}_c\left(Y_i^{\text{obs}}\right) \right|. \tag{5.8}$$

This is a more complicated test statistic than, say, the average T^{dif}. Nevertheless, because it is a scalar function of the vector of assignments and the vector of observed outcomes, it is a valid test statistic. Therefore, we use exactly the same procedure as with the simpler statistics: calculate its exact finite-sample distribution generated by the randomization and then calculate the associated exact p-value.

5.5.7 Statistics with Multiple Components

The validity of the FEP approach depends on an *a priori* (i.e., before seeing the data) commitment to a specific pair: a null hypothesis and a test statistic. The corresponding p-values are valid for each pair considered in isolation, but the p-values are not independent across pairs. Specifically, consider two possible test statistics, $T^1(\mathbf{W}, \mathbf{Y}^{\mathrm{obs}}, \mathbf{X})$ and $T^2(\mathbf{W}, \mathbf{Y}^{\mathrm{obs}}, \mathbf{X})$, with realized values $T^{1,\mathrm{obs}}$ and $T^{2,\mathrm{obs}}$. This situation may arise in a number of ways. First, it may be that there are multiple alternative hypotheses of interest. For example, under one alternative hypothesis the mean of the outcome distribution may shift (suggesting a test statistic based on the difference in means by treatment status), whereas under another alternative hypothesis the dispersion may change (suggesting a test statistic based on the ratio of sample variances by treatment status). Second, it may be that the researcher has two outcomes for each unit. In the honey study, there are, for example, measures on both cough frequency and cough severity. In that case, one statistic could be the difference in average cough frequency by treatment status and the other difference in average cough severity by treatment status. Under any sharp null hypothesis, one can calculate p-values for each of the tests, for example,

$$p_1 = \Pr(T^1 \geq T^{1,\mathrm{obs}} | \mathbf{X}, \mathbf{Y}(0), \mathbf{Y}(1), H_0) \quad \text{and} \quad p_2 = \Pr(T^2 \geq T^{2,\mathrm{obs}} | \mathbf{X}, \mathbf{Y}(0), \mathbf{Y}(1), H_0).$$

These p-values are valid for each test in isolation, but using the minimum of p_1 and p_2 as an overall p-value for the null hypothesis is not valid, nor is using the average of p_1 and p_2 for this purpose.

The simplest way to obtain a valid p-value with multiple test statistics is to combine the two (or more) test statistics into a single test statistic. One can do this directly, by defining the test statistic as a function of the two original test statistics,

$$T^{\mathrm{comb}} = g(T^1, T^2),$$

for some scalar function $g(\cdot, \cdot)$. Choices for T^{comb} could include a (weighted) average of the two statistics, or the minimum or maximum of the two statistics. Alternatively, T^{comb} could be a function of the two p-values, for example, the minimum or the average. Because T^1 and T^2 (or p_1 and p_2) are functions of $(\mathbf{W}, \mathbf{Y}, \mathbf{X})$, it follows that T^{comb} is a function of these vectors and thus a valid scalar test statistic according to our definition. Hence, its randomization distribution can be calculated, and the corresponding p-value would equal

$$p_g = \Pr(g(T^1, T^2) \geq g(T^{1,\mathrm{obs}}, T^{2,\mathrm{obs}}) | \mathbf{X}, \mathbf{Y}(0), \mathbf{Y}(1), H_0).$$

As an example, suppose we have for, each unit, two outcome measures, Y_{i1}^{obs} and Y_{i2}^{obs}. These may be distinct measurements (e.g., in the honey study, the cough frequency and cough severity, both post-treatment), or one could be a transformation of the other. For each outcome we could calculate the statistics based on the t-statistic:

$$T^{\text{t-stat},1} = \left| \frac{\overline{Y}_{t1}^{\text{obs}} - \overline{Y}_{c1}^{\text{obs}}}{\sqrt{s_{c1}^2/N_c + s_{t1}^2/N_t}} \right|, \quad \text{and} \quad T^{\text{t-stat},2} = \left| \frac{\overline{Y}_{t2}^{\text{obs}} - \overline{Y}_{c2}^{\text{obs}}}{\sqrt{s_{c2}^2/N_c + s_{t2}^2/N_t}} \right|.$$

Then we could choose for our test statistic

$$T^{\text{comb}} = \max \left(T^{\text{t-stat},1}, T^{\text{t-stat},2} \right).$$

In this case, a slightly more natural test statistic is based on Hotelling's T^2 statistic for the difference in vector of means. For $j = 1, 2$ let $\overline{Y}_{c,j}^{\text{obs}} = \sum_{i:W_i=0} Y_{i,j}^{\text{obs}}/N_c$ and $\overline{Y}_{t,j}^{\text{obs}} = \sum_{i:W_i=1} Y_{i,j}^{\text{obs}}/N_t$. Then let $\hat{V}_c/N_c + \hat{V}_t/N_t$ be an estimator for the covariance matrix of $(\overline{Y}_{t,1} - \overline{Y}_{c,1}, \overline{Y}_{t,2} - \overline{Y}_{c,2})'$, where

$$\hat{V}_c = \frac{1}{N_c - 1} \sum_{i:W_i=0} \begin{pmatrix} Y_{i,1}^{\text{obs}} - \overline{Y}_{c,1}^{\text{obs}} \\ Y_{i,2}^{\text{obs}} - \overline{Y}_{c,2}^{\text{obs}} \end{pmatrix} \cdot \begin{pmatrix} Y_{i,1}^{\text{obs}} - \overline{Y}_{c,1}^{\text{obs}} \\ Y_{i,2}^{\text{obs}} - \overline{Y}_{c,2}^{\text{obs}} \end{pmatrix}',$$

and

$$\hat{V}_t = \frac{1}{N_t - 1} \sum_{i:W_i=1} \begin{pmatrix} Y_{i,1}^{\text{obs}} - \overline{Y}_{t,1}^{\text{obs}} \\ Y_{i,2}^{\text{obs}} - \overline{Y}_{t,2}^{\text{obs}} \end{pmatrix} \cdot \begin{pmatrix} Y_{i,1}^{\text{obs}} - \overline{Y}_{t,1}^{\text{obs}} \\ Y_{i,2}^{\text{obs}} - \overline{Y}_{t,2}^{\text{obs}} \end{pmatrix}'.$$

Then a natural test statistic is

$$T^{\text{Hotelling}} = \begin{pmatrix} \overline{Y}_{t,1}^{\text{obs}} - \overline{Y}_{c,1}^{\text{obs}} \\ \overline{Y}_{t,2}^{\text{obs}} - \overline{Y}_{c,2}^{\text{obs}} \end{pmatrix}' \left(\hat{V}_c/N_c + \hat{V}_t/N_t \right)^{-1} \begin{pmatrix} \overline{Y}_{t,1}^{\text{obs}} - \overline{Y}_{c,1}^{\text{obs}} \\ \overline{Y}_{t,2}^{\text{obs}} - \overline{Y}_{c,2}^{\text{obs}} \end{pmatrix}, \tag{5.9}$$

which measures the Mahalanobis squared distance between the averages in the treatment group and the control group.

5.5.8   Choosing a Test Statistic

Given the wide variety of test statistics introduced here, let us now return to the question of how to choose one among them to calculate the one valid p-value. In principle, the choice should be governed by considering both plausible alternative hypotheses and the approximate distribution of the potential outcomes under both null and alternative hypotheses. Suppose one suspects the effect of the treatment to be multiplicative; in that case, a natural test statistic for assessing the null hypothesis of no effect would be the differences in the average logarithms of the outcomes between the treatment groups. If the null hypothesis does not hold because the effect is in fact multiplicative, such a test statistic will be more sensitive to this alternative hypothesis than the simple difference in averages, thus leading to greater power in the FEP. Similarly, if we expect the treatment to increase the dispersion of the outcomes but to leave the location unchanged, we can use the difference in or ratio of estimates of measures of dispersion, such as the sample variances or the interquartile ranges, for our test statistic. If the treatment does increase the dispersion but does not alter the location, such a test statistic will lead to more power when using the FEP than would a test statistic based on the difference in average outcomes by treatment status.

A second consideration concerns the distribution of the values of the observed potential outcomes. If the empirical distributions of the observed potential outcomes have some outliers, calculating average differences by treatment status may lead to an FEP with low power against an alternative that corresponds to a constant and additive treatment effect. In that case it may be possible to use a test statistic that measures the difference in the centers of the two observed potential outcome distributions, not affected by a few extreme values, such as the medians, trimmed means, ranks, or even maximum likelihood estimates of locations based on long-tailed distributions, such as the family of t-distributions. In practice, using the average difference in ranks is an attractive test statistic that has decent power in a wide range of settings.

5.6   A SMALL SIMULATION STUDY

To illustrate how the different statistics perform in a known setting, we conducted a small simulation study. The study was designed to see how much power various statistics had against different (e.g., additive versus multiplicative) alternatives under various distributions of the outcomes. Although we look here at multiple statistics, one must remember that the p-value retains its properties only for a single statistic: one cannot look at multiple p-values and choose the "best," as we discussed in Section 5.5.7.

In the basic simulation setting, the population distribution for $Y_i(0)$ is normal with mean zero and unit variance, $\mathcal{N}(0, 1)$. The treatment effect is τ for all units, so that $Y_i(1) = Y_i(0) + \tau \sim \mathcal{N}(\tau, 1)$. In each replication, we draw a random sample of size $N = 2000$ with $N_c = 1000$ assigned to the control group and $N_t = 1000$ assigned to the treatment group. We calculate p-values for the sharp null hypothesis that $Y_i(1) = Y_i(0)$ for all units. We carry out the calculations using three different test statistics. First, the absolute value of the simple difference in means for treated and controls, T^{dif} given in Equation (5.2). Second, we take the absolute value of the difference in medians T^{median} given in (5.4). Third, we take the absolute value of the difference in average ranks, T^{rank} given in (5.7). In all three cases, we calculate the p-value as the probability under the null hypothesis of getting a test statistic as large as the observed test statistic, or larger.

We repeat this process by repeatedly drawing random samples and approximating the corresponding p-values by simulation. We then compute the power of the tests for each test statistic as the proportion of p-values less than or equal to 0.10. We do this simulation for a range of values of $\tau > 0$. Figure 5.1 reports the proportions for the three different test statistics that generate p-values less than 0.1, as a function of τ. The solid line corresponds to the mean, the dashed line to the median, and the dotted line corresponds to the rank statistic. We see that the FEP-based rank and mean test statistics have similar performances, whereas the FEP based on the median has less power in this situation.

We then modify the basic data-generating process by changing the distribution of $Y_i(0)$. We add a binary random variable U_i to the normal components with $\Pr(U_i = 0) = 0.8$ and $\Pr(U_i = 5) = 0.2$, which leads to a distribution with 20% outliers. We again consider additive alternatives where $Y_i(1) = Y_i(0) + \tau$. In Figure 5.2 we present the

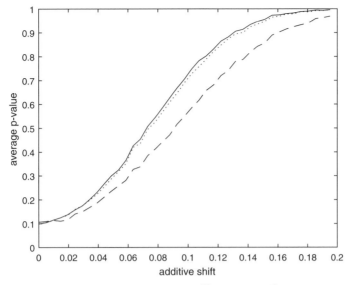

Figure 5.1.   Additive model with normal outcomes T^{dif} (solid), T^{median} (dashed), T^{rank} (dotted)

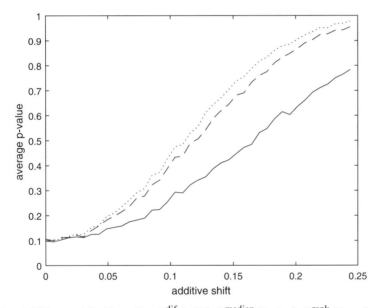

Figure 5.2.   Additive model with outliers T^{dif} (solid), T^{median} (dashed), T^{rank} (dotted)

power functions for the same three statistics. The rank-based and the median-based FEP's are superior here. The mean-based FEP has substantially worse power due to the presence of outliers.

In the third part, we change the distribution of $Y_i(0)$ so that the logarithm of $Y_i(0)$ has a normal distribution with mean zero and unit variance, and make the treatment effect multiplicative: $Y_i(1) = Y_i(0) \cdot \exp(\tau)$ for a range of values of τ. Exploiting the fact that the outcomes are positive in this case, we include a test statistic based on the difference

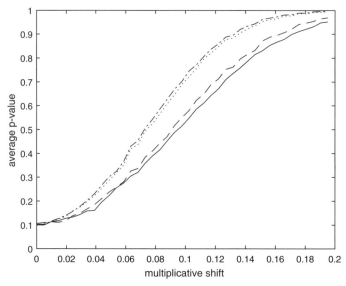

Figure 5.3.   Multiplicative model T^{dif} (solid), T^{median} (dashed), T^{rank} (dotted), T^{log} (dash-dot)

in average logarithms of the basic outcome, T^{log} given in (5.3). Figure 5.3 presents the results. Again the solid line corresponds to the mean, the dashed line to the median, and the dotted line corresponds to the rank statistic, and now the dash-dot line corresponds to the statistic based on the difference in average logarithms. The logarithm-based FEP and rank-based FEP both have superior power in this case compared to the mean-based FEP and median-based FEP.

Overall, these simulations suggest that the rank-based statistic is an attractive choice in a range of settings. It has relatively good power in all three settings considered, whereas the other choices for the test statistics performed well only in settings that play to their advantages, at the expense of relatively poor power in other settings.

5.7 INTERVAL ESTIMATES BASED ON FISHER P-VALUE CALCULATIONS

Earlier we discussed how we can use FEP calculations for null hypotheses other than that of absolutely no effect of the treatment, even if this was never considered in the original proposals by Fisher. Suppose, for example, we wish to assess the null hypothesis that for all units the effect of the treatment is an increase in test score equal to $C = 0.5$: $Y_i(1) = Y_i(0) + 0.5$. This assumption is itself a sharp null hypothesis and allows us to fill in all of the missing outcomes; Table 5.6 lists the full set of potential outcomes for the first six observations in the honey data set based on this null hypothesis. Given this complete knowledge, we can again calculate the randomization distribution of any test statistic and the corresponding p-value of any observed test statistic.

Let us now do this for a range of values of a postulated effect τ. The second column of Table 5.7 lists, for the full honey data set, the FEPs associated with a constant treatment

Table 5.6. *First Six Observations from Data from Honey Study with Missing Data in Parentheses under the Null Hypothesis of a Constant Effect of Size 0.5. Missing Potential Outcomes in Parentheses*

Unit	Potential Outcomes		Actual Treatment	Observed Outcome
	$Y_i(0)$	$Y_i(1)$		
1	(2.5)	3.0	1	3.0
2	(4.5)	5.0	1	5.0
3	(−0.5)	0.0	1	0.0
4	4.0	(4.5)	0	4.0
5	0.0	(0.5)	0	0.0
6	1.0	(1.5)	0	1.0

Note: Data based on cough frequency for first six units from honey study.

effect, C, for $C \in \{-3, -2.75, -2.50, \ldots, 1.00\}$. Here the test statistic is the absolute value of the difference in average outcomes for treated and control units minus C, and the p-value is the proportion of draws of the assignment vector leading to a test statistic at least as large as the observed value of that test statistic. From Table 5.7 we see that, for very negative values of C ($C < -1.50$) or very positive values ($C > 0.25$), the p-value is more extreme (smaller) than 0.05. Between these values there is a region where the C-based null hypothesis leads to p-values larger than 0.05. At the lower end of the range, we find that we obtain p-values less than 0.05 with a null hypothesis of a constant additive effect of -1.5, but not a constant additive effect of -1.25. The set of values where we get p-values larger than 0.05 is $[-1.44, 0.06]$, which provides a 95% "Fisher" interval for a common additive treatment effect, in the spirit of Fisher's exact p-values.

   In the third column of Table 5.7, we do the same for a rank-based test. To be clear here, let us be explicit about the calculation of the statistic and the p-value. If the null hypothesis is that the treatment effect is $Y_i(1) - Y_i(0) = C$, then we first calculate for each unit the implied value of $Y_i(0)$. For units with $W_i = 0$, we have $Y_i(0) = Y_i^{obs}$, and for units with $W_i = 1$, we have $Y_i(0) = Y_i^{obs} - C$ under the null hypothesis. Then we convert these $Y_i(0)$ to ranks R_i. Note that this rank is not the rank of Y_i^{obs}; rather it is, under the null hypothesis, the rank of $Y_i(0)$ (or, equivalently, under the null hypothesis, the rank of $Y_i(1)$). Next, we calculate the statistic as the average rank for the treated minus the average rank for the controls, $T = |\bar{R}_t - \bar{R}_c|$. Finally, we calculate the p-value for this test statistic, under the randomization distribution, as the proportion of values of the test statistic under the randomization distribution that are larger than or equal to the realized value of the test statistic. The set of values where we get p-values equal to or larger than 0.05 is $[-2.00, -0.00]$, which provides a 95% "Fisher" interval for the treatment effect.

5.8   COMPUTATION OF P-VALUES

The p-value calculations presented so far, other than those in the simulations in Section 5.6, have been exact; we have been able to calculate precisely in how many randomizations the test statistic T would be more extreme than our observed value

Table 5.7. *P-Values for Tests of Constant Treatment Effects (Full Honey Data Set from Table 5.1, with Cough Frequency as Outcome)*

Hypothesized Treatment Effect	P-Value (level)	P-Value (rank)
−3.00	0.000	0.000
−2.75	0.000	0.000
−2.50	0.000	0.000
−2.25	0.000	0.000
−2.00	0.001	0.000
−1.75	0.006	0.078
−1.50	0.037	0.078
−1.44	0.050	0.078
−1.25	0.146	0.078
−1.00	0.459	0.628
−0.75	0.897	0.428
−0.50	0.604	0.428
−0.25	0.237	0.429
0.00	0.067	0.043
0.06	0.050	0.043
0.25	0.014	0.001
0.50	0.003	0.000
0.75	0.000	0.001
1.00	0.000	0.000

Note: The level statistic is the absolute value of the difference in treated and control averages minus the hypothesized value, and the p-value is based on the proportion of statistics at least as large as the observed value. The rank-based statistic is the difference in average ranks for the treated and control units, of the value of the potential outcome under the null treatment.

of T. We could do these calculations exactly because the samples were small. In general, however, with N_t units assigned to the treatment group and N_c units assigned to the control group, the number of distinct values of the assignment vector is $\binom{N_c+N_t}{N_t}$, which, as we saw in Table 4.1 in Chapter 4, can grow very quickly with N_c and N_t. With both N_c and N_t sufficiently large, it may be infeasible to calculate the test statistic for every value of the assignment vector, even with current advances in computing. This does not mean, however, that it is difficult to calculate an accurate p-value associated with a test statistic, because we can rely on numerical approximations to the p-value.

It is typically very easy to obtain an accurate approximation of the p-value associated with a specific test statistic and null hypothesis. To do this, instead of calculating the statistic for every single value of the assignment vector $\mathbf{W} \in \mathbb{W}^+$, we calculate it for only a randomly chosen subset of possible assignment vectors. Let $T^{\text{dif,obs}}$ be the observed value of the test statistic. Then, randomly draw an N-dimensional vector with N_c zeros and N_t ones from the set of possible assignment vectors. For each draw from this set,

Table 5.8. *P-Values Estimated through Simulation for Honey Data from Table 5.1 for Null Hypothesis of Zero Effects*

Number of Simulations	P-Value	$\widehat{(s.\,e.)}$
100	0.010	(0.010)
1,000	0.044	(0.006)
10,000	0.044	(0.002)
100,000	0.042	(0.001)
1,000,000	0.043	(0.000)

Note: Statistic is absolute value of difference in average ranks of treated and control cough frequencies. P-value is proportion of draws at least as large as observed statistic.

the probability of being drawn is $1 \Big/ \binom{N_c+N_t}{N_t}$. Calculate the statistic for the first draw, say $T^{\mathrm{dif},1} = \overline{Y}_{t,1} - \overline{Y}_{c,1}$. Repeat this process $K - 1$ times, in each instance drawing a new vector of assignments and calculating the statistic $T^{\mathrm{dif},k} = \overline{Y}_{t,k} - \overline{Y}_{c,k}$, for $k = 2, \ldots, K$. We then approximate the p-value for our test statistic by the fraction of these K statistics that are as extreme as, or more extreme than, the observed value $T^{\mathrm{dif,obs}}$,

$$\hat{p} = \frac{1}{K} \sum_{k=1}^{N} \mathbf{1}_{T^{\mathrm{dif},k} \geq T^{\mathrm{dif,obs}}}.$$

If we were to draw the assignment vectors without replacement, and we sampled $\binom{N_c+N_t}{N_t}$ assignment vectors, we would have calculated the statistic for all assignment vectors, and we would obtain the exact p-value. In practice, if K is large, the p-value based on a random sample will be quite accurate. For this approximation, it does not matter whether we sample with or without replacement. The latter will lead to slightly more precise p-values for modest values of K, but both will lead to accurate p-values with K large enough because each assignment vector has the same probability of being drawn with or without replacement. The accuracy of this approximation is, therefore, entirely within the researcher's control. One can determine the number of independent draws required for a given degree of accuracy. Given a true p-value of p^*, and K draws from the set of possible assignment vectors, the large-sample standard error of the p-value is $\sqrt{p^*(1-p^*)/K}$. The maximum value for the standard error is achieved at $p^* = 1/2$, in which case the standard error of the estimated p-value is $1/(2\sqrt{K})$. Hence, if we want to estimate the p-value accurately enough that its standard error is less than 0.001, it suffices to use $K = 250,000$ draws, which is computationally entirely feasible unless the calculation of the test statistic is itself tedious (which it rarely is, although it can be, for example, when the test statistic is based on a model without closed-form estimates).

   To illustrate this approach, we now analyze the full data set from the Honey Study for which the summary statistics are presented in Table 5.1. Table 5.8 reports the p-value for the null hypothesis of no effect, and using for our approximated p-values, $K = 100$, $K = 1,000$, $K = 10,000$, $K = 100,000$, and $K = 1,000,000$. The statistic used is the absolute value of the difference between average ranks for treated and control, and the p-value reported is the proportion of assignment vectors that leads to a value for the test statistic at least as large as the observed value of the test statistic.

5.9   FISHER EXACT P-VALUES WITH COVARIATES

Thus far, all of the statistics considered have ignored the presence of any pre-treatment variables. Their presence greatly expands the set of possible test statistics. Here we discuss a few additional statistics that are feasible exploiting the presence of covariates.

First, one can use the pre-treatment variables to transform the observed outcome. For instance, if the pre-treatment variable is analogous to the outcome but measured prior to assignment to treatment or control (for instance, a pre-test score), it can be useful to subtract this variable from the potential outcomes and then carry out the test on the transformed outcomes, commonly referred to as *gain scores*. Thus, define

$$Y_i'(w) = Y_i(w) - X_i,$$

for each level of the treatment w, and define the realized transformed outcome as

$$Y_i'^{,obs} = Y_i^{obs} - X_i = \begin{cases} Y_i'(0) & \text{if } W_i = 0, \\ Y_i'(1) & \text{if } W_i = 1. \end{cases}$$

Such gain scores are often used in educational research. One should resist the temptation, though, to interpret the gain $Y_i'^{,obs}$ as a causal effect of the program for a treated unit i. Such an interpretation requires that $Y_i(0)$ is equal to X_i, which is generally not warranted.

The unit-level causal effect on the modified outcome Y' is $Y_i'(1) - Y_i'(0)$. Substituting $Y_i'(w) = Y_i(w) - X_i$ shows that this causal effect is identical to the unit-level causal effect on the original outcome Y_i, $Y_i(1) - Y_i(0)$. Hence the null hypothesis that $Y_i(0) = Y_i(1)$ for all units is identical to the null hypothesis that $Y_i'(1) = Y_i'(0)$ for all units. However, the FEP based on $Y_i'^{,obs}$ generally differs from the FEP based on Y_i^{obs}. A natural test statistic, based on average differences between treated and control units, measured in terms of the transformed outcome is

$$
\begin{aligned}
T^{\text{gain}} &= \frac{\sum_{i:W_i=1} Y_i'^{,obs}}{N_{\text{t}}} - \frac{\sum_{i:W_i=0} Y_i'^{,obs}}{N_{\text{c}}} \\[2mm]
&= \frac{\sum_{i:W_i=1} \left(Y_i^{obs} - X_i\right)}{N_{\text{t}}} - \frac{\sum_{i:W_i=0} \left(Y_i^{obs} - X_i\right)}{N_{\text{c}}} \\[2mm]
&= \overline{Y}_{\text{t}}^{obs} - \overline{Y}_{\text{c}}^{obs} - (\overline{X}_{\text{t}} - \overline{X}_{\text{c}}),
\end{aligned}
\tag{5.10}
$$

where $\overline{X}_{\text{c}} = \sum_{i:W_i=0} X_i/N_{\text{c}}$ and $\overline{X}_{\text{t}} = \sum_{i:W_i=1} X_i/N_{\text{t}}$ are the average value of the covariate in the control and treatment group respectively. Compare this test statistic with the statistic based on the simple difference in average outcomes, $T^{\text{dif}} = \overline{Y}_{\text{t}}^{obs} - \overline{Y}_{\text{c}}^{obs}$. The difference between the two statistics is equal to the difference in pre-treatment averages by treatment group, $\overline{X}_{\text{t}} - \overline{X}_{\text{c}}$. This difference is, on average (i.e., averaged over all assignment vectors), equal to zero by the randomization, but typically it is different from zero for any particular assignment vector. The distribution of the test statistic $T^{\mathrm{gain}} = \overline{Y}_{\mathrm{t}}^{\mathrm{obs}} - \overline{Y}_{\mathrm{c}}^{\mathrm{obs}} - (\overline{X}_{\mathrm{t}} - \overline{X}_{\mathrm{c}})$ will therefore generally differ from that of $T^{\mathrm{dif}} = \overline{Y}_{\mathrm{t}}^{\mathrm{obs}} - \overline{Y}_{\mathrm{c}}^{\mathrm{obs}}$, and thus so will be the associated p-value.

An alternative transformation involving the pre-test score is to use the proportional change from baseline, so that

$$Y_i''(w) = \frac{Y_i(w) - X_i}{X_i}, \quad \text{for } w = 0, 1,$$

and

$$Y_i''^{,\mathrm{obs}} = \frac{Y_i^{\mathrm{obs}} - X_i}{X_i}.$$

Here the implicit causal effect being estimated for unit i is

$$\frac{Y_i(1) - X_i}{X_i} - \frac{Y_i(0) - X_i}{X_i} = \frac{Y_i(1) - Y_i(0)}{X_i}.$$

A natural test statistic is now

$$T^{\mathrm{prop-change}} = \overline{Y''}_{\mathrm{t}} - \overline{Y''}_{\mathrm{c}} = \frac{1}{N_{\mathrm{t}}} \sum_{i:W_i=1} \frac{Y_i^{\mathrm{obs}} - X_i}{X_i} - \frac{1}{N_{\mathrm{c}}} \sum_{i:W_i=0} \frac{Y_i^{\mathrm{obs}} - X_i}{X_i}. \tag{5.11}$$

Both the gain score and the proportional change from baseline statistics are likely to lead to more powerful tests if the covariate X_i is a good proxy for $Y_i(0)$. Such a situation often arises if the covariate is a lagged value of the outcome, for example, a pre-test score in an educational testing example, or lagged earnings in a job-training example.

Both T^{gain} and $T^{\mathrm{prop-change}}$ use the covariates in a very specific way: transforming the original outcome using a known, pre-specified function. Such transformations make sense if one has a clear prior notion about the relationship between the potential outcomes and the covariate. Often, however, one may think that the covariate is highly correlated with the potential outcomes, but their scales may be different, for example, if X_i is a health index and Y_i is post-randomization medical complications for unit i. In that case, it is useful to consider a more general way to exploit the presence of covariates.

Recall that any scalar function $T = T(\mathbf{W}, \mathbf{Y}^{\mathrm{obs}}, \mathbf{X})$ can be used in the FEP framework. One possibility is to calculate a more complicated transformation that involves the values of both outcomes and pre-treatment variables for all units. For instance, let $(\hat{\beta}_0, \hat{\beta}_X, \hat{\beta}_W)$ be the least squares coefficients in a regression of Y_i^{obs} on a constant, X_i, and W_i:

$$\left(\hat{\beta}_0, \hat{\beta}_X, \hat{\beta}_W \right) = \arg \min_{\beta_0, \beta_X, \beta_W} \sum_{i=1}^{N} \left(Y_i^{\mathrm{obs}} - \beta_0 - \beta_X \cdot X_i - \beta_W \cdot W_i \right)^2.$$

These least squares coefficients are obviously functions of $(\mathbf{W}, \mathbf{Y}^{\mathrm{obs}}, \mathbf{X})$. An alternative choice for the test statistic is then

$$T^{\mathrm{reg-coef}} = \hat{\beta}_W. \tag{5.12}$$

Table 5.9. *P-Values for Honey Data from Table 5.1, for Null Hypothesis of Zero Effects Using Various Statistics*

Test Statistic	Statistic	P-Value
T^{dif}	−0.697	0.067
T^{quant} ($\delta = 0.25$)	−1.000	0.440
T^{quant} ($\delta = 0.50$)	−1.000	0.637
T^{quant} ($\delta = 0.75$)	−1.000	0.576
$T^{\text{t-stat}}$	−1.869	0.065
T^{rank}	−9.785	0.043
T^{ks}	0.304	0.021
$T^{\text{Hotelling}}$	3.499	0.182
T^{gain}	−0.967	0.006
$T^{\text{reg-coef}}$	−0.911	0.008

Note: Outcome is cough frequency (`cfa`) with the exception of $T^{\text{Hotelling}}$, which is based on cough frequency and cough severity (`cfa` and `csa`). The p-value is proportion of draws at least as large as observed statistic.

This statistic is likely to be more powerful than those based on simple differences in observed outcomes if the covariates are powerful predictors of the potential outcomes.

As before, the validity of a test based on only one such statistic does not rely on the regression model being correctly specified. However, the increases in power will be especially realized when the model provides a reasonable approximation to the distribution of values of the potential outcomes in both treatment conditions.

5.10   FISHER EXACT P-VALUES FOR THE HONEY DATA

Now we return to the full honey data set with all seventy-two observations. Table 5.9 lists ten test statistics and corresponding p-values, with the p-values estimated using 1,000,000 draws from the randomization distribution. The p-values are based on the post-treatment cough frequency (`cfa`) and the post-treatment cough severity (`csa`). Again, here we report multiple p-values, although, in theory, only one is valid, the one specified *a priori*, and in practice, one should do only one, or adjust the p-values as discussed in Section 5.5.7.

First we report the p-values when the statistic is the absolute value of the simple difference in average cough frequency by treatment status, $T^{\text{dif}} = |\overline{Y}_t^{\text{obs}} - \overline{Y}_c^{\text{obs}}|$. This leads to a p-value of 0.067. Next we report three quantile-based statistics, T^{quant} given in (5.5), for the quartiles $\delta = 0.25$, $\delta = 0.5$, and $\delta = 0.75$. Note that, due to the discrete nature of the outcome variable used here, cough frequency after the treatment, the observed values of the statistic are the same for all three choices of δ, although the implied p-values differ. The quantile-based p-values are considerably higher compared to those based on the difference-in-means statistic, illustrating that with discrete outcomes, quantile-based statistics can have low statistical power. Fifth, we use the conventional t-statistic, $T^{\text{t-stat}}$ given in (5.6). The p-value for this test is similar to that for the simple difference in means. Note that the p-value based on the normal approximation to the distribution of this statistic is 0.062, fairly close to the p-value based on the randomization distribution because the sample size is reasonably large. Next, we use the difference in average ranks, taking account of ties, using the statistic T^{rank} given in (5.7). This leads to a smaller p-value, equal to 0.042. Then we use the Kolmogorov-Smirnov-based test statistic, given in (5.8). The maximum difference observed between the cumulative distribution functions is 0.304. As can be seen from Table 5.2, this maximum difference occurs at $y = 2$, where $\hat{F}_t(2) = 0.63$ and $\hat{F}_e(2) = 0.32$. The p-value using the Kolmogorov-Smirnov-based statistic is 0.021.

The eighth p-value uses both outcomes, cough frequency and cough severity. The test statistic is based on Hotelling's T^2 statistic, $T^{\text{Hotelling}}$ in (5.9). The last two p-values involve the pre-treatment variable `cfp`. First we calculate the statistic based on the absolute value of the difference in gains scores, T^{gain}, as given in (5.10). The last test uses the estimated regression coefficient as the test statistic, $T^{\text{reg-coef}}$, as given in (5.12). Both lead to substantially lower p-values than the statistics that do not exploit the pre-treatment variables. This reflects the strong correlation between the prior cough frequency and *ex post* cough frequency (the unconditional correlation is 0.41 in the full sample).

5.11 CONCLUSION

The FEP approach is an excellent one for simple situations when one is willing to assess the premise of a sharp null hypothesis. It is also a very useful starting point, prior to any more sophisticated analysis, to investigate whether a treatment does indeed have some effect on outcomes of interest. For this purpose, an attractive approach is to use the test statistic equal to the absolute value of the difference in average ranks by treatment status, and to calculate the p-value as the probability, under the null hypothesis of absolutely no effect of the treatment, of the test statistic being as large as, or larger than, the realized value of the test statistic. In most situations, however, researchers are not solely interested in obtaining p-values for sharp null hypotheses. Simply being confident that there is some effect of the treatment for some units is not sufficient to inform policy decisions. Instead researchers often wish to obtain estimates of the average treatment effect without being concerned about variation in the effects. In such settings the FEP approach does not immediately apply. In the next chapter, we discuss a framework for inference developed by Neyman (1923) that does directly apply in such settings, at least asymptotically, while maintaining a randomization perspective.

NOTES

As stated here, what we call "Fisher interval" was not actually proposed by Fisher, but may be close to what Fisher would have called a "fiducial interval."

Extensive work on exact inference using the randomization distribution, considerably extending Fisher's work in this area, has been done by Kempthorne and in the recent literature by Rosenbaum. See among others, Kempthorne (1952, 1955), Rosenbaum (1984a, 1988, 1989b, 2002), and Imbens and Rosenbaum (2004). Rosenbaum's work also focuses on interval estimation using randomization inference. Surveys of this work include Rosenbaum (2002, 2009). Randomization tests based on residuals from regression analyses are discussed in Gail, Tian, and Piantadosi (1988). An interesting application of randomization inference to the California recall election is presented in Ho and Imai (2006).

A Bayesian approach to the analysis of randomized experiments is developed in Rubin (1978). We will discuss a closely related model-based approach in Chapter 8. Rubin (1990a) provides a general discussion of modes of inference for causal effects, relating randomization-based inference to other modes of inference, such as those discussed in Chapters 6, 7, and 8.

The Wilcoxon rank sum test was originally developed for equal-sized treatment and control groups in Wilcoxon (1945). Generalizations were developed in Mann and Whitney (1947); see also Lehman (1975) and Rosenbaum (2000).

## CHAPTER 6

Neyman's Repeated Sampling Approach to Completely Randomized Experiments

6.1 INTRODUCTION

In the last chapter we introduced the Fisher Exact P-value (FEP) approach for assessing sharp null hypotheses. As we saw, a sharp null hypothesis allowed us to fill in the values for all missing potential outcomes in the experiment. This was the basis for deriving the randomization distributions of various statistics, that is, the distributions induced by the random assignment of the treatments given fixed potential outcomes under that sharp null hypothesis. During the same period in which Fisher was developing this method, Neyman (1923, 1990) was focused on methods for the estimation of, and inference for, average treatment effects, also using the distribution induced by randomization, sometimes in combination with repeated sampling of the units in the experiment from a larger population of units. At a general level, he was interested in the long-run operating characteristics of statistical procedures under both repeated sampling from the population and randomized assignment of treatments to the units in the sample. Specifically, he attempted to find point estimators that were unbiased, and also interval estimators that had the specified nominal coverage in large samples. As noted before, his focus on average effects was different from the focus of Fisher; the average effect across a population may be equal to zero even when some, or even all, unit-level treatment effects differ from zero.

Neyman's basic questions were the following. What would the average outcome be if all units were exposed to the active treatment, $\overline{Y}(1)$ in our notation? How did that compare to the average outcome if all units were exposed to the control treatment, $\overline{Y}(0)$ in our notation? Most importantly, what is the difference between these averages, the average treatment effect $\tau_{\mathrm{fs}} = \overline{Y}(1) - \overline{Y}(0) = \sum_{i=1}^{N} (Y_i(1) - Y_i(0))/N$? (Here we use the subscript fs to be explicit about the fact that the estimand is the finite-sample average treatment effect. Later we use the notation τ_{sp} to denote the super-population average treatment effect.) Neyman's approach was to develop an estimator of the average treatment effect and derive its mean and variance under repeated sampling. By repeated sampling we refer to the sampling generated by drawing from both the population of units, and from the randomization distribution (the assignment vector $\mathbf{W}$), although Neyman never described his analysis this way. His approach is similar to Fisher's, in that both consider the distribution of statistics (functions of the observed $\mathbf{W}$ and $\mathbf{Y}^{\mathrm{obs}}$) under the randomization distribution, with all potential outcomes regarded as fixed. The similarity ends there. In Neyman's analysis, we do not start with an assumption that allows us to fill in all values of the missing potential outcomes, and so we cannot derive the exact randomization distribution of statistics of interest. However, without such an assumption we can often still obtain good estimators of aspects of this distribution, for example, first and second moments. Neyman's primary concern was whether an estimator was unbiased for the average treatment effect τ_{fs}. A secondary goal was to construct an interval estimator for the causal estimand, which he hoped to base on an unbiased estimator for the sampling variance of the average treatment effect estimator. Confidence intervals, as they were called later by Neyman (1934), are stochastic intervals that are constructed in such a way that they include the true value of the estimand with probability, over repeated draws, at least equal to some fixed value, the confidence coefficient.

The remainder of this chapter is organized as follows. In Section 6.2 we begin by describing the data that will be used to illustrate the concepts discussed in this chapter. These data are from a randomized experiment conducted by Duflo, Hanna, and Ryan (2012) to assess the effect of a teacher-incentive program on teacher performance. Next, in Section 6.3, we introduce Neyman's estimator for the average treatment effect and show that it is unbiased for the average treatment effect, given a completely randomized experiment. We then calculate, in Section 6.4, the sampling variance of this estimator and propose an estimator of this variance in Section 6.5. There are several approaches one can take in this latter step, depending on whether one assumes a constant additive treatment effect. In Section 6.6 we discuss the construction of confidence intervals. Throughout the first part of this discussion, we assume that our interest is in a finite population of size N. Because we do not attempt to infer anything about units outside this population, it does not matter how this population was selected; the entire analysis is conditional on the population itself. In Section 6.7 we relax this assumption and instead consider, as did Neyman (1923, 1990), a population of units so that we can view the sample of N units as a random sample drawn from this population. Given this shift in perspective, we reinterpret the original results, especially with respect to the choice of estimator for the sampling variance, and the associated large sample confidence interval for the average effect. In Section 6.8 we discuss the role of covariates in Neyman's approach. In the current chapter we allow only for discrete covariates. With continuous covariates the analysis is more complicated, and we discuss various methods in Chapters 7 and 8. Next, in Section 6.9, we apply Neyman's approach to the data from the Duflo-Hanna-Ryan teacher-incentive experiment. Section 6.10 concludes. Throughout the chapter we maintain the stability assumption, SUTVA.

6.2 THE DUFLO-HANNA-RYAN TEACHER-INCENTIVE EXPERIMENT DATA

To illustrate the methods discussed in this chapter, we use data from a randomized experiment conducted in rural India by Duflo, Hanna, and Ryan (2012), designed to study the effect of financial incentives on teacher performance, measured both directly by teacher absences and indirectly by educational output measures, such as average class test scores. A sample of 113 single-teacher schools was selected, and in a randomly selected subset

Table 6.1. *Summary Statistics for Duflo-Hanna-Ryan Teacher-Incentive Observed Data*

	Variable	Control ($N_c = 54$)		Treated ($N_t = 53$)			
		Average	(S.D.)	Average	(S.D.)	Min	Max
Pre-treatment	pctprewritten	0.19	(0.19)	0.16	(0.17)	0.00	0.67
Post-treatment	open	0.58	(0.19)	0.80	(0.13)	0.00	1.00
	pctpostwritten	0.47	(0.19)	0.52	(0.23)	0.05	0.92
	written	0.92	(0.45)	1.09	(0.42)	0.07	2.22
	written_all	0.46	(0.32)	0.60	(0.39)	0.04	1.43

of 57 schools, the salary structure was changed so that teachers were given a salary that was tied to their (i.e., the teachers') attendance over a month-long period, whereas in the remaining 56 schools, the salary structure was unchanged. In both treatment and control schools, the teachers were given cameras with time stamps and asked to have students take pictures of the class with the teacher, both at the beginning and at the end of every school day. In addition, there were random unannounced visits to the schools by program officials to see whether the school was open or not.

In the current chapter, to focus on Neyman's approach, we avoid complicating issues of unintended missing data, and we drop six schools with missing data and use the $N = 107$ schools with recorded values for all five key variables, in addition to the treatment indicator: four outcomes and one covariate. Out of these 107 schools/teachers, $N_t = 53$ were in the treatment group with a salary schedule tied to teacher attendance, and $N_c = 54$ were in the control sample. In our analyses, we use four outcome variables. The first is the proportion of times the school was open during a random visit (open). The second outcome is the percentage of students who completed a writing test (pctpostwritten). The third is the value of the writing test score (written), averaged over all the students in each school who took the test. Even though not all students took the test, in each class at least some students took the writing test at the end of the study. The fourth outcome variable is the average writing test score with zeros imputed for the students who did not take the test (written_all). We use one covariate in the analysis, the percentage of students who completed the written test prior to the study (pctprewritten).

Table 6.1 presents summary statistics for the data set. For all five variables (the pretreatment variables pctprewritten, and the four outcome variables open, pctpostwritten, written, and written_all), we present averages and standard deviations by treatment status, and the minimum and maximum values over the full sample.

6.3 UNBIASED ESTIMATION OF THE AVERAGE TREATMENT EFFECT

Suppose we have a population consisting of N units. As before, for each unit there exist two potential outcomes, $Y_i(0)$ and $Y_i(1)$, corresponding to the outcome under control and treatment respectively. As with the Fisher Exact P-value (FEP) approach discussed in the previous chapter, the potential outcomes are considered fixed. As a result, the only random component is the vector of treatment assignments, $\mathbf{W}$, with i^{th} element W_i, which by definition has a known distribution in a completely randomized experiment.

Neyman was interested in the population average treatment effect:

$$\tau_{\text{fs}} = \frac{1}{N} \sum_{i=1}^{N} \left(Y_i(1) - Y_i(0) \right) = \overline{Y}(1) - \overline{Y}(0),$$

where $\overline{Y}(0)$ and $\overline{Y}(1)$ are the averages of the potential control and treated outcomes respectively:

$$\overline{Y}(0) = \frac{1}{N} \sum_{i=1}^{N} Y_i(0), \quad \text{and} \quad \overline{Y}(1) = \frac{1}{N} \sum_{i=1}^{N} Y_i(1).$$

Suppose that we observe data from a completely randomized experiment in which $N_t = \sum_{i=1}^{N} W_i$ units are randomly selected to be assigned to treatment and the remaining $N_c = \sum_{i=1}^{N} (1 - W_i)$ are assigned to control. Because of the randomization, a natural estimator for the average treatment effect is the difference in the average outcomes between those assigned to treatment and those assigned to control:

$$\hat{\tau}^{\text{dif}} = \overline{Y}_t^{\text{obs}} - \overline{Y}_c^{\text{obs}},$$

where

$$\overline{Y}_c^{\text{obs}} = \frac{1}{N_c} \sum_{i:W_i=0} Y_i^{\text{obs}} \quad \text{and} \quad \overline{Y}_t^{\text{obs}} = \frac{1}{N_t} \sum_{i:W_i=1} Y_i^{\text{obs}}.$$

Theorem 6.1 *The estimator $\hat{\tau}^{\text{dif}}$ is unbiased for τ_{fs}.*

Proof of Theorem 6.1. Using the fact that $Y_i^{\text{obs}} = Y_i(1)$ if $W_i = 1$, and $Y_i^{\text{obs}} = Y_i(0)$ if $W_i = 0$, we can write the estimator $\hat{\tau}^{\text{dif}}$ as:

$$\hat{\tau}^{\text{dif}} = \frac{1}{N} \sum_{i=1}^{N} \left(\frac{W_i \cdot Y_i(1)}{N_t/N} - \frac{(1 - W_i) \cdot Y_i(0)}{N_c/N} \right).$$

Because we view the potential outcomes as fixed, the only component in this statistic that is random is the treatment assignment, W_i. Given the setup of a completely randomized experiment (N units, with N_t randomly assigned to the treatment), by Section 3.5, $\Pr_W(W_i = 1|\mathbf{Y}(0), \mathbf{Y}(1)) = \mathbb{E}_W[W_i|\mathbf{Y}(0), \mathbf{Y}(1)] = N_t/N$. (Here we index the probability and expectation, and later the variance, operators by W to stress that the probability, expectation, or variance, is taken solely over the randomization distribution, keeping fixed the potential outcomes $\mathbf{Y}(0)$ and $\mathbf{Y}(1)$, and keeping fixed the population.) Thus,

$\hat{\tau}^{\text{dif}}$ is unbiased for the average treatment effect τ_{fs}:

$$\mathbb{E}_W\left[\hat{\tau}^{\text{dif}}\,\middle|\,\mathbf{Y}(0),\mathbf{Y}(1)\right] = \frac{1}{N}\sum_{i=1}^{N}\left(\frac{\mathbb{E}_W[W_i]\cdot Y_i(1)}{N_{\text{t}}/N} - \frac{\mathbb{E}_W[1-W_i])\cdot Y_i(0)}{N_{\text{c}}/N}\right)$$

$$= \frac{1}{N}\sum_{i=1}^{N}\left(Y_i(1) - Y_i(0)\right) = \tau_{\text{fs}}.$$

$\square$

Note that the estimator is unbiased, irrespective of the share of treated and control units in the randomized experiment. This does not imply, however, that this share is irrelevant for inference; it can greatly affect the precision of the estimator, as we see in the next section.

For the teacher-incentive experiment, taking the proportion of days that the school was open (open) as the outcome of interest, this estimator for the average effect is

$$\hat{\tau}^{\text{dif}} = \overline{Y}_{\text{t}}^{\text{obs}} - \overline{Y}_{\text{c}}^{\text{obs}} = 0.80 - 0.58 = 0.22,$$

as can be seen from the numbers in Table 6.1.

6.4   THE SAMPLING VARIANCE OF THE NEYMAN ESTIMATOR

Neyman was also interested in constructing interval estimates for the average treatment effect, which he later (Neyman, 1934) termed confidence intervals. This construction involves three steps. First, derive the sampling variance of the estimator for the average treatment effect. Second, develop estimators for this sampling variance. Third, appeal to a central limit argument for the large sample normality of $\hat{\tau}$ over its randomization distribution and use its estimated sampling variance from step 2 to create a large-sample confidence interval for the average treatment effect τ_{fs}.

In this section we focus on the first step, deriving the sampling variance of the proposed estimator $\hat{\tau}^{\text{dif}} = \overline{Y}_{\text{t}}^{\text{obs}} - \overline{Y}_{\text{c}}^{\text{obs}}$. This derivation is relatively cumbersome because the assignments for different units are not independent in a completely randomized experiment. With the number of treated units fixed at N_{t}, the fact that unit i is assigned to the active treatment lowers the probability that unit i' will receive active treatment. To show how to derive the sampling variance, we start with a simple example of only two units with one unit assigned to each treatment group. We then expand our discussion to the general case with N units and N_{t} randomly assigned to active treatment.

6.4.1   The Sampling Variance of the Neyman Estimator with Two Units

Consider the simple case with one treated and one control unit. The estimand, the finite sample average treatment effect, in this case is

$$\tau_{\text{fs}} = \frac{1}{2}\cdot\left[(Y_1(1) - Y_1(0)) + (Y_2(1) - Y_2(0))\right]. \tag{6.1}$$

In a completely randomized experiment, both units cannot receive the same treatment; it follows that $W_1 = 1 - W_2$. The estimator for the average treatment effect is therefore:

$$\hat{\tau}^{\text{dif}} = W_1 \cdot \left(Y_1^{\text{obs}} - Y_2^{\text{obs}} \right) + (1 - W_1) \cdot \left(Y_2^{\text{obs}} - Y_1^{\text{obs}} \right).$$

If unit 1 receives the treatment ($W_1 = 1$), our estimate of the average treatment effect will be $\hat{\tau}^{\text{dif}} = Y_1^{\text{obs}} - Y_2^{\text{obs}} = Y_1(1) - Y_2(0)$. If on the other hand, $W_1 = 0$, the estimate will be $\hat{\tau} = Y_2^{\text{obs}} - Y_1^{\text{obs}} = Y_2(1) - Y_1(0)$, so that we can also write:

$$\hat{\tau}^{\text{dif}} = W_1 \cdot \left(Y_1(1) - Y_2(0) \right) + (1 - W_1) \cdot \left(Y_2(1) - Y_1(0) \right).$$

To simplify the following calculations of the sampling variance of this estimator, define the binary variable $D = 2 \cdot W_1 - 1$, so that $D \in \{-1, 1\}$, $W_1 = (1 + D)/2$ and $W_2 = 1 - W_1 = (1 - D)/2$. Because the expected value of the random variable W_1 is equal to $1/2$, the expected value of D, over the randomization distribution, is $\mathbb{E}_W[D] = 0$, and the variance is $\mathbb{V}_W(D) = \mathbb{E}_W[D^2] = D^2 = 1$. In terms of D and the potential outcomes, we can write the estimator $\hat{\tau}$ as:

$$\hat{\tau}^{\text{dif}} = \frac{D + 1}{2} \cdot \left(Y_1(1) - Y_2(0) \right) + \frac{1 - D}{2} \cdot \left(Y_2(1) - Y_1(0) \right),$$

which can be rewritten as:

$$\hat{\tau}^{\text{dif}} = \frac{1}{2} \cdot \left[\left(Y_1(1) - Y_1(0) \right) + \left(Y_2(1) - Y_2(0) \right) \right]$$

$$+ \frac{D}{2} \cdot \left[\left(Y_1(1) + Y_1(0) \right) - \left(Y_2(1) + Y_2(0) \right) \right]$$

$$= \tau_{\text{fs}} + \frac{D}{2} \cdot \left[\left(Y_1(1) + Y_1(0) \right) - \left(Y_2(1) + Y_2(0) \right) \right].$$

Because $\mathbb{E}_W[D] = 0$, we can see immediately that $\hat{\tau}^{\text{dif}}$ is unbiased for τ_{fs} (which we already established in Section 6.3 for the general case). However, the representation in terms of D also makes the calculation of its sampling variance straightforward:

$$\mathbb{V}_W(\hat{\tau}^{\text{dif}}) = \mathbb{V}_W\left(\tau_{\text{fs}} + \frac{D}{2} \cdot \left[\left(Y_1(1) + Y_1(0) \right) - \left(Y_2(1) + Y_2(0) \right) \right] \right)$$

$$= \frac{1}{4} \cdot \mathbb{V}_W(D) \cdot \left[\left(Y_1(1) + Y_1(0) \right) - \left(Y_2(1) + Y_2(0) \right) \right]^2,$$

because τ and the potential outcomes are fixed. Given that $\mathbb{V}_W(D) = 1$, it follows that the sampling variance of our estimator $\hat{\tau}^{\text{dif}}$ is equal to:

$$\mathbb{V}_W(\hat{\tau}^{\text{dif}}) = \frac{1}{4} \cdot \left[\left(Y_1(1) + Y_1(0) \right) - \left(Y_2(1) + Y_2(0) \right) \right]^2. \tag{6.2}$$

This representation of the sampling variance shows that this will be an awkward object to estimate, because it depends on all four potential outcomes, including products of the different potential outcomes for the same unit that are never jointly observed.

6.4.2 The Sampling Variance of the Neyman Estimator with N Units

Next, we look at the general case with $N > 2$ units, of which N_t are randomly assigned to treatment. To calculate the sampling variance of $\hat{\tau}^{\text{dif}} = \overline{Y}_t^{\text{obs}} - \overline{Y}_c^{\text{obs}}$, we need the expectations of the second and cross moments of the treatment indicators W_i for $i = 1, \ldots, N$. Because $W_i \in \{0, 1\}$, $W_i^2 = W_i$, and thus

$$\mathbb{E}_W\left[W_i^2\right] = \mathbb{E}_W\left[W_i\right] = \frac{N_t}{N}, \quad \text{and} \quad \mathbb{V}_W(W_i) = \frac{N_t}{N} \cdot \left(1 - \frac{N_t}{N}\right).$$

To calculate the cross moment in a completely randomized experiment, recall that with the number of treated units fixed at N_t, the two events – unit i being treated and unit i' being treated – are not independent. Therefore $\mathbb{E}_W\left[W_i \cdot W_{i'}\right] \neq \mathbb{E}_W\left[W_i\right] \cdot \mathbb{E}_W\left[W_{i'}\right] = (N_t/N)^2$. Rather:

$$\mathbb{E}_W[W_i \cdot W_{i'}] = \Pr_W(W_i = 1) \cdot \Pr_W(W_{i'} = 1 | W_i = 1) = \frac{N_t}{N} \cdot \frac{N_t - 1}{N - 1}, \quad \text{for } i \neq j,$$

because conditional on $W_i = 1$ there are $N_t - 1$ treated units remaining, out of a total of $N - 1$ units remaining. Given the sampling moments derived, we can infer the sampling variance and covariance of W_i and $W_{i'}$.

Theorem 6.2 *The sampling variance of $\hat{\tau}^{\text{dif}} = \overline{Y}_t^{\text{obs}} - \overline{Y}_c^{\text{obs}}$ is*

$$\mathbb{V}_W\left(\overline{Y}_t^{\text{obs}} - \overline{Y}_c^{\text{obs}}\right) = \frac{S_c^2}{N_c} + \frac{S_t^2}{N_t} - \frac{S_{ct}^2}{N}, \tag{6.3}$$

where S_c^2 and S_t^2 are the variances of $Y_i(0)$ and $Y_i(1)$ in the sample, defined as:

$$S_c^2 = \frac{1}{N - 1} \sum_{i=1}^{N} \left(Y_i(0) - \overline{Y}(0)\right)^2, \quad \text{and} \quad S_t^2 = \frac{1}{N - 1} \sum_{i=1}^{N} \left(Y_i(1) - \overline{Y}(1)\right)^2,$$

and S_{ct}^2 is the sample variance of the unit-level treatment effects, defined as:

$$S_{ct}^2 = \frac{1}{N - 1} \sum_{i=1}^{N} \left(Y_i(1) - Y_i(0) - (\overline{Y}(1) - \overline{Y}(0))\right)^2$$

$$= \frac{1}{N - 1} \sum_{i=1}^{N} \left(Y_i(1) - Y_i(0) - \tau_{fs}\right)^2.$$

Proof of Theorem 6.2. See Appendix A.

Let us consider the interpretation of the three components of this variance in turn. The first two are related to sample variances for averages of random samples. Recall that the finite-sample average treatment effect is the difference in average potential outcomes: $\tau_{fs} = \overline{Y}(1) - \overline{Y}(0)$. To estimate τ_{fs}, we first estimate $\overline{Y}(1)$, the population average potential outcome under treatment, by the average outcome for the N_t treated units, $\overline{Y}_t^{\text{obs}}$. This estimator is unbiased for $\overline{Y}(1)$. The population variance of $Y_i(1)$ is $S_t^2 = \sum_i (Y_i(1) - \overline{Y}(1))^2/(N - 1)$. Given this population variance for $Y_i(1)$, the sampling variance for an average of a random sample of size N_t would be $(S_t^2/N_t) \cdot (1 - N_t/N)$, where the last factor is the finite sample correction. The first term has this form, except for the finite sample correction. Similarly, the average outcome for the N_c units assigned to control, $\overline{Y}_c^{obs}$, is unbiased for the population average outcome under the control treatment, $\overline{Y}(0)$, and its sampling variance, ignoring the finite population correction, is S_c^2/N_c. These results follow by direct calculation, or by using standard results from the analysis of simple random samples: given a completely randomized experiment, the N_t treated units provide a simple random sample of the N values of $Y_i(1)$, and the N_c control units provide a simple random sample of the N values of $Y_i(0)$.

The third component of this sampling variance, S_{ct}^2/N, is the sample variance of the unit-level treatment effects, $Y_i(1) - Y_i(0)$. If the treatment effect is constant in the population, this third term is equal to zero. If the treatment effect is not constant, S_{ct}^2 is positive. Because it is subtracted from the sum of the first two elements in the expression for the sampling variance of $\overline{Y}_t^{obs} - \overline{Y}_c^{obs}$, Equation (6.3), the positive value for S_{ct}^2 reduces the sampling variance of this estimator for the average treatment effect.

There is an alternative representation of the sampling variance of $\hat{\tau}^{dif}$ that is useful. First we write the variance of the unit-level treatment effect as a function of ρ_{ct}, the population correlation coefficient between the potential outcomes $Y_i(1)$ and $Y_i(0)$:

$$S_{ct}^2 = S_c^2 + S_t^2 - 2 \cdot \rho_{ct} \cdot S_c \cdot S_t,$$

where

$$\rho_{ct} = \frac{1}{(N-1) \cdot S_c \cdot S_t} \sum_{i=1}^{N} \left(Y_i(1) - \overline{Y}(1) \right) \cdot \left(Y_i(0) - \overline{Y}(0) \right). \tag{6.4}$$

By definition, ρ_{ct} is a correlation coefficient and so lies in the interval $[-1, 1]$. Substituting this representation of S_{ct}^2 into Equation (6.3), the alternative expression for the sampling variance of $\hat{\tau}^{dif}$ (alternative to (6.3)) is:

$$\mathbb{V}_W \left(\overline{Y}_t^{obs} - \overline{Y}_c^{obs} \right) = \frac{N_t}{N \cdot N_c} \cdot S_c^2 + \frac{N_c}{N \cdot N_t} \cdot S_t^2 + \frac{2}{N} \cdot \rho_{ct} \cdot S_c \cdot S_t. \tag{6.5}$$

The sampling variance of our estimator is smallest when the potential outcomes are perfectly negatively correlated ($\rho_{ct} = -1$), so that

$$S_{ct}^2 = S_c^2 + S_t^2 + 2 \cdot S_c \cdot S_t,$$

and

$$\mathbb{V}_W \left(\overline{Y}_t^{obs} - \overline{Y}_c^{obs} \,\middle|\, \rho_{ct} = -1 \right) = \frac{N_t}{N \cdot N_c} \cdot S_c^2 + \frac{N_c}{N \cdot N_t} \cdot S_t^2 - \frac{2}{N} \cdot S_c \cdot S_t,$$

and largest when the two potential outcomes are perfectly positively correlated ($\rho_{ct} = +1$), so that

$$S_{ct}^2 = S_c^2 + S_t^2 - 2 \cdot S_c \cdot S_t,$$

and

$$
\mathbb{V}_W \left(\overline{Y}_t^{\text{obs}} - \overline{Y}_c^{\text{obs}} \,\middle|\, \rho_{\text{ct}} = 1 \right) = \frac{N_t}{N \cdot N_c} \cdot S_c^2 + \frac{N_c}{N \cdot N_t} \cdot S_t^2 + \frac{2}{N} \cdot S_c \cdot S_t
$$

$$
= \frac{S_c^2}{N_c} + \frac{S_t^2}{N_t} - \frac{(S_c - S_t)^2}{N}. \tag{6.6}
$$

The most notable special case of perfect correlation arises when the treatment effect is constant and additive, $Y_i(1) - Y_i(0) = \tau$ for all $i = 1, \ldots, N$. In that case,

$$
\mathbb{V}^{\text{const}} = \mathbb{V}_W \left(\overline{Y}_t^{\text{obs}} - \overline{Y}_c^{\text{obs}} \,\middle|\, \rho_{\text{ct}} = 1, S_c^2 = S_t^2 \right) = \frac{S_c^2}{N_c} + \frac{S_t^2}{N_t}. \tag{6.7}
$$

The fact that the sampling variance of $\overline{Y}_t^{\text{obs}} - \overline{Y}_c^{\text{obs}}$ is largest when the treatment effect is constant (i.e., not varying) across units may appear somewhat counterintuitive. Let us therefore return to the two-unit case and consider the form of the sampling variance there in more detail. In the two-unit case, the sampling variance, presented in Equation (6.2), is a function of the sum of the two potential outcomes for each of the two units. Consider two numerical examples. In the first example, $Y_i(0) = Y_i(1) = 10$, and $Y_2(0) = Y_2(1) = -10$, corresponding to a zero treatment effect for both units. To calculate the correlation between the two potential outcomes, we use expression (6.4) for ρ_{ct} and find the numerator of ρ_{ct} equals

$$
\frac{1}{N-1} \sum_{i=1}^{N} \left(Y_i(1) - \overline{Y}(1) \right) \cdot \left(Y_i(0) - \overline{Y}(0) \right)
$$

$$
= \left((Y_1(1) - 0) \cdot (Y_1(0) - 0) + (Y_2(1) - 0) \cdot (Y_2(0) - 0) \right) = 200,
$$

and the two components of the denominator of ρ_{ct} equal

$$
S_c^2 = \frac{1}{N-1} \sum_{i=1}^{N} \left(Y_i(0) - \overline{Y}(0) \right)^2 = \left((10 - 0)^2 + (-10 - 0)^2 \right) = 200,
$$

and

$$
S_t^2 = \frac{1}{N-1} \sum_{i=1}^{N} \left(Y_i(1) - \overline{Y}(1) \right)^2 = \left((10 - 0)^2 + (-10 - 0)^2 \right) = 200,
$$

so that the correlation between the two potential outcomes is 1. In the second example, suppose that $Y_1(0) = Y_2(1) = -10$, and $Y_1(1) = Y_2(0) = 10$. A similar calculation shows that the correlation between the two potential outcomes is now -1. In both examples the *average* treatment effect is zero, but in the first case the treatment effect is constant and thus equal to 0 for each unit, whereas in the second case the treatment effect for unit 1 is equal to 20, and for unit 2 the treatment effect is equal to -20. As a result, when estimating the average treatment effect, in the first case the two possible values of the estimator are $Y_1^{\text{obs}} - Y_2^{\text{obs}} = 20$ (if $W_1 = 1$ and $W_2 = 0$) and $Y_2^{\text{obs}} - Y_1^{\text{obs}} = -20$ (if $W_1 = 0$ and $W_2 = 1$). In contrast, in the second case the two values of the estimator are both equal to 0. Hence, the sampling variance of the estimator in the first case, with $\rho_{ct} = +1$, is positive (in fact, equal to 20^2), whereas in the second case, with $\rho_{ct} = -1$, the sampling variance is 0.

6.5   ESTIMATING THE SAMPLING VARIANCE

Now that we have derived the sampling variance of our estimator, $\hat{\tau}^{\mathrm{dif}} = \overline{Y}_t^{\mathrm{obs}} - \overline{Y}_c^{\mathrm{obs}}$, the next step is to develop an estimator for this sampling variance. To do this, we consider separately each of the three elements of the sampling variance given in Equation (6.3).

The numerator of the first term, the sample variance of the potential control outcome vector, $\mathbf{Y}(0)$, is equal to S_c^2. As shown in Appendix A, or from standard results on simple random samples, an unbiased estimator for S_c^2 is

$$s_c^2 = \frac{1}{N_c - 1} \sum_{i:W_i=0} \left(Y_i(0) - \overline{Y}_c^{\mathrm{obs}}\right)^2 = \frac{1}{N_c - 1} \sum_{i:W_i=0} \left(Y_i^{\mathrm{obs}} - \overline{Y}_c^{\mathrm{obs}}\right)^2.$$

Analogously, we can estimate S_t^2, the population variance of $Y_i(1)$, by

$$s_t^2 = \frac{1}{N_t - 1} \sum_{i:W_i=1} \left(Y_i(1) - \overline{Y}_t^{\mathrm{obs}}\right)^2 = \frac{1}{N_t - 1} \sum_{i:W_i=1} \left(Y_i^{\mathrm{obs}} - \overline{Y}_t^{\mathrm{obs}}\right)^2.$$

The third term, S_{ct}^2 (the population variance of the unit-level treatment effects), is generally impossible to estimate empirically because we never observe both $Y_i(1)$ and $Y_i(0)$ for the same unit. We therefore have no direct observations on the variation in the treatment effects across the population and therefore cannot directly estimate S_{ct}^2. As noted previously, if the treatment effects are constant and additive ($Y_i(1) - Y_i(0) = \tau_{\mathrm{fs}}$ for all units), then this component of the sampling variance is equal to zero and the third term vanishes. Thus we have proved:

Theorem 6.3 *If the treatment effect $Y_i(1) - Y_i(0)$ is constant, then an unbiased estimator for the sampling variance is*

$$\hat{\mathbb{V}}^{\mathrm{neyman}} = \frac{s_c^2}{N_c} + \frac{s_t^2}{N_t}. \tag{6.8}$$

This estimator for the sampling variance is widely used, even when the assumption of an additive treatment effect may be known to be inaccurate. There are two main reasons for the popularity of this estimator for the sampling variance. First, by implicitly setting the third element of the estimated sampling variance equal to zero, the expected value of $\hat{\mathbb{V}}^{\mathrm{neyman}}$ is at least as large as the true sampling variance of $\overline{Y}_t^{\mathrm{obs}} - \overline{Y}_c^{\mathrm{obs}}$, irrespective of the heterogeneity in the treatment effect, because the third term is non-negative. Hence, in large samples, confidence intervals generated using this estimator of the sampling variance will have coverage at least as large, but not necessarily equal to, their nominal coverage.[1] (Note that this statement still needs to be qualified by the clause "in large samples," because we rely on the central limit theorem to construct normal-distribution-based confidence intervals.) It is interesting to return to the discussion between Fisher and Neyman regarding the general interest in average treatment effects and sharp null hypotheses. Neyman's proposed estimator for the sampling variance is unbiased only in the case of a constant additive treatment effect, which is satisfied under the sharp null hypothesis of no treatment effects whatsoever, which was the case considered by Fisher. In other cases the proposed estimator of the sampling variance generally over-estimates the true sampling variance of $\overline{Y}_t^{obs} - \overline{Y}_c^{obs}$. As a result, Neyman's interval is generally statistically conservative in large samples. The second reason for using $\hat{\mathbb{V}}^{neyman}$ as an estimator for the sampling variance of $\overline{Y}_t^{obs} - \overline{Y}_c^{obs}$ is that it is always unbiased for the sampling variance of $\hat{\tau}^{dif}$ as an estimator of the infinite super-population average treatment effect; we discuss this population interpretation at greater length in Section 6.7.

In the remainder of this section, we consider two alternative estimators for the sampling variance of $\hat{\tau}^{dif}$. The first explicitly allows for treatment effect heterogeneity. Under treatment effect heterogeneity, the estimator for the sampling variance in Equation (6.8), $\hat{\mathbb{V}}^{neyman}$, provides an upwardly biased estimate: the third term, which vanishes if the treatment effect is constant, is now negative. The question arises whether we can improve upon the Neyman variance estimator without risking under coverage in large samples.

To see that there is indeed information to do so, recall our argument that an implication of constant treatment effects is that the variances S_c^2 and S_t^2 are equal. A difference between these variances, which would in large samples lead to a difference in the corresponding estimates s_c^2 and s_t^2, indicates variation in the treatment effects. To use this information to create a better estimator for the sampling variance of $\overline{Y}_t^{obs} - \overline{Y}_c^{obs}$, let us turn to the representation of the sampling variance in Equation (6.5), which incorporates ρ_{ct}, the population correlation coefficient between the potential outcomes:

$$\mathbb{V}_W\left(\overline{Y}_t^{obs} - \overline{Y}_c^{obs}\right) = S_t^2 \cdot \frac{N_t}{N \cdot N_c} + S_c^2 \cdot \frac{N_c}{N \cdot N_t} + \rho_{ct} \cdot S_c \cdot S_t \cdot \frac{2}{N}.$$

Conditional on a value for the correlation coefficient, ρ_{ct}, we can estimate this sampling variance as

$$\hat{\mathbb{V}}^{\rho_{ct}} = s_c^2 \cdot \frac{N_t}{N \cdot N_c} + s_t^2 \cdot \frac{N_c}{N \cdot N_t} + \rho_{ct} \cdot s_c \cdot s_t \cdot \frac{2}{N}. \tag{6.9}$$

This variance is again largest if the two potential outcomes are perfectly correlated, that is, $\rho_{01} = 1$. An alternative conservative estimator of the sampling variance that exploits

---

[1] This potential difference between actual and nominal coverage of confidence intervals in randomized experiments concerned Neyman, and probably with this in mind, he formally defined confidence intervals in 1934 to allow for the possibility that the actual coverage could be greater than the nominal coverage. Thus the proposed "conservative" intervals are still valid in large samples. Fisher (1934) in his discussion did not agree with the propriety of this definition.

this bound is

$$\hat{\mathbb{V}}^{\rho_{ct}=1} = s_c^2 \cdot \frac{N_t}{N \cdot N_c} + s_1^2 \cdot \frac{N_c}{N \cdot N_t} + s_c \cdot s_t \cdot \frac{2}{N}$$

$$= \frac{s_c^2}{N_c} + \frac{s_t^2}{N_t} - \frac{(s_t - s_c)^2}{N}. \tag{6.10}$$

If s_c^2 and s_t^2 are unequal, then $\hat{\mathbb{V}}^{\rho_{ct}=1}$ will be smaller than $\hat{\mathbb{V}}^{\text{neyman}}$. Using $\hat{\mathbb{V}}^{\rho_{ct}=1}$ to construct confidence intervals will result in tighter confidence intervals than using $\hat{\mathbb{V}}^{\text{neyman}}$, without compromising their large-sample validity. The intervals based on $\hat{\mathbb{V}}^{\rho_{ct}=1}$ will still be conservative in large samples, because $\hat{\mathbb{V}}^{\rho_{ct}=1}$ is still upwardly biased when the true correlation is smaller than one, although less so than $\hat{\mathbb{V}}^{\text{neyman}}$. Note, however, that with no information beyond the fact that $s_c^2 \neq s_t^2$, all choices for ρ_{ct} smaller than unity raise the possibility that we will underestimate the sampling variance and construct invalid confidence intervals.

Next consider an alternative sampling variance estimator under the additional assumption that the treatment effect is constant, $Y_i(1) - Y_i(0) = \tau$ for all i. This alternative estimator exploits the fact that under the constant treatment assumption, the population variances of the two potential outcomes, S_c^2 and S_t^2, must be equal. We can therefore define $S^2 \equiv S_c^2 = S_t^2$ and pool the outcomes for the treated and control units to estimate this common variance:

$$s^2 = \frac{1}{N-2} \cdot \left(s_c^2 \cdot (N_c - 1) + s_t^2 \cdot (N_t - 1) \right)$$

$$= \frac{1}{N-2} \cdot \left(\sum_{i:W_i=0} \left(Y_i^{\text{obs}} - \overline{Y}_c^{\text{obs}} \right)^2 + \sum_{i:W_i=1} \left(Y_i^{\text{obs}} - \overline{Y}_t^{\text{obs}} \right)^2 \right). \tag{6.11}$$

The larger sample size for this estimator (from N_c and N_t for s_c^2 and s_t^2 respectively, to N for s^2), leads to a more precise estimator for the sampling variance of $\overline{Y}_t^{\text{obs}} - \overline{Y}_c^{\text{obs}}$ if the treatment effect is constant, namely

$$\hat{\mathbb{V}}^{\text{const}} = s^2 \cdot \left(\frac{1}{N_c} + \frac{1}{N_t} \right). \tag{6.12}$$

When the treatment effects are constant this estimator is preferable to either $\hat{\mathbb{V}}^{\text{neyman}}$ or $\hat{\mathbb{V}}^{\rho_{ct}=1}$, but if not, it need not be valid. Both $\hat{\mathbb{V}}^{\text{neyman}}$ and $\hat{\mathbb{V}}^{\rho_{ct}=1}$ are valid generally and therefore may be preferred.

Let us return to the Duflo-Hanna-Ryan teacher-incentive data. The estimate for the average effect of assignment to the incentives-based salary rather than the conventional salary structure, on the probability that the school is open, is, as discussed in the previous section, equal to 0.22. Now let us consider estimators for the sampling variance. First we estimate the sample variances S_c^2, S_t^2, and the combined variance S^2; the estimates are

$$s_c^2 = 0.19^2, \quad s_t^2 = 0.13^2, \quad \text{and} \quad s^2 = 0.16^2.$$

The two sample variances s_c^2 and s_t^2 are quite different, with their ratio being larger than two. Next we use the sample variances of the potential outcomes to estimate the sampling variance for the average treatment effect estimator. The first estimate for the sampling variance, which is, in general, conservative but allows for unrestricted treatment effect heterogeneity, is

$$\hat{\mathbb{V}}^{\text{neyman}} = \frac{s_c^2}{N_c} + \frac{s_t^2}{N_t} = 0.0311^2.$$

(We report four digits after the decimal point to make explicit the small differences between the various estimators for the sampling variance, although in practice one would probably only report two or three digits.) The second estimate, still conservative, but exploiting differences in the variances of the outcome by treatment group, and again allowing for unrestricted treatment effect heterogeneity, is

$$\hat{\mathbb{V}}^{\rho_{\text{ct}}=1} = s_c^2 \cdot \frac{N_t}{N \cdot N_c} + s_t^2 \cdot \frac{N_c}{N \cdot N_t} + s_c \cdot s_t \cdot \frac{2}{N} = 0.0305^2.$$

By construction this estimator is smaller than $\hat{\mathbb{V}}^{\text{neyman}}$. However, even though the variances s_c^2 and s_t^2 differ by more than a factor of two, the difference in the estimated sampling variances $\hat{\mathbb{V}}^{\rho_{\text{ct}}=1}$ and $\hat{\mathbb{V}}^{\text{neyman}}$ is very small in this example, less than 1%. In general, the standard variance $\hat{\mathbb{V}}^{\text{neyman}}$ is unlikely to be substantially larger than $\hat{\mathbb{V}}^{\rho_{\text{ct}}=1}$, as suggested by this example. The third and final estimate of the sampling variance, which relies on a constant treatment effect for its validity, is

$$\hat{\mathbb{V}}^{\text{const}} = s^2 \cdot \left(\frac{1}{N_c} + \frac{1}{N_t} \right) = 0.0312^2,$$

slightly larger than the other estimates, but essentially the same for practical purposes.

6.6 CONFIDENCE INTERVALS AND TESTING

In the introduction to this chapter, we noted that Neyman's interest in estimating the precision of the estimator for the average treatment effect was largely driven by an interest in constructing confidence intervals. By a confidence interval with confidence coefficient $1 - \alpha$, here we mean a pair of functions $C_L(\mathbf{Y}^{\text{obs}}, \mathbf{W})$ and $C_U(\mathbf{Y}^{\text{obs}}, \mathbf{W})$, defining an interval $[C_L(\mathbf{Y}^{\text{obs}}, \mathbf{W}), C_U(\mathbf{Y}^{\text{obs}}, \mathbf{W})]$, such that

$$\Pr_W(C_L(\mathbf{Y}^{\text{obs}}, \mathbf{W}) \leq \tau \leq C_U(\mathbf{Y}^{\text{obs}}, \mathbf{W})) \geq 1 - \alpha.$$

The only reason the lower and upper bounds in this interval are random is through their dependence on $\mathbf{W}$. The distribution of the confidence limits is therefore generated by the randomization. Note that, in this expression, the probability of including the true value τ may exceed $1 - \alpha$, in which case the interval is considered valid but conservative. Here we discuss a number of ways to construct such confidence intervals and to conduct tests for hypotheses concerning the average treatment effect. We will use the Duflo-Hanna-Ryan data to illustrate the steps of Neyman's approach.

6.6.1 Confidence Intervals

Let $\hat{\mathbb{V}}$ be an estimate of the sampling variance of $\hat{\tau}^{\text{dif}}$ over its randomization distribution (in practice we recommend using $\hat{\mathbb{V}}^{\text{neyman}}$). Suppose we wish to construct a 90% confidence interval. We base the interval on a normal approximation to the randomization distribution of $\hat{\tau}^{\text{dif}}$. This approximation is somewhat intellectually inconsistent with our stress on finite-sample properties of the estimator for τ and its sampling variance, but it is driven by the common lack of empirical *a priori* information about the joint distribution of the potential outcomes. As we will see, normality is often a good approximation to the randomization distribution of standard estimates, even in fairly small samples. To further improve on this approximation, we could approximate the distribution of $\hat{\mathbb{V}}^{\text{neyman}}$ by a chi-squared distribution, and then use that to approximate the distribution of $\hat{\tau}^{\text{dif}}/\sqrt{\hat{\mathbb{V}}^{\text{neyman}}}$ by a t-distribution. For simplicity here, we use the 5th and 95th percentile of the standard normal distribution, -1.645 and 1.645, to calculate a nominal central 90% confidence interval as:

$$\text{CI}^{0.90}(\tau_{\text{fs}}) = \left(\hat{\tau}^{\text{dif}} - 1.645 \cdot \sqrt{\hat{\mathbb{V}}}, \hat{\tau}^{\text{dif}} + 1.645 \cdot \sqrt{\hat{\mathbb{V}}} \right).$$

More generally, if we wish to construct a central confidence interval with nominal confidence level $(1 - \alpha) \times 100\%$, as usual we look up the $\alpha/2$ and $1 - \alpha/2$ quantiles of the standard normal distribution, denoted by $z_{\alpha/2}$, and construct the confidence interval:

$$\text{CI}^{1-\alpha}(\tau_{\text{fs}}) = \left(\hat{\tau}^{\text{dif}} + z_{\alpha/2} \cdot \sqrt{\hat{\mathbb{V}}}, \hat{\tau}^{\text{dif}} + z_{1-\alpha/2} \cdot \sqrt{\hat{\mathbb{V}}} \right).$$

This approximation applies when using any estimate of the sampling variance, and, in large samples, the resulting intervals are valid confidence intervals under the same assumptions that make the corresponding estimator for the sampling variance an unbiased or upwardly biased estimator of the true sampling variance.

Based on the three sampling variance estimates reported in the previous section for the outcome that the school is open, we obtain the three following 90% confidence intervals. First, based on $\hat{\mathbb{V}}^{\text{neyman}} = 0.0311^2$, we find

$$\text{CI}^{0.90}_{\text{neyman}}(\tau_{\text{fs}}) = \left(\hat{\tau}^{\text{dif}} + z_{0.10/2} \cdot \sqrt{\hat{\mathbb{V}}^{\text{neyman}}}, \hat{\tau}^{\text{dif}} + z_{1-0.10/2} \cdot \sqrt{\hat{\mathbb{V}}^{\text{neyman}}} \right)$$

$$= (0.2154 - 1.645 \cdot 0.0311, 0.2154 + 1.645 \cdot 0.0311) = (0.1642, 0.2667).$$

Second, based on the sampling variance estimator assuming a constant treatment effect, $\hat{\mathbb{V}}_{\text{const}} = 0.0312^2$, we obtain a very similar interval,

$$\text{CI}^{0.90}_{\text{const}}(\tau_{\text{fs}}) = (0.1640, 0.2668).$$

Finally, based on the third sampling variance estimator, $\hat{\mathbb{V}}_{\rho_{\text{ct}}=1} = 0.0305^2$, we obtain again a fairly similar interval,

$$\text{CI}^{0.90}_{\rho_{\text{ct}}=1}(\tau_{\text{fs}}) = (0.1652, 0.2657).$$

With the estimates for the sampling variances so similar, the three 90% large-sample confidence intervals are also very similar.

6.6.2 Testing

We can also use the sampling variance estimates to carry out tests of hypotheses concerning the average treatment effect. Suppose we wish to test the null hypothesis that the average treatment effect is zero against the alternative hypothesis that the average effect differs from zero:

$$H_0^{\text{neyman}} : \quad \frac{1}{N} \sum_{i=1}^{N} (Y_i(1) - Y_i(0)) = 0, \text{ and}$$

$$H_a^{\text{neyman}} : \quad \frac{1}{N} \sum_{i=1}^{N} (Y_i(1) - Y_i(0)) \neq 0.$$

A natural test statistic to use for Neyman's "average null" is the ratio of the point estimate to the estimated standard error. For the teacher-incentive data, the point estimate is $\overline{Y}_t^{\text{obs}} - \overline{Y}_c^{\text{obs}} = 0.2154$. The estimated standard error is, using the conservative estimator for the sampling variance, $\hat{\mathbb{V}}^{\text{neyman}}$, equal to 0.0311. The resulting t-statistic is therefore

$$t = \frac{\overline{Y}_t^{\text{obs}} - \overline{Y}_c^{\text{obs}}}{\sqrt{\hat{\mathbb{V}}^{\text{neyman}}}} = -\frac{0.2154}{0.0311} = 6.9.$$

The associated p-value for a two-sided test, based on the normal approximation to the distribution of the t-statistic, is $2 \cdot (1 - \Phi(6.9)) < 0.001$. At conventional significance levels, we clearly reject the (Neyman) null hypothesis that the average treatment effect is zero.

It is interesting to compare this test, based on Neyman's approach, to the FEP approach. There are two important differences between the two approaches. First, and most important, they assess different null hypotheses, for example, a zero average effect for Neyman versus a zero effect for all units for Fisher (although Fisher's null hypothesis implies Neyman's). Second, the Neyman test relies on a large-sample normal approximation for its validity, whereas the p-values based on the FEP approach are exact.

Let us discuss both differences in more detail. First consider the difference in hypotheses. The Neyman test assesses whether the average treatment effect is zero, whereas the FEP assesses whether the treatment effect is zero for all units in the experiment. Formally, in the Fisher approach the null hypothesis is

$$H_0^{\text{fisher}} : \quad Y_i(1) - Y_i(0) = 0 \text{ for all } i = 1, \dots, N,$$

and the (implicit) alternative hypothesis is

$$H_a^{\text{fisher}} : \quad Y_i(1) - Y_i(0) \neq 0 \text{ for some } i = 1, \dots, N.$$

Depending on the implementation of the FEP approach, this difference in null hypotheses may be unimportant. If we choose to use a test statistic proportional to the average difference, we end up with a test that has virtually no power against alternatives with heterogeneous treatment effects that average out to zero. We would have power against at least some of those alternatives if we choose a different statistic. Consider as an example a population where for all units $Y_i(0) = 2$. For 1/3 of the units the treatment effect is

2. For 2/3 of the units the treatment effect is -1. In this case the Neyman null hypothesis of a zero average effect is true. The Fisher null hypothesis of no effect whatsoever is not true. Whether we can detect this violation depends on the choice of statistic. The FEP approach, with the statistic equal to the average difference in outcomes by treatment status, has no power against this alternative. However, the FEP approach, with a different statistic, based on the average difference in outcomes after transforming the outcomes by taking logarithms, does have power in this setting. In this artificial example, the expected difference in logarithms by treatment status is -0.23. The FEP based on the difference in average logarithms will detect this difference in large samples.

The second difference between the two procedures is in the approximate nature of the Neyman test, compared to the exact results for the FEP approach. We use two approximations in the Neyman approach. First, we use the *estimated* variance (e.g., $\hat{\mathbb{V}}^{\mathrm{neyman}}$) instead of the *actual* variance ($\mathbb{V}_W(\overline{Y}_{\mathrm{t}}^{\mathrm{obs}} - \overline{Y}_{\mathrm{c}}^{\mathrm{obs}})$). Second, we use a normal *approximation* for the repeated sampling distribution of the difference in averages $\overline{Y}_{\mathrm{t}}^{\mathrm{obs}} - \overline{Y}_{\mathrm{c}}^{\mathrm{obs}}$. Both approximations are justified in large samples. If the sample is reasonably large, and if there are few or no outliers, as in the application in this chapter, these approximations will likely be accurate.

6.7   INFERENCE FOR POPULATION AVERAGE TREATMENT EFFECTS

In the introduction to this chapter, we commented on the distinction between a finite population interpretation, in which the sample of size N is considered the population of interest, and a super-population perspective, in which the N observed units are viewed as a random sample from an essentially infinite population. The second argument in favor of using the sampling variance estimator $\hat{\mathbb{V}}^{\mathrm{neyman}}$ in Equation (6.8) is that, regardless of the level of heterogeneity in the unit-level treatment effect, $\hat{\mathbb{V}}^{\mathrm{neyman}}$ is unbiased for the sampling variance of the estimator $\hat{\tau}^{\mathrm{dif}}$ for the super-population, as opposed to the finite sample, average treatment effect. Here we further explore this argument, address how it affects our interpretation of the estimator of the average treatment effect, and discuss the various choices of estimators for its sampling variance.

Suppose that the population of N subjects taking part in the completely randomized experiment is itself a simple random sample from a larger population, which, for simplicity, we assume is infinite. This is a slight departure from Neyman's explicit focus on the average treatment effect for a finite population. In many cases, however, this change of focus is immaterial. Although in some agricultural experiments, farmers may be genuinely interested in which fertilizer was best for their specific fields in the year of the experiment, in most social and medical science settings, experiments are, explicitly or implicitly, conducted with a view to inform policies for a larger population of units, often assumed to have generated the N units in our sample by random sampling. However, without additional information, we cannot hope to obtain more precise estimates for the treatment effects in the super-population than for the treatment effects in the sample. In fact, the estimates for the population estimands are typically strictly less precise. Ironically it is exactly this loss in precision that enables us to obtain unbiased estimates of the sampling variance of the traditional estimator for the average treatment effect in the super-population.

Viewing our N units as a random sample of the target super-population, rather than viewing them as the population itself, induces a distribution on the two potential outcomes for each unit. The pair of potential outcome values for an observed unit i is simply one draw from the distribution in the population and is, therefore, itself stochastic. The distribution of the pair of two potential outcomes in turn induces a distribution on the unit-level treatment effects and on the average of the unit-level treatment effects within the drawn sample. To be clear about this super-population perspective, we use the subscript fs to denote the finite-sample average treatment effect and sp to denote the super-population average treatment effect:

$$\tau_{fs} = \frac{1}{N} \sum_{i=1}^{N} (Y_i(1) - Y_i(0)) \quad \text{and} \quad \tau_{sp} = \mathbb{E}_{sp} [Y_i(1) - Y_i(0)].$$

Analogously, the subscript sp on the expectations operator indicates that the expectation is taken over the distribution generated by random sampling from the super-population and not solely over the randomization distribution. Thus $\tau_{sp} = \mathbb{E}_{sp}[Y_i(1) - Y_i(0)]$ is the expected value of the unit-level treatment effect, under the distribution induced by sampling from the super-population or, equivalently, the average treatment effect in the super-population. Because of the random sampling, τ_{sp} is also equal to the expected value of the finite-sample average treatment effect,

$$\mathbb{E}_{sp} [\tau_{fs}] = \mathbb{E}_{sp} \left[\overline{Y}(1) - \overline{Y}(0) \right] = \frac{1}{N} \sum_{i=1}^{N} \mathbb{E}_{sp} [Y_i(1) - Y_i(0)] = \tau_{sp}. \tag{6.13}$$

See Appendix B for details on the super-population perspective. Let σ_{ct}^2 be the variance of the unit-level treatment effect in this super-population, $\sigma_{ct}^2 = \mathbb{V}_{sp}(Y_i(1) - Y_i(0)) = \mathbb{E}_{sp}[(Y_i(1) - Y_i(0) - \tau_{sp})^2]$, and let σ_c^2 and σ_t^2 denote the population variances of the two potential outcomes, or the super-population expectations of S_c^2 and S_t^2:

$$\sigma_c^2 = \mathbb{V}_{sp}(Y_i(0)) = \mathbb{E}_{sp} \left[(Y_i(0) - \mathbb{E}_{sp}[Y_i(0)])^2 \right],$$

and

$$\sigma_t^2 = \mathbb{V}_{sp}(Y_i(1)) = \mathbb{E}_{sp} \left[(Y_i(1) - \mathbb{E}_{sp}[Y_i(1)])^2 \right].$$

The definition of the variance of the unit-level treatment effect within the super-population, σ_{ct}^2, implies that the variance of τ_{fs} across repeated random samples is equal to

$$\mathbb{V}_{sp}(\tau_{fs}) = \mathbb{V}_{sp} \left(\overline{Y}(1) - \overline{Y}(0) \right) = \sigma_{ct}^2/N. \tag{6.14}$$

Now let us consider the sampling variance of the standard estimator for the average treatment effect, $\hat{\tau}^{dif} = \overline{Y}_t^{obs} - \overline{Y}_c^{obs}$, given this sampling from the super-population. The expectation and variance operators without subscripts denote expectations and variances taken over both the randomization distribution and the random sampling from the super-population.

We have

$$\mathbb{V}\left(\hat{\tau}^{\text{dif}}\right) = \mathbb{E}\left[\left(\overline{Y}_{\text{t}}^{\text{obs}} - \overline{Y}_{\text{c}}^{\text{obs}} - \mathbb{E}\left[\overline{Y}_{\text{t}}^{\text{obs}} - \overline{Y}_{\text{c}}^{\text{obs}}\right]\right)^2\right]$$

$$= \mathbb{E}\left[\left(\overline{Y}_{\text{t}}^{\text{obs}} - \overline{Y}_{\text{c}}^{\text{obs}} - \mathbb{E}_{\text{sp}}\left[\overline{Y}(1) - \overline{Y}(0)\right]\right)^2\right],$$

where the second equality holds because $\mathbb{E}\left[\overline{Y}_{\text{t}}^{\text{obs}} - \overline{Y}_{\text{c}}^{\text{obs}}\right] = \mathbb{E}_{\text{sp}}[\overline{Y}(1) - \overline{Y}(0)] = \tau_{\text{sp}}$, as shown above. Adding and subtracting $\overline{Y}(1) - \overline{Y}(0)$ within the expectation, this sampling variance, over both randomization and random sampling, is equal to:

$$\mathbb{V}\left(\hat{\tau}^{\text{dif}}\right)$$

$$= \mathbb{E}\left[\left(\overline{Y}_{\text{t}}^{\text{obs}} - \overline{Y}_{\text{c}}^{\text{obs}} - \left(\overline{Y}(1) - \overline{Y}(0)\right) + \left(\overline{Y}(1) - \overline{Y}(0)\right) - \mathbb{E}_{\text{sp}}\left[\overline{Y}(1) - \overline{Y}(0)\right]\right)^2\right]$$

$$= \mathbb{E}\left[\left(\overline{Y}_{\text{t}}^{\text{obs}} - \overline{Y}_{\text{c}}^{\text{obs}} - \left(\overline{Y}(1) - \overline{Y}(0)\right)\right)^2\right]$$

$$+ \mathbb{E}_{\text{sp}}\left[\left(\left(\overline{Y}(1) - \overline{Y}(0)\right) - \mathbb{E}_{\text{sp}}\left[\overline{Y}(1) - \overline{Y}(0)\right]\right)^2\right]$$

$$+ 2 \cdot \mathbb{E}\left[\left(\overline{Y}_{\text{t}}^{\text{obs}} - \overline{Y}_{\text{c}}^{\text{obs}} - \left(\overline{Y}(1) - \overline{Y}(0)\right)\right) \cdot \left(\left(\overline{Y}(1) - \overline{Y}(0)\right) - \mathbb{E}_{\text{sp}}\left[\overline{Y}(1) - \overline{Y}(0)\right]\right)\right].$$

The third term of this last expression, the covariance term, is equal to zero because the expectation of the first factor, $\overline{Y}_{\text{t}}^{\text{obs}} - \overline{Y}_{\text{c}}^{\text{obs}} - (\overline{Y}(1) - \overline{Y}(0))$, conditional on the N-vectors $\mathbf{Y}(0)$ and $\mathbf{Y}(1)$ (taking the expectation just over the randomization distribution), is zero. Hence the sampling variance reduces to:

$$\mathbb{V}\left(\overline{Y}_{\text{t}}^{\text{obs}} - \overline{Y}_{\text{c}}^{\text{obs}}\right) = \mathbb{E}\left[\left(\overline{Y}_{\text{t}}^{\text{obs}} - \overline{Y}_{\text{c}}^{\text{obs}} - \overline{Y}(1) - \overline{Y}(0)\right)^2\right]$$

$$+ \mathbb{E}_{\text{sp}}\left[\left(\overline{Y}(1) - \overline{Y}(0) - \mathbb{E}_{\text{sp}}\left[Y(1) - Y(0)\right]\right)^2\right]. \qquad (6.15)$$

Earlier we showed that $\mathbb{E}_W\left[\overline{Y}_{\text{t}}^{\text{obs}} - \overline{Y}_{\text{c}}^{\text{obs}} \middle| \mathbf{Y}(0), \mathbf{Y}(1)\right] = \tau_{\text{fs}} = \overline{Y}(1) - \overline{Y}(0)$; hence by iterated expectations, the first term on the right side is equal to the expectation of the conditional (randomization-based) variance of $\overline{Y}_{\text{t}}^{\text{obs}} - \overline{Y}_{\text{c}}^{\text{obs}}$ (conditional on the N-vector of potential outcomes $\mathbf{Y}(0)$ and $\mathbf{Y}(1)$). This conditional variance is equal to

$$\mathbb{E}_W\left[\left(\overline{Y}_{\text{t}}^{\text{obs}} - \overline{Y}_{\text{c}}^{\text{obs}} - \overline{Y}(1) - \overline{Y}(0)\right)^2 \middle| \mathbf{Y}(0), \mathbf{Y}(1)\right] = \frac{S_{\text{c}}^2}{N_{\text{c}}} + \frac{S_{\text{t}}^2}{N_{\text{t}}} - \frac{S_{\text{ct}}^2}{N}, \qquad (6.16)$$

as in Equation (6.3). Recall that these earlier calculations were made when assuming that the sample N was the population of interest and thus were conditional on $\mathbf{Y}(0)$ and $\mathbf{Y}(1)$. The expectation of (6.16) over the distribution of $\mathbf{Y}(0)$ and $\mathbf{Y}(1)$ generated by sampling from the super-population is

$$
\mathbb{E}\left[\left(\overline{Y}_t^{obs} - \overline{Y}_c^{obs} - \overline{Y}(1) - \overline{Y}(0)\right)^2\right]
$$

$$
= \mathbb{E}_{sp}\left[\mathbb{E}_W\left[\left(\overline{Y}_t^{obs} - \overline{Y}_c^{obs} - \overline{Y}(1) - \overline{Y}(0)\right)^2 \middle| \mathbf{Y}(0), \mathbf{Y}(1)\right]\right]
$$

$$
= \mathbb{E}_{sp}\left[\frac{S_c^2}{N_c} + \frac{S_t^2}{N_t} - \frac{S_{ct}^2}{N}\right] = \frac{\sigma_c^2}{N_c} + \frac{\sigma_t^2}{N_t} - \frac{\sigma_{ct}^2}{N}.
$$

The expectation of the second term on the right side of Equation (6.15) is equal to σ_{ct}^2/N, as we saw in Equation (6.14). Thus the sampling variance of $\hat{\tau}^{dif}$ over sampling from the super-population equals:

$$
\mathbb{V}_{sp} = \mathbb{V}_{sp}\left(\hat{\tau}^{dif}\right) = \frac{\sigma_c^2}{N_c} + \frac{\sigma_t^2}{N_t}, \tag{6.17}
$$

which we can estimate without bias by substituting s_c^2 and s_t^2 for σ_c^2 and σ_t^2, respectively:

$$
\hat{\mathbb{V}}^{sp} = \frac{s_c^2}{N_c} + \frac{s_t^2}{N_t}.
$$

The estimator $\hat{\mathbb{V}}^{sp}$ is identical to the previously introduced conservative estimator of the sampling variance for the finite population average treatment effect estimator, $\hat{\mathbb{V}}^{neyman}$, presented in Equation 6.8. Under simple random sampling from the super-population, the expected value of the estimator $\hat{\mathbb{V}}^{neyman}$ equals $\mathbb{V}_{sp}$. Hence, considering the N observed units as a simple random sample from an infinite super-population, the estimator in (6.8) is an unbiased estimate of the sampling variance of the estimator of the super-population average treatment effect. Neither of the alternative estimators – $\hat{\mathbb{V}}^{const}$ in Equation (6.12), which exploits the assumption of a constant treatment effect, nor $\hat{\mathbb{V}}^{\rho_{ct}=1}$ in Equation (6.10), derived through bounds on the correlation coefficient – has this attractive quality. Thus, despite the fact that $\hat{\mathbb{V}}^{const}$ may be a better estimator of the sampling variance in the finite population when the treatment effect is constant, and $\hat{\mathbb{V}}^{\rho_{ct}=1}$ may be a better estimator of $\mathbb{V}_{fs}$, $\hat{\mathbb{V}}^{neyman}$ is used almost uniformly in practice in our experience, although the logic for it appears to be rarely explicitly discussed.

6.8   NEYMAN'S APPROACH WITH COVARIATES

One can easily extend Neyman's approach for estimating average treatment effects to settings with discrete covariates. In this case, one would partition the sample into sub-samples defined by the values of the covariate and then conduct the analysis separately within these subsamples. The resulting within-subsample estimators would be unbiased for the within-subsample average treatment effect. Taking an average of these estimates, weighted by subsample sizes, gives an unbiased estimate of the overall average treatment effect. As we see in Chapter 9, we consider this method in the discussion on stratified random experiments.

It is impossible, however, in general to derive estimators that are exactly unbiased under the randomization distribution, conditional on the covariates, when there are covariate values for which we have only treated or only control units, which is likely to happen with great frequency in settings with covariates that take on many values. In such settings, building a model for the potential outcomes, and using this model to create an estimator of the average treatment effect, is a more appealing option. We turn to this topic in the next two chapters.

6.9   RESULTS FOR THE DUFLO-HANNA-RYAN TEACHER-INCENTIVE DATA

Now let us return to the teacher-incentive data and systematically look at the results based on the methods discussed in the current chapter. We analyze four outcomes in turn, plus one "pseudo-outcome." For illustrative purposes, we report here a number of point, sampling variance, and interval estimates. The first variable we analyze, as if it were an outcome, is a pre-treatment variable, and so we know *a priori* that the causal effect of the treatment on this variable is zero, both at the unit level and on average. In general, it can be useful to carry out such analyses as a check on the success of the randomization: that is, we know here that the Fisher null hypothesis of no effect whatsoever is true. The pre-treatment variable is `pctprewritten`, the percentage of students in a school that took the pre-program writing test. For this variable, we estimate, as anticipated, the average effect to be small, -0.03, with a 95% confidence interval that comfortably includes zero, $(-0.10, 0.04)$.

Now we turn to the four "real" outcomes. In Table 6.2 we report estimates of the components of the variance, and in Table 6.3 we present estimates of and confidence intervals for the average treatment effects. First we focus on the causal effect of the attendance-related salary incentives on the proportion of days that the school was open during the days it was subject to a random check. The estimated effect is 0.22, with a 95% confidence interval of $[0.15, 0.28]$. It is clear that the attendance-related salary incentives appeared to lead to a higher proportion of days with the school open. We also look at the effect on the percentage of students in the school who took the written test, `pctpostwritten`. Here the estimated treatment effect is 0.05, with a 95% confidence interval of $[-0.03, 0.13]$. The effect is not statistically significant at the 5% level, but it is at the 10% level. Next, we look at the average score on the writing test, which leads to a point estimate of 0.17, with a 95% confidence interval of $[0.00, 0.34]$. Finally, we examine the average test score, assigning zeros to students not taking the test. Now we estimate an average effect of 0.14, with a 95% confidence interval of $[0.00, 0.28]$. As with the Fisher exact p-value approach, the interpretation of nominal levels for tests and interval estimates formally holds for only one such interval. In the final analysis, we look at estimates separately for two subsamples, defined by whether the proportion of students taking the initial writing test was zero or positive, to illustrate the application of the methods developed in this chapter to subpopulations defined by covariates. Again, these analyses are for illustrative purposes only, and we do not take account of the fact that we do multiple tests. The first subpopulation (`pctprewritten`$=0$) comprises 40 schools (37%) and the second (`pctprewritten`>0) 67 schools (63%). We analyze

Table 6.2. *Estimates of Components of Variance of Estimator for the Effect of Teacher Incentives on the Proportion of Days that the School is Open; $N_c = 54$, $N_t = 53$, Duflo-Hanna-Ryan Data*

Estimated means	$\overline{Y}_c^{obs}$	0.58
	$\overline{Y}_t^{obs}$	0.80
	$\hat{\tau}$	0.22
Estimated variance components	s_c^2	0.19^2
	s_t^2	0.13^2
	s^2	0.16^2
Sampling variance estimates	$\hat{\mathbb{V}}^{neyman} = \frac{s_c^2}{N_c} + \frac{s_t^2}{N_t}$	0.03^2
	$\hat{\mathbb{V}}^{const} = s^2 \cdot \left(\frac{1}{N_c} + \frac{1}{N_t}\right)$	0.03^2
	$\hat{\mathbb{V}}^{\rho_{ct}=1} = s_c^2 \cdot \frac{N_t}{N \cdot N_c} + s_t^2 \cdot \frac{N_c}{N \cdot N_t} + s_c \cdot s_t \cdot \frac{2}{N}$	0.03^2

Table 6.3. *Estimates of, and Confidence Intervals for, Average Treatment Effects for Duflo-Hanna-Ryan Teacher-Incentive Data*

$\hat{\tau}$	$\widehat{(s.e.)}$	95% C.I.
0.22	(0.03)	(0.15,0.28)
0.05	(0.04)	(−0.03,0.13)
0.17	(0.08)	(0.00,0.34)
0.14	(0.07)	(0.00,0.28)

Table 6.4. *Estimates of, and Confidence Intervals for, Average Treatment Effects for Duflo-Hanna-Ryan Teacher-Incentive Data*

Variable	pctpre $= 0$ (N = 40)			pctprewritten > 0 (N = 67)			Difference		
	$\hat{\tau}$	$\widehat{(s.e.)}$	95% C.I.	$\hat{\tau}$	$\widehat{(s.e.)}$	95% C.I.	EST	$\widehat{(s.e.)}$	95% C.I.
open	0.23	(0.05)	(0.14,0.32)	0.21	(0.04)	(0.13,0.29)	0.02	(0.06)	(−0.10,0.14)
pctpost written	−0.004	(0.06)	(−0.16,0.07)	0.11	(0.05)	(0.01,0.21)	−0.15	(0.08)	(−0.31,0.00)
written	0.20	(0.10)	(0.00,0.40)	0.18	(0.10)	(−0.03,0.38)	0.03	(0.15)	(−0.26,0.31)
written _all	0.04	(0.07)	(−0.10,0.19)	0.22	(0.09)	(0.04,0.40)	−0.18	(0.12)	(−0.41,0.05)

separately the effect of assignment to attendance-based teacher incentives on all four outcomes. The descriptive results are reported in Table 6.4. The main substantive finding is that the effect of the incentive scheme on writing skills (written) appears lower for schools where many students entered with insufficient writing skills to take the initial test. The 95% confidence interval comfortably includes zero (−0.41, 0.05), and the 90% confidence interval is (−0.37, 0.01).

6.10   CONCLUSION

In this chapter we discuss Neyman's approach to estimation and inference in completely randomized experiments. He was interested in assessing the operating characteristics of statistical procedures under repeated sampling and random assignment of treatments. Neyman focused on the average effect of the treatment. He proposed an estimator for the average treatment effect in the finite sample, and showed that it was unbiased under repeated sampling. He also derived the sampling variance for this estimator. Finding an estimator for this sampling variance that itself is unbiased turned out to be impossible in general. Instead Neyman showed that the standard estimator for the sampling variance of this estimator is positively biased, unless the treatment effects are constant and additive, in which case it is unbiased. Like Fisher's approach, Neyman's methods have great appeal in the settings where they apply. However, again like Fisher's methods, there are many situations where we are interested in questions beyond those answered by their approaches. For example, we may want to estimate average treatment effects adjusting for differences in covariates in settings where some covariate values appear only in treatment or control groups. In the next two chapters we discuss methods that do not have the exact (finite sample) statistical properties that make the Neyman and Fisher approaches so elegant in their simplicity but that do address more complicated questions, albeit under additional assumptions or approximations.

NOTES

There was disagreement between Fisher and Neyman regarding the importance of the null hypothesis of a zero average effect versus zero effects for all units. In the reading of Neyman's 1935 paper in the *Journal of the Royal Statistical Society* on the interpretations of data from a set of agricultural experiments, the discussion became very heated:

> (Neyman) "So long as the *average* (emphasis in original) yields of any treatments are identical, the question as to whether these treatments affect *separate* yields on *single* plots seems to be uninteresting and academic. ..."
>
> (Fisher) "... It may be foolish, but that is what the z [FEP] test was designed for, and the only purpose for which it has been used. ..."
>
> (Neyman) "... I believe Professor Fisher himself described the problem of agricultural experimentation formerly not in the same manner as he does now. ..."
>
> (Fisher) "... Dr. Neyman thinks another test would be more important. I am not going to argue that point. It may be that the question which Dr. Neyman thinks should be answered is more important than the one I have proposed and attempted to answer. I suggest that before criticizing previous work it is always wise to give enough study to the subject to understand its purpose. Failing that it is surely quite unusual to claim to understand the purpose of previous work better than its author."

Given the tone of Fisher's remarks, it is all the more surprising how gracious Neyman is in later discussions, for example, the quotations in Chapter 5.

Much of the material in this chapter draws on Neyman (1923), translated as Neyman (1990). Also see Neyman (1934, 1935), with discussions, as well as the comments in Rubin (1990b) on Neyman's work in this area.

APPENDIX A SAMPLING VARIANCE CALCULATIONS

First we calculate the sampling variance of the estimator $\hat{\tau}^{\text{dif}} = \overline{Y}_{\text{t}}^{\text{obs}} - \overline{Y}_{\text{c}}^{\text{obs}}$. As before, we have N units, N_{t} receiving the treatment and N_{c} receiving the control. The average treatment effect is:

$$\tau_{\text{fs}} = \overline{Y}(1) - \overline{Y}(0) = \frac{1}{N} \sum_{i=1}^{N} (Y_i(1) - Y_i(0)).$$

The standard estimator of τ_{fs} is:

$$\hat{\tau}^{\text{dif}} = \overline{Y}_{\text{t}}^{\text{obs}} - \overline{Y}_{\text{c}}^{\text{obs}} = \frac{1}{N_{\text{t}}} \sum_{i=1}^{N} W_i \cdot Y_i^{\text{obs}} - \frac{1}{N_{\text{c}}} \sum_{i=1}^{N} (1 - W_i) \cdot Y_i^{\text{obs}}$$

$$= \frac{1}{N} \sum_{i=1}^{N} \left(\frac{N}{N_{\text{t}}} \cdot W_i \cdot Y_i(1) - \frac{N}{N_{\text{c}}} \cdot (1 - W_i) \cdot Y_i(0) \right).$$

For the variance calculations, it is useful to work with a centered treatment indicator D_i, defined as

$$D_i = W_i - \frac{N_{\text{t}}}{N} = \begin{cases} \dfrac{N_{\text{c}}}{N} & \text{if } W_i = 1 \\ -\dfrac{N_{\text{t}}}{N} & \text{if } W_i = 0. \end{cases}$$

The expectation of D_i is zero, and its variance is $\mathbb{V}(D_i) = \mathbb{E}[D_i^2] = N_{\text{c}} N_{\text{t}}/N^2$. Later we also need its cross moment, $\mathbb{E}[D_i \cdot D_j]$. For $i \neq j$ the distribution of this cross product is

$$\Pr_W (D_i \cdot D_j = d) = \begin{cases} \dfrac{N_{\text{t}} \cdot (N_{\text{t}} - 1)}{N \cdot (N - 1)} & \text{if } d = N_{\text{c}}^2/N^2 \\[2ex] 2 \cdot \dfrac{N_{\text{t}} \cdot N_{\text{c}}}{N \cdot (N - 1)} & \text{if } d = -N_{\text{t}} N_{\text{c}}/N^2 \\[2ex] \dfrac{N_{\text{c}} \cdot (N_{\text{c}} - 1)}{N \cdot (N - 1)} & \text{if } d = N_{\text{t}}^2/N^2 \\[2ex] 0 & \text{otherwise,} \end{cases}$$

thereby leading to

$$\mathbb{E}_W [D_i \cdot D_j] = \begin{cases} \dfrac{N_{\text{c}} \cdot N_{\text{t}}}{N^2} & \text{if } i = j \\[2ex] -\dfrac{N_{\text{t}} \cdot N_{\text{c}}}{N^2 \cdot (N - 1)} & \text{if } i \neq j \end{cases}.$$

In terms of D_i, our estimate of the average treatment effect is:

$$\overline{Y}_t^{obs} - \overline{Y}_c^{obs} = \frac{1}{N} \sum_{i=1}^{N} \left(\frac{N}{N_t} \cdot \left(D_i + \frac{N_t}{N} \right) \cdot Y_i(1) - \frac{N}{N_c} \cdot \left(\frac{N_c}{N} - D_i \right) \cdot Y_i(0) \right)$$

$$= \frac{1}{N} \sum_{i=1}^{N} (Y_i(1) - Y_i(0)) + \frac{1}{N} \sum_{i=1}^{N} D_i \cdot \left(\frac{N}{N_t} \cdot Y_i(1) + \frac{N}{N_c} \cdot Y_i(0) \right)$$

$$= \tau_{fs} + \frac{1}{N} \sum_{i=1}^{N} D_i \cdot \left(\frac{N}{N_t} \cdot Y_i(1) + \frac{N}{N_c} \cdot Y_i(0) \right). \tag{A.1}$$

Because $\mathbb{E}_W[D_i] = 0$ and all potential outcomes are fixed, the estimator $\overline{Y}_t^{obs} - \overline{Y}_c^{obs}$ is unbiased for the average treatment effect, $\tau_{fs} = \overline{Y}(1) - \overline{Y}(0)$.

Next, because the only random element in Equation (A.1) is D_i, the variance of $\hat{\tau} = \overline{Y}_t^{obs} - \overline{Y}_c^{obs}$ is equal to the variance of the second term in Equation (A.1). Defining $Y_i^+ = (N/N_t)Y_i(1) + (N/N_c)Y_i(0)$, the latter is equal to:

$$\mathbb{V}_W \left(\overline{Y}_t^{obs} - \overline{Y}_c^{obs} \right) = \mathbb{V}_W \left(\frac{1}{N} \sum_{i=1}^{N} D_i \cdot Y_i^+ \right) = \frac{1}{N^2} \cdot \mathbb{E}_W \left[\left(\sum_{i=1}^{N} D_i \cdot Y_i^+ \right)^2 \right].$$

$$\tag{A.2}$$

Expanding Equation (A.2), we get:

$$\mathbb{V}_W \left(\overline{Y}_t^{obs} - \overline{Y}_c^{obs} \right) = \mathbb{E}_W \left[\frac{1}{N^2} \sum_{i=1}^{N} \sum_{j=1}^{N} D_i D_j Y_i^+ Y_j^+ \right]$$

$$= \frac{1}{N^2} \sum_{i=1}^{N} (Y_i^+)^2 \cdot \mathbb{E}_W \left[D_i^2 \right] + \frac{1}{N^2} \sum_{i=1}^{N} \sum_{j \neq i} \mathbb{E}_W \left[D_i \cdot D_j \right] \cdot Y_i^+ \cdot Y_j^+$$

$$= \frac{N_c \cdot N_t}{N^4} \sum_{i=1}^{N} (Y_i^+)^2 - \frac{N_c \cdot N_t}{N^4 \cdot (N-1)} \sum_{i=1}^{N} \sum_{j \neq i} Y_i^+ \cdot Y_j^+$$

$$= \frac{N_c \cdot N_t}{N^3 \cdot (N-1)} \sum_{i=1}^{N} (Y_i^+)^2 - \frac{N_c \cdot N_t}{N^4 \cdot (N-1)} \sum_{i=1}^{N} \sum_{j=1}^{N} Y_i^+ \cdot Y_j^+$$

$$= \frac{N_c \cdot N_t}{N^3 \cdot (N-1)} \sum_{i=1}^{N} \left(Y_i^+ - \overline{Y^+} \right)^2$$

$$= \frac{N_c \cdot N_t}{N^3 \cdot (N-1)} \sum_{i=1}^{N} \left(\frac{N}{N_t} \cdot Y_i(1) + \frac{N}{N_c} \cdot Y_i(0) - \left(\frac{N}{N_t} \cdot \overline{Y}(1) + \frac{N}{N_c} \cdot \overline{Y}(0) \right) \right)^2$$

$$= \frac{N_c \cdot N_t}{N^3 \cdot (N-1)} \sum_{i=1}^{N} \left(\frac{N}{N_t} \cdot Y_i(1) - \frac{N}{N_t} \cdot \overline{Y}(1) \right)^2$$

$$+ \frac{N_c \cdot N_t}{N^3 \cdot (N-1)} \sum_{i=1}^{N} \left(\frac{N}{N_c} \cdot Y_i(0) - \frac{N}{N_c} \cdot \overline{Y}(0) \right)^2$$

$$+ \frac{2 \cdot N_c \cdot N_t}{N^3 \cdot (N-1)} \sum_{i=1}^{N} \left(\frac{N}{N_t} \cdot Y_i(1) - \frac{N}{N_t} \cdot \overline{Y}(1) \right) \cdot \left(\frac{N}{N_c} \cdot Y_i(0) - \frac{N}{N_c} \cdot \overline{Y}(0) \right)$$

$$= \frac{N_c}{N \cdot N_t \cdot (N-1)} \sum_{i=1}^{N} \left(Y_i(1) - \overline{Y}(1) \right)^2 + \frac{N_t}{N \cdot N_c \cdot (N-1)} \sum_{i=1}^{N} \left(Y_i(0) - \overline{Y}(0) \right)^2$$

$$+ \frac{2}{N \cdot (N-1)} \sum_{i=1}^{N} \left(Y_i(1) - \overline{Y}(1) \right) \cdot \left(Y_i(0) - \overline{Y}(0) \right). \tag{A.3}$$

Recall the definition of S_{ct}^2, which implies that

$$S_{ct}^2 = \frac{1}{N-1} \sum_{i=1}^{N} \left(Y_i(1) - \overline{Y}(1) - \left(Y_i(0) - \overline{Y}(0) \right) \right)^2$$

$$= \frac{1}{N-1} \sum_{i=1}^{N} \left(Y_i(1) - \overline{Y}(1) \right)^2 + \frac{1}{N-1} \sum_{i=1}^{N} \left(Y_i(0) - \overline{Y}(0) \right)^2$$

$$- \frac{2}{N-1} \sum_{i=1}^{N} \left(Y_i(1) - \overline{Y}(1) \right) \cdot \left(Y_i(0) - \overline{Y}(0) \right)$$

$$= S_t^2 + S_c^2 - \frac{2}{N-1} \sum_{i=1}^{N} \left(Y_i(1) - \overline{Y}(1) \right) \cdot \left(Y_i(0) - \overline{Y}(0) \right).$$

Hence, the expression in (A.3) is equal to

$$\mathbb{V}_W \left(\overline{Y}_t^{obs} - \overline{Y}_c^{obs} \right) = \frac{N_c}{N \cdot N_t} \cdot S_t^2 + \frac{N_t}{N \cdot N_c} \cdot S_c^2 + \frac{1}{N} \cdot \left(S_t^2 + S_c^2 - S_{ct}^2 \right)$$

$$= \frac{S_t^2}{N_t} + \frac{S_c^2}{N_c} - \frac{S_{ct}^2}{N}.$$

Now we investigate the bias of the Neyman estimator for the sampling variance, $\mathbb{V}^{neyman}$, under the assumption of a constant treatment effect. Assuming a constant treatment effect, S_{ct}^2 is equal to zero, so we need only find unbiased estimators for S_c^2 and S_t^2 to provide an unbiased estimator of the variance of $\overline{Y}_t^{obs} - \overline{Y}_c^{obs}$. Consider the estimator

$$s_t^2 = \frac{1}{N_t - 1} \sum_{i:W_i=1} \left(Y_i^{obs} - \overline{Y}_t^{obs} \right)^2.$$

The goal is to show that the expectation of s_t^2 is equal to

$$S_t^2 = \frac{1}{N-1} \sum_{i=1}^{N} \left(Y_i(1) - \overline{Y}(1) \right)^2 = \frac{N}{N-1} \left(\overline{Y^2}(1) - \left(\overline{Y}(1)\right)^2 \right).$$

First,

$$s_t^2 = \frac{1}{N_t - 1} \sum_{i=1}^{N} \mathbf{1}_{W_i=1} \cdot \left(Y_i^{\text{obs}} - \overline{Y}_t^{\text{obs}} \right)^2$$

$$= \frac{1}{N_t - 1} \sum_{i=1}^{N} \mathbf{1}_{W_i=1} \cdot \left(Y_i(1) - \overline{Y}_t^{\text{obs}} \right)^2$$

$$= \frac{1}{N_t - 1} \sum_{i=1}^{N} \mathbf{1}_{W_i=1} \cdot Y_i^2(1) - \frac{N_t}{N_t - 1} \left(\overline{Y}_t^{\text{obs}} \right)^2. \tag{A.4}$$

Consider the expectation of the two terms in (A.4) in turn. Using again $D_i = \mathbf{1}_{W_i=1} - N_t/N$, with $\mathbb{E}[D_i] = 0$, we have

$$\mathbb{E}\left[\frac{1}{N_t - 1} \sum_{i=1}^{N} \mathbf{1}_{W_i=1} \cdot Y_i^2(1) \right] = \frac{1}{N_t - 1} \sum_{i=1}^{N} \mathbb{E}\left[\left(D_i + \frac{N_t}{N} \right) \cdot Y_i^2(1) \right]$$

$$= \frac{N_t}{N_t - 1} \cdot \overline{Y^2}(1).$$

Next, the expectation of the second factor in the second term in (A.4):

$$\mathbb{E}_W\left[\left(\overline{Y}_t^{\text{obs}} \right)^2 \right] = \mathbb{E}_W\left[\frac{1}{N_t^2} \sum_{i=1}^{N} \sum_{j=1}^{N} W_i \cdot W_j \cdot Y_i^{\text{obs}} \cdot Y_j^{\text{obs}} \right]$$

$$= \mathbb{E}_W\left[\frac{1}{N_t^2} \sum_{i=1}^{N} \sum_{j=1}^{N} W_i \cdot W_j \cdot Y_i(1) \cdot Y_j(1) \right]$$

$$= \frac{1}{N_t^2} \sum_{i=1}^{N} \sum_{j=1}^{N} \mathbb{E}_W\left[\left(D_i + \frac{N_t}{N} \right) \cdot \left(D_j + \frac{N_t}{N} \right) \cdot Y_i(1) \cdot Y_j(1) \right]$$

$$= \frac{1}{N_t^2} \sum_{i=1}^{N} \sum_{j=1}^{N} Y_i(1) \cdot Y_j(1) \cdot \left(\mathbb{E}\left[D_i \cdot D_j \right] + \frac{N_t^2}{N^2} \right)$$

$$= \frac{1}{N_t^2} \sum_{i=1}^{N} Y_i^2(1) \cdot \left(\mathbb{E}_W\left[D_i^2 \right] + \frac{N_t^2}{N^2} \right)$$

$$+ \frac{1}{N_t^2} \sum_{i=1}^{N} \sum_{j \neq i} Y_i(1) \cdot Y_j(1) \cdot \left(\mathbb{E}_W\left[D_i \cdot D_j \right] + \frac{N_t^2}{N^2} \right)$$

$$= \frac{1}{N_t^2} \sum_{i=1}^{N} Y_i^2(1) \cdot \left(\frac{N_c \cdot N_t}{N^2} + \frac{N_t^2}{N^2} \right)$$

$$+ \frac{1}{N_t^2} \sum_{i=1}^{N} \sum_{j \neq i} Y_i(1) \cdot Y_j(1) \cdot \left(-\frac{N_c \cdot N_t}{N^2 \cdot (N-1)} + \frac{N_t^2}{N^2} \right)$$

$$= \frac{1}{N_t} \cdot \overline{Y^2}(1) + \frac{N_t - 1}{N \cdot (N-1) \cdot N_t} \sum_{i=1}^{N} \sum_{j \neq i} Y_i(1) \cdot Y_j(1)$$

$$= \frac{1}{N_t} \cdot \overline{Y^2}(1) - \frac{N_t - 1}{N \cdot (N-1) \cdot N_t} \sum_{i=1}^{N} Y_i^2(1) + \frac{N_t - 1}{N \cdot (N-1) \cdot N_t} \sum_{i=1}^{N} \sum_{j=1}^{N} Y_i(1) \cdot Y_j(1)$$

$$= \frac{1}{N_t} \cdot \overline{Y^2}(1) - \frac{N_t - 1}{(N-1) \cdot N_t} \cdot \overline{Y^2}(1) + \frac{(N_t - 1) \cdot N}{(N-1) \cdot N_t} \left(\overline{Y}(1) \right)^2$$

$$= \frac{N_c}{N_t \cdot (N-1)} \cdot \overline{Y^2}(1) + \frac{(N_t - 1) \cdot N}{(N-1) \cdot N_t} \left(\overline{Y}(1) \right)^2.$$

Hence, the expectation of the second term in (A.4) equals

$$-\frac{N_c}{(N_t - 1) \cdot (N-1)} \cdot \overline{Y^2}(1) + \frac{N}{(N-1)} \cdot \left(\overline{Y}(1) \right)^2,$$

and adding up the expectations of both terms in in (A.4) leads to

$$\mathbb{E}_W\left[s_t^2 \right] = \frac{N_t}{N_t - 1} \cdot \overline{Y^2}(1) - \frac{N_c}{(N_t - 1) \cdot (N-1)} \cdot \overline{Y^2}(1) - \frac{N}{(N-1)} \cdot \left(\overline{Y}(1) \right)^2$$

$$= \frac{N}{N-1} \cdot \overline{Y^2}(1) - \frac{N}{(N-1)} \cdot \left(\overline{Y}(1) \right)^2 = S_t^2.$$

Following the same argument,

$$\mathbb{E}_W\left[s_c^2 \right] = \frac{1}{N_c - 1} \cdot \mathbb{E}_W\left[\sum_{i=1}^{N} (1 - W_i) \cdot \left(Y_i^{obs} - \overline{Y}_c^{obs} \right)^2 \right] = S_c^2.$$

Hence, the estimators s_c^2 and s_t^2 are unbiased for S_c^2 and S_t^2, and can be used to create an unbiased estimator for the variance of $\overline{Y}_t^{obs} - \overline{Y}_c^{obs}$, our estimator of the average treatment effect under the constant treatment effect assumption.

APPENDIX B RANDOM SAMPLING FROM A SUPER-POPULATION

In this chapter we introduced the super-population perspective. In this appendix we provide more details of this approach and its differences from the finite population perspective. Let N_{sp} be the size of the super-population, with N_{sp} large, but countable. Each unit in this population is characterized by the pair $(Y_i(0), Y_i(1))$, for $i = 1, \ldots, N_{sp}$. Let $\mathbf{Y}_{sp}(0)$ and $\mathbf{Y}_{sp}(1)$ denote the N_{sp}-component vectors with i^{th} element equal to $Y_i(0)$ and $Y_i(1)$ respectively. We continue to view these potential outcomes as fixed. Our finite sample is a Simple Random Sample (SRS) of size N from this large super-population. We take N as fixed. Let R_i denote the sampling indicator, so that $R_i = 1$ if unit i is sampled, and $R_i = 0$ if unit i is not sampled, with $\sum_{i=1}^{N_{\mathrm{sp}}} R_i = N$. The sampling indicator is a binomial random variable with mean N/N_{sp} and variance $(N/N_{\mathrm{sp}}) \cdot (1 - N/N_{\mathrm{sp}})$. The covariance between R_i and R_j, for $i \neq j$, is $-(N/N_{\mathrm{sp}})^2$. Within the finite sample of size N, we carry out a completely randomized experiment, with N_{t} units randomly selected to receive the active treatment, and the remaining $N_{\mathrm{c}} = N - N_{\mathrm{t}}$ units assigned to receive the control treatment. For the units in the finite sample, we have $W_i = 1$ for units assigned to the treatment group, and $W_i = 0$ for units assigned to the control group. To simplify the exposition, let us assign $W_i = 0$ to all units not sampled (with $R_i = 0$).

The super-population average treatment effect is

$$\tau_{\mathrm{sp}} = \frac{1}{N_{\mathrm{sp}}} \sum_{i=1}^{N_{\mathrm{sp}}} (Y_i(1) - Y_i(0)),$$

and the variance of the treatment effect in the super-population is

$$\sigma_{\mathrm{ct}}^2 = \frac{1}{N_{\mathrm{sp}}} \sum_{i=1}^{N_{\mathrm{sp}}} \left(Y_i(1) - Y_i(0) - \tau_{\mathrm{sp}} \right)^2.$$

Now consider the finite-population average treatment effect:

$$\tau_{\mathrm{fs}} = \frac{1}{N} \sum_{i=1}^{N_{\mathrm{sp}}} R_i \cdot (Y_i(1) - Y_i(0)).$$

Viewing R_i as random, but keeping $(Y_i(0), Y_i(1))$, for $i = 1, \ldots, N_{\mathrm{sp}}$ fixed, we can take the expectation of τ_{fs} over the distribution generated by the random sampling. Indexing the expectations operator by subscript sp to be explicit about the fact that the expectation is taken over the distribution generated by the random sampling, and thus over R_i, $i = 1, \ldots, N$, we have

$$\mathbb{E}_{\mathrm{sp}} \left[\tau_{\mathrm{fs}} | \, \mathbf{Y}_{\mathrm{sp}}(0), \mathbf{Y}_{\mathrm{sp}}(1) \right] = \frac{1}{N} \sum_{i=1}^{N_{\mathrm{sp}}} \mathbb{E}_{\mathrm{sp}} \left[R_i \right] \cdot (Y_i(1) - Y_i(0))$$

$$= \frac{1}{N} \sum_{i=1}^{N_{\mathrm{sp}}} \frac{N}{N_{\mathrm{sp}}} \cdot (Y_i(1) - Y_i(0)) = \tau_{\mathrm{sp}}.$$

The variance of the finite sample average treatment effect is

$$\mathbb{V}_{\mathrm{sp}} \left(\tau_{\mathrm{fs}} | \, \mathbf{Y}_{\mathrm{sp}}(0), \mathbf{Y}_{\mathrm{sp}}(1) \right)$$

$$= \mathbb{E}_{\mathrm{sp}} \left[\left(\frac{1}{N} \sum_{i=1}^{N_{\mathrm{sp}}} R_i \cdot (Y_i(1) - Y_i(0)) - \tau_{\mathrm{sp}} \right)^2 \middle| \, \mathbf{Y}_{\mathrm{sp}}(0), \mathbf{Y}_{\mathrm{sp}}(1) \right]$$

$$= \mathbb{E}_{sp}\left[\left.\left(\frac{1}{N}\sum_{i=1}^{N_{sp}}\left(R_i - \frac{N}{N_{sp}}\right)\cdot\left(Y_i(1) - Y_i(0) - \tau_{sp}\right)\right)^2\right| \mathbf{Y}_{sp}(0), \mathbf{Y}_{sp}(1)\right]$$

$$= \frac{1}{N^2}\sum_{i=1}^{N_{sp}}\sum_{j=1}^{N_{sp}}\mathbb{E}_{sp}\left[\left(R_i - \frac{N}{N_{sp}}\right)\cdot\left(R_j - \frac{N}{N_{sp}}\right)\right.$$

$$\left.\cdot\left(Y_i(1) - Y_i(0) - \tau_{sp}\right)\cdot\left(Y_j(1) - Y_j(0) - \tau_{sp}\right)\right| \mathbf{Y}_{sp}(0), \mathbf{Y}_{sp}(1)\Big]$$

$$= \frac{1 - N/N_{sp}}{N\cdot N_{sp}}\sum_{i=1}^{N_{sp}}\left(Y_i(1) - Y_i(0) - \tau_{sp}\right)^2$$

$$- \frac{1}{N_{sp}^2}\sum_{i=1}^{N_{sp}}\sum_{j\neq i}\left(Y_i(1) - Y_i(0) - \tau_{sp}\right)\cdot\left(Y_j(1) - Y_j(0) - \tau_{sp}\right)$$

$$= \frac{\sigma_{ct}^2}{N} - \frac{\sigma_{ct}^2}{N_{sp}} - \frac{1}{N_{sp}^2}\sum_{i=1}^{N_{sp}}\sum_{j\neq i}\left(Y_i(1) - Y_i(0) - \tau_{sp}\right)\cdot\left(Y_j(1) - Y_j(0) - \tau_{sp}\right).$$

If N_{sp} is large relative to N, the last two terms are small relative to the first one, and the variance of τ_{fs} over the super-population is approximately equal to

$$\mathbb{V}_{sp}\left(\tau_{fs}| \mathbf{Y}_{sp}(0), \mathbf{Y}_{sp}(1)\right) \approx \frac{\sigma_{sp}^2}{N}.$$

Now let us consider the estimator $\hat{\tau}^{dif} = \overline{Y}_t^{obs} - \overline{Y}_c^{obs}$. We can write this in terms of the super-population as

$$\hat{\tau}^{dif} = \frac{1}{N_t}\sum_{i=1}^{N_{sp}}R_i\cdot W_i\cdot Y_i^{obs} - \frac{1}{N_c}\sum_{i=1}^{N_{sp}}R_i\cdot(1 - W_i)\cdot Y_i^{obs}.$$

We can take the expectation of this estimator, first conditional on **R** (and always conditional on $\mathbf{Y}_{sp}(1)$ and $\mathbf{Y}_{sp}(0)$), so the expectation is over the randomization distribution:

$$\mathbb{E}_W\left[\hat{\tau}^{dif}\middle| \mathbf{R}, \mathbf{Y}_{sp}(1), \mathbf{Y}_{sp}(0)\right] = \frac{1}{N_t}\sum_{i=1}^{N_{sp}}R_i\cdot \mathbb{E}_W[W_i]\cdot Y_i^{obs}$$

$$- \frac{1}{N_c}\sum_{i=1}^{N_{sp}}R_i\cdot \mathbb{E}_W[1 - W_i]\cdot Y_i^{obs}$$

$$= \frac{1}{N}\sum_{i=1}^{N_{sp}}R_i\cdot(Y_i(1) - Y_i(0)) = \tau_{fs}.$$

Thus, the expectation of $\hat{\tau}^{\text{dif}}$, over both the randomization distribution and the sampling distribution, is

$$\mathbb{E}\left[\hat{\tau}^{\text{dif}}\middle|\,\mathbf{Y}_{\text{sp}}(1),\mathbf{Y}_{\text{sp}}(0)\right]=\mathbb{E}_{\text{sp}}\left[\mathbb{E}_W\left[\hat{\tau}^{\text{dif}}\middle|\,\mathbf{R},\mathbf{Y}_{\text{sp}}(1),\mathbf{Y}_{\text{sp}}(0)\right]\middle|\,\mathbf{Y}_{\text{sp}}(1),\mathbf{Y}_{\text{sp}}(0)\right]$$

$$=\mathbb{E}_{\text{sp}}\left[\tau_{\text{fs}}|\,\mathbf{Y}_{\text{sp}}(1),\mathbf{Y}_{\text{sp}}(0)\right]=\tau_{\text{sp}}.$$

Next we calculate the sampling variance of $\hat{\tau}^{\text{dif}}$, over both the randomization distribution and the sampling distribution. By iterated expectations,

$$\mathbb{V}_{\text{sp}}=\mathbb{V}\left(\hat{\tau}^{\text{dif}}\middle|\,\mathbf{Y}_{\text{sp}}(1),\mathbf{Y}_{\text{sp}}(0)\right)$$

$$=\mathbb{E}_{\text{sp}}\left[\mathbb{V}_W\left(\hat{\tau}^{\text{dif}}\middle|\,\mathbf{R},\mathbf{Y}_{\text{sp}}(1),\mathbf{Y}_{\text{sp}}(0)\right)\middle|\,\mathbf{Y}_{\text{sp}}(1),\mathbf{Y}_{\text{sp}}(0)\right]$$

$$+\mathbb{V}_{\text{sp}}\left(\mathbb{E}_W\left[\hat{\tau}^{\text{dif}}\middle|\,\mathbf{R},\mathbf{Y}_{\text{sp}}(1),\mathbf{Y}_{\text{sp}}(0)\right]\middle|\,\mathbf{Y}_{\text{sp}}(1),\mathbf{Y}_{\text{sp}}(0)\right)$$

$$=\mathbb{E}_{\text{sp}}\left[\frac{S_c^2}{N_c}+\frac{S_t^2}{N_t}-\frac{S_{ct}^2}{N}\middle|\,\mathbf{Y}_{\text{sp}}(1),\mathbf{Y}_{\text{sp}}(0)\right]+\mathbb{V}_{\text{sp}}\left(\tau_{\text{fs}}|\,\mathbf{Y}_{\text{sp}}(1),\mathbf{Y}_{\text{sp}}(0)\right)$$

$$=\frac{\sigma_c^2}{N_c}+\frac{\sigma_t^2}{N_t}-\frac{\sigma_{ct}^2}{N}+\frac{\sigma_{ct}^2}{N}-\frac{\sigma_{ct}^2}{N_{\text{sp}}}-\frac{1}{N_{\text{sp}}^2}\sum_{i=1}^{N_{\text{sp}}}\sum_{j\neq i}\left(Y_i(1)-Y_i(0)-\tau_{\text{sp}}\right)$$

$$\cdot\left(Y_j(1)-Y_j(0)-\tau_{\text{sp}}\right)$$

$$\approx\frac{\sigma_c^2}{N_c}+\frac{\sigma_t^2}{N_t},$$

when N_{sp} is large relative to N.

## CHAPTER 7

Regression Methods for Completely Randomized Experiments

7.1 INTRODUCTION

One of the more common ways of estimating causal effects with experimental, as well as observational, data in many disciplines is based on regression methods. Typically an additive linear regression function is specified for the observed outcome as a function of a set of predictor variables. This set of predictor variables includes the indicator variable for the receipt of treatment and usually additional pre-treatment variables. The parameters of the regression equation are estimated by least squares, with the primary focus on the coefficient for the treatment indicator. Inferences, including point estimates, standard errors, tests, and confidence intervals, are based on standard least squares methods. Although popular, the use of these methods in this context is not without controversy, with some researchers arguing that experimental data should be analyzed based on randomization inference. As Freedman writes bluntly, "Experiments should be analyzed as experiments, not as observational studies" (Freedman, 2006, p. 691). It has also been pointed out that the justification for least squares methods does not follow from randomization. Again Freedman: "randomization does not justify the assumptions behind the ols [ordinary least squares] model" (Freedman, 2008a, p. 181). In this chapter we discuss in some detail the rationale for, and the interpretation and implementation of, regression methods in the setting with completely randomized experiments. This chapter can be viewed as providing a bridge between the previous chapter, which was largely focused on exact finite-sample results based on randomization, and the next chapter, which is based on fully parametric models for imputation of the unobserved potential outcomes.

The most important difference between the methods discussed in Chapters 5 and 6 and the ones discussed here is that they rely on different sampling perspectives. Both the Fisher approach discussed in Chapter 5 and the Neyman methods discussed in Chapter 6 view the potential outcomes as fixed and the treatment assignments as the sole source of randomness. In the regression analysis discussed in this chapter, the starting point is an infinite super-population of units. Properties of the estimators are assessed by resampling from that population, sometimes conditional on the predictor variables including the treatment indicator. From that perspective, the potential outcomes in the sample are random, and we can derive the bias and sampling variance of estimators over the distribution induced by this random sampling. The sampling variance of estimators derived in this approach will be seen to be very similar to the Neyman sampling variance for $\hat{\tau}^{\text{dif}}$ derived in Chapter 6, although its interpretation will be different.

There are four key features of the models considered in this chapter. First, we consider models for the observed outcomes rather than for the potential outcomes. Second, we consider models only for the conditional mean rather than for the full distribution. Third, the estimand, here always an average treatment effect, is a parameter of the statistical model. The latter implies that inferential questions can be viewed as questions of inference for parameters of a statistical model. Fourth, in the current context of completely randomized experiments, the validity of these models, that is, whether the models provide accurate descriptions of the conditional mean, is immaterial for the large-sample unbiasedness of the least squares estimator of the average treatment effect.

As the Freedman quote illustrates, the conventional justification for linear regression models, that the regression function represents the conditional expectation of the observed outcome given the predictor variables, does not follow from the randomization if there are predictors beyond the treatment indicator. Nevertheless, in the setting of a completely randomized experiment, the least squares point estimates and associated inferences can be given a causal interpretation. There is an important difference with the causal interpretation in the previous chapter, however. With the exception of the setting without additional covariates beyond the treatment indicator, where the main results are essentially identical to those discussed in the previous chapter from the Neyman approach, all results are now asymptotic (large sample) results. Specifically, exact unbiasedness no longer holds in finite samples with covariates beyond the treatment indicator because of the need to estimate additional nuisance parameters, that is, the associated regression coefficients. The possible benefit of the regression methods over the exact methods from the previous chapter is that they provide a straightforward and, for many researchers, familiar way to incorporate covariates. If these covariates are predictive of the potential outcomes, their inclusion in the regression model can result in causal inferences that are more precise than differences in observed means. This gain in precision can be substantial if the covariates are highly predictive of the potential outcomes, although in practice the gains are often modest. The disadvantage of regression models relative to the fully model-based methods that will be discussed in the next chapter is that the use of standard linear regression models often restricts the set of models considerably, and thereby restricts the set of questions that can be addressed. Thus, when using these regression models, there is often a somewhat unnatural tension between, on the one hand, models that provide a good statistical fit and have good statistical properties and, on the other hand, models that answer the substantive question of interest. This tension is not present in the full, model-based methods discussed in the next chapter.

This chapter is organized as follows. In the next section, Section 7.2, we describe the data that will be used to illustrate the techniques discussed in this chapter. The data come from a completely randomized experiment previously analyzed by Efron and Feldman (1991). Section 7.3 reviews and adds notation regarding the super-population perspective. In Section 7.4 we discuss the case with no predictor variables beyond the treatment indicator. In that case, most of the results are closely related to those from the previous chapter. In Section 7.5 we generalize the results to allow for the presence of additional predictor variables. Next, in Section 7.6, we include interactions between the predictor variables and the treatment indicator. In Section 7.7 we discuss the role of transformations of the outcome variable. The following section, Section 7.8, discusses the limits on the increases in precision that can be obtained by including covariates. In Section 7.9 we discuss testing for the presence of treatment effects. Then, in Section 7.10, we apply the methods to the Efron-Feldman data. Section 7.11 concludes.

7.2   THE LRC-CPPT CHOLESTEROL DATA

We illustrate the concepts discussed in this chapter using data from a randomized experiment, the Lipid Research Clinics Coronary Primary Prevention Trial (LRC-CPPT), designed to evaluate the effect of the drug cholestyramine on cholesterol levels. The data were previously analyzed in Efron and Feldman (1991). The data set analyzed here contains information on $N = 337$ individuals. Of these 337 individuals, $N_t = 165$ were randomly assigned to receive cholestyramine and the remaining $N_c = 172$ were assigned to the control group, which received a placebo.

For each individual, we observe two cholesterol measures recorded prior to the random assignment. The two measures differ in their timing. The first, `chol1`, was taken prior to a communication, sent to all 337 individuals in the study, about the benefits of a low-cholesterol diet, and the second, `chol2`, was taken after this suggestion, but prior to the random assignment to cholestyramine or placebo. We observe two outcomes. The primary outcome is an average of post-randomization cholesterol readings, `cholf`, averaged over two-month readings for a period of time averaging 7.3 years for all the individuals in the study. Efron and Feldman's primary outcome is the change in cholesterol level, relative to a weighted average of the two pre-treatment cholesterol levels, `cholp=` $0.25 \cdot$ `chol1` $+ 0.75 \cdot$ `chol2`. We denote this change in cholesterol levels by `chold=cholf-cholp`. The secondary outcome is a compliance measure, denoted by `comp`, the percentage of the nominally assigned dose of either cholestyramine or placebo that the individual actually took. Although individuals did not know whether they were assigned to cholestyramine or to the placebo, later we shall see that differences in side effects between the active drug and the placebo induced systematic differences in compliance behavior by treatment status. Note that all individuals, whether assigned to the treatment or the control group, were assigned the same nominal dose of the drug or placebo, for the same time period.

The availability of compliance data raises many interesting issues regarding differences between the effect of *being assigned* to the taking of cholestyramine and the effect of actually *taking* cholestyramine. We discuss some of these issues in detail in later chapters on noncompliance and instrumental variables (Chapters 23–25). Here we analyze the compliance measure solely as a secondary outcome. Note, however, that in general it is *not* appropriate to interpret either the difference in final cholesterol levels by assignment, conditional on observed compliance levels, or the difference in final cholesterol levels by actual dosage taken, as estimates of average causal effects. Such causal interpretations would require strong additional assumptions beyond randomization. For example, to validate conditioning on observed compliance levels would require that observed compliance is a proper pre-treatment variable unaffected by the assignment to treatment versus placebo. Because observed compliance reflects behavior subsequent to the assignment, it may be affected by the treatment assigned, which is an assumption. This is an assumption

Table 7.1. *Summary Statistics for PRC-CPPT Cholesterol Data*

	Variable	Control ($N_c = 172$)		Treatment ($N_t = 165$)			
		Average	Sample (S.D.)	Average	Sample (S.D.)	Min	Max
Pre-treatment	chol1	297.1	(23.1)	297.0	(20.4)	247.0	442.0
	chol2	289.2	(24.1)	287.4	(21.4)	224.0	435.0
	cholp	291.2	(23.2)	289.9	(20.4)	233.0	436.8
Post-treatment	cholf	282.7	(24.9)	256.5	(26.2)	167.0	427.0
	chold	−8.5	(10.8)	−33.4	(21.3)	−113.3	29.5
	comp	74.5	(21.0)	59.9	(24.4)	0	101.0

that can be assessed, and in the current study we can reject, at conventional significance levels, the assumption that observed compliance is a proper pretreatment variable.

In Table 7.1 we present summary statistics for the Efron-Feldman data. For the two initial cholesterol levels (chol1 and chol2), as well as the composite pre-treatment cholesterol level (cholp), the averages do not vary much by treatment status, consistent with the randomized assignment. We do see that the second pre-treatment cholesterol-level measurement, chol2, is, on average, lower than the first one, chol1. This is consistent with the fact that in between the two measurements, the individuals in the study received information about the benefits of a low cholesterol diet that may have induced them to improve their diets. For the subsequent cholesterol-level measures (cholf and chold), the averages do vary considerably by treatment status. In addition, the average level of compliance (comp) is much higher in the control group than in the treatment group. Later in this chapter we investigate the statistical precision of this difference, but here we just comment that this is consistent with relatively severe side effects of the actual drug, which are not present in the placebo. This difference signals the potential dangers of using a post-treatment variable, such as observed compliance, as a covariate.

7.3 THE SUPER-POPULATION AVERAGE TREATMENT EFFECTS

As in Section 6.7 in the previous chapter, we focus in this chapter on the average effect in the super-population, rather than in the sample. We assume that the sample of size N for which we have information can be considered a simple random sample drawn from an infinite super-population. Considering the N units in our sample as a random sample from the super-population induces a distribution on the pair of potential outcomes. The observed potential outcome and covariate values for a drawn unit are simply one draw from the joint distribution in the population and are therefore themselves stochastic. We assume that we have no information about this distribution beyond the values of the observed outcomes and covariates in our sample.

The distribution of the two potential outcomes in turn induces a distribution on the unit-level treatment effects, and thereby on the average of the unit-level treatment effect within the experimental sample. To be clear about this super-population perspective, let us, as we did in the previous chapter, index the average treatment effect τ by fs to denote the finite-sample average treatment effect and by sp to denote the super-population average treatment effect. Thus

$$\tau_{\text{fs}} = \frac{1}{N} \sum_{i=1}^{N} (Y_i(1) - Y_i(0))$$

is the average effect of the treatment in the finite sample, and

$$\tau_{\text{sp}} = \mathbb{E}_{\text{sp}} [Y_i(1) - Y_i(0)]$$

is the expected value of the unit-level treatment effect under the distribution induced by sampling from the super-population, or, equivalently, the average treatment effect in the super-population. (We index the expectations operator by sp to make explicit that the expectation is taken over the random sampling, not over the randomization distribution, as in the previous chapter.) For the discussion in this chapter, it is useful to introduce some additional notation. Define the super-population average and variance of the two potential outcomes conditional on the covariates or pre-treatment variables, e.g., $X_i = x$,

$$\mu_{\text{c}}(x) = \mathbb{E}_{\text{sp}} [Y_i(0)|X_i = x], \quad \mu_{\text{t}}(x) = \mathbb{E}_{\text{sp}} [Y_i(1)|X_i = x],$$
$$\sigma_{\text{c}}^2(x) = \mathbb{V}_{\text{sp}} (Y_i(0)|X_i = x), \quad \text{and} \quad \sigma_{\text{t}}^2 = \mathbb{V}_{\text{sp}} (Y_i(1)|X_i = x),$$

and let the mean and variance of the unit-level treatment effects at $X_i = x$ be denoted by

$$\tau(x) = \mathbb{E}_{\text{sp}}(Y_i(1) - Y_i(0)|X_i = x], \quad \text{and} \quad \sigma_{\text{ct}}^2(x) = \mathbb{V}_{\text{sp}} (Y_i(1) - Y_i(0)|X_i = x),$$

respectively. In addition, denote the marginal means and variances

$$\mu_{\text{c}} = \mathbb{E}_{\text{sp}} [Y_i(0)], \quad \mu_{\text{t}} = \mathbb{E}_{\text{sp}} [Y_i(1)],$$
$$\sigma_{\text{c}}^2 = \mathbb{V}_{\text{sp}} (Y_i(0)), \quad \text{and} \quad \sigma_{\text{t}}^2 = \mathbb{V}_{\text{sp}} (Y_i(1)).$$

Note that the two marginal means are equal to the expectation of the corresponding conditional means:

$$\mu_{\text{c}} = \mathbb{E}_{\text{sp}} [\mu_{\text{c}}(X_i)], \quad \text{and} \quad \mu_{\text{t}} = \mathbb{E}_{\text{sp}} [\mu_{\text{t}}(X_i)],$$

but, by the law of iterated expectations, the marginal variance differs from the average of the conditional variance by the variance of the conditional mean:

$$\sigma_{\text{c}}^2 = \mathbb{E}_{\text{sp}} \left[\sigma_{\text{c}}^2(X_i)\right] + \mathbb{V}_{\text{sp}} (\mu_{\text{c}}(X_i)), \quad \text{and} \quad \sigma_{\text{t}}^2 = \mathbb{E}_{\text{sp}} \left[\sigma_{\text{t}}^2(X_i)\right] + \mathbb{V}_{\text{sp}} (\mu_{\text{t}}(X_i)).$$

Finally, let

$$\mu_X = \mathbb{E}_{\text{sp}} [X_i], \quad \text{and} \quad \Omega_X = \mathbb{V}_{\text{sp}}(X_i) = \mathbb{E}_{\text{sp}} \left[(X_i - \mu_X)^T(X_i - \mu_X)\right],$$

denote the super-population mean and covariance matrix of the row vector of covariates X_i, respectively.

7.4   LINEAR REGRESSION WITH NO COVARIATES

In this section we focus on the case without covariates, that is, no predictor variables beyond the indicator W_i for the receipt of treatment. We maintain the assumption of a completely randomized experiment. We specify a linear regression function for the observed outcome Y_i^{obs} as

$$Y_i^{obs} = \alpha + \tau \cdot W_i + \varepsilon_i,$$

where the unobserved residual ε_i captures unobserved determinants of the outcome. The ordinary least squares (or ols for short) estimator for τ is based on minimizing the sum of squared residuals over α and τ,

$$(\hat{\tau}^{ols}, \hat{\alpha}^{ols}) = \arg \min_{\tau, \alpha} \sum_{i=1}^{N} \left(Y_i^{obs} - \alpha - \tau \cdot W_i \right)^2,$$

with solutions

$$\hat{\tau}^{ols} = \frac{\sum_{i=1}^{N} \left(W_i - \overline{W} \right) \cdot \left(Y_i^{obs} - \overline{Y}^{obs} \right)}{\sum_{i=1}^{N} \left(W_i - \overline{W} \right)^2}, \quad \text{and} \quad \hat{\alpha}^{ols} = \overline{Y}^{obs} - \hat{\tau}^{ols} \cdot \overline{W},$$

where

$$\overline{Y}^{obs} = \frac{1}{N} \sum_{i=1}^{N} Y_i^{obs} \quad \text{and} \quad \overline{W} = \frac{1}{N} \sum_{i=1}^{N} W_i = \frac{N_t}{N}.$$

Simple algebra shows that in this case the ols estimator $\hat{\tau}^{ols}$ is identical to the difference in average outcomes by treatment status:

$$\hat{\tau}^{ols} = \overline{Y}_t^{obs} - \overline{Y}_c^{obs} = \hat{\tau}^{dif},$$

where, as before, $\overline{Y}_t^{obs} = \sum_{i:W_i=1} Y_i^{obs}/N_t$ and $\overline{Y}_c^{obs} = \sum_{i:W_i=0} Y_i^{obs}/N_c$ are the averages of the observed outcomes in the treatment and control groups respectively.

The least squares estimate of τ is often interpreted as an estimate of the causal effect of the treatment, explicitly in randomized experiments, and sometimes implicitly in observational studies. The assumptions traditionally used in the least squares approach are that the residuals ε_i are independent of, or at least uncorrelated with, the treatment indicator W_i. This assumption is difficult to evaluate directly, as the interpretation of these residuals is rarely made explicit beyond a somewhat vague notion of capturing unobserved factors affecting the outcomes of interest. Statistical textbooks, therefore, often stress that in observational studies the regression estimate $\hat{\tau}^{ols}$ measures only the association between the two random variables W_i and Y_i^{obs} and that a causal interpretation is generally not warranted. In the current context, however, we already have a formal justification for the causal interpretation of $\hat{\tau}^{ols}$ because it is identical to $\overline{Y}_t^{obs} - \overline{Y}_c^{obs}$, which itself was shown in Chapter 6 to be unbiased for the finite-sample average treatment effect, τ_{fs}, as well as for the super-population average treatment effect, τ_{sp}. Nevertheless, it is useful to justify the causal interpretation of $\hat{\tau}^{\mathrm{ols}}$ more directly in terms of the standard justification for regression methods, using the assumptions that random sampling created the sample and a completely randomized experiment generated the observed data from that sample.

Let α be the population average outcome under the control, $\alpha = \mu_c = \mathbb{E}_{\mathrm{sp}}[Y_i(0)]$, and recall that τ_{sp} is the super-population average treatment effect, $\tau_{\mathrm{sp}} = \mu_t - \mu_c = \mathbb{E}_{\mathrm{sp}}[Y_i(1) - Y_i(0)]$. Now *define* the residual ε_i in terms of the population parameters, treatment indicator, and the potential outcomes as

$$\varepsilon_i = Y_i(0) - \alpha + W_i \cdot (Y_i(1) - Y_i(0) - \tau_{\mathrm{sp}}) = \begin{cases} Y_i^{\mathrm{obs}} - \alpha & \text{if } W_i = 0, \\ Y_i^{\mathrm{obs}} - \alpha - \tau_{\mathrm{sp}} & \text{if } W_i = 1. \end{cases}$$

Then we can write

$$\varepsilon_i = Y_i^{\mathrm{obs}} - (\alpha + \tau_{\mathrm{sp}} \cdot W_i),$$

and thus we can write the observed outcome as

$$Y_i^{\mathrm{obs}} = \alpha + \tau_{\mathrm{sp}} \cdot W_i + \varepsilon_i.$$

Random sampling allows us to view the potential outcomes as random variables. In combination with random assignment this implies that assignment is independent of the potential outcomes,

$$\mathrm{Pr}(W_i = 1 | Y_i(0), Y_i(1)) = \mathrm{Pr}(W_i = 1),$$

or in Dawid's (1979) "$\perp$" independence notation,

$$W_i \perp (Y_i(0), Y_i(1)).$$

The definition of the residual, in combination with random assignment and random sampling from a super-population, implies that the residual has mean zero conditional on the treatment indicator in the population:

$$\mathbb{E}_{\mathrm{sp}}[\varepsilon_i | W_i = 0] = \mathbb{E}_{\mathrm{sp}}[Y_i(0) - \alpha | W_i = 0] = \mathbb{E}_{\mathrm{sp}}[Y_i(0)] - \alpha = 0,$$

and

$$\begin{aligned} \mathbb{E}_{\mathrm{sp}}[\varepsilon_i | W_i = 1] &= \mathbb{E}_{\mathrm{sp}}[Y_i(1) - \alpha - \tau_{\mathrm{sp}} | W_i = 1] \\ &= \mathbb{E}_{\mathrm{sp}}[Y_i(1) - \alpha - \tau_{\mathrm{sp}} | W_i = 1] = 0, \end{aligned}$$

so that

$$\mathbb{E}_{\mathrm{sp}}[\varepsilon_i | W_i = w] = 0, \qquad \text{for } w = 0, 1.$$

The fact that the conditional mean of ε_i given W_i is zero in turn implies unbiasedness of the least squares estimator, $\hat{\tau}^{\mathrm{ols}}$ for $\tau_{\mathrm{sp}} = \mathbb{E}_{\mathrm{sp}}[Y_i(1) - Y_i(0)]$, over the distribution induced by random sampling. The above derivation shows how properties of residuals commonly asserted as assumptions in least squares analyses actually follow from random sampling and random assignment, and thus have a scientific basis in the context of a completely randomized experiment.

Another way of deriving this result, which is closer to the way we will do this for the general case with pre-treatment variables, is to consider the super-population limits of the estimators. The estimators are defined as

$$(\hat{\alpha}^{\text{ols}}, \hat{\tau}^{\text{ols}}) = \arg\min_{\alpha,\tau} \sum_{i=1}^{N} \left(Y_i^{\text{obs}} - \alpha - \tau \cdot W_i \right)^2.$$

Under some regularity conditions, these estimators converge, as the sample size goes to infinity, to the population limits (α^*, τ^*) that minimize the expected value of the sum of squares:

$$(\alpha^*, \tau^*) = \arg\min_{\alpha,\tau} \mathbb{E}_{\text{sp}} \left[\frac{1}{N} \sum_{i=1}^{N} \left(Y_i^{\text{obs}} - \alpha - \tau \cdot W_i \right)^2 \right]$$

$$= \arg\min_{\alpha,\tau} \mathbb{E}_{\text{sp}} \left[\left(Y_i^{\text{obs}} - \alpha - \tau \cdot W_i \right)^2 \right].$$

This implies that the population limit is $\tau^* = \mathbb{E}_{\text{sp}}[Y_i^{\text{obs}}|W_i = 1] - \mathbb{E}_{\text{sp}}[Y_i^{\text{obs}}|W_i = 0]$. Random assignment of W_i implies $\mathbb{E}_{\text{sp}}[Y_i^{\text{obs}}|W_i = 1] - \mathbb{E}_{\text{sp}}[Y_i^{\text{obs}}|W_i = 0] = \mathbb{E}_{\text{sp}}[Y_i(1) - Y_i(0)] = \tau_{\text{sp}}$, so that the population limit of the least squares estimator is equal to the population average treatment effect, $\tau^* = \tau_{\text{sp}}$.

Now let us analyze the least squares approach to inference (i.e., sampling variance and confidence intervals) applied to the setting of a completely randomized experiment. Let us initially assume homoskedasticity ($\sigma_{Y|W}^2 = \sigma_c^2 = \sigma_t^2$). Using least squares methods, the variance of the residuals would be estimated as

$$\hat{\sigma}_{Y|W}^2 = \frac{1}{N-2} \sum_{i=1}^{N} \hat{\varepsilon}_i^2 = \frac{1}{N-2} \sum_{i=1}^{N} \left(Y_i^{\text{obs}} - \hat{Y}_i^{\text{obs}} \right)^2,$$

where the estimated residual is $\hat{\varepsilon}_i = Y_i^{\text{obs}} - \hat{Y}_i^{\text{obs}}$, and the predicted value $\hat{Y}_i^{\text{obs}}$ is

$$\hat{Y}_i^{\text{obs}} = \begin{cases} \hat{\alpha}^{\text{ols}} & \text{if } W_i = 0, \\ \hat{\alpha}^{\text{ols}} + \hat{\tau}^{\text{ols}} & \text{if } W_i = 1. \end{cases}$$

The ols variance estimate can be rewritten as

$$\hat{\sigma}_{Y|W}^2 = \frac{1}{N-2} \left(\sum_{i:W_i=0} \left(Y_i^{\text{obs}} - \overline{Y}_c^{\text{obs}} \right)^2 + \sum_{i:W_i=1} \left(Y_i^{\text{obs}} - \overline{Y}_t^{\text{obs}} \right)^2 \right),$$

which is equivalent to our calculation of s^2, the common variance across the two potential outcome distributions, as seen in Equation (6.11) in Chapter 6. The conventional estimator for the sampling variance of $\hat{\tau}_{ols}$ is then

$$\hat{\mathbb{V}}^{homosk} = \frac{\hat{\sigma}_{Y|W}^2}{\sum_{i=1}^N (W_i - \overline{W})^2} = s^2 \cdot \left(\frac{1}{N_c} + \frac{1}{N_t} \right).$$

This expression is equal to $\hat{\mathbb{V}}^{const}$ in Equation (6.12) in Chapter 6. This result is not surprising, because the assumption of homoskedasticity in the linear model setting is implied by the assumption of a constant treatment effect.

For comparison with subsequent results, it is also useful to have the limit of the estimated sampling variance, normalized by the sample size N. Let p be the probability limit of the ratio of the number of treated units to the total number of units, $p = \text{plim}(N_t/N)$. Then, as the sample size increases, the normalized sampling variance estimator converges in probability to

$$N \cdot \hat{\mathbb{V}}^{homosk} \xrightarrow{p} \frac{\sigma_{Y|W}^2}{p \cdot (1-p)}. \tag{7.1}$$

Note, however, that the random assignment assumption we used for the causal interpretation of $\hat{\tau}^{ols}$, although it implies independence between assignments and potential outcomes, implies only zero correlation between the assignment and the residual, not necessarily full independence. Yet we rely on this independence to conclude that the variance is homoskedastic. In many cases, the homoskedasticity assumption will not be warranted, and one may wish to use an estimator for the sampling variance of $\hat{\tau}^{ols}$ that allows for heteroskedasticity. The standard robust sampling variance estimator for least squares estimators is

$$\hat{\mathbb{V}}^{hetero} = \frac{\sum_{i=1}^N \hat{\varepsilon}_i^2 \cdot (W_i - \overline{W})^2}{\left(\sum_{i=1}^N (W_i - \overline{W})^2 \right)^2}.$$

Defining, as the previous chapter,

$$s_c^2 = \frac{1}{N_c - 1} \sum_{i:W_i=0} \left(Y_i^{obs} - \overline{Y}_c^{obs} \right)^2, \quad \text{and} \quad s_t^2 = \frac{1}{N_t - 1} \sum_{i:W_i=1} \left(Y_i^{obs} - \overline{Y}_t^{obs} \right)^2,$$

we can write the variance estimator under heteroskedasticity as

$$\hat{\mathbb{V}}^{hetero} = \frac{s_c^2}{N_c} + \frac{s_t^2}{N_t}.$$

This is exactly the same estimator for the sampling variance derived from Neyman's perspective in Chapter 6 ($\hat{\mathbb{V}}^{neyman}$ in Equation (6.8)). So, in the case without additional predictors, the regression approach leads to sampling variance estimators that are familiar from the discussion in the previous chapter. It does, however, provide a different perspective on these results. First of all, it is based on a random sampling perspective. Second, this perspective allows for a natural and simple extension to the case with additional predictors.

7.5   LINEAR REGRESSION WITH ADDITIONAL COVARIATES

Now let us consider the case with additional covariates. In this section these additional covariates are included in the regression function additively. The regression function is specified as:

$$Y_i^{\text{obs}} = \alpha + \tau \cdot W_i + X_i \beta + \varepsilon_i, \tag{7.2}$$

where X_i is a row vector of covariates (i.e., pre-treatment variables). We estimate the regression coefficients again using least squares:

$$(\hat{\alpha}^{\text{ols}}, \hat{\tau}^{\text{ols}}, \hat{\beta}^{\text{ols}}) = \arg\min_{\alpha,\tau,\beta} \sum_{i=1}^{N} \left(Y_i^{\text{obs}} - \alpha - \tau \cdot W_i - X_i \beta \right)^2.$$

The first question we address in this section concerns the causal interpretation of the least squares estimate $\hat{\tau}^{\text{ols}}$ in the presence of these covariates and the associated parameters. We are not interested per se in the value of the "nuisance" parameters, β and α. In particular, we are not interested in a causal interpretation of those parameters. Moreover, we will *not* make the assumption that the regression function in (7.2) is correctly specified or that the conditional expectation of Y_i^{obs} is actually linear in X_i and W_i. However, in order to be precise about the causal interpretation of $\hat{\tau}^{\text{ols}}$, it is useful, as in Section 7.4, to define the limiting values to which the least squares estimators converge as the sample gets large. We will refer to these limiting values as the super-population values corresponding to the estimators and denote them with a superscript $*$, as in Section 7.4. Using this notation, under some regularity conditions, $(\hat{\alpha}^{\text{ols}}, \hat{\tau}^{\text{ols}}, \hat{\beta}^{\text{ols}})$ converge to $(\alpha^*, \tau^*, \beta^*)$, defined as

$$(\alpha^*, \tau^*, \beta^*) = \arg\min_{\alpha,\tau,\beta} \mathbb{E}\left[\left(Y_i^{\text{obs}} - \alpha - \tau \cdot W_i - X_i \beta \right)^2 \right].$$

These population values are generally well defined (subject, essentially, only to finite-moment conditions and positive definiteness of Ω_X, the population covariance matrix of X_i), even if the conditional expectation of the observed outcome given covariates is not linear in the covariates.

In this case with additional predictors, it is no longer true that $\hat{\tau}^{\text{ols}}$ is unbiased for τ_{sp} in finite samples. However, irrespective of whether the regression function is truly linear in the covariates in the population, the least squares estimate $\hat{\tau}^{\text{ols}}$ is unbiased in large samples for the population average treatment effect, τ_{sp}. Moreover, τ^*, the probability limit of the estimator, is equal to the population average treatment effect τ_{sp}. Finally, in large samples $\hat{\tau}^{\text{ols}}$ will be distributed approximately normally around τ_{sp}. To be precise, we state the result formally.

Theorem 7.1 *Suppose we conduct a completely randomized experiment in a sample drawn at random from an infinite population. Then,* (i)

$$\tau^* = \tau_{\text{sp}},$$

and (ii),

$$\sqrt{N} \cdot \left(\hat{\tau}^{\text{ols}} - \tau_{\text{sp}} \right) \xrightarrow{d} \mathcal{N} \left(0, \frac{\mathbb{E}\left[(W_i - p)^2 \cdot \left(Y_i^{\text{obs}} - \alpha^* - \tau_{\text{sp}} \cdot W_i - X_i \beta^* \right)^2 \right]}{p^2 \cdot (1 - p)^2} \right).$$

We will prove the first part of the result here in the body of the text. The proof of the second part, and of subsequent results, is given in the Appendix to this chapter.

Proof of Theorem 7.1(i). Consider the limiting objective function:

$$Q(\alpha, \tau, \beta) = \mathbb{E}[(Y_i^{\text{obs}} - \alpha - \tau \cdot W_i - X_i \beta)^2]$$
$$= \mathbb{E}\left[\left(Y_i^{\text{obs}} - \tilde{\alpha} - \tau \cdot W_i - (X_i - \mu_X)\beta \right)^2 \right],$$

where $\tilde{\alpha} = \alpha + \mu_X \beta$, with $\mu_X = \mathbb{E}[X_i]$. Minimizing the right-hand side over $\tilde{\alpha}$, τ, and β leads to the same values for τ and β as minimizing the left-hand side over α, τ, and β, with the least squares estimate of $\tilde{\alpha}$ equal $\hat{\alpha} + \hat{\beta}' \mu_X$. Next,

$$Q(\tilde{\alpha}, \tau, \beta) = \mathbb{E}_{\text{sp}} \left[\left(Y_i^{\text{obs}} - \tilde{\alpha} - \tau \cdot W_i - (X_i - \mu_X)\beta \right)^2 \right]$$
$$= \mathbb{E}_{\text{sp}} \left[\left(Y_i^{\text{obs}} - \tilde{\alpha} - \tau \cdot W_i \right)^2 \right] + \mathbb{E}_{\text{sp}} \left[((X_i - \mu_X)\beta)^2 \right]$$
$$\quad - 2 \cdot \mathbb{E}_{\text{sp}} \left[\left(Y_i^{\text{obs}} - \tilde{\alpha} - \tau \cdot W_i \right) \cdot (X_i - \mu_X)\beta \right]$$
$$= \mathbb{E}_{\text{sp}} \left[\left(Y_i^{\text{obs}} - \tilde{\alpha} - \tau \cdot W_i \right)^2 \right] + \mathbb{E}_{\text{sp}} \left[((X_i - \mu_X)\beta)^2 \right]$$
$$\quad - 2 \cdot \mathbb{E}_{\text{sp}} \left[Y_i^{\text{obs}} \cdot (X_i - \mu_X)\beta \right], \tag{7.3}$$

because

$$\mathbb{E}_{\text{sp}} \left[(X_i - \mu_X)\beta \right] = 0, \quad \text{and} \quad \mathbb{E}_{\text{sp}} \left[\tau \cdot W_i \cdot (X_i - \mu_X)\beta \right] = 0,$$

the first by definition, and the second because of the random sampling and the random assignment. Because the last two terms in (7.3) do not depend on $\tilde{\alpha}$ or τ, minimizing (7.3) over τ and α is equivalent to minimizing the objective function without the additional covariates,

$$\mathbb{E}_{\text{sp}} \left[\left(Y_i^{\text{obs}} - \tilde{\alpha} - \tau \cdot W_i \right)^2 \right],$$

which leads to the solutions

$$\tilde{\alpha}^* = \mathbb{E}_{\text{sp}}[Y_i^{\text{obs}} | W_i = 0] = \mathbb{E}_{\text{sp}} [Y_i(0) | W_i = 0] = \mathbb{E}_{\text{sp}} [Y_i(0)] = \mu_{\text{c}},$$

and

$$\tau^* = \mathbb{E}_{sp}[Y_i^{obs}|W_i = 1] - \mathbb{E}_{sp}[Y_i^{obs}|W_i = 0]$$
$$= \mathbb{E}_{sp}[Y_i(1)|W_i = 1] - \mathbb{E}_{sp}[Y_i(0)|W_i = 0] = \tau_{sp}.$$

Thus, the least squares estimator is consistent for the population average treatment effect τ_{sp}.   □

What is important in the first part of the result is that the consistency (large-sample unbiasedness) of the least squares estimator for τ_{sp} does *not* depend on the correctness of the specification of the regression function in a completely randomized experiment. No matter how non-linear the conditional expectations of the potential outcomes given the covariates are in the super-population, simple least square regression is consistent for estimating the population average treatment effect. The key insight into this result is that, by randomizing treatment assignment, the super-population correlation between the treatment indicator and the covariates is zero. Even though in finite samples the actual correlation may differ from zero, in large samples this correlation will vanish, and as a result the inclusion of the covariates does not matter for the limiting values of the estimator. The fact that in finite samples the correlation may differ from zero is what leads to the possibility of finite-sample bias.

Although the inclusion of the additional covariates does not matter for the limit of the corresponding estimator, it does matter for the sampling variance of the estimators. Let us interpret the sampling variance in some special cases. Suppose that, in fact, the conditional expectation of the two potential outcomes is linear in the covariates, with the same slope coefficients but different intercepts in the two treatment arms, or

$$\mathbb{E}_{sp}[Y_i(0)|X_i = x] = \alpha_c + x\beta, \quad \text{and} \quad \mathbb{E}_{sp}[Y_i(1)|X_i = x] = \alpha_t + x\beta,$$

so that, in combination with random assignment, we have

$$\mathbb{E}_{sp}\left[Y_i^{obs}\middle| X_i = x, W_i = 1\right] = \alpha_c + \tau_{sp} \cdot t + x\beta,$$

where $\tau_{sp} = \alpha_t - \alpha_c$. Suppose that, in addition, the variance of the two potential outcomes does not vary by treatment or covariates:

$$\mathbb{V}_{sp}(Y_i(w)|X_i = x) = \sigma_{Y|W,X}^2,$$

for $w = 0, 1$, and all x. Then the normalized sampling variance for the least squares estimator for τ_{sp}, given for the general case in Theorem 7.1, simplifies to

$$N \cdot \mathbb{V}_{sp}^{homosk} = \frac{\sigma_{Y|W,X}^2}{p \cdot (1-p)}. \tag{7.4}$$

This expression reveals the gain in precision from including the covariates. Instead of the unconditional variance of the potential outcomes, as in the expression for the sampling variance in the case without covariates in (7.1), we now have the conditional variance of the outcome given the covariates. If the covariates explain much of the variation in the potential outcomes, so that the conditional variance $\sigma_{Y|W,X}^2$ is substantially smaller than the marginal variance $\sigma^2_{Y|W}$, then including the covariates in the regression model will lead to a considerable increase in precision. The price paid for the increase in precision from including covariates is relatively minor. Instead of having (exact) unbiasedness of the estimator in finite samples, unbiasedness now only holds approximately, that is, in large samples.

The sampling variance for the average treatment effect can be estimated easily using standard least squares methods. Substituting averages for the expectations, and least squares estimates for the unknown parameters, we estimate the sampling variance as

$$
\hat{\mathbb{V}}^{\text{hetero}}_{\text{sp}} = \frac{1}{N\,(N-1-\dim(X_i))}
$$

$$
\cdot \frac{\sum_{i=1}^{N} (W_i - \overline{W})^2 \cdot \left(Y_i^{\text{obs}} - \hat{\alpha}^{\text{ols}} - \hat{\tau}^{\text{ols}} - X_i\hat{\beta}^{\text{ols}} \right)^2}{\left(\overline{W} \cdot (1 - \overline{W}) \right)^2}.
$$

If one wishes to impose homoskedasticity, one can still use the heteroskedasticity-consistent sampling variance estimator, but a more precise estimator of the sampling variance imposes homoskedasticity, leading to the form:

$$
\hat{\mathbb{V}}^{\text{homo}}_{\text{sp}} = \frac{1}{N\,(N-1-\dim(X_i))} \cdot \frac{\sum_{i=1}^{N} \left(Y_i^{\text{obs}} - \hat{\alpha}^{\text{ols}} - \hat{\tau}^{\text{ols}} - X_i\hat{\beta}^{\text{ols}} \right)^2}{\overline{W} \cdot (1 - \overline{W})}.
$$

7.6   LINEAR REGRESSION WITH COVARIATES AND INTERACTIONS

In this section we take the analysis of Section 7.5 one step further. In addition to including the covariates linearly, one may wish to interact the covariates with the indicator for the receipt of treatment if we expect that the association between the covariates and the outcome varies by treatment status. The motivation for this is twofold. First, adding additional covariates of any form, including those based on interactions, may further improve the precision of the estimator. Second, by interacting all such predictors with the treatment indicators, we achieve a particular form of robustness to model misspecification that we discuss in more detail later. This robustness is not particularly important in the current setting of a completely randomized experiment, but it will be important in observational studies discussed in Parts III and IV of this text. We specify the regression function as

$$
Y_i^{\text{obs}} = \alpha + \tau \cdot W_i + X_i\beta + W_i \cdot (X_i - \overline{X})\gamma + \varepsilon_i.
$$

We include the interaction of the treatment indicator with the covariates in deviations from their sample means to simplify the relationship between the population limits of the estimators for the parameters of the regression function and τ_{sp}.

Let $\hat{\alpha}^{\text{ols}}$, $\hat{\tau}^{\text{ols}}$, $\hat{\beta}^{\text{ols}}$, and $\hat{\gamma}^{\text{ols}}$ denote the least squares estimates,

$$
(\hat{\alpha}^{\text{ols}}, \hat{\tau}^{\text{ols}}, \hat{\beta}^{\text{ols}}, \hat{\gamma}^{\text{ols}}) = \arg\min_{\alpha,\tau,\beta,\gamma} \sum_{i=1}^{N} \left(Y_i^{\text{obs}} - \alpha - \tau \cdot W_i - X_i\beta - W_i \cdot (X_i - \overline{X})\gamma \right)^2,
$$

and let α^*, τ^*, β^*, and γ^* denote the corresponding population values:

$$(\alpha^*, \tau^*, \beta^*, \gamma^*) = \arg\min_{\alpha,\tau,\beta,\gamma} \mathbb{E}_{sp}\left[\left(Y_i^{obs} - \alpha - \tau \cdot W_i - X_i\beta - W_i \cdot (X_i - \mu_X)\gamma\right)^2\right].$$

Results similar to Theorem 7.1 can be obtained for this case. The least squares estimator $\hat{\tau}^{ols}$ is consistent for the average treatment effect τ_{sp}, and inference can be based on least squares methods.

Theorem 7.2 *Suppose we conduct a completely randomized experiment in a random sample from a super-population. Then (i)*

$$\tau^* = \tau_{sp},$$

and (ii),

$$\sqrt{N} \cdot \left(\hat{\tau}^{ols} - \tau_{sp}\right) \xrightarrow{d} \mathcal{N}$$

$$\left(0, \frac{\mathbb{E}_{sp}\left[(W_i - p)^2 \cdot \left(Y_i^{obs} - \alpha^* - \tau_{sp} \cdot W_i - X_i\beta^* - W_i \cdot (X_i - \mu_X)\gamma^*\right)^2\right]}{p^2 \cdot (1 - p)^2}\right).$$

The proof for this theorem is provided in the Appendix.

A slightly different interpretation of this result connects it to the imputation-based methods that are the topic of the next chapter. Suppose we take the model at face value and assume that the regression function represents the conditional expectation:

$$\mathbb{E}_{sp}\left[Y_i^{obs} \middle| X_i = x, W_i = w\right] = \alpha + \tau \cdot t + x\beta + w \cdot (x - \mu_X)\gamma. \tag{7.5}$$

In combination with the random assignment, this implies that

$$\mathbb{E}_{sp}[Y_i(0)|X_i = x] = \mathbb{E}_{sp}[Y_i(0)|X_i = x, W_i = 0]$$

$$= \mathbb{E}_{sp}\left[Y_i^{obs}\middle|X_i = x, W_i = 0\right] = \alpha + x\beta,$$

and

$$\mathbb{E}_{sp}[Y_i(1)|X_i = x] = \alpha + \tau + x\beta + (x - \mu_X)\gamma.$$

Suppose that unit i was exposed to the treatment ($W_i = 1$), so $Y_i(1)$ is observed and $Y_i(0)$ is missing. Under the model in (7.5), the predicted value for the missing potential outcome $Y_i(0)$ is

$$\hat{Y}_i(0) = \hat{\alpha}^{ols} + X_i\hat{\beta}^{ols},$$

so that for this treated unit the predicted value for the unit-level causal effect is

$$\hat{\tau}_i = Y_i(1) - \hat{Y}_i(0) = Y_i^{obs} - \left(\hat{\alpha}^{ols} + X_i\hat{\beta}^{ols}\right).$$

For a control unit i (with $W_i = 0$) the predicted value for the missing potential outcome $Y_i(1)$ is

$$\hat{Y}_i(1) = \hat{\alpha}^{\text{ols}} + \hat{\tau}^{\text{ols}} + X_i\hat{\beta}^{\text{ols}} + (X_i - \overline{X})\hat{\gamma}^{\text{ols}},$$

and the predicted value for the unit-level causal effect for this control unit i is

$$\hat{\tau}_i = \hat{Y}_i(1) - Y_i(0) = \hat{\alpha}^{\text{ols}} + \hat{\tau}^{\text{ols}} + X_i\hat{\beta}^{\text{ols}} + (X_i - \overline{X})\hat{\gamma}^{\text{ols}} - Y_i^{\text{obs}}.$$

Now we can estimate the overall average treatment effect τ_{fs} by averaging the estimates of the unit-level causal effects $\hat{\tau}_i$. Simple algebra shows that this leads to the ols estimator:

$$\frac{1}{N}\sum_{i=1}^{N}\hat{\tau}_i = \frac{1}{N}\sum_{i=1}^{N}\left\{W_i \cdot \left(Y_i(1) - \hat{Y}_i(0)\right) + (1 - W_i) \cdot \left(\hat{Y}_i(1) - Y_i(0)\right)\right\} = \hat{\tau}^{\text{ols}}.$$

Thus, the least squares estimator $\hat{\tau}^{\text{ols}}$ can be interpreted as averaging estimated unit-level causal effects in the sample, based on imputing the missing potential outcomes through a linear regression model. However, as has been stressed repeatedly, thanks to the randomization, the consistency of the ols estimator does not rely on the validity of the regression model as an approximation to the conditional expectation.

There is another important feature of the estimator based on linear regression with a full set of interactions that was alluded to at the beginning of this chapter. As the above derivation shows, the estimator essentially imputes the missing potential outcomes. The regression model with a full set of interactions does so separately for the treated and control units. When imputing the value of $Y_i(0)$ for the treated units, this procedure uses only the observed outcomes, Y_i^{obs}, for control units, without any dependence on observations on $Y_i(1)$ (and vice versa). This gives the estimator attractive robustness properties, clearly separating imputation of control and treated outcomes. This will be important in the context of observational studies.

7.7   TRANSFORMATIONS OF THE OUTCOME VARIABLE

If one is interested in the average effect of the treatment on a transformation of the outcome, one can first transform the outcome and then apply the methods discussed so far. For example, in order to estimate the average effect on the logarithm of the outcome, we can first take logarithms and then estimate the regression function

$$\ln\left(Y_i^{\text{obs}}\right) = \alpha + \tau \cdot W_i + X_i\beta + \varepsilon_i.$$

Irrespective of the form of the association between outcomes and covariates, in a completely randomized experiment, least squares estimates of τ are consistent for the average effect $\mathbb{E}[\ln(Y_i(1)) - \ln(Y_i(0))]$. This follows directly from the previous discussion. There is an important issue, though, involving such transformations that relates to the correctness of the specification of the regression function. Suppose one is interested in the average effect $\mathbb{E}[Y_i(1) - Y_i(0)]$, but suppose that one actually suspects that a model linear in logarithms provides a better fit to the distribution of Y_i^{obs} given X_i and W_i. Estimating a model linear in logarithms and transforming the estimates back to an estimate of the average effect in levels requires assumptions beyond those on the conditional expectation of the logarithm of the potential outcomes: one needs to make distributional assumptions on the unobserved component. We discuss such modeling strategies in the next chapter.

As an extreme example of this issue, consider the case where the researcher is interested in the average effect of the treatment on a binary outcome. Estimating a linear regression function by least squares will lead to a consistent estimator for the average treatment effect. However, such a linear probability model is unlikely to provide an accurate approximation of the conditional expectation of the outcome given covariates and treatment indicator. Logistic models (where $\Pr(Y_i^{\text{obs}} = 1 | W_i = w, X_i = x)$ is modeled as $\exp(\alpha + \tau \cdot w + x\beta)/(1 + \exp(\alpha + \tau \cdot w + x\beta))$), or probit models (where $\Pr(Y_i^{\text{obs}} = 1 | W_i = w, X_i = x) = \Phi(\alpha + \tau \cdot w + x\beta)$, with $\Phi(z) = \int_{-\infty}^{z} (2\pi)^{-1/2} \exp(-z^2/2)$ the normal cumulative distribution function) are more likely to lead to an accurate approximation of the conditional expectation of the outcome given the covariates and the treatment indicator. However, such a model will not generally lead to a consistent estimator for the average effect unless the model is correctly specified. Moreover, the average treatment effect cannot be expressed directly in terms of the parameters of the logistic or probit regression model.

The issue is that in the regression approach, the specification of the statistical model is closely tied to the estimand of interest. In the next chapter we separate these two issues. This separation is attractive for a number of reasons discussed in more detail in the next chapter, but it also carries a price, namely that consistency of the estimators will be tied more closely to the correct specification of the model. We do not view this as a major issue. In the setting of completely randomized experiments, the bias is unlikely to be substantial with moderate-sized samples, as flexible models are likely to have minimal bias. Moreover, this consistency property despite possible misspecification of the regression function holds only with completely randomized experiments. In observational studies, even regression models rely heavily on the correct specification for consistency of the estimator. Furthermore, large-sample results, such as consistency, are only guidelines for finite-sample properties, and as such not always reliable.

7.8 THE LIMITS ON INCREASES IN PRECISION DUE TO COVARIATES

In large samples, including covariates in the regression function will not lower, and generally will increase, the precision of the estimator for the average treatment effect. However, beyond the first few covariates, more covariates are unlikely to improve the precision substantially in modest-sized samples. Here we briefly discuss some limits to the gains in precision from including covariates in settings where the randomized assignment ensures that the covariates are not needed for bias removal.

Suppose we do not include any predictor variables in the regression beyond the indicator variable for the treatment, W_i, that is, we include no covariates. Normalized by the sample size, the sampling variance of the least squares estimator, in this case equal to the simple difference in means, is equal to

$$N \cdot \mathbb{V}_{\text{nocov}} = \frac{\sigma_c^2}{1-p} + \frac{\sigma_t^2}{p},$$

familiar in various forms from this and the previous chapter. Now suppose we have available a vector of covariates, X_i. Including these covariates, their interactions with the treatment indicator, and possibly higher-order moments of these covariates, leads to a normalized sampling variance that is bounded from below by

$$N \cdot \mathbb{V}_{\text{bound}} = \frac{\mathbb{E}_{\text{sp}}[\sigma_c^2(X_i)]}{1-p} + \frac{\mathbb{E}_{\text{sp}}[\sigma_t^2(X_i)]}{p}.$$

Instead of the marginal variances σ_c^2 and σ_t^2 in the two terms, we now take the expectation of the conditional variances $\sigma_c^2(X_i)$ and $\sigma_t^2(X_i)$. The difference between the two expressions for the sampling variance, and thus the gain from including the covariates in a flexible manner, is the sum of the sampling variances of the conditional means of $Y_i(w)$ given X_i:

$$\mathbb{V}_{\text{nocov}} - \mathbb{V}_{\text{bound}} = \left(\frac{\sigma_c^2}{1-p} + \frac{\sigma_t^2}{p} \right) - \left(\frac{\mathbb{E}_{\text{sp}}[\sigma_c^2(X_i)]}{1-p} + \frac{\mathbb{E}_{\text{sp}}\left[\sigma_t^2(X_i)\right]}{p} \right)$$

$$= \frac{\mathbb{V}_{\text{sp}}(\mu_c(X_i))}{1-p} + \frac{\mathbb{V}_{\text{sp}}(\mu_t(X_i))}{p}.$$

The more the covariates X_i help in explaining the potential outcomes, and thus the bigger the variation in $\mu_w(x)$, the bigger the gain from including them in the specification of the regression function. In the extreme case, where neither $\mu_c(x)$ nor $\mu_t(x)$ varies with the predictor variables, there is no gain from using the covariates, even in large samples. Moreover, in small samples there will actually be a loss of precision due to the estimation of coefficients, that are, in fact, zero.

7.9   TESTING FOR THE PRESENCE OF TREATMENT EFFECTS

In addition to estimating average treatment effects, the regression models discussed in this chapter have been used to test for the presence of treatment effects. In the current setting of completely randomized experiments, tests for the presence of any treatment effects are not necessarily as attractive as the Fisher exact p-value calculations discussed in Chapter 5, but their extensions to observational studies are relevant. In addition, we may be interested in testing hypotheses concerning the heterogeneity in the treatment effects that do not fit into the FEP framework because the associated null hypotheses are not sharp. As in the discussion of estimation, we focus on procedures that are valid in large samples, irrespective of the correctness of the specification of the regression model.

The most interesting setting is the one where we allow for a full set of first-order interactions with the treatment indicator and specify the regression function as

$$Y_i^{\text{obs}} = \alpha + \tau_{\text{sp}} \cdot W_i + X_i \beta + W_i \cdot (X_i - \overline{X}) \gamma + \varepsilon_i.$$

In that case we can test the null hypothesis of a zero average treatment effect by testing the null hypothesis that $\tau_{sp} = 0$. However, we can construct a different test by focusing on the deviation of either $\hat{\tau}_{sp}$ or $\hat{\gamma}$ from zero. If the regression model were correctly specified, that is, if the conditional expectation of the outcome in the population given covariates and treatment indicator were equal to

$$\mathbb{E}_{sp}\left[Y_i^{obs} \middle| X_i = x, W_i = w \right] = \alpha + \tau \cdot w + x\beta + w \cdot (x - \mu_X)\gamma',$$

this would test the null hypothesis that the average treatment effect conditional on each value of the covariates is equal to zero, or

$$H_0 : \mathbb{E}_{sp}[Y_i(1) - Y_i(0)|X_i = x] = 0, \quad \forall\, x,$$

against the alternative hypothesis

$$H_a : \mathbb{E}_{sp}[Y_i(1) - Y_i(0)|X_i = x] \neq 0, \quad \text{for some } x.$$

Without making the assumption that the regression model is correctly specified, it is still true that, if the null hypothesis that $\mathbb{E}[Y_i(1) - Y_i(0)|X_i = x] = 0$ for all x were correct, then the population values τ_{sp} and γ^* would be equal to zero. However, it is no longer true that for *all* deviations of this null hypothesis the limiting values of either τ_{sp} or γ^* differ from zero. It is possible that $\mathbb{E}[Y_i(1) - Y_i(0)|X_i = x]$ differs from zero for some values of x even though τ_{sp} and γ^* are both equal to zero.

In order to implement these tests, one can again use standard least squares methods. The normalized covariance matrix of the vector $(\hat{\tau}^{ols}, \hat{\gamma}^{ols})$ is

$$\mathbb{V}_{\tau,\gamma} = \begin{pmatrix} \mathbb{V}_\tau & \mathbb{C}_{\tau,\gamma} \\ \mathbb{C}_{\tau,\gamma}^T & \mathbb{V}_\gamma \end{pmatrix}.$$

The precise form of the components of the covariance matrix, as well as consistent estimators for these components, is given in the Appendix. In order to test the null hypothesis that the average effect of the treatment given the covariates is zero for all values of the covariates, we then use the quadratic form

$$Q_{zero} = \begin{pmatrix} \hat{\tau}^{ols} \\ \hat{\gamma}^{ols} \end{pmatrix}^T \hat{\mathbb{V}}_{\tau,\gamma}^{-1} \begin{pmatrix} \hat{\tau}^{ols} \\ \hat{\gamma}^{ols} \end{pmatrix}. \tag{7.6}$$

Note that this is not a test that fits into the Fisher exact p-value approach because it does not specify all missing potential outcomes under the null hypothesis.

The second null hypothesis we consider is that the average treatment effect is constant as a function of the covariates:

$$H_0' : \mathbb{E}_{sp}[Y_i(1) - Y_i(0)|X_i = x] = \tau_{sp}, \quad \text{for all } x,$$

against the alternative hypothesis

$$H_a' : \exists\, x_0, x_1, \text{ such that } \mathbb{E}_{sp}[Y_i(1) - Y_i(0)|X_i = x_0] \neq \mathbb{E}_{sp}[Y_i(1) - Y_i(0)|X_i = x_1].$$

This null hypothesis may be of some importance in practice. If there is evidence of heterogeneity in the effect of the treatment as a function of the covariates, one has to be more careful in extrapolating to different subpopulations. On the other hand, if there is no evidence of heterogeneity by observed characteristics, and if the distribution of these characteristics in the sample is sufficiently varied, it may be more credible to extrapolate estimates to different subpopulations. (Of course, lack of positive evidence for heterogeneity does not imply a constant treatment effect, but in cases with sufficient variation in the covariates, it does suggest that treatment-effect heterogeneity may be a second-order problem.) In order to test this null hypothesis, we can use the quadratic form

$$Q_{\mathrm{const}} = (\hat{\gamma}^{\,\mathrm{ols}})^T \hat{\mathbb{V}}_{\gamma}^{-1} \hat{\gamma}^{\,\mathrm{ols}}. \tag{7.7}$$

Theorem 7.3 *Suppose we conduct a completely randomized experiment in a random sample from a large population. If $Y_i(1) - Y_i(0) = \tau$ for some value τ and all units, then*
(i): $\gamma^* = 0$,
and (ii)

$$Q_{\mathrm{const}} \xrightarrow{d} \mathcal{X}(\dim(X_i)).$$

If $Y_i(1) - Y_i(0) = 0$ for all units, then (iii),

$$Q_{\mathrm{zero}} \xrightarrow{d} \mathcal{X}(\dim(X_i) + 1).$$

7.10   ESTIMATES FOR LRC-CPPT CHOLESTEROL DATA

Now let us return to the LRC-CPPT cholesterol data. We look at estimates for two average effects. First, the effect on post-treatment cholesterol levels, the primary outcome of interest, denoted by `cholf`. Second, partly anticipating some of the analyses in Chapters 23–25, we estimate the effect of assignment to treatment on the level of compliance, `comp`. Because compliance was far from perfect (on average, individuals assigned to the control group took 75% of the nominal dose, and individuals in the group assigned to the active treatment, on average, took 60% of the nominal dose), the estimates of the effect on post-assignment cholesterol levels should be interpreted as estimates of *intention-to-treat* (ITT) effects, that is, average effects of assignment to the drug versus assignment to the placebo, rather than as estimates of the effects of the efficacy of the drug.

For each outcome, we present four regression estimates of the average effects. First, we use a simple linear regression with only the indicator for assignment. Second, we include the composite prior cholesterol level `cholp` as a linear predictor. Third, we include both prior cholesterol-level measurements, `chol1` and `chol2`, as linear predictors. Fourth, we add interactions of the two prior cholesterol-level measurements with the assignment indicator.

Table 7.2 presents the results for these regressions. For the cholesterol-level outcome, the average effect is estimated in all cases reported to be a reduction of approximately 25–26 units, approximately an 8% reduction. Including predictors beyond the treatment

Table 7.2. *Regression Estimates for Average Treatment Effects for the PRC-CPPT Cholesterol Data from Table 7.1*

Covariates	Effect of Assignment to Treatment on			
	Post-Cholesterol Level		Compliance	
	$\hat{\tau}$	$\widehat{(\text{s. e.})}$	$\hat{\tau}$	$\widehat{(\text{s. e.})}$
No covariates	-26.22	(3.93)	-14.64	(3.51)
cholp	-25.01	(2.60)	-14.68	(3.51)
chol1, chol2	-25.02	(2.59)	-14.95	(3.50)
chol1, chol2, interacted with W	-25.04	(2.56)	-14.94	(3.49)

Table 7.3. *Regression Estimates for Average Treatment Effects on Post-Cholesterol Levels for the PRC-CPPT Cholesterol Data from Table 7.1*

Covariates	Model for Levels		Model for Logs	
	Est	$\widehat{(\text{s. e.})}$	Est	$\widehat{(\text{s. e.})}$
Assignment	-25.04	(2.56)	-0.098	(0.010)
Intercept	-3.28	(12.05)	-0.133	(0.233)
chol1	0.98	(0.04)	-0.133	(0.233)
chol2-chol1	0.61	(0.08)	0.602	(0.073)
chol1 × Assignment	-0.22	(0.09)	-0.154	(0.107)
(chol2-chol1) × Assignment	0.07	(0.14)	0.184	(0.159)
R-squared	0.63		0.57	

indicator improves the precision considerably, reducing the estimated standard error by a third. Including predictors beyond the simple composite prior cholesterol level cholp does not affect the estimated precision appreciably. For the effect of the assignment on receipt of the drug, the estimated effect is also stable across the different specifications of the regression function. For this outcome the estimated precision does not change with the inclusion of additional predictors.

The left panel of Table 7.3 presents more detailed results for the regression of the outcome on the covariates and the interaction of covariates with the treatment indicator. Although substantively the coefficients of the covariates are not of interest in the current setting, we can see from these results that the covariates do add considerable predictive power to the regression function. This predictive power is what leads to the increased precision of the estimator for the average treatment effect based on the regression with covariates relative to the regression without covariates. For the purpose of assessing the relative predictive power of different specifications, we also report, in the right panel of Table 7.3, the results for a regression after transforming all cholesterol levels to logarithms. As stressed before, this changes the estimand, and so the results are not directly comparable. It is useful to note, though, that in this case the transformation does not improve the predictive power, in the sense that the squared correlation between the observed outcomes and the covariates decreases as a result of this transformation.

Table 7.4. *P-Values for Tests for Constant and Zero Treatment Effects, Using* `chol1` *and* `chol2-chol1` *as Covariates for the PRC-CPPT Cholesterol Data from Table 7.1*

		Post-Cholesterol Level	Compliance
Zero treatment effect	$\mathcal{X}^2(3)$ approximation	<0.001	<0.001
	Fisher exact p-value	<0.001	0.001
Constant treatment effect	$\mathcal{X}^2(2)$ approximation	0.029	0.270

In Table 7.4 we report p-values for some of the tests discussed in Section 7.9. First we consider the null hypothesis that the effect of the treatment on the final cholesterol level is zero. We use the statistic Q_{zero} given in Equation (7.6), based on the regression with the two prior cholesterol levels and their interactions with the treatment as covariates. Under this null hypothesis, this statistic has, in large samples, a chi-squared distribution with three degrees of freedom. The value of the statistic in the sample is 100.48, which leads to an approximate p-value based on the chi-squared distribution with three degrees of freedom less than 0.001. We perform the same calculations using the compliance variable as the outcome of interest. Now the value of the test statistic is 19.27, again leading to an approximate p-value less than 0.001. Because under the null hypothesis of no effect whatsoever, we can apply the FEP approach, we also calculate the exact p-values. For the post-cholesterol level, the FEP calculations lead to a p-value less than 0.001. For the compliance outcome, the p-value based on the FEP approach is 0.001. The p-values under the FEP approach are similar to those based on large-sample approximations because, with the sample size used in this example, a total of 337 units, 172 in the control group and 165 in the treatment group, and the data values, the normal approximations that underlie the large-sample properties of the tests are accurate.

Next, we test the null hypothesis that the treatment effect is constant against the alternative that it varies between units, using the statistic Q_{const} given in (7.7). For the final cholesterol-level outcome, the value of the test statistic is 7.05, leading to a p-value based on the chi-squared approximation with two degrees of freedom equal to 0.029. For the compliance outcome, the value of the statistic is 2.62, leading to an approximate p-value of 0.269. Note that in this case, because of the presence of nuisance parameters (we do not restrict the level of the treatment effect, only its variance), the FEP approach is not applicable. Together the tests suggest that the evidence for the presence of treatment effects is very strong but that the evidence for heterogeneity in the treatment effect is weak.

Overall, with the caveat of the multiple testing, the message from this application supports the conclusion that including some covariates can substantially improve the estimated precision of the inferences, although including many covariates is unlikely to be helpful beyond the inclusion of the most important ones.

7.11 CONCLUSION

In this chapter we discuss regression methods for estimating causal effects in the context of a completely randomized experiment. Regression models are typically motivated by assumptions on conditional mean functions. Such assumptions are difficult to justify other than as approximations. In the context of a completely randomized experiment, however, we can use the randomization to help justify the key assumptions necessary for consistency of the least squares estimator. In contrast to the methods discussed in previous chapters, most of these results are only approximate, relying on large samples. In that sense, the regression methods can be viewed as providing a bridge from the exact results based on randomization inference to the model-based methods that will be discussed in the next chapter.

Regression methods can easily incorporate covariates into estimands and, in that sense lead to an attractive extension of Neyman's basic approach discussed in Chapter 6. In settings with completely randomized experiments, they offer a simple and widely used framework for estimating and constructing confidence intervals for average treatment effects. The main disadvantage is that they are closely tied to linearity. In completely randomized experiments, this linearity is not a particularly important concern, because the methods still lead to consistent estimators for average treatment effects. In observational studies, however, this reliance on linearity can make regression methods sensitive to minor changes in specification. In those settings, discussed in detail in Parts III and IV of this text, simple regression methods are not recommended.

NOTES

The Efron-Feldman data were also analyzed in Jin and Rubin (2008) using a principal stratification approach. In their analysis, the focus is on the causal effect of the actual dose of the drug taken, rather than on the (intention-to-treat) effect of the assignment to the drug.

Cochran (1977) and Goldberger (1991) have extensive discussions on the properties of least squares estimators in settings where the conditional expectation is not necessarily linear, and on the notion of the "best linear predictor" (Goldberger, 1991, p. 52). Gail, Wieand, and Piantadosi (1984) discuss biases in estimated treatment effects in the context of non-linear regression models with experimental data. See also Lin (2012) and Miratrix, Sekhon, and Yu (2013). Lesaffre and Senn (2003) discuss the properties of alternative covariance adjustment methods. Koch, Tangen, Jung, and Amara (1998) discuss regression methods in settings with binary and ordered discrete outcome data. Victora, Habicht, and Bryce (2004) discuss regression methods in health applications.

The discussion in Section 7.8 on the limits of the gains in precision from incorporating pre-treatment variables draws on the results in Hahn (1998). See also Robins and Rotnitzky (1995) and Hirano, Imbens, and Ridder (2003).

Freedman (2008ab) discusses the role of regression analyses in the context of randomized experiments. He suggests, as evidenced by the quotes in the introduction to this chapter, that the use of regression analysis is not always warranted, a view to which we also subscribe. Angrist and Pischke (2008) and Lin (2012) present a less critical view of the use of regression methods for causal inference.

Senn (1994) and Imai, King, and Stuart (2008) discuss the motivation for testing or not testing for baseline balance in randomized experiments.

APPENDIX

Proof of Theorem 7.1

It is convenient to reparametrize the model. Instead of (α, τ, β), we parametrize the model using $(\tilde{\alpha}, \tau, \beta)$, where $\tilde{\alpha} = \alpha - p \cdot \tau - \mathbb{E}_{sp}[X_i]\beta$. The reparametrization does not change the ols estimates for τ and β, nor their limiting values. The limiting value of the new parameter is $\tilde{\alpha}^* = \alpha^* - p \cdot \tau_{sp} - \mathbb{E}_{sp}[X_i]\beta^*$. In terms of these parameters, the objective function is

$$\sum_{i=1}^{N} \left(Y_i^{obs} - \left(\tilde{\alpha} - p \cdot \tau - \mathbb{E}_{sp}[X_i]\beta \right) - \tau \cdot W_i - X_i\beta \right)^2$$

$$= \sum_{i=1}^{N} \left(Y_i^{obs} - \tilde{\alpha} - \tau \cdot (W_i - p) - \left(X_i - \mathbb{E}_{sp}[X_i] \right) \beta \right)^2.$$

The first-order conditions for the estimators $(\hat{\tilde{\alpha}}^{ols}, \hat{\tau}^{ols}, \hat{\beta}^{ols})$ are

$$\sum_{i=1}^{N} \psi(Y_i^{obs}, W_i, X_i, \hat{\tilde{\alpha}}^{ols}, \hat{\tau}^{ols}, \hat{\beta}^{ols}) = 0,$$

where $\psi(\,\cdot\,)$ is a three-component column vector:

$$\psi(y, w, x, \alpha, \tau, \beta) = \begin{pmatrix} y - \alpha - \tau \cdot (w - p) - \left(x - \mathbb{E}_{sp}[X_i] \right) \beta \\ (w - p) \cdot \left(y - \alpha - \tau \cdot (w - p) - \left(x - \mathbb{E}_{sp}[X_i] \right) \beta \right) \\ \left(x - \mathbb{E}_{sp}[X_i] \right) \cdot \left(y - \alpha - \tau \cdot (w - p) - \left(x - \mathbb{E}_{sp}[X_i] \right) \beta \right) \end{pmatrix}.$$

Given the population values of the parameters, α^*, τ_{sp}, and β^*, standard M-estimation (or generalized method of moments) results imply that under standard regularity conditions the estimator is consistent and asymptotically normally distributed:

$$\sqrt{N} \cdot \begin{pmatrix} \hat{\tilde{\alpha}}^{ols} - \alpha^* \\ \hat{\tau}^{ols} - \tau_{sp} \\ \hat{\beta}^{ols} - \beta^* \end{pmatrix} \xrightarrow{d} \mathcal{N} \left(\begin{pmatrix} 0 \\ 0 \\ 0 \end{pmatrix}, \Gamma^{-1} \Delta (\Gamma^T)^{-1} \right),$$

where the two components of the covariance matrix are

$$\Gamma = \mathbb{E}_{sp} \left[\frac{\partial}{\partial(\alpha, \tau, \beta)} \psi(Y_i^{obs}, W_i, X_i, \alpha, \tau, \beta) \right] \Bigg|_{(\tilde{\alpha}^*, \tau_{sp}, \beta^*)}$$

$$= \mathbb{E}_{sp} \left[\begin{pmatrix} -1 & -(W_i - p) \\ -(W_i - p) & -(W_i - p)^2 \\ -(X_i - \mathbb{E}_{sp}[X_i])^T & -(W_i - p) \cdot (X_i - \mathbb{E}_{sp}[X_i])^T \end{pmatrix} \right.$$

$$\left. \begin{pmatrix} -(X_i - \mathbb{E}_{sp}[X_i]) \\ -(W_i - p) \cdot (X_i - \mathbb{E}_{sp}[X_i]) \\ -(X_i - \mathbb{E}_{sp}[X_i])^T \cdot (X_i - \mathbb{E}_{sp}[X_i]) \end{pmatrix} \right]$$

$$
= \mathbb{E}_{\mathrm{sp}} \left[\begin{pmatrix} -1 & 0 & 0 \\ 0 & -p(1-p) & 0 \\ 0 & 0 & -\mathbb{E}_{\mathrm{sp}} \left[(X_i - \mathbb{E}_{\mathrm{sp}}[X_i])^T \cdot (X_i - \mathbb{E}_{\mathrm{sp}}[X_i]) \right] \end{pmatrix} \right],
$$

and

$$
\Delta = \mathbb{E}_{\mathrm{sp}} \left[\psi(Y_i^{\mathrm{obs}}, W_i, X_i, \tilde{\alpha}^*, \tau_{\mathrm{sp}}, \beta^*) \cdot \psi(Y_i^{\mathrm{obs}}, W_i, X_i, \tilde{\alpha}^*, \tau_{\mathrm{sp}}, \beta^*)^T \right]
$$

$$
= \mathbb{E}_{\mathrm{sp}} \left[\left(Y_i^{\mathrm{obs}} - \alpha^* - \tau_{\mathrm{sp}} - X_i\beta^* \right)^2 \cdot \begin{pmatrix} 1 \\ W_i - p \\ (X_i - \mathbb{E}_{\mathrm{sp}}[X_i])^T \end{pmatrix} \begin{pmatrix} 1 \\ W_i - p \\ (X_i - \mathbb{E}_{\mathrm{sp}}[X_i])^T \end{pmatrix}^T \right].
$$

The variance of $\hat{\tau}$ is the $(2,2)$ element of the covariance matrix. Because Γ is block diagonal, the $(2,2)$ element of $\Gamma^{-1}\Delta(\Gamma^T)^{-1}$ is equal to the $(2,2)$ element of Δ divided by $(p(1-p))^2$, which is equal to

$$
\mathbb{E}_{\mathrm{sp}} \left[\left(Y_i^{\mathrm{obs}} - \alpha^* - \tau_{\mathrm{sp}} - X_i\beta^* \right)^2 \cdot (W_i - p)^2 \right].
$$

Hence the variance of $\hat{\tau}$, normalized by the sample size N, is equal to

$$
\frac{\mathbb{E}_{\mathrm{sp}} \left[\left(Y_i^{\mathrm{obs}} - \alpha^* - \tau_{\mathrm{sp}} - X_i\beta^* \right)^2 \cdot (W_i - p)^2 \right]}{p^2 \cdot (1-p)^2}.
$$

$\square$

Proof of Theorem 7.2

First we show that in this case τ^* the population value of $\hat{\tau}$, equal to

$$
(\alpha^*, \tau^*, \beta^*, \gamma^*) = \arg \min_{\alpha, \beta, \tau, \gamma} \mathbb{E}_{\mathrm{sp}} \left[\left(Y_i^{\mathrm{obs}} - \alpha - \tau \cdot W_i - X_i\beta - W_i \cdot (X_i - \mu_X)\gamma \right)^2 \right],
$$

is equal to τ_{sp}. Again it is useful to reparametrize. The new vector of parameters is

$$
\begin{pmatrix} \tilde{\alpha}_{\mathrm{c}} \\ \beta_{\mathrm{c}} \\ \tilde{\alpha}_{\mathrm{t}} \\ \beta_{\mathrm{t}} \end{pmatrix} = \begin{pmatrix} \alpha + \mu_X\beta \\ \beta \\ \alpha + \tau + \mu_X\beta \\ \gamma + \beta \end{pmatrix},
$$

with inverse

$$
\begin{pmatrix} \alpha \\ \beta \\ \tau \\ \gamma \end{pmatrix} = \begin{pmatrix} \tilde{\alpha}_{\mathrm{c}} - \mu_X\beta_{\mathrm{c}} \\ \beta_{\mathrm{c}} \\ \tilde{\alpha}_{\mathrm{t}} - \tilde{\alpha}_{\mathrm{c}} \\ \beta_{\mathrm{t}} - \beta_{\mathrm{c}} \end{pmatrix}.
$$

In terms of this parameter vector the minimization problem is

$$
\begin{aligned}
(\tilde{\alpha}_c^*, \tilde{\alpha}_t^*, \beta_c^*, \beta_t^*) & \\
= \arg \min_{\alpha_c, \alpha_t, \beta_c, \beta_t} & \mathbb{E}_{sp} \left[\left(Y_i^{obs} - \alpha_c - (\alpha_t - \alpha_c) \cdot W_i - X_i \beta_c \right. \right. \\
& \left. \left. - W_i \cdot (X_i - \mu_X)(\beta_t - \beta_c) \right)^2 \right] \\
= \arg \min_{\alpha_c, \alpha_t, \beta_c, \beta_t} & \mathbb{E}_{sp} \left[(1 - W_i) \cdot \left(Y_i^{obs} - \alpha_c - (X_i - \mu_X)\beta_c \right)^2 \right. \\
& \left. + W_i \cdot \left(Y_i^{obs} - \alpha_t - (X_i - \mu_X)\beta_t \right)^2 \right].
\end{aligned}
$$

Hence, we can solve separately

$$
(\tilde{\alpha}_c^*, \beta_c^*) = \arg \min_{\alpha_c, \beta_c} \mathbb{E}_{sp} \left[(1 - W_i) \cdot \left(Y_i^{obs} - \alpha_c - (X_i - \mu_X)\beta_c \right)^2 \right],
$$

and

$$
(\tilde{\alpha}_t^*, \beta_t^*) = \arg \min_{\alpha_t, \beta_t} \mathbb{E}_{sp} \left[W_i \cdot \left(Y_i^{obs} - \alpha_t - (X_i - \mu_X)\beta_t \right)^2 \right].
$$

Because $\mathbb{E}_{sp}[X_i|W_i = w] = \mu_X$ for $w = 0, 1$ by the randomization, this leads to the solutions

$$
\tilde{\alpha}_c^* = \mathbb{E}_{sp}[Y_i(0)], \qquad \text{and } \tilde{\alpha}_t^* = \mathbb{E}_{sp}[Y_i(1)].
$$

Hence

$$
\tau^* = \tilde{\alpha}_t^* - \tilde{\alpha}_c^* = \mathbb{E}_{sp}[Y_i(1)] - \mathbb{E}_{sp}[Y_i(0)] = \tau_{sp},
$$

proving part (*i*).

For part (*ii*) we use a different reparametrization. Let $\tilde{\alpha} = \alpha - \tau \cdot p - \mu_X \beta$, with the other parameters unchanged, so that the minimization problem becomes

$$
\begin{aligned}
(\hat{\tilde{\alpha}}^{ols}, \hat{\tau}^{ols}, \hat{\beta}^{ols}, \hat{\gamma}^{ols}) = \arg \min_{\alpha, \tau, \beta, \gamma} \frac{1}{N} \sum_{i=1}^{N} & \\
\times \left(Y_i^{obs} - \alpha - \tau \cdot (W_i - p) - \beta'(X_i - \mu_X) - \gamma'(X_i - \mu_X) \cdot W_i \right)^2. &
\end{aligned}
$$

The first-order conditions for the estimators $(\hat{\tilde{\alpha}}^{ols}, \hat{\tau}^{ols}, \hat{\beta}^{ols}, \hat{\gamma}^{ols})$ are

$$
\sum_{i=1}^{N} \psi(Y_i^{obs}, W_i, X_i, \hat{\tilde{\alpha}}^{ols}, \hat{\tau}^{ols}, \hat{\beta}^{ols}, \hat{\gamma}^{ols}) = 0,
$$

where

$$\psi(y,w,x,\alpha,\tau,\beta,\gamma) = \begin{pmatrix} y - \alpha - \tau \cdot (w-p) - (x - \mathbb{E}_{\mathrm{sp}}[X_i]) \beta - \gamma'\, (x - \mathbb{E}_{\mathrm{sp}}[X_i]) \cdot \tau \\ (w-p) \cdot \left\{ y - \alpha - \tau \cdot (w-p) - (x - \mathbb{E}_{\mathrm{sp}}[X_i]) \beta \right. \\ \left. - w \cdot (x - \mathbb{E}_{\mathrm{sp}}[X_i]) \gamma \right\} \\ (x - \mathbb{E}_{\mathrm{sp}}[X_i])^T \cdot \left\{ y - \alpha - \tau \cdot (w-p) - (x - \mathbb{E}_{\mathrm{sp}}[X_i]) \beta \right. \\ \left. - w \cdot (x - \mathbb{E}_{\mathrm{sp}}[X_i]) \gamma \right\} \\ (x - \mathbb{E}_{\mathrm{sp}}[X_i])^T \cdot w \cdot \left\{ y - \alpha - \tau \cdot (w-p) - (x - \mathbb{E}_{\mathrm{sp}}[X_i]) \beta \right. \\ \left. - w \cdot (x - \mathbb{E}_{\mathrm{sp}}[X_i]) \gamma \right\} \end{pmatrix}.$$

In large samples we have, by standard M-estimation methods,

$$\sqrt{N} \cdot \begin{pmatrix} \hat{\tilde{\alpha}}^{\mathrm{ols}} - \alpha^* \\ \hat{\tau}^{\mathrm{ols}} - \tau_{\mathrm{sp}} \\ \hat{\beta}^{\mathrm{ols}} - \beta^* \\ \hat{\gamma}^{\mathrm{ols}} - \gamma^* \end{pmatrix} \xrightarrow{d} \mathcal{N}\left(\begin{pmatrix} 0 \\ 0 \\ 0 \\ 0 \end{pmatrix}, \Gamma^{-1} \Delta (\Gamma^T)^{-1} \right), \qquad (A.1)$$

where the two components of the covariance matrix are now

$$\Gamma = \mathbb{E}_{\mathrm{sp}}\left[\left. \frac{\partial}{\partial(\alpha,\tau,\beta^T,\gamma^T)} \psi(Y_i^{\mathrm{obs}}, W_i, X_i, \alpha, \tau, \beta, \gamma) \right] \right|_{(\tilde{\alpha}^*, \tau_{\mathrm{sp}}, \beta^*, \gamma^*)}$$

$$= \mathbb{E}_{\mathrm{sp}}\left[\begin{pmatrix} -1 & -(W_i - p) \\ -(W_i - p) & -(W_i - p)^2 \\ -(X_i - \mu_X)^T & -(W_i - p)(X_i - \mu_X)^T \\ W_i (X_i - \mu_X)^T & (W_i - p) W_i (X_i - \mu_X)^T \end{pmatrix} \right.$$

$$\left. \begin{pmatrix} -(X_i - \mu_X) & W_i (X_i - \mu_X) \\ -(W_i - p)(X_i - \mu_X) & (W_i - p) W_i (X_i - \mu_X) \\ -(X_i - \mu_X)^T (X_i - \mu_X) & W_i (X_i - \mu_X)^T (X_i - \mu_X) \\ W_i (X_i - \mu_X)^T (X_i - \mu_X) & W_i^2 (X_i - \mu_X)^T (X_i - \mu_X) \end{pmatrix} \right]$$

$$= \mathbb{E}_{\mathrm{sp}}\left[\begin{pmatrix} -1 & 0 & 0 & 0 \\ 0 & -p(1-p) & 0 & 0 \\ 0 & 0 & -\Omega_X & 0 \\ 0 & 00 & & -p \cdot \Omega_X \end{pmatrix} \right],$$

and

$$\Delta = \mathbb{E}_{\mathrm{sp}}\left[\psi(Y_i^{\mathrm{obs}}, W_i, X_i, \tilde{\alpha}^*, \tau_{\mathrm{sp}}, \beta^*, \gamma^*) \cdot \psi(Y_i^{\mathrm{obs}}, W_i, X_i, \tilde{\alpha}^*, \tau_{\mathrm{sp}}, \beta^*, \gamma^*)^T \right]$$

$$= \mathbb{E}_{\mathrm{sp}}\left[(Y_i^{\mathrm{obs}} - \alpha^* - \tau_{\mathrm{sp}} - \beta^{*\prime} X_i)^2 \cdot \begin{pmatrix} 1 \\ W_i - p \\ (X_i - \mu_X)^T \\ W_i \cdot (X_i - \mu_X)^T \end{pmatrix} \begin{pmatrix} 1 \\ W_i - p \\ (X_i - \mu_X)^T \\ W_i \cdot (X_i - \mu_X)^T \end{pmatrix}^T \right].$$

The normalized variance of $\hat{\tau}^{\text{ols}} - \tau_{\text{sp}}$ is the $(2, 2)$ element of the matrix $\Gamma^{-1}\Delta(\Gamma^{T})^{-1}$, which is equal to

$$\frac{\mathbb{E}_{\text{sp}}\left[\left(Y_i^{\text{obs}} - \alpha^* - \tau_{\text{sp}} - X_i\beta^*\right)^2 \cdot (W_i - p)^2\right]}{p^2 \cdot (1-p)^2}.$$

$\square$

Proof of Theorem 7.3

We use the same reparametrization as in the first part of the proof of Theorem 7.2:

$$\begin{pmatrix} \tilde{\alpha}_{\text{c}} \\ \beta_{\text{c}} \\ \tilde{\alpha}_{\text{t}} \\ \beta_{\text{t}} \end{pmatrix} = \begin{pmatrix} \alpha + \mu_X\beta \\ \beta \\ \alpha + \tau + \mu_X\beta \\ \gamma + \beta \end{pmatrix}.$$

In terms of the new parameters, $\gamma^* = \beta_{\text{t}}^* - \beta_{\text{c}}^*$. In the proof of Theorem 7.2 it was shown that the population values for $(\tilde{\alpha}_{\text{c}}, \beta_{\text{c}})$ solve

$$(\tilde{\alpha}_{\text{c}}^*, \beta_{\text{c}}^*) = \arg\min_{\alpha_{\text{c}},\beta_{\text{c}}} \mathbb{E}_{\text{sp}}\left[(1 - W_i) \cdot \left(Y_i^{\text{obs}} - \alpha_{\text{t}} - (X_i - \mu_X)\beta_{\text{c}}\right)^2\right]$$

$$= \arg\min_{\alpha_{\text{c}},\beta_{\text{c}}} \mathbb{E}_{\text{sp}}\left[(1 - W_i) \cdot (Y_i(0) - \alpha_{\text{t}} - (X_i - \mu_X)\beta_{\text{c}})^2\right].$$

Because of the randomization, W_i is independent of $Y_i(0)$ and X_i, and so

$$(\tilde{\alpha}_{\text{c}}^*, \beta_{\text{c}}^*) = \arg\min_{\alpha_{\text{c}},\beta_{\text{c}}} (1 - p) \cdot \mathbb{E}_{\text{sp}}\left[(Y_i(0) - \alpha_{\text{c}} - (X_i - \mu_X)\beta_{\text{c}})^2\right].$$

A similar argument shows that $(\tilde{\alpha}_{\text{t}}^*, \beta_{\text{t}}^*)$ solve the same optimization problem:

$$(\tilde{\alpha}_{\text{t}}^*, \beta_{\text{t}}^*) = \arg\min_{\alpha_{\text{t}},\beta_{\text{t}}} p \cdot \mathbb{E}_{\text{sp}}\left[(Y_i(1) - \alpha_{\text{c}} - (X_i - \mu_X)\beta_{\text{t}})^2\right]$$

$$= \arg\min_{\alpha_{\text{t}},\beta_{\text{t}}} (1 - p) \cdot \mathbb{E}_{\text{sp}}\left[(Y_i(0) + \tau - \alpha_{\text{c}} - (X_i - \mu_X)\beta_{\text{t}})^2\right]$$

(because by the null hypothesis of zero effects $Y_i(1) = Y_i(0) + \tau$) and so $\gamma^* = \beta_{\text{t}}^* - \beta_{\text{c}}^* = 0$. This finishes the proof of part (i) of the theorem.

Under the null hypothesis $(Y_i(1) = Y_i(0) + \tau)$, $\gamma^* = 0$. Then $\sqrt{N}\hat{\gamma}^{\text{ols}}$ will in large samples have a normal distribution with variance V_γ, and the quadratic form Q_{const} will have a Chi-squared distribution with degrees of freedom equal to the dimension of X_i. This concludes the proof of part (ii) of the theorem.

Under the null hypothesis $(Y_i(1) = Y_i(0)$ for all units) it also follows that $\tau_{\text{sp}} = 0$. In that case $\sqrt{N}(\hat{\tau}^{\text{ols}}, \hat{\gamma}^{\text{ols}})$ are in large samples normally distributed with covariance matrix $V_{\tau,\gamma}$. Hence the quadratic form Q_{zero} will in large samples have a chi-squared distribution with degrees of freedom equal to the dimension of τ and γ, which is equal to the dimension of X_i plus one.

The covariance matrix for $(\hat{\tau}^{\text{ols}}, \hat{\gamma}^{\text{ols}})$ is most easily obtained from the parametrization in part (ii) of the proof of Theorem 7.2, in terms of $(\tilde{\alpha}, \tau, \beta, \gamma)$. The point estimates for τ and γ under this parametrization are identical to those under the parametrization $(\alpha, \tau, \beta, \gamma)$. Under the parametrization in terms of $(\tilde{\alpha}, \tau, \beta, \gamma)$ the full covariance matrix of $\sqrt{N}(\hat{\tilde{\alpha}}^{\text{ols}} - \tilde{\alpha}^{\text{ols}}, \hat{\tau}^{\text{ols}} - \tau, \hat{\beta}^{\text{ols}} - \beta, \hat{\gamma}^{\text{ols}} - \gamma)$ is given by $\Gamma^{-1}\Delta(\Gamma^{T})^{-1})$ as given in (A.1). To obtain the covariance matrix for $\sqrt{N}(\hat{\tau}^{\text{ols}} - \tau, \hat{\gamma}^{\text{ols}} - \gamma)$ partition $\Gamma^{-1}\Delta(\Gamma^{T})^{-1})$ as

$$\mathbb{V} = \Gamma^{-1}\Delta(\Gamma^{T})^{-1}) = \begin{pmatrix} \mathbb{V}_{\tilde{\alpha},\tilde{\alpha}} & \mathbb{V}_{\tilde{\alpha},\tau} & \mathbb{V}_{\tilde{\alpha},\beta^{T}} & \mathbb{V}_{\tilde{\alpha},\gamma^{T}} \\ \mathbb{V}_{\tau,\tilde{\alpha}} & \mathbb{V}_{\tau,\tau} & \mathbb{V}_{\tau,\beta^{T}} & \mathbb{V}_{\tau,\gamma^{T}} \\ \mathbb{V}_{\beta,\tilde{\alpha}} & \mathbb{V}_{\beta,\tau} & \mathbb{V}_{\beta,\beta^{T}} & \mathbb{V}_{\beta,\gamma^{T}} \\ \mathbb{V}_{\gamma,\tilde{\alpha}} & \mathbb{V}_{\gamma,\tau} & \mathbb{V}_{\gamma,\beta^{T}} & \mathbb{V}_{\gamma,\gamma^{T}} \end{pmatrix}.$$

The covariance matrix for $\sqrt{N}(\hat{\tau}^{\text{ols}} - \tau, \hat{\gamma}^{\text{ols}} - \gamma)$ is then

$$\mathbb{V}_{\tau,\gamma} = \begin{pmatrix} \mathbb{V}_{\tau,\tau} & \mathbb{V}_{\tau,\gamma^{T}} \\ \mathbb{V}_{\gamma,\tau} & \mathbb{V}_{\gamma,\gamma^{T}} \end{pmatrix}.$$

The covariance matrix for $\sqrt{N}(\hat{\gamma}^{\text{ols}} - \gamma)$ is simply $\mathbb{V}_{\gamma,\gamma^{T}}$. $\qquad\qquad\square$

## CHAPTER 8

Model-Based Inference for Completely Randomized Experiments

8.1 INTRODUCTION

As discussed in Chapters 5 and 6, both Fisher's and Neyman's approaches for assessing treatment effects in completely randomized experiments viewed the potential outcomes as fixed quantities, some observed and some missing. The randomness in the observed outcomes was generated primarily through the assignment mechanism, and sometimes also through random sampling from a population. In this chapter, as in the preceding chapter on regression methods, we consider a different approach to inference, where the potential outcomes themselves are also viewed as random variables, even in the finite sample. Because all of the potential outcomes are considered random variables, any functions of them will also be random variables. This includes any causal estimand of interest – for example, the average treatment effect or the median causal effect.

We begin by building a stochastic model for all potential outcomes that generally depends on some unknown parameters. Using the observed data to learn about these parameters, we stochastically draw the unknown parameters and use the postulated model to impute the missing potential outcomes given the observed data, and use this in turn to conduct inference for the estimand of interest. At some level, all methods for causal inference can be viewed as imputation methods, although some more explicitly than others. Because any causal estimand depends on missing potential outcomes, any estimate for such an estimand is, implicitly or explicitly, based on estimates of these missing potential outcomes. The discussion in the current chapter puts this imputation perspective front and center. Because the imputations and resulting inferences are especially straightforward from a Bayesian perspective, we primarily focus on the Bayesian approach, but we also discuss the implementation of frequentist approaches, as well as how the two differ.

This model-based approach is very flexible compared to the Fisher's exact p-value approach, Neyman's repeated sampling approach, or regression methods. For instance, this method can easily accommodate a wide variety of estimands – we may be interested not only in average treatment effects but also in quantiles, or in measures of dispersion of the distributions of potential outcomes. In general we can conduct inference in this model-based approach for any causal estimand $\tau = \tau(\mathbf{Y}(0), \mathbf{Y}(1))$, or even more generally

$$\tau = \tau(\mathbf{Y}(0), \mathbf{Y}(1), \mathbf{X}, \mathbf{W}), \tag{8.1}$$

allowing the estimand to depend on the pre-treatment variables and the vector of treatment indicators: we do restrict τ to be a row-exchangeable comparison of $\mathbf{Y}(0)$, $\mathbf{Y}(1)$, $\mathbf{X}$, and $\mathbf{W}$ on a common set of units. In addition, although we focus primarily on the finite population, the model-based approach can easily accommodate super-population estimands. And lastly, unlike Fisher's and Neyman's methods, the model-based approach can be extended readily to observational studies, where the assignment mechanism is (partially) unknown, which we study in Parts III, IV, V, and VI of this text. In such settings, although fundamentally the resulting inference may be more sensitive to the modeling assumptions, and thus less credible than in randomized experiments, the basic approach, as well as its implementation, is the same as in classical randomized experiments.

One of the practical issues in the model-based approach is the choice of a credible model for imputing the missing potential outcomes. It is important to keep in mind here that the estimand of interest need not be a particular parameter of the statistical model. In many traditional statistical analyses, the parameters themselves are taken to be the primary objects of interest. For example, in linear regression analyses for causal effects discussed in the previous chapter, the primary focus of attention was one of the slope coefficients in the regression model. In the current setting, there is no reason why the parameters should coincide with the estimands. As stressed in the introduction to this book, the estimands τ are functions of the *ex ante* observable vectors of potential outcomes $\mathbf{Y}(0)$ and $\mathbf{Y}(1)$ (and possibly $\mathbf{X}$ and $\mathbf{W}$). These potential outcomes, and thus the causal estimands, are well defined irrespective of the stochastic model for either the treatment assignment or the potential outcomes. In some cases – for example, a linear model with identical slope coefficients in treatment and control groups – the estimand of interest may happen to be equal to one of the parameters of the model. Although this can simplify matters, especially when conducting a frequentist analysis of the data, it is important to understand that any such coincidence is not of any intrinsic importance, and it should not influence the choice of estimands or models, except for pedagogical purposes; rather, the choice should be based on substantive grounds. In the current setting of a completely randomized experiment, the inferences for the estimand of interest are often relatively robust to the parametric model chosen, as long as the specification is reasonably flexible. In fact, in many cases, at least in large samples, estimates for the average treatment effect are unbiased from Neyman's repeated sampling perspective, and the resulting interval estimates have the properties of Neyman's confidence intervals. Yet in other settings, for instance in observational studies with many covariates, the specification of the model may be an inherently difficult task, and the substantive conclusions are generally sensitive to the model-specification choices made. We will return to this issue in more detail in subsequent chapters.

A final comment is that, in contrast to the discussion in the previous chapter, we focus our discussion here on simulation-based computational methods rather than on analytical methods. In principle, either can be used. We focus on computational methods in large part because they often simplify the analyses given recent advances in computational power and in computational methods, such as Markov-Chain-Monte-Carlo (MCMC) techniques. Focusing on computational methods allows us to separate the problem of drawing inferences into smaller steps, with each step often conceptually straightforward. In addition, in contrast to analytical approaches, computational methods maintain the conceptual distinction between parameters in the parametric model and the estimands of interest.

The remainder of this chapter is structured as follows. In Section 8.2 we describe the data from a randomized evaluation of a labor market training program, originally analyzed by Lalonde (1986) and subsequently by Dehejia and Wahba (1999), as well as many others. In Section 8.3, as an introduction to the ideas underlying the model-based approach, we begin with a simple example with a population of only six units and discuss two naive methods to impute the missing potential outcomes given the observed data. The first naive method ignores uncertainty altogether. The second naive method incorporates uncertainty in the value to impute but ignores uncertainty in the estimated model. In addition, both naive methods jump directly to a model of the missing potential outcomes given the observed data, rather than deriving it. But this conditional distribution is inherently a function of the two underlying primitives, the assignment mechanism and the joint distribution of the two potential outcomes, and conceptually it is attractive first to specify these primitives and then to derive the conditional distribution of missing potential outcomes given observed values from these primitives. In order to incorporate uncertainty into the model, the model-based approach starts directly from these more fundamental distributions and then derives the conditional distribution of the missing potential outcomes.

Section 8.4 is the central section in this chapter. In this section we introduce the various steps of the general structure of the model-based approach in the setting without covariates. The goal is to calculate the conditional distribution of the full vector of missing potential outcomes given observed data:

$$f(\mathbf{Y}^{\text{mis}}|\mathbf{Y}^{\text{obs}}, \mathbf{W}). \tag{8.2}$$

Once we have this conditional distribution, we can infer the distribution for any estimand of interest of the form $\tau = \tau(\mathbf{Y}(0), \mathbf{Y}(1), \mathbf{W})$ by rewriting the estimand as a function of observed and missing outcomes, and assignments, $\tau = \tau(\mathbf{Y}^{\text{mis}}, \mathbf{Y}^{\text{obs}}, \mathbf{W})$. The Bayesian approach for deriving the conditional distribution in (8.2) is implemented using two inputs. The first input is a model for the joint distribution of $(\mathbf{Y}(0), \mathbf{Y}(1))$ given a hypothetical vector of parameters θ,

$$f(\mathbf{Y}(0), \mathbf{Y}(1)|\theta). \tag{8.3}$$

By specifying this distribution in terms of a vector of unknown parameters θ, we allow for a flexible model, with essentially no loss of generality. The second input is a prior distribution for θ, representing prior beliefs about the parameter vector:

$$p(\theta). \tag{8.4}$$

In Section 8.4 we analyze the four steps taking us from the two inputs, (8.3) and (8.4), to the output, (8.2), in detail. We also discuss the choices for the model and prior distribution. To illustrate these ideas, we return to the same six units studied in Section 8.3.

In the subsequent five sections we discuss extensions of the model-based approach. First, in Section 8.5 we discuss simulation methods for approximating the distribution of τ given $\mathbf{Y}^{\mathrm{obs}}$ and $\mathbf{W}$, that is, the posterior distribution. Then, in Section 8.6, we discuss the issues concerning dependence between the two potential outcomes $(Y_i(0), Y_i(1))$ for a given unit, including the inability of the data to provide information regarding any such dependence, and the implications of that for posterior distributions. In Section 8.7 we incorporate covariates X_i into the model-based approach. Next, in Section 8.8, we discuss a super-population interpretation of the data. Up to this point, including Section 8.8, the discussion takes a Bayesian perspective, although the methods discussed in this chapter can also accommodate a frequentist (repeated sampling) approach.[1] In Section 8.9 we discuss the model-based approach from this chapter from a frequentist perspective. In contrast to the Bayesian approach, the standard frequentist approach interprets the unknown hypothetical parameters as fixed quantities and assumes that the potential outcomes (missing or observed) are random variables given these fixed parameters. In Section 8.10 we present estimates based on the Lalonde-Dehejia-Wahba data, illustrating the various methods introduced in this chapter.

8.2   THE LALONDE NSW EXPERIMENTAL JOB-TRAINING DATA

The data we use in this chapter, to illustrate the methods developed here, come from a randomized evaluation of a job training program, the National Supported Work (NSW) program, first analyzed by Lalonde (1986) and subsequently widely used in the literature on program evaluation in econometrics. The specific data set we use here is the one discussed by Dehejia and Wabha (1999), which is a subset of the Lalonde data. The population that was eligible for this program consisted of men who were substantially disadvantaged in the labor market. Most of them had very poor labor market histories with few instances of long-term employment. For each man in this subset we have data on background characteristics, including age (`age`), years of education (`education`), whether they were now or ever before married (`married`), whether they were high school dropouts (`nodegree`), and ethnicity (`black`). We also have two measures of pre-training earnings; the first is earnings in 1975 (`earn'75`), and the second is earnings thirteen to twenty-four months prior to the training, denoted by (`earn'74`) because this primarily corresponds to earnings in the calendar year 1974. We also use an indicator for zero earnings in 1975 (`earn'75=0`) and an indicator for zero earnings in the months thirteen to twenty-four prior to being randomized to training or not

---

[1] A Bayesian perspective refers to statistical analyses based on viewing all *a priori* unobserved quantities as random variables and deriving the joint conditional distribution of estimands given all observed quantities using Bayes Rule. A frequentist perspective refers to analyses of procedures in terms of their properties in repeated samples. Interestingly, Fisher's (FEP) approach is arguably closer conceptually to the Bayesian approach than to the Neyman approach (Rubin, 1984). See Appendix A for more details and references.

Table 8.1. *Summary Statistics: National Supported Work (NSW) Program Data*

Covariate	Mean	(S.D.)	Average Controls ($N_c = 260$)	Average Treated ($N_t = 185$)
age	25.37	(7.10)	25.05	25.82
education	10.20	(1.79)	10.09	10.35
married	0.17	(0.37)	0.15	0.19
nodegree	0.78	(0.41)	0.83	0.71
black	0.83	(0.37)	0.83	0.84
earn'74	2.10	(5.36)	2.11	2.10
earn'74=0	0.73	(0.44)	0.75	0.71
earn'75	1.38	(3.15)	1.27	1.53
earn'75=0	0.65	(0.48)	0.68	0.60
earn'78	5.30	(6.63)	4.56	6.35
earn'78=0	0.31	(0.46)	0.35	0.24

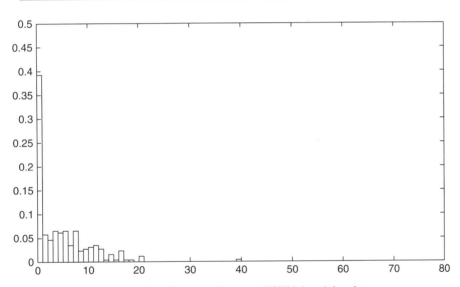

Figure 8.1.   Histogram of earnings for control group – NSW job-training data

(earn'74=0). The outcome of interest is post-program labor market experiences, earnings in 1978 (earn'78).

   Table 8.1 presents some summary statistics for the sample of $N = 445$ men, of whom $N_t = 185$ were assigned to the job training program and $N_c = 260$ were assigned to the control group. All earnings variables are in thousands of dollars. Note that annual earnings for these men are very low, even for those years; when we average only over those with positive earnings, average annual earnings in 1978 are on the order of only approximately $8,000 after the program. Prior to the program, earnings are even lower, partly because low earnings in 1978 were a component for determining eligibility. Most pre-program characteristics are reasonably well balanced between the two groups, although the lower proportion of men with zero earnings in 1975 in the treatment group might raise concerns. Figures 8.1 and 8.2 present histograms of the distribution of the outcome, earnings in 1978 in the control and treatment groups, respectively.

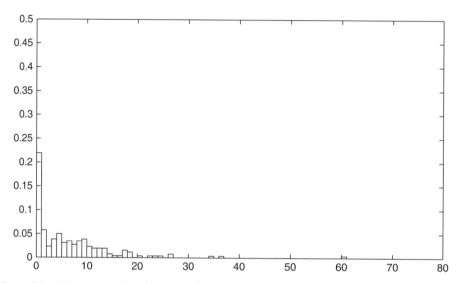

Figure 8.2.   Histogram of earnings for trainee group – NSW job-training data

8.3   A SIMPLE EXAMPLE: NAIVE AND MORE SOPHISTICATED APPROACHES TO IMPUTATION

Before we introduce the formal representation of the model-based imputation approach, we begin by working through a very simple example that introduces the key ideas underlying this approach. To illustrate this example, we use a subset of the data from the NSW evaluation. Table 8.2 lists information on six men from this data set. The first man did not go through the training program. He did not have a job in 1978, and his 1978 earnings were zero. The second man did go through the training program. He subsequently did find a job, and received earnings in 1978 equal to approximately \$9,900. There are a total of three treated and three control individuals, and thus twelve potential outcomes, six of them observed and six of them missing.

In the illustration in this section, we focus on the average treatment effect as the estimand. More general estimands can easily be accommodated in this approach, and we discuss some later. We can write the average treatment effect for this population of six men as

$$\tau_{\text{fs}} = \tau(\mathbf{Y}(0), \mathbf{Y}(1)) = \frac{1}{6} \cdot \sum_{i=1}^{6} \left(Y_i(1) - Y_i(0) \right). \tag{8.5}$$

We rely heavily on an alternative representation of the average treatment effect, in terms of observed and missing potential outcomes. To derive this representation, we use the characterization of the two potential outcomes $Y_i(0)$ and $Y_i(1)$ in terms of the missing and observed values:

$$Y_i(0) = \begin{cases} Y_i^{\text{mis}} & \text{if } W_i = 1, \\ Y_i^{\text{obs}} & \text{if } W_i = 0, \end{cases} \quad \text{and} \quad Y_i(1) = \begin{cases} Y_i^{\text{mis}} & \text{if } W_i = 0, \\ Y_i^{\text{obs}} & \text{if } W_i = 1. \end{cases} \tag{8.6}$$

Table 8.2. *First Six Observations from NSW Program Data*

Unit	Potential Outcomes		Treatment	Observed Outcome
	$Y_i(0)$	$Y_i(1)$	W_i	Y_i^{obs}
1	0	?	0	0
2	?	9.9	1	9.9
3	12.4	?	0	12.4
4	?	3.6	1	3.6
5	0	?	0	0
6	?	24.9	1	24.9

Note: Question marks represent missing potential outcomes.

Then we can write τ_{fs} in terms of observed and missing potential outcomes and treatment indicators as

$$\tau_{\text{fs}} = \tilde{\tau}(\mathbf{Y}^{\text{obs}}, \mathbf{Y}^{\text{mis}}, \mathbf{W})$$

$$= \frac{1}{6} \cdot \sum_i^N \left((W_i \cdot Y_i^{\text{obs}} + (1 - W_i) \cdot Y_i^{\text{mis}}) - ((1 - W_i) \cdot Y_i^{\text{obs}} + W_i \cdot Y_i^{\text{mis}}) \right)$$

$$= \frac{1}{6} \cdot \sum_{i=1}^N \left((2 \cdot W_i - 1) \cdot \left(Y_i^{\text{obs}} - Y_i^{\text{mis}} \right) \right). \tag{8.7}$$

We know the value of the causal estimand up to the missing potential outcome values. In the model-based approach, we estimate the average treatment effect by explicitly imputing the six missing potential outcomes, initially once, and then repeatedly to account for the uncertainty in the imputation. Let $\hat{Y}_i^{\text{mis}}$ be the imputed value for Y_i^{mis}, leading to the following estimator for the average treatment effect:

$$\hat{\tau} = \tilde{\tau}(\mathbf{Y}^{\text{obs}}, \hat{\mathbf{Y}}^{\text{mis}}, \mathbf{W}) = \frac{1}{6} \cdot \sum_{i=1}^N \left((2 \cdot W_i - 1) \cdot (Y_i^{\text{obs}} - \hat{Y}_i^{\text{mis}}) \right). \tag{8.8}$$

The key question is how to impute the missing potential outcomes $\hat{Y}_i^{\text{mis}}$, given the observed values $\mathbf{Y}^{\text{obs}}$ and the treatment assignments $\mathbf{W}$.

Let us first discuss a very simple, and naive, approach, where we impute each missing potential outcome by the average of the observed potential outcomes with that treatment level. Consider the first unit. Unit 1 received the control treatment, so we observe its potential outcome under control ($Y_1(0)$) but not its potential outcome given treatment ($Y_1(1)$). Thus $Y_1^{\text{obs}} = Y_1(0)$ and $Y_1^{\text{mis}} = Y_1(1)$. The average outcome for the three units randomly assigned to the treatment, that is, units 2, 4, and 6, is $\overline{Y}_{\text{t}}^{\text{obs}} = (Y_2(1) + Y_4(1) + Y_6(1))/3 = (9.9 + 3.6 + 24.9)/3 = 12.8$. In this illustrative example, we would therefore impute $\hat{Y}_1^{\text{mis}} = 12.8$. In contrast, Unit 2 received the treatment, thus $Y_2^{\text{mis}} = Y_2(0)$. The average observed outcome for the three randomly chosen units who did receive the control treatment is $\overline{Y}_{\text{c}}^{\text{obs}} = (Y_1(0) + Y_3(0) + Y_5(0))/3 = (0 + 12.4 + 0)/3 = 4.1$, so we impute $\hat{Y}_2^{\text{mis}} = \overline{Y}_{\text{c}}^{\text{obs}} = 4.1$. Following the same approach for the remaining

Table 8.3. *The Average Treatment Effect Using Imputation of Average Observed Outcome Values within Treatment and Control Groups for the NSW Program Data*

Unit	Potential Outcomes		Treatment	Observed Outcome
	$Y_i(0)$	$Y_i(1)$	W_i	Y_i^{obs}
1	0	(12.8)	0	0
2	(4.13)	9.9	1	9.9
3	12.4	(12.8)	0	12.4
4	(4.13)	3.6	1	3.6
5	0	(12.8)	0	0
6	(4.13)	24.9	1	24.9
Average	4.13	12.8		
Diff (ATE):		8.67		

four units, Table 8.3 presents the observed and imputed potential outcomes – the latter in parentheses – for all six units. Substituting these values in Equation (8.8) gives an average treatment effect of $\hat{\tau} = 12.8 - 4.1 = 8.7$. Notice that this is equal to the difference between the two average observed outcomes by treatment status, $\hat{\tau}^{dif} = \overline{Y}_t^{obs} - \overline{Y}_c^{obs}$. Given the imputation method, the value for the causal estimand should not be surprising, but the overall result is unsatisfying. Because we imputed the missing potential outcomes as if there were no uncertainty about their values, this method provides only a point estimate, with no sense of its precision. Yet it is clear that we are not at all certain that the missing potential outcomes $Y_1(1)$, $Y_3(1)$, and $Y_5(1)$ are all exactly equal to 12.8. In fact, for the three units with $Y_i(1)$ observed, we see that there is a fair amount of variation in the $Y_i(1)$. Even if we assume that units 1, 3, and 5 are "on average" just like the others – as we should expect, given the completely randomized experiment – we should still create imputations that reflect this variability. At most, the randomization would allow us to deduce the *distribution* of the missing potential outcomes, but almost never the exact values of the missing potential outcomes.

Let us therefore consider a second, less naive approach to imputing the missing potential outcomes. Let us again consider a unit with $W_i = w$, so that $Y_i^{mis} = Y_i(1 - w)$. Instead of setting $\hat{Y}_i^{mis}$ for such a unit equal to the corresponding average observed value $\overline{Y}_c^{obs}$ if $w = 1$ or $\overline{Y}_t^{obs}$ if $w = 0$, as we did in the first approach, let us draw Y_i^{mis} for such a unit at random from the distribution of Y_j^{obs} for those units for whom we observe $Y_j(1 - w)$, that is, units with $W_j = 1 - w$. Specifically, for Unit 1, with $Y_1^{mis} = Y_1(1)$, let us draw at random from the trinomial distribution that puts mass 1/3 on each of the three observed $Y_i(1)$ values, the observed Y_i^{obs} values for Units 2, 4, and 6, namely $Y_2(1) = 9.9$, $Y_4(1) = 3.6$, and $Y_6(1) = 24.9$. Similarly for Unit 2, impute Y_2^{mis} by drawing from the trinomial distribution with values $Y_1(0) = 0$, $Y_3(0) = 12.4$, and $Y_5(0) = 0$, each with probability equal to 1/3; because two of the values are equal, this amounts to a binomial distribution with support points 0 and 12.4, with probabilities 2/3 and 1/3, respectively. Suppose we draw 3.6 for Unit 1 and 12.4 for Unit 2, thereby imputing $\hat{Y}_1^{mis} = 3.6$ and $\hat{Y}_2^{mis} = 12.4$. For the third unit, we again draw from the distribution with values 9.9, 3.6, and 24.9; suppose we draw $\hat{Y}_3^{mis} = 9.9$. For the fourth unit, suppose we

Table 8.4. *The Average Treatment Effect Using Imputed Draws from the Empirical Distributions within Treatment and Control Groups for the First Six Units from the NSW Program Data*

Unit	Potential Outcomes		Treatment	Observed Outcome
	$Y_i(0)$	$Y_i(1)$	W_i	Y_i^{obs}
Panel A: First draw				
1	0	(3.6)	0	0
2	(12.4)	9.9	1	9.9
3	12.4	(9.9)	0	12.4
4	(12.4)	3.6	1	3.6
5	0	(9.9)	0	0
6	(0)	24.9	1	24.9
Average	6.2	10.3		
Diff (ATE):		4.1		
Panel B: Second draw				
1	0	(9.9)	0	0
2	(0)	9.9	1	9.9
3	12.4	(24.9)	0	12.4
4	(0)	3.6	1	3.6
5	0	(3.6)	0	0
6	(0)	24.9	1	24.9
Average	2.1	12.8		
Diff (ATE):		10.7		

again draw 12.4; hence $\hat{Y}_4^{mis} = \hat{Y}_2^{mis} = 12.4$. Note that because we draw with replacement, it is possible to draw the same value for more than one unit. Panel A of Table 8.4 gives these six observations with the missing values imputed in this fashion. Given the imputed and observed data, this gives an estimated average treatment effect of 4.1.

Up to this point, this process has been fairly similar to the first method: for each of the six units, we imputed the missing potential outcome and, via Equation (8.8), used those imputations to estimate the average treatment effect. Now, however, there is a crucial difference. With the current method, we can repeat this process to give a new value for the average treatment effect. Again drawing from the same assumed distributions for the missing $\mathbf{Y}(0)$ and $\mathbf{Y}(1)$, we expect to draw different values, thereby giving a different estimate for the average treatment effect. Panel B of Table 8.4 presents such a result, this time giving an estimated average treatment effect equal to 10.7.

We can repeat this procedure as many times as we wish, although at some point we will generate sets of draws identical to the ones already obtained. With six missing potential outcomes, each one drawn from a set of three possible values, there are $3^6 = 729$ different ways of imputing the data, all equally likely. Calculating the corresponding average treatment effect for each set of draws, we can then calculate the average and standard deviation of these 729 estimates. Note that not all of these will be different; the order in which the individual outcomes are imputed does not matter. Over the 729 possible vectors of imputed missing data, this leads to an average treatment effect of

8.7 and a standard deviation of 3.1. Notice that this average is again identical to the difference in average outcomes by treatment level, $\hat{\tau}^{\text{dif}} = \overline{Y}_{\text{t}}^{\text{obs}} - \overline{Y}_{\text{c}}^{\text{obs}}$. As before, this should seem intuitive, because we have calculated this value from the full set of 729 possible, equally likely, permutations. What this approach adds to the previous analysis, however, is an estimate of the entire distribution of the average treatment effect and, in particular, an estimate of the variability of the estimated average treatment effect, as reflected, for instance, in the standard deviation of this distribution.

Although this example focuses on the average treatment effect, the same procedure could be applied to any other function of the six pairs of potential outcomes. For example, one may be interested in the ratio of variances of the potential outcomes at each treatment level, or in other measures of central tendency or dispersion.

With more than six units, it quickly becomes expensive to calculate all possible imputations of the missing data. In practice one may, therefore, prefer to use a randomly selected subset of these imputations and estimate the distribution of a treatment effect as reflected by these values. Such an approach will give an accurate approximation to the distribution based on drawing all possible imputations if enough replications are made. The use of this randomization for imputing the missing potential outcomes is purely a computational device, albeit a very convenient one.

This second method for imputing the missing potential outcomes is substantially more sophisticated than the first. Nevertheless, it still does not address fully the uncertainty we face in estimating the average treatment effect. In particular, we impute the missing data as if we knew the *exact* distribution of each of the potential outcomes. Yet, in practice, we have only limited information; in this example based on six units, our information for the distributions of treatment and control outcomes comes entirely from three observations for each. For instance, we assume the distribution of $Y_i(1)$, based on the three observed values (9.9, 3.6, and 24.9), is trinomial for those three values with equal probability. If we actually observed three additional units exposed to the treatment, it is likely that their observed outcomes would differ from the first three. If we study the set of all 445 observations in the NSW data set, we see that the other treated units do have different potential outcomes from the three in Table 8.2. To take into account this additional source of uncertainty essentially requires a model for the potential outcomes – observed as well as missing – which formally addresses the uncertainty about possible values of missing potential outcomes. We turn to this next.

8.4   BAYESIAN MODEL-BASED IMPUTATION IN THE ABSENCE OF COVARIATES

Let us now formally describe the Bayesian model-based approach for inference in completely randomized experiments when no covariates are observed. The primary goal of this approach is to build a model for the missing potential outcomes, given the observed data,

$$f(\mathbf{Y}^{\text{mis}}|\mathbf{Y}^{\text{obs}}, \mathbf{W}). \tag{8.9}$$

Once we have such a model, we can derive the distribution for the estimand of interest, $\tau = \tau(\mathbf{Y}(0), \mathbf{Y}(1), \mathbf{W})$, using the fact that we can also represent the estimand in terms of observed and missing potential outcomes as $\tau = \tau(\mathbf{Y}^{\text{mis}}, \mathbf{Y}^{\text{obs}}, \mathbf{W})$.

Throughout this chapter, we are slightly informal in our use of notation, and use $f(\cdot | \cdot)$ to denote generic conditional distributions, without indexing the distribution $f(\cdot | \cdot)$ by the random variables. In each case it should be clear from the context to which random variables the distributions refer.

The previous naive approaches also build models for the missing potential outcomes but in partially unsatisfactory ways. In the first approach in Section 8.3, we specified a degenerate distribution of the missing potential outcomes for unit i as

$$\Pr\left(Y_i^{\text{mis}} = y \,\middle|\, \mathbf{Y}^{\text{obs}}, \mathbf{W}\right) = \begin{cases} 1 & \text{if } y = 12.8, \text{ and } W_i = 0, \\ 1 & \text{if } y = 4.1, \text{ and } W_i = 1, \\ 0 & \text{otherwise.} \end{cases}$$

In the second approach in Section 8.3, we specified a non-degenerate distribution of the missing potential outcomes for unit i, namely

$$\Pr\left(Y_i^{\text{mis}} = y \,\middle|\, \mathbf{Y}^{\text{obs}}, \mathbf{W}\right) = \begin{cases} 1/3 & \text{if } y \in \{3.6, 9.9, 24.9\}, \text{ and } W_i = 0, \\ 1/3 & \text{if } y = 12.4, W_i = 1, \\ 2/3 & \text{if } y = 0, W_i = 1, \\ 0 & \text{otherwise.} \end{cases}$$

Using these models, for each unit i, we predicted Y_i^{mis}, the outcome we would have observed if i had been exposed to the alternative treatment. Given these imputed missing potential outcomes, we calculated the corresponding estimand, in the specific example, the average treatment effect. These models for the missing potential outcomes were straightforward, but too simplistic, in that neither model allowed for uncertainty in the estimation of the distribution of the missing potential outcomes. In this section we consider more sophisticated methods for imputing the missing potential outcomes that allow for such uncertainty.

Although what we are ultimately interested in is simply a model for the conditional distribution of $\mathbf{Y}^{\text{mis}}$ given $(\mathbf{Y}^{\text{obs}}, \mathbf{W})$, this is not our initial focus. The reason is that it is conceptually difficult to specify directly a model for the conditional distribution of $\mathbf{Y}^{\text{mis}}$ given $\mathbf{Y}^{\text{obs}}$ and $\mathbf{W}$, and still formally conform to the distributional assumptions on the science and the assignment mechanism. The conditional distribution of $\mathbf{Y}^{\text{mis}}$ given $(\mathbf{Y}^{\text{obs}}, \mathbf{W})$ depends intricately on the joint distribution of the potential outcomes, $(\mathbf{Y}(0), \mathbf{Y}(1))$, and on the assignment mechanism. These are very different objects. Specification of the former requires scientific (e.g., subject-matter) knowledge, be it economics, biology, or some other science. In contrast, in the context of this chapter, the assignment mechanism is known by the assumption of a completely randomized experiment. In the model-based approach, we therefore step back and consider specification of the two components separately.

In the remainder of this section, we describe, at a more abstract level, the general approach for obtaining the distribution of the missing data given the observed data in settings without covariates. We separate the derivation of the posterior distribution of the causal effect of interest into four steps, laying out in detail the procedure that takes us from the specification of the joint distribution of the potential outcomes to the conditional distribution of the causal estimand given the observed data, called the posterior (meaning post-observed data) distribution of the estimand. Following the description of the general approach, we return to the six-unit example and show, in detail, how this can be implemented analytically in a very simple setting with Gaussian distributions for the potential outcomes. However, in practice there are few situations where one can derive the posterior distribution of interest analytically, and in Section 8.5 we show how simulation methods can be used to obtain draws from the posterior distribution in the same simple example. This simulation approach is much more widely applicable and often easy to implement.

8.4.1 Inputs into the Model-Based Approach

The first input for the model-based approach is a model for the joint distribution of the two potential outcomes $(\mathbf{Y}(0), \mathbf{Y}(1))$:

$$f(\mathbf{Y}(0), \mathbf{Y}(1)). \tag{8.10}$$

Under row (unit) exchangeability of the matrix $(\mathbf{Y}(0), \mathbf{Y}(1))$, and by an appeal to de Finetti's theorem, we can, with no essential loss of generality, model this joint distribution $(\mathbf{Y}(0), \mathbf{Y}(1))$ as the integral over the product of iid (independent and identically distributed) unit-level distributions,

$$f(\mathbf{Y}(0), \mathbf{Y}(1)) = \int \prod_{i=1}^{N} f(Y_i(0), Y_i(1)|\theta) \cdot p(\theta) d\theta,$$

where θ is an unknown, finite-dimensional parameter of $f(Y_i(0), Y_i(1)|\theta)$, which lies in a parameter space Θ, and $p(\theta)$ is its marginal (or prior) distribution.

Specifying the joint distribution of $(Y_i(0), Y_i(1))$ conditional on θ can be a difficult task. The joint density can involve many unknown parameters. Its specification requires subject-matter (scientific) knowledge. Although in the current setting of completely randomized experiments, inferences are often robust to different specifications, this is not necessarily true in observational studies. In the example in the next section, we use a bivariate normal distribution, but in other cases, binomial distributions or log normal distributions, or mixtures of more complicated distributions may be more appropriate.

Specifying the second input, the prior distribution of θ,

$$p(\theta), \tag{8.11}$$

can also be difficult. In many cases, however, the substantive conclusions are not particularly sensitive to this choice. In the application in this chapter we investigate this issue in more detail.

In observational studies there would be a third input into the model-based calculations: the conditional distribution of $\mathbf{W}$ given the potential outcomes, or in other words, the assignment mechanism, $f(\mathbf{W}|\mathbf{Y}(0), \mathbf{Y}(1))$. In the current setting of a completely randomized experiment with no covariate, the assignment mechanism is by definition equal to

$$\Pr(\mathbf{W}|\mathbf{Y}(0),\mathbf{Y}(1)) = \binom{N}{N_t}^{-1}, \quad \text{for all } \mathbf{W} \text{ such that } \sum_{i=1}^{N} W_i = N_t,$$

so this is an input that needs no further specification here.

8.4.2 The Four Steps of the Bayesian Approach to Model-Based Inference for Causal Effects in Completely Randomized Experiments with No Covariates

There are four steps involved in going from the two inputs to the distribution of the estimand given the observed data. The first step of the model-based approach involves deriving $f(\mathbf{Y}^{\text{mis}}|\mathbf{Y}^{\text{obs}},\mathbf{W},\theta)$. The second step involves deriving the posterior distribution for the parameter θ, that is, $f(\theta|\mathbf{Y}^{\text{obs}},\mathbf{W})$. The third step involves combining the conditional distribution $f(\mathbf{Y}^{\text{mis}}|\mathbf{Y}^{\text{obs}},\mathbf{W},\theta)$ and the posterior distribution $f(\theta|\mathbf{Y}^{\text{obs}},\mathbf{W})$ to obtain the conditional distribution of the missing data given the observed data, but without conditioning on the parameters, $f(\mathbf{Y}^{\text{mis}}|\mathbf{Y}^{\text{obs}},\mathbf{W})$, that is, integrating their product over θ. Finally, in the fourth step we use the definition of the estimand, $\tau = \tau(\mathbf{Y}(0),\mathbf{Y}(1))$, and the conditional distribution $f(\mathbf{Y}^{\text{mis}}|\mathbf{Y}^{\text{obs}},\mathbf{W})$ to obtain the conditional distribution of the estimand given the observed values, $f(\tau|\mathbf{Y}^{\text{obs}},\mathbf{W})$. We now examine these four steps in somewhat excruciating detail.

Step 1: Derivation of $f(\mathbf{Y}^{\text{mis}}|\mathbf{Y}^{\text{obs}},\mathbf{W},\theta)$ First we combine the conditional distribution of the vector of assignments given the potential outcomes, $\Pr(\mathbf{W}|\mathbf{Y}(0),\mathbf{Y}(1),\theta)$, with the model for the joint distribution of the potential outcomes given, θ, $f(\mathbf{Y}(0),\mathbf{Y}(1)|\theta)$, to get the joint distribution of $(\mathbf{W},\mathbf{Y}(0),\mathbf{Y}(1))$ given θ, as the product of these two vectors:

$$f(\mathbf{Y}(0),\mathbf{Y}(1),\mathbf{W}|\theta) = \Pr(\mathbf{W}|\mathbf{Y}(0),\mathbf{Y}(1),\theta) \cdot f(\mathbf{Y}(0),\mathbf{Y}(1)|\theta). \tag{8.12}$$

Using the joint distribution in (8.12), we derive the conditional distribution of the potential outcomes given the vector of assignments and the parameter, θ, $f(\mathbf{Y}(0),\mathbf{Y}(1)|\mathbf{W},\theta)$, for the general case as

$$f(\mathbf{Y}(0),\mathbf{Y}(1)|\mathbf{W},\theta) = \frac{f(\mathbf{Y}(0),\mathbf{Y}(1),\mathbf{W}|\theta)}{\Pr(\mathbf{W}|\theta)} = \frac{f(\mathbf{Y}(0),\mathbf{Y}(1),\mathbf{W}|\theta)}{\int f(\mathbf{y}(0),\mathbf{y}(1),\mathbf{W}|\theta)d\mathbf{y}(0)d\mathbf{y}(1)}.$$

The assumption of a completely randomized experiment implies that $\mathbf{W}$ is independent of $(\mathbf{Y}(0),\mathbf{Y}(1))$, and so that this conditional distribution is in fact equal to the marginal distribution:

$$f(\mathbf{Y}(0),\mathbf{Y}(1)|\mathbf{W},\theta) = f(\mathbf{Y}(0),\mathbf{Y}(1)|\theta).$$

This simplification more generally applies to all regular assignment mechanisms.

Next, we transform the distribution for $\mathbf{Y}(0)$ and $\mathbf{Y}(1)$ given $\mathbf{W}$ and θ into the distribution for $\mathbf{Y}^{\text{mis}}$ given $\mathbf{Y}^{\text{obs}}$, $\mathbf{W}$, and θ. Recall that we can express the pair $(Y_i^{\text{mis}}, Y_i^{\text{obs}})$ as functions of $(Y_i(0), Y_i(1), W_i)$:

$$Y_i^{\text{obs}} = \begin{cases} Y_i(0) & \text{if } W_i = 0, \\ Y_i(1) & \text{if } W_i = 1, \end{cases} \qquad Y_i^{\text{mis}} = \begin{cases} Y_i(0) & \text{if } W_i = 1, \\ Y_i(1) & \text{if } W_i = 0. \end{cases} \tag{8.13}$$

Hence $(\mathbf{Y}^{\text{mis}}, \mathbf{Y}^{\text{obs}})$ can be written as a transformation of $(\mathbf{Y}(0), \mathbf{Y}(1), \mathbf{W})$, or

$$(\mathbf{Y}^{\text{mis}}, \mathbf{Y}^{\text{obs}}) = g(\mathbf{Y}(0), \mathbf{Y}(1), \mathbf{W}).$$

We can use this transformation to obtain the distribution of $(\mathbf{Y}^{\text{mis}}, \mathbf{Y}^{\text{obs}})$ given $\mathbf{W}$ and θ,

$$f(\mathbf{Y}^{\text{mis}}, \mathbf{Y}^{\text{obs}}|\mathbf{W}, \theta). \tag{8.14}$$

This, in turn, allows us to derive:

$$f(\mathbf{Y}^{\text{mis}}|\mathbf{Y}^{\text{obs}}, \mathbf{W}, \theta) = \frac{f(\mathbf{Y}^{\text{mis}}, \mathbf{Y}^{\text{obs}}|\mathbf{W}, \theta)}{f(\mathbf{Y}^{\text{obs}}|\mathbf{W}, \theta)} = \frac{f(\mathbf{Y}^{\text{mis}}, \mathbf{Y}^{\text{obs}}|\mathbf{W}, \theta)}{\int_{\mathbf{y}^{\text{mis}}} f(\mathbf{y}^{\text{mis}}, \mathbf{Y}^{\text{obs}}|\mathbf{W}, \theta) d\mathbf{y}^{\text{mis}}}. \tag{8.15}$$

This is the conditional distribution of the missing potential outcomes given the observed values, also called the posterior predictive distribution of $\mathbf{Y}^{\text{mis}}$.

Step 2: Derivation of the Posterior Distribution of the Parameter θ, $p(\theta|\mathbf{Y}^{\text{obs}}, \mathbf{W})$ Here we combine the prior distribution on θ, $p(\theta)$, with the distribution of the observed data given θ to derive the posterior distribution of θ, $p(\theta|\mathbf{Y}^{\text{obs}}, \mathbf{W})$. In order to derive the likelihood function, which is proportional to the distribution of the observed data regarded as a function of the unknown θ, we return to our previously established joint distribution of the missing and observed potential outcomes given the parameter θ, $f(\mathbf{Y}^{\text{mis}}, \mathbf{Y}^{\text{obs}}|\mathbf{W}, \theta)$. From this, we can derive the marginal distribution of the observed outcomes given θ, that is, the likelihood function, by integrating out the missing potential outcomes,

$$\mathcal{L}(\theta|\mathbf{Y}^{\text{obs}}, \mathbf{W}) \equiv f(\mathbf{Y}^{\text{obs}}, \mathbf{W}|\theta) = \int_{\mathbf{y}^{\text{mis}}} f(\mathbf{y}^{\text{mis}}, \mathbf{Y}^{\text{obs}}, \mathbf{W}|\theta) \, d\mathbf{y}^{\text{mis}}.$$

Combining the likelihood function with the prior distribution $p(\theta)$, we obtain the posterior (that is, conditional given the observed data) distribution of the parameters:

$$p(\theta|\mathbf{Y}^{\text{obs}}, \mathbf{W}) = \frac{p(\theta) \cdot \mathcal{L}(\theta|\mathbf{Y}^{\text{obs}}, \mathbf{W})}{f(\mathbf{Y}^{\text{obs}}, \mathbf{W})}, \tag{8.16}$$

where $f(\mathbf{Y}^{\text{obs}}, \mathbf{W})$ is the marginal distribution of $(\mathbf{Y}, \mathbf{W})$ obtained by integrating over θ:

$$f(\mathbf{Y}^{\text{obs}}, \mathbf{W}) = \int_{\theta} p(\theta) \cdot \mathcal{L}(\theta|\mathbf{Y}^{\text{obs}}, \mathbf{W}) \, d\theta.$$

Step 3: Derivation of Posterior Distribution of Missing Outcomes $f(\mathbf{Y}^{\text{mis}}|\mathbf{Y}^{\text{obs}}, \mathbf{W})$ Now we combine the conditional distribution of $\mathbf{Y}^{\text{mis}}$ given $(\mathbf{Y}^{\text{obs}}, \mathbf{W}, \theta)$, given in (8.15), and the posterior distribution for θ, given in (8.16), to derive the joint distribution of $(\mathbf{Y}^{\text{mis}}, \theta)$ given $(\mathbf{Y}^{\text{obs}}, \mathbf{W})$:

$$f(\mathbf{Y}^{\text{mis}}, \theta|\mathbf{Y}^{\text{obs}}, \mathbf{W}) = f(\mathbf{Y}^{\text{mis}}|\mathbf{Y}^{\text{obs}}, \mathbf{W}, \theta) \cdot p(\theta|\mathbf{Y}^{\text{obs}}, \mathbf{W}).$$

Then we integrate over θ to derive the conditional distribution of $\mathbf{Y}^{\text{mis}}$ given $(\mathbf{Y}^{\text{obs}}, \mathbf{W})$:

$$f(\mathbf{Y}^{\text{mis}}|\mathbf{Y}^{\text{obs}}, \mathbf{W}) = \int_{\theta} f(\mathbf{Y}^{\text{mis}}, \theta|\mathbf{Y}^{\text{obs}}, \mathbf{W}) \, d\theta,$$

which gives us the conditional distribution of the missing data given the observed data.

Step 4: Derivation of Posterior Distribution of Estimand $f(\tau|\mathbf{Y}^{\text{obs}}, \mathbf{W})$ Finally, we use the conditional distribution of the missing data given the observed data $f(\mathbf{Y}^{\text{mis}}|\mathbf{Y}^{\text{obs}}, \mathbf{W})$ and the observed data $(\mathbf{Y}^{\text{obs}}, \mathbf{W})$ to obtain the distribution of the estimand of interest given the observed data. This is the first, and only, time the procedure uses the specific choice of estimand.

The general form of the estimand is $\tau = \tau(\mathbf{Y}(0), \mathbf{Y}(1), \mathbf{W})$. We can rewrite τ in terms of observed and missing potential outcomes and the treatment assignment, using (8.6):

$$(\mathbf{Y}(0), \mathbf{Y}(1)) = h(\mathbf{Y}^{\text{mis}}, \mathbf{Y}^{\text{obs}}, \mathbf{W}).$$

Thus we can write $\tilde{\tau}(\mathbf{Y}^{\text{mis}}, \mathbf{Y}^{\text{obs}}, \mathbf{W})$. Combined with the conditional distribution of $\mathbf{Y}^{\text{mis}}$ given $(\mathbf{Y}^{\text{obs}}, \mathbf{W})$, we derive the conditional distribution of τ given the observed data $(\mathbf{Y}^{\text{obs}}, \mathbf{W})$, that is, the posterior distribution of τ:

$$f(\tau|\mathbf{Y}^{\text{obs}}, \mathbf{W}).$$

Once we have this distribution, we can derive the posterior mean, standard deviation, and any other feature of the posterior distribution of the causal estimand.

We conclude this section with a general comment concerning the key differences between the formal model-based approach and the simplistic examples that opened this chapter. First, the researcher must specify a complete model for the joint distribution of the potential outcomes $\mathbf{Y}(0)$ and $\mathbf{Y}(1)$ by specifying a unit-level joint distribution, $f(Y_i(0), Y_i(1)|\theta)$, given a generally unknown parameter θ. Although this model depends on an unknown parameter, θ, and thus need not be very restrictive, at first glance this approach may seem more restrictive than the initial examples where no such model was necessary. Yet this is not necessarily correct. The earlier, naive approaches assumed that the distribution of the missing data given the observed data was known with certainty, an assumption that is more restrictive than any parametric specification. The second difference is that the model-based approach requires the researcher to choose a prior distribution for the unknown parameter θ in order to derive its posterior distribution. In practice, given a completely randomized experiment, this choice is often not critical. At least in this setting, as long as the model is reasonably flexible, the prior distribution is not too dogmatic, and the data are sufficiently informative, the substantive conclusions are typically robust. In observational studies, however, the sensitivity of conclusions to the model choice and the choice of prior distribution are typically more severe, as we see in later chapters.

8.4.3 An Analytic Example with Six Units

To illustrate the four different steps in the model-based approach, consider again the first six observations of the National Supported Work Experiment. In Appendix B we provide a more detailed derivation of the distribution of the average treatment effect in a slightly more general setting where we assume Gaussianity for both the joint distribution of the potential outcomes and a conjugate prior distribution for θ, allowing for unknown covariance matrices with non-zero correlations.

The two inputs are a model for the joint distribution of the potential outcomes, and a prior distribution for the unknown parameters of this distribution. Here, for illustrative purposes, we specify a simple normal distribution for the pair of potential outcomes with unknown means but known covariance matrix:

$$\begin{pmatrix} Y_i(0) \\ Y_i(1) \end{pmatrix} \bigg| \theta \sim \mathcal{N} \left(\begin{pmatrix} \mu_c \\ \mu_t \end{pmatrix}, \begin{pmatrix} 100 & 0 \\ 0 & 64 \end{pmatrix} \right), \tag{8.17}$$

where the parameter vector θ consists of two elements, $\theta = (\mu_c, \mu_t)$, implying

$$f(Y_i(0), Y_i(1)|\theta) = \frac{1}{2\pi \cdot \sqrt{64 \cdot 100}}$$

$$\cdot \exp \left(-\frac{1}{2 \cdot 100} (Y_i(0) - \mu_c)^2 - \frac{1}{2 \cdot 64} (Y_i(1) - \mu_t)^2 \right).$$

More generally, we may wish to relax the assumption that the covariance matrix is known; for instance, see the examples in Section 8.6 and Appendix B. We may also want to consider more flexible distributions, such as mixtures of normal distributions.

The second input is the prior distribution for the vector parameter $\theta = (\mu_c, \mu_t)$. We use here the following prior distribution:

$$\begin{pmatrix} \mu_c \\ \mu_t \end{pmatrix} \sim \mathcal{N} \left(\begin{pmatrix} 0 \\ 0 \end{pmatrix}, \begin{pmatrix} 10{,}000 & 0 \\ 0 & 10{,}000 \end{pmatrix} \right). \tag{8.18}$$

This prior distribution is relatively agnostic about the values of μ_c and μ_t over a wide range of values, relative to the data values, displayed in Table 8.2. In Appendix B we provide some calculations for a more general specification of the prior distribution, allowing for non-zero means, and a non-diagonal covariance matrix. In practice, with a reasonably sized data set and a completely randomized experiment, we would expect the results to be fairly insensitive to the choice of prior distribution.

In an observational study we would also have to specify the assignment mechanism, but here this is known to be

$$\Pr(\mathbf{W} = \mathbf{w}|\mathbf{Y}(0), \mathbf{Y}(1), \mu_c, \mu_t) = \begin{pmatrix} N \\ N_t \end{pmatrix}^{-1},$$

for all $\mathbf{w}$ with $w_i \in \{0, 1\}$ for all $i = 1, \ldots, N$, and $\sum_{i=1}^{N} w_i = N_t$, and zero elsewhere.

Step 1: Derivation of $f(\mathbf{Y}^{\mathrm{mis}}|\mathbf{Y}^{\mathrm{obs}}, \mathbf{W}, \mu_{\mathrm{c}}, \mu_{\mathrm{t}})$ Because the potential outcomes are independent across units conditional on $(\mu_{\mathrm{c}}, \mu_{\mathrm{t}})$, the specification of the joint distribution of the pair $(Y_i(0), Y_i(1))$ given θ allows us to derive the joint distribution of $\mathbf{Y}(0)$ and $\mathbf{Y}(1)$ given $\theta = (\mu_{\mathrm{c}}, \mu_{\mathrm{t}})$.

$$f(\mathbf{Y}(0), \mathbf{Y}(1)|\mu_{\mathrm{c}}, \mu_{\mathrm{t}}) = \prod_{i-1}^{N} f(Y_i(0), Y_i(1)|\mu_{\mathrm{c}}, \mu_{\mathrm{t}}).$$

Let ι_N denote the N-dimensional vector with all elements equal to one, and let I_N denote the $N \times N$ dimensional identity matrix. Then the $2N$-component vector constructed by stacking $\mathbf{Y}(0)$ and $\mathbf{Y}(1)$ is distributed, given θ, as

$$\begin{pmatrix} \mathbf{Y}(0) \\ \mathbf{Y}(1) \end{pmatrix} \bigg| \mu_{\mathrm{c}}, \mu_{\mathrm{t}} \sim \mathcal{N} \left(\begin{pmatrix} \mu_{\mathrm{c}} \cdot \iota_N \\ \mu_{\mathrm{t}} \cdot \iota_N \end{pmatrix}, \begin{pmatrix} 100 \cdot I_N & 0 \cdot I_N \\ 0 \cdot I_N & 64 \cdot I_N \end{pmatrix} \right). \tag{8.19}$$

Next we exploit the assumption that the data come from a completely randomized experiment. Therefore the distribution of $\mathbf{W}$ conditional on the potential outcomes and θ is

$$\Pr(\mathbf{W} = \mathbf{w}|\mathbf{Y}(0), \mathbf{Y}(1), \mu_{\mathrm{c}}, \mu_{\mathrm{t}}) = \binom{N}{N_{\mathrm{t}}}^{-1},$$

for all $\mathbf{w}$ such that $\sum_i w_i = N_{\mathrm{t}}$, and zero elsewhere. Deriving the conditional distribution of the potential outcomes given the assignment vector is straightforward because of the independence of $\mathbf{W}$ and $(\mathbf{Y}(0), \mathbf{Y}(1))$ given θ, so that the conditional distribution is the same as the marginal distribution given in (8.19):

$$\begin{pmatrix} \mathbf{Y}(0) \\ \mathbf{Y}(1) \end{pmatrix} \bigg| \mathbf{W}, \mu_{\mathrm{c}}, \mu_{\mathrm{t}} \sim \mathcal{N} \left(\begin{pmatrix} \mu_{\mathrm{c}} \cdot \iota_N \\ \mu_{\mathrm{t}} \cdot \iota_N \end{pmatrix}, \begin{pmatrix} 100 \cdot I_N & 0 \cdot I_N \\ 0 \cdot I_N & 64 \cdot I_N \end{pmatrix} \right). \tag{8.20}$$

Now we transform this conditional distribution to the conditional distribution of $(\mathbf{Y}^{\mathrm{mis}}, \mathbf{Y}^{\mathrm{obs}})$ given $(\mathbf{W}, \mu_{\mathrm{c}}, \mu_{\mathrm{t}})$, using the representations of Y_i^{mis} and Y_i^{obs} in terms of $Y_i(0)$, $Y_i(1)$, and W_i given in Equations (8.13). Because conditional on $(\mathbf{W}, \mu_{\mathrm{c}}, \mu_{\mathrm{t}})$ the pairs $(Y_i(0), Y_i(1))$ and $(Y_{i'}(0), Y_{i'}(1))$ are independent if $i \neq i'$, it follows that the pairs $(Y_i^{\mathrm{mis}}, Y_i^{\mathrm{obs}})$ and $(Y_{i'}^{\mathrm{mis}}, Y_{i'}^{\mathrm{obs}})$ are also independent given $(\mathbf{W}, \mu_{\mathrm{c}}, \mu_{\mathrm{t}})$ if $i \neq i'$. Hence

$$f(\mathbf{Y}^{\mathrm{mis}}, \mathbf{Y}^{\mathrm{obs}})|\mathbf{W}, \mu_{\mathrm{c}}, \mu_{\mathrm{t}}) = \prod_{i=1}^{N} f(Y_i^{\mathrm{mis}}, Y_i^{\mathrm{obs}}|\mathbf{W}, \mu_{\mathrm{c}}, \mu_{\mathrm{t}}),$$

where the joint distribution of $(Y_i^{\mathrm{mis}}, Y_i^{\mathrm{obs}})$ given $(\mathbf{W}, \mu_{\mathrm{c}}, \mu_{\mathrm{t}})$ is

$$\begin{pmatrix} Y_i^{\mathrm{mis}} \\ Y_i^{\mathrm{obs}} \end{pmatrix} \bigg| \mu_{\mathrm{c}}, \mu_{\mathrm{t}}, \mathbf{W} \sim \mathcal{N} \left(\begin{pmatrix} W_i \cdot \mu_{\mathrm{c}} + (1 - W_i) \cdot \mu_{\mathrm{t}} \\ (1 - W_i) \cdot \mu_{\mathrm{c}} + W_i \cdot \mu_{\mathrm{t}} \end{pmatrix}, \right.$$
$$\left. \begin{pmatrix} W_i \cdot 100 + (1 - W_i) \cdot 64 & 0 \\ 0 & (1 - W_i) \cdot 100 + W_i \cdot 64 \end{pmatrix} \right). \tag{8.21}$$

Because in this example Y_i^{mis} and Y_i^{obs} are uncorrelated given $(\mu_{\mathrm{c}}, \mu_{\mathrm{t}})$ – the off-diagonal elements of the covariance matrix in (8.21) are equal to zero – the conditional distribution of Y_i^{mis} given $(Y_i^{\text{obs}}, \mu_c, \mu_t)$ is simply equal to the marginal distribution of Y_i^{mis} given (μ_c, μ_t):

$$Y_i^{\text{mis}}|\mathbf{Y}^{\text{obs}}, \mathbf{W}, \mu_c, \mu_t \sim \mathcal{N}\left(W_i \cdot \mu_c + (1 - W_i) \cdot \mu_t, W_i \cdot 100 + (1 - W_i) \cdot 64\right).$$
(8.22)

Thus the joint distribution of the full N-vector $\mathbf{Y}^{\text{mis}}$ given $(\mathbf{Y}^{\text{obs}}, \mathbf{W}, \mu_c, \mu_t)$, is

$$\mathbf{Y}^{\text{mis}}|\mathbf{Y}^{\text{obs}}, \mathbf{W}, \mu_c, \mu_t \sim \mathcal{N}\left(\begin{pmatrix} W_1 \cdot \mu_c + (1 - W_1) \cdot \mu_t \\ W_2 \cdot \mu_c + (1 - W_2) \cdot \mu_t \\ \vdots \\ W_N \cdot \mu_c + (1 - W_N) \cdot \mu_t \end{pmatrix}, \right.$$

$$\left. \begin{pmatrix} W_1 \cdot 100 + (1 - W_1) \cdot 64 & 0 & \cdots & 0 \\ 0 & W_2 \cdot 100 + (1 - W_2) \cdot 64 & \cdots & 0 \\ \vdots & \vdots & \ddots & \vdots \\ 0 & 0 & \cdots & W_N \cdot 100 + (1 - W_N) \cdot 64 \end{pmatrix}\right).$$
(8.23)

For the six units in our illustrative data set, this leads to

$$\begin{pmatrix} Y_1^{\text{mis}} \\ Y_2^{\text{mis}} \\ Y_3^{\text{mis}} \\ Y_4^{\text{mis}} \\ Y_5^{\text{mis}} \\ Y_6^{\text{mis}} \end{pmatrix} \Bigg| \mathbf{Y}^{\text{obs}}, \mathbf{W}, \mu_c, \mu_t \sim \mathcal{N}\left(\begin{pmatrix} \mu_t \\ \mu_c \\ \mu_t \\ \mu_c \\ \mu_t \\ \mu_c \end{pmatrix}, \begin{pmatrix} 64 & 0 & 0 & 0 & 0 & 0 \\ 0 & 100 & 0 & 0 & 0 & 0 \\ 0 & 0 & 64 & 0 & 0 & 0 \\ 0 & 0 & 0 & 100 & 0 & 0 \\ 0 & 0 & 0 & 0 & 64 & 0 \\ 0 & 0 & 0 & 0 & 0 & 100 \end{pmatrix}\right).$$
(8.24)

Step 2: Derivation of the Posterior Distribution of the Parameter $p(\mu_c, \mu_t|\mathbf{Y}^{\text{obs}}, \mathbf{W})$
The second step consists of deriving the posterior distribution of the parameter given the observed data. The posterior distribution is proportional to the product of the prior distribution and the likelihood function:

$$p(\mu_c, \mu_t|\mathbf{Y}^{\text{obs}}, \mathbf{W}) \propto p(\mu_c, \mu_t) \cdot \mathcal{L}(\mu_c, \mu_t|\mathbf{Y}^{\text{obs}}, \mathbf{W}).$$

The prior distribution is given in (8.18), so all we need to do is derive the likelihood function. Conditional on $(\mathbf{W}, \mu_c, \mu_t)$, the distribution of the observed outcome Y_i^{obs} is

$$Y_i^{\text{obs}}|\mathbf{W}, \mu_c, \mu_t \sim \mathcal{N}\left((1 - W_i) \cdot \mu_c + W_i \cdot \mu_t, (1 - W_i) \cdot 100 + W_i \cdot 64\right). \quad (8.25)$$

Because Y_i^{obs} is independent of $Y_{i'}^{\text{obs}}$ conditional on $(\mathbf{W}, \mu_c, \mu_t)$ if $i \neq i'$, the contribution of unit i to the likelihood function is proportional to ("$\propto$")

$$\mathcal{L}_i \propto \frac{1}{\sqrt{2\pi \cdot ((1 - W_i) \cdot 100 + W_i \cdot 64)}}$$

$$\times \exp\left[-\frac{1}{2}\left(\frac{1}{(1 - W_i) \cdot 100 + W_i \cdot 64}\left(Y_i^{\text{obs}} - (1 - W_i) \cdot \mu_{\text{c}} - W_i \cdot \mu_{\text{t}}\right)^2\right)\right],$$

and the likelihood function is proportional to the product of these N factors and the probability of the assignment vector. Because the latter is a known constant, it can be ignored, and the likelihood function is proportional to

$$\mathcal{L}(\mu_{\text{c}}, \mu_{\text{t}} | \mathbf{Y}^{\text{obs}}, \mathbf{W})$$

$$\propto \prod_{i=1}^{6}\left\{\frac{1}{\sqrt{2\pi \cdot ((1 - W_i) \cdot 100 + W_i \cdot 64)}}\right.$$

$$\left. \times \exp\left[-\frac{1}{2}\left(\frac{1}{(1 - W_i) \cdot 100 + W_i \cdot 64}\left(Y_i^{\text{obs}} - (1 - W_i) \cdot \mu_{\text{c}} - W_i \cdot \mu_{\text{t}}\right)^2\right)\right]\right\}$$

$$\propto \prod_{i:W_i=0} \frac{1}{\sqrt{2\pi \cdot 100}} \exp\left[-\frac{1}{2}\left(\frac{1}{100}\left(Y_i^{\text{obs}} - \mu_{\text{c}}\right)^2\right)\right]$$

$$\times \prod_{i:W_i=1} \frac{1}{\sqrt{2\pi \cdot 64}} \exp\left[-\frac{1}{2}\left(\frac{1}{64}\left(Y_i^{\text{obs}} - \mu_{\text{t}}\right)^2\right)\right].$$

To derive the posterior distribution, we exploit the fact that both the prior distribution of μ_{c} and μ_{t}, and the likelihood function factor into a function of μ_{c} and a function of μ_{t}. This factorization leads to the following posterior distribution of $(\mu_{\text{c}}, \mu_{\text{t}})$ given the observed data:

$$p(\mu_{\text{c}}, \mu_{\text{t}} | \mathbf{Y}^{\text{obs}}, \mathbf{W}) \propto$$

$$\exp\left[-\frac{1}{2}\left(\frac{\mu_{\text{c}}^2}{10{,}000}\right)\right] \cdot \prod_{i:W_i=0} \frac{1}{\sqrt{2\pi \cdot 100}} \exp\left[-\frac{1}{2}\left(\frac{(Y_i^{\text{obs}} - \mu_{\text{c}})^2}{100}\right)\right]$$

$$\times \exp\left[-\frac{1}{2}\left(\frac{\mu_{\text{t}}^2}{10{,}000}\right)\right] \cdot \prod_{i:W_i=1} \frac{1}{\sqrt{2\pi \cdot 64}} \exp\left[-\frac{1}{2}\left(\frac{(Y_i^{\text{obs}} - \mu_{\text{t}})^2}{64}\right)\right].$$

This expression implies that

$$\begin{pmatrix} \mu_{\text{c}} \\ \mu_{\text{t}} \end{pmatrix}\Bigg| \mathbf{Y}^{\text{obs}}, \mathbf{W}$$

$$\sim \mathcal{N}\left(\begin{pmatrix} \overline{Y}_{\text{c}}^{\text{obs}} \cdot \dfrac{N_{\text{c}} \cdot 10{,}000}{N_{\text{c}} \cdot 10{,}000 + 100} \\ \overline{Y}_{\text{t}}^{\text{obs}} \cdot \dfrac{N_{\text{t}} \cdot 10{,}000}{N_{\text{t}} \cdot 10{,}000 + 64} \end{pmatrix}, \begin{pmatrix} \dfrac{1}{N_{\text{c}}/100 + 1/10{,}000} & \dfrac{1}{N_{\text{t}}/64 + 1/10{,}000} \\ 0 & \end{pmatrix}\right).$$

$$(8.26)$$

Substituting the appropriate values from the six-unit data set in Table 8.2, with $\overline{Y}_{\text{c}}^{\text{obs}} = 4.1$ and $N_{\text{c}} = 3$, we find that μ_{c} has a Gaussian posterior distribution with mean equal to 4.1 and variance equal to $33.2 = 5.8^2$. Following the same argument for μ_t, with $\overline{Y}_t^{\text{obs}} = 12.8$ and $N_t = 3$, we find that μ_t has a Gaussian posterior distribution with mean 12.8 and variance $21.3 = 4.6^2$, so that:

$$\begin{pmatrix} \mu_c \\ \mu_t \end{pmatrix} \middle| \mathbf{Y}^{\text{obs}}, \mathbf{W} \sim \mathcal{N}\left(\begin{pmatrix} 4.1 \\ 12.8 \end{pmatrix}, \begin{pmatrix} 5.8^2 & 0 \\ 0 & 4.6^2 \end{pmatrix} \right). \tag{8.27}$$

Recall our previous comment that, given a completely randomized experiment, the resulting posterior distribution is fairly insensitive to the choice of the prior distribution for μ_c, μ_t. We can see this here, where the choice of prior distribution has had little effect on any of the moments of the posterior distribution of (μ_c, μ_t). In particular, notice in (8.27) that the mean values for μ_c and μ_t are equal, up to the first significant digit, to the observed average values, $\overline{Y}_c^{\text{obs}}$ and $\overline{Y}_t^{\text{obs}}$. The posterior distribution, proportional to the product of the prior distribution for (μ_c, μ_t) and the marginal distribution of $\mathbf{Y}^{\text{obs}}$ given (μ_c, μ_t), regarded as a function of (μ_c, μ_t), puts weight on each factor proportional to their precisions, that is, the inverse of their variances. Our choice of prior distribution – with such large posited variances – implies giving almost all of the weight to the observed data, $\overline{Y}_c^{\text{obs}}$ and $\overline{Y}_t^{\text{obs}}$. This choice was made specifically to impose little structure through our assumptions, instead allowing the observed data to be the primary voice for the ultimate posterior distribution of τ.

Step 3: Derivation of Posterior Distribution of Missing Potential Outcomes $f(\mathbf{Y}^{\text{mis}}|\mathbf{Y}^{\text{obs}}, \mathbf{W})$ Now we combine the conditional distribution of $\mathbf{Y}^{\text{mis}}$ given $(\mathbf{Y}^{\text{obs}}, \mathbf{W}, \mu_c, \mu_t)$, given in (8.23), and the posterior distribution of (μ_c, μ_t) given $(\mathbf{Y}^{\text{obs}}, \mathbf{W})$, given in (8.26), to obtain the conditional distribution of $\mathbf{Y}^{\text{mis}}$ given $(\mathbf{Y}^{\text{obs}}, \mathbf{W})$. Because the distribution of $\mathbf{Y}^{\text{mis}}$ given $(\mathbf{Y}^{\text{obs}}, \mathbf{W}, \mu_c, \mu_t)$, and the distribution of (μ_c, μ_t) given $(\mathbf{Y}^{\text{obs}}, \mathbf{W})$ are Gaussian, it follows that the joint distribution of $(\mathbf{Y}^{\text{mis}}, \mu_c, \mu_t)$ given $(\mathbf{Y}^{\text{obs}}, \mathbf{W})$ is Gaussian, and thus the marginal distribution of $\mathbf{Y}^{\text{mis}}$ given $(\mathbf{Y}^{\text{obs}}, \mathbf{W})$ is Gaussian. Hence, all we need to do is derive the first two moments of this distribution in order to characterize it fully.

First consider the mean of Y_i^{mis} given $(\mathbf{Y}^{\text{obs}}, \mathbf{W})$. Conditional on $(\mathbf{Y}^{\text{obs}}, \mathbf{W}, \mu_c, \mu_t)$, we have, using (8.24):

$$\mathbb{E}\left[Y_i^{\text{mis}} \middle| \mathbf{Y}^{\text{obs}}, \mathbf{W}, \mu_c, \mu_t \right] = W_i \cdot \mu_c + (1 - W_i) \cdot \mu_t.$$

In addition, from (8.26), we have

$$\mathbb{E}\left[\begin{pmatrix} \mu_c \\ \mu_t \end{pmatrix} \middle| \mathbf{Y}^{\text{obs}}, \mathbf{W} \right] = \begin{pmatrix} \overline{Y}_c^{\text{obs}} \cdot \dfrac{N_c \cdot 10{,}000}{N_c \cdot 10{,}000 + 100} \\ \overline{Y}_t^{\text{obs}} \cdot \dfrac{N_t \cdot 10{,}000}{N_t \cdot 10{,}000 + 64} \end{pmatrix}.$$

Hence

$$\mathbb{E}\left[Y_i^{\text{mis}} | \mathbf{Y}^{\text{obs}}, \mathbf{W} \right] = W_i \cdot \left(\overline{Y}_c^{\text{obs}} \cdot \frac{N_c \cdot 10{,}000}{N_c \cdot 10{,}000 + 100} \right)$$
$$+ (1 - W_i) \cdot \left(\overline{Y}_t^{\text{obs}} \cdot \frac{N_t \cdot 10{,}000}{N_t \cdot 10{,}000 + 64} \right). \tag{8.28}$$

Next, consider the variance. By the law of iterated expectations,

$$\mathbb{V}\left(Y_i^{\mathrm{mis}}\middle|\mathbf{Y}^{\mathrm{obs}},\mathbf{W}\right) = \mathbb{E}\left[\mathbb{V}\left(Y_i^{\mathrm{mis}}\middle|\mathbf{Y}^{\mathrm{obs}},\mathbf{W},\mu_{\mathrm{c}},\mu_{\mathrm{t}}\right)\middle|\mathbf{Y}^{\mathrm{obs}},\mathbf{W}\right]$$

$$+ \mathbb{V}\left(\mathbb{E}\left[Y_i^{\mathrm{mis}}\middle|\mathbf{Y}^{\mathrm{obs}},\mathbf{W},\mu_{\mathrm{c}},\mu_{\mathrm{t}}\right]\middle|\mathbf{Y}^{\mathrm{obs}},\mathbf{W}\right)$$

$$= \mathbb{E}\left[W_i\cdot100+(1-W_i)\cdot64\middle|\mathbf{Y}^{\mathrm{obs}},\mathbf{W}\right] + \mathbb{V}\left(W_i\cdot\mu_{\mathrm{c}}+(1-W_i)\cdot\mu_{\mathrm{t}}\middle|\mathbf{Y}^{\mathrm{obs}},\mathbf{W}\right)$$

$$= W_i\cdot100+(1-W_i)\cdot64+W_i\cdot\frac{1}{N_{\mathrm{c}}/100+1/10{,}000}+(1-W_i)\cdot\frac{1}{N_{\mathrm{t}}/64+1/10{,}000}$$

$$= W_i\cdot\left(100+\frac{1}{N_{\mathrm{c}}/100+1/10{,}000}\right)+(1-W_i)\cdot\left(64+\frac{1}{N_{\mathrm{t}}/64+1/10{,}000}\right).$$

$$(8.29)$$

We also need to consider the covariance between Y_i^{mis} and $Y_{i'}^{\mathrm{mis}}$, for $i\neq i'$:

$$\mathbb{C}\left(Y_i^{\mathrm{mis}},Y_{i'}^{\mathrm{mis}}\middle|\mathbf{Y}^{\mathrm{obs}},\mathbf{W}\right) = \mathbb{E}\left[\mathbb{C}\left(Y_i^{\mathrm{mis}},Y_{i'}^{\mathrm{mis}}\middle|\mathbf{Y}^{\mathrm{obs}},\mathbf{W},\mu_{\mathrm{c}},\mu_{\mathrm{t}}\right)\middle|\mathbf{Y}^{\mathrm{obs}},\mathbf{W}\right]$$

$$+ \mathbb{C}\left(\mathbb{E}\left[Y_i^{\mathrm{mis}}\middle|\mathbf{Y}^{\mathrm{obs}},\mathbf{W},\mu_{\mathrm{c}},\mu_{\mathrm{t}}\right],\mathbb{E}\left[Y_{i'}^{\mathrm{mis}}\middle|\mathbf{Y}^{\mathrm{obs}},\mathbf{W},\mu_{\mathrm{c}},\mu_{\mathrm{t}}\right]\middle|\mathbf{Y}^{\mathrm{obs}},\mathbf{W}\right)$$

$$= 0 + \mathbb{C}\left(W_i\cdot\mu_{\mathrm{c}}+(1-W_i)\cdot\mu_{\mathrm{t}},W_{i'}\cdot\mu_{\mathrm{c}}+(1-W_{i'})\cdot\mu_{\mathrm{t}}\middle|\mathbf{Y}^{\mathrm{obs}},\mathbf{W}\right)$$

$$= W_i\cdot W_j\cdot\frac{1}{N_{\mathrm{c}}/100+1/10{,}000}+(1-W_i)\cdot(1-W_j)\cdot\frac{1}{N_{\mathrm{t}}/64+1/10{,}000}. \quad (8.30)$$

Putting this all together for the six-unit data set, we find

$$\begin{pmatrix}Y_1^{\mathrm{mis}}\\Y_2^{\mathrm{mis}}\\Y_3^{\mathrm{mis}}\\Y_4^{\mathrm{mis}}\\Y_5^{\mathrm{mis}}\\Y_6^{\mathrm{mis}}\end{pmatrix}\middle|\mathbf{Y}^{\mathrm{obs}},\mathbf{W} \sim$$

$$\mathcal{N}\left(\begin{pmatrix}12.8\\4.1\\12.8\\4.1\\12.8\\4.1\end{pmatrix},\begin{pmatrix}85.3 & 0 & 21.3 & 0 & 21.3 & 0\\0 & 133.2 & 0 & 33.2 & 0 & 33.2\\21.3 & 0 & 85.3 & 0 & 21.3 & 0\\0 & 0 & 0 & 133.2 & 0 & 33.2\\21.3 & 0 & 21.3 & 0 & 85.3 & 0\\0 & 33.2 & 0 & 33.2 & 0 & 133.2\end{pmatrix}\right). \quad (8.31)$$

Note that the missing outcomes are no longer independent. Conditional on the parameters $(\mu_{\mathrm{c}},\mu_{\mathrm{t}})$ they were independent, but the fact that they depend on common parameters introduces some dependence.

Step 4: Derivation of Posterior Distribution of Estimand, $f(\tau | \mathbf{Y}^{\text{obs}}, \mathbf{W})$ In this example, we are interested in the sample average effect of the treatment:

$$\tau_{\text{fs}} = \tau(\mathbf{Y}(0), \mathbf{Y}(1)) = \frac{1}{N} \sum_{i=1}^{N} (Y_i(1) - Y_i(0)).$$

Using (8.6) we can write this in terms of the missing and observed outcomes as

$$\tau_{\text{fs}} = \tau(\mathbf{Y}^{\text{mis}}, \mathbf{Y}^{\text{obs}}, \mathbf{W}) = \frac{1}{N} \sum_{i=1}^{N} (1 - 2 \cdot W_i) \cdot Y_i^{\text{mis}} + \frac{1}{N} \sum_{i=1}^{N} (2 \cdot W_i - 1) \cdot Y_i^{\text{obs}}.$$

Conditional on $(\mathbf{Y}^{\text{obs}}, \mathbf{W})$ the only stochastic components of this expression are the Y_i^{mis}. Because τ_{fs} is a linear function of $Y_1^{\text{mis}}, \ldots, Y_6^{\text{mis}}$, the fact that the Y_i^{mis} are jointly normally distributed implies that τ_{fs} has a normal distribution. We use the results from Step 3 to derive the first two moments of τ_{fs} given $(\mathbf{Y}^{\text{obs}}, \mathbf{W})$. The conditional mean is

$$\mathbb{E}\left[\tau_{\text{fs}} \middle| \mathbf{Y}^{\text{obs}}, \mathbf{W} \right] = \frac{1}{N} \sum_{i=1}^{N} (2 \cdot W_i - 1) \cdot Y_i^{\text{obs}} + \frac{1}{N} \sum_{i=1}^{N} (1 - 2 \cdot W_i) \cdot \mathbb{E}\left[Y_i^{\text{mis}} \middle| \mathbf{Y}^{\text{obs}}, \mathbf{W} \right]$$

$$= \frac{N_{\text{t}}}{N} \cdot \overline{Y}_{\text{t}}^{\text{obs}} - \frac{N_{\text{c}}}{N} \cdot \overline{Y}_{\text{c}}^{\text{obs}}$$

$$+ \frac{1}{N} \sum_{i=1}^{N} (1 - 2 \cdot W_i) \cdot \left(W_i \cdot \left(\overline{Y}_{\text{c}}^{\text{obs}} \cdot \frac{N_{\text{c}} \cdot 10{,}000}{N_{\text{c}} \cdot 10{,}000 + 100} \right) \right.$$

$$\left. + (1 - W_i) \cdot \left(\overline{Y}_{\text{t}}^{\text{obs}} \cdot \frac{N_{\text{t}} \cdot 10{,}000}{N_{\text{t}} \cdot 10{,}000 + 64} \right) \right)$$

$$= \overline{Y}_{\text{t}}^{\text{obs}} \cdot \frac{N_{\text{t}} \cdot 10{,}000 + 64 \cdot N_{\text{t}}/N}{N_{\text{t}} \cdot 10{,}000 + 64} - \overline{Y}_{\text{c}}^{\text{obs}} \cdot \frac{N_{\text{c}} \cdot 10{,}000 + 100 \cdot N_{\text{c}}/N}{N_{\text{c}} \cdot 10{,}000 + 100}.$$

Next, consider the conditional variance of τ_{fs}. Because τ_{fs} is a linear function of the Y_i^{mis}, the variance is a linear combination of the variances and covariances:

$$\mathbb{V}\left(\tau_{\text{fs}} \middle| \mathbf{Y}^{\text{obs}}, \mathbf{W} \right) = \frac{1}{N^2} \sum_{i=1}^{N} \mathbb{V}\left(\left(1 - 2 \cdot W_i\right) \cdot Y_i^{\text{mis}} \middle| \mathbf{Y}^{\text{obs}}, \mathbf{W} \right)$$

$$+ \frac{1}{N^2} \sum_{i=1}^{N} \sum_{i' \neq i} \mathbb{C}\left(\left(1 - 2 \cdot W_i\right) \cdot Y_i^{\text{mis}}, \left(1 - 2 \cdot W_{i'}\right) \cdot Y_{i'}^{\text{mis}} \middle| \mathbf{Y}^{\text{obs}}, \mathbf{W} \right)$$

$$= \frac{1}{N^2} \left(N_{\text{t}} \cdot \left(100 + \frac{1}{N_{\text{c}}/100 + 1/10{,}000} \right) + N_{\text{c}} \cdot \left(64 + \frac{1}{N_{\text{t}}/64 + 1/10{,}000} \right) \right)$$

$$+ \frac{1}{N^2} \left(N_{\text{t}} \cdot (N_{\text{t}} - 1) \cdot \frac{1}{N_{\text{c}}/100 + 1/10{,}000} + N_{\text{c}} \cdot (N_{\text{c}} - 1) \cdot \frac{1}{N_{\text{t}}/64 + 1/10{,}000} \right).$$

Substituting in the values for the six-unit data set ($N = 6$, $N_{\text{c}} = N_{\text{t}} = 3$), we find

$$\tau_{\text{fs}} | \mathbf{Y}^{\text{obs}}, \mathbf{W} \sim \mathcal{N}\left(8.7, 5.2^2 \right). \tag{8.32}$$

Thus, combining our assumptions on the joint distribution of $(\mathbf{Y}(0), \mathbf{Y}(1))$ given (μ_c, μ_t) and on the prior distribution of (μ_c, μ_t) with the observed data, we find that the posterior distribution of τ_{fs} given $(\mathbf{Y}^{obs}, \mathbf{W})$ is normal, with the posterior mean of the average treatment effect equal to 8.7, and the posterior standard deviation equal to 5.2. Note that our point estimate of τ_{fs} is very similar to the value we found previously in the two imputation methods in Section 8.3, namely 8.7. In contrast, the standard error estimated under the second method (the first method essentially gave a standard error of zero for the estimate) was only 2.8, much smaller than what we find using the fully model-based approach. This difference is driven by the fact that with the second method we still assumed we knew the model of $\mathbf{Y}^{mis}$ given $\mathbf{Y}^{obs}$ with certainty, whereas here we allow uncertainty via the estimation of the parameter $\theta = (\mu_c, \mu_t)$.

8.5 SIMULATION METHODS IN THE MODEL-BASED APPROACH

So far in this chapter, our calculations have all been analytical; we have derived the exact distribution of the average treatment effect, given the observed data, and given our choice of prior distribution. Unfortunately, in many settings this approach is infeasible, or at least impractical. Depending on the model for the joint distribution of the potential outcomes, the calculations required to derive the conditional distribution of the estimand τ given the observed data – in particular, the integration across the parameter space – can be quite complicated. We therefore generally rely on simulation methods for evaluating the distribution of the estimand of interest. These simulation methods intuitively link the full model-based approach back to the starting point of the chapter: the explicit imputation of the missing components of the causal estimand, that is, the missing potential outcomes.

To use simulation methods, the two key elements are the conditional distribution of the missing data given the observed data and parameters, $f(\mathbf{Y}^{mis}|\mathbf{Y}^{obs}, \mathbf{W}, \mu_c, \mu_t)$, derived in Step 1, and the posterior distribution of the parameters given the observed data, $p(\mu_c, \mu_t|\mathbf{Y}^{obs}, \mathbf{W})$, derived in Step 2. Using these distributions, we can distributionally impute the missing data – that is, we repeatedly (or multiply) impute the missing potential outcomes. In this section, we continue with the example with six individuals to illustrate these ideas. See Appendix B for a description of the simulation method with a more general example.

First, recall the posterior distribution of the parameters given data for the six units in our illustrative sample, derived in Step 2:

$$\begin{pmatrix} \mu_c \\ \mu_t \end{pmatrix} \Bigg| \mathbf{Y}^{obs}, \mathbf{W} \sim \mathcal{N}\left(\begin{pmatrix} 4.1 \\ 12.8 \end{pmatrix}, \begin{pmatrix} 5.8^2 & 0 \\ 0 & 4.6^2 \end{pmatrix} \right).$$

We draw a pair of random values (μ_c, μ_t) from this distribution. Suppose the first pair of draws is $(\mu_c^{(1)}, \mu_t^{(1)}) = (1.63, 5.09)$. Given this draw for the parameters (μ_c, μ_t), we can substitute these values into the conditional distribution of $\mathbf{Y}^{mis}$, that is, $f(\mathbf{Y}^{mis}|\mathbf{Y}^{obs}, \mathbf{W}, \mu_c, \mu_t)$ to impute, independently, all of the missing potential outcomes.

Table 8.5. *The Average Treatment Effect Using Full Model-Based Imputations for the NSW Program Data*

Unit	Potential Outcomes		Treatment	Observed Outcome
	$Y_i(0)$	$Y_i(1)$	W_i	Y_i^{obs}
Panel A: First Parameter Draw $(\mu_c^{(1)}, \mu_t^{(1)}) = (1.63, 5.09)$				
1	0	(6.1)	0	0
2	(13.5)	9.9	1	9.9
3	12.4	(7.4)	0	12.4
4	(13.5)	3.6	1	3.6
5	0	(−4.1)	0	0
6	(1.3)	24.9	1	24.9
Average	6.8	8.0		
$\tau_{\text{fs}}^{(1)}$		1.2		
Panel B: Second Parameter Draw $(\mu_c^{(2)}, \mu_t^{(2)}) = (6.01, 13.58)$				
1	0	(12.1)	0	0
2	(27.8)	9.9	1	9.9
3	12.4	(19.4)	0	12.4
4	(4.6)	3.6	1	3.6
5	0	(8.9)	0	0
6	(7.1)	24.9	1	24.9
Average	8.7	13.1		
$\tau_{\text{fs}}^{(2)}$		4.5		

Specifically, we draw $\mathbf{Y}^{\text{mis}}$ from the normal distribution

$$
\begin{pmatrix} Y_1^{\text{mis}} \\ Y_2^{\text{mis}} \\ Y_3^{\text{mis}} \\ Y_4^{\text{mis}} \\ Y_5^{\text{mis}} \\ Y_6^{\text{mis}} \end{pmatrix} \Bigg| \mathbf{Y}^{\text{obs}}, \mathbf{W}, \theta \sim \mathcal{N} \left(\begin{pmatrix} 5.09 \\ 1.63 \\ 5.09 \\ 1.63 \\ 5.09 \\ 1.63 \end{pmatrix}, \begin{pmatrix} 64 & 0 & 0 & 0 & 0 & 0 \\ 0 & 100 & 0 & 0 & 0 & 0 \\ 0 & 0 & 64 & 0 & 0 & 0 \\ 0 & 0 & 0 & 100 & 0 & 0 \\ 0 & 0 & 0 & 0 & 64 & 0 \\ 0 & 0 & 0 & 0 & 0 & 100 \end{pmatrix} \right),
$$

obtained by substituting 1.63 for μ_c and 5.09 for μ_t in Equation (8.24). Thus, the missing $Y_i(0)$ values for units 2, 4, and 6 will be drawn independently from a $\mathcal{N}(1.63, 10^2)$ distribution, and the missing $Y_i(1)$ values for units 1, 3, and 5 independently from a $\mathcal{N}(5.09, 8^2)$ distribution. Panel A of Table 8.5 shows the data with the missing potential outcomes drawn from this posterior predictive distribution. Substituting the observed and imputed missing potential outcomes into Equation (8.8) leads to an estimate for the average treatment effect of $\hat{\tau}^{(1)} = 1.2$. Notice that in this step, we impute a complete set of missing data without redrawing the unknown parameters. This is important. The alternative, drawing say Y_1^{mis} given one draw from the parameter vector and drawing Y_2^{mis} from a second draw from the parameter vector, would, in general, be incorrect.

Next we draw a new pair of parameter values. Suppose this time we draw $(\mu_c^{(2)}, \mu_t^{(2)}) = (6.01, 13.58)$. Given this draw, we again impute the full vector of missing outcomes, $\mathbf{Y}^{\text{mis}}$. The missing $Y_i(0)$ values are now drawn independently from a $\mathcal{N}(6.01, 100)$ distribution, and the missing $Y_i(1)$ values independently from a $\mathcal{N}(13.58, 64)$ distribution. Panel B of Table 8.5 shows the data with the missing outcomes drawn from these distributions, leading to a second estimate for the average treatment effect of $\hat{\tau}^{(2)} = 4.5$. To derive the full distribution for our estimate of the average treatment effect, we repeat this a number of times and calculate the average and standard deviation of the imputed estimators $\hat{\tau}^{(1)}, \hat{\tau}^{(2)}, \ldots$. Our result, based on $N_R = 10{,}000$ draws of the pair $\theta = (\mu_c, \mu_t)'$, is an average, over these 10,000 draws for $\hat{\tau}_{\text{fs}}^{(r)}$, for $r = 1, \ldots, N_R$, of 8.6 and a standard deviation of 5.3:

$$\frac{1}{N_R} \sum_{r=1}^{N_R} \tau_{\text{fs}}^{(r)} = \bar{\tau} = 8.6, \qquad \frac{1}{N_R - 1} \sum_{r=1}^{N_R} \left(\tau_{\text{fs}}^{(r)} - \bar{\tau} \right)^2 = 5.3^2.$$

Notice that the simulated mean and standard deviation are quite close to the analytically calculated mean and variance given in Equation (8.32). Hence we lose little precision by using simulation in place of the usually more complicated analytical calculation.

8.6   DEPENDENCE BETWEEN POTENTIAL OUTCOMES

As discussed in Section 8.4, usually the most critical decision in the model-based approach is the specification of the model of the joint distribution of the unit-level potential outcomes, $f(Y_i(0), Y_i(1)|\theta)$. In the six-unit example in Section 8.4, we used a joint normal distribution, where we assumed a known covariance matrix. For simplicity, we assumed no dependence between the two potential outcomes – the cross-terms of the covariance matrix were equal to zero. Typically it is more appropriate to choose a model in which the elements of the covariance matrix are also unknown. In this case, one parameter that requires special consideration is the correlation coefficient ρ or, more generally, the parameters reflecting the degree of dependence between the two potential outcomes.

Suppose, in contrast to the model we used in Section 8.4, we assume a joint distribution for the potential outcomes with unknown covariance matrix, including an unknown correlation coefficient ρ:

$$f(Y_i(0), Y_i(1)|\theta) \sim \mathcal{N} \left(\begin{pmatrix} \mu_c \\ \mu_t \end{pmatrix}, \begin{pmatrix} \sigma_c^2 & \rho \sigma_c \sigma_t \\ \rho \sigma_c \sigma_t & \sigma_t^2 \end{pmatrix} \right),$$

where now the parameter vector is $\theta = (\mu_c, \mu_t, \sigma_c^2, \sigma_t^2, \rho)'$. In this setting, the conditional distribution of Y_i^{obs} given $(\mathbf{W}, \theta)$ is

$$f(Y_i^{\text{obs}}|\mathbf{W}, \theta) = \frac{1}{\sqrt{2\pi \cdot ((1 - W_i) \cdot \sigma_c^2 + W_i \cdot \sigma_t^2)}}$$

$$\times \exp \left[-\frac{1}{2} \left(\frac{\left(Y_i^{\text{obs}} - (1 - W_i) \cdot \mu_c - W_i \cdot \mu_t \right)^2}{(1 - W_i) \cdot \sigma_c^2 + W_i \cdot \sigma_t^2} \right) \right], \qquad (8.33)$$

and the corresponding likelihood function is

$$\mathcal{L}(\mu_{\text{c}}, \mu_{\text{t}}, \sigma_{\text{c}}^2, \sigma_{\text{t}}^2, \rho | \mathbf{Y}^{\text{obs}}, \mathbf{W}) = \prod_{i=1}^{6} \frac{1}{\sqrt{2\pi \cdot ((1 - W_i) \cdot \sigma_{\text{c}}^2 + W_i \cdot \sigma_{\text{t}}^2)}}$$

$$\times \exp\left[-\frac{1}{2}\left(\frac{1}{(1 - W_i) \cdot \sigma_{\text{c}}^2 + W_i \cdot \sigma_{\text{t}}^2}\left(Y_i^{\text{obs}} - (1 - W_i) \cdot \mu_{\text{c}} - W_i \cdot \mu_{\text{t}}\right)^2\right)\right].$$

Note that the likelihood function does not depend on the correlation coefficient ρ; it is, in fact, completely unchanged from the corresponding expression in Section 8.4, other than that it replaces 100 with σ_{c}^2 and 64 with σ_{t}^2. In other words, the data contain no information about the correlation between the potential outcomes.

Suppose, in addition, that the prior distribution of the parameters θ can be factored into a function of the correlation coefficient times a function of the remaining parameters:

$$p(\theta) = p(\rho) \cdot p(\mu_{\text{c}}, \mu_{\text{t}}, \sigma_{\text{c}}^2, \sigma_{\text{t}}^2).$$

In combination with the fact that the likelihood function is free of ρ, this implies that the posterior distribution of the correlation coefficient will be identical to its prior distribution. Considering similar discussions in earlier chapters – for example, the difficulty in estimating the variance of the unit-level treatment effects in Chapter 6 – this result should not be surprising. We never simultaneously observe both potential outcomes for any unit, and thus we have no empirical information on their dependence.

To understand the implications of this change in assumptions, let us estimate the average treatment effect under the same model, except now assuming a correlation coefficient equal to 1. With the variances still known, $\sigma_{\text{c}}^2 = 100$ and $\sigma_{\text{t}}^2 = 64$, the parameter vector is again $\theta = (\mu_{\text{c}}, \mu_{\text{t}})$. The distribution of the potential outcomes is now

$$\begin{pmatrix} Y_i(0) \\ Y_i(1) \end{pmatrix} \Bigg| \theta \sim \mathcal{N}\left(\begin{pmatrix} \mu_{\text{c}} \\ \mu_{\text{t}} \end{pmatrix}, \begin{pmatrix} 100 & 80 \\ 80 & 64 \end{pmatrix}\right).$$

Using the same steps as in Section 8.4, we can derive the joint distribution of $(\mathbf{Y}^{\text{mis}}, \mathbf{Y}^{\text{obs}})$ given $(\mathbf{W}, \mu_{\text{c}}, \mu_{\text{t}})$:

$$\begin{pmatrix} Y_i^{\text{mis}} \\ Y_i^{\text{obs}} \end{pmatrix} \Bigg| \mathbf{W}, \mu_{\text{c}}, \mu_{\text{t}} \sim \mathcal{N}\left(\begin{pmatrix} W_i \cdot \mu_{\text{c}} + (1 - W_i) \cdot \mu_{\text{t}} \\ (1 - W_i) \cdot \mu_{\text{c}} + W_i \cdot \mu_{\text{t}} \end{pmatrix},\right.$$

$$\left.\begin{pmatrix} W_i \cdot 100 + (1 - W_i) \cdot 64 & 80 \\ 80 & (1 - W_i) \cdot 100 + W_i \cdot 64 \end{pmatrix}\right).$$

This distribution is almost equal to the previously calculated joint distribution for $(\mathbf{Y}^{\text{mis}}, \mathbf{Y}^{\text{obs}})$, seen in Equation (8.21), except that the cross-terms in the covariance matrix are now also non-zero.

Using this joint distribution, we can derive the conditional distribution of $\mathbf{Y}^{\text{mis}}$ given $(\mathbf{Y}^{\text{obs}}, \mathbf{W}, \mu_c, \mu_t)$:

$$Y_i^{\text{mis}} \mid \mathbf{Y}^{\text{obs}}, \mathbf{W}, \mu_c, \mu_t \sim \tag{8.34}$$

$$\sim \mathcal{N}\left(W_i \cdot \left(\mu_c + \frac{80}{64} \cdot (Y_i^{\text{obs}} - \mu_t) \right) + (1 - W_i) \cdot \left(\mu_t + \frac{80}{100} \cdot (Y_i^{\text{obs}} - \mu_c) \right), 0 \right).$$

This conditional distribution is quite different from the one derived for the case with $\rho = 0$, given in (8.22). Here the conditional variance is zero; because we assume a perfect correlation between $Y_i(0)$ and $Y_i(1)$, it follows that, given $(Y_i^{\text{obs}}, \mu_c, \mu_t)$, we know the exact value of Y_i^{mis}.

However, our interest is not in this conditional distribution. Rather, we need the distribution of $\mathbf{Y}^{\text{mis}}$ given $(\mathbf{Y}^{\text{obs}}, \mathbf{W})$ only, that is, without conditioning on (μ_c, μ_t). To derive this distribution, we need the posterior distribution of (μ_c, μ_t). Here it is key that the conditional distribution of the observed outcomes, given the assignment $\mathbf{W}$ and parameter θ, $f(\mathbf{Y}^{\text{obs}} \mid \mathbf{W}, \theta)$, is unaffected by our assumption on ρ – compare Equation (8.33), with $\sigma_t^2 = 10^2$ and $\sigma_t^2 = 8^2$, to Equation (8.25). Thus the likelihood function remains the same, and this is in fact true irrespective of the value of the correlation coefficient. If we assume the same prior distribution for θ, the posterior distributions for (μ_c, μ_t) will be the same as that derived before and given in (8.26).

Because Y_i^{mis} is a linear function of (μ_c, μ_t), normality of (μ_c, μ_t) implies normality of Y_i^{mis}. The mean and variance of Y_i^{mis} given $(\mathbf{Y}^{\text{obs}}, \mathbf{W})$ are

$$\mathbb{E}\left[Y_i^{\text{mis}} \mid \mathbf{Y}^{\text{obs}}, \mathbf{W} \right] = W_i \cdot \left\{ \overline{Y}_c^{\text{obs}} \cdot \frac{N_c \cdot 10{,}000}{N_c \cdot 10{,}000 + 100} + \frac{80}{64} \right.$$

$$\left. \cdot \left(Y_i^{\text{obs}} - \overline{Y}_t^{\text{obs}} \cdot \frac{N_t \cdot 10{,}000}{N_t \cdot 10{,}000 + 64} \right) \right\}$$

$$+ (1 - W_i) \cdot \left\{ \overline{Y}_t^{\text{obs}} \cdot \frac{N_t \cdot 10{,}000}{N_t \cdot 10{,}000 + 64} + \frac{80}{100} \cdot \left(Y_i^{\text{obs}} - \overline{Y}_c^{\text{obs}} \cdot \frac{N_c \cdot 10{,}000}{N_c \cdot 10{,}000 + 100} \right) \right\},$$

$$\mathbb{V}\left(Y_i^{\text{mis}} \mid \mathbf{Y}^{\text{obs}}, \mathbf{W} \right) = W_i \cdot \left\{ \mathbb{V}(\mu_c) + \left(\frac{80}{64} \right)^2 \cdot \mathbb{V}(\mu_t) \right\}$$

$$+ (1 - W_i) \cdot \left\{ \mathbb{V}(\mu_t) + \left(\frac{80}{100} \right)^2 \cdot \mathbb{V}(\mu_c) \right\}$$

$$= W_i \cdot \left\{ \frac{1}{N_c/100 + 1/10{,}000} + \left(\frac{80}{64} \right)^2 \cdot \frac{1}{N_t/64 + 1/10{,}000} \right\}$$

$$+ (1 - W_i) \cdot \left\{ \frac{1}{N_t/64 + 1/10{,}000} + \left(\frac{80}{100} \right)^2 \cdot \frac{1}{N_c/100 + 1/10{,}000} \right\}.$$

Finally, the covariance between Y_i^{mis} and $Y_{i'}^{\text{mis}}$, for $i \neq i'$, is

$$
\mathbb{C}\left(Y_i^{\text{mis}}, Y_{i'}^{\text{mis}} \middle| \mathbf{Y}^{\text{obs}}, \mathbf{W}\right) = W_i \cdot W_{i'}
$$

$$
\cdot \left(\frac{1}{N_c/100 + 1/10,000} + \left(\frac{80}{64}\right)^2 \cdot \frac{1}{N_t/64 + 1/10,000} \right)
$$

$$
- W_i \cdot (1 - W_{i'}) \cdot \left(\frac{80}{100} \cdot \frac{1}{N_c/100 + 1/10,000} + \frac{80}{64} \cdot \frac{1}{N_t/64 + 1/10,000} \right)
$$

$$
- (1 - W_i) \cdot W_{i'} \cdot \left(\frac{80}{100} \cdot \frac{1}{N_c/100 + 1/10,000} + \frac{80}{64} \cdot \frac{1}{N_t/64 + 1/10,000} \right)
$$

$$
+ (1 - W_i) \cdot (1 - W_{i'}) \cdot \left(\frac{1}{N_t/64 + 1/10,000} + \left(\frac{80}{100}\right)^2 \cdot \frac{1}{N_c/100 + 1/10,000} \right).
$$

Again, our ultimate interest is not in this conditional distribution, but in the conditional distribution of the estimand given $(\mathbf{Y}^{\text{obs}}, \mathbf{W})$. Using the average treatment effect as our estimand, we have

$$
\tau_{\text{fs}} = \frac{1}{N} \sum_{i=1}^{N} (2 \cdot W_i - 1) \cdot \left(Y_i^{\text{obs}} - Y_i^{\text{mis}} \right)
$$

$$
= \frac{1}{N} \sum_{i=1}^{N} (2 \cdot W_i - 1) \cdot Y_i^{\text{obs}} - \frac{1}{N} \sum_{i=1}^{N} (2 \cdot W_i - 1) \cdot Y_i^{\text{mis}}.
$$

Thus $\tau_{\text{fs}} | \mathbf{Y}^{\text{obs}}, \mathbf{W}$ has a Gaussian (normal) distribution with mean

$$
\mathbb{E}\left[\tau_{\text{fs}} \middle| \mathbf{Y}^{\text{obs}}, \mathbf{W} \right] = \frac{1}{N} \sum_{i=1}^{N} (2 \cdot W_i - 1) \cdot Y_i^{\text{obs}} + \frac{1}{N} \sum_{i=1}^{N} (1 - 2 \cdot W_i) \cdot \mathbb{E}\left[Y_i^{\text{mis}} \middle| \mathbf{Y}^{\text{obs}}, \mathbf{W} \right]
$$

$$
= \overline{Y}_t^{\text{obs}} \cdot \frac{N_t \cdot 1000 - 16 \cdot N_t/N}{N_t \cdot 1000 + 64} - \overline{Y}_c^{\text{obs}} \cdot \frac{N_c \cdot 1000 + 20 \cdot N_c/N}{N_c \cdot 1000 + 100}.
$$

and variance

$$
\mathbb{V}\left(\tau_{\text{fs}} \middle| \mathbf{Y}^{\text{obs}}, \mathbf{W} \right) = \frac{1}{N^2} \sum_{i=1}^{N} \mathbb{V}\left(Y_i^{\text{mis}} \middle| \mathbf{Y}^{\text{obs}}, \mathbf{W} \right) + \frac{1}{N^2} \sum_{i=1}^{N} \sum_{i' \neq i} \mathbb{C}\left(Y_i^{\text{mis}}, Y_{i'}^{\text{mis}} \middle| \mathbf{Y}^{\text{obs}}, \mathbf{W} \right)
$$

$$
= \frac{N_t}{N^2} \cdot \left\{ \frac{1}{N_c/100 + 1/10,000} + \left(\frac{80}{64}\right)^2 \cdot \frac{1}{N_t/64 + 1/10,000} \right\}
$$

$$
+ \frac{N_c}{N^2} \cdot \left\{ \frac{1}{N_t/64 + 1/10,000} + \left(\frac{80}{100}\right)^2 \cdot \frac{1}{N_c/100 + 1/10,000} \right\}
$$

$$
+ \frac{N_t \cdot (N_t - 1)}{N^2} \cdot \left(\frac{1}{N_c/100 + 1/10,000} + \left(\frac{80}{64}\right)^2 \cdot \frac{1}{N_t/64 + 1/10,000} \right)
$$

$$-\frac{2 \cdot N_\mathrm{c} \cdot N_\mathrm{t}}{N^2} \cdot \left(\frac{80}{100} \cdot \frac{1}{N_\mathrm{c}/100 + 1/10,000} + \frac{80}{64} \cdot \frac{1}{N_\mathrm{t}/64 + 1/10,000} \right)$$

$$+ \frac{N_\mathrm{c} \cdot (N_\mathrm{c} - 1)}{N^2} \cdot \left(\frac{1}{N_\mathrm{t}/64 + 1/10,000} + \left(\frac{80}{100} \right)^2 \cdot \frac{1}{N_\mathrm{c}/100 + 1/10,000} \right).$$

Substituting the values for the six-unit illustrative data set, we find

$$\tau_\mathrm{fs} | \mathbf{Y}^\mathrm{obs}, \mathbf{W} \sim \mathcal{N} \left(8.7, 7.7^2 \right).$$

Thus, using the same model in Section 8.4, with the sole modification of assuming a cor-relation coefficient fixed at one rather than zero, leads to an estimated average treatment effect with approximately the same mean, 8.7, but a standard deviation now equal to 7.7, somewhat larger than the standard deviation of 5.2 calculated assuming independent potential outcomes.

The main point to take from this section is that the correlation coefficient between the two potential outcomes is somewhat different from other parameters of the model because the data generally do not contain empirical information about it (more gener-ally, about the parameters governing the conditional association between $\mathbf{Y}(0)$ and $\mathbf{Y}(1)$ given $\mathbf{X}$). This leaves us with the question of how they should be modeled. Sometimes we choose to be "conservative" about this dependence and therefore assume the worst case. In terms of the posterior variance, the worst case is often the situation of perfect correlation between the two potential outcomes. Note that this mirrors our approach in Chapter 6 in the discussion of Neyman's repeated sampling approach. On the other hand, researchers often wish to avoid contamination of the imputation of the potential outcomes under the active treatment by imputed values of the potential outcomes under the control treatment, and vice versa, thus choosing to model the two potential out-come distributions as conditionally independent in an approach that is conservative in a different sense.

8.7   MODEL-BASED IMPUTATION WITH COVARIATES

The presence of covariates does not fundamentally change the underlying method for imputing the missing potential outcomes in the model-based approach. In this sense, the model-based imputation approach has a substantial advantage over Neyman's approach that was discussed in the previous chapter. In the current setting, the presence of covari-ates in principle allows for improved imputations of the missing outcomes because the covariates provide information to help predict the missing potential outcomes.

Given covariates, the first step now consists of specifying a model for the joint distri-bution of the two potential outcomes conditional on these covariates, $f(\mathbf{Y}(0), \mathbf{Y}(1) | \mathbf{X}, \theta)$. Suppose, by appealing to de Finetti's theorem, that the triples $(Y_i(0), Y_i(1), X_i)$ are mod-eled as independent and identically distributed conditional on a vector-valued parameter θ. We can always factor this distribution into two components, the joint distribu-tion of the potential outcomes given the covariates and the marginal distribution of the covariates:

$$f(Y_i(0), Y_i(1), X | \theta_{Y|X}, \theta_X) = f(Y_i(0), Y_i(1) | X, \theta_{Y|X}) \cdot f(X | \theta_X), \tag{8.35}$$

where $\theta_{Y|X}$ and θ_X are functions of θ governing the respective distributions. Often we assume that the parameters entering the marginal distribution of the covariates are distinct from those entering the conditional distribution of the potential outcomes given the covariates, and specify the prior distribution so that it factors into a function of $\theta_{Y|X}$ and a function of θ_X:

$$p(\theta_{Y|X}, \theta_X) = p(\theta_{Y|X}) \cdot p(\theta_X). \tag{8.36}$$

Although this assumption is often made in practice, it is not always innocuous. For example, when the covariates include a time series of previous measurements (prior to the intervention of the active treatment) of the same quantity as measured by the outcome, the parameters governing the distribution of the covariates could have important information about the parameters governing the outcome distribution under the control treatment. However, if (8.36) holds, the analysis simplifies. In that case we need to model only the conditional distribution of the potential outcomes given the covariates, $f(Y_i(0), Y_i(1) | X_i, \theta)$. (We drop the indexing of θ by $Y|X$ because there is only one parameter vector left.) The remaining steps are essentially unchanged. We derive the conditional distribution of the causal estimand given the observed data and parameters, now also conditional on the covariates. We also derive the posterior distribution of the parameters given the observed potential outcomes and covariates.

Let us consider an example with a scalar covariate. The models that we have studied so far have had bivariate normal distributions:

$$\begin{pmatrix} Y_i(0) \\ Y_i(1) \end{pmatrix} \sim \mathcal{N} \left(\begin{pmatrix} \mu_c \\ \mu_t \end{pmatrix}, \begin{pmatrix} \sigma_c^2 & 0 \\ 0 & \sigma_t^2 \end{pmatrix} \right). \tag{8.37}$$

One way to extend the previous model to allow for covariates is to instead model the conditional distribution of the potential outcomes conditional on the covariates as

$$\begin{pmatrix} Y_i(0) \\ Y_i(1) \end{pmatrix} \bigg| X_i, \theta \sim \mathcal{N} \left(\begin{pmatrix} X_i \beta_c \\ X_i \beta_t \end{pmatrix}, \begin{pmatrix} \sigma_c^2 & 0 \\ 0 & \sigma_t^2 \end{pmatrix} \right), \tag{8.38}$$

where we include the intercept in the vector of covariates. Thus θ now consists of the four components $\beta_c, \beta_t, \sigma_c^2$, and σ_t^2, where β_c and β_t are vectors. An alternative is to assume that the slope coefficients (the elements of β_c and β_t other than those corresponding to the intercept) are the same for both potential outcomes, although in many situations such restrictions are not supported by the data. Notice that, in model (8.38), the covariates affect only the location of the distribution, not its dispersion. This modeling assumption too can be relaxed.

Given model (8.38), the remainder of the steps in the model-based approach with covariates are very similar to those in the situation without covariates. We can derive the distribution of the average treatment effect given observed variables and parameters $\theta = (\beta_c, \beta_t, \sigma_c^2, \sigma_t^2)$. For unit i with covariate value X_i, the missing potential outcome has, given the parameter values, the distribution

$$Y_i^{\text{mis}} | \mathbf{Y}^{\text{obs}}, \mathbf{W}, \mathbf{X}, \theta \sim \mathcal{N}\left(W_i \cdot X_i \beta_{\text{c}} + (1 - W_i) \cdot X_i \beta_{\text{t}}, W_i \cdot \sigma_{\text{t}}^2 + (1 - W_i) \cdot \sigma_{\text{t}}^2 \right).$$

We combine this distribution with the posterior distribution of θ given $(\mathbf{Y}, \mathbf{W}, \mathbf{X})$ to obtain the joint posterior distribution of τ and θ, which we then use to get the marginal posterior distribution of θ. If the prior distribution for θ factors into a function of $(\alpha_{\text{c}}, \beta_{\text{c}}, \sigma_{\text{c}}^2)$ and a function of $(\alpha_{\text{t}}, \beta_{\text{t}}, \sigma_{\text{t}}^2)$, then we can factor the posterior distribution into a function of $(\alpha_{\text{c}}, \beta_{\text{c}}, \sigma_{\text{c}}^2)$ and a function of $(\alpha_{\text{t}}, \beta_{\text{t}}, \sigma_{\text{t}}^2)$, with the former depending only on the units with $W_i = 0$, and the latter depending only on units with $W_i = 1$.

In situations with covariates, analytic solutions are difficult to obtain. In practice, we use simulation methods to obtain draws from the posterior distribution of the causal estimand.

8.8   SUPER-POPULATION AVERAGE TREATMENT EFFECTS

In the discussion so far, we have focused on the average treatment effect for the sample at hand, $\tau_{\text{fs}} = \sum_{i=1}^{N} (Y_i(1) - Y_i(0))/N$. Suppose instead that we view these observations as a random sample from an infinite super-population, and that our interest lies in the average treatment effect for that super-population:

$$\tau_{\text{sp}} = \mathbb{E}_{\text{sp}}[Y_i(1) - Y_i(0)].$$

This discussion mirrors that in Chapter 6 where we used Neyman's approach with a super-population. As in that setting, we can modify the model-based approach discussed in Sections 8.1–8.6 to estimate and conduct inference for this different estimand.

Given a fully specified model for the potential outcomes, the new estimand of interest, τ_{sp}, can sometimes be expressed solely as a function of the parameters. For example, in the normal linear model we can write:

$$\tau_{\text{sp}} = \tau(\theta) = \mathbb{E}_{\text{sp}}\left[Y_i(1) - Y_i(0) | \theta \right] = \mu_{\text{t}} - \mu_{\text{c}}.$$

In general, the population average treatment effect can be defined through the model for the joint distribution of the potential outcomes as

$$\tau(\theta) = \int \int (y(1) - y(0)) f(y(1), y(0)|\theta) \, dy(1) \, dy(0).$$

If there are covariates, the estimand may depend on both the parameters and the distribution of covariates, for example,

$$\tau_{\text{sp}} = \mathbb{E}_{\text{sp}}\left[\tau(\theta, \mathbf{X}) \right], \qquad \text{where } \tau(\theta, \mathbf{X}) = \mathbb{E}_{\text{sp}}\left[Y_i(1) - Y_i(0) | \mathbf{X}, \theta \right].$$

The representation in the linear model makes inference for the population average treatment effect conceptually straightforward. As before, we draw randomly from the derived posterior distribution for θ. Then, instead of using this draw $\theta^{(1)}$ to draw from the conditional distribution of $\mathbf{Y}^{\text{mis}}$, that is, $f(\mathbf{Y}^{\text{mis}} | \mathbf{Y}^{\text{obs}}, \mathbf{W}, \theta^{(1)})$, we simply use the draw to calculate the average treatment effect directly: $\tau^{(1)} = \tau(\theta^{(1)})$. Using N_R draws from the posterior distribution of θ (given the observed data) gives us $\{\hat{\tau}_{sp}^{(r)}, r = 1, \ldots, N_R\}$. The average and sample variance of these N_R draws give us estimates of the posterior mean and variance of the population average treatment effect.

Using the same six observations, let us see how the results for the super-population average treatment effect differ from those for the sample average treatment effect. As derived in Section 8.4.3, the joint posterior distribution for $\theta = (\mu_c, \mu_t)'$ is equal to

$$\begin{pmatrix} \mu_c \\ \mu_t \end{pmatrix} \bigg| \mathbf{Y}^{obs}, \mathbf{W} \sim \mathcal{N}\left(\begin{pmatrix} 4.1 \\ 12.8 \end{pmatrix}, \begin{pmatrix} 33.2 & 0 \\ 0 & 21.3 \end{pmatrix} \right).$$

The posterior distribution for $\tau_{sp} = \mu_t - \mu_c$ is therefore

$$\mu_t - \mu_c | \mathbf{Y}^{obs}, \mathbf{W} \sim \mathcal{N}\left((12.8 - 4.1), (33.2 + 21.3 + 2 \cdot 0)\right) \sim \mathcal{N}\left(8.7, 7.4^2\right).$$

Hence the posterior mean of τ_{sp} is 8.7, identical to the posterior mean of the sample average treatment effect τ_{fs}. The posterior standard deviation for the population average treatment effect is now 7.4. For comparison, recall that when we calculated the sample average treatment effect assuming independence across the two potential outcomes (Section 8.4.3), the standard deviation was equal to 5.2; when we assumed perfect correlation (Section 8.6), it was instead 7.7. Thus the posterior standard deviation is substantially different from that derived for the sample average treatment effect under independence of the potential outcomes but close to that for the sample average treatment effect under perfect correlation. This result should not be surprising. Compared to the first task, estimating the population average treatment effect is more demanding. Even if we could observe all elements of the vectors of potential outcomes $\mathbf{Y}(0)$ and $\mathbf{Y}(1)$ in our experiment – allowing us to calculate the finite-sample average treatment effect, $\tau_{fs} = \sum_{i=1}^{N} (Y_i(1) - Y_i(0))/N$ with certainty – we would still be uncertain about the average treatment effect in the super-population from which our sample was taken. This result mirrors the discussion in Chapter 6, where we showed that using the worst-case scenario assumption of perfect correlation not only gave a "conservative" estimate of the sampling variance in a finite-population setting but also provided an unbiased estimate of the sampling variance of the point estimate in the super-population.

It is also important to note that when we are interested in the super-population average treatment effect, the value of the correlation coefficient ρ becomes unimportant: the estimand $\tau_{sp} = \mu_t - \mu_c$ does not depend on ρ at all. Because the likelihood function of the observed data does not depend on ρ either, the posterior distribution for τ will not depend on the prior distribution for ρ, when the prior distribution of θ has ρ and (μ_c, μ_t) marginally independent.

8.9   A FREQUENTIST PERSPECTIVE

In this section we consider the frequentist perspective for calculating average treatment effects via the model-based approach. So far this discussion has taken an exclusively Bayesian perspective because this is particularly convenient for the problem at hand; it treats the uncertainty in the missing potential outcomes in the same way that it treats the uncertainty in the unknown parameters. In contrast, from the standard frequentist perspective, the unknown parameters are taken as fixed quantities, always to be conditioned on, whereas the potential outcomes, missing and observed, are considered unobserved and observed random variables given parameters, respectively. Nevertheless, as in many other instances, inferences based on Bayesian and frequentist perspectives are often close in substantive terms, with Bayesian posterior intervals often having good repeated sampling coverage rates, and it is instructive to understand both perspectives. Here we therefore outline the frequentist perspective in greater detail, focusing on the case where the estimand of interest is the population average treatment effect, $\tau_{sp}(\theta)$.

Suppose, as before, we specify the joint distributions of $Y_i(0)$ and $Y_i(1)$ in terms of a parameter vector θ. As we saw in Section 8.8, the average treatment effect τ_{sp} is the difference in the two expected values, $\tau_{sp} = \mathbb{E}[Y_i(1) - Y_i(0)|\theta]$. This expectation is a function of the parameters, $\tau_{sp}(\theta)$.

Consider first the situation without covariates, where the joint distribution of the two potential outcomes is bivariate normal with means μ_c and μ_t, with both variances equal to σ^2, and the correlation coefficient equal to zero. In this case the function $\tau_{sp}(\theta)$ is simply the difference: $\tau_{sp} = \mu_t - \mu_c$. In fact, given that we are interested in the average treatment effect, we can reparameterize θ as $\tilde{\theta} = (\mu_c, \tau_{sp}, \sigma^2)$, where $\tau_{sp} = \mu_t - \mu_c$. The estimand of interest now equals one of the elements of our parameter vector, and the inferential problem is now simply one of estimating $\tilde{\theta}$ and its associated precision.

Taking this approach, we can make a direct connection to linear regression. The conditional distribution of the observed potential outcomes given the assignment and parameter vectors is now independent and identically distributed as

$$Y_i^{obs}|\mathbf{W}, \tilde{\theta} \sim \mathcal{N}(\mu_c + W_i \cdot \tau_{sp}, \sigma^2).$$

Hence we can simply estimate the population average treatment effect, τ_{sp}, by ordinary least squares (OLS), with the OLS standard errors providing the appropriate measure of uncertainty for $\hat{\tau}_{sp}$.

Although the preceding result seems appealing, it is somewhat misleading in its simplicity. Often, statistical models that are convenient for modeling the joint distribution of the potential outcomes cannot be parameterized easily in terms of the average treatment effect. In that case, τ_{sp} will generally be a more complex function of the parameter vector. Nevertheless, in general we can still obtain maximum likelihood estimates of θ, and thus of $\tau_{sp}(\theta)$, as well as estimates of the large sample precision of $\tau_{sp}(\theta)$.

To see how this works, in a slight modification of the linear model, suppose, for example, that the model is specified on the logarithm of the potential outcomes:

$$\begin{pmatrix} \ln(Y_i(0)) \\ \ln(Y_i(1)) \end{pmatrix} \Bigg| \theta \sim \mathcal{N}\left(\begin{pmatrix} \mu_c \\ \mu_t \end{pmatrix}, \begin{pmatrix} \sigma_c^2 & 0 \\ 0 & \sigma_t^2 \end{pmatrix} \right).$$

The population average treatment effect is now equal to

$$\tau_{sp} = \tau(\theta) = \exp\left(\mu_t + \frac{1}{2} \cdot \sigma_t^2 \right) - \exp\left(\mu_c + \frac{1}{2} \cdot \sigma_c^2 \right). \tag{8.39}$$

Using this model to estimate τ_{sp}, we would first obtain maximum likelihood estimates of the parameters, $\theta = (\mu_c, \mu_t, \sigma_c^2, \sigma_t^2)$. Next we would substitute these values into the transformation $\tau_{sp}(\cdot)$ to obtain point estimates $\hat{\tau}_{sp} = g(\hat{\theta})$, where $g(\cdot)$ is defined by (8.39). The potentially more complicated step is the calculation of the asymptotic precision of our estimator. This calculation requires, for example, that we first calculate the full large-sample sampling covariance matrix for the parameter vector θ (e.g., using the Fisher information matrix), followed by the application of the delta method (i.e., Taylor series approximations) to derive the asymptotic sampling variance for $\hat{\tau}_{sp}$.

In this example, the frequentist approach has been only slightly more complicated than in the simple linear model. Often when there are covariates, however, these transformations of the original parameters become quite complex. The temptation is thus to choose models for the joint distribution $f(\mathbf{Y}(0), \mathbf{Y}(1)|\mathbf{X}, \theta)$ that make this transformation as simple as possible, as in the preceding linear examples. We stress, however, that the role of the statistical model is solely to provide a good description of the joint distribution of the potential outcomes. This is conceptually different from being parameterized conveniently in terms of the estimand of interest.

The possible advantage of the frequentist approach is that it avoids the need to specify the prior distribution $p(\theta)$ for the parameters governing the joint distribution of the two potential outcomes. However, this does not come without cost. Nearly always one has to rely on large sample approximations to justify the derived frequentist confidence intervals. But in large samples, by the Bernstein–Von Mises Theorem (e.g., Van Der Vaart, 1998), the practical implications of the choice of prior distribution is limited, and the alleged benefits of the frequentist approach vanish.

8.10   MODEL-BASED ESTIMATES OF THE EFFECT OF THE NSW PROGRAM

To illustrate the methods discussed in this chapter, we return to the full data set for the National Supported Work (NSW) program introduced in Section 8.2. We focus on a couple of aspects of the modeling approach and, in particular, the sensitivity to the choice for the joint distribution of the potential outcomes. We will not discuss in detail the choice of prior distribution for the Bayesian approach. For the simple models we use here, standard diffuse prior distributions are available. They perform well and the results are not sensitive to modest deviations from them.

For each model, we report in Table 8.6 the posterior mean and posterior standard deviation for the average effect τ_{fs}, and the treatment minus control differences in quantiles by treatment status for the 0.25, 0.50, and 0.75 quantiles, $\tau_{quant,0.25}$, $\tau_{quant,0.50}$, and $\tau_{quant,0.75}$. To be precise for, say the 0.25 quantile, we report the difference between the 0.25 quantile of the N values of $Y_i(1)$, some observed and some imputed, and the 0.25 quantile of the N values of $Y_i(0)$, some observed and some imputed. This generally differs from the 0.25 quantile of the N values of the unit-level treatment effects $Y_i(1) - Y_i(0)$. The latter quantile is more difficult to estimate, because results for such an estimand are sensitive to choices for the prior distribution of the dependence structure between the two potential outcomes.

Table 8.6. *Posterior Means and Standard Deviations for Treatment Effects under Four Models for NSW Program Data*

| | | | | | Effect on Quantiles | | |
Mean Covariate Dependent	Variance Treatment Specific	Potential Outcome Independent	Two-Part Model	Mean Effect Mean (S.D.)	0.25 quant Mean (S.D.)	0.50 quant Mean (S.D.)	0.75 quant Mean (S.D.)
No	No	No	No	1.79 (0.63)	1.79 (0.63)	1.79 (0.63)	1.79 (0.63)
No	Yes	Yes	No	1.78 (0.49)	0.63 (0.35)	1.63 (0.55)	3.07 (0.64)
Yes	Yes	Yes	No	1.57 (0.50)	0.42 (0.34)	1.40 (0.55)	2.89 (0.63)
Yes	Yes	Yes	Yes	1.57 (0.74)	0.25 (0.30)	1.03 (0.53)	1.69 (0.72)

To put the model-based results in perspective, we first estimated the average effect using the simple difference in means, using Neyman's approach. The average effect of the training program on annual earnings in thousands of dollars was estimated to be $\hat{\tau}_{\mathrm{fs}} = 1.79$, with an estimated standard error of 0.63 based on $\hat{\mathbb{V}}^{\mathrm{neyman}}$. Adjusting for all ten covariates from Table 8.1 using the linear regression methods from the previous chapter, with the regression including an intercept, an indicator for the treatment, and the ten covariates, changes the estimate to 1.67 (with an estimated error equal to 0.64).

We consider four specifications for the joint distribution of the potential outcomes given covariates. The first is a joint normal distribution with the potential outcomes perfectly correlated, free from dependence on the covariates, and with identical variances in the two treatment arms:

$$\begin{pmatrix} Y_i(0) \\ Y_i(1) \end{pmatrix} \Big| X_i, \theta \sim \mathcal{N} \left(\begin{pmatrix} \mu_c \\ \mu_t \end{pmatrix}, \begin{pmatrix} \sigma^2 & \sigma^2 \\ \sigma^2 & \sigma^2 \end{pmatrix} \right). \tag{8.40}$$

To implement this model, we need to make one more decision, namely the prior distribution for the unknown parameter $\theta = (\mu_c, \mu_t, \sigma^2)$. We take the parameters to be independent *a priori*. The prior distributions for the two mean parameters, μ_c and μ_t, are normal with zero means and variances equal to 100^2, the standard deviations of 100 being large relative to the scale of the data (the earnings variables are measured in thousands of dollars and range from 0 to 60.3). The prior distribution for σ^2 is inverse gamma with parameters 1 and 0.01, respectively. The posterior mean and standard deviation for the treatment effects of interest are reported in the first row of Table 8.6. Note that, for this specification, the effect of the treatment is constant, and so the estimates of the quantile effects are all identical to that for the mean. The posterior mean of τ_{fs} is equal to 1.79, with a posterior standard deviation of 0.63.

For the results reported in the second row of Table 8.6, again we assume prior independence between the potential outcomes and allow for treatment-control differences in the conditional variances:

$$\begin{pmatrix} Y_i(0) \\ Y_i(1) \end{pmatrix} \Big| X_i, \theta \sim \mathcal{N} \left(\begin{pmatrix} \mu_c \\ \mu_t \end{pmatrix}, \begin{pmatrix} \sigma_c^2 & 0 \\ 0 & \sigma_t^2 \end{pmatrix} \right), \tag{8.41}$$

The prior distributions for the two mean parameters, μ_c and μ_t, are, as before, normal with zero means and variances equal to 100^2. The prior distributions for σ_c^2 and σ_t^2 are inverse gamma with parameters 1 and 0.01 respectively. The posterior mean for the average effect, τ_{fs}, is now 1.78, very similar to the 1.79 from before. However, the posterior standard deviation for the average effect τ_{fs} is substantially lower, 0.49. The posterior means for the quantile effects are fairly different from those reported in the first row of the table, ranging from 0.63 for the 0.25 quantile to 3.07 for the 0.75 quantile.

In the third row of Table 8.6, we allow for linear dependence of the conditional means of the potential outcomes in nine covariates:

$$\begin{pmatrix} Y_i(0) \\ Y_i(1) \end{pmatrix} \Bigg| X_i, \theta \sim \mathcal{N} \left(\begin{pmatrix} X_i\beta_c \\ X_i\beta_t \end{pmatrix}, \begin{pmatrix} \sigma_c^2 & 0 \\ 0 & \sigma_t^2 \end{pmatrix} \right). \tag{8.42}$$

For the parameters β_c and β_t, we assume prior independence from the other parameters, as well as independence from each other. The prior distributions are specified to be normal with zero means and variance equal to 100^2. The prior distributions for σ_c^2 and σ_t^2 are the same as before. The posterior mean for the average effect is now 1.57 with a posterior standard deviation equal to 0.50. The posterior means for the quantile effects range from 0.42 for the 0.25 quantile to 2.89 for the 0.75 quantile.

All three of these models implicitly assume continuity of the potential outcome distributions. These models are therefore implausible as descriptions of the distribution of the potential outcomes, considering the high proportion of zeros in the observed outcomes (equal to 31%). The fourth model is a more serious attempt to fit this conditional distribution. We model two parts of the conditional distribution. First, the probability of a positive value for $Y_i(0)$ is

$$\Pr(Y_i(0) > 0 | X_i, W_i, \theta) = \frac{\exp(X_i\gamma_c)}{1 + \exp(X_i\gamma_c)}, \tag{8.43}$$

and similarly for $Y_i(1)$:

$$\Pr(Y_i(1) > 0 | X_i, W_i, \theta) = \frac{\exp(X_i\gamma_t)}{1 + \exp(X_i\gamma_t)}.$$

Second, conditional on a positive outcome, the logarithm of the potential outcome is assumed to have a normal distribution:

$$\ln(Y_i(0)) | Y_i(0) > 0, X_i, W_i, \theta \sim \mathcal{N}\left(X_i\beta_c, \sigma_c^2\right), \tag{8.44}$$

and

$$\ln(Y_i(1)) | Y_i(1) > 0, X_i, W_i, \theta \sim \mathcal{N}\left(X_i\beta_t, \sigma_t^2\right).$$

The simulation-based results for this model are displayed in the fourth row of Table 8.6. The posterior mean for the average effect is now 1.57, with a posterior standard deviation of 0.74. The posterior mean for the 0.25 quantile is much lower in this model, equal to 0.25. These posterior distributions, especially the posterior mean for the 0.25

Table 8.7. *Posterior Distributions for Parameters for Normal/Logistic Two-Part Model – NSW Program Data*

Covariate	β_c		$\beta_t - \beta_c$		γ_0		$\gamma_1 - \gamma_0$	
	Mean	(S.D.)	Mean	(S.D.)	Mean	(S.D.)	Mean	(S.D.)
intercept	1.38	(0.84)	0.40	(1.26)	2.54	(1.49)	0.68	(2.49)
age	0.02	(0.01)	−0.02	(0.02)	−0.01	(0.02)	0.02	(0.03)
education	0.01	(0.06)	0.01	(0.09)	−0.05	(0.11)	0.02	(0.17)
married	−0.23	(0.25)	0.35	(0.35)	−0.18	(0.40)	0.91	(0.73)
nodegree	−0.01	(0.27)	−0.24	(0.39)	−0.28	(0.47)	−0.26	(0.74)
black	−0.44	(0.20)	0.37	(0.30)	−1.09	(0.44)	−0.77	(0.97)
earn'74	−0.01	(0.02)	0.01	(0.03)	0.01	(0.04)	−0.02	(0.08)
earn'74=0	0.19	(0.31)	−0.58	(0.46)	1.00	(0.56)	−3.06	(1.12)
earn'75	0.02	(0.04)	0.01	(0.05)	0.00	(0.08)	0.20	(0.17)
earn'75=0	−0.05	(0.29)	0.17	(0.40)	−0.61	(0.46)	2.13	(1.05)
$\ln(\sigma_c)$	0.02	(0.06)						
$\ln(\sigma_t)$	0.03	(0.06)						

quantile, are much more plausible given the substantial fraction of individuals who are not working in any period in the study.

In Table 8.7 we report posterior means and standard deviations for all parameter estimates in the last model. These estimates shed some light on the amount of heterogeneity in the treatment effects. We report the estimates for the parameters of the control outcomes, (β_c and γ_c), and for the differences in the parameters for the treated outcome and the control outcomes, $\beta_t - \beta_c$, and $\gamma_t - \gamma_c$.

8.11 CONCLUSION

In this chapter we outline a model-based imputation approach to estimation of and inference for causal effects. The causal effects of interest are viewed as functions of observed and missing potential outcomes. The missing potential outcomes are imputed through a statistical model for the joint distribution of the potential outcomes and a model for the assignment mechanism, which is known in the randomized experiment setting. The model for the potential outcomes is, in principle, informed by subject-matter knowledge, although in the randomized experiment setting, results tend to be relatively insensitive to modest changes in its specification. The context in this chapter is that of a completely randomized experiment, but, in principle, the general framework extends naturally to non-experimental settings.

NOTES

The data used in this chapter to illustrate the concepts introduced were first analyzed by Lalonde (1986) and used subsequently by many others, including Heckman and Hotz (1989), Dehejia and Wahba (1999), Smith and Todd (2001), Abadie and Imbens (2009), as well as others. The Lalonde study has been very influential for its conclusion that non-experimental evaluations were unable to recover experimental estimates. The data are available on Rajeev Dehejia's website, http://www.nber.org/~rdehejia/nswdata.html.

The Bayesian approach to the analysis of randomized experiments presented here was first discussed in detail in Rubin (1978). For Bayesian analyses of more complicated (non-ignorable treatment assignment) models, see Imbens and Rubin (1997b), Hirano, Imbens, Rubin, and Zhou (2000), and Zhang, Rubin, and Mealli (2009).

De Finetti's Theorem originates in de Finetti (1964, 1992). See also Hewitt and Savage (1955), Feller (1965, pp. 225–226), Rubin (1978), and for extensions to the finite N case see Diaconis (1976).

For general discussions of Bayesian methods see Box and Tiao (1973), Gelman, Carlin, Stern, and Rubin (1995), Hartigan (1983), Lancaster (2004), and Robert (1994). To implement the Bayesian analysis discussed in this chapter, it is useful to use modern numerical methods, in particular Markov-Chain-Monte-Carlo methods. For textbook discussions, in addition to the aforementioned texts on Bayesian methods, see Tanner (1996), Robert and Casella (2004), and Brooks, Gelman, Jones, and Meng (2011).

APPENDIX A POSTERIOR DISTRIBUTIONS FOR NORMAL MODELS

In this appendix, we briefly review the basic results in Bayesian inference used in the current chapter. For a fuller discussion of general Bayesian methods, see Gelman, Carlin, Stern, and Rubin (1995) and Lancaster (2004). For a discussion of the role of Bayesian methods for inference for causal effects, see Rubin (1978, 2004) and Imbens and Rubin (1997).

A.1 Prior Distributions, Likelihood Functions, and Posterior Distributions

A Bayesian formulation has two components. First we specify a "sampling" model (conditional distribution) for the data given unknown parameters. The data are denoted by $\mathbf{Z}$. Often $\mathbf{Z}$ is a matrix of dimension $N \times K$, with typical row Z_i. The parameter will be denoted by θ. The parameter lies in the set Θ. The sampling model will be denoted by $f_{\mathbf{Z}}(\mathbf{Z}|\theta)$. As a function of θ with fixed data $\mathbf{Z}$, it is known as the likelihood function: $\mathcal{L}(\theta|\mathbf{Z})$. The second component of a Bayesian formulation is the prior distribution on θ, denoted by $p(\theta)$, which is a (proper) probability (density) function, integrating to one over the parameter space Θ.

The posterior distribution of θ given the observed data $\mathbf{Z}$ is then

$$p(\theta|\mathbf{Z}) = \frac{\mathcal{L}(\theta|\mathbf{Z}) \cdot p(\theta)}{\int_{\theta \in \Theta} \mathcal{L}(\theta|\mathbf{Z}) \cdot p(\theta)d\theta}.$$

Often we write

$$p(\theta|\mathbf{Z}) \propto \mathcal{L}(\theta|\mathbf{Z}) \cdot p(\theta),$$

because the constant can be recovered using the fact that the posterior distribution integrates to one.

A.2 The Normal Distribution with Unknown Mean and Known Variance

The first special case is the normal distribution with unknown mean and known variance. Suppose $\mathbf{Z}$ is an N-vector with i^{th} component $Z_i|\mu \sim \mathcal{N}(\mu, \sigma^2)$, with σ^2 known, and all the Z_i independent given μ. We use a normal prior distribution for μ, with mean α and variance ω^2. Then the posterior distribution for μ is

$$p(\mu|\mathbf{Z}) \sim \mathcal{N}\left(\frac{\overline{Z} \cdot N/\sigma^2 + \alpha/\omega^2}{N/\sigma^2 + 1/\omega^2}, \frac{1}{N/\sigma^2 + 1/\omega^2}\right),$$

where $\overline{Z} = \sum_{i=1}^{N} Z_i/N$.

A.3 The Normal Distribution with Known Mean and Unknown Variance

Now suppose the distribution of Z_i is $\mathcal{N}(\mu, \sigma^2)$ with μ known and σ^2 unknown. We use a prior distribution for σ^2 such that, for specified S_0^2 and M, the random variable $\sigma^{-2}S_0^2/M$ has a gamma distribution with parameters $M/2$ and $1/2$ (or, equivalently, a chi-squared distribution with M degrees of freedom). Then the posterior distribution of σ^2 given $\mathbf{Z}$ is such that the distribution of $\sigma^{-2} \cdot (S_0^2 + \sum_i (Z_i - \mu)^2/(M+N)$ has a gamma distribution with parameters $(M + N)/2$ and $1/2$. Repeatedly sampling μ and σ^2, this leads to a sequence whose draws converge to a draw of (μ, σ^2) from its actual posterior distribution.

A.4 Simulation Methods for the Normal Linear Regression Model

Here we present the details for a simulation-based inference for the parameters of a normal linear regression model:

$$Y_i|\beta, \sigma^2 \sim \mathcal{N}\left(X_i\beta, \sigma^2\right), \tag{A.1}$$

with unknown β and σ. We use a normal prior distribution for β, $\mathcal{N}(\mu, \Omega)$, and prior distribution for σ^2 such that for specified S_0^2 and M, $\sigma^{-2} \cdot S_0^2/M$ has a Gamma distribution with parameters $M/2$ and $1/2$.

To draw from the posterior distribution of β and σ^2, we use Markov-Chain-Monte-Carlo (MCMC) methods where we draw sequentially from the posterior distribution of β given σ^2 and from the posterior distribution of σ^2 given β, and iterate. We initialize the chain by using the least squares estimate for β and σ^2 as the starting value.

The first step is drawing from the posterior distribution of β given σ^2. This posterior distribution is

$$p(\beta|\mathbf{Y}, \mathbf{X}, \sigma^2) \propto \mathcal{N}\left(\left(\sigma^{-2}\mathbf{X}'\mathbf{X} + \Omega^{-1}\right)^{-1}\left(\sigma^{-2}\mathbf{X}'\mathbf{Y} + \Omega^{-1}\mu\right), \left(\sigma^{-2}\mathbf{X}'\mathbf{X} + \Omega^{-1}\right)^{-1}\right).$$

It is straightforward to draw from.

The second step is drawing from a posterior distribution of σ^2 given β. This posterior distribution is such that the distribution of

$$\sigma^{-2} \cdot \sum_{i=1}^{N} (Y_i - X_i\beta)^2 / (N + M),$$

has a Gamma distribution with parameters $(N + M)/2$ and $1/2$. Repeatedly drawing β and σ^2 this way leads to a sequence whose draws converge to draws of (β, σ^2) from its actual posterior distribution.

A.5 Simulation Methods for the Logistic Regression Model

Here we discuss methods for drawing from the posterior distribution of the parameters in a logistic regression model. The model is

$$\Pr(Y_i = 1|X_i, \gamma) = \frac{\exp(X_i\gamma)}{1 + \exp(X_i\gamma)}.$$

With a sample of size N the likelihood function is

$$\mathcal{L}(\gamma|\mathbf{Y}, \mathbf{X}) = \prod_{i=1}^{N} \frac{\exp(Y_i \cdot X_i\gamma)}{1 + \exp(X_i\gamma)}.$$

We use a normal prior distribution for γ, with mean μ and covariance matrix Ω. To sample from the posterior distribution, we use the Metropolis Hastings algorithm (e.g., Gelman, Carlin, Stern, and Rubin, 2000). For the starting value we use the maximum likelihood estimates $\hat{\gamma}_{ml}$ for γ, although this may not be the best choice for assessing convergence of the chain. We can construct a chain $\gamma_0, \gamma_1, \ldots, \gamma_K$, where $\gamma_0 = \hat{\gamma}_{ml}$. Given a value γ_k we proceed as follows. We draw a candidate value γ from a normal distribution centered at $\hat{\gamma}_{ml}$ with covariance matrix $2 \cdot \hat{\mathcal{I}}^{-1}$, where $\hat{\mathcal{I}}$ is the estimated Fisher information matrix. Let $\mathcal{N}(\gamma|\mu, \Omega)$ denote the density function for a multivariate normal random variable with mean μ, covariance matrix Ω, evaluated at γ.

Given the candidate value γ, we move to this new value or stay at the current value γ_k, with probabilities

$$\Pr(\gamma_{k+1} = \gamma) = \min\left(1, \frac{\mathcal{L}(\gamma) \cdot \mathcal{N}(\gamma|\mu, \Omega) \cdot \mathcal{N}(\gamma_k|\hat{\gamma}_{ml}, 2 \cdot \hat{\mathcal{I}}^{-1})}{\mathcal{L}(\gamma_k) \cdot \mathcal{N}(\gamma_k|\mu, \Omega) \cdot \mathcal{N}(\gamma|\hat{\gamma}_{ml}, 2 \cdot \hat{\mathcal{I}}^{-1})}\right)$$

$$\Pr(\gamma_{k+1} = \gamma_k) = 1 - \Pr(\gamma_{k+1} = \gamma).$$

As with the previous method in Appendix A.3, the sequence converges to a draw from the correct posterior distribution of γ.

APPENDIX B ANALYTIC DERIVATIONS WITH KNOWN COVARIANCE MATRIX

In this appendix we derive the distribution of the average treatment effect for the case where the potential outcomes are jointly normally distributed with known covariance matrix, and the prior distribution for the parameters is also jointly normal. In this case, analytic solutions exist for the distribution of the average treatment effect, conditional on the observed data. These analytic results allow us to compare answers for various special cases, such as when the two potential outcomes are uncorrelated versus answers when they are perfectly correlated, and the finite sample versus super-population average treatment effect.

Assume N exchangeable units, indexed by $i = 1, \ldots, N$. Conditional on the parameter vector θ, we assume the potential outcomes are normally distributed:

$$\begin{pmatrix} Y_i(0) \\ Y_i(1) \end{pmatrix} \Bigg| \, \theta \overset{i.i.d.}{\sim} \mathcal{N}\left(\begin{pmatrix} \mu_c \\ \mu_t \end{pmatrix}, \begin{pmatrix} \sigma_c^2 & \rho\sigma_c\sigma_t \\ \rho\sigma_c\sigma_t & \sigma_c^2 \end{pmatrix} \right). \tag{B.1}$$

In this example the covariance matrix parameters σ_c^2, σ_c^2, and ρ are assumed known, and $\theta = (\mu_c, \mu_t)$ is the vector of unknown parameters. The distribution of the assignment vector $\mathbf{W}$ is $p(\mathbf{W})$, known by the assumption of a completely randomized experiment. Conditional on $\mathbf{W}$ and the parameters, the observed potential outcomes are independent of one another, with distribution

$$Y_i^{obs}|\mathbf{W}, \theta \sim \mathcal{N}(W_i \cdot \mu_t + (1 - W_i) \cdot \mu_c, W_i \cdot \sigma_c^2 + (1 - W_i) \cdot \sigma_c^2).$$

Thus, the likelihood function is

$$\mathcal{L}(\mu_c, \mu_t | \mathbf{Y}^{obs}, \mathbf{W}) = p(\mathbf{W}) \cdot \prod_{i=1}^{N} \frac{1}{\sqrt{2\pi \cdot ((1 - W_i) \cdot \sigma_c^2 + W_i \cdot \sigma_c^2)}} \tag{B.2}$$

$$\times \exp\left[-\frac{1}{2}\left(\frac{1}{(1 - W_i) \cdot \sigma_c^2 + W_i \cdot \sigma_c^2} (Y_i - (1 - W_i) \cdot \mu_c - W_i \cdot \mu_t)^2 \right) \right].$$

As we saw in Section 8.6, this likelihood is free of the correlation coefficient ρ.

Note that, because of the assumed normal distribution of the two potential outcomes, the average of the observed outcomes per treatment level have sampling distributions

$$\begin{pmatrix} \overline{Y}_c^{obs} \\ \overline{Y}_t^{obs} \end{pmatrix} \Bigg| \, \theta \sim \mathcal{N}\left(\begin{pmatrix} \mu_c \\ \mu_t \end{pmatrix}, \begin{pmatrix} \sigma_c^2/N_c & 0 \\ 0 & \sigma_c^2/N_t \end{pmatrix} \right), \tag{B.3}$$

where N_t is the number of treated and N_c is the number of control units. Because $(\overline{Y}_c^{obs}, \overline{Y}_t^{obs}, N_c, N_t)$ is a sufficient statistic, the likelihood function based on (B.3) is proportional to that of the likelihood function based on the full set of observed data $(\mathbf{Y}^{obs}, \mathbf{W})$. Note also that the conditional covariance (given θ) between $\overline{Y}_c^{obs}$ and $\overline{Y}_t^{obs}$ is zero, which is true irrespective of the correlation between the two potential outcomes for the same unit, because the two averages, $\overline{Y}_c^{obs}$ and $\overline{Y}_t^{obs}$, are based on different units.

To derive the conditional distribution of the missing potential outcomes given the data and the unknown parameters, first let us consider the conditional distribution of one potential outcome given the other:

$$Y_i(1)|Y_i(0), \mathbf{W}, \theta \sim \mathcal{N}\left(\mu_t + \rho \cdot \frac{\sigma_t}{\sigma_c} \cdot (Y_i(0) - \mu_c), (1 - \rho^2) \cdot \sigma_c^2\right),$$

and

$$Y_i(0)|Y_i(1), \mathbf{W}, \theta \sim \mathcal{N}\left(\mu_c + \rho \cdot \frac{\sigma_c}{\sigma_t} \cdot (Y_i(1) - \mu_t), (1 - \rho^2) \cdot \sigma_c^2\right).$$

Then, if we use Equations (8.13), the representations of Y_i^{obs} and Y_i^{mis} as functions of $Y_i(0)$ and $Y_i(1)$, the conditional distribution of Y_i^{mis} is

$$Y_i^{\text{mis}}|Y_i^{\text{obs}}, \mathbf{W}, \theta \sim \mathcal{N}\left(W_i \cdot \left(\mu_c + \rho \cdot \frac{\sigma_c}{\sigma_t} \cdot (Y_i^{\text{obs}} - \mu_t)\right)\right.$$

$$+ (1 - W_i) \cdot \left(\mu_t + \rho \cdot \frac{\sigma_t}{\sigma_c} \cdot (Y_i^{\text{obs}} - \mu_c)\right),$$

$$\left.(1 - \rho^2) \cdot ((W_i \cdot \sigma_t^2 + (1 - W_i) \cdot \sigma_c^2)\right).$$

Because of the exchangeability of the potential outcomes, Y_i^{mis} is independent of $Y_{i'}^{\text{mis}}$ if $i \neq i'$, conditional on $\mathbf{W}$ and θ.

Next we use the representation of the average treatment effect in terms of the observed and missing potential outcomes,

$$\tau_{\text{fs}} = \frac{1}{N}\sum_{i=1}^{N}(Y_i(1) - Y_i(0)) = \frac{1}{N}\sum_{i=1}^{N}\left((2W_i - 1) \cdot \left(Y_i^{\text{obs}} - Y_i^{\text{mis}}\right)\right)$$

$$= \frac{1}{N}\sum_{i=1}^{N}(2W_i - 1) \cdot Y_i^{\text{obs}} - \frac{1}{N}\sum_{i=1}^{N}(2W_i - 1) \cdot Y_i^{\text{mis}},$$

to derive the conditional distribution of τ_{fs} given $\mathbf{Y}^{\text{obs}}$, $\mathbf{W}$, and θ. The first sum is observed, and the second sum consists of N unobserved terms. Because, given $(\mathbf{Y}^{\text{obs}}, \mathbf{W})$ and θ, τ_{fs} is a linear function of normal random variables, τ_{fs} is normally distributed with mean

$$\mathbb{E}\left[\tau_{\text{fs}} \middle| \mathbf{Y}^{\text{obs}}, \mathbf{W}, \theta\right] = \frac{1}{N}\sum_{i=1}^{N}W_i \cdot \left(Y_i^{\text{obs}} - \mu_c - \rho \cdot \frac{\sigma_c}{\sigma_t} \cdot \left(Y_i^{\text{obs}} - \mu_t\right)\right) \qquad \text{(B.4)}$$

$$+ (1 - W_i) \cdot \left(\mu_t - Y_i^{\text{obs}} + \rho \cdot \frac{\sigma_t}{\sigma_c} \cdot \left(Y_i^{\text{obs}} - \mu_c\right)\right)$$

$$= \lambda_t \cdot \overline{Y}_t^{\text{obs}} + (1 - \lambda_t) \cdot \mu_t - \left(\lambda_c \cdot \overline{Y}_c^{\text{obs}} + (1 - \lambda_c) \cdot \mu_c\right),$$

where

$$\lambda_t = \frac{N_t}{N} \cdot \left(1 - \rho \cdot \frac{\sigma_c}{\sigma_t}\right), \quad \text{and} \quad \lambda_c = \frac{N_c}{N} \cdot \left(1 - \rho \cdot \frac{\sigma_t}{\sigma_c}\right),$$

and conditional variance

$$\mathbb{V}\left(\tau_{fs} \,\Big|\, \mathbf{Y}^{obs}, \mathbf{W}, \theta\right) = \frac{1 - \rho^2}{N}\left(\frac{N_t}{N} \cdot \sigma_c^2 + \frac{N_c}{N} \cdot \sigma_c^2\right). \tag{B.5}$$

Now consider inference for θ. We use a joint normal prior distribution for (μ_c, μ_t):

$$\begin{pmatrix} \mu_c \\ \mu_t \end{pmatrix} \sim \mathcal{N}\left(\begin{pmatrix} \nu_c \\ \nu_t \end{pmatrix}, \begin{pmatrix} \omega_c^2 & 0 \\ 0 & \omega_t^2 \end{pmatrix}\right), \tag{B.6}$$

where ν_c, ν_t, ω_c, and ω_t are specified constants. Combining the prior distribution in (B.6) with the (normal) likelihood function for the observed data given (μ_c, μ_t) from (B.2), leads to a conditional posterior distribution for τ_{fs} given θ that is normal with mean

$$\mu_{\theta|\mathbf{Y}^{obs}, \mathbf{W}} = \mathbb{E}\left[\begin{pmatrix} \mu_c \\ \mu_t \end{pmatrix} \Big| \mathbf{Y}^{obs}, \mathbf{W}, \theta\right] = \begin{pmatrix} \delta_c \cdot \overline{Y}_c^{obs} + (1 - \delta_c) \cdot \nu_c \\ \delta_t \cdot \overline{Y}_t^{obs} + (1 - \delta_t) \cdot \nu_t \end{pmatrix}, \tag{B.7}$$

where

$$\delta_c = \frac{N_c/\sigma_t^2}{N_c/\sigma_c^2 + 1/\omega_c^2} \quad \text{and} \quad \delta_t = \frac{N_t/\sigma_t^2}{N_t/\sigma_c^2 + 1/\omega_t^2},$$

and covariance matrix

$$\Sigma_{\theta|\mathbf{Y}^{obs}, \mathbf{W}} = \mathbb{V}\left(\begin{pmatrix} \mu_c \\ \mu_t \end{pmatrix} \Big| \mathbf{Y}^{obs}, \mathbf{W}, \theta\right) = \begin{pmatrix} \frac{1}{N_c/\sigma_t^2 + 1/\omega_c^2} & 0 \\ 0 & \frac{1}{N_t/\sigma_c^2 + 1/\omega_t^2} \end{pmatrix}. \tag{B.8}$$

Next we combine the posterior distribution for θ with the conditional posterior distribution of the average treatment effect τ_{fs} given θ to obtain the distribution of the average treatment effect conditional on only the observed data, its posterior distribution. Because both of the distributions used here are normal, with the latter linear in the parameters, the posterior distribution of τ_{fs} (i.e., marginalized over θ) will also be normal. Specifically, because $(\theta|\mathbf{Y}^{obs}, \mathbf{W}) \sim \mathcal{N}(\mu_{\theta|\mathbf{Y}^{obs}, \mathbf{W}}, \Sigma_{\theta|\mathbf{Y}^{obs}, \mathbf{W}})$, and $(\tau_{fs}|\mathbf{Y}^{obs}, \mathbf{W}, \theta) \sim \mathcal{N}(\beta_c + \beta_t'\theta, \sigma_{\tau_{fs}|\mathbf{Y}^{obs}, \mathbf{W}, \theta}^2)$ (with $\sigma_{\tau|\mathbf{Y}^{obs}, \mathbf{W}, \theta}^2$ free of θ), it follows that $(\tau_{fs}|\mathbf{Y}^{obs}, \mathbf{W}) \sim \mathcal{N}(\beta_c + \beta_t'\mu_{\theta|\mathbf{Y}^{obs}, \mathbf{W}}, \sigma_{\tau_{fs}|\mathbf{Y}^{obs}, \mathbf{W}, \theta}^2 + \beta_t'\Sigma_{\theta|\mathbf{Y}^{obs}, \mathbf{W}}\beta_t)$. Straightforward algebra then shows that $(\tau_{fs}|\mathbf{Y}^{obs}, \mathbf{W})$ is normal with mean

$$\mu_{\tau_{fs}|\mathbf{Y}^{obs}, \mathbf{W}} = \kappa_t \cdot \overline{Y}_t^{obs} + (1 - \kappa_t) \cdot \nu_t - \left(\kappa_c \cdot \overline{Y}_c^{obs} + (1 - \kappa_c) \cdot \nu_c\right), \tag{B.9}$$

where

$$\kappa_c = \lambda_c + (1 - \lambda_c) \cdot \delta_c = \frac{N_c}{N} \cdot \left(1 - \rho \cdot \frac{\sigma_t}{\sigma_c}\right) + \left(\frac{N_t}{N} + \frac{N_c}{N} \cdot \rho \cdot \frac{\sigma_t}{\sigma_c}\right) \cdot \frac{N_c/\sigma_c^2}{N_c/\sigma_c^2 + 1/\omega_c^2}$$

$$= (1 - p) \cdot \left(1 - \rho \cdot \frac{\sigma_t}{\sigma_c}\right) + \left(p + (1 - p) \cdot \rho \cdot \frac{\sigma_t}{\sigma_c}\right) \cdot \frac{(1 - p) \cdot N/\sigma_c^2}{(1 - p) \cdot N/\sigma_c^2 + 1/\omega_c^2},$$

and

$$\kappa_t = \lambda_t + (1 - \lambda_t) \cdot \delta_t = \frac{N_t}{N} \cdot \left(1 - \rho \cdot \frac{\sigma_c}{\sigma_t}\right) + \left(\frac{N_c}{N} + \frac{N_t}{N} \cdot \rho \cdot \frac{\sigma_c}{\sigma_t}\right) \cdot \frac{N_t/\sigma_c^2}{N_t/\sigma_c^2 + 1/\omega_t^2}$$

$$= p \cdot \left(1 - \rho \cdot \frac{\sigma_c}{\sigma_t}\right) + \left(1 - p + p \cdot \rho \cdot \frac{\sigma_c}{\sigma_t}\right) \cdot \frac{p \cdot N/\sigma_t^2}{p \cdot N/\sigma_c^2 + 1/\omega_t^2},$$

where $p = N_t/N$, and with posterior variance

$$\sigma_{\tau_{fs}|\mathbf{Y}^{obs},\mathbf{W}}^2 = \frac{1 - \rho^2}{N} \left(\frac{N_t}{N} \cdot \sigma_c^2 + \frac{N_c}{N} \cdot \sigma_c^2\right)$$

$$+ \left(\frac{N_t}{N} + \frac{N_c}{N} \cdot \rho \cdot \frac{\sigma_t}{\sigma_c}\right)^2 \cdot \frac{1}{N_c/\sigma_c^2 + 1/\omega_c^2}$$

$$+ \left(\frac{N_c}{N} + \frac{N_t}{N} \cdot \rho \cdot \frac{\sigma_c}{\sigma_t}\right)^2 \cdot \frac{1}{N_t/\sigma_c^2 + 1/\omega_t^2}$$

$$= \frac{1 - \rho^2}{N} \left(p \cdot \sigma_c^2 + (1 - p) \cdot \sigma_c^2\right)$$

$$+ \frac{(p + (1 - p) \cdot \rho \cdot \sigma_t/\sigma_c)^2}{(1 - p) \cdot N/\sigma_c^2 + 1/\omega_c^2} + \frac{(1 - p + p \cdot \rho \cdot \sigma_c/\sigma_t)^2}{p \cdot N/\sigma_c^2 + 1/\omega_t^2}.$$

Now let us look at some special cases. First, the large sample approximation. With N_c and N_t large, we ignore terms that are of order $o(1/N_c)$ or $o(1/N_t)$. In this case, $\kappa_c \to 1$, $\kappa_t \to 1$, and the mean and scaled variance simplify to

$$\mu_{\tau_{fs}|\mathbf{Y}^{obs},\mathbf{W},N_c,N_t \text{ large}}^2 \longrightarrow \overline{Y}_t^{obs} - \overline{Y}_c^{obs},$$

and

$$N \cdot \sigma_{\tau_{fs}|\mathbf{Y}^{obs},\mathbf{W},N_c,N_t \text{ large}}^2 \longrightarrow (1 - \rho^2) \cdot \left(p \cdot \sigma_t^2 + (1 - p) \cdot \sigma_c^2\right)$$

$$+ \left(p + (1 - p) \cdot \rho \cdot \frac{\sigma_t}{\sigma_c}\right)^2 \cdot \frac{\sigma_t^2}{1 - p} + \left((1 - p) + p \cdot \rho \cdot \frac{\sigma_c}{\sigma_t}\right)^2 \cdot \frac{\sigma_c^2}{p}.$$

For the variance, it is useful to consider the special cases with $\rho = 0$ and $\rho = 1$. In large samples,

$$N \cdot \sigma_{\tau_{fs}|\mathbf{Y}^{obs},\mathbf{W},N_c,N_t \text{ large},\rho=0}^2 \longrightarrow \sigma_t^2 \cdot \frac{p}{1 - p} + \sigma_t^2 \cdot \frac{1 - p}{p},$$

and

$$
N \cdot \sigma^2_{\tau_{\mathrm{fs}}|\mathbf{Y}^{\mathrm{obs}},\mathbf{W},N_c,N_t \ \mathrm{large},\rho=1} \longrightarrow \left(p + (1-p) \cdot \frac{\sigma_t}{\sigma_c} \right)^2 \cdot \frac{\sigma^2_c}{1-p}
$$

$$
+ \left((1-p) + p \cdot \frac{\sigma_c}{\sigma_t} \right)^2 \cdot \frac{\sigma^2_c}{p}.
$$

It is also useful to compare this to the posterior distribution for the population average treatment effect τ_{sp}. For the general prior distribution, the posterior distribution is

$$
\tau_{\mathrm{sp}}|\mathbf{Y}^{\mathrm{obs}},\mathbf{W} \sim
$$

$$
\mathcal{N}\left(\delta_t \cdot \overline{Y}^{\mathrm{obs}}_t + (1-\delta_t) \cdot \nu_t - \left(\delta_c \cdot \overline{Y}^{\mathrm{obs}}_c + (1-\delta_c) \cdot \nu_c \right), \right.
$$

$$
\left. \frac{1}{(1-p) \cdot N/\sigma^2_t + 1/\omega^2_c} + \frac{1}{p \cdot N/\sigma^2_c + 1/\omega^2_t} \right).
$$

Even in finite samples, the posterior distribution of τ_{sp} does not depend on the correlation between the potential outcomes, ρ. In large samples this simplifies to

$$
\tau_{\mathrm{sp}} \approx \mathcal{N}\left(\overline{Y}^{\mathrm{obs}}_t - \overline{Y}^{\mathrm{obs}}_c, \frac{\sigma^2_c}{(1-p) \cdot N} + \frac{\sigma^2_t}{p \cdot N} \right).
$$

Note that the difference between the normalized posterior precisions for the average effect in the sample and the population average effect does not vanish as the sample size gets large.

Finally, it is useful to derive the conditional distribution of the missing potential outcomes given the observed data, integrating out the unknown parameters θ. For this we use the conditional distribution of the missing data given the observed data and parameters, and the posterior distribution of the parameters. Again, the normality of both components ensures that the distribution of the missing data are Gaussian (normal). The mean and variance of Y^{mis}_i given $\mathbf{Y}^{\mathrm{obs}}$ and $\mathbf{W}$ are thus

$$
\mu_{Y^{\mathrm{mis}}_i|\mathbf{Y}^{\mathrm{obs}},\mathbf{W}} = W_i \cdot \left(\delta_c \cdot \overline{Y}^{\mathrm{obs}}_c + (1-\delta_c) \cdot \nu_c + \rho \cdot \frac{\sigma_t}{\sigma_c} \cdot \left(Y^{\mathrm{obs}}_i - \delta_t \cdot \overline{Y}^{\mathrm{obs}}_t + (1-\delta_t) \cdot \nu_t \right) \right)
$$

$$
+ (1-W_i) \cdot \left(\delta_t \cdot \overline{Y}^{\mathrm{obs}}_t + (1-\delta_t) \cdot \nu_t + \rho \cdot \frac{\sigma_c}{\sigma_t} \cdot \left(Y^{\mathrm{obs}}_i - \delta_c \cdot \overline{Y}^{\mathrm{obs}}_c + (1-\delta_c) \cdot \nu_c \right) \right),
$$

and

$$
\sigma^2_{Y^{\mathrm{mis}}_i|\mathbf{Y}^{\mathrm{obs}},\mathbf{W}} = W_i \cdot \left((1-\rho^2) \cdot \sigma^2_t + \frac{1}{(1-p) \cdot N/\sigma^2_t + 1/\omega^2_c} + \rho^2 \cdot \left(\frac{\sigma_c}{\sigma_t} \right)^2 \right.
$$

$$
\left. \cdot \frac{1}{p \cdot N/\sigma^2_c + 1/\omega^2_c} \right)
$$

$$+ (1 - W_i) \cdot \left((1 - \rho^2) \cdot \sigma_c^2 + \frac{1}{p \cdot N / \sigma_t^2 + 1/\omega_t^2} + \rho^2 \cdot \left(\frac{\sigma_t}{\sigma_c} \right)^2 \right.$$

$$\left. \cdot \frac{1}{((1-p) \cdot N / \sigma_c^2 + 1/\omega_c^2)} \right).$$

In this case there is also a covariance across units, through the dependence on the parameters:

$$\mathrm{Cov}(Y_i^{\mathrm{mis}}, Y_{i'}^{\mathrm{mis}} | \mathbf{Y}^{\mathrm{obs}}, \mathbf{W}) = \begin{cases} \dfrac{\rho^2 \cdot \sigma_c^2}{N_c + \sigma_c^2/\omega_c^2} + \dfrac{1}{N_t/\sigma_c^2 + 1/\omega_t^2} & \text{if } W_i = 0, W_{i'} = 0 \\[3mm] -\dfrac{\rho \cdot \sigma_t \cdot \sigma_c}{N_c + \sigma_c^2/\omega_c^2} - \dfrac{\rho \cdot \sigma_t \cdot \sigma_c}{N_t + \sigma_c^2/\omega_t^2} & \text{if } W_i = 0, W_{i'} = 1 \\[3mm] -\dfrac{\rho \cdot \sigma_t \cdot \sigma_c}{N_c + \sigma_c^2/\omega_c^2} - \dfrac{\rho \cdot \sigma_t \cdot \sigma_c}{N_t + \sigma_c^2/\omega_t^2} & \text{if } W_i = 1, W_{i'} = 0 \\[3mm] \dfrac{1}{N_c/\sigma_c^2 + 1/\omega_c^2} + \dfrac{\rho^2 \cdot \sigma_t^2}{N_t + \sigma_c^2/\omega_t^2} & \text{if } W_i = 1, W_{i'} = 1. \end{cases}$$

In large samples, these can be approximated by

$$\mu_{Y_i^{\mathrm{mis}} | \mathbf{Y}^{\mathrm{obs}}, \mathbf{W}} = W_i \cdot \left(\overline{Y}_c^{\mathrm{obs}} + \rho \cdot \frac{\sigma_t}{\sigma_c} \cdot \left(Y_i^{\mathrm{obs}} - \overline{Y}_t^{\mathrm{obs}} \right) \right)$$

$$+ (1 - W_i) \cdot \left(\overline{Y}_t^{\mathrm{obs}} + \rho \cdot \frac{\sigma_c}{\sigma_t} \cdot \left(Y_i^{\mathrm{obs}} - \overline{Y}_c^{\mathrm{obs}} \right) \right),$$

$$\sigma^2_{Y_i^{\mathrm{mis}} | \mathbf{Y}^{\mathrm{obs}}, \mathbf{W}} = W_i \cdot \sigma_c^2 \cdot \left(1 - \rho^2 + \frac{1}{(1-p) \cdot N} + \frac{\rho^2}{p \cdot N} \right)$$

$$+ (1 - W_i) \cdot \sigma_c^2 \cdot \left(1 - \rho^2 + \frac{1}{p \cdot N} + \frac{\rho^2}{((1-p) \cdot N)} \right),$$

and

$$\mathrm{Cov}(Y_i^{\mathrm{mis}}, Y_{i'}^{\mathrm{mis}} | \mathbf{Y}^{\mathrm{obs}}, \mathbf{W}) = \begin{cases} \dfrac{\rho^2 \cdot \sigma_c^2}{(1-p) \cdot N} + \dfrac{\sigma_c^2}{p \cdot N} & \text{if } W_i = 0, W_{i'} = 0, \\[3mm] -\dfrac{\rho \cdot \sigma_t \cdot \sigma_c}{(1-p) \cdot N} - \dfrac{\rho \cdot \sigma_t \cdot \sigma_c}{p \cdot N} & \text{if } W_i = 0, W_{i'} = 1, \\[3mm] -\dfrac{\rho \cdot \sigma_t \cdot \sigma_c}{(1-p) \cdot N} - \dfrac{\rho \cdot \sigma_t \cdot \sigma_c}{p \cdot N} & \text{if } W_i = 1, W_{i'} = 0, \\[3mm] \dfrac{\sigma_c^2}{(1-p) \cdot N} + \dfrac{\rho^2 \cdot \sigma_c^2}{p \cdot N} & \text{if } W_i = 1, W_{i'} = 1. \end{cases}$$

## CHAPTER 9

Stratified Randomized Experiments

9.1 INTRODUCTION

The focus in the previous chapters in Part II was on completely randomized experiments, where, in a fixed sample with N units, N_t are randomly chosen to receive the active treatment and the remaining $N_c = N - N_t$ are assigned to receive the control treatment. We considered four modes of inference: Fisher's exact p-values and associated intervals, Neyman's unbiased estimates and repeated sampling-based large-N confidence intervals, regression methods, and model-based imputation. In addition, we considered the benefits of observing covariates, that is, measurements on the units unaffected by the treatments, such as pre-treatment characteristics. In this chapter we consider the same issues for a different class of randomized experiments, stratified randomized experiments, also referred to as randomized blocks experiments to use the terminology of classical experimental design. In stratified randomized experiments, units are stratified (or grouped or blocked) according to the values of (a function of) the covariates. Within the strata, independent completely randomized experiments are conducted but possibly with different relative sizes of treatment and control groups.

Part of the motivation for considering alternative structures for randomized experiments is interest in such experiments per se. But there are other, arguably equally important reasons. In the discussion of observational studies in Parts III, IV, V, and VI of this text, we consider methods for (non-randomized) observational data that can be viewed in some way as analyzing the data as if they arose from hypothetical stratified randomized experiments. Understanding these methods in the context of randomized experiments will aid their interpretation and implementation in observational studies.

The main part of this chapter describes how the methods developed in the previous four chapters can be modified to apply in the context of stratified randomized experiments. In most cases these modifications are conceptually straightforward. We also discuss some design issues in relation to stratification. Specifically, we assess the benefits of stratification relative to complete randomization.

In the next section we describe the data used to illustrate the concepts discussed in this chapter. These data are from a randomized experiment designed to evaluate the effect of class size on academic achievement, known as Project Star. In Section 9.3 we discuss the general structure of stratified randomized experiments. In the next four sections we discuss the four approaches we described previously for completely randomized experiments: in Section 9.4 the Fisher exact p-value approach; in Section 9.5 the Neyman approach; in Section 9.6 the regression approach; and in 9.7 the model-based imputation approach. Next, in Section 9.8, we discuss design issues and specifically the common benefits of stratified randomized experiments over completely randomized experiments. Section 9.9 concludes.

9.2   THE TENNESEE PROJECT STAR DATA

We illustrate the methods for randomized block experiments using data from a random-ized evaluation of the effect of class size on test scores conducted in 1985–1986 in Tennessee called the Student/Teacher Achievement Ratio experiment, or Project Star for short. This was a very influential experiment; Mosteller (1995) calls it "one of the most important educational investigations ever carried out." In this chapter we use the kinder-garten data from schools where students and teachers were randomly assigned to small classes (13–17 students per teacher), to regular classes (22–25 students per teacher), or to regular classes with a teacher's aide. To be eligible for Project Star, a school had to have a sufficient number of students to allow the formation of at least one class of each of the three types. Once a school had been admitted to the program, a decision was made on the number of classes of each type (small, regular, regular with aide). We take as fixed the number of classes of each type in each school. The unit of analysis is the teacher or class, rather than the individual student, to help justify the no-interference part of SUTVA.

   The experiment is somewhat different from those we have discussed before, so we will be precise in its description. A school has a pool of at least 57 students, so they could support at least one small and two regular-sized classes. Two separate and inde-pendent randomizations took place. One random assignment is that of teachers to classes of different types, small, regular, or regular with aide. The second randomization is of students to classes/teachers. In our analysis, we mainly rely on the first randomization, of class-size and aides to teachers, using the teachers as the units of analysis. Irrespec-tive of the assignments of students to classes, the resulting inferences are valid for the effect on the teachers of being assigned to a particular type of class. However, the sec-ond randomization is important for the interpretation of the results. Suppose we find that assignment to a small class leads on average to better outcomes for the teacher. Without the randomization of students to classes, this could be due to systematic assignment of better students to the smaller classes. With the second randomization, this is ruled out, and systematic effects can be interpreted as the effects of class size. This type of dou-ble randomization is somewhat similar to that in "split plot" designs (Cochran and Cox, 1957), although in split plot designs two different treatments are being applied by the double randomization.

   Given the structure of the experiment, one could also focus on students as the unit of analysis, and investigate effects of class size on student-level outcomes. The con-cern, however, is that the Stable Unit Treatment Value Assumption (SUTVA) is not plausible in that case. Violations of SUTVA complicate the Neyman, regression, and imputation approaches considerably, and we therefore primarily focus on class-level (i.e., teacher-level) analyses in this chapter. As we see in Section 9.4.4, however, it remains straightforward to use the FEP approach to test the null hypothesis that assignment of students to different classes had no effect on test scores whatsoever, because SUTVA is automatically satisfied under Fisher's sharp null hypothesis of no effects of the treatment.

In the analyses in this chapter, we focus on the comparison between regular (control) and small (treated) classes, and ignore the data for regular classes with teachers' aides. We discard schools that do not have at least two classes of both the small size and the regular size. Focusing on schools with at least two regular classes and two small classes leaves us with sixteen schools, which creates sixteen strata or blocks. Most have exactly two classes of each size, but one has two regular classes and four small classes, and two other schools have three small classes and two regular-sized classes. The total number of teachers and classes in this reduced data set is $N = 68$. Out of these 68 teachers, $N_c = 32$ are assigned to regular-sized classes, and $N_t = 36$ are assigned to small classes. Outcomes are defined at the class (i.e., teacher) level. The class-level outcomes we focus on are averages of test scores over all students for their teacher. One can, however, consider other outcomes, such as median test score of the students with a specific teacher or measures of within-teacher dispersion. The specific outcome we analyze here is the class average score on a mathematics test. The individual student scores were normalized to have mean equal to zero and standard deviation equal to one across all the students in the reduced data set. These individual scores then ranged from a minimum of -4.13 to a maximum of 2.94. The averages for each of the 68 classes in our analysis are reported in Table 9.1, organized by school. Overall, the average for the regular classes is -0.13 with a standard deviation of 0.56, and the average for the small classes is 0.09 with a standard deviation of 0.61. We return to these data after introducing methods for the analysis of such studies.

9.3   THE STRUCTURE OF STRATIFIED RANDOMIZED EXPERIMENTS

In stratified randomized experiments, units are grouped together according to some pre-treatment characteristics into strata. Within each stratum, a completely randomized experiment is conducted, and thus, within each stratum, the methods discussed in Chapters 5–8 are directly applicable. However, the interest is not about hypotheses or treatment effects within a single stratum, but rather it is about hypotheses and treatment effects across all strata. Moreover, the sample sizes are often such that we cannot obtain precise estimates of typical treatment effects within any one stratum. Here we discuss how the methods developed previously can be adapted to take account of the additional structure of the experiment.

9.3.1 The Case with Two Strata

As before, we are interested both in assessing null hypotheses concerning treatment effects and in estimating typical treatment effects (usually the average). First we focus

Table 9.1. *Class Average Mathematics Scores from Project Star*

School/ Stratum	No. of Classes	Regular Classes $(W_i = 0)$	Small Classes $(W_i = 1)$
1	4	$-0.197, 0.236$	$0.165, 0.321$
2	4	$0.117, 1.190$	$0.918, -0.202$
3	5	$-0.496, 0.225$	$0.341, 0.561, -0.059$
4	4	$-1.104, -0.956$	$-0.024, -0.450$
5	4	$-0.126, 0.106$	$-0.258, -0.083$
6	4	$-0.597, -0.495$	$1.151, 0.707$
7	4	$0.685, 0.270$	$0.077, 0.371$
8	6	$-0.934, -0.633$	$-0.870, -0.496, -0.444, 0.392$
9	4	$-0.891, -0.856$	$-0.568, -1.189$
10	4	$-0.473, -0.807$	$-0.727, -0.580$
11	4	$-0.383, 0.313$	$-0.533, 0.458$
12	5	$0.474, 0.140$	$1.001, 0.102, 0.484$
13	4	$0.205, 0.296$	$0.855, 0.509$
14	4	$0.742, 0.175$	$0.618, 0.978$
15	4	$-0.434, -0.293$	$-0.545, 0.234$
16	4	$0.355, -0.130$	$-0.240, -0.150$
Average		-0.13	0.09
(S.D.)		(0.56)	(0.61)

on the case with the sample of N units divided into two subsamples, for example, females (f) and males (m), with subsample size $N(f)$ and $N(m)$, respectively, so that $N = N(f) + N(m)$. To fit the division into two subsamples into the structure developed so far, it is useful to associate with each unit a binary covariate (e.g., the unit's sex) with the membership in strata based on this covariate. Although in general in this text we use the notation X_i for the covariate for unit i, here we use the notation G_i for this particular covariate that determines stratum or group membership, with **B** denoting the N-component vector with typical element B_i. As with any other covariate, the value of G_i is not affected by the treatment. In this example G_i takes on the values f and m. Define $\tau_{\text{fs}}(f)$ and $\tau_{\text{fs}}(m)$ to be the finite-sample average treatment effects in the two strata:

$$\tau_{\text{fs}}(f) = \frac{1}{N(f)} \sum_{i:G_i=f} \left(Y_i(1) - Y_i(0)\right), \quad \text{and} \quad \tau_{\text{fs}}(m) = \frac{1}{N(m)} \sum_{i:G_i=m} \left(Y_i(1) - Y_i(0)\right).$$

Within each stratum, we conduct a completely randomized experiment with $N_t(f)$ and $N_t(m)$ units assigned to the active treatment in the two subsamples respectively, and the remaining $N_c(f) = N(f) - N_t(f)$ and $N_c(m) = N(m) - N_t(m)$ units assigned to the control treatment. It need not be the case that the proportion of treated units, the propensity score, $e(f) = N_t(f)/N(f)$ and $e(m) = N_t(m)/N(m)$ for the female and male subpopulations, respectively, is the same in both subpopulations. Let $N_t = N_t(f) + N_t(m)$ be the total number of units assigned to the treatment group, and $N_c = N_c(f) + N_c(m)$ be the total number of units assigned to the control group. Let us consider the assignment mechanism. Within the $G_i = f$ subpopulation, $N_t(f)$ units out of $N(f)$ are randomly chosen to receive the treatment. There are $\binom{N(f)}{N_t(f)}$ such allocations. For every allocation for the set of units with $G_i = m$, there are $\binom{N(m)}{N_t(m)}$ ways of choosing $N_t(m)$ units with $G_i = m$ to receive the treatment out of $N(m)$ units. All of these allocations are equally likely. Combining these two assignment vectors, the assignment mechanism for a stratified randomized experiment with two strata can be written as

$$\Pr(\mathbf{W}|\mathbf{Y}(0), \mathbf{Y}(1), \mathbf{B}) = \binom{N(f)}{N_t(f)}^{-1} \cdot \binom{N(m)}{N_t(m)}^{-1} \quad \text{for } \mathbf{W} \in \mathbb{W}^+,$$

$$\text{where } \mathbb{W}^+ = \left\{ \mathbf{W} \text{ such that } \sum_{i:G_i=f} W_i = N_t(f), \sum_{i:G_i=m} W_i = N_t(m) \right\}.$$

Compare the assignment mechanism for a stratified randomized experiment to that for a completely randomized experiment with $N_t = N_t(f) + N_t(m)$ assigned to treatment and $N_c = N(f) - N_t(f) + N(m) - N_t(m)$ assigned to control. Many assignment vectors that would have positive probability with a completely randomized experiment have probability zero with the stratified randomized experiment: all vectors with $\sum_{i=1}^N W_i = N_t(f) + N_t(m)$ but $\sum_{i:G_i=f} W_i \neq N_t(f)$ (or, equivalently, $\sum_{i:G_i=m} W_i \neq N_t(m)$). If $N_t(f)/N(f) \approx N_t(m)/N(m)$, the stratification rules out substantial imbalances in the covariate distributions in the two treatment groups that could arise by chance in a completely randomized experiment. The possible disadvantage of the stratification is that a large number of possible assignment vectors are eliminated, just as a completely randomized experiment eliminates assignment vectors that would be allowed under Bernoulli trials (where assignment for each unit is determined independently of assignment for any other unit). The advantage of a completely randomized experiment over a Bernoulli trial for drawing causal inferences was argued to be the relative lack of information on treatment effects of the eliminated assignment vectors, typically those assignment vectors with a severe imbalance between the number of controls and the number of treated.

Here the argument is similar, although not quite as obvious. If we were to partition the population randomly into strata, the assignment vectors eliminated by the stratification are in expectation as helpful as the ones included, and the stratification will not produce a more informative experiment. However, if the stratification is based on characteristics that are associated with the outcomes of interest, we shall see that stratified randomized experiments generally are more informative than completely randomized experiments. For example, in many drug trials, one may expect systematic differences in typical outcomes, both given the drug and without the drug, for men and women. In that case, conducting the experiment by stratifying the population into males and females, rather than conducting a completely randomized experiment, makes eminent sense. It can lead to more precise inferences, by eliminating the possibility of assignments with severe imbalances in sex distribution – for example, the extreme and uninformative assignment with all women exposed to the active treatment and all men exposed to the control treatment.

9.3.2 The Case with J Strata

Here we generalize the notation to the situation with multiple strata. Let J be the number of strata, and $N(j)$, $N_c(j)$, and $N_t(j)$ the total number of units, and the number of control and treated units in strata j, respectively, for $j = 1, \ldots, J$. Let $G_i \in \{1, \ldots, J\}$ denote the stratum for unit i, and let $B_i(j) = \mathbf{1}_{G_i=j}$, be the indicator that is equal to one if unit i is in stratum j, and zero otherwise. Within stratum j there are now $\binom{N(j)}{N_t(j)}$ possible assignments, so that the assignment mechanism is

$$\Pr(\mathbf{W}|\mathbf{B}, \mathbf{Y}(0), \mathbf{Y}(1)) = \prod_{j=1}^{J} \binom{N(j)}{N_t(j)}^{-1} \qquad \text{for } \mathbf{W} \in \mathbb{W}^+,$$

where $\mathbb{W}^+ = \{\mathbf{W} \in \mathbb{W} | \sum_{i=1}^{N} B_i(j) \cdot W_i = N_t(j) \text{ for } j = 1, \ldots, J\}$.

9.4   FISHER'S EXACT P-VALUES IN STRATIFIED RANDOMIZED EXPERIMENTS

In stratified randomized experiments, just as in completely randomized experiments, the assignment mechanism is completely known. Hence, given a sharp null hypothesis that specifies all unobserved potential outcomes given knowledge of the observed outcomes, we can directly apply Fisher's approach to calculate exact p-values as discussed in Chapter 5. Let us focus on Fisher's sharp null hypothesis that all treatment effects are zero: $H_0 : Y_i(0) = Y_i(1)$ for $i = 1, 2, \ldots, N$. For ease of exposition, we focus initially on the case with two strata, $G_i \in \{f, m\}$.

9.4.1 The Choice of Statistics in the FEP Approach with Two Strata

Let $\overline{Y}_c^{\text{obs}}(j)$ and $\overline{Y}_t^{\text{obs}}(j)$ be the average observed outcome for units in stratum j (currently, in the two-stratum example for $j \in \{f, m\}$, later, in the general J-stratum case for $j = 1, \ldots, J$) in the control and treatment groups, and let $e(j)$ be the propensity score:

$$\overline{Y}_c^{\text{obs}}(j) = \frac{1}{N_c(j)} \sum_{i:G_i=j} (1 - W_i) \cdot Y_i^{\text{obs}}, \quad \overline{Y}_t^{\text{obs}}(j) = \frac{1}{N_t(j)} \sum_{i:G_i=j} W_i \cdot Y_i^{\text{obs}},$$

and

$$e(j) = N_t(j)/N(j).$$

Obvious statistics are the absolute value of the difference in the average observed outcome for treated and control units in the first and in the second stratum:

$$T^{\text{dif}}(f) = \left| \overline{Y}_t^{\text{obs}}(f) - \overline{Y}_c(f)^{\text{obs}} \right| \quad \text{and} \quad T^{\text{dif}}(m) = \left| \overline{Y}_t^{\text{obs}}(m) - \overline{Y}_c^{\text{obs}}(m) \right|.$$

Neither of the statistics, $T^{\mathrm{dif}}(f)$ or $T^{\mathrm{dif}}(m)$, is particularly attractive by itself: for either one an entire stratum is ignored, and thus the test would not be sensitive to violations of the null hypothesis in the stratum that is ignored.

A more appealing statistic is based on the combination of the two within-stratum statistics, $T^{\mathrm{dif}}(f)$ and $T^{\mathrm{dif}}(m)$, for example, the absolute value of a convex combination of the two differences in averages,

$$T^{\mathrm{dif},\lambda} = \left| \lambda \cdot \left(\overline{Y}_{\mathrm{t}}^{\mathrm{obs}}(f) - \overline{Y}_{\mathrm{c}}^{\mathrm{obs}}(f) \right) + (1 - \lambda) \cdot \left(\overline{Y}_{\mathrm{t}}^{\mathrm{obs}}(m) - \overline{Y}_{\mathrm{c}}^{\mathrm{obs}}(m) \right) \right|,$$

for some $\lambda \in [0, 1]$. For any fixed value of λ, we can use the same FEP approach and find the randomized distribution of the statistic under the null hypothesis, and thus calculate the corresponding p-value. The question is what would be an attractive choice for λ? An obvious choice for λ is to weight the two differences $T^{\mathrm{dif}}(f)$ and $T^{\mathrm{dif}}(m)$ by the relative sample sizes (RSS) in the strata and choose $\lambda = \lambda_{\mathrm{RSS}} \equiv N(f)/(N(f)+N(m))$. If the relative proportions of treated and control units in each stratum, $N_{\mathrm{t}}(f)/N(f)$ and $N_{\mathrm{t}}(m)/N(m)$ respectively, are similar, then the stratification from our stratified experiment is close to the stratification from a completely randomized experiment. In that case, this choice for the weight parameter λ_{RSS} would lead to the natural statistic that is common in a completely randomized experiment,

$$T^{\mathrm{dif},\lambda_{\mathrm{RSS}}} = \left| \frac{N(f)}{N(f) + N(m)} \cdot \left(\overline{Y}_{\mathrm{t}}^{\mathrm{obs}}(f) - \overline{Y}_{\mathrm{c}}^{\mathrm{obs}}(f) \right) + \frac{N(m)}{N(f) + N(m)} \cdot \left(\overline{Y}_{\mathrm{t}}^{\mathrm{obs}}(m) - \overline{Y}_{\mathrm{c}}^{\mathrm{obs}}(m) \right) \right|.$$

If the relative proportions of treated and control units are very different, however, this choice for λ does not necessarily lead to a very powerful test statistic. Suppose, for example, that both strata contain fifty units, where in stratum f, only a single unit gets assigned to treatment, and the remaining forty-nine units get assigned to control, whereas in stratum m, the number of treated and control units is twenty-five. In that case, the test based on $T^{\mathrm{dif}}(m)$ is likely to have substantially more power than the test based on $T^{\mathrm{dif}}(f)$. Combining $T^{\mathrm{dif}}(f)$ and $T^{\mathrm{dif}}(m)$ by the relative share of the two strata in the population, thereby giving both stratum-specific average observed outcome differences $\hat{\tau}(f)$ and $\hat{\tau}(m)$ equal weight, would lead to a test statistic with poor power properties because it gives equal weight to the f stratum that is characterized by a severe imbalance in the proportions of treated and control units.

An alternative choice for λ is motivated by considering against which alternative hypotheses we would like our test statistic to have power. Often an important alternative hypothesis has a treatment effect that is constant both within and between strata. To obtain a more attractive choice for λ based on this perspective, it is useful to consider the sampling variances of the two stratum-specific statistics, $T^{\mathrm{dif}}(f)$ and $T^{\mathrm{dif}}(m)$, under Neyman's repeated sampling perspective. Applying the results from Chapter 5, we find that under the randomization distribution, the sampling variance of the two within-stratum estimates of the average treatment effects are

$$\mathbb{V}_{W}\left(\overline{Y}_{\mathrm{t}}^{\mathrm{obs}}(f) - \overline{Y}_{\mathrm{c}}^{\mathrm{obs}}(f) \right) = \frac{S_{\mathrm{t}}^{2}(f)}{N_{\mathrm{t}}(f)} + \frac{S_{\mathrm{c}}^{2}(f)}{N_{\mathrm{c}}(f)} - \frac{S_{\mathrm{ct}}(f)^{2}}{N(f)},$$

and

$$\mathbb{V}_W\left(\overline{Y}_{\text{t}}^{\text{obs}}(m) - \overline{Y}_{\text{c}}^{\text{obs}}(m)\right) = \frac{S_{\text{t}}^2(m)}{N_{\text{t}}(m)} + \frac{S_{\text{c}}^2(m)}{N_{\text{c}}(m)} - \frac{S_{\text{ct}}^2(m)}{N(m)}.$$

Suppose that, within the strata, the treatment effects are constant. In that case, $S_{\text{ct}}^2(f) = S_{\text{ct}}^2(m) = 0$, and the last term drops from both expressions. Assume, in addition, that all four variances $S_{\text{c}}^2(f)$, $S_{\text{t}}^2(f)$, $S_{\text{c}}^2(m)$, and $S_{\text{t}}^2(m)$ are equal to S^2. Then the sampling variances of the two observed differences are

$$\mathbb{V}_W\left(\overline{Y}_{\text{t}}^{\text{obs}}(f) - \overline{Y}_{\text{c}}^{\text{obs}}(f)\right) = S^2 \cdot \left(\frac{1}{N_{\text{t}}(f)} + \frac{1}{N_{\text{c}}(f)}\right),$$

and

$$\mathbb{V}_W\left(\overline{Y}_{\text{t}}(m) - \overline{Y}_{\text{c}}(m)\right) = S^2 \cdot \left(\frac{1}{N_{\text{t}}(m)} + \frac{1}{N_{\text{c}}(m)}\right).$$

In that case, a sensible choice for λ would be the value that maximizes precision by weighting the two statistics by the inverse of their sampling variances, or

$$\lambda_{\text{opt}} = \frac{1}{\frac{1}{N_{\text{t}}(f)} + \frac{1}{N_{\text{c}}(f)}} \bigg/ \left(\frac{1}{\frac{1}{N_{\text{t}}(m)} + \frac{1}{N_{\text{c}}(m)}} + \frac{1}{\frac{1}{N_{\text{c}}(m)} + \frac{1}{N_{\text{t}}(m)}}\right)$$

$$= \frac{N(f) \cdot \frac{N_{\text{t}}(f)}{N(f)} \cdot \frac{N_{\text{c}}(f)}{N(f)}}{N(f) \cdot \frac{N_{\text{t}}(f)}{N(f)} \cdot \frac{N_{\text{c}}(f)}{N(f)} + N(m) \cdot \frac{N_{\text{t}}(m)}{N(m)} \cdot \frac{N_{\text{c}}(m)}{N(m)}},$$

with the weight for each stratum proportional to the product of the stratum size and the stratum proportions of treated and control units. The statistic $T^{\text{dif},\lambda_{\text{opt}}}$ often leads to a test statistic that is more powerful against alternatives with a constant treatment effect than $T^{\text{dif},\lambda_{\text{RSS}}}$, especially in settings with substantial variation in stratum-specific proportions of treated units.

We also could have used the exact same statistics we used in Chapter 5. For example, in the setting of a completely randomized experiment, a natural statistic was the difference between average observed treated and control outcomes:

$$T^{\text{dif}} = \left|\overline{Y}_{\text{t}}^{\text{obs}} - \overline{Y}_{\text{c}}^{\text{obs}}\right|.$$

In the current setting of stratified experiments, with two strata, this statistic can be written as

$$T^{\text{dif}} = \left|\frac{1}{N_{\text{t}}(f) + N_{\text{t}}(m)} \sum_{i=1}^{N} W_i \cdot Y_i^{\text{obs}} - \frac{1}{N_{\text{c}}(f) + N_{\text{c}}(m)} \sum_{i=1}^{N} (1 - W_i) \cdot Y_i^{\text{obs}}\right|.$$

Then we can write this statistic as

$$T^{\text{dif}} = \left| \frac{N_{\text{t}}(f)}{N_{\text{t}}} \cdot \overline{Y}_{\text{t}}^{\text{obs}}(f) - \frac{N(f) - N_{\text{t}}(f)}{N_{\text{c}}} \cdot \overline{Y}_{\text{c}}^{\text{obs}}(f) + \frac{N_{\text{t}}(m)}{N_{\text{t}}} \cdot \overline{Y}_{\text{t}}^{\text{obs}}(m) - \frac{N_{\text{c}}(m)}{N_{\text{c}}} \cdot \overline{Y}_{\text{c}}^{\text{obs}}(m) \right|.$$

This statistic T^{dif} is a valid statistic for testing from the FEP perspective but somewhat unnatural in the current context. Because of Simpson's paradox, one would not always expect small values for the statistic, even when the null hypothesis holds. Suppose that the null hypothesis of zero treatment effects for all units holds and that the potential outcomes are closely associated with the covariate that determines the strata, for example, $Y_i(0) = Y_i(1) = X_i$ for all units ($Y_i(0) = Y_i(1) = 1$ for units with $X_i = 1$ and $Y_i(0) = Y_i(1) = 2$ for units with $X_i = 2$). In that case, the statistic T^{dif} is equal to

$$T^{\text{dif}} = \left| \frac{N_{\text{t}}(f)}{N_{\text{t}}} \cdot 1 - \frac{N(f) - N_{\text{t}}(f)}{N_{\text{c}}} \cdot 1 + \frac{N_{\text{t}}(m)}{N_{\text{t}}} \cdot 2 - \frac{N_{\text{c}}(m)}{N_{\text{c}}} \cdot 2 \right|.$$

If $N_f = 10$, $N_{\text{t}}(f) = 5$, $N(m) = 20$, and $N_{\text{t}}(m) = 5$, this is equal to

$$T^{\text{dif}} = \left| \frac{5}{10} \cdot 1 - \frac{5}{20} \cdot 1 + \frac{5}{10} \cdot 2 - \frac{15}{20} \cdot 2 \right| = \left| \frac{1}{2} + 1 - \frac{1}{4} - \frac{3}{2} \right| = \frac{1}{4}.$$

Under the sharp null hypothesis of no causal effects, the statistic $\overline{Y}_{\text{t}}^{\text{obs}} - \overline{Y}_{\text{c}}^{\text{obs}}$ no longer has expectation equal to zero, whereas it did have expectation zero in the completely randomized experiment. Nevertheless, T^{dif} is still a function of assignments, observed outcomes, and covariates, and as such its distribution under the null hypothesis can be tabulated, and p-values can be calculated.

Finally, let us consider rank-based statistics. In the setting with a completely randomized experiment we focused on the difference in average ranks. In that case we defined the normalized rank R_i (allowing for ties) as

$$R_i = \sum_{j=1}^{N} \mathbf{1}_{Y_j^{\text{obs}} < Y_i^{\text{obs}}} + \frac{1}{2} \left(1 + \sum_{j=1}^{N} \mathbf{1}_{Y_j^{\text{obs}} = Y_i^{\text{obs}}} \right) - \frac{N+1}{2}.$$

Given the N ranks R_i, $i = 1, \ldots, N$, an obvious test statistic is the absolute value of the difference in average ranks for treated and control units:

$$T^{\text{rank}} = \left| \overline{R}_{\text{t}} - \overline{R}_{\text{c}} \right|, \quad \text{where} \quad \overline{R}_{\text{t}} = \frac{1}{N_{\text{t}}} \sum_{i:W_i=1} R_i, \quad \text{and} \quad \overline{R}_{\text{c}} = \frac{1}{N_{\text{c}}} \sum_{i:W_i=0} R_i,$$

where $\overline{R}_{\text{t}}$ and $\overline{R}_{\text{c}}$ are the average rank in the treatment and control groups respectively. Although we can use this statistic for the FEP approach, this would not be attractive if there is substantial variation between strata. We therefore propose modifying this statistic for the setting of a stratified randomized experiment. Let R_i^{strat} be the normalized within-stratum rank of the observed outcome for unit i:

$$
R_i^{\text{strat}} = \begin{cases} \sum_{j:G_i=f} \mathbf{1}_{Y_j^{\text{obs}} < Y_i^{\text{obs}}} + \frac{1}{2}\left(1 + \sum_{j:G_i=f} \mathbf{1}_{Y_j^{\text{obs}} = Y_i^{\text{obs}}}\right) - \frac{N(f)+1}{2}, & \text{if } G_i = f, \\[3mm] \sum_{j:G_i=m} \mathbf{1}_{Y_j^{\text{obs}} < Y_i^{\text{obs}}} + \frac{1}{2}\left(1 + \sum_{j:G_i=m} \mathbf{1}_{Y_j^{\text{obs}} = Y_i^{\text{obs}}}\right) - \frac{N(m)+1}{2}, & \text{if } G_i = m. \end{cases}
$$

Then we can use the average value of the within-stratum ranks for treated and control units:

$$
T^{\text{rank,stratum}} = \left| \overline{R}_{\text{t}}^{\text{strat}} - \overline{R}_{\text{c}}^{\text{strat}} \right|,
$$

where

$$
\overline{R}_{\text{t}}^{\text{strat}} = \frac{1}{N_{\text{t}}} \sum_{i:W_i=1} R_i^{\text{strat}}, \quad \text{and} \quad \overline{R}_{\text{c}}^{\text{strat}} = \frac{1}{N_{\text{c}}} \sum_{i:W_i=0} R_i^{\text{strat}}.
$$

9.4.2 The FEP Approach with J Strata

Most of the statistics discussed in the previous section extend naturally to the case with J strata. Define for a general J-component vector λ the statistic

$$
T^{\text{dif},\lambda} = \left| \sum_{j=1}^{J} \lambda(j) \cdot \left(\overline{Y}_{\text{t}}^{\text{obs}}(j) - \overline{Y}_{\text{c}}^{\text{obs}}(j) \right) \right|.
$$

The first natural choice for λ has $\lambda(j)$ proportional to the stratum size,

$$
\lambda(j) = \frac{N(j)}{N}, \quad \text{leading to} \quad T^{\text{dif},\lambda_{\text{RSS}}} = \left| \sum_{j=1}^{J} \frac{N(j)}{N} \cdot \left(\overline{Y}_{\text{t}}^{\text{obs}}(j) - \overline{Y}_{\text{c}}^{\text{obs}}(j) \right) \right|.
$$

The second choice for λ minimizes the sampling variance of the contrast between treated and control averages under homoskedasticity, leading to

$$
\lambda^{\text{opt}}(j) = \frac{N(j) \cdot \frac{N_{\text{t}}(j)}{N(j)} \cdot \frac{N_{\text{c}}(j)}{N(j)}}{\sum_{k=1}^{J} N(k) \cdot \frac{N_{\text{t}}(k)}{N(k)} \cdot \frac{N_{\text{c}}(k)}{N(k)}},
$$

in turn leading to

$$
T^{\text{dif},\lambda_{\text{opt}}} = \left| \frac{1}{\sum_{j=1}^{J} N(j) \cdot \frac{N_{\text{t}}(j)}{N(j)} \cdot \frac{N_{\text{c}}(j)}{N(j)}} \sum_{j=1}^{J} N(j) \cdot \frac{N_{\text{t}}(j)}{N(j)} \cdot \frac{N_{\text{c}}(j)}{N(j)} \cdot \left(\overline{Y}_{\text{t}}^{\text{obs}}(j) - \overline{Y}_{\text{c}}^{\text{obs}}(j) \right) \right|.
$$

For the modified rank statistic, we define R_i^{strat} to be the normalized within-stratum rank of the observed outcome for unit i, taking account of ties:

$$R_i^{\text{strat}} = \sum_{i':G_{i'}=G_i} \mathbf{1}_{Y_{i'}^{\text{obs}}<Y_i^{\text{obs}}} + \frac{1}{2}\left(1 + \sum_{i':G_{i'}=G_i} \mathbf{1}_{Y_{i'}^{\text{obs}}=Y_i^{\text{obs}}}\right) - \frac{N(G_i)+1}{2}.$$

Then we can use the average value of the within-stratum ranks for treated and control units:

$$T^{\text{rank,stratum}} = \left|\overline{R}_t^{\text{strat}} - \overline{R}_c^{\text{strat}}\right|,$$

where, as before, $\overline{R}_t^{\text{strat}}$ and $\overline{R}_c^{\text{strat}}$ are the averages of the normalized within-stratum ranks for treated and control units.

9.4.3 The FEP Approach with Class-Level Data from Project Star

We now analyze the Project Star data using the FEP approach. Let $B_i(j)$, $i = 1, \ldots, 68$, $j = 1, \ldots, 13$ be an indicator for unit (i.e., teacher) i being from stratum (school) j. For the thirteen schools with two classes of each type, there are $\binom{4}{2} = 6$ different possible assignments. For the two schools with three small classes and two regular classes, there are $\binom{5}{2} = 10$ different possible assignments, and for the one school with four small and two regular classes, there are $\binom{6}{2} = 15$ different possible assignments. Hence, the total number of assignments of teachers to class type with positive probability is $(6^{13}) \times 10^2 \times 15 \approx 2 \times 10^{13}$. We therefore use numerical methods to approximate the p-values for the FEP approach.

We focus in this section on the null hypothesis that there is no effect of class size on the average test score that a teacher would achieve for their students,

$$H_0 : \quad Y_i(0) = Y_i(1), \quad \text{for all } i = 1, \ldots, 68,$$

in any of the sixty-eight classes. We consider four test statistics based on the stratified class-level data. (Recall that the p-value has a valid interpretation only if one statistic is specified *a priori*, and our exercise is for illustrative purposes only.) The first test statistic is the absolute value of the difference in the average mathematics scores between small (treated) and regular-sized (control) classes:

$$T^{\text{dif}} = \left|\overline{Y}_t^{\text{obs}} - \overline{Y}_c^{\text{obs}}\right|.$$

As was discussed before, this statistic, which is natural in a completely randomized experiment, is not natural in this setting because one would not necessarily expect small values even when the null hypothesis is true (especially if there is substantial variation of the shares of treated units within the strata), although the results of the test are valid. The value of the statistic in the sample is 0.224. The p-value, here calculated as the probability under the randomization distribution of finding a value of the statistic at least as large as 0.224, is $p = 0.034$, thereby suggesting that it is unlikely that the students of teachers assigned to the small classes had the same average test scores as the students of teachers assigned to large classes.

The second statistic is the average of the sixteen within-school average differences between small and regular class mathematics scores, weighted by the number of classes in the schools $N(j)$, divided by the total number of classes, $N = 68$:

$$T^{\text{dif},\lambda_{\text{RSS}}} = \left| \sum_{j=1}^{J} \frac{N(j)}{N} \cdot \left(\overline{Y}_{\text{t}}^{\text{obs}}(j) - \overline{Y}_{\text{c}}^{\text{obs}}(j) \right) \right|.$$

The realized value of the test statistic is 0.241. The p-value, now the probability under the randomization distribution of finding a value of the statistic at least as large as 0.241, is $p = 0.023$. This statistic also suggests that the teachers with smaller classes had different average test scores than teachers with regular-sized classes.

The third statistic also weights the within-school average differences, but now the weights are proportional to the product of the number of classes in each school and the proportions of treated and control classes within each school:

$$T^{\text{dif},\lambda_{\text{opt}}} = \left| \frac{1}{\sum_{j=1}^{J} \frac{N(j)}{N} \cdot \frac{N_{\text{t}}(j)}{N(j)} \cdot \frac{N_{\text{c}}(j)}{N(j)}} \sum_{j=1}^{J} \frac{N(j)}{N} \cdot \frac{N_{\text{t}}(j)}{N(j)} \cdot \frac{N_{\text{c}}(j)}{N(j)} \cdot \left(\overline{Y}_{\text{t}}^{\text{obs}}(j) - \overline{Y}_{\text{c}}^{\text{obs}}(j) \right) \right|.$$

Especially when there is considerable variation in the proportion of treated and control units between strata, this statistic is expected to be more powerful against alternative hypotheses with constant additive treatment effects. The realized value of the test statistic is 0.238, with a corresponding p-value of 0.025, leading to essentially the same substantive conclusion as that based on the previous two statistics.

In the current application, these three test-statistics lead to very similar p-values. This is partly because most of the schools have two classes of each type. If there were more dispersion in the fraction of small classes by school and in the number of classes per school, the results could well differ more for the three statistics. The value of the rank-based test $T^{\text{rank},\text{stratum}}$ is 0.48, leading to a p-value of 0.15. Because the outcomes themselves are averages (over students within the classes), there are few outliers, and in this case, the rank-based tests would not be expected to have an advantage over statistics based on simple averages.

Another interesting test statistic here is based on the variation in average mathematics scores in small and regular classes. Suppose that at the individual-student level, it makes no difference to students whether they have many or few classmates, that is, whether they are in a regular or small class. In that case, the expected value of the average mathematics score in regular and small classes should be the same. However, because in small classes the average is calculated over fewer students than in large classes, the small class averages should have a larger variance. More precisely, if the individual test scores have a mean μ and variance σ^2, then the average in a class of size K should have mean μ and variance σ^2/K. So, even if individual student scores are not affected by class size, the null hypothesis that at the teacher level the average test score is not affected by the class size need not be true. We can investigate this phenomenon by choosing a new test statistic.

Now calculate for each school and class type the difference between the highest and the lowest average score:

$$\Delta_{\rm c}(j) = \max_{i:W_i=0,G_i=j} Y_i^{\rm obs} - \min_{i:W_i=0,G_i=j} Y_i^{\rm obs},$$

and

$$\Delta_{\rm t}(j) = \max_{i:W_i=1,G_i=j} Y_i^{\rm obs} - \min_{i:W_i=1,G_i=j} Y_i^{\rm obs}.$$

(For the schools with two small classes, this amounts to the absolute value of the difference between the two small classes.) We then take, for each school, the difference between this difference for small and regular classes:

$$\Delta(j) = \Delta_{\rm t}(j) - \Delta_{\rm c}(j).$$

We then average these differences over all 16 schools, weighted by the number of classes in each school:

$$T^{\rm range} = \frac{1}{N} \sum_{j=1}^{J} N(j) \cdot \Delta(j).$$

We find that the range does, indeed, on average appear to be larger in the small classes than in the regular classes, with the realized value of the test statistic equal to 0.226. The p-value based on the FEP calculations is 0.109. Thus there is only limited evidence against the null hypothesis that the variation in average scores differs between small and regular-sized classes.

9.4.4 The FEP Approach with Student-Level Data from Project Star

Here we consider an alternative analysis of the Project Star data, using the student-level data. This analysis is specific to the FEP approach and the particular structure of the Project Star data, and is not generally applicable to stratified randomized experiments. We present it here to show the richness of the FEP approach. This section can be bypassed without loss of continuity.

The key issue is that for this analysis, the no-interference part of the stability assumption, SUTVA, is automatically satisfied. More precisely, under the null hypothesis of no effects whatsoever, the no-interference assumption holds automatically, but it need not hold under the alternative hypothesis. Recall that the experiment assigned students and teachers randomly to the classes. Without the no-interference assumption, we index potential outcomes by the assignment vector that describes the class and teacher pair for each student. The discussion in this section is relatively informal. In Appendix A we present a more formal discussion of this example, which requires substantial new notation, which is not used in the rest of the text.

First consider the data from a single stratum, in this application a school, say school j. This school has $N(j)$ students and $P(j)$ teachers and classes. These students and teachers will be randomly assigned to $P(j)$ classes, with the class size for class s equal to $M_s(j)$.

The class sizes must add to the school size, or $\sum_{s=1}^{P(j)} M_s(j) = N(j)$. The total number of ways one can select the students, given class sizes, is

$$\prod_{s=1}^{P(j)-1} \binom{N(j) - \sum_{t<s} M_t(j)}{M_s(j)}.$$

The $P(j)$ teachers can be assigned to the $P(j)$ classes in $P(j)!$ ways, so the total number of ways the students and teachers for school j can be assigned to classes is

$$\prod_{s=1}^{P(j)-1} \binom{N(j) - \sum_{t<s} M_t(j)}{M_s(j)} \cdot P(j)!.$$

For each student this is the total number of potential outcomes. The basis for the randomization distribution is this set of assignments, which are all equally likely. The total number of assignments is obtained by multiplying this for each school, across all schools:

$$\prod_{j=1}^{J} \prod_{s=1}^{S(j)-1} \binom{N(j) - \sum_{t<s} M_t(j)}{M_s(j)} \cdot P(j)!.$$

The null hypothesis we consider is that of no effect whatsoever, against the alternative hypothesis that some potential outcomes differ. The test statistic we use is the average over the schools of the average student score for students in small classes minus the average student score for students in regular-sized classes.

$$T^{\text{student}} = \left| \frac{1}{\sum_{j=1}^{J} \frac{N(j)}{N} \cdot \frac{N_c(j)}{N(j)} \cdot \frac{N_t(j)}{N(j)}} \cdot \sum_{j=1}^{J} \frac{N(j)}{N} \cdot \frac{N_c(j)}{N(j)} \cdot \frac{N_t(j)}{N(j)} \cdot \left(\overline{Y}_t(j)^{\text{obs}} - \overline{Y}_c(j)^{\text{obs}} \right) \right|,$$

with the stratum weight equal to

$$\frac{N(j)}{N} \cdot \frac{N_c(j)}{N(j)} \cdot \frac{N_t(j)}{N(j)}.$$

In the sample, the statistic is 0.242, with a p-value < 0.001. Thus we get much stronger evidence against this null hypothesis than we did for the null hypothesis using class-level data.

Now let us compare this analysis to that based on teacher-level data. If we were to maintain the no-interference assumption at the student level, the new null hypothesis requires only that changing student i's assignment from a regular to a small class does not change the outcome. In that case the student-level test score will tend to be more powerful than the class-level average test score, and the former would be preferable to the latter. However, in this application, the student-level stability assumption is a very strong and tenuous one to make. It is very plausible that there are interactions between children that would violate this assumption. Hence, even clear rejections of the null hypothesis of no differences by teacher assignment would not necessarily be credible evidence of systematic effects of class *size* – it may simply indicate the presence of effects of teachers or peers. In contrast, the teacher-level assessment does not rely on within-class, no-interference assumptions, and so clear evidence against the null hypothesis of no effect based on that assessment is more credible evidence of class-size effects.

9.5   THE ANALYSIS OF STRATIFIED RANDOMIZED EXPERIMENTS FROM NEYMAN'S REPEATED SAMPLING PERSPECTIVE

The results in Chapter 6 for a completely randomized experiment can be used to analyze data within a stratum. Specifically, within each stratum those results can be used to obtain an estimate of the average treatment effect and to obtain a conservative estimator of the repeated sampling variance of this estimator.

9.5.1 The Two-Stratum Case

Initially we focus on the simple example with two strata and apply the framework to the Project Star data in Section 9.5.2. For the first stratum, the natural unbiased estimator for the average treatment effect $\tau_{\mathrm{fs}}(f)$ is

$$\hat{\tau}^{\mathrm{dif}}(f) = \overline{Y}_{\mathrm{t}}^{\mathrm{obs}}(f) - \overline{Y}_{\mathrm{c}}^{\mathrm{obs}}(f) = \frac{1}{N_{\mathrm{t}}(f)} \sum_{i:G_i=f} W_i \cdot Y_i^{\mathrm{obs}} - \frac{1}{N_{\mathrm{c}}(f)} \sum_{i:G_i=f} (1 - W_i) \cdot Y_i^{\mathrm{obs}}.$$

The sampling variance of this estimator, under the randomization distribution, is

$$\mathbb{V}_W\left(\hat{\tau}^{\mathrm{dif}}(f)\right) = \frac{S_{\mathrm{c}}^2(f)}{N_{\mathrm{c}}(f)} + \frac{S_{\mathrm{t}}^2(f)}{N_{\mathrm{t}}(f)} - \frac{S_{\mathrm{ct}}^2(f)}{N(f)},$$

with analogous expressions for the estimator for the average treatment effect in the second stratum and its sampling variance. However, we are not necessarily interested in the two within-stratum average treatment effects. More commonly, we are interested in a weighted average of the two within-stratum average effects. A natural estimand is the finite-sample average treatment effect,

$$\tau_{\mathrm{fs}} = \frac{N(f)}{N(f) + N(m)} \cdot \tau_{\mathrm{fs}}(f) + \frac{N(m)}{N(f) + N(m)} \cdot \tau_{\mathrm{fs}}(m) = \frac{1}{N} \sum_{i=1}^N \left(Y_i(1) - Y_i(0)\right).$$

With fixed stratum sizes, unbiasedness of the two within-stratum estimators implies unbiasedness of

$$\hat{\tau}^{\mathrm{strat}} = \frac{N(f)}{N(f) + N(m)} \cdot \hat{\tau}^{\mathrm{dif}}(f) + \frac{N(m)}{N(f) + N(m)} \cdot \hat{\tau}^{\mathrm{dif}}(m),$$

for the population average treatment effect τ_{fs}. Similarly, the assumption that the randomizations in the two strata are independent, formalized in the assignment mechanism, implies that the two estimators are uncorrelated, and thus

$$
\mathbb{V}_W\left(\hat{\tau}^{\text{strat}}\right) = \left(\frac{N(f)}{N(f) + N(m)}\right)^2 \cdot \mathbb{V}_W(\hat{\tau}_f) + \left(\frac{N(m)}{N_f + N(m)}\right)^2 \cdot \mathbb{V}_W(\hat{\tau}_m)
$$

$$
= \left(\frac{N(f)}{N(f) + N(m)}\right)^2 \cdot \left(\frac{S_c(f)^2}{N_c(f)} + \frac{S_t^2(f)}{N_t(f)} - \frac{S_{ct}^2(f)}{N(f)}\right)
$$

$$
+ \left(\frac{N(m)}{N(f) + N(m)}\right)^2 \cdot \left(\frac{S_c(m)^2}{N_c(m)} + \frac{S_t^2(m)}{N_t(m)} - \frac{S_{ct}^2(m)}{N(m)}\right).
$$

The same issues that were discussed in Chapter 6 arise here in estimating this sampling variance. There is no direct way to estimate the components of this sampling variance involving the covariance of the unit-level potential outcomes, so typically those terms are ignored to obtain an estimated upper bound on the sampling variance by simply estimating the two within-stratum sampling variances:

$$
\hat{\mathbb{V}}^{\text{neyman}} = \left(\frac{N(f)}{N(f) + N(m)}\right)^2 \cdot \left(\frac{s_c^2(f)}{N_c(f)} + \frac{s_t^2(f)}{N_t(f)}\right)
$$

$$
+ \left(\frac{N(m)}{N(f) + N(m)}\right)^2 \cdot \left(\frac{s_c^2(m)}{N_c(m)} + \frac{s_t^2(m)}{N_t(m)}\right).
$$

This estimate of the sampling variance is unbiased if the within-stratum treatment effects are constant and additive, and overestimates the sampling variance in expectation otherwise. Note that we do not need to make assumptions about the variation in treatment effects between strata.

So far in this section, the discussion has focused on the estimation of the population average treatment effect, τ_{fs}. In some cases we may be interested in a different weighted average of the within-strata treatment effects. For example, we may be interested in the average effect of the treatment on the outcome for the units who received the treatment. Given the random assignment, and within the strata, this effect is equal to $\tau_{\text{fs}}(f)$ and $\tau_{\text{fs}}(m)$, respectively. Within each stratum this is, in expectation, the same as the average effect for the full stratum. However, when the proportions of treated units differ between the strata, the weights have to be adjusted to obtain an unbiased estimate of the average effect of the treatment on the units who received treatment. The appropriate weights are proportional to the fraction of treated units in each strata, leading to the estimand

$$
\tau_{\text{fs},t} = \frac{N_t(f)}{N_t(f) + N_t(m)} \cdot \tau_{\text{fs}}(f) + \frac{N_t(m)}{N_t(f) + N_t(m)} \cdot \tau_{\text{fs}}(m),
$$

and thus to the natural unbiased estimator

$$
\hat{\tau}_t^{\text{strat}} = \frac{N_t(f)}{N_t(f) + N_t(m)} \cdot \hat{\tau}^{\text{dif}}(f) + \frac{N_t(m)}{N_t(f) + N_t(m)} \cdot \hat{\tau}^{\text{dif}}(m).
$$

The sampling variance of $\hat{\tau}_t$ can be estimated in the same way as the sampling variance for the population average treatment effect, modifying the weights to reflect the new estimand:

$$\hat{\mathbb{V}}_t^{\text{neyman}} = \left(\frac{N_t(f)}{N_t(f) + N_t(m)} \right)^2 \cdot \left(\frac{s_c(f)^2}{N_c(f)} + \frac{s_t^2(f)}{N_t(f)} \right)$$

$$+ \left(\frac{N_t(m)}{N_t(f) + N_t(m)} \right)^2 \cdot \left(\frac{s_c^2(m)}{N_c(m)} + \frac{s_t^2(m)}{N_t(m)} \right).$$

More generally we can look at other weighted averages, such as the average effect for those who did not receive the treatment, but such averages are often more difficult to motivate as relevant.

Using Neyman's repeated sampling approach, we can also investigate other estimands, such as the differences between the stratum-specific average treatment effects. A natural unbiased estimator for the difference between $\tau_{\text{fs}}(m)$ and $\tau_{\text{fs}}(f)$ is

$$\hat{\tau}^{\text{dif}}(m) - \hat{\tau}^{\text{dif}}(f) = \left(\overline{Y}_t^{\text{obs}}(m) - \overline{Y}_c^{\text{obs}}(m) \right) - \left(\overline{Y}_t^{\text{obs}}(f) - \overline{Y}_c^{\text{obs}}(f) \right).$$

This estimator is unbiased for the difference in average treatment effects with sampling variance

$$\mathbb{V}_W \left(\hat{\tau}^{\text{dif}}(m) - \hat{\tau}^{\text{dif}}(f) \right) = \frac{S_c^2(f)}{N_c(f)} + \frac{S_t^2(f)}{N_t(f)} - \frac{S_{ct}^2(f)}{N(f)} + \frac{S_c^2(m)}{N_c(m)} + \frac{S_t^2(m)}{N_t(m)} - \frac{S_{ct}^2(m)}{N(m)}.$$

An estimator for the upper bound on this sampling variance is

$$\hat{\mathbb{V}}^{\text{neyman}} \left(\hat{\tau}^{\text{dif}}(m) - \hat{\tau}^{\text{dif}}(f) \right) = \frac{s_c^2(f)}{N_c(f)} + \frac{s_t^2(f)}{N_t(f)} + \frac{s_c^2(m)}{N_c(m)} + \frac{s_t^2(m)}{N_t(m)}.$$

We can use any of the estimated sampling variances and the associated unbiased estimators to construct large-sample confidence intervals for the associated estimator.

9.5.2 The Neyman Approach and Project Star

Next, let us consider point estimates and confidence intervals for the average effect of the class size based on the stratified experiment. First we present estimates that account for the stratification. For each school j, for $j = 1, \ldots, 16$, the average effect of the treatment and its corresponding sampling variance are estimated as

$$\hat{\tau}^{\text{dif}}(j) = \overline{Y}_t^{\text{obs}}(j) - \overline{Y}_c^{\text{obs}}(j), \quad \text{and} \quad \hat{\mathbb{V}}^{\text{neyman}}(j) = \frac{s_c(j)^2}{N_c(j)} + \frac{s_t(j)^2}{N_t(j)},$$

respectively. For each school, the estimated average effect and the square root of the estimated sampling variance are reported in Table 9.2. The population average effect is estimated as

$$\hat{\tau}^{\text{strat}} = \sum_{j=1}^{J} \frac{N(j)}{N} \cdot \hat{\tau}(j) = 0.241,$$

Table 9.2. *Within-School Estimates of Treatment Effect of Small Classes Relative to Regular Classes – Project Star*

School	Estimated Effect	$\widehat{(\text{s.e.})}$
1	0.223	(0.230)
2	−0.295	(0.776)
3	0.417	(0.404)
4	0.748	(0.215)
5	−0.077	(0.206)
6	1.655	(0.405)
7	−0.254	(0.255)
8	0.429	(0.306)
9	−0.006	(0.311)
10	−0.014	(0.182)
11	−0.003	(0.605)
12	0.222	(0.309)
13	0.432	(0.179)
14	0.340	(0.336)
15	0.207	(0.396)
16	−0.306	(0.245)
$\hat{\tau}^{\,\text{strat}}$	0.241	(0.092)

and its sampling variance by

$$\hat{\mathbb{V}}^{\text{neyman}} = \sum_{j=1}^{J} \left(\frac{N(j)}{N} \right)^2 \cdot \hat{\mathbb{V}}^{\text{neyman}}(j) = 0.092^2.$$

Hence the large sample 95% confidence interval for the average effect is

$$\text{CI}^{0.95}(\tau_{\text{fs}}) = \left(0.061, 0.421 \right).$$

It is interesting to compare this point estimate and its associated standard error to that based on the analysis using the (incorrect) assumption that the data arose from a completely randomized experiment. The point estimate of the average effect is then $\hat{\tau}^{\text{dif}} = \overline{Y}_{\text{t}}^{\text{obs}} - \overline{Y}_{\text{c}}^{\text{obs}} = 0.224$, with an estimated standard error of 0.141, leading to a large sample 95% confidence interval of (−0.053, 0.500). This estimator of the sampling variance is biased if there is variation in the probability of treatment between the different strata, or if there is variation in the average potential outcomes by stratum. We know the former is the case, with the probability of a small class equal to 0.5 in most schools, and equal to 0.60 and 0.67 in some schools. Assessing the latter issue is more complicated, and we shall return to this in Section 9.7.2. The fact that the point estimates differ under the assumptions of a completely randomized experiment and a stratified randomized experiment suggests that average potential outcomes also differ between strata. The estimated standard error for the stratification-based analysis is smaller than that for the completely randomized experiment, suggesting, again, that average potential outcomes differ between strata, which implies that there is a gain in precision from the stratification.

9.6 REGRESSION ANALYSIS OF STRATIFIED RANDOMIZED EXPERIMENTS

In order to interpret regression-based estimators, we take a super-population perspective with a fixed number of strata, and an infinite number of units within each stratum. Because there are few notational simplifications from considering the special case with only two strata, we look in this section immediately at the general situation with J strata.

9.6.1 The General Framework

Let $q(j) = N(j)/N$ and $e(j) = N_t(j)/N(j)$ be the proportion of each stratum in the sample from the infinite super-population, and the proportion of treated units in each stratum, or the propensity score, respectively. We consider two specifications of the regression function in this case. The first specification of the regression function treats the stratum indicators as additional regressors and includes them additively. The second specification includes a full set of interactions of the stratum indicators with the treatment indicator. We then investigate the large-sample properties of the least squares estimators of the coefficients on the treatment indicator.

Similar to the regression function specifications in Chapter 7, the first specification simply includes indicators for the strata additively in addition to the indicator for the treatment:

$$Y_i^{\text{obs}} = \tau \cdot W_i + \sum_{j=1}^{J} \beta(j) \cdot B_i(j) + \varepsilon_i, \tag{9.1}$$

where $B_i(j)$ is an indicator for unit i belonging to stratum j. Because we include, in this specification, a full set of stratum indicators $B_i(j)$, for $j = 1, \ldots, J$, we do not include an intercept in the specification of the regression function. We focus on the least squares estimator for τ,

$$(\hat{\tau}^{\text{ols}}, \hat{\beta}^{\text{ols}}) = \arg\min_{\tau, \beta} \sum_{i=1}^{N} \left(Y_i^{\text{obs}} - \tau \cdot W_i + \sum_{j=1}^{J} \beta(j) \cdot B_i(j) \right)^2. \tag{9.2}$$

As before, we define τ^* and β^* to be the population counterparts to these OLS estimators,

$$(\tau^*, \beta^*) = \arg\min_{\tau, \beta} \mathbb{E}\left[\left(Y_i^{\text{obs}} - \tau \cdot W_i + \sum_{j=1}^{J} \beta(j) \cdot B_i(j) \right)^2 \right]. \tag{9.3}$$

The first question concerns the population value τ^* corresponding to $\hat{\tau}^{\text{ols}}$. In general $\hat{\tau}^{\text{ols}}$ is not consistent for the population average treatment effect τ_{sp}. Instead, it estimates a weighted average of the within-stratum average effects, with weights proportional to the product of the fraction of observations in the stratum and the probabilities of receiving and not receiving the treatment. More specifically,

$$\omega(j) = q(j) \cdot e(j) \cdot (1 - e(j)), \quad \text{and} \quad \tau_\omega = \sum_{j=1}^{J} \omega(j) \cdot \tau_{\text{sp}}(j) \bigg/ \left(\sum_{j=1}^{J} \omega(j) \right), \quad (9.4)$$

where $\tau_{\text{sp}}(j) = \mathbb{E}[Y_i(1) - Y_i(0)|B_i(j) = 1]$. Then $\hat{\tau}^{\text{ols}}$ is consistent for τ_ω. The following theorem formalizes this result.

Theorem 9.1 *Suppose we conduct a stratified randomized experiment in a sample drawn at random from an infinite population. Then, for estimands τ^* and τ_w defined in (9.3) and (9.4), the estimator $\hat{\tau}^{\text{ols}}$ satisfies, (i)*

$$\tau^* = \tau_\omega,$$

and (ii),

$$\sqrt{N} \cdot \left(\hat{\tau}^{\text{ols}} - \tau_\omega \right) \xrightarrow{d}$$

$$\mathcal{N} \left(0, \frac{\mathbb{E}\left[\left(W_i - \sum_{j=1}^{J} q(j) \cdot B_i(j) \right)^2 \cdot \left(Y_i^{\text{obs}} - \tau^* \cdot W_i - \sum_{j=1}^{J} \beta_j^* \cdot B_i(j) \right)^2 \right]}{\left(\sum_{j=1}^{J} q(j) \cdot e(j) \cdot (1 - e(j)) \right)^2} \right).$$

The proof appears in Appendix B.

The weights $\omega(j)$ have an interesting interpretation. Suppose we estimate the within-stratum average treatment effect $\tau^{\text{dif}}(j)$ as $\hat{\tau}^{\text{dif}}(j) = \overline{Y}_{\text{t}}^{\text{obs}}(j) - \overline{Y}_{\text{c}}^{\text{obs}}(j)$. The sampling variance of $\hat{\tau}^{\text{dif}}(j)$, under the assumption of a constant treatment effect, is $(S^2/N) \cdot (q(j) \cdot e(j) \cdot (1 - e(j)))^{-1}$. Hence the weights $\omega(j)$ are proportional to the precision of natural unbiased estimators of the within-stratum treatment effects, which leads to a relatively precisely estimated weighted average effect.

The second specification of the regression function includes a full set of interactions of the stratum indicators with the indicator for the treatment W_i. In order to be able to interpret the coefficient on the treatment indicator as an average causal effect, we include the interactions with the stratum indicators relative to their share in the sample and relative to the indicator for the last stratum:

$$Y_i^{\text{obs}} = \tau \cdot W_i \cdot \frac{B_i(j)}{N(j)/N} + \sum_{j=1}^{J} \beta(j) \cdot B_i(j) + \sum_{j=1}^{J-1} \gamma(j) \cdot W_i \cdot \left(B_i(j) - B_i(J) \cdot \frac{N(j)}{N(J)} \right) + \varepsilon_i.$$

$$(9.5)$$

Note that in this specification we only include the first $J - 1$ interactions to avoid perfect collinearity in the regression function. In this case, the population value τ^*, corresponding to the large sample limit of the least squares estimator $\hat{\tau}^{\text{ols,inter}}$, is equal to the population average treatment effect τ_{sp}.

Theorem 9.2 *Suppose we conduct a stratified randomized experiment in a sample drawn at random from an infinite population. Then, for $\hat{\tau}^{\mathrm{ols,inter}}$ defined as the least squares estimator corresponding to the regression function in (9.5), and τ^* defined as the population limit corresponding to that estimator, (i)*

$$\tau^* = \tau_{\mathrm{sp}},$$

and (ii),

$$\sqrt{N} \cdot \left(\hat{\tau}^{\mathrm{ols,inter}} - \tau_{\mathrm{sp}} \right) \xrightarrow{d} \mathcal{N}\left(0, \sum_{j=1}^{J} q(j)^2 \cdot \left(\frac{\sigma_c^2(j)}{(1 - e(j)) \cdot q(j)} + \frac{\sigma_t^2(j)}{e(j) \cdot q(j)} \right) \right).$$

It is interesting to compare the sampling variance of $\hat{\tau}^{\mathrm{ols}}$ and $\hat{\tau}^{\mathrm{ols,inter}}$. In general, the sampling variance of $\hat{\tau}^{\mathrm{ols,inter}}$ is larger than that of $\hat{\tau}^{\mathrm{ols}}$.

9.6.2 Regression Analysis of Project Star

The first specification of the regression function includes the treatment indicator and the indicators for the blocks:

$$Y_i^{\mathrm{obs}} = \tau \cdot W_i + \sum_{j=1}^{J} \beta(j) \cdot B_i(j) + \varepsilon_i.$$

The point estimate and standard error for τ_{fs} are

$$\hat{\tau}^{\mathrm{ols}} = 0.238 \ (\widehat{\mathrm{s.e.}} \ 0.103).$$

Recall from the discussion in Section 9.6 that this estimator is not necessarily consistent for the average effect of the treatment in the population if there is variation in the effect of the class size by school.

The second specification of the regression function includes indicators for the strata, as well as interactions of the stratum indicators and the treatment indicator:

$$Y_i^{\mathrm{obs}} = \tau \cdot W_i \cdot \frac{B_i(J)}{N(J)/N} + \sum_{j=1}^{J} \beta_j \cdot B_i(j) + \sum_{j=1}^{J-1} \tau(j) \cdot W_i \cdot \left(B_i(j) - B_i(J) \cdot \frac{N(j)}{N(J)} \right) + \varepsilon_i.$$

The point estimate and standard error for τ, based on this specification, are

$$\hat{\tau}^{\mathrm{ols,inter}} = 0.241 \ (\widehat{\mathrm{s.e.}} \ 0.095).$$

The two estimates for the average effect are close, with similar standard errors, consistent with limited heterogeneity in the treatment effects.

9.7   MODEL-BASED ANALYSIS OF STRATIFIED RANDOMIZED EXPERIMENTS

In a model-based analysis, it is conceptually straightforward to take account of the stratification. As in the analysis of completely randomized experiments, we combine the specification of the joint distribution of the potential outcomes with the known distribution of the vector of assignment indicators to derive the posterior distribution of the causal estimand. There is one new issue that arises in this context: the link between the distributions of the potential outcomes in distinct strata.

9.7.1 General Considerations

One can choose to have distinct parameters for the distributions in different strata, that is, independent prior distributions. Alternatively the researcher may wish to link the parameters in the different strata either deterministically by imposing equality restrictions or stochastically through a dependence structure in the prior distribution, that is, for example, through a hierarchical model. In situations with few strata and many units per stratum, one may wish to pursue the first strategy and specify distinct distributions for the potential outcomes in each stratum, with independent prior distributions on the parameters of these distributions. In contrast, in settings with a substantial number of strata, and a modest number of units per stratum, one may wish to link some of the parameters. One can do so by restricting them to be equal, or by incorporating dependence into the specification of the prior distribution.

We make this more specific and illustrate the issues for the case with common and stratum-specific parameters. Suppose we specify the joint distribution of the potential outcomes in stratum j as

$$
\begin{pmatrix} Y_i(0) \\ Y_i(1) \end{pmatrix} \Bigg| B_i(j), \theta \sim \mathcal{N}\left(\begin{pmatrix} \mu_c(j) \\ \mu_t(j) \end{pmatrix}, \begin{pmatrix} \sigma_c^2(j) & 0 \\ 0 & \sigma_t^2(j) \end{pmatrix} \right),
\tag{9.6}
$$

where the means $(\mu_c(j), \mu_t(j))$ and variances $(\sigma_c^2(j), \sigma_t^2(j))$ are specific to stratum j. The full parameter vector is $\theta = (\mu_c(j), \mu_t(j), \sigma_c^2(j), \sigma_t^2(j), w = 0, 1, j = 1, \ldots, J)$.

With few strata and a substantial number of units per stratum, we may wish to use a prior distribution that makes all elements of θ *a priori* independent, for example, using normal prior distributions for the $\mu_c(j)$ and $\mu_t(j)$ and inverse chi-squared prior distributions for the $\sigma_c^2(j)$ and $\sigma_t^2(j)$.

However, if there are many strata and the number of units per stratum is modest, we may wish to specify a hierarchical prior distribution for the means to obtain more precise estimates. For example, we may wish to restrict the variances of the potential outcomes to be the same across strata, σ_c^2 and σ_t^2 for all j, and to specify the means to have a joint normal prior distribution, independent of the variances σ_c^2 and σ_t^2:

$$
\begin{pmatrix} \mu_c(1) \\ \mu_c(2) \\ \vdots \\ \mu_c(J) \\ \mu_t(1) \\ \mu_t(2) \\ \vdots \\ \mu_t(J) \end{pmatrix} \sim \mathcal{N}\left(\begin{pmatrix} \gamma_c \\ \gamma_c \\ \vdots \\ \gamma_c \\ \gamma_t \\ \gamma_t \\ \vdots \\ \gamma \end{pmatrix}, \begin{pmatrix} \eta_c^2 & 0 & \cdots & 0 & \rho\sigma_c\sigma_t & 0 & \cdots & 0 \\ 0 & \eta_c^2 & & \vdots & 0 & \rho\sigma_c\sigma_t & & \vdots \\ \vdots & & \ddots & & \vdots & & \ddots & \\ 0 & \cdots & & \eta_c^2 & 0 & \cdots & & \rho\sigma_c\sigma_t \\ \rho\sigma_c\sigma_t & 0 & \cdots & 0 & \eta_t^2 & 0 & \cdots & 0 \\ 0 & \rho\sigma_c\sigma_t & & \vdots & 0 & \eta_t^2 & & \vdots \\ \vdots & & \ddots & & \vdots & & \ddots & \\ 0 & \cdots & \rho\sigma_c\sigma_t & 0 & \cdots & & & \eta_t^2 \end{pmatrix} \right).
$$

The full parameter vector is now $\theta = (\sigma_c^2, \sigma_t^2, \gamma_c, \gamma_t, \eta_c^2, \eta_t^2, p)$.

9.7.2 A Model-Based Analysis of Project Star

We now conduct a model-based imputation analysis of the Project Star data. The model we consider for the potential outcomes is

$$\begin{pmatrix} Y_i(0) \\ Y_i(1) \end{pmatrix} \Bigg| B_i(j) = 1, \theta \sim \mathcal{N}\left(\begin{pmatrix} \mu_c(j) \\ \mu_t(j) \end{pmatrix}, \begin{pmatrix} \sigma^2 & 0 \\ 0 & \sigma^2 \end{pmatrix} \right),$$

with a common variance σ^2. In addition we assume that the pairs of stratum-specific means $(\mu_c(j), \mu_t(j))$ are independent across strata given the hyperparameters,

$$\begin{pmatrix} \mu_c(j) \\ \mu_t(j) \end{pmatrix} \Bigg| \sigma^2, \gamma_c, \gamma_t, \Sigma \sim \mathcal{N}\left(\begin{pmatrix} \gamma_c \\ \gamma_t \end{pmatrix}, \Sigma \right), \quad \begin{pmatrix} \mu_c(j) \\ \mu_t(j) \end{pmatrix} \perp\!\!\!\perp \begin{pmatrix} \mu_c(k) \\ \mu_t(k) \end{pmatrix} \Bigg| \sigma^2, \gamma_c, \gamma_t, \Sigma, j \neq k.$$

In this model, the two potential outcome means $(\mu_c(j), \mu_t(j))$ are specific to the stratum, and the variance σ^2 is common to all strata and both potential outcomes.

The full parameter vector is $\theta = (\gamma_c, \gamma_t, \Sigma, \sigma^2)$. For the prior distributions, we use conventional proper choices. For the variance parameter σ^2, we use a standard inverse Chi-squared prior distribution,

$$k_0 \cdot v_0^2 \cdot \sigma^{-2} \sim \mathcal{X}^2(k_0), \quad \text{or} \quad \sigma^2 \sim \mathcal{X}^{-2}(k_0, v_0^2),$$

using the notation from Gelman, Carlin, Stern, and Rubin (1995). Our choices for the parameters of the prior distribution are $k_0 = 2$ and $v_0^2 = 0.001$. For γ_c and γ_t, we use independent normal prior distributions,

$$\begin{pmatrix} \gamma_c \\ \gamma_t \end{pmatrix} \sim \mathcal{N}\left(\begin{pmatrix} 0 \\ 0 \end{pmatrix}, \begin{pmatrix} 100^2 & 0 \\ 0 & 100^2 \end{pmatrix} \right).$$

The prior distribution for Σ is an inverse wishart distribution,

$$\Sigma \sim \mathcal{W}^{-1}(k_1, \Gamma_1^{-1}).$$

We consider two pairs of values for (k_1, Γ_1). The first is $k_1 = 1{,}000$, $\Gamma_1 = 1{,}000 \cdot \mathcal{I}_2$, where $\mathcal{I}_k$ is the $k \times k$ identity matrix. This essentially corresponds to removing the link between the parameters in the different strata. We refer to this as the "independent" prior, corresponding to independence between the stratum-specific means. The second choice for (k_1, Γ_1) is $k_1 = 3$ and $\Gamma_1^{-1} = 0.001 \cdot k_1 \cdot \mathcal{I}_2$, which allows the hierarchical structure to influence answers. We refer to this prior distribution as the hierarchical prior.

For the independent prior distribution, the posterior mean and standard deviation are

$$\mathbb{E}[\tau_{fs}|\mathbf{Y}^{obs}, \mathbf{W}, \mathbf{B}, \text{independent}] = 0.241, \quad \mathbb{V}(\tau_{fs}|\mathbf{Y}^{obs}, \mathbf{W}, \mathbf{B}, \text{independent}) = 0.095^2.$$

Substantively it is difficult to see why one would wish to impose the *ex post* independence. Certainly, as we will see, there is strong evidence in the data to suggest that the average potential outcomes within the schools are related.

For the hierarchical prior distribution, the posterior mean and standard deviation are

$$\mathbb{E}[\tau_{fs}|\mathbf{Y}^{obs}, \mathbf{W}, \mathbf{B}, \text{hierarchical}] = 0.235, \quad \mathbb{V}(\tau_{fs}|\mathbf{Y}^{obs}, \mathbf{W}, \mathbf{B}, \text{hierarchical})^2 = 0.107^2.$$

It is also interesting to assess the evidence for variation in average potential outcomes and treatment effects by strata. In order to do so, we inspect the posterior distribution of Σ given the hierarchical prior distribution. The logarithm of the square root of the two diagonal elements corresponds to the logarithm of the standard deviation of $\mu_c(j)$ and $\mu_t(j)$ over the sixteen schools. The posterior means of logarithms of those two standard deviations are

$$\mathbb{E}\left[\ln\left(\sqrt{\Sigma_{11}}\right)\middle|\, \mathbf{Y}^{\text{obs}}, \mathbf{W}, \mathbf{B}, \text{hierarchical}\right] = -1.14,$$

$$\mathbb{V}\left(\ln\left(\sqrt{\Sigma_{11}}\right)\middle|\, \mathbf{Y}^{\text{obs}}, \mathbf{W}, \mathbf{B}, \text{hierarchical}\right) = 0.47^2,$$

and

$$\mathbb{E}\left[\ln\left(\sqrt{\Sigma_{22}}\right)\middle|\, \mathbf{Y}^{\text{obs}}, \mathbf{W}, \mathbf{B}, \text{hierarchical}\right] = -1.08,$$

$$\mathbb{V}\left(\ln\left(\sqrt{\Sigma_{22}}\right)\middle|\, \mathbf{Y}^{\text{obs}}, \mathbf{W}, \mathbf{B}, \text{hierarchical}\right) = 0.45^2.$$

There is clearly some evidence of heterogeneity in the stratum means. However, the heterogeneity is highly correlated across potential outcomes, with the posterior mean for the Fisher Z transformation of the correlation between $\beta_c(j)$ and $\beta_t(j)$ (the $(1, 2)$ element of Σ divided by the square root of the product of the $(1, 1)$ and $(2, 2)$ elements) equal to

$$\mathbb{E}\left[\frac{1}{2}\ln\left(\frac{1 + \Sigma_{12}/(\sqrt{\Sigma_{11}\Sigma_{22}})}{1 - \Sigma_{12}/(\sqrt{\Sigma_{11}\Sigma_{22}})}\right)\middle|\, \mathbf{Y}^{\text{obs}}, \mathbf{W}, \mathbf{B}, \text{hierarchical}\right] = 2.63,$$

and the posterior variance equal to

$$\mathbb{V}\left(\frac{1}{2}\ln\left(\frac{1 + \Sigma_{12}/(\sqrt{\Sigma_{11}\Sigma_{22}})}{1 - \Sigma_{12}/(\sqrt{\Sigma_{11}\Sigma_{22}})}\right)\middle|\, \mathbf{Y}^{\text{obs}}, \mathbf{W}, \mathbf{B}, \text{hierarchical}\right) = 0.67^2.$$

The posterior mean of the correlation itself is 0.96. The average treatment effect in school j is approximately $\tau(j) = \mu_t(j) - \mu_c(j)$. In terms of the parameters, the variance of the treatment effect across the sixteen schools is $(-1\ 1)\Sigma(-1\ 1)' = \Sigma_{11} - \Sigma_{12} - \Sigma_{21} + \Sigma_{22}$. We focus on the square root of this, that is, the standard deviation of the treatment effect over the schools. The posterior mean of the logarithm of the standard deviation of the treatment effect is

$$\mathbb{E}\left[\ln\left(\sqrt{\Sigma_{11} - \Sigma_{12} - \Sigma_{21} + \Sigma_{22}}\right)\middle|\, \mathbf{Y}^{\text{obs}}, \mathbf{W}, \mathbf{B}, \text{hierarchical}\right] = -2.33,$$

with posterior variance

$$\mathbb{V}\left(\ln\left(\sqrt{\Sigma_{11} - \Sigma_{12} - \Sigma_{21} + \Sigma_{22}}\right)\middle|\, \mathbf{Y}^{\text{obs}}, \mathbf{W}, \mathbf{B}, \text{hierarchical}\right) = 0.59^2.$$

Comparing the posterior mean of the standard deviation of the stratum-specific treatment effect $\tau(j)$ over the sixteen strata, (0.115), with the posterior mean of the standard deviation of the stratum-specific level under the control treatment $\mu_c(j)$ over the sixteen strata, (0.349), suggests that, although there is considerable evidence that *levels* of the average test scores vary by school, there is little evidence that average class size *effects* vary much by school. The former may be due to differences in teacher quality or to differences in student populations. This type of conclusion highlights the advantage of a fully model-based analysis, which allows for the simultaneous investigation of multiple questions.

9.8   DESIGN ISSUES: STRATIFIED VERSUS COMPLETELY RANDOMIZED EXPERIMENTS

When designing an experimental evaluation, one may often have the choice between a completely randomized experiment and a stratified randomized experiment. Here we study the implications of the choice between the different experimental designs for the expected sampling variance of the standard unbiased estimator for the average treatment effect. There is a sense in which one is never worse off stratifying on a covariate. However, to make this point precise, we need to pose the question appropriately.

We analyze the problem in a super-population setting. Each unit in this population has a binary characteristic G_i, $G_i \in \{f, m\}$. The proportion of women ($G_i = f$ types) in the population is p. We consider the following two designs. In the first design we randomly draw N units from the population. Out of this sample of size N, we randomly draw $N_t = q \cdot N$ units to receive the active treatment and $N_c = (1 - q) \cdot N$ units to receive the control treatment. Based on the randomized experiment, we estimate the average treatment effect in the super-population as

$$\hat{\tau}^{\text{dif}} = \overline{Y}_t^{\text{obs}} - \overline{Y}_c^{\text{obs}},$$

with (super-population) sampling variance

$$\mathbb{V}_{\text{sp}}(\hat{\tau}^{\text{dif}}) = \frac{\sigma_c^2}{N_c} + \frac{\sigma_t^2}{N_t}.$$

In the second design, we randomly draw $N(f) = p \cdot N$ units from the subpopulation of units who have $G_i = f$, and $N(m) = (1 - p) \cdot N$ units from the population who have $G_i = m$. In the first subsample, we randomly select $N_t(f) = p \cdot q \cdot N$ units to receive the active treatment, and the remaining $N_c(f) = p \cdot (1 - q) \cdot N$ are assigned to receive the control treatment. In the second subsample $N_t(m) = p \cdot q \cdot N$ units are randomly selected to receive the active treatment, and the remaining $N_c(m) = (1 - p) \cdot (1 - q) \cdot N$ units to receive the control treatment. Note that we assign the same proportion of units in each subpopulation to the active treatment. In this experiment, we estimate the average treatment effect within the $G_i = f$ and $G_i = m$ subpopulations as

$$\hat{\tau}^{\text{dif}}(f) = \overline{Y}_t^{\text{obs}}(f) - \overline{Y}_c^{\text{obs}}(f), \quad \text{and} \quad \hat{\tau}^{\text{dif}}(m) = \overline{Y}_t^{\text{obs}}(m) - \overline{Y}_c^{\text{obs}}(m),$$

and the overall average effect as

$$\hat{\tau}^{\text{strat}} = \frac{N(f)}{N} \cdot \hat{\tau}^{\text{dif}}(f) + \frac{N(m)}{N} \cdot \hat{\tau}^{\text{dif}}(m) = p \cdot \hat{\tau}^{\text{dif}}(f) + (1 - p) \cdot \hat{\tau}^{\text{dif}}(m).$$

The super-population variance for this estimator is

$$\mathbb{V}_{\text{sp}}(\hat{\tau}^{\text{strat}}) = \frac{p}{N} \cdot \left(\frac{\sigma_t^2(f)}{p} + \frac{\sigma_c^2(f)}{1 - p} \right) + \frac{1 - p}{N} \cdot \left(\frac{\sigma_t^2(m)}{p} + \frac{\sigma_c^2(m)}{1 - p} \right).$$

The difference between the two sampling variances, normalized by the sample size N, is

$$N \cdot \left(\mathbb{V}_{\mathrm{sp}}(\hat{\tau}^{\mathrm{dif}}) - \mathbb{V}_{\mathrm{sp}}(\hat{\tau}^{\mathrm{strat}}) \right) = p(1-p) \cdot \left((\mu_{\mathrm{c}}(f) - \mu_{\mathrm{c}}(m))^2 + (\mu_{\mathrm{t}}(f) - \mu_{\mathrm{t}}(m))^2 \right) \geq 0,$$

where $\mu_{\mathrm{c}}(f)$ is the average of $Y_i(0)$ for women, and similarly for $\mu_{\mathrm{c}}(f)$, $\mu_{\mathrm{c}}(m)$, and $\mu_{\mathrm{t}}(m)$.

Although under some conditions there is an unambiguous ranking of the population sampling variances, $\mathbb{V}_{\mathrm{sp}}(\hat{\tau}^{\mathrm{dif}})$ and $\mathbb{V}_{\mathrm{sp}}(\hat{\tau}^{\mathrm{strat}})$, the *estimated* sampling variance for the stratified experiment may be larger than for the completely randomized experiment. The natural estimator for the sampling variance of the simple unbiased estimator in a stratified randomized experiment can be larger than the natural estimators for the sampling variance in a completely randomized experiment, because of the need to estimate the within-stratum potential outcome variances.

We can assess the benefits of having the stratification for an experiment with the size of Project Star. Suppose we have J strata, each with N_{t} treated (small) and $N_{\mathrm{c}} = N_{\mathrm{t}}$ control (regular-sized) classes. Suppose that the true within-stratum variance of the potential outcomes is $\sigma^2 = 0.43^2$, which is the posterior mean for the hierarchical model estimated on the Project Star data. Suppose also that the true variance of the within-stratum average potential outcomes over the strata is $\Sigma_{11} = 0.37^2$ for the control averages $\mu_{\mathrm{c}}(j)$ and $\Sigma_{22} = 0.37^2$ for the averages given the treatment $\mu_{\mathrm{t}}(j)$, again estimated on the Project Star data. Then the ratio of the variances under a completely randomized experiment versus a stratified randomized experiment would be $(0.43^2 + 0.37^2)/0.43^2 = 1.65$. Using a stratified design reduces the variance by 40%. The stratification appears to be quite effective in Project Star.

9.9 CONCLUSION

In this chapter we discuss the analysis of stratified randomized experiments using the four approaches developed in the previous four chapters for completely randomized experiments. In general the stratification should not be ignored in design if treatment rates and potential outcomes vary systematically by stratum. All approaches can be adapted in a fairly straightforward manner to take account of the stratification. A key issue is that in the model-based analysis, a hierarchical model can be useful to take account of similarities in potential outcome distributions across strata. As we illustrate using data from the Project Star experiment on class size, stratification can increase precision of estimation when the strata are good predictors of the potential outcomes.

In the next chapter we extend these analyses to an extreme version of stratification in an experimental context, paired randomized experiments, where each stratum consists of only two units, one treated and one control.

NOTES

The Project Star data have been used by numerous researchers. For more recent research papers, see Krueger (1999), Chetty, Friedman, Hilger, Saez, Schanzenbach, and Yagan (2011) and Graham (2008). Graham (2008) looks at implications of within-class interactions on variances, as discussed in Section 9.4.3.

To implement the Bayesian analysis discussed in Sections 9.7 and 9.7.2 it is useful to use modern numerical methods, in particular Markov-Chain-Monte-Carlo methods, which we discuss in some detail in Chapter 8.

In textbook discussions of the benefits of stratification, and its extreme version, pairing versus complete randomization, it is sometimes pointed out that there are costs associated with stratification and pairing in small population settings. For example, Snedecor and Cochran (1989, p. 101) write: "If the criterion has no correlation with the response variable, a small loss in accuracy results from the pairing due to the adjustment for degrees of freedom. A substantial loss may even occur if the criterion is badly chosen so that members of a pair are negatively correlated." The possibility of negative correlation arises only if in the populations in the strata are small. For example, as discussed in Snedecor and Cochran (1967, p. 294), if the strata correspond to litters of rats, then weights within strata may well be negatively correlated. On the other hand, if the within-strata samples are drawn from large strata, in expectation the stratification can only lead to non-negative correlations.

Box, Hunter, and Hunter (2005, p. 93) also suggest that there is a trade-off in terms of accuracy or variance in the decision to stratify, writing: "Thus you would gain from the paired design only if the reduction in variance from pairing outweighed the effect of the decrease in the number of degrees of freedom of the t distribution." These comments reflect on the implications for testing and interval estimation. In expectation, with large size strata, the sampling variance of the estimated average treatment effect can only decrease as a result of stratification or pairing, not increase.

Samii and Aronow (2012) discuss comparisons between regression approaches and Neyman repeated sampling variances in this setting.

APPENDIX A: STUDENT-LEVEL ANALYSES

Here we discuss the student-level significance tests in more detail. First consider the data from a single stratum, say school j. This school has $N(j)$ students with $P(j)$ classes/teachers. The class size for class s in school j is $M_s(j)$, with $\sum_{s=1}^{P(j)} M_s(j) = N(j)$. Note that we do not require the class sizes to be the same for all small or all regular-sized classes. Even if some classes are exactly the same size, we analyze them as distinct in the sense that having a particular group of twenty students and a teacher assigned to class 1, and a second group of ten students and another teacher assigned to class 2 is a different assignment from having the first group of students and their teacher assigned to class 2 and the others to class 1. This is not necessary, but interpreting those assignments as identical would require keeping track of classes that have identical sizes versus differ by small numbers. The $N(j)$ students and the $P(j)$ teachers are assigned randomly to the $P(j)$ classes. Start with the teachers. The $P(j)$ teachers can be assigned to the $P(j)$ classes in $P(j)!$ different ways. Selecting $M_1(j)$ students for the first class can be done in $\binom{N(j)}{M_1(j)}$ different ways. Selecting the students for the next class can be done in $\binom{N(j)-M_1(j)}{M_2(j)}$ different ways, and so on, implying that the students can be assigned in

$$\prod_{s=1}^{P(j)-1} \binom{N(j) - \sum_{t<s} M_t(j)}{M_s(j)}$$

different ways. Combining this with the teachers' assignments, the total number of ways the students and teachers for school j can be assigned is

$$\prod_{s=1}^{P(j)-1} \binom{N(j) - \sum_{t<s} M_t(j)}{M_s(j)} \cdot P(j)!.$$

For each student this is the total number of potential outcomes. Thus, let $\mathbf{W}(j)$ be the $N(j)$ vector of student assignments for school j, where the ith element of $\mathbf{W}(j)$ takes on values in the set $\{1, \ldots, P(j)\}$, indicating which class student i is assigned to. In addition, $\mathbf{T}(j)$ is the $P(j)$-dimensional vector of teacher assignments in school j, again with each element of $\mathbf{T}(j)$ taking on values in the set $\{1, \ldots, P(j)\}$. Thus we can write the potential outcome for student i in school j as

$$Y_{ij}\left(\mathbf{W}(j), \mathbf{T}(j)\right).$$

The null hypothesis we consider is

$$H_0: \ Y_{ij}\left(\mathbf{W}(j), \mathbf{T}(j)\right) = Y_{ij}\left(\mathbf{W}'(j), \mathbf{T}'(j)\right) \ \text{ for all } \mathbf{W}(j), \mathbf{T}(j), \mathbf{W}'(j), \mathbf{T}'(j).$$

The basis for the randomization distribution is the full set of assignments, which are all equally likely. The total number of assignments is obtained by multiplying the number of assignments for each school:

$$\prod_{j=1}^{J} \prod_{s=1}^{P(j)-1} \binom{N(j) - \sum_{t<s} M_t(j)}{M_s(j)} \cdot P(j)!.$$

APPENDIX B: PROOFS OF THEOREMS 9.1 AND 9.2

It is convenient to reparametrize the model slightly. Instead of (τ, β), we parametrize the model as (τ, γ), where $\gamma(j) = \beta(j) - e(j) \cdot \tau$, which does not change the least squares estimate of τ. In terms of (τ, γ), the regression function is

$$Y_i^{\text{obs}} = \tau \cdot \left(W_i - \sum_{j=1}^{J} e(j) \cdot B_i(j)\right) + \sum_{j=1}^{J} \gamma(j) \cdot B_i(j) + \varepsilon_i.$$

The population values for the parameters are

$$(\tau^*, \gamma^*) = \arg\min_{\tau, \gamma} \mathbb{E}\left[\left(Y_i^{\text{obs}} - \tau \cdot \left(W_i - \sum_{j=1}^{J} e(j) \cdot B_i(j)\right) - \sum_{j=1}^{J} \gamma(j) \cdot B_i(j)\right)^2\right].$$

We can write

$$Y_i^{\text{obs}} = \sum_{j=1}^{J} \alpha(j) \cdot B_i(j) + \sum_{j=1}^{J} \tau(j) \cdot W_i \cdot B_i(j) + \eta_i,$$

where $\alpha(j) = \mathbb{E}_{\text{sp}}[Y_i(0)|B_i(j) = 1]$ and $\tau_{\text{sp}}(j) = \mathbb{E}_{\text{sp}}[Y_i(1) - Y_i(0)|B_i(j) = 1]$, and where by definition $\mathbb{E}[\eta_i|B_i(1), \ldots, B_i(J), W_i] = 0$. Therefore,

$$(\tau^*, \gamma^*) = \arg\min_{\tau, \gamma} \mathbb{E}\left[\left(\sum_{j=1}^{J} \alpha(j) \cdot B_i(j) + \sum_{j=1}^{J} \tau(j) \cdot W_i \cdot B_i(j) - \tau\right.\right.$$

$$\left.\left. \cdot \left(W_i - \sum_{j=1}^{J} e(j) \cdot B_i(j)\right) - \sum_{j=1}^{J} \gamma(j) \cdot B_i(j)\right)^2\right]$$

$$= \arg\min_{\tau, \gamma} \mathbb{E}\left[\left(\sum_{j=1}^{J} B_i(j) \cdot (\alpha(j) - \gamma(j) + \tau(j) \cdot W_i) - \tau\right.\right.$$

$$\left.\left. \cdot \left(W_i - \sum_{j=1}^{J} e(j) \cdot B_i(j)\right)\right)^2\right]$$

$$= \arg\min_{\tau, \gamma} \left\{\mathbb{E}\left[\left(\sum_{j=1}^{J} B_i(j) \cdot (\alpha(j) - \gamma(j) + \tau(j) \cdot W_i)\right)^2\right]\right.$$

$$- 2 \cdot \tau \cdot \mathbb{E}\left[\sum_{j=1}^{J} B_i(j) \cdot (\alpha(j) - \gamma(j) + \tau(j) \cdot W_i) \cdot \left(W_i - \sum_{m=1}^{J} e(j) \cdot B_i(j)\right)\right]$$

$$\left. + \tau^2 \cdot \mathbb{E}\left[\left(W_i - \sum_{j=1}^{J} e(j) \cdot B_i(j)\right)^2\right]\right\}$$

$$= \arg\min_{\tau, \gamma} \left\{\mathbb{E}\left[\left(\sum_{j=1}^{J} B_i(j) \cdot (\alpha(j) - \gamma(j) + \tau(j) \cdot W_i)\right)^2\right]\right.$$

$$- 2 \cdot \tau \cdot \mathbb{E}\left[\sum_{j=1}^{J} B_i(j) \cdot \tau(j) \cdot W_i \cdot \left(W_i - \sum_{m=1}^{J} e(j) \cdot B_i(j)\right)\right]$$

$$\left. + \tau^2 \cdot \mathbb{E}\left[\left(W_i - \sum_{j=1}^{J} e(j) \cdot B_i(j)\right)^2\right]\right\}$$

because $\mathbb{E}[W_i | B_i(1), \ldots, B_i(J)] = \sum_{j=1}^{J} e(j) \cdot B_i(j)$. Minimizing this over τ leads to

$$\tau^* = \frac{\mathbb{E}\left[\left(\sum_{j=1}^{J} B_i(j) \cdot \tau(j) \cdot W_i \cdot \left(W_i - \sum_{m=1}^{J} e(j) \cdot B_i(j) \right) \right)^2 \right]}{\mathbb{E}\left[\left(W_i - \sum_{j=1}^{J} e(j) \cdot B_i(j) \right)^2 \right]}.$$

Because $\Pr(W_i = 1) = \sum_{j=1}^{J} q(j) \cdot e(j)$, and $\Pr(B_i(j) = 1 | W_i = 1) = q(j) \cdot e(j) / \sum_{m=1}^{J} q(j) \cdot e(j)$, it follows that the numerator is equal to $\sum_{j=1}^{J} e(j) \cdot (1 - e(j)) \cdot q(j) \cdot \tau(j)$, and that the denominator is equal to $\sum_{j=1}^{J} e(j) \cdot (1 - e(j)) \cdot q(j)$, which finishes the proof of the first part of Theorem 9.1.

The first-order conditions for the estimators $(\hat{\tau}^{\text{ols}}, \hat{\gamma}^{\text{ols}})$ are

$$\sum_{i=1}^{N} \psi(Y_i^{\text{obs}}, W_i, B_i(1), \ldots, B_i(J), \hat{\tau}^{\text{ols}}, \hat{\gamma}^{\text{ols}}) = 0,$$

where

$$\psi(y, w, b(1) \ldots, b(J), \tau, \gamma)$$

$$= \begin{pmatrix} \left(w - \sum_{j=1}^{J} e(j) \cdot b(j) \right) \cdot \left(y - \tau \cdot \left(w - \sum_{j=1}^{J} e(j) \cdot b(j) \right) - \sum_{j=1}^{J} \gamma(j) \cdot b(j) \right) \\ b(j) \cdot \left(y - \tau \cdot \left(w - \sum_{j=1}^{J} e(j) \cdot b(j) \right) - \sum_{j=1}^{J} \gamma(j) \cdot b(j) \right) \end{pmatrix}.$$

Given the population values of the parameters, τ^* and γ^*, standard M-estimation (or generalized method of moments) results imply that, under standard regularity conditions, the estimator is consistent and asymptotically normally distributed:

$$\sqrt{N} \cdot \begin{pmatrix} \hat{\tau}^{\text{ols}} - \tau^* \\ \hat{\gamma}^{\text{ols}} - \gamma^* \end{pmatrix} \xrightarrow{d} \mathcal{N} \left(\begin{pmatrix} 0 \\ 0 \\ 0 \end{pmatrix}, \Gamma^{-1} \Delta (\Gamma')^{-1} \right),$$

where the two components of the covariance matrix are

$$\Gamma = \mathbb{E}\left[\frac{\partial}{\partial(\tau, \gamma')} \psi(Y_i^{\text{obs}}, W_i, B_i(1), \ldots, B_i(J), \tau, \gamma) \right]\Bigg|_{(\tau^*, \gamma^*)}$$

$$= \mathbb{E}\left[\begin{pmatrix} \sum_{j=1}^{J} e(j) \cdot (1 - e(j)) \cdot q(j) & 0 & \cdots & 0 \\ 0 & B_i(1) & \cdots & 0 \\ \vdots & \vdots & \ddots & \vdots \\ 0 & 0 & \cdots & B_i(J) \end{pmatrix} \right],$$

and

$$
\Delta = \mathbb{E}\left[\psi(Y_i^{\mathrm{obs}}, W_i, B_i(1), \ldots, B_i(J), \tau^*, \gamma^*) \cdot \psi(Y_i^{\mathrm{obs}}, W_i, B_i(1), \ldots, B_i(J), \tau^*, \gamma^*)'\right]
$$

$$
= \mathbb{E}\left[\left(\left(Y_i^{\mathrm{obs}} - \tau^* \cdot \left(W_i - \sum_{j=1}^{J} e(j) \cdot B_i(j)\right)\right) - \sum_{j=1}^{J} \gamma^*(j) \cdot B_i(j)\right)^2\right.
$$

$$
\left. \cdot \left(\frac{W_i - \sum_{j=1}^{J} e(j) \cdot B_i(j)}{B_i(j)}\right) \left(\frac{W_i - \sum_{j=1}^{J} e(j) \cdot B_i(j)}{B_i(j)}\right)'\right]
$$

$$
= \mathbb{E}\left[\left(Y_i^{\mathrm{obs}} - \tau^* \cdot W_i - \sum_{j=1}^{J} \beta^*(j) \cdot B_i(j)\right)^2\right.
$$

$$
\left. \cdot \left(\frac{W_i - \sum_{j=1}^{J} e(j) \cdot B_i(j)}{B_i(j)}\right) \left(\frac{W_i - \sum_{j=1}^{J} e(j) \cdot B_i(j)}{B_i(j)}\right)'\right].
$$

The sampling variance of $\hat{\tau}$ is the $(1, 1)$ element of the covariance matrix. Because Γ is block diagonal, the $(1, 1)$ element of $\Gamma^{-1} \Delta (\Gamma')^{-1}$ is equal to the $(1, 1)$ element of Δ divided by the square of the $(1, 1)$ element of Γ. Hence the sampling variance of $\hat{\tau}$, normalized by the sample size N, is equal to

$$
\frac{\mathbb{E}\left[\left(W_i - \sum_{j=1}^{J} q(j) \cdot B_i(j)\right)^2 \cdot \left(Y_i^{\mathrm{obs}} - \tau^* \cdot W_i - \sum_{j=1}^{J} \beta^*(j) \cdot B_i(j)\right)^2\right]}{\left(\sum_{j=1}^{J} q(j) \cdot e(j) \cdot (1 - e(j))\right)^2}.
$$

$\square$

Proof of Theorem 9.2

First write the regression function as

$$
Y_i^{\mathrm{obs}} = \sum_{j=1}^{J} \alpha(j) \cdot B_i(j) + \sum_{j=1}^{J} \tau(j) \cdot W_i \cdot B_i(j) + \varepsilon_i.
$$

Estimating the parameters of this regression function by OLS leads to

$$
\hat{\tau}^{\mathrm{ols}}(j) = \overline{Y}_{\mathrm{t}}^{\mathrm{obs}}(j) - \overline{Y}_{\mathrm{c}}^{\mathrm{obs}}(j),
$$

which is unbiased and consistent for $\tau(j)$. Then transform the parameter vector from $\tau(J)$ to $\tau = \sum_{j=1}^{J} q(j) \cdot \tau(j)$, with inverse transformation $\tau(J) = (\tau - \sum_{j'=1}^{J-1} q(j') \cdot \tau(j'))/q(J)$. In terms of the parameters $\alpha(1), \ldots, \alpha(J), \tau(1), \ldots, \tau(J-1)$ and τ, the regression function is equal to

$$
Y_i^{\mathrm{obs}} = \tau \cdot W_i \cdot \frac{B_i(J)}{q(J)} + \sum_{j=1}^{J} \alpha(j) \cdot B_i(j) + \sum_{j=1}^{J-1} \tau(j) \cdot W_i \cdot \left(B_i(j) - B_i(J) \cdot \frac{q(j)}{q(J)}\right) + \varepsilon_i.
$$

Thus $\hat{\tau}^{\mathrm{ols}}$ is identical to $\sum_{j=1}^{J} q(j) \cdot \hat{\tau}^{\mathrm{ols}}(j)$, and therefore is consistent for $\sum_{j=1}^{J} q(j) \cdot \tau(j) = \tau_{\mathrm{sp}}$.

Because the sampling variance of $\hat{\tau}^{\mathrm{ols}}(j)$ is $(\sigma_c^2(j)/((1 - e(j)) \cdot q(j)) + \sigma_t^2(j)/(e(j) \cdot q(j)))/N$, the sampling variance of $\sum_{j=1}^{J} q(j) \cdot \hat{\tau}^{\mathrm{ols}}(j)$, normalized by N, is $N \cdot \sum_{j=1}^{J} q(j)^2 \cdot \mathbb{V}(\hat{\tau}^{\mathrm{ols}}(j))$, equal to $\sum_{j=1}^{J} q(j)^2(\sigma_c^2(j)/((1 - e(j)) \cdot q(j)) + \sigma_t^2(j)/(e(j) \cdot q(j)))$. $\qquad\square$

_____

## CHAPTER 10

Pairwise Randomized Experiments

10.1 INTRODUCTION

In the previous chapter we analyzed stratified randomized experiments, where a sample of size N was partitioned into J strata, and within each stratum a completely randomized experiment was conducted. In this chapter we consider a special case of the stratified randomized experiment. Each stratum contains exactly two units, with one randomly selected to be assigned to the treatment group, and the other one assigned to the control group. Such a design is known as a *pairwise randomized experiment* or *paired comparison*. Although this can be viewed simply as a special case of a stratified randomized experiment, there are two features of this design that warrant special attention. First, the fact that there is only a single unit in each treatment group in each stratum (or pair in this case) implies that the Neyman sampling variance estimator that we discussed in the chapters on completely randomized experiments (Chapter 6) and stratified randomized experiments (Chapter 9) cannot be used; that estimator requires the presence of at least two units assigned to each treatment in each stratum. Second, each stratum has the same proportion of treated units, which allows us to analyze the within-stratum estimates symmetrically; the natural estimator for the average treatment effect weights each stratum equally.

As in the case of stratified randomized experiments, the motivation for eliminating some of the possible assignments in pairwise randomized experiments is that *a priori* those values of the assignment vectors that are eliminated are expected to lead to less informative inferences. This argument relies on the within-pair variation in potential outcomes being small relative to the between-pair variation. Often the assignment to pairs is based on covariates. Units are matched to other units based on their similarity in these covariates, with the expectation that this similarity corresponds to similarity in the potential outcomes under each treatment. Suppose, for example, that the treatment is an expensive surgical procedure for a relatively common medical condition. It may not be financially feasible to apply the treatment to many individuals. To increase the precision of an experiment, it may, in such cases, be sensible to use the following steps. First randomly draw J individuals from the target population of individuals who have the condition for which the surgery may be beneficial. Then, for each of these J individuals, find a matching individual in the same population, as similar as possible to the original unit in terms of the characteristics that may be correlated with potential outcomes and efficacy of the treatment. If the population is relatively large, it may be possible to get very close matches with respect to a large number of characteristics, thereby reducing the variation in treatment-control differences in potential outcomes. Given these J matched pairs, one can then conduct a pairwise randomized experiment by randomly selecting one member of each pair to be assigned to the active treatment.

In this chapter we discuss analyses for such pairwise randomized experiments. In particular we discuss the calculation of Fisher exact p-values and Neyman's repeated sampling perspective, as well as regression and model-based inference. We focus primarily on conceptual issues that are special to this design.

Section 10.2 describes the data set we use to illustrate the concepts discussed in this chapter, which comes from a randomized experiment conducted around 1970 to evaluate the effect of an educational children's television program on reading ability as measured through test scores. Section 10.3 discusses the structure of paired randomized experiments and introduces some additional notation. In 10.4 we discuss the application of Fisher's exact p-value calculations in the setting of paired randomized experiments. Next, in Section 10.5 we discuss the implications of pairwise randomization for the methods discussed in Chapter 6 based on Neyman's repeated sampling perspective. In Sections 10.6 and 10.7 we analyze regression and model-based imputation methods. Section 10.8 concludes.

10.2 THE CHILDREN'S TELEVISION WORKSHOP
EXPERIMENT DATA

The Children's Television Workshop experiment was designed by Ball, Bogatz, Rubin, and Beaton (1973) to evaluate *The Electric Company*, an educational television program aimed at improving reading skills for young children, somewhat similar to *Sesame Street*. The experiment was conducted in two locations, Youngstown, Ohio, and Fresno, California, where *The Electric Company* was not broadcast on local stations. In each location a number of schools was selected. Within each school, a pair of two classes was selected. Within each pair, one class was randomly assigned to be shown *The Electric Company* show during the standard reading-class period, and the other class continued with the regular reading curriculum.

Here we focus on the data from Youngstown, where two first-grade classes from each of eight schools participated in the experiment. The data for the sixteen classes for the Youngstown location from this experiment are displayed in Table 10.1, which presents values of a pre-test score, the post-test score (the primary outcome), an indicator for the pair or school to which the unit belongs, and an indicator for the treatment (one for classes that viewed *The Electric Company* program, and zero for classes in the control group).

10.3 PAIRWISE RANDOMIZED EXPERIMENTS

A pairwise randomized experiment is a special case of a stratified randomized experiment where the number of units, N, is even, the number of strata is $J = N/2$, with one

Table 10.1. *Data from Youngstown Children's Television Workshop Experiment*

Pair G_i	Treatment W_i	Pre-Test Score X_i	Post-Test Score Y_i^{obs}	Normalized Rank Post-Test Score R_i
1	0	12.9	54.6	−7.5
1	1	12.0	60.6	2.5
2	0	15.1	56.5	−4.5
2	1	12.3	55.5	5.5
3	0	16.8	75.2	0.5
3	1	17.2	84.8	4.5
4	0	15.8	75.6	1.5
4	1	18.9	101.9	7.5
5	0	13.9	55.3	−6.5
5	1	15.3	70.6	−1.5
6	0	14.5	59.3	−3.5
6	1	16.6	78.4	2.5
7	0	17.0	87.0	5.5
7	1	16.0	84.2	3.5
8	0	15.8	73.7	−0.5
8	1	20.1	108.6	7.5

treated unit and one control unit in each stratum ($N_t(j) = N_c(j) = 1$ and $N(j) = 2$ for all $j = 1, \ldots, J$), so that each stratum is a pair. Let G_i be the variable indicating the pair, with $G_i \in \{1, \ldots, N/2\}$. The pair indicator can be thought of as a function of covariates. Of course this indicator is a pre-treatment variable in the sense that it is not affected by the treatment. Within each pair there are $\binom{N(j)}{N_t(j)} = \binom{2}{1} = 2$ possible assignments, so that the probability for any assignment vector $\mathbf{W}$ is

$$p(\mathbf{W}|\mathbf{X}, \mathbf{Y}(0), \mathbf{Y}(1)) = \prod_{j=1}^{N/2} \binom{N(j)}{N_t(j)}^{-1} = \prod_{j=1}^{N/2} \frac{1}{2} = 2^{-N/2}, \quad \text{for } \mathbf{W} \in \mathbb{W}^+,$$

where

$$\mathbb{W}^+ = \left\{ \mathbf{W} \, \middle| \, \sum_{i:G_i=j} W_i = 1 \text{ for } j = 1, \ldots, N/2 \right\}.$$

Because the assignment mechanism fits into the stratified randomized experiments discussed in Chapter 9, we can directly use many of the methods discussed in that chapter. However, there is one important difference. Because all strata have the property that they contain exactly one treated and one control unit, methods that rely on the presence of multiple control or multiple treated units cannot be applied.

To facilitate the discussion of pairwise randomized experiments, it is useful to introduce some additional notation. We arbitrarily label the two units within a pair as units A and B. Then, for all pairs $j = 1, \ldots, N/2$, let $(Y_{j,A}(0), Y_{j,A}(1))$ and $(Y_{j,B}(0), Y_{j,B}(1))$ be the potential outcomes for units A and B, respectively, in pair j, and let $W_{j,A}$ and $W_{j,B}$ be

Table 10.2. *Potential Outcomes and Covariates from Children's Television Workshop Experiment, from Table 10.1*

Pair	Unit A					Unit B				
	$Y_{i,A}(0)$	$Y_{i,A}(1)$	$W_{i,A}$	$Y_{i,A}^{obs}$	$X_{i,A}$	$Y_{i,B}(0)$	$Y_{i,B}(1)$	$W_{i,B}$	$Y_{i,B}^{obs}$	$X_{i,B}$
1	54.6	?	0	54.6	12.9	?	60.6	1	60.6	12.0
2	56.5	?	0	56.5	15.1	?	55.5	1	55.5	13.9
3	75.2	?	0	75.2	16.8	?	84.8	1	84.8	17.2
4	76.6	?	0	75.6	15.8	?	101.9	1	101.9	18.9
5	55.3	?	0	55.3	13.9	?	70.6	1	70.6	15.3
6	59.3	?	0	59.3	14.5	?	78.4	1	78.4	16.6
7	87.0	?	0	87.0	17.0	?	84.2	1	84.2	16.0
8	73.7	?	0	73.7	15.8	?	108.6	1	108.6	20.1

the treatment indicators for these units. In a pairwise randomized experiment, one unit in each pair is randomly assigned to the active treatment, and the other unit is assigned to the control treatment, thus $W_{j,A} = 1 - W_{j,B}$, with $\Pr(W_{j,A} = 1|\mathbf{Y}(0), \mathbf{Y}(1), \mathbf{X}) = 1/2$. Define also

$$Y_{j,A}^{obs} = \begin{cases} Y_{j,A}(0) & \text{if } W_{j,A} = 0, \\ Y_{j,A}(1) & \text{if } W_{j,A} = 1, \end{cases} \quad \text{and} \quad Y_{j,B}^{obs} = \begin{cases} Y_{j,B}(0) & \text{if } W_{j,A} = 1, \\ Y_{j,B}(1) & \text{if } W_{ji,A} = 0. \end{cases}$$

The average treatment effect within pair j is $\tau_{\text{pair}}(j)$,

$$\tau^{\text{pair}}(j) = \frac{1}{2} \sum_{i:G_i=j} \left(Y_i(1) - Y_i(0)\right) = \frac{1}{2}\left(\left(Y_{j,A}(1) - Y_{j,A}(0)\right) + \left(Y_{j,B}(1) - Y_{j,B}(0)\right)\right).$$

The finite-sample average treatment effect is

$$\tau_{\text{fs}} = \frac{1}{N} \sum_{i=1}^{N} \left(Y_i(1) - Y_i(0)\right) = \frac{2}{N} \sum_{j=1}^{N/2} \tau^{\text{pair}}(j).$$

Also define the pair of observed variables, one treated and one control from each pair:

$$Y_{j,c}^{obs} = \begin{cases} Y_{j,A}^{obs} & \text{if } W_{i,A} = 0, \\ Y_{j,B}^{obs} & \text{if } W_{i,A} = 1, \end{cases} \quad \text{and} \quad Y_{j,t}^{obs} = \begin{cases} Y_{j,B}^{obs} & \text{if } W_{i,A} = 0, \\ Y_{j,A}^{obs} & \text{if } W_{i,A} = 1. \end{cases}$$

Table 10.2 displays some of these variables for the 16 classes in the Children's Television Workshop Experiment.

10.4 FISHER'S EXACT P-VALUES IN PAIRWISE RANDOMIZED EXPERIMENTS

The same way stratified randomization did not pose any conceptual difficulties for the calculation of Fisher Exact P-values (FEPs), pairwise randomization does not introduce any new issues. Let us focus in this discussion on the usual Fisher null hypothesis of absolutely no treatment effects for any units,

$$H_0: \quad Y_i(0) = Y_i(1), \quad \text{for all } i = 1, \ldots, N.$$

With the assignment mechanism fully known, we can, under H_0, for any fixed statistic, derive the randomization distribution and thus calculate the corresponding p-value. An obvious statistic is the average, over the $J = N/2$ pairs, of the difference between the treated and control outcomes within each pair:

$$T^{\text{dif}} = \left| \frac{1}{J} \sum_{j=1}^{J} \left(Y_{j,t}^{\text{obs}} - Y_{j,c}^{\text{obs}} \right) \right|$$

$$= \left| \frac{1}{J} \sum_{j=1}^{J} \left(W_{i,A} \cdot \left(Y_{j,A}^{\text{obs}} - Y_{j,B}^{\text{obs}} \right) + (1 - W_{i,A}) \cdot \left(Y_{j,B}^{\text{obs}} - Y_{j,A}^{\text{obs}} \right) \right) \right|.$$

Because each pair has a single treated and a single control unit, this also equals the difference between average outcomes for treated and control units, $T^{\text{dif}} = \left| \overline{Y}_t^{\text{obs}} - \overline{Y}_c^{\text{obs}} \right|$, the statistic that was the starting point of the discussion of the FEP approach in Chapter 5. However, the p-value for this statistic will be different than that calculated under the randomization distribution considered in Chapter 5 because here the randomization distribution is based on the assignment mechanism corresponding to a pairwise randomized experiment, not the assignment mechanism corresponding to a completely randomized experiment, leading to fewer elements in $\mathbb{W}^+$.

Alternative statistics include the average of within-pair differences in logarithms or other transformations of the basic outcomes, such as ranks. To calculate the rank statistic, let R_i be the rank of Y_i^{obs} among the N values $Y_1^{\text{obs}}, \ldots, Y_N^{\text{obs}}$, normalized to have mean zero, and let $R_{j,A}$ and $R_{j,B}$ be the rank of the A and B units in pair j, among all N units. For the Children's Television Workshop data, the ranks for the sixteen classes are displayed in the last column in Table 10.1. Then the rank statistic is

$$T^{\text{rank}} = \left| \overline{R}_t - \overline{R}_c \right| = \left| \frac{1}{J} \sum_{j=1}^{J} \left(W_{j,A} \cdot \left(R_{j,A} - R_{j,B} \right) + (1 - W_{j,A}) \cdot \left(R_{j,B} - R_{j,A} \right) \right) \right|.$$

Using ranks in pairwise randomized experiments has the same advantages as using ranks in completely randomized experiments, namely reducing the sensitivity to outliers. Another statistic that is specific to pairwise randomized experiments is based on the average within-pair rank of the observed outcomes. That is, for each pair we calculate an indicator for whether the observed outcome for the treated unit is larger than the observed outcome for the control unit, and an indicator whether the observed outcome for the control unit is larger than the observed outcome for the treated unit. (Using the two indicators, rather than one of the indicators alone, allows for a simpler way of dealing with within-pair ties.) We then average the difference between these indicators,

$$T^{\text{rank,pair}} = \left| \frac{2}{N} \sum_{j=1}^{N/2} \left(\mathbf{1}_{Y_{j,1}^{\text{obs}} > Y_{j,0}^{\text{obs}}} - \mathbf{1}_{Y_{j,1}^{\text{obs}} < Y_{j,0}^{\text{obs}}} \right) \right|,$$

similar to the statistic $T^{\text{rank,stratum}}$ in Chapter 9. Like the rank-based statistic, T^{rank}, this statistic is particularly insensitive to the presence of outliers in the observed potential outcomes, and when there is substantial variation in the level of the outcomes between the pairs, it has more power than the statistic T^{rank} against alternatives under which the treatment effect is constant.

We apply these Fisher exact p-value calculations to the Children's Television Workshop data, using the null hypothesis of no effect whatsoever. Although the p-value is valid only for a single statistic, for illustrative purposes we do the analysis for all three statistics. For the statistic based on the absolute value of the difference in average outcomes by treatment status, we find

$$T^{\text{dif}} = 13.4, \quad \text{p-value} = 0.031.$$

Using the rank statistic, we find

$$T^{\text{rank}} = 3.8, \quad \text{p-value} = 0.031.$$

The last statistic, based on the indicator for whether within the pair the treated outcome was larger or smaller than the control outcome, leads to

$$T^{\text{rank,pair}} = 0.5, \quad \text{p-value} = 0.145.$$

The mechanical reason that the p-value for the within-pair rank statistic is less significant than for the other statistics is that for the two pairs where the outcome for the treated unit is less than the outcome for the control unit in the pair, the difference in outcomes is small. These small differences do not affect the average difference much, but they do affect the within-pair rank statistic. The other two p-values suggest that the television program did affect reading ability at conventional significance levels.

10.5 THE ANALYSIS OF PAIRWISE RANDOMIZED EXPERIMENTS FROM NEYMAN'S REPEATED SAMPLING PERSPECTIVE

Consider first the analysis of the average treatment effect in a single pair. The obvious estimator for the average treatment effect in pair j, $\tau^{\text{pair}}(j)$, is

$$\hat{\tau}^{\text{pair}}(j) = Y_{j,t}^{\text{obs}} - Y_{j,c}^{\text{obs}} = \sum_{i:G_i=j} (2 \cdot W_i - 1) \cdot Y_i^{\text{obs}}.$$

The values of $\hat{\tau}^{\text{pair}}(j)$ for the eight pairs in the Children's Television Workshop data are displayed in Table 10.3.

Table 10.3. *Observed Outcome Data from Children's Television Workshop Experiment by Pair*

Pair	Outcome for Control Unit	Outcome for Treated Unit	Difference
1	54.6	60.6	6.0
2	56.5	55.5	−1.0
3	75.2	84.8	9.6
4	75.6	101.9	26.3
5	55.3	70.6	15.3
6	59.3	78.4	19.1
7	87.0	84.2	−2.8
8	73.7	108.6	34.9
Mean	67.2	80.6	13.4
(S.D.)	(12.2)	(18.6)	(13.1)

Next, let us consider inference, first for the within-pair average treatment effect $\tau^{\text{pair}}(j)$. For each pair we have a completely randomized experiment with two units of which one unit is assigned to active treatment. From the results in Chapter 6 on Neyman's repeated sampling approach, it follows that the estimator $\hat{\tau}^{\text{pair}}(j)$ is unbiased for the average treatment effect $\tau^{\text{pair}}(j)$ within this pair and that its sampling variance, based on the randomization distribution, is equal to

$$\mathbb{V}_W(\hat{\tau}^{\text{pair}}(j)) = \frac{S_c(j)^2}{N_c(j)} + \frac{S_t^2(j)}{N_t(j)} - \frac{S_{ct}(j)^2}{N(j)}.$$

With $N(j) = 2$ and $N_c(j) = N_t(j) = 1$, this expression simplifies to

$$\mathbb{V}_W(\hat{\tau}^{\text{pair}}(j)) = S_c(j)^2 + S_t^2(j) - \frac{S_{ct}(j)^2}{2}.$$

The within-pair variances can be written as

$$S_c^2(j) = \sum_{i:G_i=j} \left(Y_i(0) - \overline{Y}_j(0)\right)^2 = \frac{1}{2} \cdot \left(Y_{j,A}(0) - Y_{j,B}(0)\right)^2,$$

$$S_t^2(j) = \sum_{i:P_i=j} \left(Y_i(1) - \overline{Y}_j(1)\right)^2 = \frac{1}{2} \cdot \left(Y_{j,A}(1) - Y_{j,B}(1)\right)^2,$$

and

$$S_{ct}^2(j) = \frac{1}{2} \cdot \left(\left(Y_{j,A}(1) - Y_{j,A}(0)\right) - \left(Y_{j,B}(1) - Y_{j,B}(0)\right)\right)^2,$$

where

$$\overline{Y}_j(0) = \frac{1}{2} \cdot \left(Y_{j,A}(0) + Y_{j,B}(0)\right) \quad \text{and} \quad \overline{Y}_j(1) = \frac{1}{2} \cdot \left(Y_{j,A}(1) + Y_{j,B}(1)\right).$$

If the primary interest is in the finite-sample average treatment effect, τ_{fs}, that is, the within-pair average treatment effect averaged over the $N/2$ pairs,

$$\tau_{fs} = \frac{1}{N/2} \sum_{j=1}^{N/2} \tau^{pair}(j),$$

the natural estimator is

$$\hat{\tau}^{dif} = \frac{1}{N/2} \sum_{j=1}^{N/2} \hat{\tau}^{pair}(j) = \overline{Y}_t^{obs} - \overline{Y}_c^{obs}. \tag{10.1}$$

By unbiasedness of the within-pair estimators, $\hat{\tau}$ is unbiased for the sample average treatment effect, τ_S. Its sampling variance over the randomization distribution is

$$\mathbb{V}_W\left(\hat{\tau}^{dif}\right) = \frac{1}{(N/2)^2} \sum_{j=1}^{N/2} \left(S_c^2(j) + S_t^2(j) - \frac{S_{ct}^2(j)}{2} \right).$$

So far the discussion is exactly analogous to the discussion for stratified randomized experiments in the previous chapter. However, one of the special features of pairwise randomized experiments, alluded to in the introduction to this chapter, creates a complication for the estimation of the sampling variance. In a completely randomized experiment (and similarly, within a stratum in the stratified randomized experiment), the standard estimator for the sampling variance for the observed difference in treatment and control averages is

$$\hat{\mathbb{V}}^{neyman}\left(\overline{Y}_t^{obs} - \overline{Y}_c^{obs}\right) = \frac{s_c^2}{N_c} + \frac{s_t^2}{N_t},$$

with

$$s_c^2 = \frac{1}{N_c - 1} \sum_{i:W_i=0} \left(Y_i(0) - \overline{Y}_c^{obs}\right)^2 = \frac{1}{N_c - 1} \sum_{i:W_i=0} \left(Y_i^{obs} - \overline{Y}_c^{obs}\right)^2,$$

and analogously

$$s_t^2 = \frac{1}{N_t - 1} \sum_{i:W_i=1} \left(Y_i^{obs} - \overline{Y}_t^{obs}\right)^2.$$

Because within each stratum (or pair in this case) the numbers of control and treated units are $N_c = N_t = 1$, these estimators, s_c^2 and s_t^2, cannot be used, and the standard estimator for the sampling variance of the estimated overall average effect is not feasible.

One solution to this problem is to assume that the treatment effect is constant and additive, not only within pairs but also across pairs. Because of the assumption of a constant treatment effect within pairs, it follows that the within-pair sampling variance is

$$\mathbb{V}_W(\hat{\tau}^{pair}(j)) = 2 \cdot S^2(j), \quad \text{where } S^2(j) = S_c^2(j) = S_t^2(j).$$

Moreover, if the treatment effect is constant across pairs, $\tau^{\text{pair}}(j) = \tau_S$ for all j, the within-pair variances are constant, $S^2(j) = S^2$ for all j, and

$$
\mathbb{V}_W\left(\hat{\tau}^{\text{dif}}\right) = \frac{1}{(N/2)^2} \sum_{j=1}^{N/2}\left(S_c^2(j) + S_t^2(j) - \frac{S_{ct}^2(j)}{2}\right) = \frac{4}{N}\cdot S^2,
$$

which can be estimated by calculating the sample variance of the pair-level treatment effect estimates:

$$
\hat{\mathbb{V}}^{\text{pair}}\left(\hat{\tau}^{\text{dif}}\right) = \frac{4}{N\cdot(N-2)}\cdot\sum_{j=1}^{N/2}\left(\hat{\tau}^{\text{pair}}(j) - \hat{\tau}^{\text{dif}}\right)^2.
$$

If there is heterogeneity in the treatment effects, then this sampling variance estimator is upwardly biased, and the corresponding confidence intervals will be conservative in the usual statistical sense.

Theorem 10.1 *Suppose we have J pairs of units, and randomly assign one unit from each pair to the active treatment and the other unit to the control treatment. Then (i) $\hat{\tau}^{\text{dif}}$ is unbiased for τ_{fs}, (ii) the sampling variance of $\hat{\tau}^{\text{dif}}$ is*

$$
\mathbb{V}_W\left(\hat{\tau}^{\text{dif}}\right) = \frac{1}{N^2}\sum_{j=1}^{N/2}(Y_{j,A}(0) + Y_{j,A}(1) - (Y_{j,B}(0) + Y_{j,B}(1)))^2,
$$

and (iii) the estimator for the sampling variance

$$
\hat{\mathbb{V}}^{\text{pair}}\left(\hat{\tau}^{\text{dif}}\right) = \frac{4}{N\cdot(N-2)}\cdot\sum_{j=1}^{N/2}\left(\hat{\tau}^{\text{pair}}(j) - \hat{\tau}^{\text{dif}}\right)^2,
$$

satisfies

$$
\mathbb{E}\left[\hat{\mathbb{V}}^{\text{pair}}\left(\hat{\tau}^{\text{dif}}\right)\right] = \mathbb{V}_W(\hat{\tau}^{\text{dif}}) + \frac{4}{N\cdot(N-2)}\cdot\sum_{j=1}^{N/2}\left(\tau^{\text{pair}}(j) - \tau\right)^2,
$$

with the expected value equal to $\mathbb{V}_W(\hat{\tau}^{\text{dif}})$ if the treatment effect is constant across and within pairs.

Proof of Theorem 10.1: See Appendix.

Let us return to the data from the Children's Television Workshop experiment. The within-pair differences $\hat{\tau}^{\text{pair}}(j)$ are displayed in Table 10.3. Their average is

$$
\hat{\tau}^{\text{dif}} = \frac{1}{8}\cdot\sum_{j=1}^{8}\hat{\tau}^{\text{pair}}(j) = 13.4,
$$

and its estimated sampling variance is

$$\hat{\mathbb{V}}^{\text{pair}}\left(\hat{\tau}^{\text{dif}}\right) = \frac{1}{8 \cdot (8-1)} \cdot \sum_{j=1}^{8}\left(\hat{\tau}^{\text{pair}}(j) - \hat{\tau}^{\text{dif}}\right)^2 = 4.6^2.$$

The standard, Gaussian-distribution-based asymptotic 95% confidence interval is

$$\text{CI}^{0.95}(\tau_{\text{fs}}) = \left(\hat{\tau} - 1.96 \times \sqrt{\hat{\mathbb{V}}^{\text{pair}}\left(\hat{\tau}^{\text{dif}}\right)}, \hat{\tau} + 1.96 \times \sqrt{\hat{\mathbb{V}}^{\text{pair}}\left(\hat{\tau}^{\text{dif}}\right)}\right) = (4.3, 22.5).$$

$$(10.2)$$

Because we have only eight pairs of classes, one may wish to use a confidence interval based on the t-distribution with degrees of freedom equal to $N/2 - 1 = 7$, with 0.975 quantile equal to 2.365, leading to a slightly wider confidence interval

$$\text{CI}_{t(7)}^{0.95}(\tau_{\text{fs}}) = \left(\hat{\tau} - 2.365 \times \sqrt{\hat{\mathbb{V}}^{\text{pair}}\left(\hat{\tau}^{\text{dif}}\right)}, \hat{\tau} + 2.365 \times \sqrt{\hat{\mathbb{V}}^{\text{pair}}\left(\hat{\tau}^{\text{dif}}\right)}\right)$$

$$= (2.5, 24.3).$$

$$(10.3)$$

Let us now illustrate the benefits of doing a pairwise randomized experiment instead of a completely randomized experiment. Suppose we had done a completely randomized experiment and had the same assignment vector. In that case we would have the same point estimate for the average treatment effect, namely $\hat{\tau}^{\text{dif}} = \overline{Y}_t^{\text{obs}} - \overline{Y}_c^{\text{obs}} = 13.4$. However, we would have a different estimate of the sampling variance. Using the standard Neyman estimated sampling variance discussed in Chapter 6, we would have estimated the sampling variance of the two potential outcomes as

$$s_c^2 = \frac{1}{N_c - 1}\sum_{i:W_i=0}\left(Y_i^{\text{obs}} - \overline{Y}_c^{\text{obs}}\right)^2 = 18.5^2, \quad \text{and} \quad s_t^2 = 12.2^2,$$

leading to an estimate for the sampling variance of the estimated average effect of

$$\hat{\mathbb{V}}^{\text{neyman}} = \frac{s_c^2}{8} + \frac{s_t^2}{8} = 7.8^2.$$

This sampling variance estimate is substantially larger than the estimate based on the pairwise randomization, $\hat{\mathbb{V}}^{\text{pair}} = 4.6^2$, because the observed variance of potential outcome within pairs is substantially smaller than it would be if units were randomly assigned to pairs. In other words, in this application, the assignment to pairs is effective, in the sense that it is based on factors that make the within-pair units substantially more similar than randomly selected units, probably leading to substantially more precise estimates.

10.6 REGRESSION-BASED ANALYSIS OF PAIRWISE RANDOMIZED EXPERIMENTS

In this section the second special feature of pairwise randomized experiments, alluded to in the introduction of this chapter, motivates an analysis that is different from that discussed for stratified randomized experiments. In the discussions of regression-based analyses in completely and stratified randomized experiments, the basic outcome in the analysis was Y_i^{obs}, the observed outcome for unit i. Here, instead, we use as the primary outcome in the regression analysis the within-pair difference in observed outcomes of the treated and the control unit in the pair,

$$\hat{\tau}^{\text{pair}}(j) = Y_{j,t}^{\text{obs}} - Y_{j,c}^{\text{obs}},$$

with the pair serving as the unit of analysis. We take a super-population perspective, where the pairs of units are drawn randomly from a large population, and one member of each pair is randomly assigned to the treatment group, and the other to the control group. The population average treatment effect is $\tau_{\text{sp}} = \mathbb{E}_{\text{sp}}[\tau^{\text{pair}}(j)]$, with the expectation taken over the random sampling of the pairs.

The standard estimator for the average treatment effect in a pairwise randomized experiment is the simple average of the within-pair differences,

$$\hat{\tau}^{\text{dif}} = \frac{2}{N} \sum_{j=1}^{N/2} \hat{\tau}^{\text{pair}}(j).$$

This estimator can also be interpreted as a regression estimator, where the regression function is specified simply as a constant:

$$\hat{\tau}^{\text{pair}}(j) = \tau_{\text{sp}} + \varepsilon_j.$$

The more interesting question is how to include additional covariates, beyond the implicit use of the pair indicators, into the regression function. As before, because of the randomization, we do not *need* to include additional covariates in order to remove bias, because the estimator $\hat{\tau}$ is unbiased over the randomization distribution without including covariates. The goal when including additional covariates is to improve the precision of the estimator in cases where the covariates are strongly correlated with the treatment-control differences in potential outcomes. Before discussing particular specifications, we first define $X_{j,A}$ and $X_{j,B}$ to be the covariate values for units A and B respectively within pair j. Then we define the within-pair observed difference in covariates between the treated and control units,

$$\Delta_{Xj} = \left(W_{j,A} \cdot \left(X_{j,A} - X_{j,B} \right) + (1 - W_{j,A}) \cdot \left(X_{j,B} - X_{j,A} \right) \right),$$

and the average covariate value within the pair,

$$\overline{X}_j = \left(X_{j,A} + X_{j,B} \right) / 2.$$

There are two leading approaches to including the covariates in the regression analysis. First, we can include them in the form of the within-pair difference $\Delta_{X,j}$. This is an attractive option if one thinks the conditional expectation given covariates of the pairwise difference of potential outcomes is additive and linear in the treatment minus control difference in covariates. In other words, the inclusion of $\Delta_{X,j}$ in the regression function makes sense if the covariate X_i is associated with both potential outcomes $Y_i(0)$ and $Y_i(1)$ to approximately equal degrees. Second, we can include the average value of the covariates $\overline{X}_j$. This is a natural specification if one thinks the treatment effect, the difference in potential treated and control outcomes, rather than the level of the potential outcomes, is linear in the covariates. The most general version of the regression function we consider includes the covariates both as within-pair differences and pair averages, where the latter is in deviations from the overall covariate mean $\overline{X}$:

$$\hat{\tau}^{\text{pair}}(j) = \tau + \beta \cdot \Delta_{X,j} + \gamma \cdot (\overline{X}_j - \overline{X}) + \varepsilon_j.$$

Let $(\tau^*, \beta^*, \gamma^*)$ be the population values, defined analogously to the way they were defined in Chapter 7:

$$\left(\tau^*, \beta^*, \gamma^*\right) = \arg \min_{\tau, \beta, \gamma} \mathbb{E}\left[\left(\hat{\tau}^{\text{pair}}(j) - \tau - \beta \cdot \Delta_{X,j} - \gamma \cdot (\overline{X}_j - \mu_X)\right)^2\right],$$

where $\mu_X = \mathbb{E}_{\text{sp}}(X)$ is the super-population mean of X_i. Here we use again the convention that the expectation operator without subscript is both over the randomization distribution and over the distribution induced by the random sampling from the super-population. Also let $(\hat{\tau}^{\text{ols}}, \hat{\beta}^{\text{ols}}, \hat{\gamma}^{\text{ols}})$ be the least squares estimators,

$$(\hat{\tau}^{\text{ols}}, \hat{\beta}^{\text{ols}}, \hat{\gamma}^{\text{ols}}) = \arg \min_{\tau, \beta, \gamma} \sum_{i=1}^{N} \left(\hat{\tau}^{\text{pair}}(j) - \tau - \beta \cdot \Delta_{X,j} - \gamma \cdot (\overline{X}_j - \overline{X})\right)^2.$$

Theorem 10.2 *Suppose we conduct a pairwise randomized experiment in a sample of pairs drawn at random from the super-population. Then, (i),*

$$\tau^* = \tau_{\text{sp}},$$

and (ii),

$$\sqrt{N} \cdot \left(\hat{\tau}^{\text{ols}} - \tau_{\text{sp}}\right) \xrightarrow{d} \mathcal{N}\left(0, \mathbb{E}_{\text{sp}}\left[\left(\hat{\tau}^{\text{pair}}(j) - \tau^* - \beta^* \cdot \Delta_{X,j} - \gamma^* \cdot (\overline{X}_j - \mu_X)\right)^2\right]\right).$$

Proof of Theorem 10.2 See Appendix.

Now let us estimate the average treatment effect using four different specifications for the regression function. First, for the regression model with only a constant, the least squares estimator for τ is

$$\hat{\tau}^{\text{ols}} = \frac{2}{N} \sum_{j=1}^{N/2} \hat{\tau}^{\text{pair}}(j) = \hat{\tau}^{\text{dif}},$$

equal to the estimator in Equation (10.1). Note that we do not directly include the treatment indicator, because the unit of the least squares analysis here is the pair, not the individual unit. Applying this to the Children's Television experiment data leads to

$$\hat{\tau}^{\text{ols}} = 13.4 \quad (\widehat{\text{s.e.}}\ 4.3)$$

(standard errors in brackets). The next specification for the regression function includes the within-pair difference $\Delta_{X,j}$:

$$\hat{\tau}^{\text{pair}}(j) = \tau + \beta \cdot \Delta_{X,j} + \varepsilon_j.$$

With the Children's Television Workshop data, this specification leads to

$$\hat{\tau}^{\text{pair}}(j) = \begin{array}{cc} 9.0 & + & 5.4 & \times & \Delta_{X,j}, \\ (1.5) & & (0.6) & & \end{array}$$

with a substantially smaller standard error for $\hat{\tau}^{\text{ols}}$, 1.5 instead of 4.3, because the covariate $\Delta_{X,j}$ is a strong predictor of the observed within-pair difference in outcomes. The next specification includes $\overline{X}_j$ as an additional regressor.

$$\hat{\tau}^{\text{pair}}(j) = \tau + \gamma \cdot \overline{X}_j + \varepsilon_j.$$

This leads to

$$\hat{\tau}^{\text{pair}}(j) = \begin{array}{cc} 13.4 & + & 3.9 & \times & \overline{X}_j. \\ (3.5) & & (1.7) & & \end{array}$$

Whereas including $\Delta_{X,j}$ in the regression reduced the standard error of the estimator of the average treatment effect from 4.3 to 1.5, including $\overline{X}_j$ instead of $\Delta_{X,j}$ gives a standard error of 3.5. The final specification includes both $\Delta_{X,j}$ and $\overline{X}_j$, leading to

$$\hat{\tau}^{\text{pair}}(j) = \begin{array}{cc} 8.5 & + & 5.9 & \times & \Delta_{X,j} & -1.0 & \times & \overline{X}_j, \\ (1.5) & & (0.8) & & & (0.7) & & \end{array}$$

with again a substantial reduction of the standard error, to 1.5, relative to that using the specification without covariates, but basically the same as the specification that includes only $\Delta_{X,j}$ but not $\overline{X}_j$.

10.7   MODEL-BASED ANALYSIS OF PAIRWISE RANDOMIZED EXPERIMENTS

In principle the model-based imputation approach to the analysis of pairwise randomized experiments is little different from that for the case of stratified randomized experiments. In both cases we can carry out the analysis using the covariate that indicates pair or stratum membership, G_i. In practice, the fact that each pair contains only two units implies that we cannot be as flexible regarding the specification of the joint distribution of the potential outcomes within pairs as would be possible within strata in the stratified case where we have a larger number of units in each stratum. More appropriate is an analysis with some structure on the variance within pairs, such as a hierarchical structure.

The starting point is, as in the chapter on the model-based approach to completely randomized experiments, a model for the joint distribution of the potential outcomes given the covariates, including the pair indicators, in terms of an unknown vector parameter θ:

$$f(\mathbf{Y}(0), \mathbf{Y}(1)|\mathbf{X}, \mathbf{G}, \theta),$$

in combination with a prior distribution on θ, $p(\theta)$. These two components, in combination with the known assignment mechanism, allow us to obtain the joint distribution of the missing potential outcomes $\mathbf{Y}^{\mathrm{mis}}$ given the observed data $(\mathbf{X}, \mathbf{G}, \mathbf{Y}^{\mathrm{obs}}, \mathbf{W})$, and thus allow us to obtain the posterior distribution of the estimand of interest (e.g., the average effect of the treatment).

First we assume that, conditional on $(\mathbf{X}, \mathbf{G}, \mathbf{W})$ and the parameter θ, the potential outcomes are independent by the usual appeal to de Finetti's theorem:

$$f(\mathbf{Y}(0), \mathbf{Y}(1)|\mathbf{X}, \mathbf{G}, \mathbf{W}, \theta) = \prod_{i=1}^{N} f(Y_i(0), Y_i(1)|X_i, G_i, \theta),$$

where we implicitly assume that the parameters governing the marginal distribution of (X_i, G_i) are distinct from θ. The specific model we consider has a hierarchical structure, with pair-specific mean parameters μ_j, for $j = 1, \ldots, J$. Conditional on pair indicators, covariates, and parameters,

$$\left. \begin{pmatrix} Y_i(0) \\ Y_i(1) \end{pmatrix} \right| G_i = j, X_i = x, \mu(1), \ldots, \mu(N/2), \gamma, \beta, \sigma_c^2, \sigma_t^2$$

$$\sim \mathcal{N} \left(\begin{pmatrix} \mu(j) + x \cdot \beta \\ \mu(j) + \gamma + x \cdot \beta \end{pmatrix}, \begin{pmatrix} \sigma_c^2 & 0 \\ 0 & \sigma_t^2 \end{pmatrix} \right).$$

Conditional on pair-specific mean parameters μ_j, and common parameters γ and β, we assume that the mean of the two potential outcomes is linear in x. We assume the variances are constant across pairs but allow them to differ between potential outcomes. This model is similar in spirit to the regression model where the difference in within-pair observed outcomes was modeled as linear in the difference in within-pair covariate values. Note that given this model, the parameter γ corresponds to the super-population average treatment effect, τ_{sp}. However, in this discussion we focus on inference for the finite-sample average treatment effect, τ_{fs}, by multiply imputing the missing potential outcomes. For that reason, the interpretation of the parameters in the statistical model is incidental.

Next, we specify a model for the pair-specific means μ_j:

$$\left. \begin{pmatrix} \mu(1) \\ \vdots \\ \mu(N/2) \end{pmatrix} \right| \mathbf{G}, \mathbf{X}, \mathbf{W}, \gamma, \beta, \sigma_c^2, \sigma_t^2, \mu \sim \mathcal{N} \left(\begin{pmatrix} \mu \\ \vdots \\ \mu \end{pmatrix}, \begin{pmatrix} \sigma_\mu^2 & \cdots & 0 \\ \vdots & \ddots & \vdots \\ 0 & \cdots & \sigma_\mu^2 \end{pmatrix} \right).$$

Just as in the previous chapter, using simulation methods is generally essential here for the purpose of doing inference. Even in simple cases, there are no analytic expressions

Table 10.4. *Posterior Moments and Quantiles for Youngstown Children's Television Workshop Experiment Data from Table 10.1*

Parameter	Mean	(S.D.)	Quantiles	
			0.025	0.975
γ	8.6	(1.6)	5.1	11.7
β	5.9	(0.6)	4.8	7.0
$\ln(\sigma_c)$	1.1	(0.5)	-0.3	1.9
$\ln(\sigma_t)$	0.5	(0.7)	-0.8	1.7
μ	-9.2	(2.2)	-13.6	-4.7
$\ln(\sigma_\mu)$	1.5	(0.4)	0.4	2.2

for the posterior distributions for estimands of interest in such hierarchical models. However, as we discussed in Chapter 8, this is of no intrinsic importance. Modern Bayesian simulation methods offer efficient algorithms for drawing from the posterior distribution of the estimands given the data. We provide some details in the Appendix for this specific case.

We now implement this model on the Children's Television Workshop data. The single covariate X_i is the pre-test score. We specify independent prior distributions for μ, σ_μ^2, σ_c^2, σ_t^2, γ, and β. For the mean parameters (μ, γ, β), we use normal prior distributions centered at zero, with variance 100^2. For the three variance parameters $(\sigma_\mu^2, \sigma_c^2, \sigma_t^2)$, we use, as we did in Chapter 8, inverse Chi-squared distributions, here with parameters 1 and 1. Based on the Children's Television Workshop data, the posterior mean and variance for the average treatment effect are

$$\mathbb{E}[\tau_{fs}|\mathbf{Y}^{obs}, \mathbf{W}, \mathbf{X}, \mathbf{G}] = 8.4, \qquad \mathbb{V}(\tau_{fs}|\mathbf{Y}^{obs}, \mathbf{W}, \mathbf{X}, \mathbf{G}) = 1.7^2.$$

These estimates are quite similar to those for the regression model with the covariate equal to differences in pre-treatment variables, where we estimated the average effect to be 9.0 with a standard error of 1.5. In Table 10.4 we report posterior means, standard deviations, as well as upper and lower limits for 95% posterior intervals for all parameters.

10.8   CONCLUSION

In this chapter we analyze a special case of stratified randomized experiments: paired randomized experiments. In this special case, each of the strata, now called pairs, contains two units, one assigned to the treatment group and one assigned to the control group. This simplifies some analyses and complicates others. The Fisher exact p-value approach is conceptually not affected by the restrictions on the set of assignments. The Neyman and model-based analyses are modified to take account of the special features of this design. Within each pair there is a natural estimator for the treatment effect, namely the difference in observed outcomes for the treated unit in the pair and the control unit in the same pair. Estimation of the sampling variance for estimators is more complicated in the pairwise randomized experiment because we cannot estimate the sampling variance within each pair separately the way we could estimate the sampling variance within each stratum in the previous chapter on randomized block designs. In the Neyman analysis, we therefore focus on a statistically conservative estimator for the overall sampling variance, based on the sample variance of the within-pair differences. In the regression analyses, the differences between the stratified randomized experiment case and the pairwise randomized experiment case are reflected by the focus on the within-pair difference in outcomes as the dependent variable and the pair as the unit of analysis. Finally, just like in the randomized block design, in the model-based analyses the difference between a completely randomized and a pairwise randomized experiment is reflected by the utility of a hierarchical structure for the latter case.

NOTES

The Children's Television Workshop experiment is discussed in detail in Ball, Bogatz, Rubin, and Beaton (1973). See also Gelman and Hill (2006).

The analysis of pairwise randomized experiments is discussed in detail in standard references on classical experimental design: Hinkelmann and Kempthorne (2005, 2008), Cox and Reid (2000), Cox (1958), and Snedecor and Cochran (1967, 1989). To address the issue of the variance estimation, Lynn and McCulloch (1992) suggest estimating the variance assuming homoskedasticity, ignoring the paired design. See also Donner (1987), Diehr, Martin, Koepsell, and Cheadle (1995). Shipley, Smith, and Dramaix (1989) discuss power calculations for pairwise randomized experiments. Rosenbaum (1989b) analyzes optimal matching strategies to construct matched samples that can then be analyzed using the methods for pairwise randomized experiments discussed in this chapter.

Imai (2008) obtains the same expression for the statistically conservative estimator of the sampling variance as we do in Theorem 10.1.

APPENDIX: PROOFS

Proof of Theorem 10.1
Within each pair we have a completely randomized experiment. Therefore we can use the results on the sampling variance from Chapter 6. This directly implies unbiasedness of $\hat{\tau}^{\text{pair}}(j)$ for $\tau^{\text{pair}}(j)$, and thus unbiasedness of $\hat{\tau}$ for τ_{fs}. This proves part (i) of the theorem.

Next consider part (ii). The sampling variance expression from Chapter 6 implies

$$\mathbb{V}_W(\hat{\tau}^{\text{pair}}(j)) = \frac{S_c(j)^2}{N_c(j)} + \frac{S_t^2(j)}{N_t(j)} - \frac{S_{ct}^2(j)}{N(j)}.$$

With $N(j) = 2$ and $N_c(j) = N_t(j) = 1$, this expression simplifies to

$$\mathbb{V}_W(\hat{\tau}^{\text{pair}}(j)) = S_c(j)^2 + S_t^2(j) - \frac{S_{ct}^2(j)}{2}.$$

The within-pair variances can be written as

$$S_c^2(j) = \sum_{i:G_i=j} \left(Y_i(0) - \overline{Y}_j(0)\right)^2,$$

$$S_t^2(j) = \sum_{i:G_i=j} \left(Y_i(1) - \bar{Y}_j(1)\right)^2,$$

and

$$S_{ct}^2(j) = \sum_{i:G_i=j} \left(Y_i(1) - Y_i(0) - \tau^{\text{pair}}(j)\right)^2,$$

where

$$\bar{Y}_j(0) = \frac{1}{2} \cdot \sum_{i:G_i=j} Y_i(0) = \frac{1}{2} \cdot \left(Y_{j,A}(0) + Y_{j,B}(0)\right),$$

and

$$\bar{Y}_j(1) = \frac{1}{2} \cdot \sum_{i:G_i=j} Y_i(1) = \frac{1}{2} \cdot \left(Y_{j,A}(1) + Y_{j,B}(1)\right).$$

Because pair j comprises two units, indexed by A and B, we can rewrite these expressions as

$$S_c^2(j) = \frac{1}{2} \cdot \left(Y_{j,A}(0) - Y_{j,B}(0)\right)^2, \qquad S_t^2(j) = \frac{1}{2} \cdot \left(Y_{j,A}(1) - Y_{j,B}(1)\right)^2,$$

and

$$S_{ct}^2(j) = \frac{1}{2} \cdot \left(\left(Y_{j,A}(1) - Y_{j,A}(0)\right) - \left(Y_{j,B}(1) - Y_{j,B}(0)\right)\right)^2.$$

Hence the sampling variance of $\hat{\tau}^{\text{dif}} = (2/N) \sum_{j=1}^{N/2} \hat{\tau}^{\text{pair}}(j)$ is

$$\mathbb{V}_W(\hat{\tau}^{\text{dif}}) = \frac{4}{N^2} \sum_{j=1}^{N/2} \mathbb{V}_W(\hat{\tau}^{\text{pair}}(j)) = \frac{4}{N^2} \sum_{j=1}^{N/2} \left(S_c(j)^2 + S_t^2(j) - \frac{S_{ct}^2(j)}{2}\right).$$

Substituting for $S_c^2(j)$, $S_t^2(j)$, and $S_{ct}^2(j)$ leads to

$$\mathbb{V}_W(\hat{\tau}^{\text{dif}}) = \frac{1}{N^2} \sum_{j=1}^{N/2} \Big(2 \cdot \left(Y_{j,A}(0) - Y_{j,B}(0)\right)^2 + 2 \cdot \left(Y_{j,A}(1) - Y_{j,B}(1)\right)^2$$

$$- \left(\left(Y_{j,A}(1) - Y_{j,A}(0)\right) - \left(Y_{j,B}(1) - Y_{j,B}(0)\right)\right)^2\Big),$$

which simplifies to

$$\mathbb{V}_W(\hat{\tau}^{\text{dif}}) = \frac{1}{N^2} \sum_{j=1}^{N/2} \left(Y_{j,A}(0) + Y_{j,A}(1) - \left(Y_{j,B}(0) + Y_{j,B}(1)\right)\right)^2.$$

Finally, consider part *(iii)*. If the treatment effect is constant, then $Y_{j,A}(1) = Y_{j,A}(0) + \tau$ and $Y_{j,B}(1) = Y_{j,B}(0) + \tau$ for all j. Hence the expression for the sampling variance simplifies to

$$\mathbb{V}_W(\hat{\tau}) = \frac{1}{N^2} \sum_{j=1}^{N/2} \left(Y_{j,A}(0) + Y_{j,A}(1) - \left(Y_{j,B}(0) + Y_{j,B}(1) \right) \right)^2$$

$$= \frac{1}{N^2} \sum_{j=1}^{N/2} \left(2 \cdot Y_{j,A}(0) + \tau - \left(2 \cdot Y_{j,B}(0) + \tau \right) \right)^2$$

$$= \frac{4}{N^2} \sum_{j=1}^{N/2} \left(Y_{j,A}(0) - Y_{j,B}(0) \right)^2.$$

Now consider the variance estimator $\hat{\mathbb{V}}^{\mathrm{pair}}$,

$$\hat{\mathbb{V}}^{\mathrm{pair}} = \frac{4}{N \cdot (N-2)} \cdot \sum_{j=1}^{N/2} \left(\hat{\tau}^{\mathrm{pair}}(j) - \hat{\tau} \right)^2.$$

We calculate the expectation of $\hat{\mathbb{V}}^{\mathrm{pair}}$. Note that

$$\mathbb{E}_W \left[\hat{\tau}^{\mathrm{pair}}(j) \right] = \tau^{\mathrm{pair}}(j),$$

and

$$\mathbb{E}_W \left[\hat{\tau}^{\mathrm{pair}}(j) \cdot \hat{\tau}^{\mathrm{pair}}(k) \right] =$$

$$\begin{cases} \tau^{\mathrm{pair}}(j) \cdot \tau^{\mathrm{pair}}(k) & \text{if } j \neq k, \\ \tau^{\mathrm{pair}}(j)^2 + \frac{1}{4} \cdot \left(Y_{j,A}(0) + Y_{j,A}(1) - \left(Y_{j,B}(0) + Y_{j,B}(1) \right) \right)^2 & \text{if } j = k. \end{cases}$$

Then:

$$\mathbb{E} \left[\sum_{j=1}^{N/2} \left(\hat{\tau}^{\mathrm{pair}}(j) - \hat{\tau}^{\mathrm{dif}} \right)^2 \right] = \mathbb{E} \left[\sum_{j=1}^{N/2} \left(\hat{\tau}^{\mathrm{pair}}(j) - \frac{2}{N} \cdot \sum_{k=1}^{N/2} \hat{\tau}^{\mathrm{pair}}(k) \right)^2 \right]$$

$$= \mathbb{E} \left[\sum_{j=1}^{N/2} \hat{\tau}^{\mathrm{pair}}(j)^2 - \frac{4}{N} \cdot \sum_{j=1}^{N/2} \sum_{k=1}^{N/2} \hat{\tau}^{\mathrm{pair}}(j) \cdot \hat{\tau}^{\mathrm{pair}}(k) + \frac{2}{N} \left(\sum_{k=1}^{N/2} \hat{\tau}^{\mathrm{pair}}(k) \right)^2 \right]$$

$$= \mathbb{E} \left[\sum_{j=1}^{N/2} \hat{\tau}^{\mathrm{pair}}(j)^2 - \frac{4}{N} \cdot \sum_{j=1}^{N/2} \sum_{k=1}^{N/2} \hat{\tau}^{\mathrm{pair}}(j) \cdot \hat{\tau}^{\mathrm{pair}}(k) + \frac{2}{N} \sum_{j=1}^{N/2} \sum_{k=1}^{N/2} \hat{\tau}^{\mathrm{pair}}(j) \cdot \hat{\tau}^{\mathrm{pair}}(k) \right]$$

$$= \mathbb{E} \left[\sum_{j=1}^{N/2} \hat{\tau}^{\mathrm{pair}}(j)^2 - \frac{4}{N} \cdot \sum_{j=1}^{N/2} \hat{\tau}^{\mathrm{pair}}(j)^2 - \frac{4}{N} \cdot \sum_{j=1} \sum_{k \neq j} \hat{\tau}^{\mathrm{pair}}(j) \cdot \hat{\tau}^{\mathrm{pair}}(k) \right.$$

$$\left. + \frac{2}{N} \sum_{j=1}^{N/2} \hat{\tau}^{\mathrm{pair}}(j)^2 + \frac{2}{N} \sum_{j=1}^{N/2} \sum_{k \neq j} \hat{\tau}^{\mathrm{pair}}(j) \cdot \hat{\tau}^{\mathrm{pair}}(k) \right]$$

$$
= \frac{N-2}{N} \cdot \mathbb{E}\left[\sum_{j=1}^{N/2} \hat{\tau}^{\text{pair}}(j)^2\right] - \frac{2}{N} \cdot \mathbb{E}\left[\sum_{j=1}^{N/2}\sum_{k \neq j} \hat{\tau}^{\text{pair}}(j) \cdot \hat{\tau}^{\text{pair}}(k)\right]
$$

$$
= \frac{N-2}{N} \cdot \sum_{j=1}^{N/2} \tau^{\text{pair}}(j)^2 - \frac{2}{N} \cdot \sum_{j=1}^{N/2}\sum_{k \neq j} \tau^{\text{pair}}(j) \cdot \tau^{\text{pair}}(k)
$$

$$
+ \frac{N-2}{4 \cdot N} \cdot \sum_{j=1}^{N/2} \left(Y_{j,A}(0) + Y_{j,A}(1) - \left(Y_{j,B}(0) + Y_{j,B}(1)\right)\right)^2
$$

$$
= \sum_{j=1}^{N/2} \left(\tau^{\text{pair}}(j) - \tau_S\right)^2 + \frac{N \cdot (N-2)}{4} \cdot \mathbb{V}_W\left(\hat{\tau}^{\text{dif}}\right).
$$

Thus,

$$
\mathbb{E}\left[\hat{\mathbb{V}}^{\text{pair}}\right] = \mathbb{E}\left[\frac{4}{N \cdot (N-2)} \cdot \sum_{j=1}^{N/2} \left(\hat{\tau}^{\text{pair}}(j) - \hat{\tau}^{\text{dif}}\right)^2\right]
$$

$$
= \frac{4}{N \cdot (N-2)} \cdot \left(\sum_{j=1}^{N/2} \left(\tau^{\text{pair}}(j) - \tau_{\text{fs}}\right)^2 + \frac{N \cdot (N-2)}{4} \cdot \mathbb{V}_W\left(\hat{\tau}^{\text{dif}}\right)\right)
$$

$$
= \mathbb{V}_W\left(\hat{\tau}^{\text{dif}}\right) + \frac{4}{N \cdot (N-2)} \cdot \sum_{j=1}^{N/2} \left(\tau^{\text{pair}}(j) - \tau_{\text{fs}}\right)^2.
$$

Proof of Theorem 10.2 □

First let us expand the expectation:

$$
\mathbb{E}\left[\left(\hat{\tau}^{\text{pair}}(j) - \tau - \beta \cdot \Delta_{X,j} - \gamma \cdot (\overline{X}_j - \mu_X)\right)^2\right] \tag{A.1}
$$

$$
= \mathbb{E}_{\text{sp}}\left[\mathbb{E}_W\left[\left(\hat{\tau}^{\text{pair}}(j) - \tau - \beta \cdot \Delta_{X,j} - \gamma \cdot (\overline{X}_j - \mu_X)\right)^2\right]\right]
$$

$$
= \mathbb{E}_{\text{sp}}\left[\mathbb{E}_W\left[\left(\tau^{\text{pair}}(j) - \tau - \beta \cdot \Delta_{X,j} - \gamma \cdot (\overline{X}_j - \mu_X)\right)^2\right]\right] \tag{A.2}
$$

$$
+ \mathbb{E}_{\text{sp}}\left[\mathbb{E}_W\left[\left(\hat{\tau}^{\text{pair}}(j) - \tau^{\text{pair}}(j)\right)^2\right]\right]
$$

$$
+ 2 \cdot \mathbb{E}_{\text{sp}}\left[\mathbb{E}_W\left[\left(\hat{\tau}^{\text{pair}}(j) - \tau^{\text{pair}}(j)\right) \cdot \left(\tau^{\text{pair}}(j) - \tau - \beta \cdot \Delta_{X,j} - \gamma \cdot (\overline{X}_j - \mu_X)\right)\right]\right].
$$

Consider the three terms separately.

The first term equals

$$
\mathbb{E}_{\text{sp}}\left[\mathbb{E}_W\left[\left(\tau^{\text{pair}}(j) - \tau - \beta \cdot \Delta_{X,j} - \gamma \cdot (\overline{X}_j - \mu_X)\right)^2\right]\right]
$$

$$
= \mathbb{E}_{\text{sp}}\left[\mathbb{E}_W\left[\left(\tau^{\text{pair}}(j) - \tau - \gamma \cdot (\overline{X}_j - \mu_X)\right)^2\right]\right] + \mathbb{E}_{\text{sp}}\left[\mathbb{E}_W\left[\left(\beta \cdot \Delta_{X,j}\right)^2\right]\right]
$$

$$
- 2 \cdot \beta \cdot \mathbb{E}_{\text{sp}}\left[\mathbb{E}_W\left[\Delta_{X,j}\left(\tau^{\text{pair}}(j) - \tau - \gamma \cdot (\overline{X}_j - \mu_X)\right)\right]\right]
$$

$$= \mathbb{E}_{sp}\left[\left(\tau^{\text{pair}}(j) - \tau - \gamma \cdot (\overline{X}_j - \mu_X)\right)^2\right] + \beta^2 \cdot \mathbb{E}\left[\left(\Delta_{X,j}\right)^2\right]$$

$$= \mathbb{E}_{sp}\left[\left(\tau^{\text{pair}}(j) - \tau\right)^2\right] + \mathbb{E}_{sp}\left[\left(\gamma \cdot (\overline{X}_j - \mu_X)\right)^2\right]$$

$$\quad - 2 \cdot \mathbb{E}_{sp}\left[\left(\tau^{\text{pair}}(j) - \tau\right) \cdot \left(\gamma \cdot (\overline{X}_j - \mu_X)\right)\right] + \beta^2 \cdot \mathbb{E}\left[\left(\Delta_{X,j}\right)^2\right]$$

$$= \mathbb{E}_{sp}\left[\left(\tau^{\text{pair}}(j) - \tau\right)^2\right] + \gamma^2 \cdot \mathbb{E}_{sp}\left[\left(\overline{X}_j - \mu_X\right)^2\right]$$

$$\quad - 2 \cdot \mathbb{E}_{sp}\left[\tau^{\text{pair}}(j) \cdot \left(\gamma \cdot (\overline{X}_j - \mu_X)\right)\right] + \beta^2 \cdot \mathbb{E}\left[\left(\Delta_{X,j}\right)^2\right].$$

The second term equals

$$\mathbb{E}_{sp}\left[\mathbb{E}_W\left[\left(\hat{\tau}^{\text{pair}}(j) - \tau^{\text{pair}}(j)\right)^2\right]\right]$$

$$= \mathbb{E}_{sp}\left[\frac{1}{4} \cdot \left(Y_{j,A}(0) + Y_{j,A}(1) - \left(Y_{j,B}(0) + Y_{j,B}(1)\right)\right)^2\right],$$

which does not depend on the parameters (τ, β, γ), and therefore can be ignored for the purpose of determining the minimand of the objective function (A.1).

The third term equals

$$2 \cdot \mathbb{E}_{sp}\left[\mathbb{E}_W\left[\left(\hat{\tau}^{\text{pair}}(j) - \tau^{\text{pair}}(j)\right) \cdot \left(\tau^{\text{pair}}(j) - \tau - \beta \cdot \Delta_{X,j} - \gamma \cdot (\overline{X}_j - \mu_X)\right)\right]\right]$$

$$= -2 \cdot \beta \cdot \mathbb{E}_{sp}\left[\mathbb{E}_W\left[\left(\hat{\tau}^{\text{pair}}(j) - \tau^{\text{pair}}(j)\right) \cdot \Delta_{X,j}\right]\right]$$

$$= -2 \cdot \beta \cdot \mathbb{E}\left[\left(\hat{\tau}^{\text{pair}}(j) - \tau^{\text{pair}}(j)\right) \cdot \Delta_{X,j}\right].$$

Collecting the terms that depend on (τ, β, γ) leads to

$$= \mathbb{E}_{sp}\left[\left(\tau^{\text{pair}}(j) - \tau\right)^2\right] + \gamma^2 \cdot \mathbb{E}_{sp}\left[\left(\overline{X}_j - \mu_X\right)^2\right]$$

$$\quad - 2 \cdot \gamma \cdot \mathbb{E}_{sp}\left[\tau^{\text{pair}}(j) \cdot \left((\overline{X}_j - \mu_X)\right)\right] + \beta^2 \cdot \mathbb{E}\left[\left(\Delta_{X,j}\right)^2\right]$$

$$\quad - 2 \cdot \beta \cdot \mathbb{E}\left[\left(\hat{\tau}^{\text{pair}}(j) - \tau^{\text{pair}}(j)\right) \cdot \Delta_{X,j}\right].$$

Minimizing this over (τ, β, γ) leads to

$$\tau^* = \mathbb{E}_{sp}\left[\tau^{\text{pair}}(j)\right] = \tau_{sp},$$

$$\gamma^* = \frac{\mathbb{E}_{sp}\left[\tau^{\text{pair}}(j) \cdot \left((\overline{X}_j - \mu_X)\right)\right]}{\mathbb{E}_{sp}\left[\left(\overline{X}_j - \mu_X\right)^2\right]}, \quad \text{and} \quad \beta^* = \frac{\mathbb{E}_{sp}\left[\left(\hat{\tau}^{\text{pair}}(j) - \tau^{\text{pair}}(j)\right) \cdot \Delta_{X,j}\right]}{\mathbb{E}_{sp}\left[\left(\Delta_{X,j}\right)^2\right]}.$$

Next, consider part (ii) of the theorem:

$$(\hat{\tau}^{\text{ols}}, \hat{\beta}^{\text{ols}}, \hat{\gamma}^{\text{ols}}) = \arg\min_{\tau, \beta, \gamma} \sum_{i=1}^{N} \left(\hat{\tau}^{\text{pair}}(j) - \tau - \beta \cdot \Delta_{X,j} - \gamma \cdot (\overline{X}_j - \overline{X})\right)^2. \qquad \text{(A.3)}$$

Define $\Delta_{Y,j} = \hat{\tau}^{\mathrm{pair}}(j)$ and $\hat{\mu} = \overline{X}$. The first-order conditions for the estimators $(\hat{\tau}^{\mathrm{ols}}, \hat{\beta}^{\mathrm{ols}}, \hat{\gamma}^{\mathrm{ols}}, \hat{\mu}^{\mathrm{ols}})$ in the minimization problem (A.3) are

$$\sum_{j=1}^{N/2} \psi(\Delta_{Y,j}, \Delta_{X,j}, \overline{X}_j, \hat{\tau}^{\mathrm{ols}}, \hat{\beta}^{\mathrm{ols}}, \hat{\gamma}^{\mathrm{ols}}, \hat{\mu}^{\mathrm{ols}}) = 0,$$

where

$$\psi(\Delta_y, \Delta_x, \overline{x}, \tau, \beta, \gamma, \mu) = \begin{pmatrix} \Delta_y - \tau - \beta \cdot \Delta_x - \gamma \cdot (\overline{x} - \mu) \\ \Delta_x \cdot \left(\Delta_y - \tau - \beta \cdot \Delta_x - \gamma \cdot (\overline{x} - \mu) \right) \\ (\overline{x} - \mu) \cdot \left(\Delta_y - \tau - \beta \cdot \Delta_x - \gamma \cdot (\overline{x} - \mu) \right) \\ \overline{x} - \mu \end{pmatrix}.$$

By the same arguments as used in the proofs in Chapter 7,

$$\sqrt{N} \cdot \begin{pmatrix} \hat{\tau}^{\mathrm{ols}} - \tau_{\mathrm{sp}} \\ \hat{\beta}^{\mathrm{ols}} - \beta^* \\ \hat{\gamma}^{\mathrm{ols}} - \gamma^* \\ \hat{\mu}^{\mathrm{ols}} - \mu_X \end{pmatrix} \xrightarrow{d} \mathcal{N}\left(\begin{pmatrix} 0 \\ 0 \\ 0 \\ 0 \end{pmatrix}, \Gamma^{-1} \Delta (\Gamma')^{-1} \right),$$

where the two components of the covariance matrix are

$$\Gamma = \mathbb{E}\left[\frac{\partial}{\partial(\tau, \beta, \gamma, \mu)} \psi\left(\Delta_{Y,j}, \Delta_{X,j}, \overline{X}_j, \tau, \beta, \gamma, \mu \right) \right]\bigg|_{(\tau_{\mathrm{sp}}, \beta^*, \gamma^*, \mu_X)}$$

and

$$\Delta = \mathbb{E}\left[\psi\left(\Delta_{Y,j}, \Delta_{X,j}, \overline{X}_j, \tau_{\mathrm{sp}}, \beta^*, \gamma^*, \mu_X \right) \cdot \psi\left(\Delta_{Y,j}, \Delta_{X,j}, \overline{X}_j, \tau_{\mathrm{sp}}, \beta^*, \gamma^*, \mu_X \right)' \right].$$

$$\Gamma = \begin{pmatrix} -1 & -\mathbb{E}\left[\Delta_{X,j} \right] & -\mathbb{E}\left[\overline{X} - \mu_X \right] & \gamma^* \\ -\mathbb{E}\left[\Delta_{X,j} \right] & -\mathbb{E}\left[\Delta_{X,j}^2 \right] & -\mathbb{E}\left[\Delta_{X,j} \cdot (\overline{X} - \mu_X) \right] & \gamma^* \cdot \mathbb{E}\left[\Delta_{X,j} \right] \\ -\mathbb{E}\left[\overline{X} - \mu_X \right] & -\mathbb{E}\left[\Delta_{X,j} \cdot (\overline{X} - \mu_X) \right] & -\mathbb{E}\left[(\overline{X} - \mu_X)^2 \right] & 2 \cdot \gamma^* \cdot \mathbb{E}\left[\overline{X}_j - \mu_X \right] \\ 0 & 0 & 0 & -1 \end{pmatrix}$$

$$= \begin{pmatrix} -1 & 0 & 0 & \gamma^* \\ 0 & -\mathbb{E}\left[\Delta_{X,j}^2 \right] & 0 & 0 \\ 0 & 0 & -\mathbb{E}\left[(\overline{X}_j - \mu_X)^2 \right] & 0 \\ 0 & 0 & 0 & -1 \end{pmatrix}.$$

Thus $\mathbb{V}(\hat{\tau}^{\mathrm{ols}})$, the $(1, 1)$ element of $\Gamma^{-1} \Delta (\Gamma')^{-1}$, is equal to $\Delta_{11} - \gamma^* \cdot \Delta_{14}$, where Δ_{km} is the (k, m) element of Δ. Because

$$\Delta_{14} = \mathbb{E}\left[\left(\Delta_{Y,j} - \tau_{\mathrm{sp}} - \beta^* \cdot \Delta_{X,j} - \gamma^* \cdot (\overline{X}_j - \mu_X) \right) \cdot (\overline{X}_j - \mu_X) \right] = 0,$$

it follows that the $(1, 1)$ element of $\Gamma^{-1} \Delta (\Gamma')^{-1}$ is equal to

$$\mathbb{V}_{\mathrm{sp}}(\hat{\tau}^{\mathrm{ols}}) = \Delta_{11} = \mathbb{E}\left[\left(\Delta_{Y,j} - \tau_{\mathrm{sp}} - \beta^* \cdot \Delta_{X,j} - \gamma^* \cdot (\overline{x} - \mu_X) \right)^2 \right].$$

## CHAPTER 11

Case Study: An Experimental Evaluation of a Labor Market Program

11.1 INTRODUCTION

In this chapter we illustrate some of the methods discussed in the previous chapters in an application. The application involves a social program designed to improve labor market outcomes for individuals with relatively poor skills and labor market histories: the Saturation Work Initiative Model (SWIM) program in San Diego, evaluated during the period 1985–1987. As is typical, a substantial amount of background information on the individuals in the program was collected, including demographics and recent labor market histories, allowing us to investigate heterogeneity in the effects of the program. The outcomes of interest, post-program earnings and employment records, are either discrete or mixed discrete-continuous, suggesting that constant additive treatment-effect assumptions are typically not plausible.

Using these data we will calculate Fisher exact p-values for sharp null hypotheses and construct Neyman large-sample confidence intervals. We will also discuss, in detail, regression and model-based inferences for various average treatment effects, using the covariates to increase precision as well as to estimate treatment effects for subpopulations. We emphasize the model selection choices and the various other decisions faced by researchers.

11.2 THE SAN DIEGO SWIM PROGRAM DATA

SWIM primarily targeted women who were eligible for Aid to Families with Dependent Children (AFDC), with children at least six years old (although, as the summary statistics show, there was a substantial proportion of women with younger children, a small number of men, and some individuals with no children). It was a mandatory program, with fairly strong participation enforcement, and provided a sequence of group job search, unpaid work experience, education, and job skills training. Compared to similar programs in other locations, it had broad coverage, with the intention to reach a wide range of individuals eligible for AFDC, including those who may not have participated in such assistance programs. The average cost of participating in this program was $919 per trainee, paid for by the local authorities. The participants faced no direct

Table 11.1. *Summary Statistics San Diego SWIM Data*

Variable		All (N = 3211)		Controls (N_c = 1607)		Treated (N_t = 1604)	
		Mean	(S.D.)	Mean	(S.D.)	Mean	(S.D.)
Pre-treatment variables							
female	female	0.91	(0.28)	0.92	(0.28)	0.91	(0.28)
agege35	(age ≥ 35)	0.46	(0.50)	0.46	(0.50)	0.46	(0.50)
hsdip	(high school diploma)	0.56	(0.50)	0.56	(0.50)	0.56	(0.50)
nevmar	(never married)	0.30	(0.46)	0.30	(0.46)	0.30	(0.46)
divwid	(divorced or widowed)	0.37	(0.48)	0.37	(0.48)	0.36	(0.48)
numchild	(number of children)	1.76	(1.08)	1.76	(1.07)	1.76	(1.10)
chldlt6	(children younger than 6)	0.10	(0.30)	0.10	(0.31)	0.10	(0.29)
af-amer	(African-American)	0.42	(0.49)	0.43	(0.49)	0.42	(0.49)
hisp	(Hispanic)	0.25	(0.44)	0.25	(0.43)	0.26	(0.44)
earnyrm1	(earnings year minus 1)	1.57	(3.54)	1.60	(3.56)	1.53	(3.51)
empyrm1	(positive earnings year minus 1)	0.39	(0.49)	0.40	(0.49)	0.39	(0.49)
Outcomes variables							
earnyr1	(earnings year 1)	1.85	(3.78)	1.69	(3.76)	2.02	(3.80)
empyr1	(positive earnings year 1)	0.46	(0.50)	0.40	(0.49)	0.52	(0.50)
earnyr2	(earnings year 2)	2.57	(5.08)	2.26	(4.68)	2.89	(5.44)
empyr2	(positive earnings year 2)	0.45	(0.50)	0.40	(0.49)	0.49	(0.50)

expenses for the program, although there are likely to have been indirect costs, such as child care and travel expenses. The evaluation started in 1985. Eligible individuals enrolled in the study were randomized to receive training or not. The randomization did use demographics and labor market histories. This program is typical of many labor market programs in the 1980s and 1990s, a substantial number of which were evaluated using randomized experiments. The general emphasis on experimental evaluations around this time was motivated by research (most notably a paper by Lalonde published in 1986, whose data we use in other chapters) that had concluded that non-experimental evaluations (in practice with analyses limited to linear covariance adjustment or regression methods) were often unable to replicate experimental results, and therefore claimed that non-experimental evaluations were not credible in these settings. See the notes at the end of this chapter for more discussion on this topic.

Table 11.1 presents some summary statistics for this data set. We have information on $N = 3{,}211$ individuals, with $N_t = 1{,}604$ randomly assigned to receive the training, and the remaining $N_c = 1{,}607$ assigned to the control group, which was not to receive any training as part of the SWIM program. Individuals in the control group had no access to SWIM program services but may have had access to other, possibly similar, services outside of the SWIM program. This is a common problem with social programs, where individuals assigned to the control group often have access to related programs. This feature implies that the effects should be interpreted as the effect of participating in the program versus being denied access to this particular program, rather than as the effect of participating versus not participating in any job-training program.

There are two sets of pre-treatment variables. First there are some covariates measuring individual-level background characteristics. These pre-treatment variables include whether the individual had a high school diploma, was female (`female`), was at least 35 years old (`agege35`), had a high school diploma (`hsdip`), had never married (`nevmar`), and was divorced or widowed (`divwid`), the number of children (`numchild`); whether any children were present in the household who were younger than six (`chldlt6`); and whether the individual was African-American (`af-amer`) or Hispanic (`hisp`). Second, there are records for earnings for the year prior to the randomization. We use both the actual earnings measure (`earnyrm1`) and an indicator for positive earnings in this pre-randomization year (`empyrm1`). The outcome variables of interest are total earnings in the first and second year post-randomization (`earnyr1` and `earnyr2`) and indicators for these earnings being positive. For these covariates and the outcome variables, means and standard deviations for the entire sample, as well as means and standard deviations by treatment status, are displayed in Table 11.1. Notice that approximately 60% of the participants have no earning the year prior to the assignment, suggesting that simple gain scores may not be particularly helpful. All earnings variables are yearly earnings, measured in thousands of dollars.

11.3   FISHER'S EXACT P-VALUES

First we analyze the experimental data using Fisher's exact p-value approach discussed in Chapter 5. We focus on tests of the null hypothesis that there is no effect of the program for any individual:

$$H_0: \ Y_i(0) = Y_i(1), \quad \text{for } i = 1, \ldots, N.$$

We calculate the p-values for tests of this null hypothesis for a variety of test statistics using the first and second year post-program earnings (`empyr1` and `empyr2`) as the outcomes. We analyze the full sample and, separately, the subsamples created by whether individuals had graduated from high school. Table 11.2 contains all the p-values discussed in the text. Although for illustrative purposes we calculate a large number of p-values, we should note that the formal interpretation of each holds for one p-value at a time.

Our primary p-value is based on the difference in ranks in first year post-program earnings. As before, we define the normalized rank as:

$$R_i = \sum_{i'=1}^{N} \mathbf{1}_{Y_{i'}^{\text{obs}} < Y_i^{\text{obs}}} + \frac{1}{2}\left(1 + \sum_{i'=1}^{N} \mathbf{1}_{Y_{i'}^{\text{obs}} = Y_i^{\text{obs}}}\right) - \frac{N+1}{2}.$$

Then the rank-based test statistic is

$$T^{\text{rank}} = \left|\overline{R}_{\text{t}} - \overline{R}_{\text{c}}\right|,$$

where $\overline{R}_{\text{t}}$ and $\overline{R}_{\text{c}}$ are the average ranks in the treatment and control groups respectively. The average rank is higher for individuals in the treatment group than for individuals in

Table 11.2. *P-Values for Fisher Exact Tests on San Diego SWIM Data (based on 1,000,000 draws from randomization distribution)*

Post-Program Earnings	Statistic	All (3,211)	No High School (1,409)	High School (1,802)
Year 1	T^{rank}	< 0.0001	< 0.0001	0.0014
	$T^{\text{rank}-\text{gain}}$	< 0.0001	< 0.0001	0.0001
	T^{dif}	0.0131	0.0051	0.1967
Year 2	T^{rank}	< 0.0001	0.0017	< 0.0001
	$T^{\text{rank}-\text{gain}}$	< 0.0001	0.0020	0.0002
	T^{dif}	0.0004	0.0980	0.0018

the control group, leading to a p-value less than 0.0001, strong evidence against the null hypothesis of no effect of the treatment.

For comparison purposes, we report p-values for two other statistics. The first of these exploits the additional information in the form of the covariates. Specifically, because we have values for earnings prior to the program, we may wish to base the test statistic on the rank of the gains, rather than the rank of the level of earnings. Let X_i denote the level of prior earnings. Then the rank of the gains is defined as

$$R'_i = \sum_{i'=1}^{N} \mathbf{1}_{Y^{\text{obs}}_{i'}-X_{i'} < Y^{\text{obs}}_{i}-X_i} + \frac{1}{2}\left(1 + \sum_{i'=1}^{N} \mathbf{1}_{Y^{\text{obs}}_{i'}-X_{i'} = Y^{\text{obs}}_{i}-X_i}\right) - \frac{N+1}{2}.$$

Then the rank-based test statistic is

$$T^{\text{rank,gain}} = \left|\overline{R'}_t - \overline{R'}_c\right|,$$

where $\overline{R'}_t$ and $\overline{R'}_c$ are the average ranks of the gain in the treatment and control groups respectively. The p-values based on this statistic are similar to those based on the simple rank statistic. In both cases the evidence against the null is strong for the full sample and for the subsamples based on whether the individuals have a high school degree or not.

The third statistic is the widely (perhaps too widely) used difference in means of the observed outcomes:

$$T^{\text{dif}} = \left|\overline{Y}^{\text{obs}}_t - \overline{Y}^{\text{obs}}_c\right|.$$

Here the evidence against the null hypothesis is statistically significant at conventional levels in most cases, although not quite as strong as for the rank-based tests. The reason appears to be that the distribution of the outcome is heavily skewed. About 50% of the individuals have positive earnings in either Year 1 or Year 2 post-treatment. Figures 11.1 and 11.2 present histograms of the level of earnings and its logarithm, for those with positive earnings. For such distributions, rank-based tests tend to be more sensitive to violations of the null hypothesis of no effect of the treatment than tests based on averages of the levels.

In principle, we can also use sequences of Fisher tests to create Fisher intervals as described in Chapter 5. Such Fisher intervals require specification of the treatment effect for each unit. In most cases we would implement this by considering the set of values c

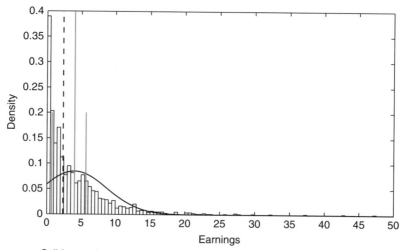

Solid curve is normal approximation, vertical tall line is mean, vertical dashed
line is median, short vertical solid lines are 0.25 and 0.75 quantiles.

Figure 11.1. Histogram-based estimate of the distribution of Year 1 earnings, for those with positive earnings, San Diego SWIM program data

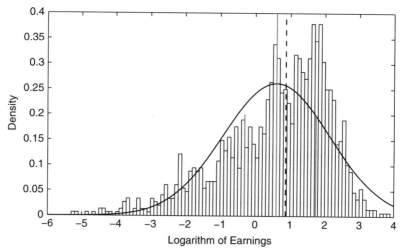

Solid curve is normal approximation, vertical tall line is mean, vertical dashed
line is median, short vertical solid lines are 0.25 and 0.75 quantiles.

Figure 11.2. Histogram-based estimate of the distribution of the logarithm of year 1 earnings, for those with positive earnings, San Diego SWIM program data

such that we cannot reject the null hypothesis of a constant treatment effect equal to c. In this data set, such an approach is possible, but it is not attractive. Many individuals have earnings equal to zero in some year, because they do not have a job in that year. It is difficult to imagine that the training program would move all these individuals to some positive amount of earnings. On substantive grounds it is therefore extremely unlikely that there is a constant treatment effect, even after considering transformations of the outcome. We will therefore not pursue this strategy.

11.4   NEYMAN'S REPEATED SAMPLING-BASED POINT ESTIMATES AND LARGE-SAMPLE CONFIDENCE INTERVALS

In this section we apply Neyman's repeated sampling approach. For the full sample, as well as various subsamples, we estimate the average treatment effect on earnings in the first year after the program, and construct confidence intervals for this average effect. The results for these analyses are displayed in Table 11.3.

First we consider the full sample. The simple difference in average treatment and control outcomes is

$$\hat{\tau}^{\text{dif}} = \overline{Y}_t^{\text{obs}} - \overline{Y}_c^{\text{obs}} = 2.02 - 1.69 = 0.33, \tag{11.1}$$

with sampling variance

$$\mathbb{V}_W \left(\hat{\tau}^{\text{dif}} \right) = \mathbb{E} \left[\left(\overline{Y}_t^{\text{obs}} - \overline{Y}_c^{\text{obs}} - \tau_{\text{fs}} \right)^2 \right] = \frac{S_c^2}{N_c} + \frac{S_t^2}{N_t} - \frac{S_{ct}^2}{N}.$$

Using the standard estimator for this sampling variance discussed in Chapter 6, we find

$$\hat{\mathbb{V}}^{\text{neyman}} = \frac{s_c^2}{N_c} + \frac{s_t^2}{N_t} = \frac{3.76^2}{1607} + \frac{3.80^2}{1604} = 0.13^2.$$

The implied large sample 95% confidence interval is

$$\text{CI}^{0.95}(\tau_{\text{fs}}) = \left(\hat{\tau}^{\text{dif}} - 1.96 \cdot \sqrt{\frac{s_c^2}{N_c} + \frac{s_t^2}{N_t}}, \hat{\tau}^{\text{dif}} + 1.96 \cdot \sqrt{\frac{s_c^2}{N_c} + \frac{s_t^2}{N_t}} \right) = (0.07, 0.59).$$

$$\tag{11.2}$$

Next, we carry out the same calculations for some subpopulations. This serves two purposes. First, we may be interested in average treatment effects by subpopulations. Second, it may lead to more precise estimates of the overall average treatment effect. We begin by partitioning the sample into those at least thirty-five years old and those younger than thirty-five. The subsample of older individuals consists of 1,473 individuals, and the younger subsample consists of 1,738 individuals. For the older group we find

$$\hat{\tau}^{\text{dif}}(\text{old}) = 0.50 \ (\widehat{\text{s. e.}} \ 0.21), \qquad \text{CI}^{0.95}(\tau_{\text{fs}}(\text{old})) = (0.09, 0.91).$$

For the younger group the estimated average treatment effect is

$$\hat{\tau}^{\text{dif}}(\text{young}) = 0.19 \ (\widehat{\text{s. e.}} \ 0.17), \qquad \text{CI}^{0.95}(\tau_{\text{fs}}(\text{young})) = (-0.14, 0.51).$$

Next we partition the sample into those with no employment experience during the pre-program period, as indicated by zero earnings in the pre-program year (empyrm1 equal to zero, which holds for 1,949 individuals) versus those with positive experience (1,262 individuals with empyrm1 equal to one). For the first group, the estimated effect and associated estimated standard error are

Table 11.3. *Estimates for Average Treatment Effects on Year 1 Earnings Based on Neyman's Repeated Sampling Approach, San Diego SWIM Program Data*

Post-Program Earnings		All (3,211)	Young (1,738)	Old (1,473)	Unemployed (1,949)	Employed (1,262)	No HS (1,409)	HS (1,802)
Year 1	Est	0.33	0.19	0.50	0.34	0.38	0.41	0.27
	(s. e.)	(0.13)	(0.17)	(0.21)	(0.13)	(0.25)	(0.15)	(0.21)
Year 2	Est	0.63	0.52	0.76	0.58	0.77	0.31	0.87
	(s. e.)	(0.18)	(0.24)	(0.27)	(0.19)	(0.33)	(0.19)	(0.28)

$$\hat{\tau}^{\text{dif}}(\text{unempl}) = 0.34 \ (\widehat{\text{s.e.}} \ 0.13), \qquad \text{CI}^{0.95}(\tau_{\text{fs}}(\text{unempl})) = (0.08, 0.601).$$

For the second group, the estimated average treatment effect is

$$\hat{\tau}^{\text{dif}}(\text{empl}) = 0.38 \ (\widehat{\text{s.e.}} \ 0.25), \qquad \text{CI}^{0.95}(\tau_{\text{fs}}(\text{emp})) = (-0.12, 0.87).$$

We can also combine these to obtain an estimate of the overall average treatment effect τ_{fs} that is possibly more precise than $\hat{\tau}^{\text{dif}}$. We implement this by weighting the two estimates, $\hat{\tau}^{\text{dif}}(\text{empl})$ for the employed and $\hat{\tau}^{\text{dif}}(\text{unempl})$ for the unemployed, by their shares in the full sample. These shares are $1{,}262/(1{,}262 + 1{,}949) = 0.39$ for those with positive earnings and 0.61 for those with zero earnings in the year prior to the program. The weighted estimated average treatment effect, or employment-adjusted estimate is

$$\hat{\tau}^{\text{strat}} = \frac{N(\text{empl})}{N(\text{empl}) + N(\text{unempl})} \cdot \hat{\tau}^{\text{dif}}(\text{empl}) + \frac{N(\text{unempl})}{N(\text{empl}) + N(\text{unempl})} \cdot \hat{\tau}^{\text{dif}}(\text{unempl})$$

$$= \frac{1262}{1262 + 1949} \cdot 0.38 + \frac{1949}{1262 + 1949} \cdot 0.34 = 0.36 \ (\widehat{\text{s.e.}} \ 0.15),$$

with the large sample 95% confidence interval equal to

$$\text{CI}^{0.95}_{\text{combined}}(\tau_{\text{fs}}) = (0.11, 0.61).$$

Note that this point estimate differs slightly from $\hat{\tau}$ in (11.1) where we took the simple difference in average outcomes by treatment status, which reflects a small imbalance in the proportion of treated and control units among those with positive and zero earnings. More specifically, among those with positive earnings, 49.2% were assigned to the active treatment and 50.8% were assigned to the control treatment; and among those with zero earnings, 50.4% were assigned to the active treatment and 49.6% were assigned to the control treatment. This does not mean the randomization was compromised, merely that there is some random variation in these proportions because the randomization was not stratified on initial employment status.

The estimated sampling variance of the average treatment effect is also affected by the post-stratification on prior employment. If the treatment effect varies by covariates, then estimating the average effects within relatively homogeneous subpopulations, and then averaging over them will often reduce the sampling variance and lead to more precise inferences. Here, the change in estimated precision is fairly small.

Finally, we partition the sample into those with no high school diploma (1,409 individuals) and those with a high school diploma (1,802 individuals). For the high school dropouts, we find

$$\hat{\tau}^{\text{dif}}(\text{no-hs}) = 0.41 \ (\widehat{\text{s.e.}} \ 0.15), \quad \text{CI}^{0.95}(\tau_{\text{fs}}(\text{no-hs})) = (0.12, 0.70).$$

For the high school graduates, the estimated average treatment effect is

$$\hat{\tau}^{\text{dif}}(\text{hs}) = 0.27 \ (\widehat{\text{s.e.}} \ 0.21), \quad \text{CI}^{0.95}(\tau_{\text{fs}}(\text{hs})) = (-0.14, 0.68).$$

11.5 REGRESSION-BASED ESTIMATES

We now consider regression-based estimates of the average effect of the treatment, on the earnings in both the first and the second year after the program started. We consider specifications of the regression function that include the set of eleven pre-treatment variables listed in Table 11.1, indicators for being female (female), being at least 35 years old (agege35), having a high school diploma (hsdip), never having been married (nevmar), being divorced or widowed (divwid), having children younger than six years (chldlt6), being African-American (af-amer), being Hispanic (hisp), the discrete variable giving the number of children (numchild), and the lagged outcome, earnings in the year preceding the training program (earnyrm1), and an indicator for earnings being positive in that prior year (empyrm1). Denoting the row vector of these eleven pre-treatment variables by X_i, the basic specification of the regression function we estimate includes an intercept, the indicator for the treatment, the vector of pre-treatment variables, and the interaction of the two:

$$Y_i^{\text{obs}} = \alpha + \tau \cdot W_i + (X_i - \overline{X})\beta + W_i \cdot (X_i - \overline{X})\gamma + \varepsilon_i.$$

The covariates are included in deviations from the sample average, so that the estimated coefficient on the treatment indicator, τ, can be interpreted as an estimator for the average effect of the treatment in the population. Implicitly this specification allows for separate slope coefficients for treated and control regression functions. For comparison, we also include least squares estimates of the regression function without pre-treatment variables:

$$Y_i^{\text{obs}} = \alpha + \tau \cdot W_i + \varepsilon_i,$$

which gives the least squares estimate for τ equal to the difference in average outcomes by treatment status,

$$\hat{\tau}^{\text{ols}} = \hat{\tau}^{\text{dif}} = \overline{Y}_{\text{t}}^{\text{obs}} - \overline{Y}_{\text{c}}^{\text{obs}} = 2.02 - 1.69 = 0.33.$$

The estimates of the average effect of the treatment do not change much with the inclusion of the eleven pre-treatment variables. For the first year earnings, the point estimate increases from 0.33 (in thousands of dollars) to 0.36, and in the second year, the estimate increases from 0.63 to 0.66. This is not unexpected: the fact that the randomization was done without regard to the pre-treatment variables implies that, on average, the

Table 11.4. *Regression Estimates for Average Treatment Effects on Earnings, for the San Diego Swim Data*

Covariates	Earnings Year 1				Earnings Year 2			
	Est	(s.e.)	Est	(s.e.)	Est	(s.e.)	Est	(s.e.)
Treat	0.33	(0.13)	0.36	(0.12)	0.63	(0.18)	0.66	(0.17)
Intercept	1.69	(0.09)	1.68	(0.09)	2.26	(0.12)	2.25	(0.11)
Covariates								
female			0.35	(0.29)			−0.03	(0.39)
agege35			−0.09	(0.17)			−0.01	(0.23)
hsdip			0.79	(0.20)			0.86	(0.25)
nevmar			0.38	(0.21)			0.47	(0.29)
divwid			0.32	(0.20)			0.41	(0.26)
numchild			0.10	(0.08)			0.03	(0.11)
chdlt6			−0.46	(0.25)			−0.20	(0.36)
af-amer			−0.22	(0.22)			−0.54	(0.28)
hisp			0.05	(0.23)			−0.25	(0.30)
earnyrm1			0.33	(0.08)			0.33	(0.09)
empyrm1			0.75	(0.30)			0.78	(0.34)
Interactions with treatment indicator								
treat×female			−0.01	(0.43)			0.48	(0.59)
treat×age 35			0.17	(0.25)			0.18	(0.36)
treat×high school dip			−0.15	(0.27)			0.54	(0.36)
treat×never married			−0.40	(0.29)			−0.33	(0.41)
treat×divorced/widowed			0.34	(0.29)			0.36	(0.41)
treat×number of children			−0.18	(0.11)			−0.29	(0.15)
treatchdlt6			0.42	(0.39)			1.15	(0.60)
treat×african-american			−0.29	(0.31)			−0.14	(0.42)
treat×hispanic			−0.26	(0.34)			0.31	(0.48)
treatearnyrm1			0.09	(0.10)			0.22	(0.13)
treatempyrm1			−0.30	(0.40)			−0.72	(0.50)
R-squared	0.002		0.190		0.004		0.151	

pre-treatment variables should be approximately the same in treatment group and control group and that their inclusion or omission usually should not change point estimates of treatment effects as a result of the linear predictive power. The estimated standard error does not change much either. They decrease slightly, as a result of the predictive power of the covariates, but because this predictive power is fairly modest, the reduction in estimated standard error is small.

The main interest in the regression estimates is that they provide some evidence regarding heterogeneity in the effect of the program, which can be seen directly by inspecting the least squares estimates of the coefficients of the interactions of the pre-treatment variables with the treatment indicator, as reported in Table 11.4. In addition to these estimates, we also report tests of hypotheses about the coefficients in the linear

Table 11.5. *P-Values for Tests of Constant and Zero Treatment Effects Assumptions, for San Diego SWIM Data*

Null Hypothesis		Earnings Year 1	Earnings Year 2
Zero effect	$\mathcal{X}^2(12)$ approximation	0.018	<0.001
	Fisher exact p-value	0.157	0.014
Constant effect	$\mathcal{X}^2(11)$ approximation	0.122	0.002

regression model. Specifically we consider two null hypotheses. First, consider the null hypotheses that all least squares coefficients involving the treatment indicator are equal to zero. Formally,

$$H_0: \quad \tau = 0 \text{ and } \gamma = 0,$$

against the alternative that either τ or some components of γ differ from zero,

$$H_a: \quad \tau \neq 0 \text{ or } \gamma \neq 0,$$

where 0 denotes a vector of zeros. The results from an F-test on the least squares coefficients are reported in Table 11.5. The value of the F-statistic using the first-year earnings as the outcome variable is 2.11, leading to a p-value of 0.018 based on the asymptotic approximation using the F-distribution with 12 degrees of freedom. We also carried out a different version of this test, where we used the F-statistic in a Fisher-exact-p-value calculation, under the null of no effect of the treatment whatsoever. This led to a considerably less significant p-value of 0.157. The results for the p-value are also reported in Table 11.5. For the second-year earnings outcome, the F-statistic is 3.78, leading to a p-value based on the F-distribution less than 0.001, and a p-value based on the randomization distribution equal to 0.014. Next, we considered the null hypothesis of no treatment effect heterogeneity by pre-treatment variables. In terms of the least squares coefficients, this corresponds to testing the null hypothesis

$$H_0: \quad \gamma = 0,$$

against the alternative that some components of γ differ from zero,

$$H_a: \quad \gamma \neq 0.$$

We find somewhat different results for the first- and second-year earnings. For the first year we find an F-statistic equal to 1.50, leading to a p-value of 0.122. This suggests little evidence for heterogeneity of the treatment effect. The F-statistic for second-year earnings is 2.68, leading to a p-value of 0.002, suggesting clear evidence that the treatment effect on second-year earnings varies by the values of the pre-treatment variables.

11.6 MODEL-BASED POINT ESTIMATES

Now let us consider the model-based approach. To avoid reporting a large number of estimates, we focus first on estimating the average treatment effect for earnings in the second year.

A simple strategy is to specify a joint normal distribution for the two potential outcomes with unit correlation. If we use a normal prior distribution for the mean parameters and inverse Chi-squared distributions for the two variance parameters, we return to the case analyzed in Chapter 8. With the number of observations as large as in the SWIM program, the choice of prior distribution is unlikely to matter much. We estimate two versions of the normal model. First, a model with no covariates; for the mean parameters, we use normal prior distributions centered at zero with prior variances equal to 100^2. For the variance parameters, we use inverse Chi-squared distributions with parameters equal to $1/2$ and 0.0005. The posterior mean for τ_{fs} is 0.33, and the posterior standard deviation is equal to 0.09. Next we include the eleven covariates in the model, assuming they enter linearly for the mean. Now the posterior mean for τ_{fs} is 0.36 and the posterior standard deviation is 0.08. Although the covariates are moderately strongly associated with the potential outcomes, including the covariates does not affect the posterior distribution for the average effect of interest very much. These results are very similar to those obtained through the Neyman approach, which is not surprising because the sample size implies that, using versions of the central limit theorem, normal distributions are likely to give accurate approximations to both the sampling and the posterior distributions.

It is clear, however, that the model used in this first attempt is not an appropriate one. The distributions are far from normal, with 54% of individuals having zero earnings one year after the program started, as the summary statistics in Table 11.1 show. A more plausible approximation to the distribution of earnings in each treatment regime is therefore a mixed discrete-continuous distribution. We use the following model with one parameter governing the probability of the point mass at zero and a normal distribution for the continuous component (which led to a better fit than a log normal distribution for the continuous part),

$$\Pr(Y_i(0) > 0|X_i, \theta) = \frac{\exp(\gamma_c)}{1 + \exp(\gamma_c)}, \quad \left(Y_i(0)|Y_i(0) > 0, X_i, \theta\right) \sim \mathcal{N}(\mu_c, \sigma_c^2),$$

$$\Pr(Y_i(1) > 0|X_i, \theta) = \frac{\exp(\gamma_t)}{1 + \exp(\gamma_t)}, \quad \left(Y_i(1)|Y_i(1) > 0, X_i, \theta\right) \sim \mathcal{N}(\mu_t, \sigma_t^2),$$

and assume independence between the potential outcomes. For this specification, it is difficult to derive an analytic expression for the posterior distribution of the average treatment effect in terms of the observed data for most prior distributions. We focus, therefore, on simulation methods.

We use independent prior distributions for the six elements of the parameter vector $\theta = (\gamma_c, \gamma_t, \mu_c, \mu_t, \sigma_c^2, \sigma_t^2)$. For γ_c, γ_t, μ_c, and μ_t, we use normal prior distributions centered at zero and with variance equal to 100^2. The prior distributions for the variance parameters are inverse Chi-squared, with parameters $1/2$ and $\sigma_c^2/2$ and $\sigma_t^2/2$, respectively. The mean and standard deviation of the posterior distribution for τ are 0.33 and

Table 11.6. *Posterior Means and Standard Deviations for Model-Based Imputation Estimates, Year 1 Earnings, for San Diego SWIM Data*

	Linear Model No Covariates		Linear Model Covariates		Two-Part Model No Covariates				Two-Part Model Covariates			
					Logit		Normal		Logit		Normal	
	Mean	(S.D.)	Mean	(S.D.)	Mean	(S.D.)	Mean	(S.D.)	Mean	(S.D.)	Mean	(S.D.)
Control Outcome												
Intercept	1.69	(0.09)	−0.13	(0.40)	−0.39	(0.05)	4.17	(0.20)	−1.56	(0.27)	1.77	(0.86)
female			0.35	(0.32)					0.04	(0.21)	0.55	(0.68)
agege35			−0.09	(0.19)					−0.28	(0.13)	0.19	(0.41)
hsdip			0.78	(0.19)					0.46	(0.12)	1.39	(0.42)
nevmar			0.38	(0.24)					0.26	(0.16)	0.52	(0.52)
divwid			0.32	(0.21)					0.13	(0.14)	0.68	(0.46)
numchild			0.10	(0.09)					0.06	(0.06)	0.14	(0.19)
chldlt6			−0.47	(0.29)					−0.13	(0.19)	−0.89	(0.63)
af-amer			−0.22	(0.21)					−0.05	(0.14)	−0.58	(0.45)
hisp			0.05	(0.24)					0.04	(0.16)	0.13	(0.53)
earnyrm1			0.33	(0.03)					0.10	(0.02)	0.32	(0.05)
empyrm1			0.75	(0.21)					1.49	(0.14)	−0.63	(0.44)
σ_c	3.76	(0.07)	3.45	(0.06)			4.97	(0.14)			4.72	(0.13)
Treated Outcome												
Intercept	2.02	(0.09)	0.69	(0.38)	0.06	(0.05)	3.92	(0.16)	−0.62	(0.24)	2.67	(0.67)
female			0.34	(0.30)					0.09	(0.20)	0.08	(0.51)
agege35			0.08	(0.18)					−0.17	(0.12)	0.31	(0.32)
hsdip			0.64	(0.18)					0.22	(0.11)	0.91	(0.32)
nevmar			−0.02	(0.23)					0.23	(0.15)	−0.39	(0.41)
divwid			0.66	(0.21)					0.51	(0.13)	0.59	(0.35)
numchild			−0.08	(0.09)					−0.08	(0.05)	−0.07	(0.15)
chldlt6			−0.04	(0.30)					−0.23	(0.18)	0.40	(0.53)
af-amer			−0.51	(0.21)					−0.14	(0.13)	−0.80	(0.34)
hisp			−0.21	(0.24)					−0.09	(0.15)	−0.29	(0.40)
earnyrm1			0.42	(0.03)					0.09	(0.03)	0.43	(0.04)
empyrm1			0.45	(0.21)					1.13	(0.14)	−0.37	(0.34)
σ_t	3.80	(0.07)	3.38	(0.06)			4.53	(0.11)			4.10	(0.10)
τ_{fs}	0.33	(0.09)	0.36	(0.08)			0.33	(0.09)			0.36	(0.08)

0.09, respectively. The posterior means and standard deviations for all elements of θ are presented in Tables 11.6 (year 1 earnings) and 11.7 (year 2 earnings).

Next, we consider a similar mixed discrete-continuous model with covariates, often called a "two-part" model. Let X_i denote the vector of covariates reported in Table 11.1. The model is now

$$\Pr(Y_i(0) > 0 | X_i = x, \theta) = \frac{\exp(x\gamma_c)}{1 + \exp(x\gamma_c)}, \quad \left(Y_i(0) | X_i = x, Y_i(0) > 0, \theta\right) \sim \mathcal{N}(x\beta_c, \sigma_c^2),$$

$$\Pr(Y_i(1) > 0 | X_i = x, \theta) = \frac{\exp(x\gamma_t)}{1 + \exp(x\gamma_t)} \quad \text{and} \quad \left(Y_i(1) | X_i = x, Y_i(1) > 0, \theta\right) \sim \mathcal{N}(x\beta_t, \sigma_t^2).$$

Table 11.7. *Posterior Means and Standard Deviations for Model-Based Imputation Estimates, Year 2 Earnings, for San Diego SWIM Data*

	Linear Model No Covariates		Linear Model Covariates		Two-Part Model No Covariates				Two-Part Model Covariates			
					Logit		Normal		Logit		Normal	
	Mean	(S.D.)	Mean	(S.D.)	Mean	(S.D.)	Mean	(S.D.)	Mean	(S.D.)	Mean	(S.D.)
Control Outcome												
Intercept	2.26	(0.12)	0.96	(0.50)	−0.40	(0.05)	5.62	(0.23)	−1.03	(0.25)	4.04	(1.01)
female			−0.06	(0.40)					−0.12	(0.20)	−0.26	(0.82)
agege35				(0.23)					−0.18	(0.12)	0.35	(0.51)
hsdip			0.88	(0.24)					0.08	(0.12)	2.18	(0.49)
nevmar			0.46	(0.31)					0.29	(0.15)	0.70	(0.64)
divwid			0.40	(0.27)					0.30	(0.13)	0.40	(0.56)
numchild			0.03	(0.11)						(0.05)	0.16	(0.24)
chldlt6			−0.22	(0.38)					−0.02	(0.17)	−0.55	(0.77)
af-amer			−0.52	(0.26)					0.05	(0.12)	−1.59	(0.55)
hisp			−0.24	(0.31)					0.06	(0.14)	−0.83	(0.62)
earnyrm1			0.33	(0.04)					0.06	(0.02)	0.38	(0.06)
empyrm1			0.76	(0.27)					1.06	(0.14)	−0.61	(0.51)
σ_c	4.68	(0.08)	4.42	(0.08)			5.97	(0.17)			5.65	(0.16)
Treated Outcome												
Intercept	2.89	(0.13)	1.05	(0.55)	−0.03	(0.05)	5.86	(0.24)	−0.73	(0.24)	4.06	(0.98)
female			0.43	(0.43)					0.10	(0.18)	0.02	(0.75)
agege35			0.18	(0.28)					0.01	(0.11)	0.36	(0.46)
hsdip			1.39	(0.28)					0.36	(0.12)	2.09	(0.49)
nevmar			0.15	(0.34)					0.13	(0.14)	0.10	(0.59)
divwid			0.78	(0.31)					0.33	(0.14)	0.87	(0.51)
numchild			−0.26	(0.12)					−0.12	(0.06)	−0.22	(0.24)
chldlt6			0.96	(0.45)					0.26	(0.18)	1.17	(0.72)
af-amer			−0.65	(0.30)					−0.20	(0.12)	−0.96	(0.51)
hisp			0.06	(0.36)					0.33	(0.14)	−0.61	(0.57)
earnyrm1			0.55	(0.04)					0.09	(0.02)	0.59	(0.06)
empyrm1			0.06	(0.31)					0.77	(0.13)	−1.23	(0.52)
σ_t	5.44	(0.10)	4.97	(0.09)			6.53	(0.16)			5.97	(0.15)
τ_{fs}	0.64	(0.13)	0.66	(0.12)			0.63	(0.13)			0.67	(0.12)

The posterior mean for τ_{fs} given this model is 0.36 with a posterior standard deviation equal to 0.08 (Table 11.8) The posterior means and standard deviations for all other elements of θ are again presented in Tables 11.6 and 11.7.

One major advantage of the model-based imputation approach is that we can easily accommodate different estimands. Suppose that instead of focusing on the average effect of the treatment, we are interested in the effect of the training program on the probability that individuals who were not working before now have jobs paying more than $5,000. Within the context of the imputations, this is a straightforward calculation. The imputation procedure is exactly as before. Now to calculate the posterior distribution of the

Table 11.8. *Summary Statistics Posterior Distribution for Finite-Sample Average Treatment Effect, for San Diego SWIM Data*

Post-Program Earnings	Model	Covariates	Mean	(S. D.)	Posterior Quantiles				
					0.025	0.25	0.5	0.75	0.975
Year 1	Linear	No	0.33	(0.09)	0.14	0.27	0.33	0.40	0.51
Year 1	Linear	Yes	0.36	(0.08)	0.19	0.30	0.36	0.41	0.52
Year 1	Two-part	No	0.33	(0.09)	0.14	0.27	0.33	0.39	0.51
Year 1	Two-part	Yes	0.37	(0.09)	0.20	0.31	0.37	0.42	0.53
Year 2	Linear	No	0.63	(0.13)	0.38	0.54	0.63	0.71	0.88
Year 2	Linear	Yes	0.66	(0.12)	0.43	0.58	0.66	0.74	0.89
Year 2	Two-part	No	0.63	(0.13)	0.38	0.54	0.63	0.71	0.87
Year 2	Two-part	Yes	0.67	(0.12)	0.44	0.59	0.67	0.75	0.90

estimand, we simply calculate the fraction, among individuals who had zero earnings before, of individuals who now have earnings more than \$5,000. Using the two-part model with covariates, the posterior mean and standard deviation for this probability are 0.029 and 0.009. Another advantage is that it is straightforward to report results on the posterior distribution of the estimands beyond moments, for example posterior quantiles.

Table 11.8 reports posterior quantiles for the average effect of the treatment on post-program earnings.

11.7   CONCLUSION

In this chapter we illustrate the four basic methods for analyzing classical randomized experiments discussed in the second part of the text. Taking as the example a randomized experiment of a job-training program, we illustrate the calculation of Fisher exact p-values, the construction of confidence intervals based on Neyman's repeated sampling approach, regression analyses, and model-based analyses. The methods generally agree here: there is strong evidence of an effect of the program, and we can estimate its average effects precisely. Ultimately the choice of methods here is somewhat subtle: the randomization ensures that the point estimates tend to be similar, the estimated precisions are similar because the covariates are only moderately predictive of the potential outcomes, and the methods differ mostly in the precise questions they ask. In the next parts of the book, where we address observational studies, these differences are often amplified, and the choices become more consequential.

NOTES

For more detail on the San Diego SWIM program and similar labor market training programs, see Friedlander and Robbins (1995), Friedlander and Gueron (1995), Hotz, Imbens, and Mortimer (2005), and Hotz, Imbens, and Klerman (2001).

Research that concluded that non-experimental evaluations were not credible in social sciences led to a renewed interest in experimental evaluations. Important papers in this literature are Lalonde (1986), Fraker and Maynard (1987), and Friedlander and Robbins (1995). The central thesis in this literature was the claim that non-experimental methods led to a wide range of results, with no reliable methods for choosing among these results. Later research cast some doubt on these claims. Dehejia and Wahba (1999) showed that methods based on the propensity score were considerably more successful in replicating experimental results than the regression-based methods considered by Lalonde (1986).

Regular Assignment Mechanisms: Design

## CHAPTER 12

Unconfounded Treatment Assignment

12.1 INTRODUCTION

In Part III of this text we leave the conceptually straightforward world of perfect randomized experiments and move toward the more common setting of observational studies. Although in simple situations we can still directly apply the tools from randomized experiments and exploit the exact results that accompany them, quickly we will be forced to make approximations in our inferences. No longer will estimators be exactly unbiased as in Chapter 6, nor will we be able to calculate exact p-values of the type considered in Chapter 5.

The first step toward addressing observational studies is to relax the classical randomized experiment assumption that the probability of treatment assignment is a known function. We do maintain, however, in this part of the text, the *unconfoundedness* assumption that states that assignment is free from dependence on the potential outcomes. Moreover, we continue to assume that the assignment mechanism is *individualistic*, so that the probability for unit i is essentially a function of the pre-treatment variables for unit i only, free of dependence on the values of pre-treatment variables for other units. We also maintain the assumption that the assignment mechanism is *probabilistic*, so that the probability of receiving any level of the treatment is strictly between zero and one for all units.

The implication of these assumptions is that the assignment mechanism can be interpreted as if, within subpopulations of units with the same value for the covariates, a completely randomized experiment of the type discussed in Chapters 5–8 was conducted, although an experiment with unknown assignment probabilities for the units. Thus, under these assumptions, we can analyze data from a subsample with the same value of the covariates as if it came from such an experiment. Although we do not know *a priori* the assignment probabilities for each of these units, we know these probabilities are identical because their covariate values are identical, and hence, conditional on the number of treated and control units composing such a subpopulation, the probability of receiving the treatment, the propensity score, is equal to $e(x) = N_t(x)/(N_c(x) + N_t(x))$ for all units with $X_i = x$; here $N_c(x)$ and $N_t(x)$ are the number of units in the control and treatment groups respectively with pre-treatment value $X_i = x$. In practice, this insight alone is of limited value, as typically there are too many distinct values of the covariates in the sample to partition the sample in this way without having either $N_c(x)$ or $N_t(x)$ equal to zero in some strata. Nevertheless, this insight has an important implication that suggests feasible alternatives for analyses.

In this chapter we discuss some general aspects of the unconfoundedness assumption, including the broad strategies we recommend in settings where unconfoundedness is viewed as an appropriate assumption, and we provide a road map for the third and fourth parts of the text. In Section 12.2 we discuss the assumption itself, its implications, and why we think the setting with unconfoundedness is an important case deserving special attention. In Section 12.3 we further explore a particular implication of unconfoundedness related to the propensity score. Even if a large set of covariates is used to ensure unconfoundedness, it is generally sufficient, in a certain sense, to adjust for a scalar function of the covariates, namely the propensity score. We discuss the balancing property of the propensity score, and what other functions of the covariates share this property. Next, in Section 12.4 we outline broad strategies for estimation and inference under regular assignment mechanisms. We discuss the general merits of the various strategies and describe methods that we discuss in more detail in the subsequent chapters. In Section 12.5, we discuss preliminary analyses not involving the outcome data that we recommend as part of what we call the *design* stage of the observational study. In Section 12.6 we outline how, in some settings, one can do additional analyses that help the researcher assess the plausibility of the unconfoundedness assumption, even though in general unconfoundedness is not testable. Section 12.7 concludes.

12.2   REGULAR ASSIGNMENT MECHANISMS

In this section we revisit the properties of a regular assignment mechanism, the implications of these properties, and why we view this as a central class of assignment mechanisms to consider in observational studies.

12.2.1 The Implications of a Regular Assignment Mechanism

As discussed in Chapter 3, a regular assignment mechanism satisfies three conditions. First, the assignment mechanism must be *probabilistic*, requiring that the unit-level assignment probabilities are strictly between zero and one:

$$0 < p_i\left(\mathbf{X}, \mathbf{Y}(0), \mathbf{Y}(1)\right) < 1, \quad \text{for } i = 1, \dots, N.$$

Second, it must be *individualistic*, requiring that (*i*) the unit level assignment probabilities can be written as a common function of that unit's potential outcomes and covariates,

$$p_i\left(\mathbf{X}, \mathbf{Y}(0), \mathbf{Y}(1)\right) = q(X_i, Y_i(0), Y_i(1)), \quad \text{for } i = 1, \dots, N,$$

and (*ii*) that

$$\Pr(\mathbf{W} \,|\, \mathbf{X}, \mathbf{Y}(0), \mathbf{Y}(1)) = c \cdot \prod_{i=1}^{N} q(X_i, Y_i(0), Y_i(1))^{W_i} \cdot (1 - q(X_i, Y_i(0), Y_i(1)))^{1-W_i},$$

for some constant c, for $\mathbf{W} \in \mathbb{W}^+$, and zero elsewhere. Third, it must be *unconfounded*, requiring that all the assignment probabilities $\Pr(\mathbf{W} \,|\, \mathbf{X}, \mathbf{Y}(0), \mathbf{Y}(1))$ are free from dependence on the potential outcomes. In combination with individualistic assignment, this implies that we can write the assignment mechanism as

$$\Pr(\mathbf{W} \,|\, \mathbf{X}, \mathbf{Y}(0), \mathbf{Y}(1)) = c \cdot \prod_{i=1}^{N} e(X_i)^{W_i} \cdot (1 - e(X_i))^{1-W_i},$$

where $e(x)$ is the propensity score. This defines the basic framework we use in Parts III and IV of this text.

Under the assumptions for a regular assignment mechanism, we can give a causal interpretation to the comparison of observed outcomes for treated and control units within subpopulations defined by values of the pre-treatment variables. Specifically, suppose we look at the subpopulation of all units with $X_i = x$; within this subpopulation the difference in the distributions of the observed outcomes, between treated and control units, fairly represent the effects of the treatment in this subpopulation, because, within this subpopulation, the treated and control units are both random samples from that subpopulation. For example, the difference in average observed outcomes is unbiased for the average effect of the treatment at $X_i = x$.

Let us first consider the case with a single binary covariate (e.g., sex), so that $X_i \in \{f, m\}$. Within the subsamples of women and men, the average finite sample treatment effects are, respectively,

$$\tau_{\text{fs}}(f) = \frac{1}{N(f)} \sum_{i:X_i=f} \left(Y_i(1) - Y_i(0)\right), \quad \text{and} \quad \tau_{\text{fs}}(m) = \frac{1}{N(m)} \sum_{i:X_i=m} \left(Y_i(1) - Y_i(0)\right),$$

where $N(f)$ and $N(m)$ are the number of women and men, respectively, in the sample. Within each of these subsamples, estimation and inference are entirely standard. We can directly use the methods from, for example, Chapter 6 in Part II of this text on Neyman's repeated sampling perspective in completely randomized experiments. The fact that we do not know *a priori* the probability of assignment to the treatment is irrelevant here: we can use the results for the analysis of completely randomized experiments by conditioning on the number of treated women and treated men. If, instead of being interested in $\tau(f)$ and $\tau(m)$ separately, we are interested in the overall average effect

$$\tau_{\text{fs}} = \frac{N(f)}{N(f) + N(m)} \cdot \tau_{\text{fs}}(f) + \frac{N(m)}{N(f) + N(m)} \cdot \tau_{\text{fs}}(m),$$

we can simply use the methods for stratified randomized experiments discussed in Chapter 9.

This approach of partitioning the population into strata by values of the pre-treatment variables extends, in principle, to all settings with discrete-valued pre-treatment variables. However, with pre-treatment variables taking on many distinct values in the sample, there may be a substantial number of strata with only treated or with only control units. For such strata, we cannot estimate the stratum-specific treatment effects using this approach, and thus we cannot estimate overall treatment effects following this strategy. This setting is of great practical relevance, and it is the primary focus of the chapters in Parts III and IV of this text, and indeed of much of the theoretical literature on estimation of, and inference for, causal effects in statistics and related disciplines. In this case, we compare outcomes for treated and control units with "similar" but not identical values for the pre-treatment variables. For such comparisons to be appropriate, we require smoothness and modeling assumptions, and decisions regarding tradeoffs between differences in one covariate versus another. How we make such trade-offs, and what are sensible approaches to find estimators and inferential procedures that lead to robust and credible results, are central topics in Parts III and IV of this text. Beyond depending on substantive insights regarding the association of particular pre-treatment variables with treatment status and potential outcomes, and related assessments of the unconfoundedness assumption, evaluating the various approaches to estimation and inference also requires statistical expertise.

12.2.2 A Super-Population Perspective

For the purpose of discussing various frequentist approaches to estimation and inference under unconfoundedness, it is useful to take a super-population perspective. Moreover, it is helpful to view the covariates X_i as having been randomly drawn from an approximately continuous distribution. If, instead, we view the covariates as having a discrete distribution with finite support, the implication of unconfoundedness is simply that one should stratify by the values of the covariates. In that case there will be, with high probability, in sufficiently large samples, both treated and control units with the exact same values of the covariates. In this way we can immediately remove all biases arising from differences between covariates, and many adjustment methods will give similar, or even identical, answers. However, as we stated before, this case rarely occurs in practice. In many applications it is not feasible to stratify fully on all covariates, because too many strata would have only a single unit. The differences between various adjustment methods arise precisely in such settings where it is not feasible to stratify on all values of the covariates, and mathematically these differences are most easily analyzed in settings with random samples from large populations using effectively continuous distributions for the covariates.

In the super-population, unconfoundedness implies a restriction on the joint distribution of $(Y_i(0), Y_i(1), W_i, X_i)$, namely

$$\Pr(W_i = 1 | Y_i(0), Y_i(1), X_i) = \Pr(W_i = 1 | X_i) = e(X_i), \tag{12.1}$$

or, in the Dawid (1979) conditional independence notation,

$$W_i \perp\!\!\!\perp \left(Y_i(0), Y_i(1) \right) \,\Big|\, X_i,$$

where we leave implicit the conditioning on the parameters governing the distributions, as in Section 3.5. Probabilistic assignment now requires that

$$0 < e(x) < 1,$$

for all x in the support of X_i, where we ignore measure-theoretic details.

12.2.3 Unconfoundedness Is Not Testable

A key feature of the unconfoundedness assumption is that it has no directly testable implications, even in settings with a large number of units. There is no information in the data that can tell us that unconfoundedness does not hold. Of course this does not mean that unconfoundedness actually holds, or even that it is plausible, but it implies that any assertion that it does *not* hold must rely on additional, substantive, information beyond the assessment of assumptions of probabilistic and individualistic assignment.

To gain further insight into this feature of the unconfoundedness assumption, it is useful to look at this assumption in a setting with a large sample, where we can estimate the joint distribution of $(Y_i^{\text{obs}}, W_i, X_i)$.

Theorem 12.1 (Super-Population Unconfoundedness) *Super-population unconfoundedness implies two restrictions on the conditional distributions of the potential outcomes. First,*

$$\left(Y_i(0) \,\Big|\, W_i = 1, X_i \right) \sim \left(Y_i(0) \,\Big|\, W_i = 0, X_i \right), \quad \text{for } i = 1, \dots, N, \tag{12.2}$$

and, second,

$$\left(Y_i(1) \,\Big|\, W_i = 0, X_i \right) \sim \left(Y_i(1) \,\Big|\, W_i = 1, X_i \right), \quad \text{for } i = 1, \dots, N. \tag{12.3}$$

(Here "$\sim$" denotes equality in distribution.)

Proof. By super-population unconfoundedness, defined in Chapter 3, Section 5, W_i is independent of $(Y_i(0), Y_i(1))$ given X_i. Hence $Y_i(0)$ is independent of W_i given X_i, implying the first claim in Theorem 12.1. The second claim follows by an analogous argument. $\square$

The first restriction states that the conditional distribution of $Y_i(0)$ given $W_i = 1$ and the pre-treatment variables X_i is the same as the conditional distribution of $Y_i(0)$ given $W_i = 0$ and X_i. It is useful to restate this, and (12.3), in terms of missing and observed outcomes:

$$\left(Y_i^{\text{mis}} \,\Big|\, W_i = w, X_i \right) \sim \left(Y_i^{\text{mis}} \,\Big|\, W_i = 1 - w, X_i \right), \quad \text{for } i = 1, \dots, N.$$

Now it becomes clear that the unconfoundedness assumption implies the equality of the distribution of a missing potential outcome (a distribution about which the data are not directly informative) to the distribution of an observable outcome (about which the data are informative). In large samples we can infer the conditional distribution of Y_i^{obs} given W_i and X_i, but no amount of observable data will allow us to infer the distribution of Y_i^{mis} given W_i and X_i.

Although unconfoundedness is not testable, there are in some cases analyses one may be able to carry out that assist the researcher when assessing the plausibility of this critical assumption. These supporting analyses rely on more restrictive assumptions that *do* generate testable consequences. In Chapter 21 we discuss such analyses in detail.

12.2.4 Why Is Unconfoundedness an Important Assumption?

Before discussing specific methods for estimation and inference based on regular assignment mechanisms, it is useful to discuss why we view this assumption as so important that we devote a large part of this text to methods assuming it.

Of the three assumptions required for regularity of the assignment mechanism, probabilistic assignment is the easiest to motivate. If a particular subpopulation has zero probability of being in one of the treatment groups, then estimates of treatment effects for this subpopulation must, by necessity, rely on extrapolation. There is often little basis for such extrapolation, and we may simply have to put such subpopulations aside. For example, suppose we are interested in evaluating a new drug, and suppose the sample studied contains both women and men, $X_i \in \{f, m\}$. However, suppose that the treatment group contains only women, so that $e(m) = \Pr(W_i = 1 | X_i = m) = 0$. In that case it would clearly require strong, possibly implausible, assumptions to estimate the effect of the treatment for men – or, for that matter, for the entire population. It would appear more reasonable to estimate the effect for women and then separately discuss the plausibility of extrapolating that estimate for women to men. Even more prevalent is the case where the probabilistic assumption is close to being violated, without the probabilities being exactly equal to zero or one, which can severely impact our ability to obtain precise estimates of the causal estimands. This raises a number of issues, which we discuss in detail in Chapters 15 and 16.

In practice, the second assumption, individualistic assignment, is rarely controversial. Although formally it is possible that there is dependence in the assignment indicators beyond that allowed through, for example, stratification on covariates, there are no practical examples we are aware of, other than sequential assignment mechanisms (which we do not discuss in this text), where this is plausibly violated.

Next, let us comment on some aspects of what is, typically, the most controversial component of the three requirements for a regular assignment mechanism: the assumption of unconfoundedness. First of all, the assumption is extremely widely used. Although this is obviously not in itself an argument for its validity, it should be noted that, by a wide margin, most analyses involving observational studies fundamentally rely on unconfoundedness, often implicitly, and often in combination with other assumptions, in order to estimate causal effects. It is not always immediately transparent that such an assumption is employed, as it is often formulated in combination with functional form or distributional assumptions, but in many such applied examples, the implication of the assumptions is that differences in outcomes for units with the same values for some set of observed pre-treatment variables, but with different levels of the treatment, can be interpreted as credible estimates of causal effects.

Let us give an example of such an assumption. In many empirical studies in social sciences, causal effects are estimated through linear regression, where, typically it is implicitly assumed that in the super-population,

$$\mathbb{E}\left[Y_i(w)|X_i\right] = \alpha + \tau_{\text{sp}} \cdot w + X_i\beta,$$

for some values of the three unknown parameters α, τ_{sp}, and β, where $\tau_{\text{sp}} = \mathbb{E}_{\text{sp}}[Y_i(1) - Y_i(0)]$. Defining $\varepsilon_i = Y_i^{\text{obs}} - \tau_{\text{sp}} \cdot W_i - X_i\beta$, so that we can write

$$Y_i^{\text{obs}} = \alpha + \tau_{\text{sp}} \cdot W_i + X_i\beta + \varepsilon_i, \tag{12.4}$$

it is then assumed that

$$\varepsilon_i \perp\!\!\!\perp W_i, X_i.$$

This assumption is often referred to as *exogeneity* of the treatment (and the pre-treatment variables) in the econometrics literature. The regression function (12.4) is interpreted as a causal relation, in our sense of the term "causal," namely that if we manipulate the treatment W_i, then the outcome would change in expectation by an amount τ_{sp}. Hence, in the potential outcome formulation, we have

$$Y_i(0) = \alpha + X_i\beta + \varepsilon_i, \quad \text{and} \quad Y_i(1) = Y_i(0) + \tau_{\text{sp}}.$$

Then, because ε_i is a function of $Y_i(0)$ and X_i given the parameters,

$$\Pr(W_i = 1|Y_i(0), Y_i(1), X_i) = \Pr(W_i|\varepsilon_i, X_i),$$

and by exogeneity of the treatment indicator, we have

$$\Pr(W_i|\varepsilon_i, X_i) = \Pr(W_i|X_i),$$

and thus unconfoundedness holds. However, the exogeneity assumption combines unconfoundedness with functional form and constant treatment effect assumptions that are quite strong, and arguably unnecessary. Therefore we focus here on the cleaner, functional-form-free unconfoundedness assumption.

A second motivation for the unconfoundedness assumption is based on a comparison with alternative assumptions. Unconfoundedness implies that one should compare units similar in terms of pre-treatment variables, that is, one should compare "like with like." This has great intuitive appeal, and underlies many informal, as well as formal, causal inferences. Without this assumption, and without additional assumptions to replace it, we would no longer have guidance on which control units would make good comparisons for particular treated units (and the other way around). In the absence of unconfoundedness, one could still conduct a sensitivity analysis or, in an extreme version, calculate ranges of values for the causal estimands consistent with the data. We discuss such approaches in Chapter 22. However, any alternative approach that would provide specific guidance on which treated units to compare with which control units would have to compare units that differ in terms of observed pre-treatment variables. As Rubin (2006) writes concerning the example of the causal effect of smoking versus not smoking, "it would make little sense to compare disease rates in well-educated non-smokers and poorly educated smokers" (page 3). To be specific, suppose we are interested in the causal effect of a job-training program. Now suppose there is a forty-year-old man who has been unemployed for six months, and who was continuously employed for eighteen months prior to that in the automobile industry, with a high school education, who is going through this training program. Assuming unconfoundedness implies that in order to estimate the causal effect of this program for him, we should look for a man with the same pre-training characteristics, who did not go through the training program. Any plausible alternative strategy would still involve looking for a person, or combination of persons, who did not go through the training program. But, in order to be different from the strategy under unconfoundedness, any alternative must imply looking for a person, or combination of persons, who are systematically different from the forty-year-old male high school graduate with six months of unemployment and eighteen months of employment in the automobile industry. In other words, an alternative to unconfoundedness must involve looking for a comparison person who is systematically *different* in terms of observed pre-treatment variables from the person who went through the training. In many cases it would appear implausible that individuals who differ in terms of pre-treatment characteristics would be more suitable comparisons. Of course, it may be that individuals who differ in terms of two or more pre-treatment variables may have offsetting unobserved differences such that ultimately they provide a better comparison, but it would appear to be difficult to improve systematically comparisons in this manner. Note that the claim is *not* that unconfoundedness is always plausible per se. The claim is the much weaker statement, that allowing for systematic differences in such pre-treatment characteristics is unlikely to improve comparisons in general practice.

Let us expand on this argument in an example to be clearer. Suppose that a researcher is concerned that the unconfoundedness assumption may be violated, because typically individuals who enrolled in this job market program may be more interested in finding jobs, that is, more motivated, than the individuals who did not enroll. Such a concern is common in the analysis of job-training programs in settings with voluntary enrollment. Let us suppose, for expositional reasons, that motivation is a permanent characteristic of individuals, not affected by the training program. It is plausible that more highly motivated individuals are, typically, better at finding employment conditional on their observed treatment status. Unconfoundedness may in this case be a reasonable assumption *if* motivation were observed. If motivation is not observed, however, the implication is that the potential outcomes would be correlated with the treatment indicator, and thus unconfoundedness would be violated. However, it is not clear that, in such a scenario, using a control person who *differs* in terms of observed pre-treatment characteristics as the comparison would improve the credibility of the causal interpretation. In order to improve the comparison, one would have to be able to trade off observed pre-treatment characteristics against the unobserved motivation, without direct information on the latter. It would appear often difficult to do so in a credible manner.

A third aspect of our motivation for focusing special attention on the setting with unconfoundedness concerns the interpretation of assignment processes that lead to differences in treatment levels for units who are identical in terms of observed pre-treatment characteristics. In randomized experiments the differences in treatment levels are due to randomization. In observational studies it is less clear why such similar units should receive different treatment assignments. Especially in settings where the units are individuals and the assignment mechanism is based on individual choices, one might be concerned that individuals who look *ex ante* identical (i.e., identical in terms of pre-treatment characteristics) but who make different choices must be different in unobserved ways that invalidates a causal interpretation of differences in their outcomes. Examples of such settings include those where individuals choose to enroll in labor market assistance programs, based on their assessment of the costs and benefits of such programs, and those where medical treatment decisions are made by physicians, in consultation with patients, choosing treatments based on their perceived costs and benefits. However, in such cases, the unobserved differences that lead to differences in treatments need not lead to violations of unconfoundedness. If the unobserved differences that led the individuals to make different choices, are independent of the potential outcomes, conditional on observed covariates, unconfoundedness still holds. This may arise, for example, in settings where unobserved differences in terms of the costs associated with exposure to the treatment are unrelated to the potential outcomes.

Let us make this argument slightly more specific using an example. Suppose two patients with a particular medical condition have identical symptoms. Suppose they also share the same physician. This physician, in consultation with these patients, faces the choice between two treatments, say drug A and drug B. Suppose drug A is expensive relative to drug B. Furthermore, suppose that as a result of differing health insurance plans, the incremental cost of taking drug A relative to drug B is higher for one patient than for the other. This cost difference may well affect the choice of drug, and as a result one may have data on individuals with similar medical conditions exposed to different treatments without violating unconfoundedness (if we assume that the choice of insurance plan is not related to outcomes given exposure to drug A or drug B, especially after conditioning on observed covariates such as sex or age).

12.2.5   Selecting Pre-Treatment Variables for Conditioning

So far, the only requirement we have imposed on the pre-treatment variables is that they precede the treatment, or that they are not themselves affected by the treatment. Variables that are possibly affected by the treatment, such as intermediate outcomes, should not be included in this set, and correctly adjusting for differences in such variables is generally difficult.

Given this set of proper pre-treatment variables, one generally wants to control for as many as possible, or all of them. If we are interested in, for example, the evaluation of a labor market training program on individuals disadvantaged in the labor market, one would like to include detailed labor market histories and individual characteristics of the individuals to eliminate such characteristics as alternative explanations for differences in outcomes between trainees and control individuals. There are some exceptions to this general advice. In some cases there is additional prior information regarding the dependence of potential outcomes on pre-treatment variables that suggests alternative estimation strategies that do not remove differences in all observed pre-treatment variables. An important case is *instrumental variables* discussed in more detail in Chapters 23–25. In practice, however, such cases are typically easy to recognize and rarely lead to confusion. Variables that are truly instrumental variables are relatively rare, and when they exist, it is even more rare that they are mistakenly used as covariates for adjustment.

12.3 BALANCING SCORES AND THE PROPENSITY SCORE

Now let us return to the theoretical discussion, using a super-population perspective. Under unconfoundedness, we can remove all biases in comparisons between treated and control units by adjusting for differences in observed covariates. Although feasible in principle, in practice this will be difficult to implement with a large number of covariates. The idea of balancing scores is to find lower-dimensional functions of the covariates that suffice for removing the bias associated with differences in the pre-treatment variables. Formally, a balancing score is a function of the covariates such that the probability (in the super-population) of receiving the active treatment given the covariates is free of dependence on the covariates given the balancing score.

Definition 12.1 (Balancing Scores)
A balancing score $b(x)$ is a function of the covariates such that

$$W_i \perp\!\!\!\perp X_i \mid b(X_i).$$

(Here we continue to leave the conditioning on parameters implicit in the super-population context.) Balancing scores are not unique. By definition, the vector of covariates X_i itself is a balancing score, and any one-to-one function of a balancing score is also a balancing score. We are most interested in low-dimensional balancing scores. One scalar balancing score is the propensity score, the conditional probability of receiving the treatment given $X_i = x$ (or any one-to-one transformation of the propensity score, such as the linearized propensity score or log odds ratio, $\ell(x) = \ln(e(x)/(1-e(x)))$). First, we show that the propensity score is indeed a balancing score:

Lemma 12.1 (Balancing Property of the Propensity Score)
The propensity score is a balancing score.

Proof. We show that

$$W_i \perp\!\!\!\perp X_i \mid e(X_i),$$

or, equivalently,

$$\Pr(W_i = 1 | X_i, e(X_i)) = \Pr(W_i = 1 | e(X_i)),$$

implying that W_i is independent of X_i given the propensity score. First, consider the left-hand side:

$$\Pr(W_i = 1 | X_i, e(X_i)) = \Pr(W_i = 1 | X_i) = e(X_i),$$

where the first equality follows because the propensity score is a function of X_i and the second is by the definition of the propensity score. Second, consider the right-hand side.

By the definition of probability and iterated expectations,

$$\Pr(W_i = 1|e(X_i)) = \mathbb{E}[W_i|e(X_i)] = \mathbb{E}\left[\mathbb{E}[W_i|X_i, e(X_i)]|e(X_i)\right] = \mathbb{E}[e(X_i)|e(X_i)] = e(X_i).$$

□

Balancing scores have an important property: if assignment to treatment is unconfounded given the full set of covariates, then assignment is also unconfounded conditioning only on a balancing score:

Lemma 12.2 (Unconfoundedness Given a Balancing Score)
Suppose assignment to treatment is unconfounded. Then assignment is unconfounded given any balancing score:

$$W_i \perp\!\!\!\perp Y_i(0), Y_i(1) \mid b(X_i).$$

Proof. We show that

$$\Pr_W(W_i = 1|Y_i(0), Y_i(1), b(X_i)) = \Pr_W(W_i = 1|b(X_i)),$$

which is equivalent to the statement in the lemma. By iterated expectations we can write

$$\Pr_W(W_i = 1|Y_i(0), Y_i(1), b(X_i)) = \mathbb{E}_W\left[W_i|Y_i(0), Y_i(1), b(X_i)\right]$$

$$= \mathbb{E}\left[\mathbb{E}_W\left[W_i|Y_i(0), Y_i(1), X_i, b(X_i)\right]\middle|Y_i(0), Y_i(1), b(X_i)\right].$$

By unconfoundedness, the inner expectation is equal to $\mathbb{E}[W_i|X_i, b(X_i)]$ and by the definition of balancing scores, this is equal to $\mathbb{E}[W_i|b(X_i)]$. Hence the last expression is equal to

$$\mathbb{E}\left[\mathbb{E}_W[W_i|b(X_i)]\middle|Y_i(0), Y_i(1), b(X_i)\right] = \mathbb{E}[W_i|b(X_i)] = \Pr(W_i = 1|b(X_i)),$$

which is equal to the right-hand side.   □

The first implication of Lemma 12.2 is that, given a vector of covariates that ensure unconfoundedness, adjustment for treatment-control differences in balancing scores suffices for removing all biases associated with differences in the covariates. The intuition is that, conditional on a balancing score, the treatment assignment is independent of the covariates. Hence, even if a covariate is associated with the potential outcomes, differences in covariates between treated and control units do not lead to bias because they cancel out by averaging over all units with the same value for the balancing score. The situation is analogous to that in a completely randomized experiment, where the distribution of covariates is the same in both treatment arms. Even though the covariates may differ between specific treated and control units with the same value for the balancing score, they have the same *distribution* of values in the treatment and control groups.

Because the propensity score is a balancing score, Lemma 12.2 implies that, conditional on the propensity score, assignment to treatment is unconfounded. But within the class of balancing scores, the propensity score has a special place, formally described in the following lemma:

Lemma 12.3 (Coarseness of Balancing Scores)
The propensity score is the coarsest balancing score. That is, the propensity score is a function of every balancing score.

Proof. Let $b(x)$ be a balancing score. Suppose that we can *not* write the propensity score as a function of the balancing score. Then it must be the case that for two values x and x' we have $b(x) = b(x')$, and at the same time $e(x) \neq e(x')$. Then, $\Pr(W_i = 1 | X_i = x) = e(x) \neq e(x') = \Pr(W_i = 1 | X_i = x')$, and so W_i and X_i are not independent given $b(X_i) = b(x)$, which violates the definition of a balancing score. $\square$

Because the propensity score is the coarsest possible balancing score, it provides the biggest benefit in terms of reducing the number of variables we need to adjust for. An important difficulty though arises from the complication that we do not know the value of the propensity score for all units, and thus we cannot directly exploit this result.

12.4   ESTIMATION AND INFERENCE

In this section we discuss general issues regarding estimation and inference for causal effects in regular assignment mechanisms. In subsequent chapters we go into more detail for some of our preferred methods, but here we provide a general overview and discuss the merits of various approaches.

12.4.1 Efficiency Bounds

Before discussing some of the specific approaches to estimation, it is useful to examine how well these methods can work. An important tool for this purpose is the *semiparametric efficiency bound*. This is a generalization of the Cramér-Rao sampling variance bound for unbiased estimators.

In order to formulate the variance bound, some additional notation is helpful. Define

$$\mu_c(x) = \mathbb{E}_{sp} [Y_i(0) | X_i = x], \qquad \mu_t(x) = \mathbb{E}_{sp} [Y_i(1) | X_i = x],$$

$$\sigma_c^2(x) = \mathbb{V}_{sp} (Y_i(0) | X_i = x), \quad \text{and} \quad \sigma_t^2(x) = \mathbb{V}_{sp} (Y_i(1) | X_i = x),$$

to be the conditional expectation and conditional variance of the potential outcomes, respectively. These expectations are with respect to the distribution generated by random sampling from the super-population. Furthermore, let τ_{sp} be the super-population average treatment effect defined as

$$\tau_{sp} = \mathbb{E}_{sp} [Y_i(1) - Y_i(0)] = \mathbb{E}_{sp} \left[\tau_{sp}(X_i) \right],$$

where

$$\tau_{sp}(x) = \mu_t(x) - \mu_c(x) = \mathbb{E}_{sp}[Y_i(1) - Y_i(0) | X_i = x].$$

It is useful to distinguish τ_{sp} from two other average treatment effects, first, the average effect of the treatment for the sample of N units at hand, or the *finite-sample average treatment effect* τ_{fs},

$$\tau_{fs} = \frac{1}{N} \sum_{i=1}^{N} \left(Y_i(1) - Y_i(0) \right),$$

and, second, the finite-sample average effect conditional on the values of the pre-treatment variables in the finite sample, the *conditional average treatment effect*,

$$\tau_{cond} = \frac{1}{N} \sum_{i=1}^{N} \tau_{sp}(X_i).$$

In the current setting, under unconfoundedness and probabilistic assignment, and without additional functional form restrictions beyond smoothness, the sampling variance bound for estimators for τ_{sp}, normalized by the sample size, is

$$\mathbb{V}_{sp}^{eff} = \mathbb{E}_{sp} \left[\frac{\sigma_c^2(X_i)}{1 - e(X_i)} + \frac{\sigma_t^2(X_i)}{e(X_i)} + (\tau_{sp}(X_i) - \tau_{sp})^2 \right]. \tag{12.5}$$

Details and references for this result are provided in the notes at the end of this chapter. This result implies that for any *regular* estimator (see again the notes for more details), its asymptotic sampling variance, after normalizing by the square root of the sample size, cannot be smaller than $\mathbb{V}_{sp}^{eff}$. The sampling variance bound consists of three terms. The first term shows that it is more difficult to estimate the average treatment effect if there is a substantial number of units with propensity score values close to one, in the sense that any estimator will have a high sampling variance in such cases. Similarly, the second term shows that it is more difficult to estimate the average treatment effect if there is a substantial number of units with propensity score values close to zero. The third term is the variance of the treatment effect conditional on the pre-treatment variables. This term is zero if the treatment effect is constant. Overall the variance expression (12.5) shows that, if the population distribution of covariates is unbalanced between treated and control units, the sampling variance of any estimator will be large. This will be important for analyses, and we return to this issue in Chapters 15 and 16.

If instead of focusing on the population average effect τ_{sp}, we focus on τ_{cond}, the efficiency bound changes to

$$\mathbb{V}_{cond}^{eff} = \mathbb{E}_{sp} \left[\frac{\sigma_c^2(X_i)}{1 - e(X_i)} + \frac{\sigma_t^2(X_i)}{e(X_i)} \right].$$

We can, at least in principle, estimate τ_{cond} more accurately than τ_{sp} because the latter also reflects the difference between the distribution of the covariates in the sample and the population. The intuition for this is easily presented in terms of a simple example. Suppose there is a single binary covariate, with unknown marginal distribution in the super-population, $X_i \in \{f, m\}$, with $\Pr(X_i = f) = p$ unknown. Suppose we can estimate the average effects $\tau_{sp}(f)$ and $\tau_{sp}(m)$ accurately for both subpopulations separately because the conditional variances are small, and suppose these average effects differ substantially. Then it follows that we can estimate τ_{cond} accurately because it is a known function of $\tau_{sp}(f)$ and $\tau_{sp}(m)$. However, because p is unknown, we would not be able to estimate τ_{sp} as accurately.

The implication is that it is important for inference to be precise about the estimand. If we focus on τ_{fs} or τ_{cond}, we need to use a different estimator for the sampling variance than if we focus on τ_{sp}.

12.4.2 Strategies for Estimation

We discuss five broad classes of strategies for estimation, with some overlap between them. These four strategies are model-based imputation, weighting, blocking, and matching methods. These four basic approaches differ in their focus on the unknown components of the joint distribution of the potential outcomes, assignment process, and covariates. In this section, we briefly describe these four general approaches, as well as a fifth class of estimators that combines aspects of some of these strategies. Variations of all five of these strategies have been used extensively in empirical work, although we do not recommend all of them. In Chapters 17 and 18 in Part IV, we discuss in more detail the implementation for two specific strategies that we view as particularly attractive in practice. These two strategies are blocking (i.e., subclassification) on the propensity score, in combination with covariance adjustment within the blocks (Chapter 17), and matching, again in combination with covariance adjustment, possibly within the matched pairs (Chapter 18). We view these two approaches as relatively attractive because of the robustness properties that stem from the combination of methods that ensure approximate comparability, either through blocking or matching, with additional bias removal and precision increases through covariance adjustment.

Although all four general approaches aim at estimating the same treatment effects, there are fundamental differences among them. One important difference between the model-based imputations and the other three (weighting, blocking, and matching methods) is that the first requires building models for the potential outcomes, whereas for the other three all decisions regarding the implementation of the estimators without covariate adjustment can be made before seeing any outcome data. This difference is important because not having outcome data prevents the researcher from adapting the model to make it fit prior notions about the treatment effects of interest. Although the researcher does have to make a number of important decisions when using weighting, blocking, and matching methods, these can be implemented in a way that does not introduce bias in the estimates for treatment effects and so have arguably more credibility.

Model-Based Imputation

The first strategy relies on imputing the missing potential outcomes by building a model for the missing outcomes and using this model to predict what would have happened to a specific unit had this unit been subject to the treatment to which it was not exposed. We discussed this approach for completely randomized experiments in Chapter 8, and the discussion here is closely related. Following the exposition from Chapter 8, we need a model for

$$\mathbf{Y}^{\text{mis}} \mid \mathbf{Y}^{\text{obs}}, \mathbf{X}, \mathbf{W}.$$

Given such a model, we can impute the missing data by drawing from the conditional distribution of $\mathbf{Y}^{\text{mis}}$ given $\mathbf{Y}^{\text{obs}}$, $\mathbf{W}$, and $\mathbf{X}$. Suppose we specify a model for the joint distribution of the two vectors of potential outcomes given the covariates, now explicitly in terms of an unknown parameter θ:

$$\mathbf{Y}(0), \mathbf{Y}(1) \mid \mathbf{X}, \theta. \tag{12.6}$$

Because of unconfoundedness, $\mathbf{W}$ is independent of $(\mathbf{Y}(0), \mathbf{Y}(1))$ given $\mathbf{X}$, and the specification of (12.6) implies the distribution

$$\mathbf{Y}(0), \mathbf{Y}(1) \mid \mathbf{W}, \mathbf{X}, \theta, \tag{12.7}$$

which in turns allows us to derive the conditional distribution of the missing data given the observed data following the argument in Chapter 8. We therefore focus on specifying a model for $(\mathbf{Y}(0), \mathbf{Y}(1))$ given $\mathbf{X}$. Given exchangeability of the units and an appeal to De Finetti's Theorem, all we need to specify is the joint distribution of

$$(Y_i(0), Y_i(1)) \mid X_i, \theta,$$

for some parameter vector θ. Given such a distribution, we can, following the same approach as in Chapter 8, impute the missing potential outcomes and use the observed and imputed potential outcomes to estimate the treatment effects of interest.

The critical part of this approach is the specification of the joint distribution of $(Y_i(0), Y_i(1))$ given X_i and parameter θ. With no covariates – or, more generally, a low-dimensional set of covariates – it is relatively easy to specify a flexible functional form for this conditional distribution. If there are many covariates, however, such a specification is more difficult, and the results can be sensitive to alternative choices. This situation is qualitatively different from the randomized experiment setting in Chapter 8, where such sensitivity will often be minor because the covariate distributions in treatment and control groups are similar. Because this approach treats the problem essentially as a prediction one, it is particularly amenable to Bayesian methods with their focus on treating unobserved quantities, including both the missing potential outcomes and unknown parameters, as unobserved random variables.

In this approach, often there is no need to specify a parametric model for the conditional distribution of the treatment indicator given the covariates, the super-population assignment mechanism,

$$p(\mathbf{W}|\mathbf{X}; \phi),$$

because, if ϕ and θ are distinct parameters, inference for causal effects is not affected by the functional form of the specification of this assignment mechanism. However, it is important for this argument that ϕ and θ are distinct parameters.

The Concern with Regression Estimators

In practice, however, this approach is often used with standard "off-the-shelf" methods, where typically linear models are postulated for average outcomes, without a full specification of the conditional joint potential outcome distribution. Let us briefly consider the linear regression approach here. Suppose we model the potential outcome distributions as normally distributed with treatment-specific parameters governing the conditional means and variances of the potential outcomes:

$$\left(\begin{array}{c} Y_i(0) \\ Y_i(1) \end{array} \right) \Big| X_i, \theta \sim \mathcal{N} \left(\left(\begin{array}{c} X_i \beta_c \\ X_i \beta_t \end{array} \right), \left(\begin{array}{cc} \sigma_c^2 & \sigma_c \cdot \sigma_t \\ \sigma_c \cdot \sigma_t & \sigma_t^2 \end{array} \right) \right),$$

where $\theta = (\beta_c, \beta_t, \sigma_c^2, \sigma_t^2)$. (Note that the vector of covariates X_i is assumed to include a constant term.) Then we can estimate β_c and β_t by least squares methods:

$$\hat{\beta}_c^{ols} = \arg \min_\beta \sum_{i:W_i=0} (Y_i - X_i \beta)^2, \quad \text{and} \quad \hat{\beta}_t^{ols} = \arg \min_\beta \sum_{i:W_i=1} (Y_i - X_i \beta)^2.$$

The population and sample average treatment effects are then estimated as

$$\hat{\tau}^{ols} = \frac{1}{N} \sum_{i=1}^N \left(W_i \cdot (Y_i^{obs} - X_i \hat{\beta}_c^{ols}) + (1 - W_i) \cdot (X_i \hat{\beta}_t^{ols} - Y_i^{obs}) \right).$$

We do not recommend this approach, introduced in Chapter 7, in the context of completely randomized experiments, without substantial modifications. The concern with the simple application of this approach is that, in many situations outside randomized experiments, it can rely heavily on extrapolation. To see this, it is useful to rewrite the estimator as

$$\hat{\tau}^{ols} = \frac{N_t}{N_t + N_c} \cdot \hat{\tau}_t^{ols} + \frac{N_c}{N_t + N_c} \cdot \hat{\tau}_c^{ols},$$

where $\hat{\tau}_c^{ols}$ and $\hat{\tau}_t^{ols}$ are estimators for the population average effect of the treatment for the control and treated units, respectively:

$$\hat{\tau}_c^{ols} = \frac{1}{N_c} \sum_{i:W_i=0} \left(X_i \hat{\beta}_t - Y_i^{obs} \right), \quad \text{and} \quad \hat{\tau}_t^{ols} = \frac{1}{N_t} \sum_{i:W_i=1} \left(Y_i^{obs} - X_i \hat{\beta}_c \right).$$

Furthermore, because of the presence of a constant term in X_i, we can write $\hat{\tau}_t$ as

$$\hat{\tau}_t^{ols} = \overline{Y}_t^{obs} - \overline{X}_t \hat{\beta}_c^{ols} = \overline{Y}_t^{obs} - \overline{Y}_c^{obs} - (\overline{X}_t - \overline{X}_c) \hat{\beta}_c^{ols}, \tag{12.8}$$

and similarly

$$\hat{\tau}_c^{ols} = \overline{X}_c \hat{\beta}_t^{ols} - \overline{Y}_c^{obs} = \overline{Y}_t^{obs} - \overline{Y}_c^{obs} - (\overline{X}_t - \overline{X}_c) \hat{\beta}_t^{ols}. \tag{12.9}$$

The last terms in expressions (12.8) and (12.9), $(\overline{X}_t - \overline{X}_c)\hat{\beta}_c^{ols}$ and $(\overline{X}_t - \overline{X}_c)\hat{\beta}_t^{ols}$, are at the core of the concern. If the two covariate distributions are substantially apart, the difference $\overline{X}_t - \overline{X}_c$ is substantial. Then the "adjustment" terms $(\overline{X}_t - \overline{X}_c)\hat{\beta}_c^{ols}$ and $(\overline{X}_t - \overline{X}_c)\hat{\beta}_t^{ols}$ will be sensitive to details of the specification of the regression function. In the context of completely randomized experiments, this was less of an issue, because the randomization ensured that, at least in expectation, the covariate distributions were balanced, with $\mathbb{E}_W\left[\overline{X}_t - \overline{X}_c\right] = 0$, with the expectation taken over the randomization distribution. Here, in contrast, the covariate distributions can be far apart even under unconfoundedness. Prior to using regression methods or other modeling approaches, therefore, one has to ensure that there is balance in the two covariate distributions. We return to this issue in Section 12.5 and in more detail in Chapters 14 and 15.

Weighting Estimators That Use the Propensity Score

Whereas the first strategy focused on estimating the two conditional outcome distributions, or at least the two conditional regression functions, the second strategy focuses on estimating the propensity score. Given knowledge of the propensity score, one can directly use some of the strategies that apply to the analysis of randomized experiments with variation in assignment probabilities. Such possible strategies include weighting, subclassification (similar to stratification in the case of randomized experiments), and matching. The key difference between these and the general imputation strategy is that the former three focus on modeling and estimating the conditional probability of assignment, whereas an imputation strategy models the conditional outcome distributions. The issues in implementing any of these three methods therefore are related to estimation of the propensity score. One approach is to treat the estimation of the propensity score as a standard problem of estimating an unknown regression function with a binary outcome and exploit the relevant literature. An alternative approach, more widely used in the evaluation literature, focuses on the essential property of the propensity score, that of balancing the covariates between treated and control groups. In this approach a specification is sought for the propensity score such that, within blocks with similar values of the propensity score, the first few (cross) moments of the covariates are balanced between treatment groups.

The first method involving the propensity score is weighting. Weighting exploits the two equalities

$$\mathbb{E}\left[\frac{Y_i^{\text{obs}} \cdot W_i}{e(X_i)}\right] = \mathbb{E}_{\text{sp}}\left[Y_i(1)\right], \quad \text{and} \quad \mathbb{E}\left[\frac{Y_i^{\text{obs}} \cdot (1 - W_i)}{1 - e(X_i)}\right] = \mathbb{E}_{\text{sp}}\left[Y_i(0)\right].$$

(Here we again index expectations by sp if they are over the distribution generated by random sampling from the super-population and by W if they are over the randomization distribution. Expectations without a subscript are over both the randomization and the random sampling from the super-population.) These equalities follow by taking iterated expectations, and exploiting unconfoundedness, for example,

$$\mathbb{E}\left[\frac{Y_i^{\text{obs}} \cdot W_i}{e(X_i)}\right] = \mathbb{E}_{\text{sp}}\left[\mathbb{E}\left[\frac{Y_i^{\text{obs}} \cdot W_i}{e(X_i)} \,\middle|\, X_i\right]\right]$$

$$= \mathbb{E}_{\text{sp}}\left[\mathbb{E}\left[\frac{Y_i(1) \cdot W_i}{e(X_i)} \,\middle|\, X_i\right]\right]$$

$$= \mathbb{E}_{\mathrm{sp}} \left[\frac{\mathbb{E}_{\mathrm{sp}}[Y_i(1)|X_i] \cdot \mathbb{E}_W[W_i|X_i]}{e(X_i)} \right]$$

$$= \mathbb{E}_{\mathrm{sp}} \left[\mathbb{E}_{\mathrm{sp}}[Y_i(1)|X_i] \right] = \mathbb{E}_{\mathrm{sp}} [Y_i(1)] ,$$

and similarly for the second equality. One can exploit these equalities by estimating the average treatment effect as

$$\hat{\tau}^{\mathrm{ht}} = \frac{1}{N} \sum_{i=1}^{N} \frac{W_i \cdot Y_i^{\mathrm{obs}}}{e(X_i)} - \frac{1}{N} \sum_{i=1}^{N} \frac{(1 - W_i) \cdot Y_i^{\mathrm{obs}}}{1 - e(X_i)}$$

$$= \frac{1}{N} \sum_{i:W_i=1} \lambda_i \cdot Y_i^{\mathrm{obs}} - \frac{1}{N} \sum_{i:W_i=0} \lambda_i \cdot Y_i^{\mathrm{obs}},$$

where

$$\lambda_i = \frac{1}{e(X_i)^{W_i} \cdot (1 - e(X_i))^{1-W_i}} = \begin{cases} 1/(1 - e(X_i)) & \text{if } W_i = 0, \\ 1/e(X_i) & \text{if } W_i = 1. \end{cases}$$

The superscript "ht" here stands for Horvitz and Thompson (1952) who introduced, in a somewhat different setting, the weighting by the inverse of the selection probability. In practice typically we do not know the true population propensity score, and we have to use an estimate of the propensity score, $\hat{e}(x)$ in place of $e(x)$, for the corresponding estimated weights. In addition, instead of using the weights λ_i directly, one can adjust the weights, so that they add up to the sample size for each treatment group, that is, use $\hat{\lambda}_i$, where

$$\hat{\lambda}_i = \begin{cases} N \cdot (1 - \hat{e}(X_i))^{-1} / \sum_{j:W_j=0} (1 - \hat{e}(X_i))^{-1} & \text{if } W_i = 0, \\ N \cdot \hat{e}(X_i)^{-1} / \sum_{j:W_j=1} \hat{e}(X_i)^{-1} & \text{if } W_i = 1. \end{cases}$$

Just like we do not recommend the simple regression estimator, we do not recommend this type of estimator in settings with a substantial difference in the covariate distributions by treatment status. In a completely randomized experiment, the propensity score would be constant, and even when the propensity score is estimated, the weights are likely to be similar for all treated and for all control units. In contrast, when the covariate distributions are far apart, the estimated propensity score will be close to zero or one for some units, and the weights, proportional to $1/\hat{e}(X_i)$ or $1/(1 - \hat{e}(X_i))$, can be large. As a result, in such settings estimators can be sensitive to minor changes in the specification of the model for the propensity score.

Blocking Estimators That Use the Propensity Score

A more robust approach involving the propensity score is to coarsen it through blocking (i.e., subclassification). In this third approach, the sample is partitioned into subclasses, based on the value of the estimated propensity score. Within each subclass, the data can be analyzed as if they arose from a completely randomized experiment. Let b_j, $j = 0, 1, \ldots, J$ denote the subclass boundaries, with $b_0 = 0$ and $b_J = 1$, and let $B_i(j)$ be a binary indicator, equal to 1 if $b_{j-1} < \hat{e}(X_i) < b_j$, and zero otherwise. Then we estimate the finite-sample average effect in subclass j, $\tau_{\text{fs}}(j)$, by $\hat{\tau}^{\text{dif}}(j)$, the difference in the average outcome for treated and control units in this subclass:

$$\hat{\tau}^{\text{dif}}(j) = \frac{\sum_{i:B_i(j)=1} Y_i \cdot W_i}{\sum_{i:B_i(j)=1} W_i} - \frac{\sum_{i:B_i(j)=1} Y_i \cdot (1 - W_i)}{\sum_{i:B_i(j)=1} (1 - W_i)}.$$

To estimate the overall finite-sample average effect of the treatment, τ_{fs}, we average these within-block differences $\hat{\tau}^{\text{dif}}(j)$,

$$\hat{\tau}^{\text{strat}} = \sum_{j=1}^{J} \frac{N(j)}{N} \cdot \hat{\tau}^{\text{dif}}(j),$$

where $N(j) = \sum_{i=1}^{N} B_i(j)$, and the label "strat" is used to stress the connection with the estimators used in the stratified randomized experiments discussed in Chapter 9. Although this method is more robust than the weighting estimator to the presence of units with extreme values of the estimated propensity score, we still do not recommend it without some modifications. In particular, we recommend reducing the bias and increasing the precision further by using covariance adjustment within the subclasses. In Chapter 17 we describe our specific approach to combining subclassification and covariance adjustment in detail.

Matching Estimators

Unlike model-based imputation and weighting and blocking methods, the fourth approach, matching, does not always rely on estimating an unknown function. Instead it relies on finding direct comparisons, that is, matches, for each unit. For a given treated unit with a particular set of values for the covariates, one looks for a control unit with as similar a set of covariates as possible. This approach has great intuitive appeal. Suppose we wish to assess the effect of a job-training program on the labor market outcomes for a particular person, say a thirty-year-old woman with two children under the age of six, with a high school education and four months of work experience in the past twelve months, who went through this training program. In the matching approach we look for a thirty-year-old woman with two children under the age of six, with a high school education and four months of work experience in the past twelve months, who did *not* attend the training program. If exact matches can be found, this is a particularly attractive and simple strategy. If no exact matches can be found, which is typically the case if the number of covariates is large compared to the number of units, this approach becomes more unwieldy. In that case one needs to assess the trade-offs of different violations of exact matching. Who should we use as a match for the thirty-year-old woman with two children and four months of work experiments who went through the training program? One possibility may be a woman from the control group who is four years older, with two months more work experience. A second possibility might be a woman who is two years younger with only one child and two months fewer work experience in the past twelve months. Assessing the relative merits of such matches requires careful inspection of the joint distribution of the covariates and substantive knowledge of the relative importance of the different characteristics for predicting outcomes. Clearly, as soon as such compromises need to be made, matching is more difficult to implement. Difficulties in dealing with many covariates show up here in a different form than in the model-based imputation methods, but they do not disappear. With many covariates, the quality of the matching, measured by some metric of the typical distance between covariates of units and the covariates of their matches, decreases. To implement the matching approach, one needs to be able to assess the trade-offs in choosing between different controls, and this requires a distance metric. We discuss in Chapter 18 some of the choices that have been used in the literature.

Mixed Estimators

In addition to the four basic approaches, there are a number of estimation methods that combine features of two or more of these basic methods in an attempt to combine the benefits of each of them. Regression (i.e., covariance adjustment), for example, is a powerful and effective method for adjusting for modest between-group differences, but it is less effective when the covariate distributions differ substantially between treatment and control groups. Using regression, not globally, but only within blocks with similar covariate distributions for treated and control units – for example, defined by the estimated propensity score – may therefore combine attractive properties of regression adjustment in relatively well-balanced samples with the robustness of subclassification methods across different distributions. Similarly one can combine matching with regression, again exploiting the strengths of both methods. We view these two combinations, subclassification with covariate adjustment within subclasses, and matching with covariance adjustment, as two of the more attractive methods in practice for estimating treatment effects with regular assignment mechanisms, especially when flexibly implemented. We discuss these approaches, and specific methods for implementing them, in more detail in Chapters 17 and 18.

12.5   DESIGN PHASE

Prior to implementing any of the methods for estimating causal effects in settings with regular assignment mechanisms, it is important to conduct what we call the *design phase* of an observational study. In this stage, we recommend investigating the extent of overlap in the covariate distributions. This, in turn, may lead to the construction of a subsample more suitable for estimating causal estimands, in the sense of being better balanced in terms of covariate distributions. There is one important feature of this initial analysis: this stage does not involve the outcome data, which need not be available at this stage, or even collected yet. As a result, this analysis cannot be "contaminated" by knowledge of estimated outcome distributions, or by preferences, conscious or unconscious, for particular results.

12.5.1 Assessing Balance

The first part of the design stage is to assess the degree of balance in the covariate distributions between treated and control units, which involves comparing the distributions of covariates in the treated and control samples. We focus on a couple of specific statistics that are useful in assessing the imbalance. First is the difference in average covariate values by treatment status, scaled by their sample standard deviation. This provides a scale-free way to assess the differences. As a rule-of-thumb, when treatment groups have important covariates that are more than one-quarter or one-half of a standard deviation apart, simple regression methods are unreliable for removing biases associated with differences in covariates, a message that goes back to the early 1970s but is often ignored.

Beyond looking at simple differences in average covariate values, we focus on the distributions of the propensity score. If the super-population covariate distributions are identical in the two treatment groups, then the true propensity score must be constant, and vice versa. Variation in the estimated propensity score is therefore a simple way to assess differences between two multivariate distributions. In practice we rarely know the propensity score *ex ante*, and so we typically have to estimate it, which involves choosing a specification for the propensity score and estimating the unknown parameters of that specification. In Chapter 13 we discuss flexible methods for doing so.

We discuss the specific methods for comparing covariate distributions and assessing balance in detail in Chapter 14.

12.5.2 Subsample Selection Using Matching on the Propensity Score

If the basic sample exhibits a substantial amount of imbalance, we may wish to construct a subsample that is characterized by better balance. Such a subsample leads to more robust and thus more credible causal inferences. In Chapter 15 we provide details for one method of implementing this approach that relies on having a relatively large number of controls and is appropriate for settings where we are interested in the effect of the treatment on the subpopulation of treated units. The proposed procedure consists of two steps. First we estimate the propensity score. Then we sequentially match each treated unit to the closest control unit in terms of the estimated propensity score, typically with the treated units ordered by decreasing estimated propensity score, although the order rarely matters much in practice. We match here without replacement, leading to matched samples with an equal number of treated and control units. We do not simply estimate the average effect of the treatment by taking the difference in average outcomes for the matched sample. Rather, within this matched sample, we apply some of the adjustment methods introduced previously, including those that allow for estimation of more general causal estimands than average effects, with the expectation that, because this sample has better covariate balance, the estimators for the matched sample will be more robust than the corresponding estimators applied to the original, full sample.

12.5.3 Subsample Selection through Trimming Using the Propensity Score

In Chapter 16 of the text, we discuss in more detail a second method for constructing balanced samples that also uses the estimated propensity score. The idea here is that for units with covariate values such that the propensity score is close to zero or one, it is difficult to obtain precise estimates of the typical effect of the treatment because, for such units, there are few controls relative to the number of treated units, or the other way around. We therefore propose putting aside such units and focusing on estimating causal effects in the subpopulation of units with propensity score values bounded away from zero and one. More precisely, we discard all units with estimated propensity scores outside an interval, and we propose a specific way to choose the interval.

12.6   ASSESSING UNCONFOUNDEDNESS

In Chapter 21, in Part V of the text, we discuss methods for assessing the unconfoundedness assumption. We purposely use the term "assess" here rather than "test," because unconfoundedness has no directly testable implications. Nevertheless, there are a number of statistical analyses that we can conduct that can shed light on its plausibility. Some of these analyses, like the analyses assessing balance, do not involve the outcome data, and so are part of the design stage. The conclusion from such analyses can be that one may deem unconfoundedness an unattractive assumption for the specific data at hand and decide not to pursue further analyses with the outcome data; or it can be that one decides that unconfoundedness is plausible, and analyses based on this assumption are credible. Here we briefly introduce three of these analyses.

12.6.1   Estimating the Effect of the Treatment on an Unaffected Outcome

The first set of assessments focuses on estimating the causal effect of the treatment on a variable that is known *a priori* not to be affected by the treatment, typically because its value is determined prior to the treatment itself. Such a variable can be a time-invariant covariate, but the most interesting case is where this is a lagged outcome. In this case, one uses all the covariates except the single covariate that is being assessed, say the lagged outcome. One estimates the pseudo-treatment effects on the lagged outcome. If these estimated effects are near zero, it is deemed more plausible that the unconfoundedness assumption holds than if the estimated effects are large. Of course, the assessment is not directly testing the unconfoundedness assumption, and so, no matter what the p-value of the null hypothesis of no effect, it does not directly reflect on the assumption of interest, unconfoundedness. Nevertheless, if the variables used in this proxy test are closely related to the outcome of interest, the assessment has arguably more force than if the variables are unrelated to the outcome of interest. For these analyses, it is clearly helpful to have a number of lagged outcomes. This approach is a *design* approach, not using any outcome data.

12.6.2   Estimating the Effect of a Pseudo-Treatment on the Outcome

The second set of assessments focuses on estimating the causal effect of a different treatment on the original outcome, and in particular a pseudo-treatment that is known *a priori* not to have an effect. This approach relies on the presence of multiple control groups and uses actual outcome data, but only for the control units. Suppose one has two possible control groups. One interpretation of the assessment is that one compares estimated treatment effects calculated using one control with average treatment effects calculated using the other control group. This procedure can also be interpreted as estimating an average treatment effect using only the two control groups, with the treatment indicator redefined as an indicator for one of the two control groups. In that case, the pseudo-treatment effect is known to be zero, and statistical evidence of a non-zero estimated treatment effect suggests that, for at least one of the control groups, the unconfoundedness assumption is violated. Again, failure to reject this "test" does not mean the unconfoundedness assumption is valid because it could be that both control groups have similar biases, but non-rejection in the case where the two control groups are *a priori* likely to have different biases makes it more plausible that the unconfoundedness assumption holds. The key for the value of this assessment is to have control groups that are likely to have different biases, if at all. One may use different geographic control groups, for example on either side of the treatment group. This approach is a *semi-design* approach, using only outcome data for the control units.

12.6.3 Assessing Sensitivity of Estimates to the Choice of Pre-Treatment Variables

The last approach for assessing the unconfoundedness assumption uses outcome data for all units. The idea is to partition the covariates again into two parts. Now the assessment involves comparing estimates for treatment effects using only a subset of the covariates to those for the full set of covariates. Substantial differences suggest that either unconfoundedness relies critically on all covariates, or it does not hold. Because this approach uses outcome data for all units, it is not a (semi-)design approach.

12.7 CONCLUSION

In this chapter we discuss the assumptions underlying regular assignment mechanisms and provide a brief overview of Parts III through V of this text. We focus primarily on the generally most controversial of these assumptions, unconfoundedness, and provide motivation for the central role this assumption plays in the third and fourth parts of this book. We then describe briefly how estimation and inference may proceed with regular assignment mechanisms. In settings where the pre-treatment variables take on few distinct values in the sample, the analysis is simple and follows exactly the same path as that under stratified randomized experiments. The more challenging setting is that where the covariates take on too many distinct values in the sample to allow for exact stratification on the covariates with each stratum having both treated and control units. It is this setting that is the focus of a large theoretical literature in statistics and related disciplines. In Chapters 13–22 we provide details on the methods we view as most promising in practice in this setting.

NOTES

The term "unconfoundedness" was introduced in Rubin (1990a, p. 284). Other terms have been used to describe the same, or closely related, assumptions. Rosenbaum and Rubin (1983a) refer to the combination of unconfoundedness and the assumption that assignment is probabilistic as "strong ignorability." Lechner (1999) and Angrist and Pischke (2008) use the term "conditional independence assumption" for the unconfoundedness assumption. The concept of unconfoundedness is closely related to what in the econometrics literature is called "exogeneity." There are no widely agreed upon definitions of exogeneity, although some authors do view it as synonymous with unconfoundedness. Manski, Sandefur, McLanahan, and Powers (1992, p. 28) describe the treatment indicator in this setting as "'exogenous,' or synonymously, 'strongly ignorable.'" Imbens (2004) discusses the link with definitions of exogeneity in parametric regression models. Following the work by Barnow, Cain, and Goldberger (1980) in a regression setting, it is also referred to as "selection on observables." For a standard discussion of exogeneity in the econometric literature, see Engle, Hendry, and Richard (1974). For general discussions of unconfoundedness in the econometrics literature, with different perspectives, see Blundell and Costa-Dias (2000, 2002), Imbens (2004), and Heckman and Vytlacil (2007ab)

Hirano and Imbens (2001), Huber, Lechner, and Wunsch (2012), and Belloni, Chernozhukov, and Hansen (2014) discuss methods for variable selection in the context of estimating the propensity score. Rosenbaum (1984b) discusses the concerns when adjusting for covariates that are affected by the treatment.

Early applications in economics include Ashenfelter (1978), Ashenfelter and Card (1985), and Card and Sullivan (1988). The semiparametric efficiency bound for τ_{sp} is derived in Hahn (1998). See also Hirano, Imbens, and Ridder (2003).

The merits of and concerns with regression (covariance) adjustments in settings where the covariate distributions differ substantially between treatment and control groups are discussed in Cochran (1965, 1968), Rubin (1973b, 1979, 2006), and Cochran and Rubin (1973).

Rosenbaum (2009) and Rubin (2007, 2008) discuss the importance of the design stage of an observational study. The discussion in Section 12.6.2 is closely related to Rosenbaum's (1987) notion of multiple control groups. An early application of these ideas is in Lalonde (1986).

There is also a literature concerned with the difficulties of adjusting for many covariates. See Angrist and Hahn (2004), Robins and Ritov (1997), Robins and Rotnitzky (1995), and Belloni, Chernozhukov, and Hansen (2014).

There is now much software available for implementing these methods. Software includes STATA programs by Becker and Ichino (2002), Abadie, Drukker, Herr, and Imbens (2003), and Sianesi (2001), and R-programs by Sekhon (2004–2013) and Hansen (2006).

## CHAPTER 13

Estimating the Propensity Score

13.1 INTRODUCTION

Many of the procedures for estimating and assessing causal effects under unconfoundedness involve the propensity score. In practice it is rare that we know the propensity score *a priori* in settings other than those involving randomized experiments. Such practical settings could have complex designs where the unit-level probabilities differ in known ways. An example is the allocation of admissions to students applying for medical school in The Netherlands in the 1980s and 1990s. Based on high school grades, applicants would be assigned a priority score that determined their *probability* of getting admitted to medical school. The actual admission to medical school was then based on a (random) lottery. Such settings are rare, however, and a more common situation is where, given the pre-treatment variables available, a researcher views unconfoundedness as a reasonable approximation to the actual assignment mechanism, with only vague *a priori* information about the form of the dependence of the propensity score on the observed pre-treatment variables. For example, in many medical settings, decisions are based on a set of clinically relevant patient characteristics observed by doctors and entered in patients' medical records. However, there is typically no explicit rule that requires physicians to choose a specific treatment based on particular values of the pre-treatment variables. In light of this degree of physician discretion, there is no explicitly known form for the propensity score. In such cases, for at least some of the methods for estimating and assessing treatment effects discussed in this part of the book, the researcher needs to estimate the propensity score. In this chapter we discuss some specific methods for doing so.

It is important to note that the various methods that will be discussed in the chapters following this one, specifically Chapters 14–17, use the propensity score in different ways. Some of these methods rely more heavily than others on an accurate approximation of the true propensity score by the estimated propensity score. As a consequence, estimators for the treatment effects may be more or less sensitive to the decisions made in the specification of the propensity score. For example, one way in which we can use the propensity score is to construct strata or subclasses, within which further adjustment methods can be used. In that case, the exact specification will likely matter less than when using methods where we rely solely on weighting by the inverse of the estimated propensity score to eliminate all biases in estimated treatment effects arising from differences in covariates distributions. Such "Horvitz-Thompson" type weighting methods, briefly discussed in Chapter 12, are therefore not emphasized in this text.

In the basic problem we study in this chapter, we have a sample of N units, viewed as a random sample from an infinite super-population. Each unit in this super-population is either exposed to, or not exposed to, the treatment. In the sample, N_c units are exposed to the control treatment and N_t units are exposed to the active treatment, with $N = N_c + N_t$. As usual, the observed treatment indicator is denoted by $W_i \in \{0, 1\}$ for unit i. For each unit in the sample, we also observe a K-component row vector of pre-treatment variables, denoted by X_i for unit i. Although many of the uses for the propensity score described in later chapters are motivated by the assumption of unconfoundedness, we do not explicitly use this assumption in the current chapter. In this chapter, the sole focus is on the statistical problem of estimating the conditional probability of receiving the treatment given the observed covariates,

$$\Pr(W_i = 1 | X_i = x) = \mathbb{E}\left[W_i | X_i = x\right], \tag{13.1}$$

which is equal to the super-population propensity score, $e(x)$, and we will use that notation here. (Here, for ease of notation we continue to omit the conditioning on the parameters governing these distributions.) If the covariate X_i is a binary scalar, or more generally takes on only a few values, the statistical problem of estimating the propensity score is straightforward: we can simply partition the sample into subsamples that are homogeneous in the values of the covariates, and estimate the propensity score for each subsample as the proportion of treated units in that subsample. Using such a fully saturated model is not feasible in many realistic settings. Often we find that many strata defined by unique values of the covariates in the sample contain only a single unit, so that the proportion of treated units within the stratum is either zero or one. Such an occurrence makes many of the methods that rely on the estimated propensity score discussed in this text infeasible, and therefore we explicitly focus in this chapter on settings where the covariates take on too many values to allow for a fully saturated model, so that some form of smoothing is essential.

The goal is to obtain estimates of the propensity score that balance the covariates between treated and control subsamples. More precisely, we would like to have an estimate of the propensity score such that, within subsamples with similar values of the estimated propensity score, the distribution of covariates among the treated units is similar to the distribution of covariates among the control units. This criterion is somewhat vague, and we elaborate on its implementation later. First, it is important to note, however, that the goal is *not* simply to get the best estimate of the propensity score in terms of mean-integrated-squared-error, or a similar criterion based on minimizing the difference between the estimated and true propensity score. Such a criterion would always suggest that using the true propensity score is preferable to using an estimated propensity score. In contrast, for our purposes, it is often preferable to use the estimated propensity score. The reason is that using the estimated score may lead to superior covariate balance in the sample compared to that achieved when using the true super-population propensity score. For example, in a completely randomized experiment with a single binary covariate (but the assignment probability free of dependence on that covariate), using the estimated propensity score to stratify units would lead to perfect within-stratum balance on the covariates in the sample, whereas using the true propensity score generally would not. The difficulty is that our criterion, in-sample balance in the covariates given the (estimated) propensity score, is not as easy to formalize and operationalize as some of the conventional goodness-of-fit measures,

There are two parts to the proposed algorithm for specifying the propensity score. First we specify an initial model, motivated by substantive knowledge. Second, we assess the statistical adequacy of an estimate of that initial model, by checking whether the covariates are balanced within strata defined by the estimated propensity score. In principle, one can iterate back and forth between these two stages, specification of the model and assessment of that model, each time refining the specification of the model. In this chapter we describe an automatic procedure (i.e., an algorithm) for selecting a specification that can, at the very least, provide a useful starting point for such an iterative procedure, and in many cases will lead to a fairly flexible specification with good balancing properties. The specific procedure selects a subset of the covariates to enter linearly into specification of the propensity score, as well as a subset of all second-order interactions of the basic set of linearly included covariates. Although, in principle, one can also include third- and higher-order terms, in our practical experience it is rare that such higher-order terms substantially improve balance for the sample sizes and data configurations commonly encountered in practice. Of course, what is "linear" and what is "higher order" depends on what initial transformation of the covariates has been applied. If one wishes to allow for the inclusion of third- and higher-order terms, or have functions of the covariates such as logarithms, or indicators for regions of the covariate space, one can easily do so by selecting them following largely the same procedure that we discuss for selecting second-order terms.

Three general comments are in order. First, it is important to keep in mind that during this entire process, and in fact in this entire chapter, we do not use the outcome data, and there is, therefore, no way of deliberately biasing the final estimation results for the treatment effects. Consequently, there is no concern regarding the statistical properties of the ultimate estimates of the average treatment effects obtained from iterating back and forth between (i) the specification of the propensity score, and (ii) balance assessments of the estimated propensity score, until an adequate specification is found.

A second point is that, in general, it is difficult to give a fully automatic procedure for specifying the propensity score in a way that leads to a specification that passes all the tests and diagnostics that we may subject that specification to in the second stage. The specification may be much improved by incorporating subject-matter knowledge regarding the role of the covariates in the treatment assignment decision *and* the outcome process. We therefore emphatically recommend against relying solely and routinely on automatic procedures. Nevertheless, we do present some automatic procedures that lead to flexible specifications of the propensity score, specifications that are increasingly flexible as the sample size grows. Such automatic procedures can provide useful starting points, as well as benchmarks for comparisons against more sophisticated and scientifically motivated specifications. Our procedure is likely to be an improvement over commonly used approaches, such as simply including all pre-treatment variables linearly in a logistic model specification. We should also note that there are many other algorithms one could use for specifying models for the propensity score, and we provide references to some of them in the notes to this chapter.

A final point to emphasize is that the primary goal is to find an adequate specification of the propensity score, in the sense of a specification that achieves statistical balance in the covariates. We are *not* directly interested in a structural, behavioral, or causal interpretation of the propensity score, although inspecting and assessing the strength and nature of the dependence of the propensity score on the covariates may be helpful when assessing the plausibility of the unconfoundedness assumption. Finding an adequate specification is, therefore, in essence, a statistical problem that relies less on subject-matter knowledge than other aspects of the modeling of causal effects. The goal is simply to find a specification for the propensity score that leads to adequate balance between covariate distributions in treatment and control groups in our sample.

The remainder of this chapter is organized as follows. The next section describes the data used in this chapter, which come from a study of the effect of barbiturate exposure on cognitive outcomes. In Section 13.3 we discuss methods for choosing the specification of the propensity score, that is, selecting the covariates for inclusion in the specification of the propensity score. Although for purposes of obtaining balanced samples a simple linear specification for the propensity score may well be adequate, we follow a conventional approach in the literature and use logistic regression models. In Section 13.4 we illustrate our proposed covariate selection procedure with the barbiturate data. In the remainder of this chapter we discuss methods for assessing the adequacy of the specification of the propensity score. We do so by assessing whether, conditional on values of the estimated propensity score, the covariates are uncorrelated with the treatment indicator, that is, whether the mean covariate values for the controls are approximately equal, conditional on the estimated propensity score. We implement this idea by first constructing strata (i.e., subclasses or blocks) within which the estimated propensity score is almost constant. In Section 13.5 we discuss an automatic method for constructing such blocks. In Section 13.6 we illustrate this method with the barbiturate data. In Section 13.7 we discuss assessing within-block balance in the covariates. In Section 13.8 we illustrate this, again using the barbiturate data. Section 13.9 concludes.

13.2 THE REINISCH ET AL. BARBITURATE EXPOSURE DATA

The data we use to illustrate the methods in this chapter come from a study of the effect of prenatal exposure to barbiturates (Reinisch, Sanders, Mortenson, and Rubin, 1995). The data set contains information on $N = 7{,}943$ men and women born between 1959 and 1961 in Copenhagen, Denmark. Of these 7,943 individuals, $N_t = 745$ men and women had been exposed *in utero* to substantial amounts of barbiturates due to maternal medical conditions. The comparison group consists of $N_c = 7{,}198$ individuals from the same birth cohort who were not exposed *in utero* to barbiturates. The substantive interest is in the effect of the barbiturate exposure on cognitive development measured many years later, although we do not access the outcome information in this chapter. The data set contains information on seventeen covariates that are potentially related to both the outcomes of interest, reflecting cognitive development, and the likelihood of having been

Table 13.1. *Summary Statistics Reinisch Data Set*

Label	Variable Description	Controls (N_c =7198)		Treated (N_t =745)		t-Stat
		Mean	(S.D.)	Mean	(S.D.)	Difference
sex	Sex of child (female is 0)	0.51	(0.50)	0.50	(0.50)	−0.3
antih	Exposure to antihistamine	0.10	(0.30)	0.17	(0.37)	4.5
hormone	Exposure to hormone treatment	0.01	(0.10)	0.03	(0.16)	2.5
chemo	Exposure to chemotherapy agents	0.08	(0.27)	0.11	(0.32)	2.5
cage	Calendar time of birth	−0.00	(1.01)	0.03	(0.97)	0.7
cigar	Mother smoked cigarettes	0.54	(0.50)	0.48	(0.50)	−3.0
lgest	Length of gestation (10 ordered categories)	5.24	(1.16)	5.23	(0.98)	−0.3
lmotage	Log of mother's age	−0.04	(0.99)	0.48	(0.99)	13.8
lpbc415	First pregnancy complication index	0.00	(0.99)	0.05	(1.04)	1.2
lpbc420	Second pregnancy complication index	−0.12	(0.96)	1.17	(0.56)	55.2
motht	Mother's height	3.77	(0.78)	3.79	(0.80)	0.7
motwt	Mother's weight	3.91	(1.20)	4.01	(1.22)	2.0
mbirth	Multiple births	0.03	(0.17)	0.02	(0.14)	−1.9
psydrug	Exposure to psychotherapy drugs	0.07	(0.25)	0.21	(0.41)	9.1
respir	Respiratory illness	0.03	(0.18)	0.04	(0.19)	0.7
ses	Socioeconomic status (10 ordered categories)	−0.03	(0.99)	0.25	(1.05)	7.0
sib	If sibling equal to 1, otherwise 0	0.55	(0.50)	0.52	(0.50)	−1.6

prescribed and taking, barbiturates. Many of the covariates relate to the mother's physical and socioeconomic situation and thus are plausibly related to children's subsequent cognitive development.

Table 13.1 presents summary statistics for the data, including averages and standard deviations for the two groups, and t-statistics assessing the test of the null hypothesis of equality of means of the covariates in the control and treatment groups. It is clear that the two groups differ substantially in the distribution of their background characteristics. The subsample of individuals exposed *in utero* to barbiturates has, on average, higher socio-economic status, older mothers, and a higher prevalence of pregnancy complications (in particular, the second composite pregnancy complication index lpbc420). Such differences may bias a simple comparison of outcomes by treatment status and suggest that, at the very least, adjustments for pre-treatment differences are required to obtain credible inferences for the causal effect of barbiturate exposure, on, say, cognitive development outcomes.

13.3 SELECTING THE COVARIATES AND INTERACTIONS

In many empirical studies, the number of covariates can be large relative to the number of units. As a result, it is not always feasible simply to include all covariates in a model for the propensity score. Moreover, for some of the most important covariates, it may not be sufficient to include them only linearly, and we may wish to include functions, such as logarithms, and higher-order terms, such as quadratic terms, or interactions between the basic covariates. Here we describe a stepwise procedure for selecting the covariates and higher-order terms for inclusion in the propensity score. In the notes to this chapter, there are references to alternative flexible methods for finding a suitable specification for the propensity score, where again "suitable" refers to obtaining balance on the important covariates.

We focus here on logistic regression models where the log odds ratio of receiving the treatment is modeled as linear in a number of (functions of) the basic covariates, with unknown coefficients. We estimate the coefficients by maximum likelihood; see the Appendix for details. The main question now concerns the selection of the functions of the basic covariates to include in the specification.

The approach starts with the K-component vector of covariates X_i. We select a subset of these K covariates to be included linearly when estimating the log odds ratio of the propensity score, as well as a subset of all $K \cdot (K+1)/2$ second-order terms (both quadratic and interactions terms). This leads to a potential set of included predictors equal to $K + K \cdot (K+1)/2 = K \cdot (K+3)/2$. We do not directly compare all possible subsets of this set because this might be too large for commonly encountered values of K (the number of such subsets is $2^{K \cdot (K+3)/2}$). Instead we follow a stepwise procedure with three stages.

In the first stage, we select a set of K_B basic covariates to be included in the propensity score, regardless of their statistical association with the treatment indicator, because they are viewed as important on substantive grounds. These substantive grounds may be based on *a priori* expected associations with the assignment process, or *a priori* expected associations with the outcome. In the second stage, we decide which of the remaining $K - K_B$ covariates will also be included linearly to estimate the log odds ratio. At the conclusion of this step, we have a total of K_L covariates entering linearly in the log odds ratio. In the third stage we decide which of the $K_L \cdot (K_L+1)/2$ interactions and quadratic terms involving the K_L selected covariates to include. This stage will lead to the selection of K_Q second-order terms, leaving us with a vector of covariates with $K_L + K_Q$ components to be included linearly in the specification of the log odds ratio.

Now let us consider each of these three stages in more detail.

Step 1: Basic Covariates

In the first step we decide to include K_B basic covariates on substantive grounds, which may include covariates that are *a priori* viewed as important for explaining the assignment and plausibly related to some outcome measures. It may also be that $K_B = 0$ if the researcher has little substantive knowledge regarding the relative importance of the covariates. In evaluations of labor market programs, this step might lead to including covariates that are viewed as important for the decision of the individual to participate, such as recent labor market experiences. The set of covariates selected at this stage may also include covariates that are *a priori* viewed as likely to be strongly associated with the outcomes. Again, in the setting of labor market programs, this could include proxies for human capital, such as prior earnings or education levels. In the barbiturate exposure example analyzed in this chapter, this set includes three pre-treatment variables: mother's age (lmotage), which is plausibly related to cognitive outcomes for the child; socio-economic status (ses), which is strongly related to the number of physician visits during pregnancies and thus exposes the mother to greater risk of barbiturate prescriptions;

and, finally, sex of the child (sex), which may be associated with measures of cognitive outcomes.

Step 2: Additional Linear Terms

In the second step we select some of the remaining covariates for inclusion in the specification of the propensity score. There are $K - K_B$ covariates not included yet. We only consider at most $(K - K_B)$ of the $2^{K - K_B}$ different subsets involving these covariates. Exactly how many and which of the subsets we consider depends on the configuration of the data. We consider one of the remaining covariates at a time, each time checking whether we wish to add it. More specifically, suppose that at some point in the covariate selection process, we have selected $\tilde{K}_L$ linear terms, including the K_B terms selected in the first step. At that point we are faced with the decision whether to include an additional covariate from the set of $K - \tilde{K}_L$ covariates, and if so, which one. This decision is based on the results of $K - \tilde{K}_L$ additional logistic regression models. In each of these $K - \tilde{K}_L$ additional logistic regression models, we add to the basic specification with $\tilde{K}_L$ covariates and an intercept, a single one of the remaining $K - \tilde{K}_L$ covariates. For each of these $K - \tilde{K}_L$ specifications, we calculate the likelihood ratio statistic assessing the null hypothesis that the newly included covariate has a zero coefficient. If all the likelihood ratio statistics are less than some pre-set constant C_L, we stop, and we include only the $\tilde{K}_L$ covariates linearly. If at least one of the likelihood ratio test statistics is greater than C_L, we add the covariate with the largest likelihood ratio statistic. We now have $\tilde{K}_L + 1$ covariates, and check whether any of the remaining $K - \tilde{K}_L - 1$ covariates should be included by calculating likelihood ratio statistics for each of them. We continue this process until none of the remaining likelihood ratio statistics exceeds C_L. This second stage leads to the addition of $K_L - K_B$ covariates to the K_B covariates already selected for inclusion in the linear set in the first stage, for a total of K_L covariates.

Step 3: Quadratic and Interaction Terms

In the third step we decide which of the interactions and quadratic terms to include in the specification of the propensity score. Given that we have selected $K_L \leq K$ covariates in the linear stage, we now decide which of the $K_L \cdot (K_L + 1)/2$ quadratic and interaction terms involving these K_L covariates to include. (If some of the covariates are binary, some of these $K_L \cdot (K_L + 1)/2$ quadratic terms would be identical to some of the linear terms and thus known not to improve the specification, and so the effective set of possible second-order terms may be smaller than $K_L \cdot (K_L + 1)/2$.) Note that with this approach, we include only higher-order terms involving the K_L covariates selected for inclusion in the linear part. We follow essentially the same procedure as for the linear stage. Suppose at some point we have added $\tilde{K}_Q$ of the $K_L \cdot (K_L + 1)/2$ possible interactions. We then estimate $K_L \cdot (K_L + 1)/2 - \tilde{K}_Q$ logistic regressions, each of which includes the intercept, the K_L linear terms (including the K_B basic ones), the $\tilde{K}_Q$ second-order terms already selected, and one of the remaining $K_L \cdot (K_L + 1)/2 - \tilde{K}_Q$ terms. For each of these $K_L \cdot (K_L + 1)/2 - \tilde{K}_Q$ logistic regressions, we calculate the likelihood ratio statistic for the null hypothesis that the most recently added second-order term has a coefficient of zero. If the largest likelihood ratio statistic is greater than some pre-determined constant C_Q, we include that interaction term in the model. Then we re-calculate the likelihood ratio statistics for the remaining $K_L \cdot (K_L + 1)/2 - \tilde{K}_Q - 1$ interaction terms, and we keep including the term with the largest likelihood ratio statistic until all of the remaining likelihood ratio statistics are less than C_Q.

This algorithm leaves us with a selection of K_L linear covariates and a selection of K_Q second-order terms (plus an intercept). We estimate the propensity score using this vector of $1 + K_L + K_Q$ terms. To illustrate the implementation of this strategy, we use the threshold value for the likelihood ratio statistic of $C_L = 1$ and $C_Q = 2.71$, corresponding implicitly to z-statistics of 1 and 1.645, respectively.

13.4 CHOOSING THE SPECIFICATION OF THE PROPENSITY SCORE FOR THE BARBITURATE DATA

Here we illustrate the implementation of the covariate selection procedure on the barbiturate data. The ultimate interest in this application is in the effect of *in utero* barbiturate exposure on cognitive outcomes for young adults, although in this chapter we do not look at the outcome data. Based on the substantive argument in the original papers using these data, it was argued that the child's sex, the mother's age, and mother's socio-economic status (sex, lmotage, and ses respectively) are particularly important covariates, the first two because they are likely to be associated with the outcomes of interest, and the last two because they are likely to be related to barbiturate exposure. We therefore include these three basic covariates in the specification of the propensity score, irrespective of the strength of their statistical association with barbiturate exposure (i.e., $K_B = 3$).

As the first step toward deciding which other covariates to include linearly, we estimate the baseline model with an intercept and the three previously selected covariates, sex, lmotage, and ses. The results for this model are in Table 13.2. Both lmotage and ses are statistically significantly (at the 0.05 level) associated with *in utero* exposure to barbiturates.

Next we estimate fourteen logistic regression models, each including an intercept, sex, lmotage, and ses, and one of the fourteen remaining covariates. For each specification, we calculate the likelihood ratio statistic for the test of the null hypothesis that the coefficient on the additional covariate is equal to zero. For example, for the covariate lpbc420, the second pregnancy complication index, the results are reported in Table 13.3. The likelihood ratio statistic (twice the difference between the unrestricted and restricted log likelihood values), is equal to 1308.0. We do this for each of the fourteen remaining covariates (seventeen covariates minus the three pre-selected). We report the fourteen likelihood ratio statistics in the first column of Table 13.4. We find that the covariate that leads to the biggest increase in the likelihood function is lpbc420. The likelihood ratio statistic for that covariate is 1308.0. Because this value exceeds our threshold of $C_L = 1$, we include the second pregnancy complication index lpbc420 in the specification of the propensity score.

Next we estimate thirteen logistic regression models where we always include an intercept, sex, lmotage, ses, and lpbc420, and additionally include, one at a time, the remaining thirteen covariates. The likelihood ratio statistics for the inclusion of these thirteen covariates are reported in the second column of Table 13.5. Now mbirth, the indicator for multiple births, is the most important covariate in terms of increasing

Table 13.2. *Estimated Parameters of Propensity Score: Baseline Case; Barbiturate Data*

Variable	EST	$(\widehat{s.e.})$	t-Stat
Intercept	−2.38	(0.06)	−41.0
sex	−0.01	(0.08)	−0.2
lmotage	0.48	(0.04)	11.7
ses	0.10	(0.04)	2.6

Table 13.3. *Estimated Parameters of Propensity Score: Baseline Case with* `lpbc420` *Added; Barbiturate Data*

Variable	EST	$(\widehat{s.e.})$	t-Stat
Intercept	−3.71	(0.10)	−36.3
sex	0.07	(0.09)	0.8
lmotage	0.22	(0.05)	4.7
ses	0.15	(0.05)	3.3
lpbc420	2.11	(0.08)	27.2
LR statistic	1308.0		

Table 13.4. *Likelihood Ratio Statistics for Sequential Selection of Covariates to Enter Linearly; Barbiturate Data*

Covariate					Step →						
sex	–	–	–	–	–	–	–	–	–	–	–
antih	17.5	0.5	1.6	1.3	2.1	1.8	1.6	1.6	1.7	1.3	–
hormone	3.9	0.3	0.7	0.7	0.4	0.8	0.7	0.7	0.7	0.8	0.9
chemo	10.0	36.6	41.9	–	–	–	–	–	–	–	–
cage	0.8	5.8	6.4	7.2	7.6	7.9	–	–	–	–	–
cigar	4.3	2.3	3.5	3.7	3.0	2.1	2.1	1.7	2.1	–	–
lgest	0.4	11.1	5.0	6.4	7.3	5.5	5.6	–	–	–	–
lmotage	–	–	–	–	–	–	–	–	–	–	–
lpbc415	0.6	0.0	0.2	0.2	0.0	0.0	0.1	0.1	0.0	0.0	0.0
lpbc420	1308.0	–	–	–	–	–	–	–	–	–	–
motht	0.1	0.1	0.0	0.0	0.0	0.0	0.0	0.0	0.0	0.0	0.0
motwt	6.1	1.5	0.6	1.2	2.5	2.7	2.4	3.4	–	–	–
mbirth	4.6	66.1	–	–	–	–	–	–	–	–	–
psydrug	93.1	29.8	38.9	46.8	–	–	–	–	–	–	–
respir	0.1	0.0	0.0	0.1	0.0	0.0	0.0	0.0	0.0	0.0	0.0
ses	–	–	–	–	–	–	–	–	–	–	–
sib	21.0	13.8	12.5	15.0	15.7	–	–	–	–	–	–

Table 13.5. *Estimated Parameters of Propensity Score: Baseline Case with* `lpbc420` *and* `mbirth` *Added; Barbiturate Data*

Variable	EST	$(\widehat{s.e.})$	t-Stat
Intercept	−3.73	(0.10)	−35.9
sex	0.08	(0.09)	0.9
lmotage	0.21	(0.05)	4.5
ses	0.16	(0.05)	3.4
lpbc420	2.21	(0.08)	27.5
mbirth	−1.96	(0.30)	−6.6
LR statistic	66.1		

the likelihood function, and because the likelihood ratio statistic for the inclusion of `mbirth`, 66.1, exceeds the threshold of $C_L = 1$, `mbirth` is added to the specification.

We keep checking whether there is any covariate that, when added to the baseline model, increases the likelihood function sufficiently, and if so, we include it in the specification of the propensity score. Proceeding this way leads to the inclusion, in the second step, after the three covariates `sex`, `lmotage`, and `ses`, which were selected in the first step, ten additional covariates. In the order they were added to the specification, these are, `lpbc420`, `mbirth`, `chemo`, `psydrug`, `sib`, `cage`, `lgest`, `motwt`, `cigar`, and `antih`. The likelihood ratio statistics are reported in Table 13.4. Once we have a model with these thirteen covariates and an intercept, none of the remaining four covariates satisfied our criterion to warrant inclusion in the specification of the propensity score.

Next we consider quadratic terms and interactions. With the thirteen covariates selected in the previous two steps for inclusion in the linear part of the propensity score, there are potentially $13 \times (13 + 1)/2 = 91$ second-order terms. Not all 91 potential second-order terms are feasible, because some of the thirteen covariates selected in the first two steps are binary indicator variables, so that the corresponding quadratic terms are identical to the linear terms. We select a subset of the non-trivial second-order terms in the same way we selected the linear terms, with the only difference being that the threshold for the likelihood ratio statistic is now 2.71, which corresponds to nominal statistical significance at the 10% level. Following this procedure, adding one second-order term at a time, leads to the inclusion of seventeen second-order terms.

Table 13.6 reports the parameter estimates for the propensity score with all the linear and second-order terms selected, with the variables in the order in which they were selected for inclusion in the specification of the propensity score.

13.5 CONSTRUCTING PROPENSITY-SCORE STRATA

The specification for the propensity score, with estimates for the unknown parameters in that specification, leads to an estimated propensity score at each value x of the covariates, denoted by $\hat{e}(x)$. Next we wish to assess the adequacy of that specification by exploiting a

Table 13.6. *Estimated Parameters of Propensity
Score: Final Specification; Barbiturate Data*

Variable	EST	$(\widehat{s.e.})$	t-Stat
Intercept	−5.67	(0.23)	−24.4
Linear terms			
sex	0.12	(0.09)	1.3
lmotage	0.52	(0.11)	4.7
ses	0.06	(0.09)	0.6
lpbc420	2.37	(0.36)	6.6
mbirth	−2.11	(0.36)	−5.9
chemo	−3.51	(0.67)	−5.2
psydrug	−3.37	(0.55)	−6.1
sib	−0.24	(0.22)	−1.1
cage	−0.56	(0.26)	−2.2
lgest	0.57	(0.23)	2.5
motwt	0.49	(0.17)	2.9
cigar	−0.15	(0.10)	−1.5
antih	0.17	(0.13)	1.3
Second-order terms			
lpbc420 × sib	0.60	(0.19)	3.1
motwt × motwt	−0.10	(0.02)	−4.5
lpbc420 × psydrug	1.88	(0.39)	4.8
ses × sib	−0.22	(0.10)	−2.2
cage × antih	−0.39	(0.14)	−2.8
lpbc420 × chemo	1.97	(0.49)	4.0
lpbc420 × lpbc420	−0.46	(0.14)	−3.3
cage × lgest	0.15	(0.05)	3.0
lmotage × lpbc420	−0.24	(0.10)	−2.5
mbirth × cage	−0.88	(0.39)	−2.3
lgest × lgest	−0.04	(0.02)	−2.0
ses × cigar	0.20	(0.09)	2.2
lpbc420 × motwt	0.15	(0.07)	2.0
chemo × psydrug	−0.93	(0.46)	−2.0
lmotage × ses	0.10	(0.05)	1.9
cage × cage	−0.10	(0.05)	−1.8
mbirth × chemo	−∞	(0.00)	−∞

property of the true propensity score, namely the independence of the treatment indicator and the vector of covariates given the true super-population propensity score,

$$W_i \perp\!\!\!\perp X_i \mid e(X_i). \tag{13.2}$$

We substitute the estimated propensity score for the true propensity score and investigate whether, at least approximately,

$$W_i \perp\!\!\!\perp X_i \mid \hat{e}(X_i), \tag{13.3}$$

that is, whether, conditional on the estimated propensity score, the covariates and the treatment indicator are independent. Ideally we would do this by stratifying the sample into subsamples or blocks within each of which all units would have the exact same value of $\hat{e}(x)$, and then assessing whether W_i and X_i within each resulting block are independent. This plan is feasible only if the estimated propensity score takes on a relatively small number of values, and thus if the covariates jointly only take on a relatively small number of values in the sample. Typically, in practice, that is not the case, and so we coarsen the estimated propensity score by constructing blocks (i.e., strata or subclasses) within which the estimated propensity scores vary only little. For a set of boundary points, $0 = b_0 < b_1 < \ldots < b_{J-1} < b_J = 1$, define the block indicator $B_i(j)$, for the ith unit, as

$$B_i(j) = \begin{cases} 1 & \text{if } b_{j-1} \leq \hat{e}(X_i) < b_j, \\ 0 & \text{otherwise}, \end{cases}$$

for $j = 1, \ldots, J$. (Here we ignore the possibility that there are units with $\hat{e}(X_i)$ exactly equal to $B_i(J) = 1$.) Then we assess adequacy of the estimated propensity score by assessing whether

$$W_i \perp\!\!\!\perp X_i \mid B_i(1), \ldots, B_i(J). \tag{13.4}$$

We operationalize the assessment of independence by examining whether the treatment indicator and the covariates are uncorrelated within each of these blocks:

$$\mathbb{E}[X_i | W_i = 1, B_i(j) = 1] = \mathbb{E}[X_i | W_i = 0, B_i(j) = 1], \tag{13.5}$$

for all blocks $j = 1, \ldots, J$.

The first step in implementing this procedure is the choice of boundary values b_j, for $j = 0, \ldots, J$. We want to choose the boundary values in such a way that within each stratum the variation in the estimated propensity score is modest. The reason is that, if the propensity score itself varies substantially within a stratum, then any evidence that the covariates are correlated with the treatment indicator within that same stratum is not compelling evidence of misspecification of the estimated propensity score. Thus, we choose the boundary values in such a way that, within any stratum, the indicator of receiving the treatment appears statistically unrelated to the estimated propensity score.

We implement the selection of boundary points by an iterative procedure as follows. First we drop from this analysis all control units with an estimated propensity score less than the smallest value of the estimated propensity score among the treated units,

$$\underline{e}_t = \min_{i:W_i=1} \hat{e}(X_i),$$

as well as all treated units with an estimated propensity score greater than the largest value of the estimated propensity score among the control units,

$$\bar{e}_c = \max_{i:W_i=0} \hat{e}(X_i).$$

This trimming ensures some overlap between the groups: among units i with estimated propensity score values $\hat{e}(X_i)$ such that $\hat{e}(X_i) < \underline{e}_t$ or $\hat{e}(X_i) > \bar{e}_c$, there are no comparisons between treated and control units, without at least some extrapolation. We then start with a single block: $J = 1$, with boundaries equal $b_0 = \underline{e}_t$ and $b_1 = b_J = \bar{e}_c$. With these starting values, we iterate through the following two steps.

1. Assessment of Adequacy of Blocks
In the first step, we check whether the current number of blocks, at this step in the algorithm equal to J, is adequate. In this procedure we use the estimated linearized propensity score (or log odds ratio), defined as

$$\hat{\ell}(x) = \ln\left(\frac{\hat{e}(x)}{1 - \hat{e}(x)}\right).$$

The main reason to focus on the linearized propensity score rather than the propensity score itself is that, compared to the propensity score, the linearized propensity score is more likely to have a distribution that is well approximated by a normal distribution. Using the linearized propensity scores, we check the following two conditions for each block $j = 1, \ldots, J$.

1.A Independence Is the estimated linearized propensity score within the block approximately uncorrelated with the treatment indicator? We assess this by calculating a t-statistic. Let $N_c(j)$ and $N_t(j)$ denote the subsample sizes for controls and treated in block j,

$$N_c(j) = \sum_{i=1}^{N}(1 - W_i) \cdot B_i(j), \quad \text{and} \quad N_t(j) = \sum_{i=1}^{N} W_i \cdot B_i(j),$$

and let $\bar{\ell}_c(j)$ and $\bar{\ell}_t(j)$ denote the average values for the estimated linearized propensity score, by treatment status and block,

$$\bar{\ell}_c(j) = \frac{1}{N_c(j)} \sum_{i=1}^{N}(1 - W_i) \cdot B_i(j) \cdot \hat{\ell}(X_i), \quad \bar{\ell}_t(j) = \frac{1}{N_t(j)} \sum_{i=1}^{N} W_i \cdot B_i(j) \cdot \hat{\ell}(X_i),$$

and finally, let S_{ℓ}^2 denote the sample variance of the linearized propensity score within block j,

$$S_{\ell}^2(j) = \frac{1}{N(j) - 2} \times \left(\sum_{i:B_i(j)=1}(1 - W_i) \cdot \left(\hat{\ell}(X_i) - \bar{\ell}_c(j)\right)^2 + \sum_{i:B_i(j)=1} W_i \cdot \left(\hat{\ell}(X_i) - \bar{\ell}_t(j)\right)^2\right).$$

The t-statistic for block j is then defined as

$$t_j = \frac{\bar{\ell}_t(j) - \bar{\ell}_c(j)}{\sqrt{s_\ell^2(j) \cdot (1/N_c(j) + 1/N_t(j))}}. \tag{13.6}$$

We compare this t-statistic for each stratum to a threshold value, which we fix at t_{max}, e.g., $t_{max} = 1$. If the t-statistic is less than or equal to t_{max}, we assess the estimated propensity score as varying little within the block, and if the t-statistic exceeds t_{max}, we assess the block as exhibiting substantial variation in the propensity score.

1.B New Strata Size If we were to split the current j^{th} stratum into two substrata, what would the new boundary value be, and how many observations would fall in each of the new substrata? We compute the median value of the propensity score among the $N_c(j) + N_t(j)$ units with an estimated propensity score in the interval (b_{j-1}, b_j). Denote this median by b'_j. (To be precise, if the current number of units in the stratum, $N_c(j) + N_t(j)$, is odd, the median is the middle value, and if the number of units in the stratum is even, the median is defined as the average of the two middle values.) Then, with the superscripts l and u denoting the low and high substratum respectively, let

$$N_c^l(j) = \sum_{i=1}^{N} (1 - W_i) \cdot B_i(j) \cdot \mathbf{1}_{\hat{e}(X_i) < b'_j}, \quad N_c^u(j) = \sum_{i=1}^{N} (1 - W_i) \cdot B_i(j) \cdot \mathbf{1}_{\hat{e}(X_i) \geq b'_j},$$

$$N_t^l(j) = \sum_{i=1}^{N} W_i \cdot B_i(j) \cdot \mathbf{1}_{\hat{e}(X_i) < b'_j}, \quad \text{and} \quad N_t^u(j) = \sum_{i=1}^{N} W_i \cdot B_i(j) \cdot \mathbf{1}_{\hat{e}(X_i) \geq b'_j},$$

be the number of control and treated units with estimated propensity scores in the lower subinterval (b_{j-1}, b'_j) and in the upper subinterval (b'_j, b_j) respectively.

The current block j is assessed to be inadequately balanced if the t-statistic is too high, $|t_j| > t_{max}$, and amenable to splitting if the number of units in each new block of each treatment type is sufficiently large to allow for a split at the median, $\min(N_c^l(j), N_t^l(j), N_c^u(j), N_t^u(j)) \geq 3$, and $\min(N_c^l(j) + N_t^l(j), N_c^u(j) + N_t^u(j)) \geq K + 2$, where K is the number of pre-treatment variables. We choose these numbers so that we can compare mean covariate values within blocks, and so that later we can do at least some adjustment for remaining covariate differences within blocks.

2. Split Blocks That Are Both Inadequately Balanced and Amenable to Splitting If block j is assessed to be inadequately balanced and amenable to splitting, then this block is split into two new blocks, corresponding to propensity score values in $[b_{j-1}, b'_j)$ and in (b'_j, b_j), and the number of strata is increased by one. We iterate between the assessment step (1) and the splitting step (2) until all blocks are assessed to be either adequately balanced or too small to split.

13.6   CHOOSING STRATA FOR THE BARBITURATE DATA

For the specification of the propensity score obtained in Section 13.4, we implement the strata selection procedure discussed in the previous section.

We start with a single block, $J = 1$, with the lower and upper boundaries equal to $b_0 = \underline{e}_t = \min_{i:W_i=1} \hat{e}(X_i) = 0.0080$, and $b_1 = \bar{e}_c = \max_{i:W_i=0} \hat{e}(X_i) = 0.9252$ respectively. Out of the 7,198 individuals who were not exposed to barbiturates *in utero*, 2,737 have estimated propensity scores less than $b_0 = \underline{e}_t$, and out of the 745 individuals who were exposed to barbiturates before birth, 3 have estimated propensity scores exceeding $b_1 = \bar{e}_c$. We discard at this stage both the 2,737 control individuals with estimated propensity scores less than b_0, and the 3 exposed individuals with estimated propensity scores exceeding b_1. Hence, in this first stratum we have $N_c(1) = 4,461$ controls and $N_t(1) = 742$ treated individuals left with estimated propensity scores between $b_0 = 0.0080$ and $b_1 = 0.9252$. For this first block (i.e., subclass), we calculate the t-statistic, t_1, for the test of the null hypothesis that the estimated linearized propensity score has the same mean in the treated and control subsamples, using the expression in (13.7). This leads to a t-statistic of $t_1 = 36.3$, which exceeds by a substantial amount the threshold of $t_{\max} = 1$. Morever, if we split the block at the median of the estimated propensity scores within this stratum (equal to 0.06), there will be a sufficient number of observations in each sub-stratum: $N_c^l(1) = 2,540$, $N_t^l(1) = 61$, $N_c^u(1) = 1,921$, and $N_t^u(1) = 681$. Therefore the current single-block subclassification is deemed inadequate, and the single block is split into two new blocks, with the new boundary equal to the median in the original subclass, equal to 0.06. These results are in the first panel of Table 13.7.

In the new stratification with two blocks, the first block with boundaries 0.01 and 0.06 has $N_c(1) = 2,540$ individuals in the control group and $N_t(1) = 61$ individuals in the treatment group. The t-statistic for the test of the null hypothesis of equality of the average estimated linearized propensity scores by treatment status for this block is 3.2. If we split the block into two parts at the median value of the propensity score (equal to 0.02), we find 1,280 control and 20 treated units in the first sub-block, and 1,260 control and 41 treated units in the second sub-block. The number of units in each subclass is sufficiently large, and therefore the original block will be split into two new blocks, at the median value of 0.02. For the second block with boundary values 0.06 and 0.9252, we again find that the stratification is inadequate, with a t-statistic of 23.7. These results are in the second panel of Table 13.7. As a result, we split both blocks, leading to four new blocks.

When we continue this procedure with the four new blocks, we find that the second of the four new blocks was sufficiently balanced in terms of the linearized propensity score. The remaining three new blocks were not well balanced and should be split again, leading to a total of seven blocks in the next round. See the third panel of Table 13.7.

We continue checking the adequacy of the blocks until either all the *t*-statistics are below the threshold value of one or splitting a block would lead to a new block that would contain an insufficient number of units of one treatment type or another. This algorithm leads to ten blocks, with the block boundaries, block widths, and the number of units of each type in the block presented in the last panel of Table 13.7. In the last column of this table, we also present the t-statistics. One can see that most of the blocks are well balanced in the linearized propensity score, with only two blocks somewhat unbalanced with t-statistics exceeding the threshold of $t_{\max} = 1$. For example, the second block is not particularly well balanced in the linearized propensity score, with a t-statistic of 1.7, but splitting it would lead to a new block with no treated units, and therefore this block is not split further.

Table 13.7. *Determination of the Number of Blocks and Their Boundaries; Barbiturate Data*

Step	Block	Lower Bound	Upper Bound	Width	# Controls	# Treated	t-Stat
1	1	0.00	0.94	0.94	4462	742	36.3
2	1	0.00	0.06	0.06	2540	61	3.2
	2	0.06	0.94	0.88	1922	681	23.7
3	1	0.00	0.02	0.01	1280	20	2.2
	2	0.02	0.06	0.05	1260	41	0.5
	3	0.06	0.20	0.14	1163	138	3.9
	4	0.20	0.94	0.74	759	543	10.9
4	1	0.00	0.01	0.00	644	6	−0.0
	2	0.01	0.02	0.01	636	14	1.7
	3	0.02	0.06	0.05	1260	41	0.5
	4	0.06	0.11	0.05	604	46	−0.3
	5	0.11	0.20	0.09	559	92	1.0
	6	0.20	0.37	0.17	458	192	1.2
	7	0.37	0.94	0.57	301	351	5.6
5	1	0.00	0.01	0.00	644	6	−0.0
	2	0.01	0.02	0.01	636	14	1.7
	3	0.02	0.06	0.05	1260	41	0.5
	4	0.06	0.11	0.05	604	46	−0.3
	5	0.11	0.20	0.09	559	92	1.0
	6	0.20	0.37	0.17	458	192	1.2
	7	0.37	0.50	0.13	181	144	2.5
	8	0.50	0.94	0.44	120	207	2.3
6	1	0.00	0.01	0.00	644	6	−0.0
	2	0.01	0.02	0.01	636	14	1.7
	3	0.02	0.06	0.05	1260	41	0.5
	4	0.06	0.11	0.05	604	46	−0.3
	5	0.11	0.20	0.09	559	92	1.0
	6	0.20	0.37	0.17	458	192	1.2
	7	0.37	0.42	0.05	101	61	0.3
	8	0.42	0.50	0.08	80	83	0.7
	9	0.50	0.61	0.11	73	90	0.8
	10	0.61	0.94	0.34	47	117	−0.3

Note: Boldface block numbers indicate blocks that were split at this step.

13.7 ASSESSING BALANCE CONDITIONAL ON THE ESTIMATED PROPENSITY SCORE

Here we discuss assessing the within-block equality of means of the covariates across the treatment groups. One problem when conducting this assessment is the large amount of relevant information. We may have a large number of covariates (in the barbiturate study, there are seventeen covariates), and a substantial number of blocks (ten in our application). Even if we were to have data from a randomized experiment, where the covariates would be balanced perfectly in expectation, in any finite sample one would expect some covariates, in at least some strata, to be sufficiently correlated with treatment status that some statistical tests ignoring the multiplicity of comparisons would suggest statistical significance of some comparisons at conventional single-test levels. Here we propose a method for assessing the overall balance for a particular specification of the propensity score, and a given set of strata, that allows for comparisons of balance across specifications of the propensity score and across strata definitions.

As before, let the block or stratum indicators be denoted by $B_i(j)$, and let $N_c(j)$ and $N_t(j)$ be the number of control and treated units in block j, for $j = 1, \ldots, J$. Let us also define $\overline{X}_{c,k}(j)$ and $\overline{X}_{t,k}(j)$ to be the average of the kth component of the K-component covariate vector X_i, for control and treated units within stratum j,

$$\overline{X}_{c,k}(j) = \frac{1}{N_c(j)} \sum_{i:W_i=0} B_i(j) \cdot X_{ik}, \quad \text{and} \quad \overline{X}_{t,k}(j) = \frac{1}{N_t(j)} \sum_{i:W_i=1} B_i(j) \cdot X_{ik},$$

respectively, for $k = 1, \ldots, K$, and $j = 1, \ldots, J$.

We are interested in assessing

$$W_i \perp\!\!\!\perp X_i \,\Big|\, B_i(1), \ldots, B_i(J),$$

implemented through an assessment of the equality,

$$\mathbb{E}\left[X_i | W_i = 1, B_i(j) = 1\right] = \mathbb{E}\left[X_i | W_i = 0, B_i(j) = 1\right], \quad \text{for } j = 1, \ldots, J.$$

We discuss three sets of tests for each covariate. The first two are based on single statistics: first, a test for each covariate based on the average of the within-block average differences by treatment status; second, a test based on all within-strata correlations with W_i; and third, a set of tests based on separate within-stratum comparisons.

13.7.1 Assessing Global Balance for Each Covariate across Strata

For the first set of tests, we analyze the data as if they arose from a stratified randomized experiment. Each of the K covariates X_{ik}, $k = 1, \ldots, K$, is taken in turn as if it were the outcome, and the pseudo-average effect of the treatment on this pseudo-outcome, denoted by τ_k^X, is estimated using the Neyman-style methods discussed in Chapter 9 on stratified randomized experiments. Alternatively we could have used Fisher exact p-values. Take the kth component of the vector covariate X_i, X_{ik}. In stratum j the pseudo-average causal effect of the treatment on this covariate can be estimated by

$$\hat{\tau}_k^X(j) = \overline{X}_{t,k}(j) - \overline{X}_{c,k}(j),$$

The sampling variance of this estimator $\hat{\tau}_k^X(j)$ is estimated as

$$\hat{\mathbb{V}}_k^X(j) = s_k^2(j) \cdot \left(\frac{1}{N_c(j)} + \frac{1}{N_t(j)} \right),$$

where

$$s_k^2(j) = \frac{1}{N(j) - 2} \left(\sum_{i:B_i(j)=1}^{N} (1 - W_i) \cdot \left(X_{ik} - \overline{X}_{c,k}(j)\right)^2 + \sum_{i:B_i(j)=1}^{N} W_i \cdot \left(X_{ik} - \overline{X}_{t,k}(j)\right)^2 \right).$$

The estimate of the pseudo-average causal effect is then the weighted average of these within-block estimates,

$$\hat{\tau}_k^X = \sum_{j=1}^{J} \frac{N_c(j) + N_t(j)}{N} \cdot \hat{\tau}_k^X(j),$$

with estimated sampling variance

$$\hat{\mathbb{V}}_k^X = \sum_{j=1}^{J} \left(\frac{N_c(j) + N_t(j)}{N} \right)^2 \cdot \hat{\mathbb{V}}_k^X(j).$$

Finally we convert these into a z-value for the (two-sided) test of the null hypothesis that the pseudo-average causal effect τ_k^X is equal to zero, against the alternative hypothesis that it differs from zero,

$$z_k = \frac{\hat{\tau}_k^X}{\sqrt{\hat{\mathbb{V}}_k^X}}.$$

We then assess the distribution of these K correlated z-values, one for each covariate, based on a normal reference distribution. If we find that the z-values are substantially larger in absolute values than one would expect if they were drawn independently from a normal distribution, we would conclude that the stratification does not lead to satisfactory balance in the covariates, suggesting the specification of the propensity score is not adequate.

13.7.2 Assessing Balance for Each Covariate within All Blocks

The average pseudo-causal effects τ_k^X may be zero, even if some of the stratum-specific pseudo-causal effects $\tau_k^X(j)$ are not. Next we therefore assess overall balance by calculating F-statistics across all strata, one covariate at a time. Treating the kth covariate as a pseudo-outcome, we use a two-way Analysis of Variance (ANOVA) procedure to test the null hypothesis that its mean for the treated subpopulation is identical to that of the mean of the control subpopulation in each of the J strata. One way to calculate the F-statistic is through a linear regression of the form

$$\mathbb{E}\left[X_{ik} \mid W_i, B_i(1), \ldots, B_i(J)\right] = \sum_{j=1}^{J} \alpha_{kj} \cdot B_i(j) + \sum_{j=1}^{J} \tau_k^X(j) \cdot B_i(j) \cdot W_i.$$

First we estimate the unrestricted estimates $(\hat{a}^{\text{ur}}, \hat{\tau}^X)$ by minimizing

$$(\hat{a}^{\text{ur}}, \hat{\tau}^X) = \arg\min_{a, \tau} \sum_{i=1}^{N} \left(X_{ik} - \sum_{j=1}^{J} \alpha_{kj} \cdot B_i(j) - \sum_{j=1}^{J} \tau_k^X(j) \cdot B_i(j) \cdot W_i \right)^2,$$

which leads to

$$\hat{a}_{kj}^{\text{ur}} = \overline{X}_{c,k}(j), \quad \text{and} \quad \hat{\tau}_k^X(j) = \overline{X}_{t,k}(j) - \overline{X}_{c,k}(j).$$

Next we estimate the restricted estimates $\hat{a}^{\text{r}}$ (under the restriction that all the $\tau_k^X(j) = 0$) by minimizing

$$\hat{a}^{\text{r}} = \arg\min_{a} \sum_{i=1}^{N} \left(X_{ik} - \sum_{j=1}^{J} \alpha_{kj} \cdot B_i(j) \right)^2,$$

leading to

$$\hat{a}_{kj}^{\text{r}} = \frac{N_c(j)}{N_c(j) + N_t(j)} \cdot \overline{X}_{c,k}(j) + \frac{N_t(j)}{N_c(j) + N_t(j)} \cdot \overline{X}_{t,k}(j).$$

The F-test of interest is then the statistic for testing the null hypothesis that all $\tau_k^X(j) = 0$, for $j = 1, \ldots, J$. The form of the F-statistic for covariate X_{ik} is

$$F_k = \frac{(\text{SSR}_k^{\text{r}} - \text{SSR}_k^{\text{ur}})/J}{\text{SSR}_k^{\text{ur}}/(N - 2J)},$$

where the restricted sum of squared residuals is

$$\text{SSR}_k^{\text{r}} = \sum_{i=1}^{N} \left(X_{ik} - \sum_{j=1}^{J} \hat{a}_{kj}^{\text{r}} \cdot B_i(j) \right)^2,$$

and the unrestricted sum of squares is

$$\text{SSR}_k^{\text{ur}} = \sum_{i=1}^{N} \left(X_{ik} - \sum_{j=1}^{J} \hat{a}_{kj}^{\text{ur}} \cdot B_{ij} - \sum_{j=1}^{J} \hat{\tau}_k^X(j) \cdot B_i(j) \cdot W_i \right)^2.$$

We then convert the p-value associated with this F-statistic, under normality of the covariates nominally from an F-distribution with J and $N - 2 \cdot J$ degrees of freedom, to a z-value. Following this procedure for each of the K covariates X_{ik}, we obtain a set of K z-values, one for each of the K covariates. Label these K z-values z_k, $k = 1, \ldots, K$. If the covariates are well balanced between treatment and control groups conditional on the propensity score, we would expect to find the z-values to be concentrated toward smaller (more negative) values relative to a normal distribution (suggesting less evidence against the null hypothesis of no difference between the two groups). Finding large positive values suggests that the covariates are not balanced within the strata.

13.7.3 Assessing Balance within Strata for Each Covariate

The third approach for assessing balance focuses on a single covariate in a single stratum at a time. For each covariate X_{ik}, for $k = 1, \ldots, K$, and for each stratum $j = 1, \ldots, J$, we test the null hypothesis

$$\mathbb{E}\left[X_i | W_i = 1, B_i(j) = 1\right] = \mathbb{E}\left[X_i | W_i = 0, B_i(j) = 1\right] \quad \text{for } j = 1, \ldots, J$$

against the alternative hypothesis that the two averages differ. For the kth covariate, and for this stratum j, we calculate a z-value z_{jk}, analogous to the t-statistics we calculated before. With the stratum-specific sample variances $s^2)k(j)$ define before, the z-value is

$$z_{jk} = \frac{\overline{X}_{\text{t},k}(j) - \overline{X}_{\text{c},k}(j)}{\sqrt{s_k^2(j) \cdot (1/N_{\text{c}}(j) + 1/N_{\text{t}}(j))}}. \tag{13.7}$$

If the covariates are well balanced, we would expect to find the absolute values of the z-values to be concentrated toward smaller (less significant) values relative to a normal distribution. To summarize the $K \times J$ z-values it is useful to present Q-Q plots, comparing the z-values against their expected values under independent draws from the normal distribution. If the covariates are well balanced, we would expect the Q-Q plots to be flatter than a 45° line.

13.8   ASSESSING COVARIATE BALANCE FOR THE BARBITURATE DATA

Given the stratification for the barbiturate data obtained in Section 13.6, using the covariate selection methods outlined in Section 13.3, we estimate the propensity score. We then construct the blocks using the methods from Section 13.5, leading to ten blocks as discussed in Section 13.6. Given these ten blocks, and given the estimated propensity score, we calculate a number of statistics to assess the adequacy of the propensity score specification. First, following the discussion in Section 13.7.1 we calculate a t-statistic for the null hypothesis that the block-adjusted average difference in average covariate values is equal to zero for each covariate. This leads to 17 t-statistics or z-values. Next, as discussed in Section 13.7.2, we calculate the F-statistic for assessing the null hypothesis that the difference in average covariate values is zero in each block. We do this separately for each covariate and convert the p-value for the F-statistic to a z-value. Small values here indicate small F-statistics, and so we are concerned only with the presence of large z-values. Next, following the discussion in Section 13.7.3, we calculate t-statistics for each stratum and each covariate separately, leading to $K \times J = 170$ z-values for the stratum-covariate specific t-tests. The results are presented in Table 13.8, with the rows corresponding to the seventeen covariates, and the columns corresponding to the ten blocks. In addition, there are two columns for the two overall tests, and one for the z-value of the test of equality of (unadjusted) average covariate values for treatment and control groups, and one for the test of the stratum-adjusted average covariate values for

Table 13.8. *z-Values for Balancing Tests: Final Propensity Score Specification; Barbiturate Data*

| Covariate | Within Blocks | | | | | | | | | | Overall | | 1-Block |
	1	2	3	4	5	6	7	8	9	10	t-Test	F-Test (z-Value)	t-Test
sex	−0.05	−2.27	1.97	0.81	0.89	−1.28	0.04	−0.39	−1.42	1.14	0.13	1.22	−0.73
antih	−0.67	−0.47	0.67	0.03	0.37	−0.25	0.38	−0.53	−0.11	0.27	−0.17	−2.88	3.21
hormone	−0.14	−0.42	−0.65	−1.00	0.25	0.71	−0.22	−1.05	−1.10	0.21	−0.99	−0.66	1.66
chemo	0.55	−0.39	−0.78	−0.75	−1.17	1.47	−0.94	0.61	0.66	0.29	−0.27	−0.61	1.76
cage	−1.41	−0.29	−1.04	−0.46	2.11	0.28	0.20	0.46	−1.48	−0.74	−1.38	0.34	1.15
cigar	−0.37	0.55	0.58	1.50	0.31	−0.93	0.21	−0.99	0.25	−0.39	0.52	−1.17	−3.13
lgest	0.90	0.58	−0.07	−0.82	0.79	−0.36	0.05	−0.33	−1.14	1.21	0.71	−1.48	0.12
lmotage	−2.20	−1.37	0.56	1.64	0.95	0.60	−0.96	−1.73	−1.47	0.36	−1.26	1.45	8.56
lpbc415	−0.48	−1.84	−1.00	−0.34	0.59	0.44	−0.20	−0.16	1.07	−0.10	−1.49	−0.82	0.75
lpbc420	1.04	0.84	−0.67	−0.86	−1.61	1.80	−0.39	1.62	1.14	−1.80	0.51	0.59	32.04
motht	−0.84	0.45	−0.67	0.75	0.64	0.09	0.30	−1.37	−0.60	−0.13	−0.50	−1.37	0.90
motwt	1.23	1.14	0.12	−1.23	−0.05	−0.45	−0.32	1.94	−0.01	−0.47	1.08	−0.18	1.44
mbirth	−0.44	−0.80	−1.54	−0.37	1.80	0.20	0.00	2.25	−1.58	−1.60	−1.28	1.00	−2.93
psydrug	−0.66	−1.01	1.05	−0.15	−0.78	0.06	−0.18	0.08	0.09	0.89	−0.29	−1.40	6.32
respir	−0.49	0.53	−0.21	0.98	1.38	0.24	−0.78	−1.51	0.22	−0.28	0.24	−0.49	0.19
ses	−0.60	−0.31	−0.74	1.16	0.82	−0.08	−0.03	−0.82	−0.91	0.36	−0.56	−1.37	5.19
sib	1.42	2.37	−1.09	−1.58	−1.53	0.11	0.63	1.63	1.19	0.23	0.98	1.64	1.48

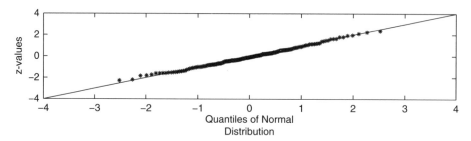

Figure 13.1.   Balance in covariates: QQ-Plot based on $C_L = 1$, $C_Q = 2.78$, barbiturate data

treatment and control groups. Finally, for comparison purposes, we also present the t-statistic for the null hypothesis that the overall average covariate values are equal in the two treatment groups, not adjusted for the blocks.

Starting with the last column, the z-value for the test of equality of unadjusted average covariate values, we find that many covariates have unconditional means that differ significantly between treatment and control groups, which is not surprising because assignment was not randomized. It is also not very informative, merely telling us that some adjustment for covariate differences is necessary and that simply comparing average outcomes for treated and control units would not lead to credible estimates of causal effects of barbiturate exposure. Out of the 170 z-values, only two exceed 2.0 in absolute value. Next, consider the column with the heading "t-test," presenting z-values for the test of zero average pseudo-causal effects for each of the seventeen covariates after stratification on the estimated propensity score. The largest of the absolute values of the seventeen t-statistics is 1.49, suggesting excellent balance. An alternative test is based on comparing each of the within-stratum pseudo-causal effects to zero using an F-test. For the F-test based on this null hypothesis, converted to a z-value, we find that the largest value is 1.64, with all the others below 1.50, again suggesting excellent balance conditional on the propensity score. Note that for these z-values large negative values suggest good balance, and we are concerned only with large positive values.

The first ten columns of the table give the z-values separately for each block and each of the seventeen covariates. The largest of these 170 z-values is 2.37. To facilitate the overall assessment of these z-values we construct a Q-Q plot, where we plot the ordered z-values, against the corresponding quantiles of the normal distribution. The Q-Q plot is presented in Figure 13.1. The Q-Q plot closely follows the 45° line. It shows that there are, if anything, slightly fewer large negative values and fewer large positive values than one would expect to see if the z-values were independent draws from a normal distribution.

From these balance assessments, we conclude that the specification of the propensity score is adequate in the sense that it leads to somewhat better balance than one would expect to see if assignment were randomized within blocks. If we had found that the balance was poor, we might have attempted to improve balance by changing the specification for the propensity score. We propose no general algorithm to improve balance beyond providing some general guidelines. For example, if one finds that many of the t-statistics for a particular covariate are large in absolute value, one may wish to include

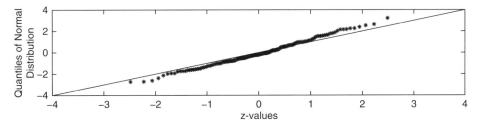

Figure 13.2.   Balance in covariates: QQ-plot based on $C_L = 1, C_Q = \infty$ barbiturate data

more flexible functional forms for that covariate, possibly piecewise linear components, or indicator variables for particularly important regions of its values.

To put the extent of the covariate balance given our preferred specification in perspective, we consider two alternative specifications of the propensity score.

In the first alternative specification, we include all seventeen linear terms but no second-order terms. Within our algorithm this corresponds to $C_L = 0$, $C_Q = \infty$. This specification appears to be common in empirical work, where researchers often simply include all covariates linearly in the propensity score without investigating whether that specification of the propensity score leads to adequate balance in the covariates. Constructing the blocks with this specification of the propensity score leads to nine blocks. Table 13.9 displays the z-values corresponding to this specification. We find that fifteen out of 153 z-values exceed 2.0, compared to only two out of 170 with our preferred specification of the propensity score. In Figure 13.2 we present the Q-Q plot for the 153 z-values based on the nine blocks and seventeen covariates. Comparing Figure 13.2 to Figure 13.1, it is clear that including some second-order terms leads to substantially better balance in the covariates, supporting the importance of doing a careful assessment of the adequacy of the propensity score specification by inspecting covariate balance.

In the second alternative specification we use lasso methods to select among all seventeen linear terms and 153 second-order terms. We use ten-fold cross-validation to select the penalty term. The lasso procedure selects fourteen covariates, three linear ones (chemo, lpbc420, and mbirth), and eleven second-order terms. Table 13.10 displays the z-values corresponding to this specification. We find that there are now fourteen out of 204 z-values exceeding 2.0, again, compared to two out of 170 with our preferred specification of the propensity score. In Figure 13.3 we present the Q-Q plot for the 153 z-values based on the nine blocks and seventeen covariates. Comparing Figure 13.3 to Figure 13.1, it appears that the lasso does not lead to as good an in-sample fit as our proposed specification, possibly due to its focus on out-of-sample prediction.

The correlation between the linearized propensity score based on our proposed specification and the linear specification is 0.95, between the proposed specification and the lasso specification the correlation is 0.96, and the correlation between the linear and the lasso specification is 0.98. The log likelihood values for the three specifications are $-1{,}556.3$ for the proposed specification, $-1{,}627.7$ for the linear specification, and $-1{,}614.7$ for the lasso specification.

Table 13.9. *z-Values for Balancing Tests: Simple Linear Propensity Score Specification; Barbiturate Data*

Covariate	Within Blocks									Overall		1-Block
	1	2	3	4	5	6	7	8	9	t-Test	F-Test (z-Value)	t-Test
sex	1.68	0.41	-0.39	0.09	-0.25	-0.51	0.78	-0.63	-0.20	1.47	-1.16	-0.87
antih	-0.98	1.75	0.17	0.29	-1.11	0.60	-0.51	-0.07	0.68	-0.18	-0.54	3.43
hormone	-0.34	-0.75	-0.45	1.23	-1.38	0.73	1.23	0.22	-0.54	-0.58	-0.16	1.78
chemo	-1.00	-2.37	-0.37	-0.90	-1.44	-1.22	2.36	1.88	0.51	-2.03	2.41	-0.02
cage	-2.54	0.38	-1.40	1.08	0.60	-0.71	1.76	-0.59	-0.07	-2.07	1.11	0.86
cigar	-0.41	0.61	-0.36	0.95	2.21	-1.16	-0.87	-1.59	0.67	0.04	0.70	-2.96
lgest	-0.06	-0.81	1.06	1.88	-0.63	1.18	-0.92	-1.86	1.19	-0.01	0.80	-0.31
lmotage	0.50	1.66	1.86	1.30	2.04	-0.10	-1.34	-2.57	-0.63	1.58	2.26	10.74
lpbc415	-1.10	-1.10	-1.53	0.42	0.91	0.46	0.40	0.48	-0.03	-1.34	-0.58	0.98
lpbc420	1.69	-1.93	0.73	-1.97	-1.93	0.17	2.63	2.52	1.82	0.77	3.09	36.35
motht	-1.94	0.61	0.19	-0.27	1.02	-0.48	-0.15	0.27	-0.59	-1.35	-0.70	0.57
motwt	-0.92	0.34	-0.70	-1.59	-0.94	0.30	0.06	-0.07	1.43	-1.01	-0.29	1.31
mbirth	-0.65	-0.91	2.95	-1.22	-1.22	3.24	1.35	-0.85	-1.65	-0.62	2.33	-3.26
psydrug	-0.25	-1.37	-0.02	-0.72	-1.50	-1.94	0.63	0.45	2.76	-1.30	3.09	7.20
respir	-0.63	-0.60	1.97	-1.00	1.27	0.49	0.08	-0.39	-0.59	-0.30	0.05	0.19
ses	-0.30	1.62	1.52	0.03	0.87	-0.12	-1.92	-1.40	1.14	0.63	0.97	5.61
sib	-2.24	-1.00	-2.24	-1.67	-2.80	0.25	1.58	2.21	2.18	-2.93	3.09	-0.78

Table 13.10. *z-Values for Balancing Tests: Lasso Propensity Score Specification; Barbiturate Data*

| | Within Blocks | | | | | | | | | | | | Overall | | 1-Block |
	1	2	3	4	5	6	7	8	9	10	11	12	t-Test	F-Test (z-Value)	t-Test
Covariate															
sex	−0.16	0.76	0.87	−0.44	1.21	0.81	1.11	−0.49	0.80	−0.22	−0.15	0.87	0.98	−1.19	−0.31
antih	−1.22	2.02	1.61	−0.09	−0.98	0.20	0.68	−0.48	1.05	−0.34	1.28	0.84	0.89	0.36	3.32
hormone	−0.59	−0.57	−0.49	1.37	−0.69	−0.49	−0.29	0.00	−1.37	0.76	1.50	−2.18	−1.07	1.94	1.76
chemo	−1.37	−1.71	−1.09	−1.74	−0.66	−0.83	−1.03	−0.94	−0.19	1.90	1.53	0.96	−2.27	1.37	−0.49
cage	−0.31	0.01	0.82	1.86	0.75	0.07	0.73	0.14	−0.36	3.54	−0.22	1.33	1.35	1.41	1.76
cigar	−0.42	0.12	0.29	0.61	2.09	−0.51	−0.91	−0.33	0.19	−2.21	−0.87	−1.10	−0.39	0.37	−3.03
lgest	0.16	0.76	1.11	0.39	0.81	1.22	−0.29	0.79	−0.60	1.19	−2.62	0.66	1.11	0.26	0.87
lmotage	−1.11	−0.91	2.81	−0.22	2.13	0.88	0.34	−0.48	0.12	0.04	−0.82	−1.24	0.16	1.29	7.71
lpbc415	−1.03	−2.33	−1.27	0.44	1.75	−0.29	−0.84	−0.69	0.05	1.33	−0.09	−0.42	−1.78	0.65	0.81
lpbc420	0.06	0.11	−2.39	0.90	−2.13	0.25	−0.63	−0.32	−0.51	1.99	2.58	0.32	−0.45	1.26	29.15
motht	−0.94	−0.37	1.20	1.49	−0.11	0.45	0.73	−0.31	−0.41	−1.24	0.27	−0.93	−0.30	−0.72	0.70
motwt	−0.93	0.63	−1.03	−0.49	−1.11	−1.46	−0.47	0.14	−0.91	−0.20	−1.92	0.64	−1.64	0.10	0.52
mbirth	−1.11	0.27	−0.92	1.74	−1.10	2.53	−0.41	0.99	−0.59	0.07	0.00	−0.67	−0.61	1.43	−1.74
psydrug	−1.01	−0.24	−1.54	0.07	−1.43	−0.99	0.00	−1.08	−1.25	1.01	1.94	0.89	−1.41	1.45	6.86
respir	−0.28	−0.91	1.72	−0.80	0.06	1.13	−0.29	−0.52	0.46	0.30	−1.24	−0.47	0.00	−0.63	−0.11
ses	−0.57	1.65	1.41	−1.65	−0.11	−0.20	1.15	0.70	−0.16	−0.91	−0.29	−0.57	0.29	−0.17	4.72
sib	0.20	0.64	−1.61	−1.65	−3.50	−0.17	−0.91	−0.10	−0.25	0.78	1.58	0.70	−0.91	1.81	1.43

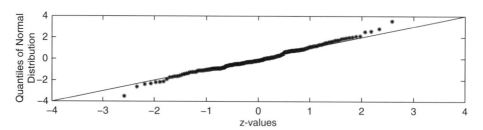

Figure 13.3.   Balance in covariates: QQ-plot based on lasso, barbiturate data

13.9   CONCLUSION

In this chapter we discuss methods for estimating the propensity score and for creating subclasses based on the estimated propensity score. There are two key points. One is that none of the analyses in this chapter involves the outcome data. There is therefore no concern with introducing biases for estimated causal effects through specification searches and pre-testing. A second key point is that the goal in this chapter is to obtain an estimated propensity score that balances the covariates within subclasses, rather than one that simply estimates the hypothetical true propensity score as accurately as possible. As has been noted in the literature, using the estimated propensity score often leads to better balance than using the true propensity score.

We propose a specific data-driven algorithm for choosing the specification of the propensity score. Although there, undoubtedly, will be situations where our proposed algorithm does not lead to adequate balance, in our limited experience it often performs adequately. We also discuss methods for assessing covariate balance, which show, for our particular application, that the specification of the propensity score and selection of subclasses lead to excellent covariate balance, better than one would expect in a randomized blocks experiment, and also better than the balance achieved by a specification for the propensity score that simply includes all covariates linearly. The algorithm uses two tuning parameters, which define cutoff values for inclusion of covariates linearly and for inclusion of second-order terms.

NOTES

The problem of estimating the propensity score is essentially one of nonparametric estimation of a regression function. There are numerous statistical procedures for doing so. Some are based on kernel smoothing. Such methods tend not to perform well in settings with a substantial number of covariates. Other methods are based on selecting subsets of the covariates for inclusion in the specification. These include subset selection (Breiman and Spector, 1992) and the lasso and related methods (Tibshirani, 1996; Bühlmann and Van Der Geer, 2011; Belloni, Chernozhukov, and Hansen, 2014). We are agnostic about what is the "best" procedure. The key is whether a proposed method leads to adequate balance. Bayesian methods are discussed in Clogg, Rubin, Schenker, Schultz, and Weidman (1991).

The point that using the estimated propensity score rather than the true propensity score leads to better balance and better estimators for causal effects has been made in Rubin and Thomas (1992a, 1992b, 1996, 2000) and Hirano, Imbens, and Ridder (2003).

Ketel, Leuven, Oosterbeek, and VanderKlaauw (2013) analyze data from the Dutch medical school admission lotteries mentioned in the introduction to this chapter to estimate the causal effect of becoming a doctor on earnings.

APPENDIX: LOGISTIC REGRESSION

The basic strategy in this chapter uses logistic regression models. Here we describe briefly how to obtain maximum likelihood estimates of the parameters of such models. Let X be the K-vector of covariates with support $\mathbb{X}$. Then for a known L-component row vector of functions $h : \mathbb{X} \mapsto \mathbb{R}^L$ we model the probability of receiving the active treatment in the super-population as

$$\Pr(W_i = 1 | X_i = x; \phi) = \frac{\exp(h(x)\phi)}{1 + \exp(h(x)\phi)}, \quad (13.8)$$

where ϕ is an unknown parameter, local to this appendix. A simple case would correspond to choosing $h(x) = x$ and estimating

$$\Pr(W_i = 1 | X_i = x; \phi) = \frac{\exp(x\phi)}{1 + \exp(x\phi)}. \quad (13.9)$$

More generally, in our algorithm, the function $h(\cdot)$ may consist of only a subset of the covariates, and additionally may include higher-order terms or transformations of the basic covariates.

The likelihood function can be written as

$$\mathcal{L}(\phi | \mathbf{Y}^{\text{obs}}, \mathbf{W}, \mathbf{X}) = \prod_{i=1}^{N} \Pr(W_i = 1 | X_i; \phi)^{W_i} \cdot (1 - \Pr(W_i = 1 | X_i; \phi))^{1-W_i} = \prod_{i=1}^{N} \frac{\exp(W_i \cdot X_i \phi)}{1 + \exp(X_i \phi)},$$

so that he logarithm of the likelihood function is

$$L(\phi | \mathbf{Y}^{\text{obs}}, \mathbf{W}, \mathbf{X}) = \sum_{i=1}^{N} W_i \cdot X_i \phi - \ln(1 + \exp(X_i \phi)).$$

The maximum likelihood estimator is

$$\hat{\phi}_{\text{ml}} = \arg \max_{\phi} L(\phi | \mathbf{Y}^{\text{obs}}, \mathbf{W}, \mathbf{X}).$$

The log likelihood function is straightforward to maximize because it is globally concave if the matrix $\sum_{i=1}^{N} h(X_i)^T \cdot h(X_i)$ is positive definite. As a result, a simple Newton-Raphson algorithm can be effective for finding the maximum likelihood estimates. If the function of covariates, $h(x)$, includes an intercept and has the form $h(x) = (1 \; h_1(x))$, a useful starting vector of starting values is $\phi^0 = (\ln(N_t/N_c), 0^T)^T$, with updating rule

$$\phi^{k+1} = \phi^k - \left(\frac{\partial^2}{\partial\phi\partial\phi^T}L(\phi^k)\right)^{-1}\frac{\partial}{\partial\phi}L(\phi^k).$$

As $k \longrightarrow \infty$, ϕ^k generally converges to $\hat{\phi}_{ml}$, again provided $\sum_{i=1}^{N}h(X_i)^T \cdot h(X_i)$ is positive definite. Given the maximum likelihood estimates $\hat{\phi}$, the standard errors are estimated as the square roots of the diagonal elements of inverse of the estimated information matrix

$$\hat{\mathbb{V}}\left(\hat{\phi}_{ml}\right) = -\left(\frac{\partial^2}{\partial\phi\partial\phi^T}L(\hat{\phi})\right)^{-1}.$$

An alternative to the logit function for the link function is to use the normal distribution function, leading to the probit model with

$$\Pr(W_i = 1 | X_i = x) = \Phi(h(x)\phi),$$

where $\Phi(a) = \int_{-\infty}^{a}(1/\sqrt{2\pi})\exp(-z^2/2)\,dz$ is the cumulative normal distribution function. A third possibility, called the robit model where the "r" stands for robust, uses the cumulative distribution for the t-distribution as a link function (Liu, 2004). If the degrees of freedom are approximately seven, this is close to the logit model, and with a large number for the degrees of freedom, this is close to the probit model. Low values for the degrees of freedom parameter correspond to more robust choices. There is little practical experience with these models to suggest whether they make a substantial difference relative to the logit model.

## CHAPTER 14

Assessing Overlap in Covariate Distributions

14.1 INTRODUCTION

When a researcher wishes to proceed to estimate causal effects under the assumption of unconfoundedness, there are various statistical methods that can be used to attempt to adjust for differences in covariate distributions. These methods include simple linear regressions, which is adequate in simple situations. They also include more sophisticated methods involving subclassification on the propensity score and matching, the latter two possibly in combination with model-based imputation methods, which can work well even in complicated situations. In order to decide on the appropriate methods, it is important first to assess the severity of the statistical challenge to adjust for the differences in covariates. In other words, it is useful to assess how different the covariate distributions are in the treatment and control groups. If the covariate distributions are similar, as they would be, in expectation, in the setting of a completely randomized experiment, there is less reason to be concerned about the sensitivity of estimates to the specific method chosen than if these distributions are substantially different. On the other hand, even if unconfoundedness holds, it may be that there are regions of the covariate space with relatively few treated units or relatively few control units, and, as a result, inferences for such regions rely largely on extrapolation and are therefore less credible than inferences for regions with substantial overlap in covariate distributions.

In this chapter we address the problem of assessing the degree of overlap in the covariate distributions – or, in other words, the *covariate balance* between the treated and control samples prior to any analyses to adjust for these differences. These assessments do not involve the outcome data and therefore do not introduce any systematic biases in subsequent analyses. In principle we are interested in the comparison of two multivariate distributions, the distributions of the covariates in the treated and control subsamples. We wish to explore how different the measures of central tendency are, and how much overlap there is in the tails of the distributions. There are two aspects of these differences in relation to the statistical challenges faced when adjusting for covariates. First, we ask how different are the two covariate distributions by treatment status. Partly for technical reasons, this part of the discussion focuses initially on assessing differences in population distributions. We then implement these concepts in finite samples. The answer to this first question is important for the choice of methods used to adjust for covariate differences. Some methods are more robust to substantial differences in the covariate distributions than others. The second part of the discussion focuses on the question concerning whether there exist, for most units in the sample, similar units with the opposite level of the treatment. Unlike the answers to the first question, the answer to this question depends partly on the sample sizes for the two subsamples: even if the moments of two distributions differ substantially, if the range of values is similar, then at least in large samples one should be able to find close matches for most units. The answer to this second question bears on the ability of *any* method to adjust credibly for covariate differences.

To focus ideas, in Section 14.2 we initially look at the case with only a single covariate, that is, a scalar X_i, where we compare two univariate distributions. We focus on differences in location, differences in measures of dispersion, and two direct measures of overlap. We then look in Section 14.3 at direct comparisons of multivariate distributions. Next, in Section 14.4, we look at the role the propensity score can play when assessing overlap in covariate distributions in settings with unconfoundedness. In Section 14.5 we assess the ability to adjust for differences in covariates by treatment status, taking into account the sample sizes in the two treatment groups. We illustrate the methods discussed in this chapter in Section 14.6 using four different data sets. These data sets range from one obtained from an experimental evaluation with a high degree of overlap to one from an observational study where covariate distributions exhibit extremely limited overlap.

14.2   ASSESSING BALANCE IN UNIVARIATE DISTRIBUTIONS

Let us first think about measuring the difference between two known univariate population distributions. We denote these probability distributions by $f_c(x)$ and $f_t(x)$, for the (conditional) covariate distribution for the controls and treated subpopulations respectively, with $F_c(x)$ and $F_t(x)$ denoting the cumulative distribution functions. Although we are ultimately interested in differences between covariate distributions, rather than between the population covariate distributions, it is useful for technical reasons to focus initially on the differences between the population distributions. We propose four summary measures of the differences between two distributions. Let $\mu_c = \mathbb{E}[X_i|W_i = 0]$ and $\mu_t = \mathbb{E}[X_i|W_i = 1]$ denote the population means for the two distributions, and let $\sigma_c^2 = \mathbb{V}(X_i|W_i = 0)$ and $\sigma_t^2 = \mathbb{V}(X_i|W_i = 1)$ denote the population variances for the two distributions. A natural measure of the difference between the locations of the distributions is what we call the *normalized difference*,

$$\Delta_{ct} = \frac{\mu_t - \mu_c}{\sqrt{(\sigma_t^2 + \sigma_c^2)/2}}, \tag{14.1}$$

which is a scale-free (affinely invariant) measure of the difference in locations, equal to the difference in means, scaled by the square root of the average of the two within-group variances.

To estimate this measure, Δ_{ct}, of the difference in covariate distributions, let $\overline{X}_c$ and $\overline{X}_t$ denote the sample averages of the covariate values for the control and treatment group respectively:

$$\overline{X}_c = \frac{1}{N_c} \sum_{i:W_i=0} X_i, \quad \text{and} \quad \overline{X}_t = \frac{1}{N_t} \sum_{i:W_i=1} X_i,$$

where, as before, N_c is the number of control units, and N_t is the number of treated units. Also, let s_c^2 and s_t^2 denote the conditional within-group sample variances of the covariate:

$$s_c^2 = \frac{1}{N_c - 1} \sum_{i:W_i=0} (X_i - \overline{X}_c)^2 \quad \text{and} \quad s_t^2 = \frac{1}{N_t - 1} \sum_{i:W_i=1} (X_i - \overline{X}_t)^2.$$

Then the empirical counterpart to Δ_{ct} is the difference in average covariate values, normalized by the square root of the average of the two within-treatment group sample variances:

$$\hat{\Delta}_{ct} = \frac{\overline{X}_t - \overline{X}_c}{\sqrt{(s_c^2 + s_t^2)/2}}. \tag{14.2}$$

It is useful to relate the normalized difference to a different statistic that is often reported in causal analyses, the t-statistic for the test of the null hypothesis that $\mu_c = \mu_t$, against the alternative hypothesis that $\mu_c \neq \mu_t$. When σ_c^2 is thought to differ from σ_t^2, this t-statistic is equal to

$$T_{ct} = \frac{\overline{X}_t - \overline{X}_c}{\sqrt{s_c^2/N_c + s_t^2/N_t}}. \tag{14.3}$$

This t-statistic serves a very different purpose and is less relevant for the problem of assessing the adequacy of simple adjustment methods than the normalized difference. Our aim is *not* to test whether the data contain sufficient information to support the claim that the two covariate means in the different treatment regimes are different. One typically suspects that the population means are, in fact, different, and whether the sample size is sufficiently large to detect this, or the significance level at which we may be able to reject the null hypothesis of no difference, is not of great importance. Rather, the goal is, at least at this point, to assess whether the differences between the two distributions are so large that simple adjustment methods, such as linear covariance (i.e., regression) adjustment, are unlikely to be adequate to remove most biases in estimated treatment/control average differences associated with differences in covariates.

Another way to see why the t-statistic T_{ct} is less relevant for assessing the difference between the two distributions than the normalized difference $\hat{\Delta}_{ct}$, consider what would happen if, for a given pair of distributions $f_c(x)$ and $f_t(x)$, we quadruple the sample size N. In expectation, the t-statistic would double in value, whereas the normalized difference would, in expectation, remain unchanged. Clearly, the statistical challenge of adjusting for differences in the covariates would be simpler rather than more difficult if we had available four times as many units: more observations drawn from the same distributions will ease the task of finding good comparisons in the treatment and control groups.

In addition to comparing the differences in location in the two distributions, one may wish to compare measures of dispersion in the two distributions. For two population distributions, a natural measure of the difference in dispersion, and one that is invariant to scale, is the logarithm of the ratio of standard deviations:

$$\Gamma_{ct} = \ln\left(\frac{\sigma_t}{\sigma_c}\right) = \ln(\sigma_t) - \ln(\sigma_c). \tag{14.4}$$

The sample analogue of this population difference is the difference in the logarithms of the two sample standard deviations:

$$\hat{\Gamma}_{ct} = \ln(s_t) - \ln(s_c). \tag{14.5}$$

We use the difference in logarithms because it is typically more normally distributed than the difference in their standard deviations or their ratio.

As a second approach to comparing the population distributions, one can investigate what fraction of the treated (control) units have covariate values that are in the tails of the distribution of the covariate values for the controls (treated). In the case with known distributions, one may wish to calculate, for example, for a fixed value α (e.g., $\alpha = 0.05$), the probability mass of the covariate distribution for the treated that is outside the $1 - \alpha/2$ and the $\alpha/2$ quantiles of the covariate distribution for the controls:

$$\pi_t^\alpha = \left(1 - F_t\big(F_c^{-1}(1 - \alpha/2)\big)\right) + F_t\big(F_c^{-1}(\alpha/2)\big),$$

and the analogous quantity for the control distribution:

$$\pi_c^\alpha = \left(1 - F_c\big(F_t^{-1}(1 - \alpha/2)\big)\right) + F_c\big(F_t^{-1}(\alpha/2)\big).$$

The idea is that, for values of x in between the quantiles $F_c^{-1}(\alpha/2)$ and $F_c^{-1}(1 - \alpha/2)$, missing control outcomes $Y_i(0)$ for the treated units are relatively easy to impute, because there are relatively many control observations in this part of the covariate space. On the other hand, for values of x less than $F_c^{-1}(\alpha/2)$, or for values of x greater than $F_c^{-1}(1 - \alpha/2)$, it will be relatively more difficult to impute $Y_i(0)$ for treated units because there are relatively few control observations in this part of the covariate space. If the proportion of such treated units, π_t^α, is high, it will be relatively difficult to predict missing potential outcomes for the treated units. Note that in a completely randomized experiment, at least in expectation, $\pi_c^\alpha = \pi_t^\alpha = \alpha$, and only $\alpha \times 100\%$ of the units have covariate values that make the prediction of the missing potential outcomes relatively difficult.

To implement this approach given the sample, let $\hat{F}_c(\cdot)$ and $\hat{F}_t(\cdot)$ be the empirical distribution function of X_i in the control and treated subsamples, respectively,

$$\hat{F}_c(x) = \frac{1}{N_c}\sum_{i:W_i=0} \mathbf{1}_{X_i \leq x}, \quad \text{and} \quad \hat{F}_t(x) = \frac{1}{N_t}\sum_{i:W_i=1} \mathbf{1}_{X_i \leq x},$$

and let $\hat{F}_c^{-1}(q)$ and $\hat{F}_t^{-1}(q)$ denote the inverse of these distributions:

$$\hat{F}_c^{-1}(q) = \min_{-\infty < x < \infty}\{x : \hat{F}_c(x) \geq q\}, \quad \text{and} \quad \hat{F}_t^{-1}(q) = \min_{-\infty < x < \infty}\{x : \hat{F}_t(x) \geq q\}.$$

Now let us pick $\alpha = 0.05$. Then $\hat{\pi}_c$ and $\hat{\pi}_t$ are the proportion of control and treated units with covariate values outside the 0.025 and 0.975 quantiles of the empirical distribution of the covariate values among the treated and control units:

$$\hat{\pi}_c^{0.05} = \left(1 - \left(\hat{F}_c\left(\hat{F}_t^{-1}(0.975)\right)\right) + \hat{F}_c\left(\hat{F}_t^{-1}(0.025)\right)\right), \tag{14.6}$$

and

$$\hat{\pi}_t^{0.05} = \left(1 - \left(\hat{F}_t\left(\hat{F}_c^{-1}(0.975)\right)\right) + \hat{F}_t\left(\hat{F}_c^{-1}(0.025)\right)\right). \tag{14.7}$$

An advantage of these last two overlap measures is that they separately indicate the difficulty when predicting missing potential outcomes for the treated and for the control units. It is possible that the data are such that predicting the missing potential outcomes for the treated units is relatively easy, with the control units sufficiently dispersed that there are close comparisons for all covariate values that are observed among the treated. Yet, for the same data set, it may be difficult to find good comparisons for some of the control units if the distribution of the covariates among the treated is less dispersed than among the control units. In that case it may be difficult to estimate, for example, the overall average effect of the treatment, τ_{fs}, but it may be possible to estimate well the average effect of the treatment for the treated units, $\tau_{fs,t} = \sum_{i:W_i=1} (Y_i(1) - Y_i(0))/N_t$.

These four measures, the standardized difference in averages, the logarithm of the ratio of standard deviations, and the two sets of coverage frequencies, give good summary measures of the balance of a scalar covariate when the distributions are symmetric. More generally, one may wish to inspect normalized differences for higher-order moments of the covariates, or of functions of the covariates (logarithms, or indicators of covariates belonging to subsets of the covariate space). In practice, however, assessing balance simply by inspecting these four measures should provide a good initial sense of possible important differences in the univariate distributions. Finally, it may be useful to construct histograms of the distribution of a covariate in both treatment arms to detect visually subtle differences not captured by differences in means and variances, especially for covariates that are *a priori* believed to be highly associated with the outcomes.

14.3   DIRECT ASSESSMENT OF BALANCE IN MULTIVARIATE DISTRIBUTIONS

Now consider the case with multiple covariates. Let K be the number of covariates, the number of components of the vector of pre-treatment variables X_i. We may wish to start by looking at each of the K covariates separately using the methods discussed in Section 14.2, but it can also be useful to have a single measure of the difference between the distributions. As before, we look initially at the population distribution of the difference between the covariate values of a random draw from the treated and control distributions. The means of those distributions are the K-vectors μ_c and μ_t, respectively, and the $K \times K$

covariance matrices are Σ_c and Σ_t. An overall summary measure of the difference in locations between the two population distributions is

$$\Delta_{ct}^{mv} = \sqrt{(\mu_t - \mu_c)' \left(\frac{\Sigma_c + \Sigma_t}{2} \right)^{-1} (\mu_t - \mu_c)}, \tag{14.8}$$

the Mahalanobis distance between the means with respect to the $((\Sigma_c + \Sigma_t)/2)^{-1}$ inner product. For the sample equivalent of this measure, we use the sample averages $\overline{X}_c$ and $\overline{X}_t$ and the following estimators for the covariance matrices,

$$\hat{\Sigma}_c = \frac{1}{N_c - 1} \sum_{i:W_i=0} (X_i - \overline{X}_c) \cdot (X_i - \overline{X}_c)', \quad \text{and} \quad \hat{\Sigma}_t = \frac{1}{N_t - 1} \sum_{i:W_i=1} (X_i - \overline{X}_t) \cdot (X_i - \overline{X}_t)',$$

leading to an estimated measure of the multivariate difference in covariate distributions:

$$\hat{\Delta}_{ct}^{mv} = \sqrt{(\overline{X}_t - \overline{X}_c)' \left(\frac{\hat{\Sigma}_c + \hat{\Sigma}_t}{2} \right)^{-1} (\overline{X}_t - \overline{X}_c)}. \tag{14.9}$$

14.4   ASSESSING BALANCE IN MULTIVARIATE DISTRIBUTIONS USING THE PROPENSITY SCORE

A complementary way to assess the overall difference in the covariate distributions is to use the propensity score. The propensity score plays a number of key roles in our discussion of causal analyses under unconfoundedness, and one of these is for assessing balance in covariate distributions. The main reason is that *any* imbalance in the population covariate distributions, whether in expectation, in dispersion, or in the shape of the distributions, leads to a difference in the population distributions of the true propensity scores by treatment status. As a result, it is theoretically sufficient to assess (e.g., visualize) differences in the distribution of the (true) propensity score in order to assess overlap in the full, joint, covariate distributions. This is very useful because it is easier to assess (e.g., visualize) differences between two univariate distributions than between two multivariate distributions. Moreover, any difference in covariate distributions by treatment status leads to a difference in the population *averages* of the true propensity scores for the treatment and control groups. There is therefore, in principle, no need to look beyond a mean difference in the true propensity scores by treatment status. In fact, given that there can be dispersion in the marginal (unconditional) distribution of the true propensity score only if the average values of the propensity scores for treated and controls differ, it is, in fact, also sufficient to assess the amount of dispersion in the marginal distribution of the propensity score: a non-zero variance of the marginal propensity score implies, and is implied by, differences in the covariate distributions by treatment status.

To state some formal results, let us initially focus on the case where the propensity score is known, which is why the previous paragraph kept emphasizing the "true" propensity score. We assume that the assignment mechanism is unconfounded, individualistic, and probabilistic (see Chapter 3 for formal definitions). Let $e(x)$ denote the true propensity score, and let $\ell(x)$ denote the linearized propensity score or log odds ratio of being in the treatment group versus the control group given covariate value $X_i = x$,

$$\ell(x) = \ln\left(\frac{e(x)}{1 - e(x)}\right).$$

We can simply look at the normalized difference in means for the propensity score or, better, the linearized propensity score, the same way we did for univariate X_i. Define $\overline{\ell}_c$ and $\overline{\ell}_t$ to be the average values for the linearized propensity scores for control and treated units,

$$\overline{\ell}_c = \frac{1}{N_c} \sum_{i:W_i=0} \ell(X_i), \quad \text{and} \quad \overline{\ell}_t = \frac{1}{N_t} \sum_{i:W_i=1} \ell(X_i),$$

and $s_{\ell,c}^2$ and $s_{\ell,c}^2$ to be the sample variances of the linearized propensity scores,

$$s_{\ell,c}^2 = \frac{1}{N_c - 1} \sum_{i:W_i=0} \left(\ell(X_i) - \overline{\ell}_c\right)^2, \quad \text{and} \quad s_{\ell,t}^2 = \frac{1}{N_t - 1} \sum_{i:W_i=1} \left(\ell(X_i) - \overline{\ell}_t\right)^2.$$

Then the estimated difference in average linearized propensity scores, scaled by the square root of the average squared within-treatment-group standard deviations is

$$\hat{\Delta}_{ct}^{\ell} = \frac{\overline{\ell}_t - \overline{\ell}_c}{\sqrt{\left(s_{\ell,c}^2 + s_{\ell,t}^2\right)/2}}. \tag{14.10}$$

There is not as much need to normalize this difference, $\overline{\ell}_t - \overline{\ell}_c$, by the square root of the average squared within-treatment-group standard deviations of the linearized propensity score as there was for the original covariates, because the propensity score, and thus any function of the propensity score, is scale-invariant.

The discussion so far is very similar to the discussion where we assessed balance in a single covariate. There are, however, two important differences that make inspection of the difference in average estimated propensity score values by treatment status particularly salient. The first is that differences in the super-population covariate distributions by treatment status imply, and are implied by, variation in the true propensity score. In other words, either the super-population distribution of the true propensity score values is degenerate and the super-population covariate distributions are identical in the two treatment arms, or the super-population distribution of propensity score values is non-degenerate and the super-population covariate distributions in treatment and control groups differ. Second, if the super-population distributions of the covariates in the two treatment groups differ, then it must be the case that the expected value (in the super-population) of the propensity score in the treatment group is larger than the expected value (in the super-population) of the propensity score in the control group. The key implication of these two results is that differences in covariate distributions by treatment status imply, and are implied by, differences in the average value of the propensity score by treatment status. Thus, differences in the average propensity score, or differences in averages of strictly monotone functions of the propensity score, such as the linearized propensity score, are scalar measures of the degree of overlap in covariate distributions.

Let us formalize the two claims above. Let $f_c(x)$ and $f_t(x)$ denote the conditional covariate distributions in the control and treated subpopulations respectively, and let p be the expected value of the propensity score, $p = \mathbb{E}[W_i] = \mathbb{E}[e(X_i)]$.

Theorem 14.1 (Propensity Score and Covariate Balance) *Suppose the assignment mechanism is unconfounded and individualistic. Then, (i) the variance of the true propensity score satisfies*

$$\mathbb{V}(e(X_i)) = \mathbb{E}\left[\left(\frac{f_t(X_i) - f_c(X_i)}{f_t(X_i) \cdot p + f_c(X_i) \cdot (1-p)}\right)^2\right] \cdot p^2 \cdot (1-p)^2, \tag{14.11}$$

and (ii) the expected difference in propensity scores by treatment status satisfies

$$\mathbb{E}[e(X_i)|W_i = 1] - \mathbb{E}[e(X_i)|W_i = 0] = \frac{\mathbb{V}(e(X_i))}{p \cdot (1-p)}. \tag{14.12}$$

Proof. Under unconfoundedness, and individualistic assignment, we can write the propensity score as

$$e(x) = \Pr(W_i = 1 | X_i = x) = \frac{f_t(x) \cdot p}{f_t(x) \cdot p + f_c(x) \cdot (1-p)}. \tag{14.13}$$

Using (14.13) we can write the deviation of the propensity score $e(x)$ from its population mean p as

$$e(x) - p = \frac{f_t(x) - f_c(x)}{f_t(x) \cdot p + f_c(x) \cdot (1-p)} \cdot p \cdot (1-p).$$

Hence the population variance of the propensity score is

$$\mathbb{V}(e(X_i)) = \mathbb{E}\left[(e(x) - p)^2\right] = \mathbb{E}\left[\left(\frac{f_t(X_i) - f_c(X_i)}{f_t(X_i) \cdot p + f_c(X_i) \cdot (1-p)}\right)^2\right] \cdot p^2 \cdot (1-p)^2,$$

demonstrating part (*i*) of the theorem.

Let us consider part (*ii*) of the theorem. Let $f^E(e)$ be the marginal distribution of the propensity score $e(X_i)$ in the population, let $f_c^E(e)$ and $f_t^E(e)$ denote the conditional distribution of the propensity score in the two treatment arms:

$$f_t^E(e) = \frac{f^E(e) \cdot \Pr(W_i = 1 | e(X_i) = e)}{\Pr(W_i = 1)} = \frac{f^E(e) \cdot e}{p} \quad \text{and} \quad f_c^E(e) = \frac{f^E(e) \cdot (1-e)}{1-p}.$$

The two conditional means of the propensity score by treatment status are

$$\mathbb{E}[e(X_i)|W_i = 1] = \int e f_t^E(e) de = \int e^2 f^E(e) de / p = \frac{\mathbb{V}(e(X_i))}{p} + p,$$

and

$$\mathbb{E}[e(X_i)|W_i = 0] = (\mathbb{E}[e(X_i)] - \mathbb{E}[e(X_i)|W_i = 1] \cdot p) / (1 - p) = p - \frac{\mathbb{V}(e(X_i))}{1 - p}.$$

The difference in means for the treatment and control group propensity scores is then:

$$\mathbb{E}[e(X_i)|W_i = 1] - \mathbb{E}[e(X_i)|W_i = 0] = \frac{\mathbb{V}(e(X_i))}{p \cdot (1 - p)}.$$

Hence, unless the distribution of the true propensity score is degenerate with $Pr(e(X_i) = p) = 1$ (so that the marginal variance of the propensity score, $\mathbb{V}(e(X_i))$, is equal to zero), there will be a difference in expected true propensity score values between treatment and control groups. Thus a zero difference between expected true propensity scores for treatment and control groups is equivalent to perfect expected balance.

Even though there can be no differences in the distribution of the true propensity score by treatment status unless there is a difference in the conditional expectation of the true propensity score by treatment status, it can be useful to inspect a histogram of the sample distributions of the estimated propensity scores in both groups to get a sense of the full distribution. When the number of covariates is large, it may be impractical to inspect histograms for each of the covariates separately, and inspecting the histogram of the estimated propensity score is a useful way to visualize a summary of the differences between the two distributions.

This discussion highlights the importance of assessing balance in the propensity score. The key insight is that differences in the expected distribution of the covariates lead to differences in expected values of the true propensity scores by treatment group, and that, therefore, inspecting the estimated propensity score distributions by treatment status should be a useful tool for assessing differences in covariate distributions. Although the formal results are based on differences in the population distributions of the true propensity score by treatment status, the practical implication is that it may be useful to assess differences in the sample distributions of the estimated propensity score.

14.5 ASSESSING THE ABILITY TO ADJUST FOR DIFFERENCES IN COVARIATES BY TREATMENT STATUS

In the previous sections we focused on differences between the covariate and estimated propensity score distributions by treatment status. If these differences are substantial, simple methods will likely not be adequate to obtain credible and robust estimates of the causal effects of interest. These measures of distributional differences considered so far do not depend on the sample sizes. The sample sizes by treatment group, however, are important determinants of whether even sophisticated methods will be adequate for obtaining credible and robust estimates. In this section we explore this question further. Specifically, we focus on the question whether for proportions of the samples there are close comparisons in the other treatment group. We do this separately by treatment group.

Consider a unit i, with treatment status W_i. We ask the question whether, for this unit, there is any other unit i' with the opposite treatment, $W_{i'} = 1 - W_i$, such that the difference in linearized propensity scores, $\ell(X_i) - \ell(X_{i'})$ is, in absolute value, less than or equal to, a threshold ℓ^u. In the current discussion, we focus on a threshold of $\ell^u = 0.1$, implying that the difference in propensity scores is approximately less than 10%. For units for whom there are units with the other treatment with differences in propensity scores less than 10%, we may be able to obtain credible (in the sense of close to unbiased), estimates of the causal effects without extrapolation. For units for whom there are no similar units with the opposite treatment level, it will be more difficult to obtain credible estimates of causal effects, irrespective of the methods used. If there are many such units, we may wish to trim the sample to improve balance using some of the methods discussed in the next two chapters.

First define, for each unit i, the indicator ς_i that takes on the value one if there is at least one unit i' with $W_{i'} = 1 - W_i$ that has a similar value for the linearized propensity score and zero otherwise:

$$\varsigma_i = \begin{cases} 1 & \text{if } \sum_{i':W_{i'}\neq W_i} \mathbf{1}_{|\hat{\ell}(X_{i'})-\hat{\ell}(X_i)|\leq \ell^u} \geq 1, \\ \\ 0 & \text{otherwise.} \end{cases}$$

Then our two overlap measures are the proportion of units in each treatment group with close comparisons,

$$q_c = \frac{1}{N_c} \sum_{i:W_i=0} \varsigma_i \quad \text{and} \quad q_t = \frac{1}{N_t} \sum_{i:W_i=1} \varsigma_i.$$

14.6   ASSESSING BALANCE: FOUR ILLUSTRATIONS

In this section we illustrate the methods discussed in this chapter. We apply these methods to four data sets, thereby illustrating a range of possible findings arising from the inspection of covariate balance. These four data sets range from a completely randomized experiment with, at least in expectation, identical covariate distributions, to an observational study with covariate distributions exhibiting very limited overlap, as well as two observational data sets with moderate amounts of overlap. In each case, we first estimate the propensity score using the methods from the previous chapter. We follow the algorithm described in that chapter to select, from K covariates X_i, some covariates to enter linearly and, in addition, some second-order terms. The tuning parameters for the algorithm were set, as proposed in Chapter 13, at $C_L = 1$ and $C_Q = 2.71$. In each case some covariates are always included in the propensity score, again as described in general terms in that chapter. We also present the graphical evidence for the adequacy of the estimated propensity score. Finally, we present, for each of the four data sets, the four covariate balance measures: normalized differences in means, log ratio of standard deviations, the two coverage measures, and the proportions of units with close comparisons.

14.6.1 Assessing Balance: The Barbiturate Data

The first application of the methods discussed in this chapter is based on the Reinisch barbiturate data set that was introduced in Chapter 13. These data contain information on 7,943 individuals, 745 of whom were exposed *in utero* to barbiturates, and 7,198 individuals in the control group, who were not exposed to barbiturates while *in utero*. We have seventeen covariates, sex, antih, hormone, chemo, cage, cigar, lgest, lmotage, lpbc415, lpbc420, motht, motwt, mbirth, psydrug, respir, ses, and sib. For a more detailed description of the data, the reader is referred to Chapter 13, where we discussed a method for specifying the propensity score. Starting with the automatic inclusion of three pre-treatment variables, sex (sex of the child), lmotage (mother's age), and ses (parents' socio-economic status), the specific method led to the inclusion of all covariates other than lpbc415, motht, and respir, in the linear part of the propensity score and, in addition, led to the inclusion of nineteen second-order terms, as detailed in the previous chapter. In this chapter we continue to utilize that specification of the propensity score and the resulting estimates.

We start by presenting, in Table 14.1, the summary statistics for the barbiturate data. For each of the seventeen covariates, as well as for the propensity score and the linearized propensity score, we report averages and sample standard deviations by treatment group. In addition, we report four measures of overlap for each covariate: $\hat{\Delta}_{ct}$, the difference in means by treatment group, normalized by the square root of the average within-group squared standard deviation; $\hat{\Gamma}_{ct}$, the log of the ratio of the sample standard deviations; and $\hat{\pi}_c^{0.05}$ $\hat{\pi}_t^{0.05}$, and the proportions of control units and treated outside the 0.025 and 0.975 quantiles of the covariate distributions for both the control and treated units, respectively. These four measures are reported in the last four columns of Table 14.1. The specification of the propensity score, selected in Chapter 13, led to the inclusion of the interaction between the indicator for chemotherapy (chemo) and the indicator for multiple births (mbirth). There was a small set of seventeen individuals who had been exposed to chemotherapy and who had experienced multiple births. These seventeen individuals were all in the control group, so we estimated the propensity score to be equal to zero for these individuals. In the calculation of the average linearized propensity score (lps) by treatment group, in the last row of Table 14.1, these seventeen individuals were excluded from further analyses.

Table 14.1 reveals that there is one covariate that is particularly unbalanced: lpbc420, a constructed index of pregnancy complications; it is highly predictive of exposure to barbiturates, with more than a full standard deviation difference in means. This is also the only variable for which the $\pi^{0.05}$ overlap measure suggests that there are substantial proportions of both the treated and control units with covariate values that are outside the central 0.95 part of the distribution for the other treatment group. A full 48% of the control units have values for lpbc420 outside the 0.025 and 0.975 quantiles of the distribution of lpbc420 among the treated units, and similarly 28% of the treated units have values for lpbc420 outside the 0.025 and 0.975 quantiles of the distribution among the control units. To further investigate the imbalance of lpbc420, Figures 14.1a and 14.1b present histograms of its distribution by treatment status. These figures show that the range of values for lpbc420 is substantially different for the two treatment groups. In the control group, the value of this variable ranges from -2.41 to 2.59, with a mean of -0.12 and a standard deviation of 0.96. In the treatment group, the range

Table 14.1. *Balance between Treated and Controls for Barbiturate Data*

	Controls ($N_c = 7{,}198$)		Treated ($N_t = 745$)		Overlap Measures		$\pi^{0.05}$	
	Mean	(S.D.)	Mean	(S.D.)	Nor Dif	Log Ratio of STD	Controls	Treated
sex	0.51	(0.50)	0.50	(0.50)	−0.01	0.00	0.00	0.00
antih	0.10	(0.30)	0.17	(0.37)	0.19	0.20	0.00	0.00
hormone	0.01	(0.10)	0.03	(0.16)	0.11	0.43	0.00	0.03
chemo	0.08	(0.27)	0.11	(0.32)	0.10	0.14	0.00	0.00
cage	0.00	(1.01)	0.03	(0.97)	0.03	−0.04	0.07	0.03
cigar	0.54	(0.50)	0.48	(0.50)	−0.12	0.00	0.00	0.00
lgest	5.24	(1.16)	5.23	(0.98)	−0.01	−0.17	0.05	0.02
lmotage	−0.04	(0.99)	0.48	(0.99)	0.53	0.00	0.07	0.07
lpbc415	0.00	(0.99)	0.05	(1.04)	0.05	0.06	0.01	0.03
lpbc420	−0.12	(0.96)	1.17	(0.56)	1.63	−0.55	0.48	0.28
motht	3.77	(0.78)	3.79	(0.80)	0.03	0.03	0.00	0.00
motwt	3.91	(1.20)	4.01	(1.22)	0.08	0.02	0.00	0.00
mbirth	0.03	(0.17)	0.02	(0.14)	−0.07	−0.21	0.03	0.00
psydrug	0.07	(0.25)	0.21	(0.41)	0.41	0.47	0.00	0.00
respir	0.03	(0.18)	0.04	(0.19)	0.03	0.07	0.00	0.00
ses	−0.03	(0.99)	0.25	(1.05)	0.28	0.06	0.00	0.00
sib	0.55	(0.50)	0.52	(0.50)	−0.06	0.00	0.00	0.00
Multivariate measure					1.78			
pscore	0.07	(0.12)	0.37	(0.22)	1.67	0.62	0.44	0.63
linearized pscore	−5.12	(3.40)	−0.77	(1.35)	1.68	−0.93	0.45	0.63

is from −0.24 to 2.50, with a mean of 1.17 and a standard deviation of 0.56. In the control group, 2,914 out of 7,198 individuals (approximately 40%) have a value for lpbc420 that is smaller than −0.2440, the smallest value observed in the treatment group. This suggests that differences in the value for this variable will be difficult to adjust reliably using simple covariance adjustment methods and that we should pay close attention to the balance for this variable using some of the design methods discussed in the next two chapters. The remaining covariates are substantially better balanced, with the largest standardized difference in means for lmotage, equal to 0.53 standard deviations. We also find that the logarithm of the ratio of standard deviations is far from zero for some of the covariates, suggesting that the dispersion varies between treatment groups. The multivariate measure is $\hat{\Delta}_{ct}^{mv} = 1.78$, suggesting that overall the two groups are substantially apart.

Next, we present, in Figures 14.2a and 14.2b, histogram estimates of the distribution of the linearized propensity score by treatment group. These figures reveal considerable imbalance between the two groups, further supporting the evidence from Table 14.1, where we found that the difference in estimated propensity scores by treatment status was more than a standard deviation. Figure 14.3a displays graphically the balance property of the propensity score. As discussed in the previous chapter, this is a Q-Q plot for the

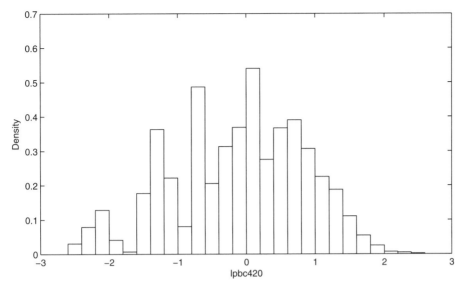

Figure 14.1a. Histogram-based estimate of the distribution of lpbc420 for control group, for barbiturate data

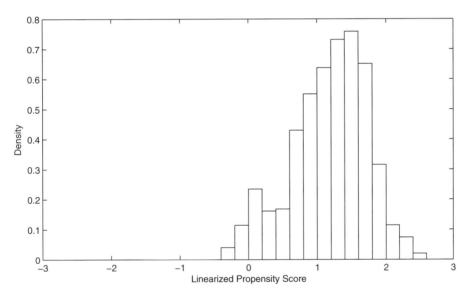

Figure 14.1b. Histogram-based estimate of the distribution of lpbc420 for treatment group, for barbiturate data

z-values, measuring within-block equality of the covariate means. The algorithm discussed in the previous chapter led to 10 blocks for the barbiturate data. As discussed in Chapter 13, this figure suggests that the specification of the propensity score is adequate.

Finally, we present in the first numerical column of Table 14.2 the matching statistics q_c and q_t. For the barbiturate data we find that $q_c = 0.60$, and $q_t = 0.98$, which suggests that it will be challenging to estimate causal effects for a substantial number of control units under unconfoundedness. In contrast, because $q_t = 0.98$, we can find comparable units for almost all treated units, suggesting that we can credibly estimate causal effects

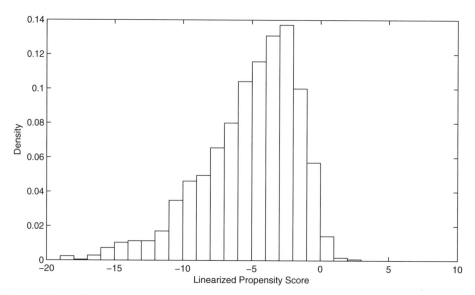

Figure 14.2a. Histogram-based estimate of the distribution of linearized propensity score for control group, for barbiturate data

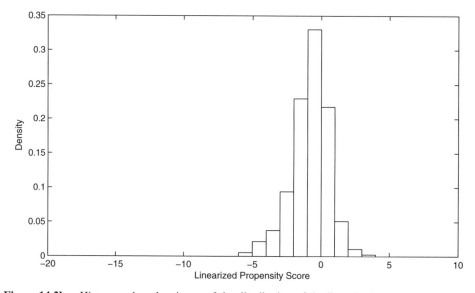

Figure 14.2b. Histogram-based estimate of the distribution of the linearized propensity score for treatment group, for barbiturate data

for the treated subpopulation. In this application, that is the natural population of interest, so the fact that we cannot credibly estimate causal effects for many of the control units need not be a concern.

14.6.2 Assessing Balance: The Lottery Data

Next, we use a data set collected by Imbens, Rubin, and Sacerdote (2001), who were interested in estimating the effect of unearned income on economic behavior, including

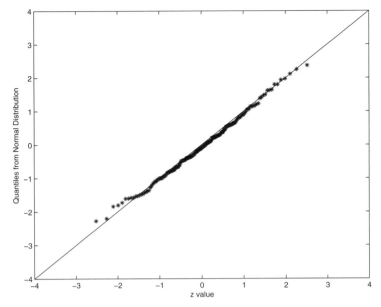

Figure 14.3a.   Q-Q plot for covariate balance conditional on propensity score for barbiturate data

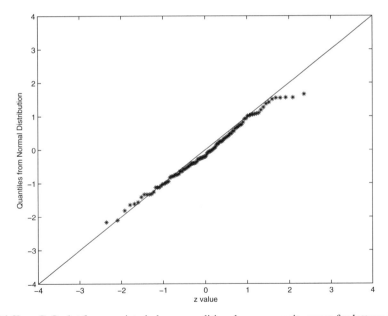

Figure 14.3b.   Q-Q plot for covariate balance conditional on propensity score for lottery data

labor supply, consumption, and savings. In order to study this question, they surveyed individuals who had played and won large sums of money in the Massachusetts lottery (the "winners"). For a comparison group, they collected data on a second set of

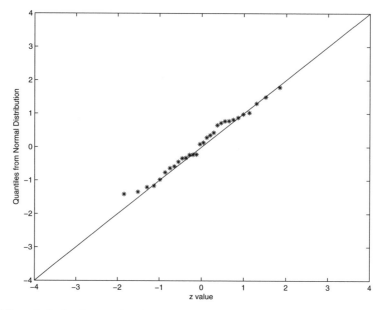

Figure 14.3c. Q-Q plot for covariate balance conditional on propensity score for Lalonde experimental data

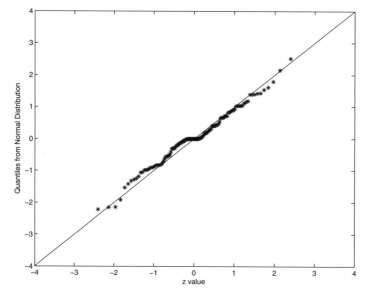

Figure 14.3d. Q-Q plot for covariate balance conditional on propensity score for Lalonde non-experimental data

individuals who also played the lottery but who had won only small prizes, referred to here as "losers." Constructing a comparison group of lottery players who did not win anything was not feasible because the Lottery Commission did not have contact information for such individuals. Although Imbens et al. analyze differences within the winners group by the amount of the prize won, here we focus only on the second comparison of winners versus losers. Specifically, here we analyze a subset of the data with $N_t = 259$

Table 14.2. *Proportion of Units with Match Discrepancy in Terms of Linearized Propensity Score Less Than 0.10*

	Barbiturate	Lottery	Lalonde Experimental	Lalonde Non-Experimental Data
q_c	0.60	0.75	0.98	0.21
q_t	0.98	0.69	0.97	0.97

winners and $N_c = 237$ losers in the sample of $N = 496$ lottery players. We know the year these individuals won or played the lottery (Year Won), the number of tickets they typically bought (Tickets Bought), their age in the year they won (Age), an indicator for being male (Male), education (Years of Schooling), whether they were working during the year they won (Working Then), and their social security earnings for the six years preceding the year they won (Earnings Year -6 to Earnings Year -1), and six indicators for each of these earnings being positive (Pos Earn Year -6 to Pos Earn Year -1).

We return to a more complete analysis of these data, involving the outcome variables, in Chapter 17. Here we only mention that the outcome we focus on in subsequent analyses is annual labor income, averaged over the first six years after playing the lottery.

We first estimate the propensity score for these data. We use the method discussed in Chapter 13 for selecting the specification, with, as before, cutoff values for the linear and second-order terms equal to $C_L = 1$ and $C_Q = 2.71$, respectively. The four covariates Tickets Bought, Years of Schooling, Working Then, and Earnings Year -1 were selected *a priori* to be included in the propensity score, partly based on *a priori* beliefs that they would be highly associated with winning the lottery (Tickets Bought), or highly associated with post-lottery earnings (Years of Schooling, Working Then, and Earnings Year -1). The algorithm then led to the inclusion of four additional covariates, for a total of eight out of the eighteen covariates entering the propensity score linearly, and ten second-order terms. The parameter estimates for this specification, with the covariates listed in the order they were selected for inclusion in the propensity score, are given in Table 14.3. Figure 14.3b suggests that the specification of the propensity score is adequate, in the sense that conditional on the propensity score, the covariates are balanced.

In Table 14.4 we present the balance statistics for the lottery data, which reveal that there are substantial differences between the covariate distributions in the two groups. Most important for post-treatment comparisons of economic behavior, we find that, prior to winning the lottery, the winners were earning significantly less than losers, with differences in all six of the pre-winning years statistically different from zero at conventional significance levels, and also large in substantial terms (on the order of 30% of average annual earnings). We also find that these differences are large relative to their variances, with the normalized differences for many variables on the order of 0.3, with some as high as 0.9 (for Tickets Bought). This suggests that simple regression methods will not reliably remove the biases associated with the differences in covariates. At the same time, the overlap statistics, $\hat{\pi}_c^{0.05}$ and $\hat{\pi}_t^{0.05}$, suggest that there is substantial overlap in the central ranges of the covariate distributions, suggesting that more sophisticated methods for adjustment may lead to credible results.

Table 14.3. *Estimated Parameters of Propensity Score for the Lottery Data*

Variable	EST	$(\widehat{s.e.})$	t-Stat
Intercept	30.24	(0.13)	231.8
Linear terms			
Tickets Bought	0.56	(0.38)	1.5
Years of Schooling	0.87	(0.62)	1.4
Working Then	1.71	(0.55)	3.1
Earnings Year -1	−0.37	(0.09)	−4.0
Age	−0.27	(0.08)	−3.4
Year Won	−6.93	(1.41)	−4.9
Pos Earnings Year -5	0.83	(0.36)	2.3
Male	−4.01	(1.71)	−2.3
Second-order terms			
Year Won × Year Won	0.50	(0.11)	4.7
Earnings Year -1 × Male	0.06	(0.02)	2.7
Tickets Bought × Tickets Bought	−0.05	(0.02)	−2.6
Tickets Bought × Working Then	−0.33	(0.13)	−2.5
Years of Schooling × Years of Schooling	−0.07	(0.02)	−2.7
Years of Schooling × Earnings Year -1	0.01	(0.00)	2.8
Tickets Bought × Years of Schooling	0.05	(0.02)	2.2
Earnings Year -1 × Age	0.00	(0.00)	2.3
Age × Age	0.00	(0.00)	2.2
Year Won × Male	0.44	(0.25)	1.7

The estimates for the propensity score also suggest that there are substantial differences between the two covariate distributions. These differences are revealed in the coverage proportions for the treated and controls, $\hat{\pi}_c$ and $\hat{\pi}_t$, which are 0.39 and 0.36 for the propensity score, even though these coverage proportions are below 0.10 for each of the covariates separately. Figures 14.4a and 14.4b present histograms estimates of the estimated propensity score.

The values for the overlap statistics, $q_c = 0.75$ and $q_t = 0.69$, suggest that, given the sample size, there are a substantial number of units for whom we will not be able to find close counterparts in the other treatment group, which indicates that we may have to trim the sample in order to focus on a subsample with better overlap. We will discuss specific methods for doing so in Chapters 15 and 16.

14.6.3 Assessing Balance: The Lalonde Experimental Data

These data were previously used and discussed in Chapter 8. Here the four earnings pre-treatment variables, earn'74, earn'74= 0, earn'75, and earn'75= 0, were selected *a priori* to be included in the propensity score. With these data, the algorithm for the specification of the propensity score leads to the inclusion of three additional pre-treatment variables as linear terms and to the inclusion of three second-order terms. Even if the randomization had been carried out correctly, and there were no missing data,

Table 14.4. *Balance between Winners and Losers for Lottery Data*

	Losers ($N_c = 259$)		Winners ($N_t = 237$)		Nor Dif	Log Ratio of STD	π^α	
	Mean	(S.D.)	Mean	(S.D.)			Controls	Treated
Year Won	6.38	(1.04)	6.06	(1.29)	−0.27	0.22	0.00	0.15
Tickets Bought	2.19	(1.77)	4.57	(3.28)	0.90	0.62	0.03	0.00
Age	53.21	(12.90)	46.95	(13.80)	−0.47	0.07	0.06	0.12
Male	0.67	(0.47)	0.58	(0.49)	−0.19	0.05	0.00	0.00
Years of Schooling	14.43	(1.97)	12.97	(2.19)	−0.70	0.11	0.01	0.09
Working Then	0.77	(0.42)	0.80	(0.40)	0.08	−0.06	0.00	0.00
Earnings Year −6	15.56	(14.46)	11.97	(11.79)	−0.27	−0.20	0.03	0.00
Earnings Year −5	15.96	(14.98)	12.12	(11.99)	−0.28	−0.22	0.10	0.00
Earnings Year −4	16.20	(15.40)	12.04	(12.08)	−0.30	−0.24	0.10	0.00
Earnings Year −3	16.62	(16.28)	12.82	(12.65)	−0.26	−0.25	0.03	0.00
Earnings Year −2	17.58	(16.90)	13.48	(12.96)	−0.27	−0.26	0.10	0.00
Earnings Year −1	18.00	(17.24)	14.47	(13.62)	−0.23	−0.24	0.03	0.00
Pos Earn Year −6	0.69	(0.46)	0.70	(0.46)	0.03	−0.01	0.00	0.00
Pos Earn Year −5	0.68	(0.47)	0.74	(0.44)	0.14	−0.07	0.00	0.00
Pos Earn Year −4	0.69	(0.46)	0.73	(0.44)	0.10	−0.04	0.00	0.00
Pos Earn Year −3	0.68	(0.47)	0.73	(0.44)	0.13	−0.06	0.00	0.00
Pos Earn Year −2	0.68	(0.47)	0.74	(0.44)	0.15	−0.07	0.00	0.00
Pos Earn Year −1	0.69	(0.46)	0.74	(0.44)	0.10	−0.05	0.00	0.00
Multivariate measure					1.49			
pscore	0.25	(0.24)	0.73	(0.26)	1.91	0.10	0.39	0.36
linearized pscore	−1.57	(1.67)	1.70	(2.10)	1.73	0.23	0.39	0.36

one would expect that the algorithm would select some covariates for inclusion in the specification of the propensity score despite the fact that the true propensity score would be constant. In reality, there are missing data, and the data set used here consists only of the records for individuals for whom all the relevant information is observed, strengthening the case for a non-degenerate specification of the true propensity score. Table 14.5 presents the estimated parameters of the propensity score. Figure 14.3c presents the balancing properties of the estimated propensity score.

Table 14.6 presents the balance statistics for the experimental Lalonde data. Not surprisingly, the summary statistics suggest that the balance in the covariate distributions is excellent, by all four measures, and for all ten pre-treatment variables, as well as for the two overlap statistics q_c and q_t. Across the ten pre-treatment variables, the maximum value of the normalized difference in covariate means is 0.30, and for the propensity score, the normalized difference is 0.54. The coverage proportion is above 0.91 for all covariates as well as for the propensity score. Figures 14.5a and 14.5b present histogram estimates of the estimated propensity score. These again suggest excellent balance, and thus simple covariance adjustment methods may be reliable here. The overlap statistics are $q_c = 0.98$ and $q_t = 0.97$, indicating that we can hope to estimate causal effects credibly for most units without extrapolation.

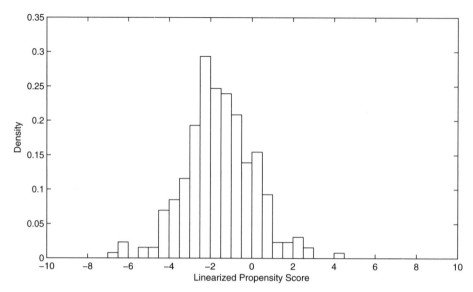

Figure 14.4a. Histogram-based estimate of the distribution of the linearized propensity score for control group, for lottery data

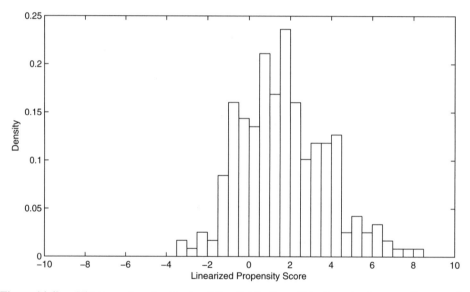

Figure 14.4b. Histogram-based estimate of the distribution of the linearized propensity score for treatment group, for lottery data

14.6.4 Assessing Balance: The Lalonde Non-Experimental Data

The primary focus of Lalonde's (1986) original paper was to examine the ability of statistical methods for non-experimental evaluations to obtain credible estimates of average causal effects. The idea was to investigate the accuracy of the estimates obtained by then correct and standard non-experimental methods by comparing them to estimates from a randomized experiment. Taking the experimental evaluation of the National Supported Work (NSW) program, Lalonde set aside the experimental control group, and

Table 14.5. *Estimated   Parameters   of   Propensity Score for the Lalonde Experimental Data*

Variable	EST	$(\widehat{s.e.})$	t-Stat
Intercept	−3.48	(0.10)	−34.6
Linear terms			
earn '74	0.03	(0.05)	0.7
unempl '74	−0.24	(0.39)	−0.6
earn '75	0.06	(0.05)	1.1
unempl '75	−3.48	(1.65)	−2.1
nodegree	7.33	(4.25)	1.7
hispanic	−0.65	(0.39)	−1.7
education	0.29	(0.37)	0.8
Second-order terms			
nodegree × education	−0.67	(0.35)	−1.9
earn '74 × nodegree	−0.13	(0.06)	−2.3
unempl '75 × education	0.30	(0.16)	1.9

Table 14.6. *Balance   between   Trainees   and   Experimental   Controls   for   Lalonde Experimental Data*

	Controls (N_c =260)		Trainees (N_t =185)		Nor	Log Ratio	$\pi^{0.05}$	
	Mean	(S.D.)	Mean	(S.D.)	Dif	of STD	Controls	Treated
black	0.83	(0.38)	0.84	(0.36)	0.04	−0.04	0.00	0.00
hispanic	0.11	(0.31)	0.06	(0.24)	−0.17	−0.27	0.00	0.00
age	25.05	(7.06)	25.82	(7.16)	0.11	0.01	0.01	0.03
married	0.15	(0.36)	0.19	(0.39)	0.09	0.08	0.00	0.00
nodegree	0.83	(0.37)	0.71	(0.46)	−0.30	0.20	0.00	0.00
education	10.09	(1.61)	10.35	(2.01)	0.14	0.22	0.01	0.08
earn '74	2.11	(5.69)	2.10	(4.89)	−0.00	−0.15	0.04	0.01
unempl '74	0.75	(0.43)	0.71	(0.46)	−0.09	0.05	0.00	0.00
earn '75	1.27	(3.10)	1.53	(3.22)	0.08	0.04	0.02	0.03
unempl '75	0.68	(0.47)	0.60	(0.49)	−0.18	0.05	0.00	0.00
Multivariate measure					0.44			
pscore	0.39	(0.11)	0.46	(0.14)	0.54	0.21	0.06	0.09
linearized pscore	−0.49	(0.53)	−0.18	(0.63)	0.53	0.17	0.06	0.09

to replace it, he constructed a comparison group from the Current Population Survey (CPS). (Lalonde also constructed a comparison group from the Panel Study of Income Dynamics, PSID, but we do not analyze these data here.) For this group, he observed the same variables as for the experimental sample. He then attempted to use the non-experimental CPS comparison group, in combination with the experimental trainees, to estimate the average causal effect of the training on the trainees. Here we focus on

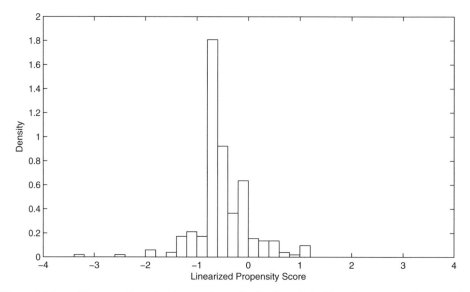

Figure 14.5a. Histogram-based estimate of the distribution of the linearized propensity score for control group, for Lalonde experimental data

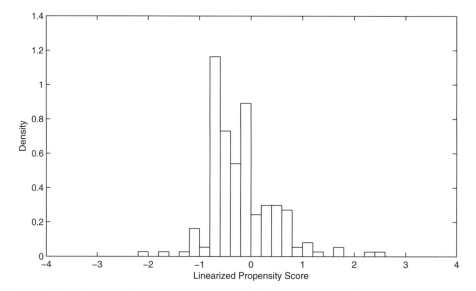

Figure 14.5b. Histogram-based estimate of the distribution of the linearized propensity score for treatment group, for Lalonde experimental data

the covariate balance between the experimental trainees and the CPS comparison group. The treatment group consists of the same set of 185 individuals who received job training that was used in the discussion in Section 14.6.3. The CPS comparison group consists of 15,992 individuals who did not receive the specific NSW training, but these individuals might, of course, have participated in other training programs. This does not affect the analysis but implies that the interpretation of the causal effect being estimated is the net effect of receiving the training associated with the NSW program, beyond any other services these individuals might receive. As in Section 14.6.3, we select the four earning

Table 14.7. *Estimated Parameters of Propensity Score*
for the Lalonde Non-Experimental Data

Variable	EST	$(\widehat{s.e.})$	t-Stat
Intercept	−16.20	(0.69)	−23.4
Linear terms			
earn '74	0.41	(0.11)	3.7
unempl '74	0.42	(0.41)	1.0
earn '75	−0.33	(0.06)	−5.5
unempl '75	−2.44	(0.77)	−3.2
black	4.00	(0.26)	15.1
married	−1.84	(0.30)	−6.1
nodegree	1.60	(0.22)	7.2
hispanic	1.61	(0.41)	3.9
age	0.73	(0.09)	7.8
Second-order terms			
age × age	−0.01	(0.00)	−7.5
unempl '74 × unempl '75	3.41	(0.85)	4.0
earn '74 × age	−0.01	(0.00)	−3.3
earn '75 × married	0.15	(0.06)	2.6
unempl '74 × earn '75	0.22	(0.08)	2.6

pre-treatment variables (earn'74, earn'74= 0, earn'75, and earn'75= 0) for prior inclusion in the propensity score. With the non-experimental Lalonde data set, the algorithm for the specification of the propensity score leads to the inclusion of five additional covariates as linear terms (excluding only education (years of education), but including the closely related variable nodegree, indicating whether an individual received at least a high school degree), and to the inclusion of five second-order terms. It is not surprising that the algorithm favors including substantially more covariates in the non-experimental case than it did in the experimental case discussed in Section 14.6.3. Table 14.7 presents the parameter estimates for the specification of the propensity score selected by the algorithm in this non-experimental case. Figure 14.3d presents the conditional balancing property of the estimated propensity score. Conditional on the propensity score, the covariates are again well balanced, suggesting that the algorithm used to select the specification of the propensity score performed well.

Table 14.8 presents the balance statistics for the non-experimental Lalonde data, and Figures 14.6a and 14.6b present histogram estimates of the estimated propensity score. For these data the balance is very poor. For a number of the covariates, the means by treatment status differ by more than a standard deviation. Consider earnings in 1975 (earn '75). Figures 14.7a and 14.7b present histograms for this covariate by treatment status. If we focus on post-program earnings as the primary outcome, as we will do in a later analysis of this program, it is clear that such large differences between the two groups in a variable such as earn '75, which is expected to be highly correlated with the outcome, could well lead to substantial biases in our estimates unless carefully controlled. All these measures suggest that, in order to estimate causal effects reliably,

Table 14.8. *Balance between Trainees and CPS Controls for Lalonde Non-experimental Data*

	Controls (N_c =15,992)		Trainees (N_c =185)		Nor Dif	Log Ratio of STD	$\pi^{0.05}$	
	Mean	(S.D.)	Mean	(S.D.)			Controls	Treated
black	0.07	(0.26)	0.84	(0.36)	2.43	0.33	0.00	0.00
hispanic	0.07	(0.26)	0.06	(0.24)	−0.05	−0.09	0.00	0.00
age	33.23	(11.05)	25.82	(7.16)	−0.80	−0.43	0.21	0.00
married	0.71	(0.45)	0.19	(0.39)	−1.23	−0.14	0.00	0.00
nodegree	0.30	(0.46)	0.71	(0.46)	0.90	−0.00	0.00	0.00
education	12.03	(2.87)	10.35	(2.01)	−0.68	−0.36	0.19	0.04
earn '74	14.02	(9.57)	2.10	(4.89)	−1.57	−0.67	0.51	0.01
unempl '74	0.12	(0.32)	0.71	(0.46)	1.49	0.34	0.00	0.00
earn '75	13.65	(9.27)	1.53	(3.22)	−1.75	−1.06	0.60	0.00
unempl '75	0.11	(0.31)	0.60	(0.49)	1.19	0.45	0.00	0.00
Multivariate measure					3.29			
pscore	0.01	(0.04)	0.41	(0.29)	1.94	1.93	0.86	0.85
linearized pscore	−10.04	(4.37)	−0.76	(2.08)	2.71	−0.74	0.86	0.85

we need to adjust for covariate differences in a sophisticated manner and, in particular, that simple regression methods are unlikely to be adequate.

It is interesting here to inspect the two overlap statistics, q_c and q_t. We find $q_c = 0.21$ and $q_t = 0.97$, indicating that we cannot hope to estimate credibly, for example, the average effect of the training program for the control group consisting of individuals surveyed in the Current Population Survey, even if we are willing to assume unconfoundedness. On the other hand, the fact that $q_t = 0.97$ suggests that there is hope of credibly estimating causal effects of the training program for the subpopulation of treated units.

14.6.5 Assessing Balance: Conclusions from the Illustrations

Figures 14.3a through 14.3d show that the algorithm for specifying the propensity score performs well in terms of generating balance in the covariates conditional on the propensity score. For each of the four specifications, the conditional balance is better than what one would expect in a randomized experiment. Unconditionally, however, the balance varies widely. This suggests that, in applications similar to the ones examined here, simple linear covariance adjustment methods are unlikely to lead to reliable estimates. Moreover, these differences suggest that we may wish to create more balanced subsamples, as well as use more sophisticated methods, to adjust for such differences.

14.7 SENSITIVITY OF REGRESSION ESTIMATES TO LACK OF OVERLAP

Here we present a simple illustration of the pitfalls that the lack of balance can lead to, especially in the context of naive adjustment methods such as linear regression. We

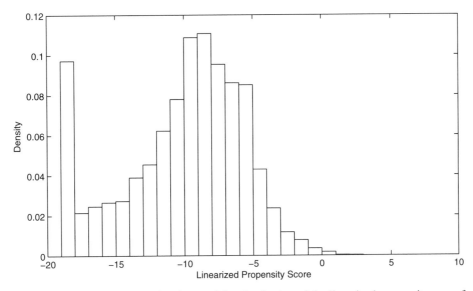

Figure 14.6a.   Histogram-based estimate of the distribution of the linearized propensity score for control group, for Lalonde non-experimental data

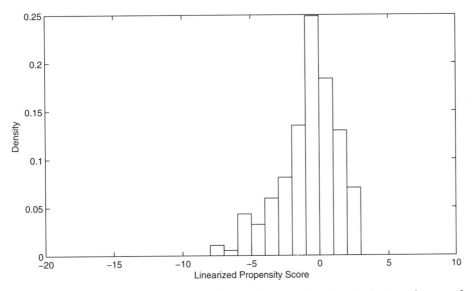

Figure 14.6b.   Histogram-based estimate of the distribution of the linearized propensity score for treatment group, for Lalonde non-experimental data

alluded to these issues at a more abstract level in Chapter 12, Section 4.2. Suppose we are interested in the average effect of the treatment on the subpopulation of treated units,

$$
\tau_{\text{fs},t} = \frac{1}{N_t} \sum_{i:W_i=1} \left(Y_i(1) - Y_i(0) \right) = \overline{Y}_t^{\text{obs}} - \frac{1}{N_t} \sum_{i:W_i=1} Y_i(0).
$$

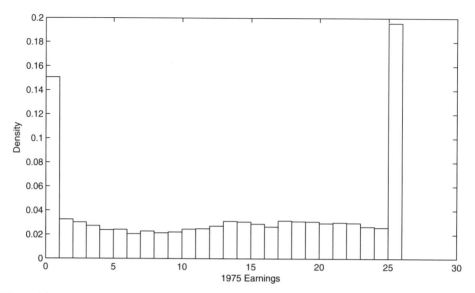

Figure 14.7a. Histogram-based estimate of the distribution of the linearized propensity score for control group, for Lalonde non-experimental data

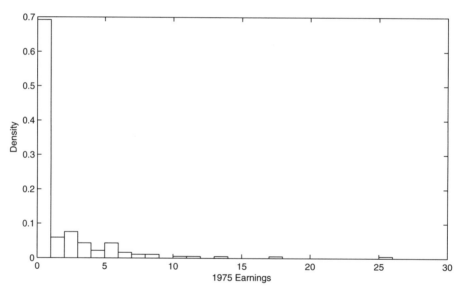

Figure 14.7b. Histogram-based estimate of the distribution of the linearized propensity score for treatment group, for Lalonde non-experimental data

In order to estimate $\tau_{\mathrm{fs},t}$, we need to impute, essentially, the missing potential outcomes, $Y_i(0)$ for all treated units, given the covariates X_i. We compare predictions based on the experimental data in Section 14.6.3, with predictions based on the non-experimental data in Section 14.6.4, using earnings in 1975 as the only covariate. We compare seven different linear regression models. These models are all of the polynomial form

$$\mathbb{E}[Y_i(0)|X_i = x] = \sum_{m=0}^{M} \beta_m \cdot x^m,$$

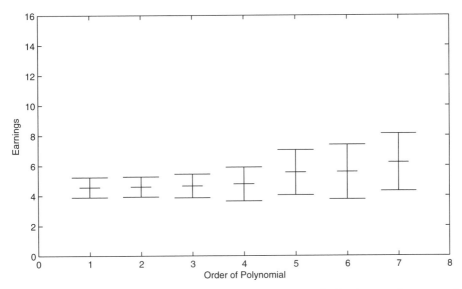

Figure 14.8a.   Intervals for predicted average earnings for trainees in the absence of treatment, for Lalonde experimental data

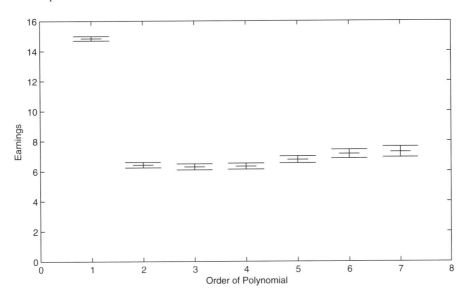

Figure 14.8b.   Intervals for predicted average earnings for trainees in the absence of treatment, for Lalonde non-experimental data

with the difference in the specification of the regression functions corresponding to the degree of the polynomial approximation. To illustrate, we use seven different models, corresponding to $M = 0, 1, \ldots, 6$, to predict the outcome, that is, 1978 earnings, for a hypothetical trainee at the average value of 1975 earnings, which is \$1,532 ($X_i = 1.532$).

Figures 14.8a and 14.8b give the 95% nominal intervals for the predicted average of 1978 earnings for trainees with 1975 earnings equal to \$1,532, in the absence of the training, in thousands of dollars. The results based on the experimental data are in Figure 14.8a, and the results based on the CPS comparison group are in Figure 14.8b.

It is clear that with the experimental data the choice of M, that is, the number of terms in the polynomial, does not matter much: as we increase the number of terms the estimated precision decreases somewhat, but the point estimates do not change much. With the non-experimental data, however, there is substantial sensitivity to the order of the polynomial. Even if we ignore the very substantial change in the results based on the specifications with no covariates, the sensitivity to higher-order terms is striking. With a third-order (cubic) approximation, the 95% nominal interval for $\mathbb{E}[Y_i(0)|X_i = 1.532]$ is $[6.13, 6.53]$, whereas with a fifth-order polynomial the 95% nominal interval is $[6.85, 7.43]$, which does not even overlap with the 95% nominal interval for the cubic approximation to the regression function. The difficulty when *a priori* choosing the order of the polynomial makes it impossible to arrive at a credible estimator based on simple regression methods in this setting.

14.8   CONCLUSION

In this chapter we develop methods for assessing covariate balance in treatment and control groups. If there is considerable balance, simple adjustment methods may well suffice to obtain credible estimates of the causal effects of interest. However, in cases where overlap is limited, such simple methods are likely to be sensitive to minor changes in the methods used, as illustrated in Section 14.7. In the following chapters, we explore two approaches for taking these issues into account. First, we develop methods for constructing subsamples with improved balance in covariate distributions between treatment groups. Second, we discuss methods for adjusting for differences in covariate distributions between treatment and control groups that are more sophisticated than linear adjustment methods. Ultimately we advocate combining both approaches to obtain more credible estimates of the causal estimands: balancing covariate distributions by matching or subclassification, and model-based adjustment.

NOTES

The importance of inspecting covariate balance and the dangers of simple linear regression adjustment goes back a long time (e.g., Cochran and Rubin, 1973; Rubin, 1973ab, 1979). This advice has not always been followed, however, and in empirical studies researchers often focus simply on t-statistics for testing the null hypotheses of no difference in average values between treatment and control groups. More recent publications stressing the importance of assessing balance compared to simply testing for equality of means include Imbens (2004, 2015), Imai, King, and Stuart (2008), Austin (2008), and Rubin (2006, 2008).

## CHAPTER 15

Matching to Improve Balance in Covariate Distributions

15.1 INTRODUCTION

In observational studies, the researcher has no control over the assignment of the treatment to units. This lack of control makes such studies inherently more sensitive and controversial than evaluations based on randomized assignment, where biases can be eliminated automatically, at least in expectation, through design, and as a result, for example, p-values can be assigned to sharp null hypotheses without relying on additional assumptions. Nevertheless, even in observational studies, one can carry out what we like to call a *design* phase during which researchers can construct a sample such that, within this selected sample, inferences are more robust and credible. We refer to this as a design phase because, just like in the design phase of a randomized study, it precedes the phase of the study during which the outcome data are analyzed. In this design phase, researchers can select a sample where the treatment and control samples are more balanced than in the original full sample. Balance here refers to the similarity of the *marginal* (generally multivariate) covariate distributions in the two treatment arms. This balance is not to be confused with the covariate balance *conditional* on the true propensity score that we discussed in the previous chapter. The latter holds, in expectation, by definition.

An extreme case of imbalance occurs when the ranges of data values of the two covariate distributions by treatment differ, and as a result there are regions of covariate values that are observed in only one of the two treatment arms. More typical, even if the ranges of data values of the covariate distributions in the two treatment arms are identical, there may be substantial differences in the shapes of the covariate distributions by treatment status. In a completely randomized experiment, the two covariate distributions are exactly balanced, in expectation. In that case, many different estimators – for example, simple treatment-control average differences, covariance-adjusted average differences, as well as many different model-based methods – tend to give similar point estimates of causal effects when sample sizes are at least moderately large. In contrast, in observational studies we often find substantial differences between covariate distributions in the two treatment arms. Such lack of covariate balance creates two problems. First, it can make subsequent inferences sensitive to ostensibly minor changes in the methods and specifications used. For example, adding an interaction or quadratic term to a linear regression specification can change the estimated average treatment effect substantially when the covariate distributions are far apart. Second, lack of balance can make the inferences imprecise. For covariate values with either few treated or few controls, it may be difficult to obtain precise estimates for treatment effects, and this, in turn, may make the estimates of overall treatment effects imprecise. In this chapter we discuss one systematic way to address these issues. In the next chapter we discuss an alternative.

In the approach to improving balance discussed in the current chapter, we focus on a setting characterized by a modest number of treated units, and a relatively large pool of possible controls. We are interested in estimating causal effects for the subpopulation of treated units. For example, consider designing an evaluation of a voluntary job-training program, where we are interested in the average effect of the training on those who completed the training program. The population of treated participants is typically well defined. The set of possible controls may include all individuals who are potentially comparable to the participants, which may well be a much larger set of individuals than the set of individuals sampled from the participants in the program. Prior to collecting the data on the outcomes for all individuals in this study, we have to select a set of individuals to serve as a control group. There is no harm in having data available on all possible control individuals, even if some are almost entirely irrelevant for the analysis. However, in practice, there may be trade-offs in terms of costs associated with collecting detailed information on a small set of units, versus those associated with collecting a limited amount of information on more units. With that trade-off in mind, it may be useful to select a subset of the full set of possible controls, based on covariate or pre-treatment information, for which we eventually collect the outcome data. Thus, the specific problem we study in this approach becomes one of selecting this subset, using solely covariate information, in order to create an informative sample for subsequent analyses. These subsequent analyses are likely to involve model-based imputation of the missing potential outcomes, matching, or propensity-score-based methods, all designed to adjust comparisons between treated and control units for remaining differences in covariate distributions. Details of the specific adjustment methods are discussed in subsequent chapters. The focus in this chapter is on selecting a control sample that is more balanced with respect to the treated sample than a random sample from the full population of possible controls. This selection will serve the purpose of making any subsequent analyses, irrespective of the choice of method, more robust, and thus more credible. Here we discuss both some practical and some theoretical issues concerning the selection of the control sample.

In this discussion we consider the set of treated units to be fixed *a priori*. We discuss two specific matching methods where, in each case, we construct the control sample by matching one or more distinct controls to each treated unit. We consider first Mahalanobis metric matching, where the distance between units is measured using all covariates, and second propensity score matching, where the distance is measured solely in terms of the difference in the estimated propensity score (or, more typically, a monotone transformation of the propensity score such as the linearized propensity score, the logarithm of the odds ratio). We then discuss the theoretical properties of these two matching methods and their relative merits, as well as methods that combine features of both.

This chapter is organized as follows. In the next section we discuss the Reinisch barbiturate data used in this chapter. In Section 15.3 we develop the mechanics of matching without replacement. Next, in Section 15.4, we illustrate the methods developed so far using a small subsample with seven units from the Reinisch barbiturate data. In Section 15.5 we discuss some theoretical issues related to matching. In Section 15.6 we apply the methods discussed in this chapter to the Reinisch barbiturate data. Section 15.7 concludes.

15.2   THE REINISCH ET AL. BARBITURATE EXPOSURE DATA

We illustrate the issues discussed in this chapter using the same barbiturate data, originally analyzed by Reinisch et al., that were previously used in Chapters 13 and 14. The barbiturate data contain information on 745 individuals exposed to barbiturates while *in utero*, as well as on 7,198 individuals who were not exposed to barbiturates *in utero* but born in the same group of hospitals as the exposed individuals. The averages and standard deviations by treatment status are presented for these data in Table 15.1, which repeats some of the information from Table 14.1. The last four columns in this table present measures of the degree of overlap introduced in Chapter 12. For each of the covariates, the propensity score, and the linearized propensity score, we present the normalized difference,

$$\hat{\Delta}_{ct} = \frac{\overline{X}_t - \overline{X}_c}{\sqrt{(s_c^2 + s_t^2)/2}},$$

the logarithm of the ratio of the standard deviations by treatment status,

$$\hat{\Gamma}_{ct} = \ln\left(\frac{s_t}{s_c}\right),$$

and the overlap probabilities for control and treated units, defined as

$$\hat{\pi}_c^\alpha = 1 - \hat{F}_c\left(\hat{F}_t^{-1}(1 - \alpha/2)\right) + \hat{F}_c\left(\hat{F}_t^{-1}(\alpha/2)\right),$$

where $\hat{F}_c(\cdot)$ and $\hat{F}_c^{-1}(\cdot)$ are the empirical distribution function and its inverse in the control subsample, and

$$\hat{F}_c(x) = \frac{1}{N_c} \sum_{i:W_i=0} \mathbf{1}_{X_i \leq x}, \quad \text{and} \quad \hat{F}_c^{-1}(q) = \min_{-\infty < x < \infty} \{x : \hat{F}_c(x) \geq q\},$$

with analogous definitions for $\hat{F}_t(\cdot)$ and $\hat{F}_t^{-1}(\cdot)$. We report $\hat{\pi}_c^\alpha$ and $\hat{\pi}_t^\alpha$ for $\alpha = 0.05$.

15.3   SELECTING A SUBSAMPLE OF CONTROLS THROUGH MATCHING TO IMPROVE BALANCE

In this section we discuss matching as a method for creating a subsample that has more balance in the covariates. First we put some structure on the problem, and then we discuss two specific matching methods: the Mahalanobis metric matching, which attempts to balance all covariates directly; and propensity score matching, which matches only on a

Table 15.1. *Summary Statistics for the Reinisch et al. Barbiturate Data*

	Controls ($N = 7{,}198$)		Treated ($N = 745$)		Nor	Log Ratio	$\pi^{0.05}$	
	Mean	(S.D.)	Mean	(S.D.)	Dif	of STD	Controls	Treated
sex	0.51	(0.50)	0.50	(0.50)	−0.01	0.00	0.00	0.00
antih	0.10	(0.30)	0.17	(0.37)	0.19	0.20	0.00	0.00
hormone	0.01	(0.10)	0.03	(0.16)	0.11	0.43	0.00	0.03
chemo	0.08	(0.27)	0.11	(0.32)	0.10	0.14	0.00	0.00
cage	−0.00	(1.01)	0.03	(0.97)	0.03	−0.04	0.07	0.03
cigar	0.54	(0.50)	0.48	(0.50)	−0.12	0.00	0.00	0.00
lgest	5.24	(1.16)	5.23	(0.98)	−0.01	−0.17	0.05	0.02
lmotage	−0.04	(0.99)	0.48	(0.99)	0.53	0.00	0.07	0.07
lpbc415	0.00	(0.99)	0.05	(1.04)	0.05	0.06	0.01	0.03
lpbc420	−0.12	(0.96)	1.17	(0.56)	1.63	−0.55	0.48	0.28
motht	3.77	(0.78)	3.79	(0.80)	0.03	0.03	0.00	0.00
motwt	3.91	(1.20)	4.01	(1.22)	0.08	0.02	0.00	0.00
mbirth	0.03	(0.17)	0.02	(0.14)	−0.07	−0.21	0.03	0.00
psydrug	0.07	(0.25)	0.21	(0.41)	0.41	0.47	0.00	0.00
respir	0.03	(0.18)	0.04	(0.19)	0.03	0.07	0.00	0.00
ses	−0.03	(0.99)	0.25	(1.05)	0.28	0.06	0.00	0.00
sib	0.55	(0.50)	0.52	(0.50)	−0.06	0.00	0.00	0.00
Multivariate measure					1.78			
pscore	0.07	(0.12)	0.37	(0.22)	1.67	0.62	0.44	0.63
linearized pscore	−5.12	(3.40)	−0.77	(1.35)	1.68	−0.93	0.45	0.63

The header "Overlap Measures[a]" spans the columns Nor Dif, Log Ratio of STD, and $\pi^{0.05}$.

[a] $\pi_t^{0.05}$ measures the proportion of treated units with a covariate value that is either below the 0.025 quantile of the covariate values or above the 0.975 quantile of the covariate values for the controls, and similarly for $\pi_c^{0.05}$.

scalar function of the covariates, created to balance all covariates in an attempt to mimic randomization.

15.3.1 Setup

Suppose we have N_t treated units, indexed by $i = 1, \ldots, N_t$. In addition, we have a pool of possible controls, of size N_c', larger than N_t. We wish to select $N_c < N_c'$ units from this set to construct a sample of size $N = N_c + N_t$ of units that will be used to estimate treatment effects. Let $\mathbb{I}_c'$ denote the set of indices for the set of possible controls, $\mathbb{I}_c' = \{N_t + 1, \ldots, N_t + N_c'\}$. We focus on the problem of choosing a subset $\mathbb{I}_c$ of the full set of controls, $\mathbb{I}_c \subseteq \mathbb{I}_c'$, that has better balance with respect to the treated units than a random sample of the full set of possible controls. We would like the covariates of the units included in $\mathbb{I}_c$ to be well balanced in terms of covariates relative to the set of treated units and, at the same time, the cardinality of the set $\mathbb{I}_c$ to be sufficiently large to allow precise causal inferences whenever possible and, also, no larger than necessary to minimize costs associated with collecting outcome data for units in $\mathbb{I}_c$.

In principle this is a decision problem, and we could set it up that way by explicitly defining the cost of data collection, the disutility associated with lack of balance and that associated with lack of precision. These costs may in practice be difficult to specify, especially *a priori*, and so we simplify the problem by fixing $N_c = N_t$, the number of treated units. Using exactly the same procedures, we could also select a number of matches for each treated unit. We focus on the case with $M = 1$ here for ease of exposition. Fixing $N_c = N_t$ may be a reasonable choice if we consider the effect of N_c on the sampling variance of estimators for causal effects. In a randomized experiment, the sampling variance of the usual estimator for the average treatment effect under homoskedasticity and constant treatment effects, is $\sigma^2 \cdot (1/N_t + 1/N_c)$. In that case, this variance tends to be dominated by the sample size of the smaller of the treatment and control groups. Adding many more controls than treated units therefore does not improve the precision much in this simple situation, whereas with fewer controls than treated units, the sampling variance is sensitive to the number of controls. This sampling variance calculation does not directly apply to the unconfoundedness setting we are studying in this part of the book, but the intuition is still correct that the sampling variance of the estimated treatment effect is dominated by the sample size of the smaller of the treatment and control groups. Choosing $N_c = N_t$ is also a convenient choice because some of the specific methods we discuss for selecting a set of controls rely on assigning a fixed number of controls to each treated unit.

Given this restriction, the decision problem becomes one of selecting a set of N_t controls from the set $\mathbb{I}_c'$ to optimize balance. We operationalize this objective by ordering the treated units and then sequentially selecting control units that are closest to each treated unit. Let $\mathbb{I}_t = \{1, \ldots, N_t\}$ denote the ordered set of indices for the treated units. Suppose for convenience that the treated units are ordered based on the value of the propensity score, with the units with the highest value of the estimated propensity score to be matched first, which corresponds to matching the units that are *a priori* the most difficult to match first. The choice of ordering can alter the results, although in practice the results tend to be fairly robust to this choice. Let $d(x, x')$ denote some measure of the "distance" between two vectors of covariates (formally not necessarily a distance because we allow $d(x, x')$ to be zero even if the vectors are not identical). Later we discuss various choices for the measure. Given the choice of the metric, let $\mathcal{M}_i^c \subset \mathbb{I}_c'$ denote the set of matched controls for treated unit i. At the moment this set is a singleton, $\mathcal{M}_i^c = \{m_i\}$, where m_i is the index of the control unit that is matched to treated unit i, but later we allow for more general matching strategies. For the first treated unit, $i = 1$, the set containing the closest match is

$$\mathcal{M}_1^c = \left\{ j \in \mathbb{I}_c' \,\middle|\, d(X_1, X_j) = \min_{j' \in \mathbb{I}_c'} d(X_1, X_{j'}) \right\}.$$

For the ith treated unit, this set is

$$\mathcal{M}_i^c = \left\{ j \in \mathbb{I}_c' - \cup_{i'=1}^{i-1} \mathcal{M}_{i'}^c \,\middle|\, d(X_i, X_j) = \min_{j' \in \mathbb{I}_c' - \cup_{i'=1}^{i-1} \mathcal{M}_{i'}^c} d(X_i, X_{j'}) \right\},$$

where $\mathbb{I}'_c - \cup_{i'=1}^{i-1} \mathcal{M}_{i'}^c$ is the subset of $\mathbb{I}'_c$ excluding the set of all the control units previously used as matches, $\cup_{i'=1}^{i-1} \mathcal{M}_{i'}^c$. Following this approach for all treated units, $i = 1, \ldots, N_t$, leads to a set of matches $\mathbb{I}_c = \cup_{i=1}^{N_t} \mathcal{M}_i^c$ with N_t distinct elements.

The remaining issue is the choice of distance metric $d(x, x')$. In the next two subsections we discuss two of the leading choices.

15.3.2 Mahalanobis Metric Matching

The first choice for the distance measure is the Mahalanobis metric, where the distance between units with covariate values x and x' is defined to be

$$d_M(x, x') = (x - x') \left(\frac{N_c \cdot \hat{\Sigma}_c + N_t \cdot \hat{\Sigma}_t}{N_c + N_t} \right)^{-1} (x - x')^T,$$

where, as previously,

$$\hat{\Sigma}_c = \frac{1}{N_c} \sum_{i:W_i=0} (X_i - \overline{X}_c)^T \cdot (X_i - \overline{X}_c) \quad \text{and} \quad \hat{\Sigma}_t = \frac{1}{N_t} \sum_{i:W_i=1} (X_i - \overline{X}_t)^T \cdot (X_i - \overline{X}_t),$$

are the within-group sample covariance matrices of the covariates, and, as previously,

$$\overline{X}_c = \frac{1}{N_c} \sum_{i:W_i=0} X_i \quad \text{and} \quad \overline{X}_t = \frac{1}{N_t} \sum_{i:W_i=1} X_i,$$

are the within-group averages of the covariates. This metric amounts to normalizing the covariates so that under the assumption $\Sigma_c \propto \Sigma_t$, they have the identity matrix as the within-group covariance matrix, and then defining the distance as the sum of squared differences. An important property of the Mahalanobis metric is that the resulting set of matches is invariant to affine transformations of the covariates.

15.3.3 Propensity Score Matching

The second distance measure considers only differences in a scalar function of the covariates, namely the estimated propensity score (or a monotone transformation thereof). The motivation for this choice is twofold. First, the motivation relies on the result, discussed in Chapter 12, that adjusting for differences in the propensity score between treated and control groups eliminates all systematic biases associated with differences in observed covariates. Second, it is simpler to find close matches on a scalar (function of the) covariate(s), than it is to find close matches on all covariates jointly. Let $e(x)$ be the propensity score, and $\ell(x) = \ln(e(x)/(1 - e(x)))$ be the linearized propensity score (lps), or the logarithm of the odds ratio. To make this specific, we use as the metric the squared difference in the lps:

$$d_\ell(x, x') = \left(\ell(x) - \ell(x') \right)^2 = \left(\ln \left(\frac{e(x)}{1 - e(x)} \right) - \ln \left(\frac{e(x')}{1 - e(x')} \right) \right)^2.$$

It is convenient to use differences in the lps rather than differences in the propensity score itself because typically this transformation takes account of the fact that typically the difference in propensity scores of 0.10 and 0.05 is larger in substantive effects on outcomes than the difference between propensity scores of 0.55 and 0.50. Put differently, the potential outcomes are more likely to be approximately linear in the lps than in the propensity score. For example, if the potential outcomes are linear in the covariates, the covariates are jointly normal, and the propensity score follows a logistic form, then the potential outcomes are linear in the lps.

In practice we typically do not know the propensity score. In that case we use an estimated version of it to construct the matches. Formally, with the estimated propensity score denoted by $\hat{e}(x)$, we define

$$d_\ell(x, x') = \left(\hat{\ell}(x) - \hat{\ell}(x')\right)^2 = \left(\ln\left(\frac{\hat{e}(x)}{1 - \hat{e}(x)}\right) - \ln\left(\frac{\hat{e}(x')}{1 - \hat{e}(x')}\right)\right)^2.$$

The use of an estimated function of the covariates for matching raises two issues. First, the estimated propensity score may actually improve the quality of the matches over using the true propensity core, a theme mentioned earlier and one that we return to later. Here, we just note that matching on the estimated propensity score rather than the true propensity score can adjust for random imbalances between covariate distributions, such as those that can arise in a randomized experiment. A second issue is that the model for the propensity score may be misspecified. In that case the balance in covariates conditional on the estimated propensity score may not hold, and the credibility of subsequent inferences may be compromised. In the current setting where we use the propensity score for creating a more balanced sample through matches this is not as likely to be an important concern as it would be if we used the estimated propensity score for weighting or blocking, because the matching is just the first step in the analysis, with subsequent steps consisting of adjustments for remaining differences in covariates.

15.3.4 Hybrid Matching Methods

In some cases, one may wish to ensure that the matched sample is perfectly balanced in some key covariates that are viewed *a priori* as possibly highly associated with the outcomes. For example, one may wish to ensure that the proportions of men and women are the same in the treatment and control groups. One can achieve this by a simple modification of the previously discussed method. Specifically, one can in such cases partition the samples by values of these covariates, and then match, within the partitioned samples, on the estimated propensity score.

15.3.5 Rejecting Matches of Poor Quality

In some cases, even the closest match may not be close enough. If one finds that the closest match for a particular treated unit is substantially different, as measured by the distance $d(x, x')$, it may be appropriate to drop the treated unit from the analysis entirely. We discuss a general approach to select the sample based on the estimated propensity score in the next chapter, but here we discuss a simple modification to address this issue in the context of matching methods.

A simple rule would be to drop treated units if the distance between a treated unit and its closest control match is larger than a fixed threshold. For example, we could drop all matches where the estimated linearized propensity score exceeds $d_{\max}$,

$$\left| \hat{\ell}(X_i) - \hat{\ell}(X_{m_i}) \right| > d_{\max},$$

for some pre-specified $d_{\max}$, say $d_{\max} = 0.1$. In practice, this rule will often eliminate only treated units with propensity score values close to one, because, with a reasonably sized set of possible controls, it is likely that there will be sufficiently close control matches for treated units with propensity scores away from one.

15.3.6 Caliper Matching Methods

The two matching methods discussed earlier, Mahalanobis matching and propensity score matching, both assign one control unit to each treated unit, but more generally the method could allow for two or more matches. An alternative strategy is to assign to each treated unit all controls that are within some distance from that treated unit. Given a distance function $d(x, x')$, we could assign to treated unit $i = 1$ all control units $j \in \mathbb{I}'_c$ such that

$$d\left(X_1, X_j \right) \le d_{\mathrm{cal}}$$

for some pre-set number d_{cal}. Let $\mathcal{M}^c_1 \subset \mathbb{I}'_c$ be the set of labels for these units. After matching treated unit $i = 1$, we seek to match the second treated unit $i = 2$ to all control units from the set of potential controls excluding the ones matched to treated unit $i = 1$, $\mathbb{I}'_c - \mathcal{M}^c_1$, with distance $d\left(X_2, X_j \right)$ less than d_{cal}, and so on, with the set of control units matched to treated unit i defined analogously.

   The advantage of the caliper-matching method is that more control units are used in the analysis, and thus potentially more information is used to estimate the missing control potential outcomes for the treated units. Its disadvantage is that the sample that results from this approach is not necessarily very well balanced. It may be that for some treated units there are many control units within the caliper, whereas for other treated units there are only one or two control units. Especially if we match without replacement, the order in which we match the treated units can be important because the method can lead to difficulties in finding good matches for some treated units if other treated units have already been matched with a large number of control units.

15.4   AN ILLUSTRATION OF PROPENSITY SCORE MATCHING WITH SIX OBSERVATIONS

Here we illustrate some of the methods discussed so far using a subset of the Reinisch barbiturate data. We use observations on seven units, two with *in utero* exposure to barbiturates, and five from the control group. The values for the estimated propensity score and lps are reported in Table 15.2. (Note that the propensity score is estimated on the full sample of $N = 7{,}643$ units.) In terms of the notation introduced in Section 15.3, $\mathbb{I}_t = \{1, 2\}$, $\mathbb{I}'_c = \{3, 4, 5, 6, 7\}$. We order the two treated units by the decreasing value of their estimated propensity scores.

Table 15.2. *Seven   Units from   the   Reinisch   et   al. Barbiturate Data Set*

Unit	W_i	$\hat{e}(X_i)$	$\hat{\ell}(X_i)$
1	1	0.577	0.310
2	1	0.032	−3.398
3	0	0.136	−1.846
4	0	0.003	−5.913
5	0	0.310	−0.798
6	0	0.000	−9.424
7	0	0.262	−1.033

First let us consider matching on the (estimated) lps. The closest match for unit 1, with an estimated lps equal to 0.310, is control unit 5, with an estimated lps equal to −0.798. For the second treated unit, with an lps equal to −3.398, the closest control unit in $\mathbb{I}'_c - \{5\} = \{3, 4, 6, 7\}$ is unit 3, with an estimated lps equal to −1.846. Control units 4, 6, and 7 are not used as matches, so that $\mathbb{I}_c = \{3, 5\}$.

Note that the order of the matching is irrelevant here. Had we started with the second treated unit, the matches would have been identical. It is important here, though, that we match on the lps. If we match on the propensity score itself, the closest match for treated unit 2 would be control unit 4 instead of control unit 3, so that in that case $\mathbb{I}_c$ would be $\{4, 5\}$.

15.5   THEORETICAL PROPERTIES OF MATCHING PROCEDURES

In this section we discuss some of the theoretical properties of the matching procedures discussed in the previous section. This section is more technical than others, and a full understanding of it is not essential for implementing the methods. It is primarily intended to provide additional understanding of the way these methods work, and in particular to provide insights into the differences between matching on the propensity score, Mahalanobis matching, and other matching methods. Most of the section deals with special cases where more-precise properties can be derived. In these special cases we assume that the vectors of covariates in both treatment arms have a normal distribution with mean vectors μ_c and μ_t, indexed by the treatment status, and common covariance matrix Σ. The results can be generalized to allow for ellipsoidally symmetric distributions with proportional inner product matrices.

We are primarily concerned with differences in covariate distributions in the matched samples relative to the original sample. This is somewhat of a simplification, because it is likely that one will not simply compare outcomes for treated and control units in the matched or original sample. Instead, it is likely that one will analyze the matched sample using additional methods of the type discussed in Chapters 17 and 18 to adjust for biases associated with remaining differences in covariate distributions. Nevertheless, the stated comparison will provide a good indication of the efficacy of matching for removing differences in covariates. Specifically, we are here concerned with biases in estimators for the super-population average treatment effect for the treated, $\tau_{\mathrm{sp,t}} = \mathbb{E}[Y_i(1) - Y_i(0)|W_i = 1]$. Moreover, here we consider only estimators based on the difference in average outcomes for treated and (matched) controls. Without matching, the estimator is $\hat{\tau}^{\mathrm{dif}} = \overline{Y}_{\mathrm{t}}^{\mathrm{obs}} - \overline{Y}_{\mathrm{c}}^{\mathrm{obs}}$, with bias

$$\mathbb{E}\left[\overline{Y}_{\mathrm{t}}^{\mathrm{obs}} - \overline{Y}_{\mathrm{c}}^{\mathrm{obs}} - \tau_{\mathrm{sp,t}}\right] = \mathbb{E}\left[Y_i(0)\big|W_i = 1\right] - \mathbb{E}\left[Y_i(0)\big|W_i = 0\right]$$

$$= \mathbb{E}\left[\mathbb{E}\left[Y_i(0)\big|X_i\right]\big|W_i = 1\right] - \mathbb{E}\left[\mathbb{E}\left[Y_i(0)\big|X_i\right]\big|W_i = 0\right],$$

with the second equality following by unconfoundedness. This bias depends on the relation between the outcomes and the covariates, $\mathbb{E}[Y_i(0)|X_i]$, and on the distributions of the covariates in the two treatment groups. We do not know this relationship, or this distribution at this stage, and in general do not wish to rely overly on knowledge about it for choosing the matching method. We therefore focus on biases in terms of general linear combinations of the covariates. Let us assume that in the super-population the conditional mean of $Y_i(0)$ given the covariates is $\mathbb{E}[Y_i(0)|X_i = x] = x\beta$, where for normalization we assume $\beta^T\beta = 1$. We do not really believe that the relationship between the outcomes and the covariates is linear. In fact, if we were confident about the linearity of the conditional mean, we could simply estimate this relationship by linear regression, which would eliminate all biases associated with differences in covariate distributions if the conditional mean were truly linear. However, the goal here is to find a meaningful comparison between different matching methods, and for that purpose, it is enlightening to focus on the effect of these matching methods on biases assuming a linear relationship between outcomes and covariates.

In combination with the notation μ_{c} and μ_{t} for the population mean of the covariate values in the control and treatment groups, the linearity for the conditional mean of $Y_i(0)$ given X_i implies that the bias for the simple average difference estimator, $\hat{\tau}^{\mathrm{dif}} = \overline{Y}_{\mathrm{t}}^{\mathrm{obs}} - \overline{Y}_{\mathrm{c}}^{\mathrm{obs}}$, is

$$\mathbb{E}\left[\overline{Y}_{\mathrm{t}}^{\mathrm{obs}} - \overline{Y}_{\mathrm{c}}^{\mathrm{obs}} - \tau_{\mathrm{sp,t}}\right] = \mathbb{E}\left[\mathbb{E}\left[Y_i(0)\big|X_i\right]\big|W_i = 1\right]$$
$$- \mathbb{E}\left[\mathbb{E}\left[Y_i(0)\big|X_i\right]\big|W_i = 0\right] = (\mu_{\mathrm{t}} - \mu_{\mathrm{c}})\beta.$$

Suppose that a generic matching method M, in expectation, changes the mean of the vector covariates for the N_{t} matched controls from μ_{c} to μ_{c}^M. This changes the bias for the simple average difference estimator from $(\mu_{\mathrm{t}} - \mu_{\mathrm{c}})\beta$ to $(\mu_{\mathrm{t}} - \mu_{\mathrm{c}}^M)\beta$. The *percentage bias reduction*, or pbr, is

$$\mathrm{pbr}(\gamma) = 100 \times \frac{(\mu_{\mathrm{t}} - \mu_{\mathrm{c}}^M)\beta}{(\mu_{\mathrm{t}} - \mu_{\mathrm{c}})\beta}. \tag{15.1}$$

In general the percentage bias reduction will depend on the value of β. Some matching methods have the feature that the percentage bias reduction is the same for all linear combinations β, so that for all β we have, for some constant c_M,

$$(\mu_{\mathrm{t}} - \mu_{\mathrm{c}}^M)\beta = c_M \cdot (\mu_{\mathrm{t}} - \mu_{\mathrm{c}})\beta.$$

Such methods are called *equal percentage bias reducing* or epbr methods. Within the context of our special case assuming normality (or, more generally, ellipsoidal symmetry and proportional inner products), this property is shared by Mahalanobis metric and propensity score matching. We shall argue that epbr is an attractive property, even though at first it may not appear to be an important property. As long as a particular matching method reduces the bias for each covariate, it might appear not to be a major concern that it reduces the bias more for some covariates than it does for others. However, if a matching method is *not* epbr, it reduces bias for some linear combinations of covariates but increase bias for others, and in fact to an infinite degree. The key insight is that if a matching method is not epbr, then there are linear combinations of the covariates (actually, an infinite number) such that the bias in the matched sample is non-zero, whereas the bias for that linear combination in the original sample was zero. Hence the matching makes the bias infinitely worse for that particular linear combination. The implication is that only epbr matching methods improve the bias for *every* linear combination.

Let us discuss this property of epbr methods in more detail. First, let us decompose the inverse of the $K \times K$ covariance matrix of the covariates Σ^{-1} (assumed proportional in both treatment groups) as GG^T, where G is a lower triangular matrix, so that $\Sigma = (G^T)^{-1}G^{-1}$. In addition, let H be any orthonormal matrix with the first column equal to $H_1 = G^T(\mu_t - \mu_c)^T/((\mu_t - \mu_c)GG^T(\mu_t - \mu_c)^T)$, so that $H^T G^T(\mu_t - \mu_c)^T/((\mu_t - \mu_c)GG^T(\mu_t - \mu_c)^T) = \mathbf{1}_K$, where $\mathbf{1}_K$ is the K-component vector with the kth element equal to one and the others equal to zero (where K is the dimension of the covariate vector). Because H is orthonormal, it follows that $HH^T = I_K$, and thus $GHH^TG^T = GG^T = \Sigma^{-1}$. By construction, G and H are invertible, and thus GH is invertible. In terms of the basis defined by the columns of $(H^T G^T)^{-1}$, the difference in covariate vectors $\mu_t - \mu_c$ is

$$H^T G^T(\mu_t - \mu_c)^T = \delta \cdot \mathbf{1}_K,$$

where the constant of proportionality δ is $\delta = \left((\mu_t - \mu_c)GG^T(\mu_t - \mu_c)^T\right)^{-1}$. Thus, the bias of the original sample is, for a linear combination ξ, measured in the basis defined by the columns of $(H^T G^T)^{-1}$, equal to

$$(\mu_t - \mu_c)GH\xi = \delta \cdot \xi^T \begin{pmatrix} 1 \\ 0 \\ \vdots \\ 0 \end{pmatrix} = \delta \cdot \xi_1,$$

where ξ_1 is the first element of ξ.

Now let us compare two matching methods, matching method A, which is epbr, and matching method B, which is not. Because matching method A is epbr, it follows that the expectation of the average of the covariates for the matched controls, μ_c^A, satisfies, for some scalar constant c_A, $(\mu_t - \mu_c^A)\gamma = c_A \cdot (\mu_t - \mu_c)\gamma$ for all linear combinations β. Choose $\beta = GH\xi$, so that

$$(\mu_t - \mu_c^A)\gamma = c_A \cdot (\mu_t - \mu_c)\gamma = c_A \cdot (\mu_t - \mu_c)GH\xi = c_A \cdot \delta \cdot \xi_1.$$

Because matching method B is not epbr, there is no scalar constant c_B such that $(\mu_1 - \mu_c^B) = c_B \cdot (\mu_t - \mu_c)$. Hence by invertibility of $H^T G^T$, it follows that there is no c_B such that

$$H^T G^T (\mu_t - \mu_0^B)^T = c_B \cdot H^T G^T (\mu_1 - \mu_c)^T.$$

Because $H^T G^T (\mu_{ct} - \mu_c)^T = \delta \cdot \mathbf{1}_K$, it follows that there is no c_B such that

$$H^T G^T (\mu_t - \mu_0^B)^T = c_B \cdot \delta \cdot \mathbf{1}_K$$

Thus it follows that some element of $H^T G^T (\mu_t - \mu_c^B)$, other than the first element, must differ from zero. Suppose that one such element is the jth one, $j \neq 1$. Let β be the jth column of HG. Then $(\mu_t - \mu_c^B)\beta$ differs from zero (so the bias after matching is non-zero), whereas the bias before matching was $(\mu_t - \mu_c)\beta = 0$. Hence matching method B has made the bias for this linear combination infinitely worse.

Second, consider propensity score and Mahalanobis matching in our special case where the covariates in both treatment arms have normal distributions with means μ_w for $w = 0, 1$ and covariance matrix Σ. First transform the covariates from X to $Z = H^T G^T (X - \mu_c)$. For both Mahalanobis and propensity score matching, the matching results are invariant to affine linear transformations of the covariates, so whether we match on X_i or Z_i is irrelevant. After the transformation from X_i to Z_i, we have in the original sample, $Z_i|W_i = 0 \sim \mathcal{N}(0, c_0 \cdot \mathbf{1}_K, I_K)$, and $Z_i|W_i = 1 \sim \mathcal{N}(c_0 \cdot \mathbf{1}_K, I_K)$ for some constant c_0, where, as before, $\mathbf{1}_K$ is the K-vector with the first element equal to one and the others equal to zero. The transformed covariates are uncorrelated and thus, because of the normality, statistically independent. In terms of Z the bias in the original sample is $c_0 \cdot \mathbf{1}_K$, concentrated in the first element. In terms of the transformed covariates, the propensity score is a function of the first element Z_{i1} only. Now consider matching on (a function of) Z_{i1}, which includes matching on the propensity score or matching on the lps. Because, under normality, the other components of Z_i are independent of Z_{i1}, matching on (a function of) Z_{i1} does not affect the other component's distributions in the two treatment arms. Combined with the fact that there is no bias in the original sample orthogonal to Z_{i1}, this fact implies that there will be no bias in the matched samples orthogonal to Z_{i1}. The matching can affect only the difference in distributions for the first covariate that is being used in the matching, Z_{i1}, and therefore $\mu_t - \mu_c^M = c_1 \cdot \mathbf{1}_K = (c_1/c_0) \cdot (\mu_t - \mu_c)$ and thus all matching methods that match only on (functions of) Z_{i1} are epbr.

Before considering the properties of Mahalanobis matching, consider matching on a K-vector Z_i such that in the original sample $Z_i|W_i = w \sim \mathcal{N}(0, I_K)$ for both $w = 0, 1$. In that case, there is no bias in the original sample. Matching on all these (for the bias irrelevant) covariates leaves the difference in means unchanged, or $\mu_t - \mu_c^M = \mu_t - \mu_c = 0$, and so there is no bias in the matched samples, and Mahalanobis matching is epbr in this case. Now consider the case of interest, where $\mu_t - \mu_c = c_0 \cdot \mathbf{1}_1$. In that case there is a bias, coming from the difference in the first element of Z. Matching on all the covariates does not introduce any bias in the other elements of Z, and so $\mu_t - \mu_c^M = c_M \cdot \mathbf{1}_K$, and Mahalanobis matching is epbr.

Note that both propensity score and Mahalanobis matching methods are epbr, where bias is defined in terms of the *average* difference between covariates. This does *not* mean that they also reduce differences in other aspects of the distribution. In fact, they may introduce bias in terms of other moments, even when there was none to begin with. It is easy to see that this can happen. Suppose we are matching on a single covariate X_i, with the same $\mathcal{N}(0, 1)$ distribution in both treatment arms. In the matched samples the variance of the covariate in the control distribution will be less than one, and thus there will be a difference in the distribution of the covariates in the two treatment arms, despite there being no such difference in the original sample. To be precise, consider a treated unit with $X_i = x < 0$. Because the probability density function for X_i is increasing in x for $x < 0$, there will tend to be slightly more control units j, with X_j close to x and $X_j > x$ than control units with X_j close to x and $X_j < x$. Thus, the expected value of X_j for a control unit matched to a treated unit with $X_i < 0$ will be larger than X_i, and the opposite for control units matched to treated units with $X_i > 0$.

The preceding discussion under normality also illustrates an important aspect of the difference between Mahalanobis and propensity score matching. The latter matches only on the scalar covariate whose distribution differs between treatment and control groups. The former matches in addition on a set of covariates whose distributions are identical in both the treatment and control groups, as well as independent of the key (function of the) covariates whose distribution differs between treatment arms. In this simplified setting with normally distributed covariates, it is clear that Mahalanobis matching is "wasteful" in terms of bias reduction in the sense that it puts much emphasis on matching covariates whose distributions are already perfectly matched in expectation. Putting any emphasis on covariates that are already balanced is disadvantageous for two reasons. First, it may lead to less bias reduction for the covariates that are not balanced in the original sample. Especially when there are many covariates, attempting to match on all of them using Mahalanobis matching may substantially erode the effectiveness for reducing bias in the function of the covariates that matters most, that is, the propensity score. Second, by matching on the covariates that are already balanced, Mahalanobis matching may compromise the balance that is already there in the distribution. On the other hand, even if a covariate is balanced in expectation, as in a randomized experiment, it may still be beneficial in terms of precision to match on such a covariate to eliminate random variation. In addition, a key advantage of Mahalanobis matching is that it has good robustness properties. Outside the special case with normally or, more generally, ellipsoidally distributed covariates, Mahalanobis matching will still balance all covariates with large enough control samples, where estimated propensity score matching may fail to do so, for example, when the model for the propensity score is misspecified.

15.6   CREATING MATCHED SAMPLES FOR THE BARBITURATE DATA

In this section we apply matching methods to the Reinisch barbiturate data. We compare results obtained using Mahalanobis metric matching and matching on the estimated lps, which we refer to as propensity score matching, in a slight abuse of language. In both cases, we match each of the 745 individuals who had been exposed *in utero* to barbiturates to a single control individual, selected from the pool of 7,198 individuals with no history of prenatal barbiturate exposure. Table 15.1 presents summary statistics for the full sample. The propensity score was estimated using the algorithm described in Chapter 13, with

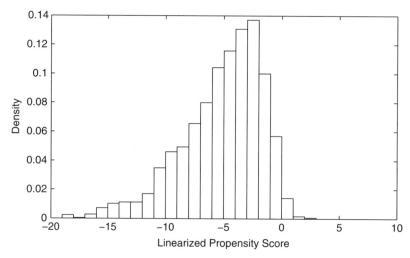

Figure 15.1a. Histogram-based estimate of the distribution of linearized propensity score for control group, for Reinisch barbiturate data

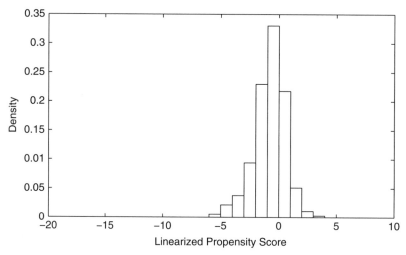

Figure 15.1b. Histogram-based estimate of the distribution of linearized propensity score for treatment group, for Reinisch barbiturate data

fourteen linear terms and nineteen second-order terms selected into the specification of the propensity score. See Table 13.6 in Chapter 13 for details on the parameter estimates for the estimated propensity score. Figures 15.1a and 15.1b, which are analogous to Figures 14.2a and 14.2b in Chapter 14, present histogram estimates of the distribution of the estimated lps for the treated and control subsamples for the Reinisch barbiturate data.

For both matching methods (Mahalanobis and lps), we report in Table 15.3 the average covariate differences between treated and control units' matched sample, scaled by the standard deviation of the covariate in the matched sample. For comparison purposes, we include a column with the normalized differences in means in the full sample. We scale all comparisons by the standard deviation in the full sample to make the columns comparable. We also report the results for the balance on the propensity score and the

Table 15.3. Between Treated and Control Units before and after Matching for the Reinisch Barbiturate Data

| | Full Sample | | | | Matched Samples | | | | | | | |
| | | | | | Mahalanobis | | | | Propensity Score | | | |
	Nor Dif	Log Rat of STD	$\pi^{0.05}$ Controls	$\pi^{0.05}$ Treated	Nor Dif	Log Rat of STD	$\pi^{0.05}$ Controls	$\pi^{0.05}$ Treated	Nor Dif	Log Rat of STD	$\pi^{0.05}$ Controls	$\pi^{0.05}$ Treated
sex	-0.01	0.00	1.00	1.00	0.00	-0.00	1.00	1.00	-0.03	0.00	1.00	1.00
antih	0.19	0.20	1.00	1.00	0.02	0.01	1.00	1.00	-0.03	-0.02	1.00	1.00
hormone	0.11	0.43	1.00	0.97	0.00	0.00	1.00	1.00	0.01	0.03	1.00	0.97
chemo	0.10	0.14	1.00	1.00	0.00	0.00	1.00	1.00	0.08	0.10	1.00	1.00
cage	0.03	-0.04	0.93	0.97	-0.03	0.03	0.96	0.95	-0.01	-0.00	0.95	0.95
cigar	-0.12	0.00	1.00	1.00	-0.01	-0.00	1.00	1.00	-0.01	-0.00	1.00	1.00
lgest	-0.01	-0.17	0.95	0.98	-0.02	0.13	0.98	0.97	0.00	0.01	0.98	0.97
lmotage	0.53	0.00	0.93	0.93	0.13	0.02	0.97	0.95	0.02	-0.01	0.95	0.97
lpbc415	0.05	0.06	0.99	0.97	0.03	0.06	0.98	0.99	0.07	-0.06	0.99	0.97
lpbc420	1.63	-0.55	0.52	0.72	0.59	-0.01	0.90	0.86	0.10	0.09	0.96	0.94
motht	0.03	0.03	1.00	1.00	-0.03	0.15	1.00	1.00	-0.03	0.03	1.00	1.00
motwt	0.08	0.02	1.00	1.00	0.02	0.09	1.00	1.00	0.05	-0.02	1.00	1.00
mbirth	-0.07	-0.21	0.97	1.00	0.00	0.00	0.98	0.98	0.03	0.12	0.99	0.98
psydrug	0.41	0.47	1.00	1.00	0.00	0.00	1.00	1.00	0.13	0.09	1.00	1.00
respir	0.03	0.07	1.00	1.00	0.00	0.00	1.00	1.00	0.03	0.07	1.00	1.00
ses	0.28	0.06	1.00	1.00	0.03	0.08	0.99	0.96	-0.04	0.02	0.99	0.96
sib	-0.06	0.00	1.00	1.00	0.03	-0.00	1.00	1.00	0.04	-0.00	1.00	1.00
Multivariate measure	0.43				0.24				0.05			
pscore	1.67	0.62	0.44	0.63	0.08	0.08	0.83	0.82	0.08	0.11	0.96	0.93
linearized pscore	1.65	-0.96	0.44	0.63	0.45	0.11	0.83	0.82	0.02	0.11	0.96	0.93

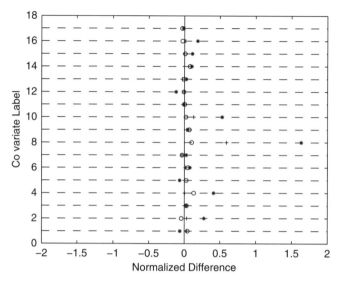

Figure 15.2. Covariate balance before (*) and after (+) lps and after Mahalanobis (o) matching, for the Reinisch barbiturate data

lps. The results show that the matching leads to a substantial improvement in balance. In the full sample, the normalized difference for one of the key covariates, lpbc420, is 1.63. Mahalanobis matching reduces this to 0.59, and propensity score matching reduces it further, to 0.10. In fact, after propensity score matching, none of the normalized differences exceeds 0.13, a degree of balance comparable to what one might expect in a completely randomized experiment. Figure 15.2 shows graphically how the normalized differences have decreased as a result of the matching. In this figure, the stars denote the original normalized differences before matching, the circles denote the normalized differences after lps matching, and the plus signs denote the normalized differences after Mahalanobis matching.

The improvement in balance can be shown graphically by comparing the distributions of the lps by treatment status in the full and matched samples. In order to do so, we re-estimate the propensity score in the matched samples, using the same algorithm as described in Chapter 13. The three covariates sex, lmotage, and ses are automatically selected for inclusion in the propensity score. First, consider the propensity score matched sample. The algorithm now selects six linear terms and one second-order term, compared to the thirty-three terms selected in the full sample. The fact that the algorithm selects fewer terms already indicates the improved balance. The parameter estimates for the propensity score are presented in Table 15.4. Second, consider the Mahalanobis matched sample. The algorithm for estimating the propensity score now selects six additional linear and six second-order terms. The results are in Table 15.5. Figures 15.1a and 15.1b present the distribution of the lps by treatment status in the full sample. Figures 15.3a and 15.3b present the distribution of the (newly estimated) lps in the lps matched samples, and Figures 15.4a and 15.4b present the distributions of the (newly estimated) lps in the Mahalanobis matched sample.

Figure 15.5 shows the distribution of differences in lps within the 745 matches after propensity score matching. This figure shows that about half the matches have

Table 15.4. *Estimated Parameters of Propensity Score for LPS Matched Sample Using the Algorithm from Chapter 13*

Variable	Est	$(\widehat{s.e.})$	t-Stat
Intercept	0.03	(0.05)	0.63
Linear terms			
sex	−0.04	(0.10)	−0.38
lmotage	0.03	(0.06)	0.45
ses	−0.04	(0.05)	−0.78
lpbc420	−0.61	(0.29)	−2.09
psydrug	0.05	(0.15)	0.32
Second-order terms			
lpbc420 × lpbc420	0.43	(0.14)	3.07

Table 15.5. *Estimated Parameters of Propensity Score for Mahalanobis Matched Sample for Barbiturate Data Using Algorithm from Chapter 13*

Variable	EST	$(\widehat{s.e.})$	t-Stat
Intercept	0.03	(0.06)	0.49
Linear terms			
sex	0.13	(0.12)	1.05
lmotage	0.27	(0.13)	2.12
ses	−0.12	(0.08)	−1.49
lpbc420	1.17	(0.28)	4.21
psydrug	−2.98	(0.67)	−4.46
chemo	−1.04	(0.21)	−5.06
mbirth	−1.68	(0.53)	−3.17
motwt	−0.11	(0.05)	−2.15
lgest	−0.69	(0.35)	−1.98
Second-order terms			
lpbc420× lpbc420	0.61	(0.17)	3.52
ses×ses	0.20	(0.06)	3.51
lgest×lgest	0.08	(0.03)	2.40
lpbc420×psydrug	1.15	(0.49)	2.35
lmotage×lpbc420	−0.24	(0.12)	−2.09
lmotage×motwt	1.12	(0.63)	1.77

differences in the lps less than 0.03, with the remainder spread out over the range 0.02 to 0.7.

To gain insight into the differences between propensity score and Mahalanobis matching, it is useful to consider the columns in Table 15.3 corresponding to the two matching

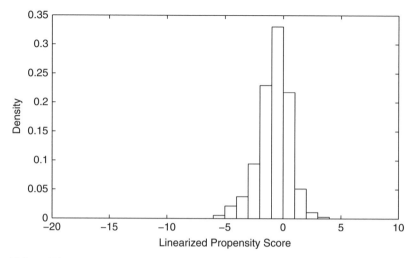

Figure 15.3a. Histogram-based estimate of the distribution of linearized propensity score after lps matching for the treatment group, for the Reinisch barbiturate data

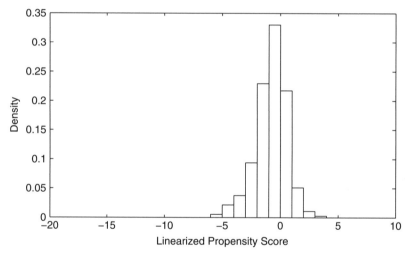

Figure 15.3b. Histogram-based estimate of the distribution of linearized propensity score after lps matching for the control group, for the Reinisch barbiturate data

methods in more detail. For most of the covariates for which there is a substantial difference in average values after matching, Mahalanobis matching leads to less balance than propensity score matching. For example, for lpbc420 (a pregnancy complication index), the normalized difference in averages is 0.59 for Mahalanobis matching and 0.10 for lps matching. For lmotage (logarithm of mother's age), the numbers are 0.09 and −0.02 for Mahalanobis and lps matching respectively. It may seem surprising that propensity score matching, which considers only one particular linear combination of the covariates for determining the match, does better in terms of generating balance

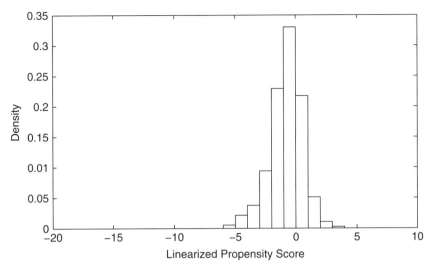

Figure 15.4a.   Histogram-based estimate of the distribution of linearized propensity score after Mahalanobis matching for the treatment group, for the Reinisch barbiturate data

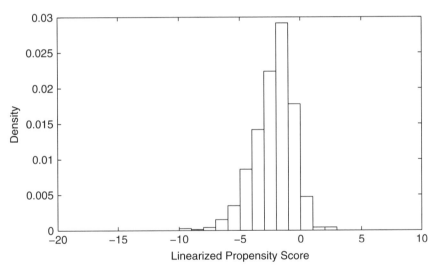

Figure 15.4b.   Histogram-based estimate of the distribution of linearized propensity score after Mahalanobis matching for the control group, for the Reinisch barbiturate data

on the individual covariates than Mahalanobis matching, which directly focuses on all the covariates. However, part of this comparison is misleading. Mahalanobis matching is designed to minimize differences in all covariates *within* matches, not to minimize differences in average covariates *across* all matched pairs. Suppose we look, for each covariate separately, at the square root of the average of the squares of within-pair differences, normalized by the square root of the sum of the squares of the sample standard deviations:

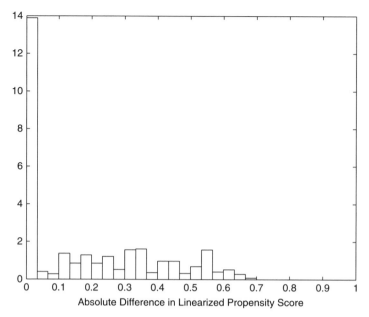

Figure 15.5. Histogram-based estimate of the distribution of the absolute difference in linearized propensity score for matches, for the Reinisch barbiturate data

$$\Delta_k = \frac{\sqrt{\frac{1}{N_t}\sum_{i=1}^{N_t}\left(X_{i,k} - X_{m_i,k}\right)^2}}{\sqrt{s_{c,k}^2 + s_{t,k}^2}}, k = 1, \ldots, K.$$

By this measure, Mahalanobis matching does considerably better than propensity score matching. For example, for `lmotage`, the two measures are 0.42 and 0.97 for Mahalanobis and lps matching respectively. Only for the pregnancy complication index, `lpbc420`, which given its importance in the propensity score, is essentially what propensity score matching is matching on in this data set, do we see a different comparison, with the numbers equal to 0.85 and 0.59 for Mahalanobis and propensity score matching, respectively. In general, propensity score matching leads to better overall balance, but Mahalanobis matching leads to smaller average differences within the matches.

It is also interesting to look at specific matches. In Table 15.6 the covariate values for three matches are presented, for both Mahalanobis matching and propensity score matching: first, the match for the treated unit with the largest value for the propensity score (0.97); second, the match for the treated unit with the median value of the propensity score (0.36); and, finally, the match for the treated unit with the smallest value of the propensity score (0.00). When we inspect the covariate values for the match for the treated unit with the largest value of the estimated propensity score, we see that propensity score matching leads to a good match in terms of `lpbc420`, the covariate that enters most prominently in the propensity score. Mahalanobis matching leads to a considerably worse match in terms of this covariate. In comparison, Mahalanobis matching leads to better match quality for some of the covariates that do not enter in the propensity score, such as `cage`.

Because the goal in the current chapter is not to create matches for specific units but to create a sample with substantial overlap in covariate distributions, matching on the lps is

Table 15.6. *Three Treated Units and Their Matches Based on Mahalanobis and Linearized Propensity Score Matching Algorithm, for the Reinisch Barbiturate Data*

| Covariate | Obs 1 (Max Pscore) | | | Obs 373 (Med Pscore) | | | Obs 745 (Min Pscore) | | |
| | Treated | Match | | Treated | Match | | Treated | Match | |
		Maha	LPS		Maha	LPS		Maha	LPS
sex	0.00	0.00	0.00	1.00	1.00	1.00	1.00	1.00	1.00
antih	1.00	1.00	0.00	0.00	0.00	0.00	0.00	0.00	0.00
hormone	0.00	0.00	0.00	0.00	0.00	0.00	0.00	0.00	0.00
chemo	0.00	0.00	1.00	0.00	0.00	0.00	0.00	0.00	0.00
cage	−0.68	−0.88	−1.23	−1.40	−1.34	0.27	−1.00	−1.47	−0.84
cigar	1.00	1.00	1.00	0.00	0.00	0.00	1.00	1.00	1.00
lgest	5.00	4.00	5.00	6.00	6.00	5.00	7.00	7.00	2.00
lmotage	0.27	0.57	0.57	1.64	1.85	−1.71	−0.82	−0.82	−0.09
lpbc415	0.26	0.26	0.26	0.74	0.44	0.93	−0.26	−0.26	0.74
lpbc420	2.50	1.41	2.45	1.21	0.85	0.98	−0.20	0.06	−0.35
motht	2.00	3.00	3.00	4.00	3.00	4.00	4.00	4.00	4.00
motwt	6.00	4.00	4.00	4.00	4.00	4.00	5.00	4.00	4.00
mbirth	0.00	0.00	0.00	0.00	0.00	0.00	0.00	0.00	0.00
psydrug	1.00	1.00	0.00	0.00	0.00	0.00	0.00	0.00	0.00
respir	0.00	0.00	1.00	0.00	0.00	0.00	0.00	0.00	0.00
ses	0.48	1.29	−1.15	0.48	0.07	−1.15	−0.34	−0.34	−1.15
sib	1.00	0.00	1.00	0.00	0.00	1.00	0.00	0.00	0.00
pscore	0.97	0.40	0.94	0.36	0.24	0.33	0.00	0.01	0.00
lps	3.48	−0.40	2.83	−0.59	−1.14	−0.70	−5.59	−4.68	−5.59

Note: Treated observations with the largest value for the estimated propensity score, the median value for the propensity score, and the smallest value for the propensity score.

Table 15.7. *Five Worst Matches for LPS Matching in Terms of LPS Distance, for the Reinisch Barbiturate Data*

| P-Score | | LPS | | Dif in LPS |
Treated	Control	Treated	Control	
0.79	0.66	1.34	0.64	0.69
0.79	0.66	1.34	0.67	0.68
0.81	0.69	1.45	0.79	0.66
0.81	0.69	1.45	0.80	0.65
0.97	0.94	3.48	2.83	0.64

clearly preferable to matching on all covariates through Mahalanobis matching, and we recommend it for this purpose, when there are more than a few covariates being matched.

Next, let us inspect, for the propensity score matched sample, the quality of the worst matches (in terms of the distance between the treated units and their matches). Table 15.7 presents, for the five worst matches, the value of the propensity score for the treated unit and the control unit, the lps, and the difference in lps. Even for these poorest of the matches, the discrepancies are modest. It is interesting to note that the worst matches are not simply for the units with the largest value of the propensity score. In this case there is little reason to discard any of the matches because of their poor quality.

15.7   CONCLUSION

In this chapter we discuss one approach to the design phase in an analysis of observational data. In this part of the analysis we select the sample for which we subsequently attempt to estimate causal effects. We attempt to construct a sample where the covariate distributions are well balanced, motivated by the fact that lack of balance can make any subsequent analysis imprecise, as well as sensitive to minor changes in the specification of the model for the outcomes given the covariates. The methods discussed in the current chapter use matching to create a control sample, selected from a larger donor pool of possible controls, in such a way that the covariate distribution in the matched control group is similar to the covariate distribution in the treated sample. In the application in this chapter, propensity score matching is effective in greatly reducing the imbalance between the covariate distributions, with the normalized differences between covariates reduced, from a maximum value of 1.63 in the full sample to a maximum value of 0.13 in the propensity score matched sample.

An important aspect of the analysis in this chapter is that it is entirely based on the covariate and treatment data, and never uses the outcome data. As such, it cannot intentionally introduce biases in the subsequent analyses.

NOTES

The formal results in this chapter on bias reduction for matching methods draw heavily on Rubin and Thomas (1992ab, 1996, 2000). Generalizing earlier ones in Rubin (1973ab, 1976) and Cochran and Rubin (1973), the results in the Rubin and Thomas work and extensions in Rubin and Stuart (2006) are more general than the ones reported in the current chapter, allowing for ellipsoidal distributions, of which normal distributions discussed here are a special case. For ease of exposition, we focus in the current chapter on cases with normal distributions. The chapter also borrows extensively from the discussion in Rosenbaum and Rubin (1984). See also Rubin (2006).

Gu and Rosenbaum (1993) distinguish between two goals of matching: minimizing distance between units within matched pairs and maximizing balance. In this chapter the goal of the matching is the latter: improving balance in covariate distributions between the two treatment groups.

Many applied papers use either Mahalanobis or propensity score matching methods to construct estimators. We discuss some of these methods in Chapter 18. Here, however, we focus on matching solely as a strategy to create more balanced samples rather than to create estimators. Subsequently we discuss various methods for estimating causal effects, all of which will generally be more effective in balanced samples. See also Ho, Imai, King, and Stuart (2007), Rosenbaum and Rubin (1985), and Pattanayak, Rubin, and Zell (2011).

## CHAPTER 16

Trimming to Improve Balance in Covariate Distributions

16.1 INTRODUCTION

The propensity score matching approach discussed in the previous chapter was aimed primarily at settings where the focus is on estimating treatment effects for the subset of treated units. The specific plan was to select a set of controls with a joint distribution of covariates similar to that for the treated units and discard the remaining controls. In the current chapter, we discuss a different approach to improving covariate balance. Starting with observations on covariates and treatment status for a sample of units with only limited overlap in terms of covariates, we construct a subsample that has a more substantial degree of overlap. We do so by discarding some units in the treatment group and some in the control group. For the resulting trimmed sample, we focus on estimating causal effects of the treatment versus control. By trimming the sample, this method generally alters the estimand, by changing the reference population. In that sense, this method sacrifices some *external validity* – the eventual estimators are less likely to be valid for typical (e.g., average) treatment effects in the original sample. The advantage is that the *internal validity* may be improved because estimators for causal effects in the trimmed sample are likely to be more credible and accurate than estimators for causal effects in the original, full sample. This primacy of internal validity, at the expense of external validity, is a general theme in this book as well as in the literature on design of randomized experiments. In studies of causal effects, there is often a trade-off between internal and external validity, with typically more focus on internal validity: given a well-defined population of interest, having a credible and precise answer for a subpopulation is often considered more important than a controversial (in the sense of relying on dubious assumptions) or imprecise answer for the full (original target) population.

The key to the trimming is the propensity score, the conditional probability of receiving the treatment given the pre-treatment variables. This role emerges naturally, rather than being imposed, as a consequence of a mathematical objective function to be minimized that does not itself involve the propensity score. If, for some units, the true propensity score is exactly equal to zero or one, it follows that for such units there are no counterparts with the alternative treatment. Thus, we cannot credibly and accurately estimate the effect of the treatment for such units without relying heavily on extrapolation. In practice, we often set aside such units, acknowledging that estimates for treatment effects for such units are not credible because of the extrapolation. The practical issue is what to do with units with values for the estimated propensity score close, but not exactly equal, to zero or one. In this chapter we argue that, in some situations, we may still wish to put aside such units, and estimate treatment effects for the set of units with estimated propensity scores substantially away from zero or one. To provide further motivation for this approach, consider units with the true value of the propensity score equal to $e(X_i) = 0.999$. Conditional on such a value for the propensity score, the probability that a unit is in the treatment group is, by definition, $e(X_i) = 0.999$. Hence, among units with $e(X_i) = 0.999$, there are almost 1,000 times as many treated units as control units. To estimate, say, the average effect of the treatment for such units using simple methods, we would either have to put a very large weight on the few control units with such propensity score values (and for this to even be feasible, we would obviously need a very large data set, large enough that there are in fact control units with such propensity score values), or we would need to extrapolate from control units with possibly quite different values for the propensity score. Neither using large weights nor relying on extrapolation is attractive: the first leads to a large sampling variance for the estimator, and the second one may lead to substantial bias.

In this chapter we discuss a principled and systematic way of selecting units with propensity score values away from zero and one, which involves choosing a threshold to assess whether the estimated propensity score is too close to zero or one. The criterion we use is based on the joint distribution of treatment indicators and pre-treatment variables and, importantly, does not involve data on the outcome variables, and therefore is a design-stage activity. It relies on the asymptotic sampling variance of estimators for average treatment effects and leads to a covariate-and-treatment-indicator-dependent criterion for determining a threshold, denoted by α, such that all units with estimated propensity score values in the intervals $[0, \alpha]$ and $[1 - \alpha, 1]$ are discarded, and causal effects are estimated only for units with values for the estimated propensity score in the interval $[\alpha, 1 - \alpha]$. In terms of motivating the threshold, we will take an infinite super-population perspective, where the sample at hand is viewed as a random sample from this super-population as introduced in Chapter 3, Section 3.5, and used in earlier chapters in this part of the text.

In practice one may wish to use the methods discussed in this chapter as a starting point for trimming the sample to achieve sufficient balance, in combination with scientific judgments. In our examples, however, we illustrate the methods using a rigid rule.

The chapter is organized as follows. In the next section we describe the data used in this chapter to illustrate the concepts and methods, which come from a study by Murphy and Cluff (1990) to investigate the effect of right heart catheterization on survival. In Section 16.3 we discuss, in detail, the intuition behind our approach in the context of a stylized example with a single binary covariate. In Section 16.4 we present results for the general case with multiple and multi-valued covariates. In Section 16.5 we return to the Catheterization Data to illustrate the general concepts developed in this chapter. Section 16.6 concludes.

16.2   THE RIGHT HEART CATHETERIZATION DATA

Murphy and Cluff (1990) studied the effectiveness of right heart catheterization in an observational setting, using data from the "Study to Understand Prognoses and Preferences

Table 16.1. *Summary Statistics for Selected Pre-Treatment Variables, for Right Heart Catherization Data*

Variable	Controls ($N_c = 3{,}551$)		Treated ($N_t = 2{,}184$)		Normalized Difference
	Mean	(S.D.)	Mean	(S.D.)	
cat1_copd	0.11	(0.32)	0.03	(0.16)	−0.32
cat2_lung	0.004	(0.060)	0.001	(0.03)	−0.05
neuro	0.16	(0.37)	0.05	(0.23)	−0.33
aps1	51	(19)	61	(20)	0.49
meanbp1	85	(39)	68	(34)	−0.44
pafi1	241	(117)	192	(106)	−0.42

for Outcomes and Risks of Treatments." Right heart catheterization is a diagnostic procedure used for critically ill patients. Their study collected data on hospitalized adult patients at five medical centers in the United States. Based on information from a panel of experts, a rich set of forty-nine covariates (recoded as seventy-two pre-treatment variables) relating to the decision to perform right heart catheterization was collected, as was detailed outcome data. Connors et al. (1996) used a one-to-one propensity score matching approach to study the same data set. Detailed information about the study and the nature of the variables can be found in Murphy and Cluff (1990) and Connors et al. (1996). Connors et al. (1996) found that, based on an analysis assuming unconfounded treatment assignment, right heart catheterization appeared to lead to adverse outcomes, namely lower survival rates. This conclusion contradicted the popular perception among practitioners that right heart catheterization was beneficial to critically ill patients.

The data set from the Connors et al. (1996) study that we use in this chapter consists of observations on $N = 5{,}735$ individuals, $N_t = 2{,}184$ of them in the treatment group and the remaining $N_c = 3{,}551$ in the control group. For each individual, we observe treatment status W_i, equal to one if right heart catheterization was applied within twenty-four hours of admission, and zero otherwise; seventy-two covariates; and eventually the outcome, which is an indicator for survival at thirty days. Hirano and Imbens (2001) present a table containing summary statistics for all seventy-two covariates. In Table 16.1 we present summary statistics for some selected covariates. Note that the cat1_copd (chronic obstructive pulmonary disease) is a fairly rare condition that differs considerably in its prevalence among treated and control units. We focus on the normalized differences,

$$\hat{\Delta}_{ct} = \frac{\overline{X}_t - \overline{X}_c}{\sqrt{(s_t^2 + s_c^2)/2}}.$$

With this many covariates, inspecting all normalized differences in means separately is cumbersome. In Figure 16.1, we present a histogram estimate of the distribution of the absolute values of the normalized differences. From this figure one can see that many of the covariates are fairly well balanced, although a number of them have substantially different distributions in the two treatment groups. For example, aps1 (Apache score) has a normalized difference of 0.49, and meanbp1 (mean blood pressure) has a normalized difference of 0.44. The mean and standard deviation of the seventy-two absolute values of the normalized differences in the full sample are 0.14 and 0.11, with 51% of the normalized

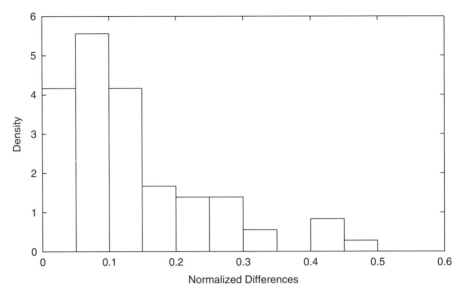

Figure 16.1. Histogram-based estimate of the distribution of the absolute values of the normalized differences for full sample, for Connors RHC data

differences exceeding 0.1, and 15% exceeding 0.25. Such differences suggest that simple methods, such as regression analysis, are unlikely to lead to effective and credible adjustments for pre-treatment differences and thereby reliable estimates of treatment effects. In this case, trimming the sample by removing units with extreme values of the estimated propensity score to improve overlap should lead to more robust inferences at the subsequent analysis stage.

16.3   AN EXAMPLE WITH A SINGLE BINARY COVARIATE

To set the stage for the issues discussed in this chapter, consider an example with a single pre-treatment variable X_i taking on two values, say, for illustrative purposes, f and m (female and male). We have a random sample of size N from an infinite super-population. Let $N(x)$ be the sample size for the subsample with $X_i = x$, with $x \in \{f, m\}$, so that $N = N(f) + N(m)$ is the total sample size. Also let q be the super-population share of $X_i = m$ units, $q = \mathbb{E}_{sp}[N(m)/N]$. Let the population average treatment effect conditional on $X_i = x$ be equal to $\tau_{sp}(x) = \mathbb{E}_{sp}[Y_i(1) - Y_i(0)|X_i = x]$. The super-population average treatment effect is

$$\tau_{sp} = \mathbb{E}_{sp}[Y_i(1) - Y_i(0)] = (1 - q) \cdot \tau_{sp}(f) + q \cdot \tau_{sp}(m).$$

Let

$$N_c(x) = \sum_{i:X_i=x} (1 - W_i) \quad \text{and} \quad N_t(x) = \sum_{i:X_i=x} W_i,$$

be the number of control and treated units with covariate value $X_i = x$, and let $e(x) = N_t(x)/N(x)$ be the propensity score at x. Finally, let

$$\overline{Y}_c^{obs}(x) = \frac{1}{N_c(x)} \sum_{i:X_i=x} Y_i^{obs} \cdot (1 - W_i) \quad \text{and} \quad \overline{Y}_t^{obs}(x) = \frac{1}{N_t(x)} \sum_{i:X_i=x} Y_i^{obs} \cdot W_i,$$

for $x = f, m$ be the average outcome within each of the four subpopulations defined by treatment status and covariate value. Assume, for ease of exposition, that the super-population variance of $Y_i(w)$ given $X_i = x$ is σ^2 for all x and w.

Natural estimators for the average treatment effects for each of the two subpopulations, $X_i = f, m$, are the simple differences in averages by treatment status for each of the two covariate values:

$$\hat{\tau}^{dif}(f) = \overline{Y}_t^{obs}(f) - \overline{Y}_c^{obs}(f), \quad \text{and} \quad \hat{\tau}^{dif}(m) = \overline{Y}_t^{obs}(m) - \overline{Y}_c^{obs}(m).$$

The sampling variances for these estimators derived from Neyman's repeated sampling perspective follow from calculations in earlier chapters. Here it is convenient to work with the approximate, asymptotic, sampling variances, the large-sample approximations to the exact variances normalized by the overall sample size N, denoted by $\mathbb{AV}(\hat{\tau})$ for a generic estimator $\hat{\tau}$. Then, the asymptotic sampling variance, defined here simply as the probability limit of the sampling variance normalized by the sample size, equals:

$$N \cdot \mathbb{V}\left(\hat{\tau}^{dif}(f)\right) = N \cdot \sigma^2 \cdot \left(\frac{1}{N_c(f)} + \frac{1}{N_t(f)}\right)$$

$$\longrightarrow \frac{\sigma^2}{(1-q)} \cdot \frac{1}{e(f) \cdot (1 - e(f))} = \mathbb{AV}\left(\hat{\tau}^{dif}(f)\right),$$

and

$$N \cdot \mathbb{V}\left(\hat{\tau}^{dif}(m)\right) = N \cdot \sigma^2 \cdot \left(\frac{1}{N_c(m)} + \frac{1}{N_t(m)}\right)$$

$$\longrightarrow \frac{\sigma^2}{q} \cdot \frac{1}{e(m) \cdot (1 - e(m))} = \mathbb{AV}\left(\hat{\tau}^{dif}(m)\right).$$

The natural estimator for the population average treatment effect, $\tau_{sp} = \mathbb{E}_{sp}[Y_i(1) - Y_i(0)]$, is

$$\hat{\tau}^{strat} = \frac{N(f)}{N(f) + N(m)} \cdot \hat{\tau}^{dif}(f) + \frac{N(m)}{N(f) + N(m)} \cdot \hat{\tau}^{dif}(m).$$

Because the two estimates $\hat{\tau}^{dif}(f)$ and $\hat{\tau}^{dif}(m)$ are independent, the sampling variance of the population average treatment effect is simply the weighted average of the two sampling variances:

$$\mathbb{V}\left(\hat{\tau}^{strat}\right) = \left(\frac{N(f)}{N(f) + N(m)}\right)^2 \cdot \mathbb{V}\left(\hat{\tau}^{dif}(m)\right) + \left(\frac{N(m)}{N(f) + N(m)}\right)^2 \cdot \mathbb{V}\left(\hat{\tau}^{dif}(m)\right).$$

Thus, the normalized sampling variance for $\hat{\tau}$ converges to

$$N \cdot \mathbb{V}\left(\hat{\tau}^{\text{strat}}\right) \longrightarrow \sigma^2 \cdot \left(\frac{q}{e(m) \cdot (1 - e(m))} + \frac{1-q}{e(f) \cdot (1 - e(f))} \right) = \mathbb{AV}(\hat{\tau}^{\text{strat}}).$$

Let us now consider the three asymptotic sampling variances, $\mathbb{AV}(\hat{\tau}^{\text{strat}})$, $\mathbb{AV}(\hat{\tau}^{\text{dif}}(f))$, and $\mathbb{AV}(\hat{\tau}^{\text{dif}}(m))$. If $e(f)$ is close to zero or one, it is difficult to estimate $\tau_{\text{sp}}(f)$ precisely. For a given total sample size N, the asymptotic variance increases without limit as $e(f)$ approaches zero or one. The extreme case where $e(f)$ is equal to zero or one implies that neither the estimator nor the sampling variance of the estimator exists in the sense of being finite. If $e(f)$ approaches zero or one, the sampling variance of $\hat{\tau}^{\text{strat}}$ will also increase, unless q is close to one (and consequently there are few $X_i = f$ units). However, given fixed N, the precision with which we can estimate $\tau_{\text{sp}}(m)$ is *not* affected by $e(f)$. Therefore, and this is the key insight, if $e(f)$ is close to zero or one, the researcher may choose to put aside all the women (the $X_i = f$ units) and focus on estimating solely the average effect for men, $\tau_{\text{sp}}(m)$.

Now let us pursue this idea more formally. Consider again the three normalized asymptotic variances $\mathbb{AV}(\hat{\tau}^{\text{strat}})$, $\mathbb{AV}(\hat{\tau}^{\text{dif}}(f))$, and $\mathbb{AV}(\hat{\tau}^{\text{dif}}(m))$. Suppose that

$$\frac{e(m) \cdot (1 - e(m))}{e(f) \cdot (1 - e(f))} \leq \frac{1-q}{1 - 2 \cdot q}. \tag{16.1}$$

Then

$$\mathbb{AV}(\hat{\tau}^{\text{dif}}(f)) \leq \mathbb{AV}(\hat{\tau}^{\text{strat}}) \leq \mathbb{AV}(\hat{\tau}^{\text{dif}}(m)).$$

Hence, under condition (16.1), it is "easier" to estimate $\tau_{\text{sp}}(f)$ than it is to estimate either $\tau(m)$ or τ_{sp}. (Here, "easier" refers to the precision of these estimators.) If, on the other hand,

$$\frac{1+q}{q} \leq \frac{e(m) \cdot (1 - e(m))}{e(f) \cdot (1 - e(f))}, \tag{16.2}$$

then

$$\mathbb{AV}(\hat{\tau}^{\text{dif}}(m)) \leq \mathbb{AV}(\hat{\tau}^{\text{strat}}) \leq \mathbb{AV}(\hat{\tau}^{\text{dif}}(f)),$$

and then $\tau_{\text{sp}}(m)$ is more precisely estimable than either $\tau_{\text{sp}}(f)$ or τ_{sp}. If neither condition (16.1) nor condition (16.2) holds, and thus

$$\frac{1-q}{2-q} \leq \frac{e(m)(1 - e(m))}{e(f)(1 - e(f))} \leq \frac{1+q}{q}, \tag{16.3}$$

then

$$\mathbb{AV}(\hat{\tau}^{\text{strat}}) \leq \min \left(\mathbb{AV}(\hat{\tau}^{\text{dif}}(m)), \mathbb{AV}(\hat{\tau}^{\text{dif}}(f)) \right).$$

The general idea behind the trimming approach in this chapter is based on the estimation of average effects for a subpopulation of units with $X_i \in \mathbb{C}$, or

$$\tau_{\mathbb{C}} = \mathbb{E}[Y_i(1) - Y_i(0)|X_i \in \mathbb{C}],$$

for a subset of the covariate space, $\mathbb{C} \subset \mathbb{X}$. We look for an "optimal" subset $\mathbb{C}^\star$ of the covariate space $\mathbb{X}$ where the average treatment effect is most precisely estimable. In this example with a single binary covariate, and covariate space $\mathbb{X} = \{f, m\}$, the set of possible subsets of $\mathbb{X}$ is $\{\{f, m\}, \{f\}, \{m\}, \emptyset\}$. We choose the subset $\mathbb{C}^\star$ of the covariate space as

$$\mathbb{C}^\star = \begin{cases} \{f\} & \text{if } \dfrac{e(m) \cdot (1 - e(m))}{e(f) \cdot (1 - e(f))} < \dfrac{1 - q}{1 - 2 \cdot q} \\[3mm] \{m\} & \text{if } \dfrac{1 + q}{q} \leq \dfrac{e(m) \cdot (1 - e(m))}{e(f) \cdot (1 - e(f))} \\[3mm] \{f, m\} & \text{otherwise.} \end{cases}$$

We then discard all units with $X_i \notin \mathbb{C}^\star$, and thus focus on estimating

$$\tau_{\mathbb{C}^\star} = \mathbb{E}_{\mathrm{sp}}\left[Y_i(1) - Y_i(0)\big|X_i \in \mathbb{C}^\star\right],$$

based solely on the subsample of units with $X_i \in \mathbb{C}^\star$. In that subsample there are few units with the propensity score close to zero or one, and thus there is, in that sense, substantial overlap for all covariate values in that subsample, making estimators generally more robust to the precise specification of the models used.

Let us make two general points about the trimming approach in the context of this binary example. First, this approach largely ignores external validity, focusing exclusively on internal validity. The binary covariate example reveals what the main issues are. The key is the product of the propensity score and one minus the propensity score, $e(x) \cdot (1 - e(x))$. If the propensity score for units with $X_i = f$ is close to zero or one, we cannot estimate the average treatment effect for this subpopulation precisely. In that case, we may be able to estimate the average treatment effect for the $X_i = m$ subpopulation more accurately than for the population as a whole, even though we might lose a substantial number of observations by discarding units with $X_i = f$. Similarly, if the propensity score for the $X_i = m$ subpopulation is close to zero or one, we may still be able to estimate the average treatment effect for the $X_i = f$ subpopulation more accurately than for the population as a whole. If neither $e(f) \cdot (1 - e(f))$ nor $e(m) \cdot (1 - e(m))$ is close to zero, we can estimate the average effect for the population as a whole more accurately than for either of the two subpopulations.

A second point is that the choice of the subset $\mathbb{C}$, or equivalently, the amount of trimming, is not tied to a specific estimator. Although in this example we compared the asymptotic variance of specific estimators for average treatment effects for a given subset $\mathbb{C}$, in general we will compare asymptotic efficiency bounds (in other words, the asymptotic sampling variance for the "best" estimator in a certain sense) for average treatment effects for different subsets $\mathbb{C}$.

16.4 SELECTING A SUBSAMPLE BASED ON THE PROPENSITY SCORE

Now let us look at the general case, which allows for multi-component and continuous covariates, where we cannot simply list all subsets of the covariate space (i.e., the power set of the covariate space) and compare within-subset sampling variances, because there are infinitely many such subsets. In fact, for a given subset, we cannot even calculate the exact sampling variance the way we did for the binary covariate case. Instead we focus on the *asymptotic* sampling variance for the efficient estimator for the average treatment effect for each subset. Under some regularity conditions (mainly concerning smoothness of the various distributions) and as discussed in Chapter 12, the asymptotic sampling variance for the efficient estimator, ignoring any model-based adjustments, for the finite-sample average treatment effect τ_{fs}, normalized by the sample size, is

$$\mathbb{AV}_{\text{fs}}^{\text{eff}} = \mathbb{E}_{\text{sp}} \left[\frac{\sigma_t^2(X_i)}{e(X_i)} + \frac{\sigma_c^2(X_i)}{1 - e(X_i)} \right]. \tag{16.4}$$

Inspection of this variance bound gives some insight into the problem. If, for a substantial part of the sample, the propensity score is close to zero or one, the sampling variance bound will be relatively large. On the other hand, if the propensity score is far from zero or one for most units, the sampling variance bound will be relatively small. Dropping units for which the propensity score is close to zero or one may, therefore, improve our ability to estimate average treatment effects.

Now suppose we focus on the average treatment effect given that the covariate value X is in some subset $\mathbb{C}$ of the covariate space, $\tau_{\mathbb{C}}$, defined as

$$\tau_{\mathbb{C}} = \mathbb{E}_{\text{sp}} \left[\tau(X_i) | X_i \in \mathbb{C} \right]. \tag{16.5}$$

The asymptotic sampling variance of the efficient estimator for this average treatment effect is, with the original sample size N for the normalization,

$$\mathbb{AV}_{\text{fs}}^{\text{eff}}(\mathbb{C}) = \frac{1}{q(\mathbb{C})} \cdot \mathbb{E}_{\text{sp}} \left[\frac{\sigma_t^2(X_i)}{e(X_i)} + \frac{\sigma_c^2(X_i)}{1 - e(X_i)} \bigg| X \in \mathbb{C} \right], \tag{16.6}$$

where

$$q(\mathbb{C}) = \text{Pr}_{\text{sp}}(X_i \in \mathbb{C}),$$

is the probability of the covariate being in the subset $\mathbb{C}$ in the super-population. If we compare (16.4) and (16.6), there are two competing effects on the asymptotic sampling variance. The first effect is that making the subset $\mathbb{C}$ smaller decreases the effective sample size, as measured by $q(\mathbb{C})$, and thus increases the asymptotic sampling variance. In fact, if the propensity score were constant $e(x) = c$, and the potential outcomes were homoskedastic, $\sigma_t^2(x) = \sigma_c^2(x) = \sigma^2$ for all x, the asymptotic sampling variance would be proportional to $1/q(\mathbb{C})$, that is, proportional to the inverse of the effective sample size. The second effect relies on variation in $e(x)$, $\sigma_c^2(x)$, and $\sigma_t^2(x)$. Choosing $\mathbb{C}$ such that $\sigma_t^2(x)/e(x)$ and $\sigma_c^2(x)/(1 - e(x))$ are relatively small lowers the asymptotic sampling variance. The question now is how to balance these two effects, that is, how to minimize Equation (16.6).

If we assume homoskedasticity, $\mathbb{V}(Y_i(w)|X_i = x) = \sigma^2$, for all w and x, the optimal sampling variance simplifies to

$$\mathrm{AV}_{\mathrm{fs}}^{\mathrm{eff}}(\mathbb{C}) = \frac{\sigma^2}{q(\mathbb{C})} \cdot \mathbb{E}_{\mathrm{sp}}\left[\frac{1}{e(X_i)} + \frac{1}{1 - e(X_i)} \,\bigg|\, X_i \in \mathbb{C} \right]. \tag{16.7}$$

Now we look for the optimal $\mathbb{C}$, denoted by $\mathbb{C}^\star$, that is, the set $\mathbb{C}$ that minimizes the asymptotic sampling variance (16.7) among all subsets $\mathbb{C}$ of $\mathbb{X}$, ignoring possible subsequent model-based adjustments. There are two possibilities. If

$$\sup_{x \in \mathbb{X}} \frac{1}{e(x) \cdot (1 - e(x))} \leq 2 \cdot \mathbb{E}_{\mathrm{sp}}\left[\frac{1}{e(X_i) \cdot (1 - e(X_i))} \right],$$

then the optimal $\mathbb{C}$ is equal to the entire covariate space, $\mathbb{C}^\star = \mathbb{X}$. Otherwise, the optimal set $\mathbb{C}^\star$ has the form

$$\mathbb{C}^\star = \{ x \in \mathbb{X} \,|\, \alpha \leq e(x) \leq 1 - \alpha \},$$

where the threshold α is equal to

$$\alpha = \frac{1}{2} - \sqrt{\frac{1}{4} - \frac{1}{\gamma}},$$

where γ is a solution to

$$\gamma = 2 \cdot \mathbb{E}_{\mathrm{sp}}\left[\frac{1}{e(X_i) \cdot (1 - e(X_i))} \,\bigg|\, \frac{1}{e(X_i) \cdot (1 - e(X_i))} \leq \gamma \right]. \tag{16.8}$$

It is interesting to note that the value of α depends solely on the marginal distribution of the propensity score. In general there will be a unique solution to the equation characterizing γ, (16.8), and we can simply estimate the threshold point, α, for the propensity score to provide guidance about trimming.

To implement this procedure we conduct the following calculations. First we estimate the propensity score using the methods discussed in Chapter 13. Given the estimated propensity score $\hat{e}(x)$, we check whether

$$\max_{i=1,\dots,N} \frac{1}{\hat{e}(X_i) \cdot (1 - \hat{e}(X_i))} \leq 2 \cdot \frac{1}{N} \sum_{i=1}^{N} \frac{1}{\hat{e}(X_i) \cdot (1 - \hat{e}(X_i))}. \tag{16.9}$$

If this inequality holds, then $\hat{\mathbb{C}} = \mathbb{X}$. If the inequality in (16.9) does not hold, then we solve for a value of γ satisfying

$$\frac{\gamma}{N} \sum_{i=1}^{N} \mathbf{1}_{(\hat{e}(X_i)\cdot(1-\hat{e}(X_i)))^{-1} \leq \gamma} = \frac{2}{N} \sum_{i=1}^{N} \frac{1}{\hat{e}(X_i) \cdot (1 - \hat{e}(X_i))} \cdot \mathbf{1}_{(\hat{e}(X_i)\cdot(1-\hat{e}(X_i)))^{-1} \leq \gamma}. \tag{16.10}$$

In general there will not be an exact solution for γ. However, if the inequality does not hold, it is the case that for very large values of γ the left-hand side of (16.10) exceeds the right-hand side. If $\gamma = \min_i (\hat{e}(X_i)(1 - \hat{e}(X_i))^{-1}$, then the left-hand side is smaller than the right-hand side. Hence there will be a largest value of γ such that the left-hand side is smaller than the right-hand side. We focus on this value for γ, denoted by $\hat{\gamma}$. Then we calculate $\hat{\alpha} = 1/2 - \sqrt{1/4 - 1/\hat{\gamma}}$, and finally

$$\hat{\mathbb{C}} = \left\{ x \in \mathbb{X} \,\middle|\, \hat{\alpha} \le \hat{e}(x) \le 1 - \hat{\alpha} \right\}.$$

We exclude units i with $\hat{e}(X_i)$ outside $\hat{\mathbb{C}}$, and focus on balance and estimation of treatment effects for the subset of units with $X_i \in \hat{\mathbb{C}}$.

16.5   THE OPTIMAL SUBSAMPLE FOR THE RIGHT HEART CATHETERIZATION DATA

We start by estimating the propensity score in the full sample. We use the two-stage selection procedure for choosing the pre-treatment variables or covariates that enter linearly and the interactions in the specification of the propensity score discussed in detail in Chapter 13. The thresholds we use for the likelihood ratio statistics are 1 for the inclusion of linear terms and 2.71 for the inclusion of interaction terms. We do not select any of the 72 covariates *a priori* to be included irrespective of their correlation with the treatment indicator because we assume that we have no substantive information beyond the inclusion of the 72 covariates into the set of potentially important covariates. The procedure from Chapter 13 selects 49 covariates out of the collection of 72 for inclusion in the linear part of the propensity score. The second stage leads to the inclusion of 116 interactions of these 49 covariates, out of a total of 1,225 second-order terms, for a total of 165 pre-treatment variables included in the specification of the propensity score.

Before calculating the threshold for the trimming procedure, let us inspect the distribution of the values of the estimated propensity score in the two treatment arms. Table 16.2 displays some summary statistics and some of the extreme values of the propensity score. It is clear that, although there is generally reasonable balance, there are some units without good counterparts in the other treatment group. In fact, for some control units, we estimate the propensity score to be equal to zero, and for some treated units, we estimate the propensity score to be equal to one, so that Inequality (16.9) does not hold. To eliminate systematically units with propensity score values for whom there are no good counterparts, we estimate the threshold value α. Given the estimated propensity score, we find $\hat{\alpha} = 0.0976$. There are 1,336 units with estimated propensity scores less than 0.0976 (mainly control units), and 280 units with estimated propensity scores exceeding $1 - \hat{\alpha} = 0.9024$ (mainly treated units), which leaves 4,119 units in the trimmed sample. Table 16.3 displays the subsample sizes by treatment group and propensity score value. For the trimmed sample with 4,119 units, we re-calculate the summary statistics, including the normalized differences. The results for a few selected covariates are displayed in Table 16.4. We also include the means of the covariates for units with propensity score values less than $\hat{\alpha}$ and propensity score values exceeding $1 - \hat{\alpha}$ to improve our understanding of the part of the sample that is discarded. In Figure 16.2 we present a histogram of the distribution of the absolute values

Table 16.2. *Estimated Propensity Scores for Full Sample, Connors Heart Catheterization Data*

	Controls	Treated
Mean	0.2399	0.6099
0.05 quantile	0.0057	0.1455
0.25 quantile	0.0548	0.4257
0.50 quantile	0.1702	0.6508
0.75 quantile	0.3654	0.8154
0.95 quantile	0.6963	0.9532
Ten smallest values		
1	0.0000	0.0162
2	0.0000	0.0187
3	0.0000	0.0219
4	0.0000	0.0231
5	0.0000	0.0256
6	0.0000	0.0261
7	0.0000	0.0280
8	0.0000	0.0301
9	0.0000	0.0323
10	0.0000	0.0351
Ten largest values		
10	0.9198	0.9981
9	0.9217	0.9991
8	0.9238	0.9991
7	0.9253	0.9996
6	0.9320	1.0000
5	0.9469	1.0000
4	0.9473	1.0000
3	0.9473	1.0000
2	0.9520	1.0000
1	0.9560	1.0000

Table 16.3. *Sample Sizes for Trimming Based on Estimated Propensity Score ($\alpha = 0.0976$), Connors Right Heart Catherization Data*

	$\hat{e}(X_i) < \alpha$	$\hat{\alpha} < \hat{e}(X_i) < 1 - \alpha$	$1 - \alpha < \hat{e}(X_i)$	All
Controls	1,282	2,252	17	3,551
Treated	54	1,867	263	2,184
All	1,336	4,119	280	5,735

Table 16.4. *Summary Statistics for Selected Pre-Treatment Variables for Trimmed Sample, for Connors Right Heart Catherization Data*

| Variable | Controls ($N_c = 2,252$) | | Treated ($N_t = 1,867$) | | Normalized Difference | Discarded (1,616) | |
	Mean	(S.D.)	Mean	(S.D.)		$\hat{e}(X_i) < \alpha$ Mean	$\hat{e}(X_i) > 1 - \alpha$ Mean
cat1_copd	0.05	(0.22)	0.03	(0.16)	−0.13	0.22	0.01
cat2_lung	0.000	(0.000)	0.000	(0.000)	0.000	0.010	0.007
neuro	0.09	(0.29)	0.05	(0.23)	−0.15	0.28	0.03
aps1	54.7	(18.8)	59.1	(19.5)	0.23	44.1	75.2
meanbp1	78.0	(36.5)	69.5	(34.0)	−0.24	97.6	52.3
pafi1	221	(111)	196	(105)	−0.23	278	151

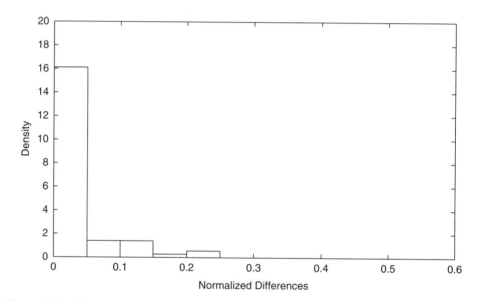

Figure 16.2. Histogram-based estimate of the distribution of the absolute values of the normalized differences for trimmed sample, for Connors RHC data

of the normalized differences. Here one can see that the normalized differences are substantially smaller in the trimmed sample than they are in the full sample. The average and standard deviation of the absolute value of the normalized differences are 0.07 and 0.06, with only 20% exceeding 0.10, and none of the absolute values of the normalized differences exceed 0.25. For comparison, in the full sample the average and standard deviation were 0.14 and 0.11, with 51% of the normalized differences exceeding 0.1, and 15% exceeding 0.25 in absolute value.

One can also see that the discarded units tend to have relatively extreme values for some of the covariates, e.g., pafi1, or meanbp1. As a result, the trimmed sample is more likely to lead to robust and credible estimates for causal estimands. Interestingly, the value of the pre-treatment variable cat2_lung (lung cancer) is zero for all units in the trimmed sample. In the full sample there are fifteen individuals who have this condition (out of the full sample of 5,735). Only two of these fifteen (15%) are in the

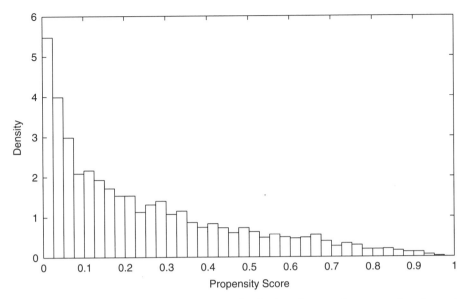

Figure 16.3a.   Histogram-based estimate of the distribution of propensity score values for control units in full sample, for Connors RHC data

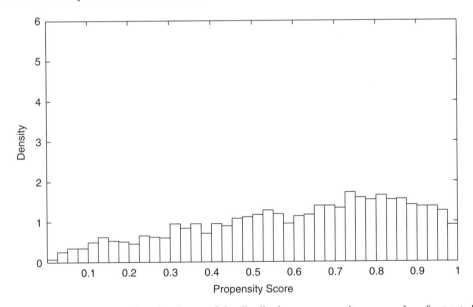

Figure 16.3b.   Histogram-based estimate of the distribution on propensity score values for treated units in full sample, for Connors RHC data

treatment group. Clearly it would be difficult to estimate the effect of the treatment for such individuals, and our automatic trimming procedure eliminates these individuals from the sample.

We re-estimate the propensity score on this trimmed sample, following the same procedure for selecting linear and interaction terms. Figures 16.3a and 16.3b present histogram estimates of the distributions of propensity score values for control and treated units in the full sample. Although in the original sample all units with propensity score

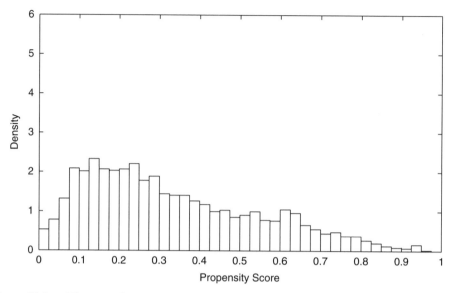

Figure 16.4a. Histogram-based estimate of the distribution of propensity score values for control units in trimmed sample, for Connors RHC data

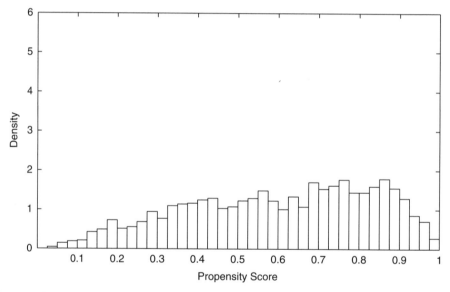

Figure 16.4b. Histogram-based estimate of the distribution on propensity score values for treated units in trimmed sample, for Connors RHC data

values below $\alpha = 0.0976$ or above $1 - \hat{\alpha} = 0.9024$ are dropped, after we re-estimate the propensity score on the trimmed sample, there are a few units with values of the estimated propensity score below 0.0976 and above 0.9024, but the number of such units is relatively small as one can see from Figures 16.4a and 16.4b. One could trim the sample again using the procedures discussed in this chapter if one felt the covariate distributions were not sufficiently balanced. Table 16.5 presents summary statistics for the propensity score values by treatment group for the trimmed sample.

Table 16.5. *Estimated Propensity Scores for Trimmed Sample, Connors Right Heart Catherization Data*

	Controls	Treated
Mean	0.3328	0.5983
0.05 quantile	0.0634	0.1906
0.25 quantile	0.1611	0.4201
0.50 quantile	0.2849	0.6241
0.75 quantile	0.4793	0.7931
0.95 quantile	0.7307	0.9234
Ten smallest values		
1	0.0000	0.0433
2	0.0000	0.0438
3	0.0000	0.0519
4	0.0000	0.0600
5	0.0000	0.0607
6	0.0000	0.0654
7	0.0014	0.0655
8	0.0028	0.0688
9	0.0044	0.0711
10	0.0048	0.0782
Ten largest values		
10	0.9258	0.9895
9	0.9279	0.9905
8	0.9302	0.9905
7	0.9330	0.9911
6	0.9385	0.9970
5	0.9412	0.9976
4	0.9460	0.9983
3	0.9466	0.9990
2	0.9474	1.0000
1	0.9530	1.0000

16.6 CONCLUSION

In this chapter we discuss our second approach to the design phase in an analysis of observational data. In this second approach, we select a subsample of the full sample for which we subsequently attempt to estimate causal effects. We attempt to construct a subsample where the covariate distributions are well balanced, motivated by the fact that lack of balance can make any subsequent analysis both imprecise and sensitive to minor changes in the specifications. The approach in this chapter is to trim the sample by discarding units with propensity score values close to zero or one, with the exact threshold determined by the joint distribution of covariates and treatment status in order to optimize asymptotic precision. The automatic trimming that we propose is simply guidance and need not be followed religiously. One should use scientific judgment when applying these rules to the initial samples and to subsequent trimmed samples with a re-estimated propensity score.

An important aspect of the analysis in this chapter, shared with the matching approach in the previous chapter, is that it is entirely based on the covariate and treatment data, and never uses the outcome data. As such it cannot intentionally introduce systematic biases in the subsequent analyses for causal effects on outcomes.

NOTES

The trimming approach discussed in this chapter is based on Crump, Hotz, Imbens, and Mitnik (2009) where formal arguments for deriving the optimal threshold are provided. Previously researchers appear to have used more *ad hoc* methods for trimming the sample to eliminate units with values for the covariates for whom there were no suitable counterparts with the opposite treatment. Dehejia and Wahba (1999, 2002), for example, drop all control units with a value for the estimated propensity score less than the smallest value for the estimated propensity score among the treated units. Lechner (2008) suggests an alternative three-step procedure to drop units with extreme values for the estimated propensity score.

There are many discussions regarding the relative importance of internal versus external validity. See Shadish, Cook, and Campbell (2002), Imbens (2010), Deaton (2010), and Manski (2013) for recent discussions and Fisher (1935), Cochran (1965), and Rubin (1978) for older arguments.

## PART IV

Regular Assignment Mechanisms: Analysis

## CHAPTER 17

Subclassification on the Propensity Score

17.1   INTRODUCTION

In this chapter we discuss a method for estimating causal effects given a regular assignment mechanism, based on *subclassification* on the estimated propensity score. We also refer to this method as *blocking* or *stratification*.

Given the assumptions of individualistic assignment and unconfoundedness, the definition of the propensity score in Chapter 3 implies that the super-population propensity score equals the conditional probability of receiving the treatment given the observed covariates. As shown in Chapter 12, the propensity score is a member of a class of functions of the covariates, collectively called *balancing scores*, that share an important property: within subpopulations with the same value of a balancing score, the super-population distribution of the covariates is identical in the treated and control subpopulations. This, in turn, was shown to imply that, under the assumption of super-population unconfoundedness, systematic biases in comparisons of outcomes for treated and control units associated with observed covariates can be eliminated entirely by adjusting solely for differences between treated and control units on a balancing score. The practical relevance of this result stems from the fact that a balancing score may be of lower dimension than the original covariates. (By definition, the covariates themselves form a balancing score, but one that has no dimension reduction.) When a balancing score is of lower dimension than the full set of covariates, adjustments for differences in this balancing score may be easier to implement than adjusting for differences in all covariates, because it avoids high-dimensional considerations. Within the class of balancing scores, the propensity score, as well as strictly monotonic transformations of it (such as the linearized propensity score or log odds ratio), have a special place. All balancing scores $b(x)$ satisfy the property that if for two covariate values x and x', $b(x) = b(x')$, then it must be the case that $e(x) = e(x')$.

In this chapter we examine a leading approach to estimating causal effects that relies on blocking, subclassification, or stratification on the estimated propensity score. The sample is partitioned into subclasses (also referred to as strata or blocks), based on the values of the estimated propensity scores, so that within the subclasses, the estimated propensity scores are approximately constant. We then can estimate causal effects within each subclass as if assignment was completely at random within each subclass, using either the Neyman-based methods for completely randomized experiments from Chapter 6, or the regression and model-based methods from Chapters 7 and 8. To estimate, for example, the overall average treatment effect, we could average the within-subclass estimated treatment effects, weighted by the subclass sizes. We can estimate other estimands, as discussed in more detail in Chapter 21, using, for example, the model-based methods from Chapter 8. Two important practical issues arise in the implementation of subclassification. First, the choice of the number of subclasses or blocks, and, second, the choice of boundary values for the blocks.

As just mentioned, we can combine subclassification with further adjustments for covariates, and in fact we generally recommend doing so. Such further adjustments have two objectives. First, because blocking typically does not eliminate all biases associated with differences in the covariates (because the estimated propensity score is typically not constant within the blocks), regression or model-based adjustments can further reduce bias of estimates. Second, these adjustments can improve the precision of estimators for causal effects even if the estimated propensity scores were constant within the blocks, similar to the way adjusting for covariates can improve efficiency even in completely randomized experiments. There is an important difference, though, between the covariance adjustment in this setting, within blocks defined by a balancing score, and its use in the full sample in observational studies. In the latter case there is generally concern that the implicit imputations of the missing potential outcomes through model-based methods rely, possibly heavily, on extrapolation. Here, by the construction of the strata, the differences in covariate distributions within each stratum are small, the extrapolation in the estimators is therefore more limited, and, as a result, the estimators are more robust to violations of the assumptions in model-based approaches, such as non-linearities in the conditional expectations, than these estimators would be without the stratification.

In the next section we return to the Imbens-Rubin-Sacerdote lottery data, previously used in Chapter 14, which is also used here to illustrate the concepts discussed in this chapter. After that, we return to theoretical issues. In Section 17.3 we discuss the construction of subclasses and the bias reduction properties of these methods. In Section 17.4 we implement subclassification methods with the lottery data. In Sections 17.5 and 17.6, we develop simple estimators for causal effects based on subclassification. These methods are then implemented on the lottery data in Section 17.7. In Section 17.8 we discuss the relation to Horvitz-Thompson style weighting methods. We conclude in Section 17.9.

17.2 THE IMBENS-RUBIN-SACERDOTE LOTTERY DATA

In this chapter we use the lottery data set originally collected by Imbens, Rubin, and Sacerdote (2001) that we used as one of the illustrations in Chapter 14. In Chapter 14 we assessed the overlap in covariate distributions for the lottery data and found that overlap was substantial, although there were subsets of covariate values with little overlap. The second column in Table 17.1 presents the normalized differences for the full sample. Note that the normalized difference for the covariate # Tickets (number of tickets bought in a typical week) is 0.64, suggesting that simple linear regression may not be adequate to remove reliably biases associated with differences in this covariate. To address these concerns with overlap in covariate distributions, we apply the methods discussed in Chapter 16 designed to improve the overlap by discarding units with values

Table 17.1. *Normalized Differences in Covariates after Subclassification for the IRS Lottery Data*

Variable	Full Sample		Trimmed Sample			
	One Block	Horvitz-Thompson	One Block	Two Blocks	Five Blocks	Horvitz-Thompson
Year Won	−0.26	0.10	−0.06	−0.03	0.07	0.07
# Tickets	0.91	0.10	0.51	0.17	0.07	−0.04
Age	−0.50	−0.30	−0.09	−0.03	0.05	0.05
Male	−0.19	0.09	−0.11	−0.10	−0.14	−0.13
Education	−0.70	0.48	−0.51	−0.18	−0.10	−0.01
Work Then	0.09	0.05	0.03	0.03	0.01	0.00
Earn Year -6	−0.32	0.01	−0.18	−0.10	−0.03	0.06
Earn Year -5	−0.28	0.01	−0.19	−0.07	−0.00	0.09
Earn Year -4	−0.29	−0.01	−0.23	−0.09	−0.01	0.06
Earn Year -3	−0.26	0.05	−0.18	−0.03	0.03	0.10
Earn Year -2	−0.31	0.06	−0.19	−0.03	0.01	0.09
Earn Year -1	−0.23	0.11	−0.17	−0.01	0.00	0.06
Pos Earn Year -6	0.03	0.16	−0.00	−0.09	−0.09	−0.01
Pos Earn Year -5	0.14	−0.14	0.10	0.01	−0.01	0.06
Pos Earn Year -4	0.10	−0.19	0.06	−0.00	−0.01	0.03
Pos Earn Year -3	0.13	−0.17	0.03	−0.04	−0.05	−0.00
Pos Earn Year -2	0.14	−0.17	0.06	0.00	−0.04	0.01
Pos Earn Year -1	0.10	0.17	−0.01	−0.04	−0.07	−0.01

Table 17.2. *Number of Units within Selected Subsamples Defined by the Estimated Propensity Score for the IRS Lottery Data*

	Low	Middle	High	All
	$\hat{e}(X_i) < 0.0891$	$0.0891 \leq \hat{e}(X_i) \leq 0.9109$	$0.9109 < \hat{e}(X_i)$	
Losers	82	172	5	259
Winners	4	151	82	237
All	86	323	87	496

of their estimated propensity scores close to zero or one. Following the specific recommendations from that chapter suggests dropping units with estimated propensity scores outside the interval $[0.0891, 0.9009]$. Table 17.2 presents the subsample sizes in the various propensity score strata. Out of the 496 units in the full sample, 259 losers and 237 winners, there are $N = 323$ with estimated propensity scores in the interval $[0.0891, 0.9009]$, of whom $N_c = 172$ are losers and $N_t = 151$ are winners. There are eighty-six units discarded because of small estimated propensity score values (less than 0.0891), eighty-two losers and four winners, and eighty-seven units discarded because of large estimated propensity score values (larger than 0.9009), five losers and eighty-two winners. This trimmed sample with 323 units is the sample we focus on in this chapter.

The fourth column in Table 17.1 presents the normalized differences for the trimmed sample. To facilitate the comparison with the normalized differences in the full sample

Table 17.3. *Estimates of Propensity Score in Trimmed Sample for the IRS Lottery Data*

Covariate	Est	$(\widehat{s.e.})$	t-Stat
Intercept	21.77	(0.13)	164.8
Linear terms			
# Tickets	−0.08	(0.46)	−0.2
Education	−0.45	(0.08)	−5.7
Working Then	3.32	(1.95)	1.7
Earnings Year -1	−0.02	(0.01)	−1.4
Age	−0.05	(0.01)	−3.7
Pos Earnings Year -5	1.27	(0.42)	3.0
Year Won	−4.84	(1.53)	−3.2
Earnings Year -5	−0.04	(0.02)	−2.1
Quadratic terms			
Year Won × Year Won	0.37	(0.12)	3.2
Tickets Bought × Year Won	0.14	(0.06)	2.2
Tickets Bought × Tickets Bought	−0.04	(0.02)	−1.8
Working Then × Year Won	−0.49	(0.30)	−1.6

presented in the second column, we normalize the difference in average covariate values in both columns by the square root of the average of the sample variances in the full sample. The results in the table show that trimming substantially improves the covariate balance. For example, the normalized difference for the Year Won pre-treatment variable decreases from −0.26 in the full sample to −0.06 in the trimmed sample.

On this trimmed sample, we re-estimate the propensity score using the algorithm discussed in Chapter 13 for selecting linear and second-order terms. Starting with the four variables selected for automatic inclusion, # Tickets, Education, Working Then, and Earnings Year -1, the algorithm selects four additional linear terms, Age, Pos Earnings Year -5, Year Won, and Earnings Year -5. In addition the application of the algorithm selects four second-order terms, Year Won × Year Won, Tickets Bought × Year Won, Tickets Bought × Tickets Bought, and Working Then × Year Won. Table 17.3 presents the parameter estimates for the logistic specification chosen. This is the estimated propensity score that we use for the purpose of subclassification. Note that when we used the same algorithm on the full sample, we included more terms, eight linear terms and ten second-order terms (see Table 14.3 in Chapter 14); the substantially improved covariate balance after trimming leads to this algorithm selecting fewer terms for the specification of the propensity score.

17.3 SUBCLASSIFICATION ON THE PROPENSITY SCORE AND BIAS REDUCTION

In Chapter 12 we showed that, if the assignment mechanism is regular, to eliminate biases in comparisons between treated and control units associated with covariates, it is sufficient to adjust for differences in the true propensity score, or, in fact, for differences in any balancing score. Here we classify or stratify units by a coarsened version of the estimated propensity score, similar to the way we used propensity score strata in Chapter 13 to evaluate the specification of the model for the propensity score. Note that the construction of strata based directly on the full set of covariates would be infeasible with a large number of covariates, because the number of subclasses that would be required to make the variation in eleven covariates within subclasses modest would generally be very large. For example, with the eighteen covariates in the lottery example, even if we defined subclasses in terms of just two (ranges of) values of each of the covariates, this would lead to an infeasibly large number of subclasses, namely $2^{18} = 262{,}144$, substantially larger than the original sample size of 496 (or 323 in the trimmed sample).

17.3.1 Subclassification

Following the discussion in Chapter 13, let us partition the range of the propensity score into J blocks, that is, intervals of the type $[b_{j-1}, b_j)$, where $b_0 = 0$ and $b_J = 1$ so that $\cup_{j=1}^{J}[b_{j-1}, b_j) = [0, 1)$. We intend to analyze the data as if they arose from a stratified randomized experiment. Initially this means that we analyze units with propensity scores within an interval $[b_{j-1}, b_j)$ *as if* they have identical propensity scores. For large J, and choices for the boundary values of the intervals so that $\max_{j=1,\ldots,J} |b_j - b_{j-1}|$ is at least moderately small, this may be a reasonable approximation.

Recall the notation from Chapter 13: for $i = 1, \ldots, N$, and for $j = 1, \ldots, J$, the binary stratum indicators $B_i(j)$ are

$$B_i(j) = \begin{cases} 1 & \text{if } b_{j-1} \leq \hat{e}(X_i) < b_j, \\ 0 & \text{otherwise.} \end{cases}$$

(Here we ignore the possibility that there are units with $\hat{e}(X_i)$ exactly equal to 1, in which case we would have to modify the definition for the last stratum.) To keep the notation consistent with the interpretation of the blocks as covariates, let the number of units of each treatment type in each strata be denoted by

$$N_c(j) = \sum_{i=1}^{N}(1 - W_i) \cdot B_i(j), \quad N_t(j) = \sum_{i=1}^{N} W_i \cdot B_i(j), \quad N(j) = N_c(j) + N_t(j),$$

for $j = 1, \ldots, J$. Let $q(j)$ be the fraction of units in stratum j:

$$q(j) = \frac{N(j)}{N}, \quad \text{for } j = 1, \ldots, J.$$

We implement the selection of boundary points using the iterative procedure introduced in Chapter 13. We start with a single block: $J = 1$, with boundaries equal to $b_0 = 0$ and $b_j = b_1 = 1$. We then cycle through the following two steps. In the first step we assess the adequacy of the current number of blocks. This assessment involves calculating, for each stratum, a t-statistic for the null hypothesis that the average value of the estimated linearized propensity score is the same for treated and control units in that stratum. The specific t-statistic used is

$$t_\ell(j) = \frac{\overline{\ell}_t(j) - \overline{\ell}_c(j)}{\sqrt{s_\ell^2(j) \cdot (1/N_c(j) + 1/N_t(j))}},$$

where

$$\overline{\ell}_c(j) = \frac{1}{N_c(j)} \sum_{i=1}^{N} (1 - W_i) \cdot B_i(j) \cdot \hat{\ell}(X_i), \qquad \overline{\ell}_t(j) = \frac{1}{N_t(j)} \sum_{i=1}^{N} W_i \cdot B_i(j) \cdot \hat{\ell}(X_i),$$

and

$$s_\ell^2(j) = \frac{1}{N_t(j) + N_c(j) - 2}$$

$$\times \left(\sum_{i=1}^{N} (1 - W_i) \cdot B_i(j) \cdot \left(\hat{\ell}(X_i) - \overline{\ell}_c(j) \right)^2 + \sum_{i=1}^{N} W_i \cdot B_i(j) \cdot \left(\hat{\ell}(X_i) - \overline{\ell}_t(j) \right)^2 \right).$$

In addition we find, within each of the current strata, the number of treated and control units left in each substratum after a subsequent split, at the median value of the estimated propensity score. Specifically, we check whether the number of controls and treated, $N_c(j)$ and $N_t(j)$, and the total number of units, $N(j)$, in each new stratum, would be greater than some minimum. If at least one of the strata is not adequately balanced, and if splitting that stratum would lead to two new strata each with a sufficient number of units, that stratum is split and the new strata are assessed for adequacy. In order to implement this algorithm, we need to specify three parameters: the maximum acceptable t-statistic (t_{max}); the minimum number of treated or control units in a stratum, $\min(N_c(j), N_t(j)) \geq N_{min,1}$; and the minimum number of units in a new stratum, $N(j) \geq N_{min,2}$. Here we choose $t_{max} = 1.96$, $N_{min,1} = 3$, and $N_{min,2} = K + 2$, where K is the number of components of the covariate vector X_i for which we want to apply further adjustments. The latter choice is motivated by the fact that we may wish to do additional modeling of potential outcome distributions, conditional on covariates, within the strata.

17.3.2 The Subclassification Estimator for the Average Treatment Effect

The first estimator for the average causal effect we consider is the simple blocking estimator. Within block j we estimate the block-specific average effect of the treatment as

$$\hat{\tau}^{dif}(j) = \overline{Y}_t^{obs}(j) - \overline{Y}_c^{obs}(j),$$

where

$$\overline{Y}_t^{obs}(j) = \frac{1}{N_t(j)} \sum_{i=1}^{N} W_i \cdot B_i(j) \cdot Y_i^{obs} \quad \text{and} \quad \overline{Y}_c^{obs}(j) = \frac{1}{N_c(j)} \sum_{i=1}^{N} (1 - W_i) \cdot B_i(j) \cdot Y_i^{obs}.$$

We then estimate the overall average treatment effect by averaging these estimates over the blocks, weighted by the relative block sizes:

$$\hat{\tau}^{\text{strat}} = \sum_{j=1}^{J} q(j) \cdot \hat{\tau}^{\text{dif}}(j).$$

Later we will modify this estimator by introducing additional adjustments based on some of the covariates, but first we explore some of the properties of this simple subclassification estimator.

17.3.3 Subclassification and Bias Reduction

To gain insights into the properties of estimators based on subclassification, we investigate here some implications for bias reduction. In this discussion we build on the theoretical analysis of the bias-reducing properties of matching presented in Chapter 15. We initially assume, but do not necessarily believe, that, in the super-population, the conditional expectations of the two potential outcomes, conditional on the covariates, are linear in the covariates, with identical slope coefficients under both treatment conditions:

$$\mathbb{E}_{\text{sp}}\left[Y_i(w)\,|\,X_i = x\right] = \alpha + \tau_{\text{sp}} \cdot w + \beta' x,$$

for $w = 0, 1$. As in most of Part III of the text, the expectation here is taken over the distribution induced by random sampling from an infinite super-population. As before, we do not believe this linearity assumption is necessarily a good approximation (in fact, if the assumption were true, one could simply remove all biases associated with the covariates by simple covariance adjustment), but linearity provides a useful approximation to assess the bias-reducing properties of subclassification.

Now consider estimating the average effect of the treatment on the full sample. Let $\overline{X}_c$, $\overline{X}_t$, and $\overline{X}$ be the average values of the covariates in the control, treated, and full samples respectively,

$$\overline{X}_c = \frac{1}{N_c} \sum_{i:W_i=0} X_i, \quad \overline{X}_t = \frac{1}{N_t} \sum_{i:W_i=1} X_i, \quad \text{and} \quad \overline{X} = \frac{1}{N} \sum_{i=1}^{N} X_i = \frac{N_c}{N} \cdot \overline{X}_c + \frac{N_t}{N} \cdot \overline{X}_t.$$

In addition, let $\overline{X}_c(j)$, $\overline{X}_t(j)$, and $\overline{X}(j)$ denote the analogous covariate averages within stratum j,

$$\overline{X}_c(j) = \frac{1}{N_c(j)} \sum_{i=1}^{N} (1 - W_i) \cdot B_i(j) \cdot X_i, \quad \overline{X}_t(j) = \frac{1}{N_t(j)} \sum_{i=1}^{N} W_i \cdot B_i(j) \cdot X_i,$$

and

$$\overline{X}(j) = \frac{1}{N(j)} \sum_{i=1}^{N} B_i(j) \cdot X_i,$$

for $j = 1, \ldots, J$. First we consider the estimator with no adjustment for differences in the covariates at all, where we simply estimate the average treatment in the full sample, without subclassification, by differencing the average outcomes for treated and control units. Alternatively, this can be viewed as the subclassification estimator with only a single stratum. We find

$$\hat{\tau}^{\text{dif}} = \overline{Y}_{\text{t}}^{\text{obs}} - \overline{Y}_{\text{c}}^{\text{obs}} = \frac{1}{N_{\text{t}}} \sum_{i:W_i=1} Y_i^{\text{obs}} - \frac{1}{N_{\text{c}}} \sum_{i:W_i=0} Y_i^{\text{obs}}.$$

The bias of $\hat{\tau}^{\text{dif}}$, conditional on the covariates,

$$\mathbb{E}_{\text{sp}}\left[\hat{\tau}^{\text{dif}}\,\middle|\,\mathbf{X}\right] - \frac{1}{N} \sum_{i=1}^{N} \mathbb{E}_{\text{sp}}[Y_i(1) - Y_i(0)|X_i],$$

arises from two sources. First, we estimate the average treatment potential outcomes for the treatment for the N_{c} control units, in expectation equal to $\mathbb{E}[Y_i(1)|W_i = 0]$ by $\overline{Y}_{\text{t}}^{\text{obs}}$; this estimator is equal to the average outcome for the N_{t} treated units, which, in expectation, equals $\mathbb{E}[Y_i(1)|W_i = 1]$. The second source of bias of $\hat{\tau}^{\text{dif}}$ arises from the difference between the expected control potential outcome for the N_{t} treated units, $\mathbb{E}[Y_i(0)|W_i = 1]$, and the expected value of its estimator, $\overline{Y}_{\text{c}}^{\text{obs}}$, which equals the expectation of the control outcomes for the control units, $\mathbb{E}[Y_i(0)|W_i = 0]$. Hence the conditional bias of $\hat{\tau}^{\text{dif}}$ is, under the linear model specification for the regression function, equal to:

$$
\begin{aligned}
\mathbb{E}\left[\hat{\tau}^{\text{dif}} - \tau_{\text{fs}}\,\middle|\,\mathbf{X},\mathbf{W}\right] &= \frac{N_{\text{c}}}{N} \cdot (\mathbb{E}[Y_i(1)|W_i = 1, X_i] - \mathbb{E}[Y_i(1)|W_i = 0, X_i]) \\
&\quad - \frac{N_{\text{t}}}{N} \cdot (\mathbb{E}[Y_i(0)|W_i = 1, X_i] - \mathbb{E}[Y_i(0)|W_i = 0, X_i]) \\
&= \frac{N_{\text{c}}}{N} \cdot (\overline{X}_{\text{t}} - \overline{X}_{\text{c}})\,\beta - \frac{N_{\text{t}}}{N} \cdot (\overline{X}_{\text{c}} - \overline{X}_{\text{t}})\,\beta \\
&= (\overline{X}_{\text{t}} - \overline{X}_{\text{c}})\,\beta.
\end{aligned}
$$

Now consider estimating the average treatment τ_{fs} by the subclassification estimator $\hat{\tau}^{\text{strat}}$ with J strata, with no further covariance adjustment within the strata (i.e., subclasses). In stratum j the bias is, using the same argument as for the overall bias,

$$\mathbb{E}\left[\hat{\tau}^{\text{dif}}(j) - \tau_{\text{fs}}(j)\,\middle|\,\mathbf{X},\mathbf{W}\right] = (\overline{X}_{\text{t}}(j) - \overline{X}_{\text{c}}(j))\,\beta.$$

The overall bias for the subclassification estimator is the weighted average of the within-block biases,

$$\mathbb{E}\left[\hat{\tau}^{\text{strat}} - \tau_{\text{fs}}|\mathbf{X},\mathbf{W}\right] = \left(\sum_{j=1}^{J} q(j) \cdot (\overline{X}_{\text{t}}(j) - \overline{X}_{\text{c}}(j))\right)\beta.$$

As a result of the subclassification, the bias that can be attributed to differences in $X_{i,k}$, the kth element of the covariate vector X_i, is reduced, under our simple linear model, from

$$\left(\overline{X}_{\text{t},k} - \overline{X}_{\text{c},k}\right) \cdot \beta_k \quad \text{to} \quad \left(\sum_{j=1}^{J} q(j) \cdot \left(\overline{X}_{\text{t},k}(j) - \overline{X}_{\text{c},k}(j)\right)\right) \cdot \beta_k,$$

where $\overline{X}_{\text{c},k}(j)$ and $\overline{X}_{\text{t},k}(j)$ are the kth elements of $\overline{X}_\text{c}(j)$ and $\overline{X}_\text{t}(j)$ respectively. Thus, the bias attributable to the kth covariate is reduced by a factor

$$\gamma_k = \sum_{j=1}^{J} q(j) \cdot \left(\overline{X}_{\text{t},k}(j) - \overline{X}_{\text{c},k}(j)\right) \Big/ \left(\overline{X}_{\text{t},k} - \overline{X}_{\text{c},k}\right). \tag{17.1}$$

We can calculate these ratios γ_k for any particular subclassification, for each covariate, to assess the bias reduction from the subclassification in a particular application.

17.4 SUBCLASSIFICATION AND THE LOTTERY DATA

Here we return to the lottery data and determine the number of subclasses (or strata) according to the algorithm described in Section 17.3. We use the cutoff values $t_{\max} = 1.96$, and $N_{\min,1} = 3$, and $N_{\min,2} = K + 2$, where K, the number of covariates possibly used for model-based adjustments, is here 18, so that $N_{\min,2} = 20$. These choices for the tuning parameters lead to five blocks. The details for the five blocks, including the cutoff values for the propensity score, the number of units by treatment status in each block, and the t-statistics for the null hypothesis of a zero difference in average propensity scores between treated and control units in the block, are presented in Table 17.4. For example, the first stratum contains 67 control and 13 treated units, with the propensity scores ranging from 0.03 to 0.24. The t-statistic for the null hypothesis of no difference in average linearized propensity score values between the two treatment groups within this stratum is -0.1, so there is actually very little difference in average linearized propensity scores between the two groups within the first block. For comparison purposes Table 17.5 presents results based on only two blocks, where the blocks' boundary is the median value of the propensity score, 0.44. Here the treatment and control groups are substantially less balanced.

Next, we investigate for these two specifications of the blocks the extent of the bias reduction based on a simple linear specification of the regression function. Columns two and four of Table 17.1 present, for both the full and trimmed samples, the average difference in covariates, $\overline{X}_{\text{t},k} - \overline{X}_{\text{c},k}$, normalized by the square root of the average of the sample variances for treated and controls, $\sqrt{(s_{\text{c},k}^2 + s_{\text{t},k}^2)/2}$ (with the latter calculated on the full sample for the second column and on the selected sample for the fourth column). For the trimmed sample, based on the subclassifications with two or five subclasses, we also present, in Columns five and six,

$$\hat{\Delta}_{\text{ct}} = \sum_{j=1}^{J} q(j) \cdot \frac{\overline{X}_{\text{t},k}(j) - \overline{X}_{\text{c},k}(j)}{\sqrt{\left(s_{\text{c},k}^2 + s_{\text{t},k}^2\right)/2}}$$

Table 17.4. *Final Subclassification for the IRS Lottery Data*

Subclass	Min P-Score	Max P-Score	# Controls	# Treated	t-Stat
1	0.03	0.24	67	13	−0.1
2	0.24	0.32	32	8	0.9
3	0.32	0.44	24	17	1.7
4	0.44	0.69	34	47	2.0
5	0.69	0.99	15	66	1.6

Table 17.5. *Subclassification with Two Subclasses, Split at Median Propensity Score for the IRS Lottery Data*

Subclass	Min P-Score	Max P-Score	# Controls	# Treated	t-Stat
1	0.03	0.44	123	38	2.8
2	0.44	0.99	49	113	3.8

(normalized by the same function of the standard deviations in the trimmed sample so that the normalized differences are directly comparable to those in Column four). The ratios of the fifth and sixth columns to the fourth column show how much the subclassifications reduce the bias arising from linear effects of the covariates in the trimmed sample, that is, the γ_k in Equation (17.1). We see that the covariates exhibiting substantial differences between the treated and control groups in the full sample show much smaller differences after trimming, and even smaller differences after subsequent subclassifications. For example, consider the covariate # tickets. In the full sample there is a normalized difference of 0.91, whereas trimming the sample reduces that to 0.51. Subclassification with only two blocks reduces that further to 0.17, and five subclasses reduces this to 0.07, or about 6% of the original 0.91. For the covariate with the second biggest normalized difference in the full sample, education, which exhibited a normalized difference of 70% in the full sample, we similarly get a reduction to about 7% in the trimmed sample. For covariates with small initial differences, the reduction is not as dramatic, but with five subclasses, the largest of the normalized differences is 0.14 (for male). Subclassification has clearly been effective in removing most of the mean differences for all eighteen covariates in this data set.

17.5 ESTIMATION BASED ON SUBCLASSIFICATION WITH ADDITIONAL BIAS REDUCTION

The simple estimator for the average treatment effect based on subclassification is

$$\hat{\tau}^{\text{strat}} = \sum_{j=1}^{J} q(j) \cdot \hat{\tau}^{\text{dif}}(j),$$

where $\hat{\tau}^{\text{dif}}(j) = \overline{Y}_{\text{t}}^{\text{obs}}(j) - \overline{Y}_{\text{c}}^{\text{obs}}(j)$. This simple estimator is not necessarily very attractive. Even when the propensity score is known, the differences $\overline{Y}_{\text{t}}^{\text{obs}}(j) - \overline{Y}_{\text{c}}^{\text{obs}}(j)$ will likely be biased for the average treatment effects within the blocks because the propensity score is only approximately constant within the blocks. We therefore may wish to attempt to reduce further any remaining bias by modifying the basic estimator. Two leading alternatives are to use regression (covariance) adjustment or model-based imputation within the blocks, which raises an important issue regarding the choice of blocks. With many blocks, typically some will contain relatively few units, and so it may be difficult to estimate even simple linear regression functions precisely within each block. Therefore, if one intends to combine subclassification with regression or model-based adjustment, one may wish to ensure a relatively large number of units in each stratum, or appropriately smooth models across blocks, or both.

Here we further discuss the least squares regression approach. It is useful to start by re-interpreting the within-block difference in average treatment and control outcomes $\hat{\tau}^{\text{dif}}(j)$ as the least squares estimator of the average causal effect in stratum j, $\tau(j)$, using the regression function

$$Y_i^{\text{obs}} = \alpha(j) + \tau(j) \cdot W_i + \varepsilon_i. \tag{17.2}$$

We estimate the parameters of this regression function using only the $N(j)$ observations in the jth stratum (i.e., the jth block). We can then generalize this estimator to allow for covariates by specifying within block j the regression function

$$Y_i^{\text{obs}} = \alpha(j) + \tau(j) \cdot W_i + X_i \beta(j) + \varepsilon_i, \tag{17.3}$$

again using only the $N(j)$ observations in block j. If the balancing on the estimated propensity score created perfect expected balance on the true propensity score, the population correlation between the covariates and the treatment indicator within a block would be zero. In that case the inclusion of the covariates in this regression is intended to improve precision (actual and estimated), the same way using covariates in the analysis of a completely randomized experiment can improve precision – even though on average the estimator based on (17.2) would be the same as the estimator based on (17.3). When using an estimated propensity score, however, does not eliminate all correlations within blocks between treatment indicator and covariates, the role of the regression adjustment in (17.3) is threefold. In addition to improving actual and estimated precision, it also can help to reduce any remaining conditional bias arising from imbalances in covariate distributions between treated and controls within the blocks. It is important to note that conceptually the use of regression adjustment is quite different here from using regression methods on the full sample. Within each block there is less concern about using the regression function to extrapolate out of sample, because the blocking has already ensured that the covariate distributions within blocks are similar. In practice the use of regression methods at this stage is more like its use in randomized experiments where the similarity of the covariate distributions greatly reduces the sensitivity to the specification of the regression function.

Mechanically the analysis now estimates the average treatment effects within the blocks using linear regression:

$$\left(\hat{\alpha}(j), \hat{\tau}^{\text{adj}}(j), \hat{\beta}(j)\right) = \arg\min_{\alpha, \tau, \beta} \sum_{i=1}^{N} B_i(j) \cdot \left(Y_i^{\text{obs}} - \alpha - \tau \cdot W_i - X_i\beta\right)^2, \qquad (17.4)$$

based on the $N(j)$ units within stratum j. Within each block, the procedure is the same as that for analyzing completely randomized experiments with regression adjustment discussed in Chapter 6. These within-block least squares estimates, $\hat{\tau}^{\text{adj}}(j)$, are then averaged to obtain an estimator for the overall average treatment effect,

$$\hat{\tau}^{\text{strat,adj}} = \sum_{j=1}^{J} q(j) \cdot \hat{\tau}^{\text{adj}}(j),$$

with the stratum weights still equal to the stratum shares $q(j) = N(j)/N$.

17.6   NEYMANIAN INFERENCE

For the simple subclassification estimator with no further covariance adjustment, we can directly apply the Neyman analysis for completely randomized experiments. Using the results from Chapter 9 on Neyman's repeated sampling perspective, applied in the context of stratified randomized experiments, the sampling variance of $\hat{\tau}^{\text{dif}}(j)$ is

$$\mathbb{V}\left(\hat{\tau}^{\text{dif}}(j)\right) = \frac{S_c^2(j)}{N_c(j)} + \frac{S_t^2(j)}{N_t(j)} - \frac{S_{ct}^2(j)}{N(j)},$$

where,

$$S_c^2(j) = \frac{1}{N(j) - 1} \sum_{i=1}^{N} B_i(j) \cdot \left(Y_i(0) - \overline{Y}(0,j)\right)^2,$$

$$S_t^2(j) = \frac{1}{N(j) - 1} \sum_{i=1}^{N} B_i(j) \cdot \left(Y_i(1) - \overline{Y}(1,j)\right)^2,$$

$$S_{ct}^2(j) = \frac{1}{N - 1} \sum_{i=1}^{N} B_i(j) \cdot (Y_i(1) - Y_i(0) - \tau(j))^2,$$

and

$$\overline{Y}(w,j) = \frac{1}{N(j)} \sum_{i=1}^{N} B_i(j) \cdot Y_i(w).$$

To obtain a statistically conservative estimate of the sampling variance $\mathbb{V}(\hat{\tau}^{\text{dif}}(j))$, we substitute

$$s_{\text{c}}^2(j) = \frac{1}{N_{\text{c}}(j) - 1} \sum_{i=1}^{N} (1 - W_i) \cdot B_i(j) \cdot \left(Y_i^{\text{obs}} - \overline{Y}_{\text{c}}^{\text{obs}}(j) \right)^2,$$

and

$$s_{\text{t}}^2(j) = \frac{1}{N_{\text{t}}(j) - 1} \sum_{i=1}^{N} W_i \cdot B_i(j) \cdot \left(Y_i^{\text{obs}} - \overline{Y}_{\text{t}}^{\text{obs}}(j) \right)^2,$$

for $S_{\text{c}}^2(j)$ and $S_{\text{t}}^2(j)$ respectively, and $s_{\text{ct}}^2(j) = 0$ for $S_{\text{ct}}^2(j)$ to obtain the following estimator,

$$\hat{\mathbb{V}}\left(\hat{\tau}^{\text{dif}}(j)\right) = \frac{1}{N_{\text{c}}(j) \cdot (N_{\text{c}}(j) - 1)} \sum_{i:W_i=0} B_i(j) \cdot \left(Y_i^{\text{obs}} - \overline{Y}_{\text{c}}^{\text{obs}}(j) \right)^2$$
$$+ \frac{1}{N_{\text{t}}(j)(N_{\text{t}}(j) - 1)} \sum_{i:W_i=1} B_i(j) \cdot \left(Y_i^{\text{obs}} - \overline{Y}_{\text{t}}^{\text{obs}}(j) \right)^2.$$

Because, conditional on **X**, the within-stratum estimator $\hat{\tau}^{\text{dif}}(j)$ is independent of $\hat{\tau}^{\text{dif}}(j')$ when $j \neq j'$, we can estimate the sampling variance of $\hat{\tau}^{\text{strat}} = \sum_{j=1}^{J} q(j) \cdot \hat{\tau}(j)$ by adding the within-block estimated sampling variances, multiplied by the square of the block proportions:

$$\hat{\mathbb{V}}(\hat{\tau}^{\text{strat}}) = \sum_{j=1}^{J} \hat{\mathbb{V}}(\hat{\tau}^{\text{dif}}(j)) \cdot q(j)^2 = \sum_{j=1}^{J} \hat{\mathbb{V}}(\hat{\tau}^{\text{dif}}(j)) \cdot \left(\frac{N(j)}{N}\right)^2.$$

In practice, however, we typically do further covariance adjustment to reduce the remaining bias. Here we focus on the specific estimator discussed in the previous subsection, where we use linear regression within the blocks, with identical slopes in the treatment and control subsamples, because of possibly small block sizes $N_{\text{c,j}}$ and $N_{\text{t,j}}$. We use the standard robust estimated sampling variance for ols estimators, robust to general heteroskedasticity. Let $\left(\hat{\alpha}(j), \hat{\tau}^{\text{adj}}(j), \hat{\beta}(j)\right)$ be the ordinary least squares estimates defined in Equation (17.4). Then define the matrices $\hat{\Delta}$ and $\hat{\Gamma}$ as

$$\hat{\Gamma}(j) = \frac{1}{N(j)} \sum_{i=1}^{N} B_i(j) \begin{pmatrix} 1 & W_i & X_i' \\ W_i & W_i & W_i \cdot X_i' \\ X_i & W_i \cdot X_i & X_i \cdot X_i' \end{pmatrix},$$

and

$$\hat{\Delta}(j) = \frac{1}{N(j)} \sum_{i=1}^{N} B_i(j) \left(Y_i - \hat{\alpha}(j) - \hat{\tau}^{\text{adj}}(j)W_i - X_i\hat{\beta}(j) \right)^2 \cdot \begin{pmatrix} 1 & W_i & X_i' \\ W_i & W_i & W_i \cdot X_i' \\ X_i & W_i \cdot X_i & X_i \cdot X_i' \end{pmatrix}.$$

Then the robust estimator for the sampling variance of $\hat{\tau}^{\mathrm{adj}}(j)$ is $\hat{\mathbb{V}}(\hat{\tau}^{\mathrm{adj}}(j))$, the natural generalization of the Neyman sampling variance estimator, is

$$\hat{\mathbb{V}}\left(\hat{\tau}^{\mathrm{adj}}(j)\right) = \frac{1}{N(j)} \left(\hat{\Gamma}(j)\hat{\Delta}(j)^{-1}\hat{\Gamma}(j)\right)^{-1}_{(2,2)},$$

the $(2,2)$ element of the $(K+2) \times (K+2)$ dimensional matrix $\left(\hat{\Gamma}(j)\hat{\Delta}(j)^{-1}\hat{\Gamma}(j)\right)^{-1}/N(j)$. We then combine the within-block variances the same way we did before:

$$\hat{\mathbb{V}}\left(\hat{\tau}^{\mathrm{strat,adj}}\right) = \sum_{j=1}^{J} \hat{\mathbb{V}}\left(\hat{\tau}^{\mathrm{adj}}_j\right) \cdot q(j)^2, \tag{17.5}$$

which is the estimated variance we use in the calculations in the next section.

If we are interested in the average treatment effect for the treated subsample, we do not need to modify the within-block estimates $\hat{\tau}^{\mathrm{adj}}(j)$ or estimated sampling variances $\hat{\mathbb{V}}(\hat{\tau}^{\mathrm{adj}}(j))$. Because we analyze the data within the blocks as if assignment is completely random, the average effect for the subsample of treated units within the block is identical to the average effect for all units within the block. In order to estimate the average effect for the treated for the entire sample, however, we do modify the block weights to reflect the proportions of treated units in the different blocks. Instead of using the sample proportions $q(j) = N(j)/N$, the appropriate weights are now equal to the proportion of treated units in each block, $N_{\mathrm{t}}(j)/N_{\mathrm{t}}$, leading to

$$\hat{\tau}^{\mathrm{strat,adj}}_{\mathrm{t}} = \sum_{j=1}^{J} \hat{\tau}^{\mathrm{adj}}(j) \cdot \frac{N_{\mathrm{t}}(j)}{N_{\mathrm{t}}}.$$

Similarly for the estimated sampling variance, we sum the within-block estimated sampling variances, multiplied by the square of the block proportions of treated:

$$\hat{\mathbb{V}}\left(\hat{\tau}^{\mathrm{strat,adj}}_{\mathrm{t}}\right) = \sum_{j=1}^{J} \hat{\mathbb{V}}\left(\hat{\tau}^{\mathrm{adj}}(j)\right) \cdot \left(\frac{N_{\mathrm{t}}(j)}{N_{\mathrm{t}}}\right)^2.$$

17.7 AVERAGE TREATMENT EFFECTS FOR THE LOTTERY DATA

Now let us return to the lottery data. The algorithm for choosing the number of blocks led to five blocks. Within each of these five blocks we estimate the average treatment effect either (*i*) using no further adjustment, (*ii*) using linear regression with four covariates (the same four covariates that are always included in the specification of the propensity score, # Tickets, Education, Working Then, and Earnings Year -1, based on substantive arguments), or (*iii*) using linear regression with the full set of eighteen covariates.

Table 17.6 presents results for the parameter estimates from the least squares regression for the five blocks with no covariates and with the limited set of four covariates. Although the parameter estimates are of only limited interest here, we note that we see

Table 17.6. *Independent Least Squares Regressions within Blocks, with Common Slope Coefficients for Treated and Controls within Blocks for the IRS Lottery Data*

Covariates	Block 1 (N = 80)		Block 2 (N = 40)		Block 3 (N = 41)		Block 4 (N = 81)		Block 5 (N = 81)	
	Est	(s.̂e.)	Est	(s.̂e.)	Est	(s.̂e.)	Est	(s.̂e.)	Est	(s.̂e.)
No covariates										
Intercept	20.02	(2.25)	12.70	(2.67)	15.59	(3.07)	19.69	(2.76)	12.75	(3.26)
Treatment	−10.82	(4.70)	2.07	(5.10)	−1.17	(4.97)	−9.43	(3.23)	−2.89	(3.59)
Limited covariates										
Intercept	−20.04	(10.66)	4.47	(9.80)	−9.91	(10.87)	−8.65	(5.58)	−6.70	(5.21)
Treatment	−6.21	(4.01)	−6.51	(3.86)	−4.81	(3.87)	−5.88	(1.82)	−2.56	(2.39)
# Tickets	−3.48	(1.39)	1.17	(1.26)	1.85	(1.24)	−0.48	(0.34)	−0.20	(0.37)
Education	2.03	(0.87)	−0.37	(0.81)	0.48	(0.93)	1.17	(0.49)	0.59	(0.42)
Work Then	−2.66	(2.96)	−0.51	(1.84)	5.98	(4.35)	1.16	(2.18)	5.30	(2.52)
Earn Year -1	0.84	(0.06)	0.83	(0.09)	0.60	(0.15)	0.76	(0.07)	0.62	(0.10)

Table 17.7. *Estimated Average Treatment Effects with Final Subclassification for the IRS Lottery Data (regression estimates as in Table 17.6)*

Covariates	Full Sample 1 Block		Trimmed Sample 1 Block		Trimmed Sample 2 Blocks		Trimmed Sample 5 Blocks	
	Est	(s.̂e.)	Est	(s.̂e.)	Est	(s.̂e.)	Est	(s.̂e.)
None	−6.2	(1.4)	−6.6	(1.7)	−6.0	(1.9)	−5.7	(2.0)
# Tickets, Education, Work Then, Earn Year-1	−2.8	(0.9)	−4.0	(1.1)	−5.6	(1.2)	−5.1	(1.2)
All	−5.1	(1.0)	−5.3	(1.1)	−6.4	(1.1)	−5.7	(1.1)

some evidence that the covariates do affect the outcomes and also that there is sufficient difference in the covariate distributions within the blocks that the adjustment alters the estimates of the effect of the treatment within the blocks.

Table 17.7 presents the estimates of the overall treatment effect on average annual post-lottery earnings based on the full sample, the trimmed sample with no subclassification, the trimmed sample with two blocks, and the trimmed sample with five blocks as selected by the algorithm. In each case, we present the estimates without covariance adjustment, covariance adjustment with the limited set of four covariates, and covariance adjustment based on the full set of eighteen covariates. The key observation is that both trimming and subclassification greatly reduce the sensitivity to the inclusion of covariates in the regression specification. In the full sample, the estimates range from −6.2 to −2.8 (in terms of thousands of dollars) in reduced labor earnings as a result of winning the lottery, a range of 3.4. In the trimmed sample, the estimates range from −6.6 to −4.0, a range of 2.6. In the trimmed sample with two blocks the range is only 0.8, and with five blocks, the range is down to 0.6. The conclusion is that, at least for this data set, trimming and subclassification greatly reduce the sensitivity to the specific least squares regression method used and thus lead to more credible estimates of causal effects.

17.8 WEIGHTING ESTIMATORS AND SUBCLASSIFICATION

There is an alternative way to use the propensity score that is, at first sight, quite different from subclassification. Closer inspection, however, reveals a close conceptual connection. In this approach, related to the work by Horvitz and Thompson (1952) in survey research, the inverse of the estimated propensity score is used to weight the units in order to eliminate biases associated with differences in observed covariates. We discuss this approach to estimation in this section partly because understanding it provides additional insight into the properties and benefits of our preferred method of subclassification.

17.8.1 Weighting Estimators

The Horvitz-Thompson estimator exploits the following two equalities, which follow from super-population unconfoundedness:

$$\mathbb{E}\left[\frac{W_i \cdot Y_i^{\text{obs}}}{e(X_i)}\right] = \mathbb{E}_{\text{sp}}\left[Y_i(1)\right] \quad \text{and} \quad \mathbb{E}\left[\frac{(1 - W_i) \cdot Y_i^{\text{obs}}}{1 - e(X_i)}\right] = \mathbb{E}_{\text{sp}}\left[Y_i(0)\right]. \qquad (17.6)$$

These inequalities can be derived as follows. Because Y_i^{obs} is $Y_i(1)$ when $W_i = 1$, it follows that

$$\mathbb{E}\left[\frac{W_i \cdot Y_i^{\text{obs}}}{e(X_i)}\right] = \mathbb{E}\left[\frac{W_i \cdot Y_i(1)}{e(X_i)}\right].$$

By iterated expectations, we can write this as

$$\mathbb{E}\left[\frac{W_i \cdot Y_i(1)}{e(X_i)}\right] = \mathbb{E}\left[\mathbb{E}\left[\frac{W_i \cdot Y_i(1)}{e(X_i)}\middle| X_i\right]\right].$$

By super-population unconfoundedness W_i is independent of $Y_i(1)$ conditional on X_i, so that the expectation of the product $W_i \cdot Y_i(1)$ given X_i is the product of the conditional expectations,

$$\mathbb{E}\left[\frac{W_i \cdot Y_i(1)}{e(X_i)}\middle| X_i\right] = \frac{\mathbb{E}_W\left[W_i| X_i\right] \cdot \mathbb{E}_{\text{sp}}\left[Y_i(1)| X_i\right]}{e(X_i)} = \frac{e(X_i) \cdot \mathbb{E}_{\text{sp}}\left[Y_i(1)| X_i\right]}{e(X_i)}$$

$$= \mathbb{E}_{\text{sp}}\left[Y_i(1)| X_i\right],$$

and thus

$$\mathbb{E}\left[\frac{W_i \cdot Y_i(1)}{e(X_i)}\right] = \mathbb{E}_{\text{sp}}\left[\mathbb{E}_{\text{sp}}\left[Y_i(1)| X_i\right]\right] = \mathbb{E}_{\text{sp}}\left[Y_i(1)\right].$$

The same argument leads to the second equality in (17.6) for the average control potential outcome.

The two equalities in (17.6) suggest estimating $\mathbb{E}[Y_i(1)]$ and $\mathbb{E}[Y_i(0)]$ as

$$\widehat{\mathbb{E}_{sp}[Y_i(1)]} = \frac{1}{N} \sum_{i=1}^{N} \frac{W_i \cdot Y_i^{obs}}{e(X_i)} \quad \text{and} \quad \widehat{\mathbb{E}_{sp}[Y_i(0)]} = \frac{1}{N} \sum_{i=1}^{N} \frac{(1 - W_i) \cdot Y_i^{obs}}{1 - e(X_i)},$$

and thus estimating the average treatment effect $\tau_{sp} = \mathbb{E}_{sp}[Y_i(1) - Y_i(0)]$ as a Horvitz-Thompson estimator,

$$\tilde{\tau}^{ht} = \frac{1}{N} \sum_{i=1}^{N} \left(\frac{W_i \cdot Y_i^{obs}}{e(X_i)} - \frac{(1 - W_i) \cdot Y_i^{obs}}{1 - e(X_i)} \right) = \frac{1}{N} \sum_{i=1}^{N} \left(\frac{(W_i - e(X_i)) \cdot Y_i^{obs}}{e(X_i) \cdot (1 - e(X_i))} \right).$$

(17.7)

In practice we rarely know the propensity score, so we rarely can use the estimator in (17.7) directly. Instead we can weight using the estimated propensity score $\hat{e}(X_i)$, and use the estimator

$$\hat{\tau}^{ht} = \sum_{i=1}^{N} \frac{W_i \cdot Y_i^{obs}}{\hat{e}(X_i)} \bigg/ \sum_{i=1}^{N} \frac{W_i}{\hat{e}(X_i)} - \sum_{i=1}^{N} \frac{(1 - W_i) \cdot Y_i^{obs}}{1 - \hat{e}(X_i)} \bigg/ \sum_{i=1}^{N} \frac{1 - W_i}{1 - \hat{e}(X_i)}.$$

(17.8)

(Normalizing the weights to one in finite samples rather than merely in expectation typically improves the mean-squared-error properties of the estimator.) The basic Horvitz-Thompson estimator can be modified easily to incorporate covariates. For this purpose, it is useful to write the weighting estimator as a weighted regression estimator. Consider the regression function

$$Y_i^{obs} = \alpha + \tau \cdot W_i + \varepsilon_i,$$

estimated by weighted least squares with estimated weights $\hat{\lambda}_i^{ht}$, where

$$\hat{\lambda}_i^{ht} = \frac{1}{(1 - \hat{e}(X_i))^{1 - W_i} \cdot e(X_i)_i^{W}} = \begin{cases} \frac{1}{1 - \hat{e}(X_i)} & \text{if } W_i = 0, \\ \frac{1}{\hat{e}(X_i)} & \text{if } W_i = 1. \end{cases}$$

This weighted regression estimator for τ is identical to $\hat{\tau}^{ht}$ as defined in (17.8). With this weighted regression version, it is straightforward to include covariates. Instead of estimating the regression function with only an intercept and an indicator for the treatment, one can estimate a regression function that includes additional covariates,

$$Y_i^{obs} = \alpha + \tau \cdot W_i + X_i \beta + \varepsilon_i,$$

using the same weights $\hat{\lambda}_{ht}^i$. The weighted regression estimator is consistent for τ_{fs} as long as either the specification of the propensity score is correct, or the specification of the regression function is correct, a property referred to as "double-robustness," although it is not necessarily robust in the standard usage of the term "robustness."

It is useful to see how this Horvitz-Thompson estimator relates to the subclassification estimator. The basic subclassification estimator, with no further adjustment for covariates, has the form

$$\hat{\tau}^{\text{strat}} = \sum_{j=1}^{J} q(j) \cdot \hat{\tau}^{\text{dif}}(j) = \sum_{j=1}^{J} q(j) \cdot \left(\overline{Y}_t(j) - \overline{Y}_c(j) \right),$$

which can be written as

$$\hat{\tau}^{\text{strat}} = \frac{1}{N} \sum_{i=1}^{N} W_i \cdot Y_i^{\text{obs}} \cdot \lambda_i^{\text{strat}} - \frac{1}{N} \sum_{i=1}^{N} (1 - W_i) \cdot Y_i^{\text{obs}} \cdot \lambda_i^{\text{strat}},$$

where the weights λ_i^{strat} satisfy

$$\lambda_i^{\text{strat}} = \sum_{j=1}^{J} B_i(j) \cdot \left(\frac{1 - W_i}{N_c(j)/N(j)} + \frac{W_i}{N_t(j)/N(j)} \right)$$

$$= \begin{cases} \sum_{j=1}^{J} B_i(j) \cdot \frac{N(j)}{N_c(j)} & \text{if } W_i = 0, \\ \sum_{j=1}^{J} B_i(j) \cdot \frac{N(j)}{N_t(j)} & \text{if } W_i = 1. \end{cases}$$

Thus the basic subclassification estimator can be interpreted as a weighting estimator where the weights are based on the block-based coarsened propensity score. Instead of using the original estimator for the propensity score, $\hat{e}(X_i)$, the blocking estimator implicitly uses as an estimate of the propensity score the fraction of treated units within the propensity score stratum to which the unit belongs:

$$\tilde{e}(X_i) = \sum_{j=1}^{J} B_i(j) \cdot \frac{N_t(j)}{N(j)}.$$

Thus, it coarsens the propensity score, approximately averaging it within the subclasses. This modification to the estimated propensity score increases very small values of the propensity score and decreases very large values, and thus it lowers extreme values for the weights in the weighted-average interpretation of the estimator.

What are the relative merits of the subclassification estimator versus the Horvitz-Thompson estimator? We discuss three issues. Ultimately we prefer the subclassification estimator and see little reason to use the estimator based on weighting by the estimated propensity score. However, in many cases this choice is not important, because it will not make much difference whether one uses the Horvitz-Thompson or subclassification weights. If the number of blocks is large, so that the dispersion of the propensity score within the strata is limited, then the weights according to the blocking estimator will be close to those according to the Horvitz-Thompson estimator, which is also true if there is only limited variation in the propensity score overall, and if there are few extreme values for the propensity score. The weights will be different only if, in at least some blocks, there is substantial variation in the propensity score, which is most likely to happen in blocks with propensity score values close to zero and one. In fact, the similarity between the estimators turns to equality in simple cases where the model for the propensity score is fully saturated and the number of blocks is sufficiently large that within a block there is no variation in the propensity score.

Now consider bias properties of the two estimators. If one uses the inverse of the true propensity score, the Horvitz-Thompson estimator is exactly unbiased. If one does not know the propensity score, it might then seem that using the best possible estimate of the propensity score (in the sense of minimizing expected mean squared error), rather than an estimator that is further smoothed, may be sensible. This appears the most powerful argument in favor of the Horvitz-Thompson estimator, but it is not particularly persuasive though. Although weighting by the inverse of the true propensity score leads to unbiased estimators for the average treatment effect, weighting using the inverse of a noisy, unbiased, estimator for the propensity score may generate considerable bias because the estimated propensity score enters in the denominator of the weights. Smoothing the weights by essentially averaging them within blocks, as the subclassification estimator does, may remove some of this bias. Moreover, in practice the propensity score is likely to be misspecified, which may affect the performance of the Horvitz-Thompson estimator more than the subclassification estimator. More specifically, suppose a particular covariate $X_{i,k}$ is omitted from the propensity score specification. If this covariate is correlated with the potential outcomes, any (small) bias from omitting it may be increased by the larger weights used in the Horvitz-Thompson approach.

The second point concerns the estimated sampling variance. Here the relative merits are clear. By smoothing over the extreme weights from the Horvitz-Thompson estimator, the subclassification estimator tends to have a smaller sampling variance, which also may make the Horvitz-Thompson estimator less robust than the blocking estimator because the large weights also tend to be the ones that are relatively imprecisely estimated or affected by misspecification of the propensity score model. For that reason, shrinking them to their mean within subclasses, as subclassification does, can improve the properties of the resulting estimator.

A final issue concerns modifications of the Horvitz-Thompson and blocking estimator involving additional covariance adjustment. The covariance adjustment version of the Horvitz-Thompson estimator uses a single set of parameters to model the dependence of the outcome on the covariates. In other words, it uses a global approximation to the regression function. Such a global approximation can lead to poor approximations to the two regression functions for some values of the covariates. An analogous procedure given the subclassification would be to restrict the slope coefficients on the covariates to be the same across all blocks. This is not what is typically done, or what we discussed in the previous sections. Instead, the slope coefficients are unrestricted between the blocks, allowing the estimated regression function to provide a better approximation to the conditional mean.

17.8.2 Weighting Estimators and the Lottery Data

To illustrate the Horvitz-Thompson estimator let us return to the lottery data. We look both at the full sample with 496 units and at the trimmed sample with 323 units. In both cases we calculate the weights according to the propensity score estimated through the algorithm described in Section 17.3. Based on the estimated propensity score, we

Table 17.8. *Some Descriptive Statistics for Weights for Horvitz-Thompson and Subclassification Estimators for the IRS Lottery Data*

| | Full Sample | | Trimmed Sample | |
	Horvitz-Thompson	Subclass	Horvitz-Thompson	Subclass
Minimum	0.92	1.06	1.00	1.19
Maximum	79.79	17.71	18.18	6.15
Standard deviation	4.20	2.63	1.69	1.35

normalize the weights within each treatment group to ensure they sum to N. We then estimate the implicit weights in the blocking estimator, again for both the full sample and for the trimmed sample.

Table 17.8 presents summary statistics for the weights. Within each data set there is a substantial difference between ranges of the Horvitz-Thompson and subclassification weights. In the full sample, the correlation coefficient between the Horvitz-Thompson and subclassification weights is only 0.64. In the trimmed sample the correlation is higher, namely 0.82. The second observation is that the weights are considerably more extreme for the Horvitz-Thompson estimator. In the full sample the largest of the weights is almost 80 for the Horvitz-Thompson estimator, compared to 17.8 for the subclassification estimator. With the smallest weights around one (the smallest weight would be at least equal to one if it was not for the normalization to ensure that the weights add to the sample size), the weight for this unit is eighty times that for the low-weight unit, making any estimates overly sensitive to the outcome for this unit; for example, increasing the outcome for this individual by one standard deviation (i.e., increasing average post-lottery earnings by $15,000), would lead to a change in the estimated average treatment effect of $(80/496) \times 15,000 = 2,500$, which is substantial, given the variation in our subclassification estimates in Table 17.7. The sensitivity of the estimates to the outcome for this unit in the subclassification estimator is less because its weight is only a fifth as large. In the trimmed sample, the largest weights are 18.2 and 6.2 for the Horvitz-Thompson and subclassification estimators respectively, so now changing the outcome for any single unit by a standard deviation leads to a change in the subclassification estimated average effect of at most $(6.2/323) \times 15,000 = 300$. In particular, for subclassification in the trimmed sample, the ratio of largest to smallest weight is 5.2, ensuring that no single unit is unduly affecting the estimates. The third observation is that the trimming greatly reduces the variation in the weights, and lowers the largest weights, by improving the balance and shrinking the propensity score toward average values. In general, subclassification smooths the weights, avoiding excessively large weights.

Suppose, as we have done before in illustrative calculations, that the conditional expectation of the potential outcomes is linear in the covariates:

$$\mathbb{E}_{\mathrm{sp}}\left[Y_i(w)|X_i\right] = \alpha + \tau \cdot W_i + X_i\beta,$$

with constant variance:

$$\mathbb{V}_{\mathrm{sp}}\left(Y_i(w)|X_i\right) = \sigma^2.$$

Table 17.9. *Least Squares Regression Estimates for the IRS Lottery Data*

Covariate	Full Sample		Trimmed Sample	
	Est	$(\widehat{s.e.})$	Est	$(\widehat{s.e.})$
Intercept	21.20	(4.80)	22.76	(6.49)
Treatment Indicator	−5.08	(0.95)	−5.34	(1.08)
Year Won	−0.64	(0.34)	−0.34	(0.44)
# Tickets	0.06	(0.15)	0.31	(0.21)
Age	−0.26	(0.04)	−0.29	(0.05)
Male	−0.58	(0.89)	0.44	(1.17)
Education	0.04	(0.20)	−0.12	(0.27)
Work Then	0.93	(1.12)	1.30	(1.45)
Earn Year -6	−0.00	(0.11)	0.01	(0.14)
Earn Year -5	−0.02	(0.13)	−0.02	(0.17)
Earn Year -4	0.02	(0.12)	0.01	(0.14)
Earn Year -3	0.29	(0.12)	0.36	(0.15)
Earn Year -2	0.04	(0.11)	−0.20	(0.16)
Earn Year -1	0.48	(0.08)	0.64	(0.11)
Pos Earn Year -6	0.19	(1.66)	0.05	(2.18)
Pos Earn Year -5	1.78	(2.10)	1.44	(2.72)
Pos Earn Year -4	−1.04	(1.99)	−0.28	(2.45)
Pos Earn Year -3	−1.60	(1.90)	−2.65	(2.50)
Pos Earn Year -2	−1.08	(2.01)	0.30	(2.98)
Pos Earn Year -1	−0.36	(1.79)	−2.52	(2.65)
Residual σ^2	8.45^2		8.59^2	

Table 17.10. *Estimated Bias and Estimated Sampling Variance for Horvitz-Thompson and Subclassification Estimators under Linear Model for the IRS Lottery Data*

	Full Sample		Trimmed Sample	
	Horvitz-Thompson	Subclass	Horvitz-Thompson	Subclass
Bias	4.34	2.68	1.29	0.30
Variance	2.59^2	0.83^2	1.29^2	1.15^2
Bias^2+Variance	5.06^2	2.81^2	1.83^2	1.19^2

If this linearity assumption were actually true, we could simply estimate τ by least squares. We present the relevant least squares estimates in Table 17.9. However, the point here is not to get an estimate of the average treatment effect under this assumption but rather to compare the Horvitz-Thompson estimate versus the subclassification estimate, under this assumption.

An estimator of the form

$$\hat{\tau}_\lambda = \frac{1}{N} \sum_{i=1}^{N} \left(W_i \cdot Y_i^{\mathrm{obs}} \cdot \lambda_i - (1 - W_i) \cdot Y_i^{\mathrm{obs}} \cdot \lambda_i \right),$$

with the weights $\lambda_i = \lambda(W_i, X_i, \mathbf{W}_{(i)}, \mathbf{X}_{(i)})$, has, conditional on $\mathbf{W}$ and $\mathbf{X}$, the following bias and sampling variance:

$$\mathrm{Bias}_\lambda = \mathbb{E}_{\mathrm{sp}} \left[\hat{\tau}_\lambda - \tau_{\mathrm{fs}} \middle| \mathbf{W}, \mathbf{X} \right] = \frac{1}{N} \sum_{i=1}^{N} \left(W_i \cdot \mu_{\mathrm{t}}(X_i) \cdot \lambda_i - (1 - W_i) \cdot \mu_{\mathrm{c}}(X_i) \cdot \lambda_i \right) - \tau,$$

and sampling variance

$$\mathbb{V}_{\mathrm{sp}} \left(\hat{\tau}_\lambda \middle| \mathbf{W}, \mathbf{X} \right) = \frac{1}{N^2} \sum_{i=1}^{N} \lambda_i^2 \cdot \left(W_i \cdot \sigma_{\mathrm{t}}^2(X_i) + (1 - W_i) \cdot \sigma_{\mathrm{c}}^2(X_i) \right).$$

Under our linear homoskedastic model assumptions, the bias simplifies to

$$\mathrm{Bias}_\lambda = \frac{1}{N} \sum_{i=1}^{N} (2 \cdot W_i - 1) \cdot X_i \beta \cdot \lambda_i = \left(\overline{X}_{\mathrm{t,weighted}} - \overline{X}_{\mathrm{c,weighted}} \right) \beta,$$

where

$$\overline{X}_{\mathrm{c,weighted}} = \sum_{i:W_i=0} X_i \cdot \lambda_i \middle/ \sum_{i:W_i=0} \lambda_i, \quad \text{and} \quad \overline{X}_{\mathrm{t,weighted}} = \sum_{i:W_i=1} X_i \cdot \lambda_i \middle/ \sum_{i:W_i=1} \lambda_i,$$

the weighted average of the control and treated covariates, respectively. Under homoskedasticity, the sampling variance simplifies to

$$\mathbb{V}_{\mathrm{sp}} \left(\hat{\tau}_\lambda \middle| \mathbf{W}, \mathbf{X} \right) = \frac{\sigma^2}{N^2} \cdot \sum_{i=1}^{N} \lambda_i^2.$$

We can estimate these two objects, Bias_λ and Var_λ, as well as the sum of the sampling variances and the square of the bias, that is, the expected-mean-squared-error, for our particular data set, leading to

$$\widehat{\mathrm{MSE}} = \left(\left(\overline{X}_{\mathrm{t,weighted}} - \overline{X}_{\mathrm{c,weighted}} \right) \hat{\beta} \right)^2 + \frac{\sigma^2}{N^2} \cdot \sum_{i=1}^{N} \lambda_i^2.$$

The results are reported in Tables 17.6 and 17.10. In Table 17.6 we report the least squares estimates of the regression function, for both the full and the trimmed samples. In the third and seventh columns of Table 17.1, we report the average difference in covariates, weighted according to the Horvitz-Thompson estimator and normalized by the square root of the sum of the standard deviations

$$\frac{\overline{X}_{\mathrm{t,weighted}} - \overline{X}_{\mathrm{c,weighted}}}{\sqrt{\left(s_{\mathrm{c}}^2 + s_{\mathrm{t}}^2 \right) / 2}}.$$

If the Horvitz-Thompson estimator were based on the true propensity scores, the average difference in covariates should be zero, at least in expectation. They are not, due in part to sampling variation and due in part to misspecification of the propensity score. We see that, for most covariates, the Horvitz-Thompson estimator has approximately the same normalized differences as the subclassification estimator. Sometimes the Horvitz-Thompson differences are larger, as for the important (in the sense of being, *a priori*, likely to be correlated with the potential outcomes) lagged earnings variables, and sometimes smaller, as for education and some of the employment indicators. The larger normalized differences are largely due to the presence of extreme weights in the Horvitz-Thompson approach.

Table 17.10 presents the components of the estimated expected-mean-squared-error. It is not surprising that, for both the full and the trimmed samples, the estimated sampling variance is smaller for the subclassification estimator, which is a direct consequence of the smoothed weights of the subclassification estimator. Possibly more surprising is the fact that, for both the full and the trimmed samples, the estimated bias is actually considerably larger for the Horvitz-Thompson estimator than for the subclassification estimator. Not surprising is that the estimated bias and the estimated sampling variance are substantially smaller in the trimmed sample than in the full sample (with the exception of the estimated sampling variance for the subclassification estimator, which is slightly smaller in the full sample than in the trimmed sample).

17.9   CONCLUSION

In this chapter we discuss one of the leading classes of estimators for average treatment effects under unconfoundedness. This subclassification estimator uses the propensity score to construct strata within which the covariates are well balanced. Within the strata, the average treatment effect is estimated by simply differencing average outcomes for treated and control units, or, in our preferred version, by further adjusting for some remaining covariate differences through linear regression. The subclassification estimator with further adjustment is similar conceptually to weighting estimators, although less variable in settings with units with propensity score values close to zero or one. We illustrate the practical value of this estimator using the lottery data.

NOTES

Subclassificiation as a method for estimating treatment effects in the presence of observed confounders has a long tradition in statistics. Early discussions can be found in Cochran (1965, 1968). See also Rosenbaum and Rubin (1983a, 1984). There are many recent applications, including Dehejia and Wahba (1999) and Rubin (2001).

The estimator that combines weighting with regression has been developed by Robins, Rotnitzky, and Zhao (1995). They show that the weighted regression estimator is consistent as long as either the specification of the propensity score is correct, or the specification of the regression function is correct, a property Robins and coauthors refer to as "double-robustness." See Hirano and Imbens (2001), Kang and Shafer (2007) and Waernbaum (2012) for some discussion on the properties of doubly-robust estimators and for some simulation studies of blocking.

See Hirano, Imbens, and Ridder (2003) on formal properties of the Horvitz-Thompson estimator with a discussion of the implications of using the estimated versus the true population propensity score to construct the weights for the precision of the resulting estimators.

An interesting extension of the equalities in Equation (17.6) is the following equality, which holds under unconfoundedness:

$$\mathbb{E}\left[Y_i^{\mathrm{obs}} \cdot \frac{W_i - e(X_i)}{e(X_i) \cdot (1 - e(X_i))} \middle| X_i = x\right] = \tau_{\mathrm{sp}}(x).$$

Thus the conditional expectation of the transformed outcome $Y_i^* = Y_i^{\mathrm{obs}} \cdot (W_i - e(X_i)/(e(X_i) \cdot (1 - e(X_i))))$, conditioning on X_i but not on W_i, is equal to $\tau(X_i)$. Athey and Imbens (2014) exploit this equality to adapt machine learning algorithms developed for prediction problems to the problem of estimating conditional average treatment effects.

## CHAPTER 18

Matching Estimators

18.1 INTRODUCTION

Following the discussion of subclassification (i.e., blocking, or stratification) in the previous chapter, we discuss in this chapter a second general approach to estimation of treatment effects in regular designs, namely matching. As earlier, we mainly focus on average effects, although the methods readily extend to estimating other causal estimands, for example, the difference in the median or other quantiles by treatment status, or differences in variances. Many of the specific techniques in this chapter are similar to the methods discussed in Chapter 15, but the aim is different. In Chapter 15 we were interested in constructing a sample with improved balance in the covariates. Here we take the sample as given, and focus on estimating treatment effects. In this chapter we consider both methods where only the treated units are matched (and where the focus is on the effects of the treatment for the treated), and methods are matched in order to estimate the effects of the treatment for the full sample.

Matching estimators – based on direct comparisons of outcomes for observationally equivalent "matched" units that received different levels of a treatment – are among the most intuitive estimators for treatment effects. Informal assessments of causality often rely implicitly on matching: "This unemployed individual found a job because of the skills acquired in a job-training program." Typically the case for or against such a claim is made by a comparison to an individual who did not participate in the training program but who is similar with respect to observed background characteristics. If we maintain the unconfoundedness assumption – that the probability of receipt of treatment is free of dependence on the potential outcomes, once observed pre-treatment characteristics are held constant – such comparisons between treated and control units with the same covariate values have a causal interpretation. The matching approach estimates average treatment effects by pairing such similar units and averaging the within-pair differences in observed outcomes.

Moreover, in many observational studies there exists no systematically better approach for estimating the effect of a treatment on an individual unit than by finding a control unit identical on all observable aspects except on the treatment received and then comparing their outcomes. For example, suppose we wish to evaluate the effect of a job-training program on a 30-year-old single mother with two children, ages 4 and 6, who had been employed for eighteen months before being out of work for the last six months, who participated in the program, and about whom we have no additional information. Lacking a randomized design for the evaluation of the training program, it appears most credible to assess the benefits of this program by comparing the labor market outcomes for this woman to those of another 30-year-old single mother with two children, aged 4 and 6, with a similar recent labor market history, in the same geographic location, but who did *not* go through the job-training program. This is exactly what matching aims to do: it attempts to find the control unit most comparable to the treated unit in all possible pre-treatment characteristics. Although making units comparable along observable dimensions need not be sufficient for obtaining credible causal comparisons, it is often a prerequisite for doing so.

To provide additional intuition for matching, consider the analysis of paired randomized experiments discussed in Chapter 10. Matching can be interpreted as reorganizing the data from an observational study in such a way that the assumptions from a paired randomized experiment hold, at least approximately. There are, however, two important differences between paired randomized experiments and matching in observational studies. The first difference is that in the latter case, unconfoundedness must be assumed – it is not guaranteed to be satisfied by design, as it is in a randomized experiment. Even when treated and control observations can be matched exactly on the observed covariates, there may exist, in observational studies, unobservable factors that vary systematically between members of the pairs, affecting both their underlying probabilities of receiving the treatment and their potential outcomes, and therefore creating biases. Thus, inferences based on matched observational data are inherently less credible than those based on data from a paired randomized experiment. The second difference from paired randomized experiments is that matching is often inexact, so that systematic differences in pre-treatment variables across the matched pairs may remain. In contrast, the within-pair randomization guarantees that the assignment probabilities are identical within pairs, and so no systematic biases can arise. Hence the assumptions from a paired randomized experiment do not generally apply, even if unconfoundedness holds, when the matching is not exact.

In this chapter we discuss matching estimators in more detail. In Section 18.2 we introduce the data set that will be used to illustrate the methods discussed in this chapter. They come from an influential study by Card and Krueger evaluating the effect of a change in the minimum wage in New Jersey in 1993. Next, in Section 18.3, we discuss the simplest form of matching estimators where we match each treated unit to a single control unit, with exactly the same values of the covariates, using each control unit at most once as a control. This matching may have been the result of the design strategy in Chapter 15. The resulting pairs of treated and control units can be analyzed using the methods developed for paired randomized experiments in Chapter 10. The natural estimator for the average treatment effect for the treated units is, in this case, simply the average difference within the pairs, and one can estimate the sampling variance by the sample variance of the within-pair differences divided by the number of pairs. This setting is too simple to cover many cases of interest, and in the remainder of the chapter we discuss extensions to more complex and realistic cases, as well as modifications of the simple matching estimator to improve its properties in those more complex situations.

These extensions and complications fall into two broad categories. The first involves dealing with the inability to find exact matches in the finite sample at hand. This category includes the issues raised by the choice between various close, but inexact, matches, as well as options to reduce biases from inexact matches. The second category involves departures from the distinct-pair setup, where each pair consists of a single unique treated and a single unique control unit, with distinct units across distinct pairs. This second category includes extensions where units are used more than once as a match, or where multiple matches are used. We now briefly describe the various specific extensions and complications.

The first three complications fit into the first category. In Section 18.4 we address the possibility that there are some treated units for whom we cannot find a control unit that is identical in terms of covariates. In that case, one option is to include matches that are not exact in terms of covariates, which in turn may lead to situations where the order in which the observations are matched affects the specific composition of the pairs, which suggests either choosing a systematic or random ordering of the units, or using a more complicated matching algorithm that takes into account the effect of one choice of match on the remaining pool of control units. A second complication is that, once the matching is inexact, we need to specify a distance measure to operationalize the notion of the "closest" match. Especially when we match on multiple covariates, the choice of metric can be important: that is, with multiple covariates, we often need to choose whether to trade off a difference in, for example, age against a difference in the number of children, or against a difference in previous labor market experience. We discuss some leading choices for such distance measures in Section 18.5. If the matching is inexact, one may be concerned that the quality of some of the matches is not adequate, in the sense that the differences in covariate values within matches is substantial. In Section 18.6 we illustrate these concepts using a small sample from the Card-Krueger data. In Section 18.7 we discuss the biases that may result from this inexact matching. There are several techniques available that attempt to reduce these biases, and we discuss some in Section 18.8. These techniques provide somewhat of a bridge between the matching estimators, discussed in this chapter, and the regression and model-based methods discussed in the context of randomized experiments in Chapters 7 and 8.

Next we discuss three extensions that fit into the second category of techniques. In Section 18.9 we discuss matching with replacement, where we allow a control unit to be used as a match more than once to increase the set of potential matches for each treated unit. Allowing a control unit to be used as a match for more than one treated unit can therefore improve the quality of the matches in the sense that it reduces the expected distance between the treated unit and its control match by expanding the potential set of control matches. Another advantage of matching with replacement is that it removes the dependence on the ordering of the units to be matched, or the need for more sophisticated matching methods that take account of the effect an early matching choice has on future possible matches. A disadvantage of such matching is that it can increase the sampling variance of the matching estimator by decreasing the number of matched controls.

In Section 18.10 we discuss the extension to multiple matches for each treated unit. Often only a single unit is used as a match. However, if multiple high-quality matches are available, one can improve the precision of the matching estimator without substantially increasing its bias. We discuss the potential gain in terms of precision as well as the potential cost in terms of bias. In Section 18.11 we discuss using matching to estimate treatment effects for the control units, rather than just for the treated units, and for the full sample.

In Section 18.12 we turn to the full data set from the Card and Krueger study to compare the estimates of the average treatment effect using various matching approaches. In addition, we compare these results to ordinary least squares estimates of the average treatment effect, calculated with, and without, using some or all of the matching variables in the regression model. This example illustrates that, regardless of the number of matches, the distance metric, or bias-adjustment approach used, all of the matching estimates can be fairly tightly clustered. In contrast, as anticipated, the ordinary least squares (regression) results can be sensitive to the specification chosen.

18.2 THE CARD-KRUEGER NEW JERSEY AND PENNSYLVANIA MINIMUM WAGE DATA

The data used in this chapter to illustrate matching methods are from an influential study by Card and Krueger (1995). They were interested in evaluating the effect of raising the state minimum wage in New Jersey in 1993 and collected data on employment at fast-food restaurants in New Jersey and in the neighboring state of Pennsylvania. The unit of analysis here is a restaurant. In addition to the number of employees measured prior to the raise in the minimum wage in New Jersey (initial empl), Card and Krueger collected for each restaurant information on starting wages (initial wage), average time until first raise (time until raise), and the identity of the chain (burger king, kfc, roys, or wendys). The outcome is employment after the raise in the minimum wage (final empl). Here we analyze the data as if they arose from a regular design, which includes the unconfoundedness assumption that, conditional on these covariates, the probability of being exposed to the new minimum wage (i.e., being from New Jersey rather than Pennsylvania) does not depend on the potential outcomes.

Table 18.1 presents summary statistics for the data set. There are 347 restaurants in the data set, 279 in New Jersey (the treated units) and 68 in Pennsylvania (the control units). For the purposes of this discussion we view the New Jersey restaurants as "treated" restaurants (subject to the intervention of the higher minimum wage), and the Pennsylvania restaurants as the "control" restaurants. A quick look at the overlap statistics suggests that the data are fairly well balanced. The largest of the normalized differences, calculated for each covariate as $(\overline{X}_t - \overline{X}_c)/\sqrt{(s_t^2 + s_c^2)/2}$, is equal to 0.28, for the initial employment variable, initial empl.

We estimate the propensity score, using the methods discussed in Chapter 13, as summarized in Table 18.2. The only covariate we pre-select for inclusion in the propensity score is the initial level of employment, initial empl. The algorithm does not select any other covariate to enter linearly and also does not select any second-order term. Had we not pre-selected initial employment, the algorithm would have selected it in any case, so the results are not sensitive to this choice. The estimated propensity score ranges from

Table 18.1. *The Card-Krueger New Jersey and Pennsylvania Minimum Wage Data*

	($N = 347$)		($N_t = 279$) (treated)		($N_c = 68$) (controls)		Nor Dif	Log Ratio of STD	$\pi^{0.05}$ Controls	Treated
	Mean	(S.D.)	Mean	(S.D.)	Mean	(S.D.)				
initial empl	17.84	(9.62)	20.17	(11.96)	17.27	(8.89)	−0.28	−0.30	0.10	0.03
burger king	0.42	(0.49)	0.43	(0.50)	0.42	(0.49)	−0.02	−0.01	0.00	0.00
kfc	0.19	(0.40)	0.13	(0.34)	0.21	(0.41)	0.20	0.17	0.00	0.00
roys	0.25	(0.43)	0.25	(0.44)	0.25	(0.43)	0.00	−0.00	0.00	0.00
wendys	0.14	(0.35)	0.19	(0.40)	0.13	(0.33)	−0.18	−0.18	0.00	0.00
initial wage	4.61	(0.34)	4.62	(0.35)	4.60	(0.34)	−0.05	−0.02	0.03	0.01
time until raise	17.96	(11.01)	19.05	(13.46)	17.69	(10.34)	−0.11	−0.26	0.10	0.03
pscore	0.80	(0.05)	0.79	(0.06)	0.81	(0.04)	0.28	−0.35	0.10	0.03
final empl	17.37	(8.39)	17.54	(7.73)	17.32	(8.55)				

Table 18.2. *Estimated Parameters of Propensity Score for the Card-Krueger New Jersey and Pennsylvania Minimum Wage Data*

Variable	Est	$(\widehat{s.\,e.})$	t-Stat
Intercept	1.93	(0.14)	14.05
Linear terms			
initial empl	−0.03	(0.01)	−2.17

0.4247 to 0.8638, again suggesting there is no need to trim part of the sample for lack of overlap.

In some of the initial discussions, we use a small subset of the Card-Krueger data to illustrate in detail some of the specific methods. For this purpose we selected twenty restaurants, five from New Jersey and fifteen from Pennsylvania, for which selected variables are presented in Table 18.3. This subset includes only burger king and kfc restaurants, and we use only initial employment (initial empl) and restaurant chain (burger king or kfc) as pre-treatment variables for this small sample.

18.3 EXACT MATCHING WITHOUT REPLACEMENT

In this section we discuss the simplest case of matching, exact matching without replacement. Initially we focus on the case where only treated units are matched, each to a unique single control. Initially we make the, generally unrealistic, assumption that there is a sufficiently large number of control units such that exact matches exist for each treated unit without the need to use the same control more than once. This may be more plausible after discarding some units using the design methods developed in Chapters 15

Table 18.3. *20 Units from the Card-Krueger New Jersey and Pennsylvania Minimum Wage Data*

Unit i	State W_i	chain X_{i1}	initial empl X_{i2}	final empl Y_i^{obs}
1	NJ	BK	22.5	40.0
2	NJ	KFC	14.0	12.5
3	NJ	BK	37.5	20.0
4	NJ	KFC	9.0	3.5
5	NJ	KFC	8.0	5.5
6	PA	BK	10.5	15.0
7	PA	KFC	13.8	17.0
8	PA	KFC	8.5	10.5
9	PA	BK	25.5	18.5
10	PA	BK	17.0	12.5
11	PA	BK	20.0	19.5
12	PA	BK	13.5	21.0
13	PA	BK	19.0	11.0
14	PA	BK	12.0	17.0
15	PA	BK	32.5	22.5
16	PA	BK	16.0	20.0
17	PA	KFC	11.0	14.0
18	PA	KFC	4.5	6.5
19	PA	BK	12.5	31.5
20	PA	BK	8.0	8.0

and 16. If there are multiple control units that are exact matches for a particular treated unit, we choose one element from this set randomly.

To be precise, and in order to deal with some of the subsequent extensions, let us introduce some notation. As before, we have a sample with N units, indexed by $i = 1, \ldots, N$. Let $\mathbb{I}_t = \{1, \ldots, N_t\}$ be the set of indices for the N_t treated units and $\mathbb{I}_c = \{N_t + 1, \ldots, N_t + N_c\}$ the set of indices for the N_c controls. Because (by assumption) distinct exact matches exist for each treated unit, we will obtain a set of N_t pairs. Let $\mathcal{M}_i^c \subset \mathbb{I}_c$ be the set of control indices containing the matches for treated unit i. Because we use a single match, $\mathcal{M}_i^c$ is a singleton, $\mathcal{M}_i^c = \{m_i^c\}$, where m_i^c is the index of the unit with the closest covariate values among the units with the opposite treatment to that of unit i. Because the matches are all distinct, it follows that if $i \neq i'$, then $\mathcal{M}_i^c \cap \mathcal{M}_{i'}^c = \emptyset$, and because the matching is exact, $X_i = X_{m_i^c}$ for all $i = 1, \ldots, N_t$. The superscript "c" on the set $\mathcal{M}_i^c$ indicates that the matches are control matches; later, when we also match control units, the set of their matches will be denoted by $\mathcal{M}_i^t$.

To be clear, suppose we have five units in the population, with units 1 and 2 treated units and 3, 4, and 5 control units. In that case, we have $\mathbb{I}_t = \{1, 2\}$, $\mathbb{I}_c = \{3, 4, 5\}$; $N_t = 2$ so that we construct two pairs. One possible pair of matches is to have the first pair equal to $(1, 3)$ and the second pair equal to $(2, 5)$ – for example, if $X_1 = X_3$, and $X_2 = X_5$.

For such a matching scheme to be at all possible, we obviously need $N_c \geq N_t$, and in practice we may need the reservoir of possible control units to be much larger than the number of treated units. We ignore these practical issues for now, but later we discuss such issues in some detail (see Section 18.9).

Now consider the ith matched pair, (i, m_i^c), with covariate values $X_i = X_{m_i^c} = x$. Because of super-population unconfoundedness, the probability is $1/2$ that, of these two units, it is unit i rather than unit m_i that received the treatment, conditional on the covariate value x and conditional on the pair of potential outcomes for each element of the pair. Given unconfoundedness, these N_{t} matched pairs, therefore, can be considered as comprising data from a paired randomized experiment and can be analyzed using the methods discussed in Chapter 10. A key implication, from the results in Chapter 10, is that the matched pair difference for the ith pair,

$$\hat{\tau}_i^{\text{match}} = Y_i^{\text{obs}} - Y_{m_i^c}^{\text{obs}},$$

is an unbiased estimator of the causal effect at $X = X_i$ for both units in the pair, and thus

$$\hat{\tau}_{\text{t}}^{\text{match}} = \frac{1}{N_{\text{t}}} \sum_{i:W_i=1} \hat{\tau}_i^{\text{match}} = \frac{1}{N_{\text{t}}} \sum_{i:W_i=1} \left(Y_i^{\text{obs}} - Y_{m_i^c}^{\text{obs}}\right) = \frac{1}{N_{\text{t}}} \sum_{i:W_i=1} \left(Y_i(1) - Y_{m_i^c}(0)\right),$$

(18.1)

is an unbiased estimator for the average treatment effect for the units in $\mathbb{I}_{\text{t}}$. The second implication is that

$$\hat{\mathbb{V}}\left(\hat{\tau}_{\text{t}}^{\text{match}}\right) = \frac{1}{N_{\text{t}}} \sum_{i:W_i=1} \left(Y_i^{\text{obs}} - Y_{m_i^c}^{\text{obs}} - \hat{\tau}_{\text{t}}^{\text{match}}\right)^2,$$

(18.2)

is a statistically conservative estimator of the sampling variance of the unbiased estimator in (18.1). We can also calculate exact p-values based on Fisher's approach, conditional on the N_{t} pairs. In both approaches, the analysis is entirely standard based on the results for the paired randomized experiment discussed in Chapter 10.

In practice, such an exact matching scheme is rarely feasible. The first impediment is that exact matching is typically impossible, and we must instead rely on "close" rather than exact matches, with a host of attendant complications. The second issue is that the pool of potential matches is often too small to ignore the conflicts that may arise when the same control is the best match for more than one treated unit. There are three general options to address this latter complication. One can explicitly match in such a way that the N_{t} matches remain distinct – matching without replacement. An alternative is to pick a particular order of the units and match the units in that order. A third possibility is to allow for duplication in the use of controls in the pairs (matching with replacement). In the remainder of this chapter, we discuss such methods and their attendant complications, as well as provide a number of practical ways to implement matching.

18.4 INEXACT MATCHING WITHOUT REPLACEMENT

In this section we discuss the conventional matching estimator, where we continue to match only the treated units without replacement of chosen controls (assuming $N_{\text{t}} <$

N_c), but now without assuming the existence of perfect matches for all units. For each of the N_t treated units we attempt to find the "closest" match within the set of all controls, $\mathbb{I}_c$, with respect to the covariates, thereby leading to N_t pairs. We would like to match the ith treated unit, with covariate values X_i, to control unit m_i^c, that is, the control unit that solves

$$m_i^c = \mathrm{argmin}_{i' \in \mathbb{I}_c} \|X_i - X_j\|, \tag{18.3}$$

where $\|x\|$ denotes a generic metric or distance function.[1] The solution to this minimization problem is control unit j that is the closest match to the treated unit being considered. When multiple controls are equally close matches, we could choose randomly one of them.

Even with a specified metric, there remains an issue with this approach. Solving Equation (18.3) for each treated unit separately may lead to the same control unit being selected as a match more than once. In other words, it may be that control unit $j \in \mathbb{I}_c$ is not only the best match for treated unit i but also for treated unit i'. Because at this stage we rule out matching with replacement, we cannot use control unit j as a match for both. There are two ways we can address this. The first is to attempt to match all units simultaneously to obtain the "optimal" allocation of matches across the full population $\mathbb{I}_t$. Formally, we can do this by minimizing an aggregate measure of the matching distances such as their sum. This amounts to simultaneously choosing the N_t indices $m_1, \ldots, m_{N_t} \in \mathbb{I}_c$ that solve

$$\mathrm{argmin}_{m_1^c, \ldots, m_{N_t}^c \in \mathbb{I}_c} \sum_{i=1}^{N_t} \|X_i - X_{m_i^c}\|, \qquad \text{subject to } m_i \neq m_{i'}, \text{ for } i \neq i'. \tag{18.4}$$

Although this "optimal matching" problem is straightforward to solve in settings with few units, it can become a demanding task computationally if the sample size is moderately large. Researchers therefore often follow an alternative approach by matching units sequentially, using what is often called a "greedy" or "nearest available matching" algorithm. In the first step, the first treated unit, $i = 1$, is matched to its closest control unit – ignoring the effect this choice has on subsequent matches – by solving

$$m_1^c = \mathrm{argmin}_{m_1^c \in \mathbb{I}_c} \|X_1 - X_j\|.$$

The second treated unit, $i = 2$, is then matched by searching over the remaining controls:

$$m_2^c = \mathrm{argmin}_{i' \in \mathbb{I}_c - \mathcal{M}_1^c} \|X_2 - X_j\|,$$

where the notation $\mathbb{I}_c - \mathcal{M}_1^c$ denotes the set of control units excluding the control unit matched to treated unit 1, $\mathcal{M}_1^c = \{m_1^c\}$. The ith treated unit is then matched to the closest

---

[1] We will discuss a number of choices for the distance metric in Section 18.5. For now it may be useful to think of the generic distance measure, where, for a K-dimensional vector x, $\|x - x'\| = \|x - x\|_V = ((x - x')V^{-1}(x - x')^T)^{1/2}$ for some positive semi-definite matrix V. This metric may not be a formal distance because $\|x - x'\|$ may be zero even when $x \neq x'$.

control unit in the set of all control units, excluding the first $i-1$ sets of control matches, leading to

$$m_i^c = \text{argmin}_{i' \in \mathbb{I}_c - \left(\cup_{i'=1}^{i-1} \mathcal{M}_{i'}^c\right)} \|X_i - X_j\|,$$

and so on, until all N_t treated units are matched.

It is important to realize that the result of this matching is now dependent on the ordering of the treated units. Rather than assigning this order randomly, researchers sometimes match first those units that are *a priori* most likely to be difficult to match. One such order is based on the estimated propensity score, the estimated probability of receiving treatment. Control units have, in expectation, a smaller estimated propensity score than treated units, and thus treated units with a larger value for their estimated propensity score tend to be more difficult to match. A common approach is therefore to match treated units based on the rank of their estimated propensity scores, starting with those with the highest value for the propensity score. Such a greedy matching algorithm is easier to implement than an optimal one, and the loss in terms of the criterion in (18.4) is often small. In fact, the chosen set of controls tends to be very similar across such matching orderings.

The result of the matching so far is, again, a set of pairs (i, m_i^c), for $i = 1, \ldots, N_t$, now with approximately – rather than exactly – the same values for all covariates. Hence, even under the assumption of unconfoundedness, the probability of assignment to the treatment may be now only approximately the same for both units in each pair. If we ignore this inexactness, we can once again rely on the paired randomized experiment results to obtain an approximately unbiased estimator for the average treatment effect on the treated, and its sampling variance, given in (18.1) and (18.2), respectively.

When searching for the best match for treated unit i, there may be two or more equally close control units. There are several ways one can deal with this issue. First, one can use the average of the outcomes for this set of tied matches as the estimate of the control potential outcome for treated unit i, $\sum_{j \in \mathcal{M}_i^c} Y_{i'}(0)/M_i$, where M_i is the cardinality of the set $\mathcal{M}_i^c$. Or, instead, one can use some mechanism for choosing among this set of potential matches, potentially by random selection. The first choice has the advantage of reducing the sampling variance of the resulting estimator for the treatment effect at X_i. It is also more systematic than randomly choosing among the set of potential matches. Yet it has the disadvantage of removing more units from the pool of possible control units available for subsequent matches. If the overall pool of possible control matches is relatively small, and if there are many ties, this method of using all potential matches may lead to poor-quality matches for the remaining treated units compared to randomly selecting one of the possible control matches.

Inference based on matching estimators that match without replacement is typically still based on the sampling variance estimator for paired randomized experiments given by Equation (18.2). Even though there is a potential bias in the estimator for the average treatment effect (formally, the expectation of the estimator conditional on the covariates is not exactly equal to the estimand), in practice this is ignored, which can be justified by appealing to special large-sample results where the size of $\mathbb{I}_c$ is much larger than the size of $\mathbb{I}_t$. See the notes at the end of the chapter for more details and formal results.

18.5   DISTANCE MEASURES

Before we can implement these ideas in practice, we must discuss how to operationalize the notion of "closeness" in practical situations when exact matching is not possible. Consider the case of a single covariate. In that case, one may, for example, choose between defining distance in terms of differences in levels or logarithms. Consider matching an individual who is 20 years old, with two potential matches, one individual age 15 and one age 26. In terms of levels, the first match is closer, with a difference of only 5 years rather than 6 years. However, if one considers the logarithm of age, so that the difference corresponds approximately to the percentage difference, the first match (between individuals age 20 and 15) corresponds to a difference of 0.29 versus a difference of only 0.26 for the second match (between individuals age 20 and 26). Hence the latter would be considered a closer match if closeness is measured on a logarithmic scale.

This problem of scaling, or transforming, the covariates is particularly relevant if one matches not on the original covariate but on some bounded function of it, such as the propensity score. In substantive terms, the difference between a probability of 0.01 and 0.06 (a sixfold increase) is often much larger than the difference between a probability of 0.06 and 0.11 (less than doubling), even though in both cases the difference in levels of the propensity score is equal to 0.05. In that case, an often more attractive metric is based on the linearized propensity score or log odds ratio, obtained by transforming the probability $e(x)$ into $\ell(x) = \ln(e(x)/(1 - e(x)))$, which would make the difference between probabilities of 0.01 and 0.06 equal to $|-4.60 - (-2.75)| = 1.84$, much bigger than the difference in terms of the linearized propensity score between probabilities of 0.06 and 0.11, namely $|-2.75 - (-2.09)| = 0.66$.

This problem of the choice of metric is compounded by the presence of multiple covariates, each of which can be continuous, discrete, or a simple indicator variable. A first, commonly used principle when choosing among possible distance metrics is that many covariates have no natural scale, and therefore one should use a metric that is invariant to their scale. Hence, after a transformation is chosen (e.g., logarithm versus level) for a covariate, researchers typically should normalize all covariates to a common variance before matching. However, even choosing a transformation and normalizing the result does not solve all issues with the choice of the metric. In settings with inexact matching and multiple covariates, there is a fundamental problem involving trading off the various covariates. In terms of the Card-Krueger example, if we want a match for a Burger King restaurant in New Jersey with 20 initial employees, should we prefer (as a control from the set of Pennsylvania restaurants) a Burger King with 23 initial employees, or a Kentucky Fried Chicken with 21 initial employees?

We consider distance metrics of the form $d_V(x, x') = (x'V^{-1}x)^{1/2}$ for a positive definite weight matrix V. A common choice for distance is the *Mahalanobis metric*, where the weight matrix is based on the average of the within-treatment-group sample covariance matrices:

$$V_M = \frac{1}{2} \cdot \left(\frac{1}{N_c} \sum_{i:W_i=0} (X_i - \overline{X}_c)^T \cdot (X_i - \overline{X}_c) + \frac{1}{N_t} \sum_{i:W_i=1} (X_i - \overline{X}_t)^T \cdot (X_i - \overline{X}_t) \right).$$

This metric takes account of correlations across covariates and leads to matches that are invariant to affine transformations of the covariates.[2] This is a particularly attractive property if most of the pre-treatment variables have no natural scale. The second choice we consider is what we call the *Euclidean metric*,

$$V_E = \text{diag}(V_M),$$

the diagonal matrix with variances on the diagonal ignoring the covariances. An even simpler metric is the sum of squared differences, without normalizing, which we do not recommend in general but use purely for illustrative purposes in Section 18.6.

Using the affinely invariant Mahalanobis metric can have possibly unexpected consequences. Consider the case where one matches on two highly correlated covariates X_1 and X_2 with equal variances. To be specific, assume that the correlation coefficient is equal to $\rho = 0.9$, and both variances are equal to $\sigma_X^2 = 1$. Suppose that we wish to find a match for a treated unit i, with $(X_{i1}, X_{i2}) = (0, 0)$. The two potential matches are control unit j with $(X_{j1}, X_{j2}) = (5, 5)$ and control unit j' with $(X_{j'1}, X_{j'2}) = (4, 0)$. The differences in covariates for the two matches are the vectors $X_i - X_j = (5, 5)$ and $X_i - X_{j'} = (4, 0)$, respectively. Some intuition suggests that the second match is better: it is strictly closer to the treated unit with respect to both covariates. Using the Euclidean metric, which sets the off-diagonal elements of V_M equal to zero, this is in fact true; the distance between the second potential match and the treated unit is $\|X_i - X_{j'}\|_{V_E} = 4$, considerably smaller than the distance to the first, $\|X_i - X_j\|_{V_E} = \sqrt{50} \approx 7.07$. By comparison, using the Mahalanobis metric, the distance to the first match is $\|X_i - X_j\|_{V_M} = \sqrt{5/0.19} \approx 5.13$, whereas the distance to the second is a much larger $\|X_i - X_{j'}\|_{V_M} = \sqrt{16/0.19} \approx 9.18$. Because of the correlation between the covariates in the sample, the difference in covariate values between the matches is interpreted differently by the two metrics.

To see why this situation arises, and to see the role of affine transformations, consider the artificial regressor $X_3 = (X_1 - \rho \cdot X_2)/\sqrt{1 - \rho^2} \approx (X_1 - 0.9 \cdot X_2)/\sqrt{0.19}$. Like X_1 and X_2, the third covariate has variance $\sigma_X^2 \cdot (1 - \rho^2)/0.19 = 1$. The pair of covariates (X_2, X_3) are an affine transformation of the pair of covariates (X_1, X_2). The transformation is chosen, however, so that X_2 and X_3 have zero correlation. Because the transformation is affine, the ranking of the matches does not change after the transformation according to the Mahalanobis distance, which is not true for the Euclidean distance. More precisely, the values of the X_3 regressor for the three units in the example are $X_{i3} = 0$, $X_{j3} = 0.5/\sqrt{0.19} \approx 1.15$, and $X_{j'3} = 4/\sqrt{0.19} \approx 9.18$. Thus, in terms of X_3, unit j is a better match for unit i than is unit j'. This is also true if we calculate the Euclidean and Mahalanobis distance based on covariates X_2 and X_3. Define $\tilde{X} = (X_2, X_3)'$. Based on the pair of covariates (X_2, X_3), the Euclidean distance between unit i and unit j is $\|\tilde{X}_i - \tilde{X}_j\|_{V_E} = \sqrt{25 + 16/0.19} \approx 10.45$. The Euclidean distance between unit i and unit j' is $\|\tilde{X}_i - \tilde{X}_{j'}\|_{V_E} \approx 5.13$. Because the correlation between X_2 and X_3 is zero, the Mahalanobis distance is identical to the Euclidean distance, and $\|\tilde{X}_i - \tilde{X}_j\|_{V_M} \approx 10.45$ and $\|\tilde{X}_i - \tilde{X}_{j'}\|_{V_M} \approx 5.13$. A choice between the Euclidean and Mahalanobis metrics corresponds implicitly to a stance on what the appropriate match would be in a case such as this. The choice of the Euclidean distance versus the Mahalanobis metric makes little

---

[2]   An affine transformation is a transformation of the form $x' = a + Bx$.

difference for estimating treatment effects in situations with low correlations between the covariates, as we will see in Section 18.12 when we calculate various matching estimates of the treatment effect of a minimum wage increase on employment levels.

One may wish to impose additional structure on the distance metric. For example, a particular indicator variable may be considered especially important so that the researcher may insist that it be matched exactly. In the evaluation of a medical treatment, for example, one may wish to impose that women exposed to the new treatment be matched solely to women exposed to the control treatment, and that men be matched solely to men, irrespective of differences in other characteristics. Similarly, in the example discussed here, one may require that restaurants subject to the new minimum wage law be matched only to restaurants in the same chain. More generally, one can choose a distance metric that assigns more weight to covariates that are considered more important *a priori* by increasing the relevant element of the matrix V^{-1} to increase its weight when building the scalar distance measure. Notice that "importance" here refers to the loss of credibility resulting from inexact matching on that particular component of X.

Ideally, when considering alternative distance metrics in the pursuit of estimating treatment effects for treated units, the intermediate goal is to obtain a metric that creates matched pairs (i, m_i^c) with $X_i = x$ and $X_{m_i^c} = x'$ such that the expected control outcomes at the covariate values, $\mathbb{E}_{sp}[Y_i(0)|X_i = x]$ and $\mathbb{E}_{sp}[Y_i(0)|X_i = x']$, are identical, or at least very similar. To achieve this objective, however, one would need to know the relationship between $Y_i(0)$ and X_i. In some situations it is possible to estimate this relation and use that information to choose between metrics. However, it is, in our view, unattractive to base the matching metric on a relation between potential outcomes and the covariates estimated on the same data set. Suppose, for example, that we estimate the conditional expectation $\mathbb{E}[Y_i(0)|X]$ based on a parsimonious model for the control potential outcomes in terms of the covariates. Matching units based on $\hat{\mathbb{E}}[Y_i(0)|X]$ can lead to results that are sensitive to the specification chosen. Remember that much of the appeal of the matching approach is precisely its lack of reliance on modeling the relationship between the potential outcomes and covariates in the data set at hand. Hence, making the construction of a matched sample depend on an initial estimation step that involves outcome data generally detracts from the general appeal of this approach. Moreover, matching is often used to create estimates of causal effects for more than one outcome variable.

18.6   MATCHING AND THE CARD-KRUEGER DATA

Initially we look at a small subset of these data, five restaurants in New Jersey and fifteen in Pennsylvania (listed in Table 18.3). The covariates used are the initial employment level (`initial empl`), measured prior to the minimum wage change (although not prior to its announcement, which could in principle create problems for this analysis), and the restaurant chain identity (`burger king` or `kfc`). Initial employment is a more or less continuous variable (not necessarily an integer because part-time workers are counted as fractions).

Suppose we want to match without replacement these five treated observations using a greedy algorithm. Consider the first, a New Jersey BK with 22.5 employees prior to the minimum wage increase (unit 1 in Table 18.3). Now let us look for the best match for this restaurant, that is, the most similar unit from Pennsylvania. Among the fifteen Pennsylvania restaurants in our sample, there are eleven BKs and four KFCs. In terms of initial employment, the closest restaurants are one with 25.5 employees (unit 9) and one with 20 (unit 11). Both are BKs, so it is clear that the closest match will be one of these. In terms of the absolute difference, unit 11 is clearly closer. In terms of logs, the initial employment value for unit 1 is 3.11, for unit 9 it is 3.24, and for 11 it is 3.00. Thus, unit 11, the closest match both in levels and in logarithms, seems to be the natural match.[3]

Skipping units 2 through 4 for the moment, consider matching next the fifth treated observation, a KFC with an initial employment of eight workers. There are four KFCs in the control (Pennsylvania) sample, although none with an employment level of exactly eight. There is also one BK with exactly eight employees (unit 20). The Pennsylvania KFC with employment closest to that of unit 5 is unit 8, with 8.5 initial employees. We therefore face a choice: Is it more important to match exactly on the initial number of employees, or to match exactly on the restaurant chain? In this case, we may think that a difference of half an employee (e.g., a single part-time worker) out of a total of eight is less important than matching exactly on chain. But suppose the nearest KFC restaurant had an initial employment that differed from that of unit 5 (eight employees) by more than three or four employees. At what point would we decide that the better match would be the BK restaurant with exactly eight initial employees?

As we discussed in Section 18.5 on distance metrics, it is clear that the choice of metric establishes a systematic trade-off between matching discrepancies in one variable versus the other. To do so, we first convert the indicator variable into a numerical measure. Suppose we code BK as "0" and KFC as "1." Now for each control we can calculate the covariate difference between itself and the treated unit being matched and convert this into a distance. Suppose we simply square the differences and sum them. In practice we would typically start by normalizing the covariate values, but to simplify the illustrative calculations here we omit this step. Then the distance between unit 5 and the two potential matches, units 8 and 20, is $1/4$ and 1, respectively. According to this criterion, unit 8 is closer. However, suppose we had instead coded the chains as "0" and "1/3." In that case the order would be reversed, with the distances now $1/4$ and $1/9$. When there is no particular reason to assign the indicator variable a difference of 1 across our two types, it is recommended to normalize the data to make the matching results invariant to such choices.

Thus far we have had to make two decisions, first the choice of matching order, and second the choice of distance metric. The three panels of Table 18.4 list the results of matching the five New Jersey restaurants varying the match order and the distance metric used. In each we match without replacement using a greedy algorithm.

In the first panel the treated units are matched in their original order and, for illustrative purposes, the metric used is the sum of the squared differences. Notice that unit 5 is *not* matched to unit 8 (the KFC with 8.5 employees discussed earlier), because unit 8 has

---

[3] Note, however, that it is easy to find strictly monotone transformations of numbers of employees such that unit 9 is closer to unit 1 than is unit 11.

Table 18.4. *The Roles of Match Order and Distance Metric, for the 20 Units from the Card and Krueger Fast-Food Restaurant Employment Data*

Match Order = 1,2,3,4,5; Metric = $x_1^2 + x_2^2$

i	m_i^c	Y_i^{obs}	$Y_{m_i^c}^{obs}$	$\hat{\tau}_i^{match}$
1	11	40.0	19.5	20.5
2	7	12.5	17	−4.5
3	15	20.0	22.5	−2.5
4	8	3.5	10.5	−7
5	20	5.5	8.0	−2.5
$\hat{\tau}_t^{match}$				+0.8

Match Order = 1,2,3,5,4; Metric = $x_1^2 + x_2^2$

i	m_i^c	Y_i^{obs}	$Y_{m_i^c}^{obs}$	$\hat{\tau}_i^{match}$
1	11	40.0	19.5	20.5
2	7	12.5	17.0	−4.5
3	15	20.0	22.5	−2.5
5	8	5.5	10.5	−5
4	20	3.5	8.0	−4.5
$\hat{\tau}_t^{match}$				+0.8

Match Order = 1,2,3,4,5; Metric = $100 \cdot x_1^2 + x_2^2$

i	m_i^c	Y_i^{obs}	$Y_{m_i^c}^{obs}$	$\hat{\tau}_i^{match}$
1	11	40.0	19.5	20.5
2	7	12.5	17.0	−4.5
3	15	20.0	22.5	−2.5
4	8	3.5	10.5	−7
5	17	5.5	14.0	−8.5
$\hat{\tau}_t^{match}$				−0.4

already been "used up" in matching unit 4. Hence, because we are matching without replacement, we are forced to settle for a lower-quality match. For each matched pair, we then estimate the unit-level treatment effect, $\hat{\tau}_i^{match} = Y_i^{obs} - Y_{m_i^c}^{obs} = Y_i(1) - Y_{m_i^c}(0)$. Across the five pairs, this process gives an estimated average treatment effect for the treated of +0.8 employees. (It may come as somewhat of a surprise to find a positive estimate, because all else being equal, standard economic theory predicts that a rise in the minimum wage will lower employment levels. But remember that this estimate is based on only five matched pairs.)

In the second panel, the metric remains the same, but the order changes: unit 5 is now matched before unit 4. This leads to a change in the matches: whereas in the first scheme unit 4 was matched to unit 8, and unit 5 was matched to unit 20, these matches are now reversed. Notice, however, that the estimator of the average treatment effect remains the same. Because the same set of five controls is being used, regardless of *which* treated units are being matched, the average post-treatment employment difference across the five pairs is unchanged.

In the third panel we return to the original order but change the distance metric effectively to require exact matching on the chain identity. In practice, this was done by adjusting the standard metric to multiply the square of the difference in chain by 100. Whereas before unit 5 (a New Jersey KFC with initial employment of 8) was matched to unit 20 (a Pennsylvania Burger King with equal initial employment), it is now matched to unit 17 (a Pennsylvania KFC with initial employment of 11). This adjustment in matches changes the estimate of the average treatment effect for the treated from $+0.8$ to -0.4.

18.7   THE BIAS OF MATCHING ESTIMATORS

We now return to the issue of the potential bias created by discrepancies between the pre-treatment covariates of the units within a matched pair. Consider the ith matched pair (i, m_i^c), where i indexes the treated unit. The unit-level treatment effect for the treated unit (i.e., the unit to be matched, as opposed to the unit used as a match) is equal to $\tau_i = Y_i(1) - Y_i(0)$. Because we can never simultaneously observe both potential outcomes for a given unit, we estimate this causal effect using the difference in observed outcomes for the two units of the matched pair:

$$\hat{\tau}_i^{\text{match}} = Y_i^{\text{obs}} - Y_{m_i^c}^{\text{obs}} = Y_i(1) - Y_{m_i^c}(0).$$

When the match is perfect, both units of this pair have covariate values equal to that for the matched unit, that is, $X_i = X_{m_i^c}$. With inexact matching, however, $X_i \neq X_{m_i^c}$. We call the difference in covariate values between the matched treated unit and its control match the *matching discrepancy*:

$$D_i = X_i - X_{m_i^c}.$$

Taking the super-population perspective, let

$$\mu_c(x) = \mathbb{E}_{\text{sp}}[Y_i(0)|X_i = x], \quad \text{and} \quad \mu_t(x) = \mathbb{E}_{\text{sp}}[Y_i(1)|X_i = x],$$

denote the super-population means for each potential outcome at covariate value $X = x$. If the matching discrepancy is equal to zero – an exact match – the expected difference in outcomes within the pair is equal to the average treatment effect conditional on $X_i = x$. That is, if $D_i = 0$, then the expected difference between outcomes within the pair is equal to the super-population average treatment effect for units with $X_i = x$:

$$\mathbb{E}_{\text{sp}}\left[Y_i^{\text{obs}} - Y_{m_i^c}\,\middle|\, W_i = 1, X_i = X_{m_i^c} = x\right] = \mathbb{E}_{\text{sp}}\left[Y_i(1) - Y_{m_i^c}(0)\,\middle|\, X_i = X_{m_i^c} = x\right]$$

$$= \mu_t(x) - \mu_c(x) = \tau(x).$$

In general, with a non-zero matching discrepancy, the expectation of the matching estimator of the unit-level treatment effect, which is the difference in observed outcomes in the matched pair, will be equal to

$$\mathbb{E}_{sp}\left[\hat{\tau}_i^{match}\middle|W_i=1,X_i,X_{m_i^c}\right]=\mathbb{E}_{sp}\left[Y_i(1)-Y_{m_i^c}(0)\middle|X_i,X_{m_i^c}\right]=\mu_t(X_i)-\mu_c(X_{m_i^c})$$
$$=\tau(X_i)+(\mu_c(X_i)-\mu_c(X_{m_i^c})).$$

We refer to the last term of this expression,

$$B_i=\mu_c(X_i)-\mu_c(X_{m_i^c}),$$

as the *unit-level bias* of the matching estimator.

A matching discrepancy D_i can lead to different levels of bias depending on the conditional expectation of the control outcome, $\mu_c(x)$. If this regression function does not depend on X, then clearly there is no discrepancy in these covariates that can introduce a bias. In general, the larger the absolute correlation between the covariates and the potential outcomes, the more bias a fixed matching discrepancy D_i can introduce.

In practice it will be easier to find good matches if the distributions of the covariates in the treatment and control groups are similar, that is, if there is much overlap between the two distributions. In contrast, if the propensity scores are concentrated near the endpoints – for the treated units near a propensity score of 1 and for the control units near a propensity score of 0 – it will be difficult to find close matches.

18.8   BIAS-CORRECTED MATCHING ESTIMATORS

In cases where matching is imperfect, there are several model-based approaches, all involving observed outcome data, one can use to attempt to reduce the unit-level bias created by the matching discrepancies. Each of these methods uses the within-pair pre-treatment covariate values X_i and $X_{m_i^c}$, combined with additional model-based adjustments, in an attempt to further reduce biases associated with differences in covariates. Here we introduce a general approach to bias adjustment and discuss its justification. In Sections 18.8.1 through 18.8.3, we then discuss three specific methods for applying this adjustment to the matching estimator.

Again consider a matched pair (i,m_i^c) where i indexes the treated unit, $i=1,\ldots,N_t$. As discussed earlier, the unadjusted estimator of the unit-level treatment effect is equal to $\hat{\tau}_i^{match}=Y_i^{obs}-Y_{m_i^c}^{obs}$, with expected value for this estimator, conditional on covariates and treatment indicators, equal to $\mathbb{E}_{sp}[\hat{\tau}_i^{match}|\mathbf{X},\mathbf{W}]=\mu_t(X_i)-\mu_c(X_{m_i^c})$. However, conditional on $\mathbf{X}$ and $\mathbf{W}$, the super-population expected treatment effect for the matched unit (the treated unit i) is $\tau(X_i)=\mu_t(X_i)-\mu_c(X_i)$. The difference is the unit-level bias for matched pair i:

$$B_i=\mathbb{E}_{sp}[Y_i(1)-Y_{m_i^c}(0)|\mathbf{X},\mathbf{W}]-\tau(X_i)=\mu_c(X_i)-\mu_c(X_{m_i^c}). \tag{18.5}$$

Three simple approaches have been proposed to reduce this bias, which modify the unadjusted unit-level estimate for the treatment effect, $\hat{\tau}_i^{match}$, by subtracting an estimate of the bias B_i in (18.5). Thus, instead of estimating the control outcome $Y_i(0)$ by the realized outcome for its match, $Y_{m_i^c}(0)$, we use

$$\hat{Y}_i(0) = Y_{m_i^c}(0) + \hat{B}_i,$$

which leads to the following bias-adjusted estimate of the average treatment effect:

$$\hat{\tau}_t^{adj} = \frac{1}{N_t} \sum_{i:W_i=1} \left(Y_i(1) - \hat{Y}_i(0) \right) = \frac{1}{N_t} \sum_{i:W_i=1} \left(Y_i(1) - Y_{m_i^c}(0) - \hat{B}_i \right).$$

Although it is conceptually straightforward to use more general functional forms, in practice, and in all three methods discussed in the following sections, the bias adjustment is based on a simple linear regression estimate of the conditional bias B_i.[4] Suppose the conditional mean of the potential outcome under the control treatment, $\mu_c(x) = \mathbb{E}_{sp}[Y_i(0)|X_i = x]$, is linear in the covariates:

$$\mu_c(x) = \alpha_c + x\beta_c. \tag{18.6}$$

For the subsequent discussion, it will be useful to specify an analogous equation for the conditional expectation of the potential outcomes given treatment, possibly with different parameters:

$$\mu_t(x) = \alpha_t + x\beta_t. \tag{18.7}$$

If Equation (18.6) holds, then the unit-level bias is $B_i = (X_i - X_{m_i^c})\beta_c = D_i\beta_c$, where $D_i = X_i - X_{m_i^c}$, the matching discrepancy. More generally, this approach can be thought of as approximating the difference $\mu_c(X_i) - \mu_c(X_{m_i^c})$ by a function linear in $X_i - X_{m_i^c}$. The three model-based approaches discussed here differ in the way they estimate the regression coefficients in this linear regression adjustment.

It is important to note that this approximation is conceptually distinct from the general regression approach discussed in Chapter 7. In that case we also approximate the regression function $\mu_c(x)$ by a linear function. However, there we relied on this approximation not just locally but across the full covariate space. We therefore were concerned about the sensitivity of the results to the specification chosen (e.g., the linearity of the regression function) because the distributions of the covariates may differ substantially between the two treatment levels. The current setting is different. Through matching, we have created a subsample in which the distributions of the covariates are likely to be well balanced between the two treatments. Hence, whereas with the full sample the regression function may be used to predict relatively far out of sample, here it is only used locally, and the corresponding results should be less sensitive to minor changes in the specification of the regression function. This statement does not suggest that the specification no longer matters at all, just that it is likely to matter less.

---

[4] It may be useful to use a more local estimate, for example, within strata defined by the covariates or by the propensity score.

18.8.1 Regression on the Matching Discrepancy

In the first bias-adjustment approach, we assume that the regression functions (18.6) and (18.7) are parallel:

$$\mu_c(x) = \alpha_d + x\beta_d, \qquad \text{and} \qquad \mu_t(x) = \tau + \mu_c(x) = \tau + \alpha_d + x\beta_d. \qquad (18.8)$$

We exploit this assumption by estimating the bias-adjustment coefficient β_d through a least squares regression of the within-pair difference in outcomes, $\hat{\tau}_i^{\text{match}} = Y_i^{\text{obs}} - Y_{m_i^c}^{\text{obs}}$ on the matching discrepancy, the within-pair difference in pre-treatment values, $D_i = X_i - X_{m_i^c}$.

To see why this works, consider the difference in observed outcomes, which for each pair is our unadjusted estimate of the unit-level treatment effect, $\hat{\tau}_i^{\text{unadj}} = Y_i^{\text{obs}} - Y_{m_i^c}^{\text{obs}}$. Using the notation introduced in (18.8), we can write this difference as

$$Y_i^{\text{obs}} - Y_{m_i^c}^{\text{obs}} = \tau_i(X_i) \qquad (18.9)$$

$$+ \left(\mu_c(X_i) - \mu_c(X_{m_i^c}) \right) \qquad (18.10)$$

$$+ (Y_i(1) - \mu_t(X_i)) - \left(Y_{m_i^c}(0) - \mu_c(X_{m_i^c}) \right). \qquad (18.11)$$

This equation states that $Y_i - Y_{m_i^c}$ is equal to the average treatment effect (18.9), plus the bias due to the matching discrepancy (18.10), plus, for each member of the pair, the difference between the observed outcome and its expected value, (18.11). Now let us define the residual

$$\nu_i = (Y_i(1) - \mu_t(X_i)) - \left(Y_{m_i^c}(0) - \mu_c(X_{m_i^c}) \right),$$

where ν_i is equal to the sum (18.10) and (18.11). We can then write the within-pair difference in observed outcomes, under the linear specification in (18.8), as

$$Y_i^{\text{obs}} - Y_{m_i^c}^{\text{obs}} = \tau + \left(X_i - X_{m_i^c} \right) \beta_d + \nu_i = \tau + D_i \beta_d + \nu_i. \qquad (18.12)$$

By definition, ν_i will have zero mean conditional on $\mathbf{X}$ and $\mathbf{W}$. Furthermore, because $D_i = X_i - X_{m_i^c}$ is a function of $\mathbf{X}$ and $\mathbf{W}$, it follows that ν_i also has mean zero conditional on D_i, for $i = 1, \ldots, N_t$. Hence we can use ordinary least squares to estimate the regression function in Equation (18.12) by regressing the within-pair outcome difference on the matching discrepancy, D_i, which leads to the following coefficient estimates for the slope parameters:

$$\hat{\beta}_d = \left(\sum_{i:W_i=1} (D_i - \overline{D})^T \cdot (D_i - \overline{D}) \right)^{-1} \left(\sum_{i:W_i=1} (D_i - \overline{D})^T \cdot (Y_i^{\text{obs}} - Y_{m_i^c}^{\text{obs}}) \right),$$

where $\overline{D} = \sum_{i:W_i=1} D_i / N_t$.

We then use $\hat{\beta}_d$ to adjust the outcome for the match within each pair, $Y_{m_i^c}(0)$:

$$\hat{Y}_i(0) = Y_{m_i^c}(0) + \hat{B}_i = Y_{m_i^c}(0) + \left(X_i - X_{m_i^c} \right) \hat{\beta}_d.$$

To calculate the bias-adjusted estimate of the average treatment effect, we then use these adjusted values $\hat{Y}_i(0)$ in place of the observed values $Y_{m_i^c}(0)$ in the standard equation for the estimated treatment effect:

$$\hat{\tau}_t^{\text{adj,d}} = \frac{1}{N_t} \sum_{i:W_i=1} \left(Y_i(1) - \hat{Y}_i(0) \right)$$

$$= \frac{1}{N_t} \sum_{i:W_i=1} \left(Y_i(1) - Y_{m_i^c}(0) - (X_i - X_{m_i^c})\hat{\beta}_d \right)$$

$$= \frac{1}{N_t} \sum_{i:W_i=1} \left(Y_i - Y_{m_i^c} - D_i\hat{\beta}_d \right) = \hat{\tau}_t^{\text{match}} - \overline{D}\hat{\beta}_d. \tag{18.13}$$

18.8.2 Control Regression on the Covariates

In the second bias-adjustment approach, we estimate the regression function (18.6) using all control units within the matched sample. We then use these regression coefficients to adjust the observed outcome for the match in a direction toward the expected outcome if the unit and its match had equal covariate values X_i. Specifically, in this approach we estimate the regression function

$$Y_{m_i^c} = \alpha_c + X_{m_i^c}\beta_c + \nu_{ci}, \tag{18.14}$$

where $\nu_{ci} = Y_{m_i^c} - \mu_0(X_{m_i^c})$. We estimate the regression using the control units in each of the N_t pairs. Thus, using the N_t controls, with outcomes $Y_{m_1^c}, \ldots, Y_{m_{N_t}^c}$ and covariate values $X_{m_1^c}, \ldots, X_{m_{N_t}^c}$, we estimate α_c and β_c as

$$\hat{\beta}_c = \left(\sum_{i:W_i=1} (X_{m_i^c} - \overline{X}_m^c)^T \cdot (X_{m_i^c} - \overline{X}_m^c) \right)^{-1} \left(\sum_{i:W_i=1} (X_{m_i^c} - \overline{X}_m^c) \cdot Y_{m_i^c} \right),$$

and

$$\hat{\alpha}_c = \overline{Y}_m^c - \overline{X}_m^c \hat{\beta}_c,$$

where $\overline{X}_m^c = \sum_{i:W_i=1} X_{m_i^c}/N_t$, and $\overline{Y}_m^c = \sum_{i:W_i=1} Y_{m_i^c}^{\text{obs}}/N_t$.

We use the estimated regression functions to adjust the potential outcomes for the matches within each pair. The adjusted potential control outcome is equal to

$$\hat{Y}_i(0) = Y_{m_i^c}(0) + (X_i - X_{m_i^c})\hat{\beta}_c.$$

Note that we do *not* replace the match control outcome by its value predicted by the regression function, $\hat{Y}_{m_i^c}(0) = \hat{\alpha}_c + X_{m_i^c}\hat{\beta}_c$. Instead, we simply adjust the observed outcome for the match by a relatively small amount $(X_i - X_{m_i^c})\hat{\beta}_c$.[5] The implied estimate

---

[5] Note that this is a small adjustment whenever unit i is fairly well matched, that is, whenever the matching discrepancy $X_i - X_{m_i^c}$ is small.

for the bias-adjusted average treatment effect is thus

$$\hat{\tau}_t^{\text{adj},c} = \frac{1}{N_t} \sum_{i:W_i=1} \left(Y_i(1) - \hat{Y}_i(0) \right)$$

$$= \frac{1}{N_t} \sum_{i:W_i=1} \left(Y_i^{\text{obs}} - Y_{m_i^c}^{\text{obs}} - (X_i - X_{m_i^c})\hat{\beta}_c \right) = \hat{\tau}_t^{\text{match}} - \overline{D}\hat{\beta}_c. \qquad (18.15)$$

The difference with the expression in (18.13) is in the estimator $\hat{\beta}_c$ in (18.15) versus $\hat{\beta}_d$ in (18.13).

18.8.3 Parallel Regressions on Covariates

Like the first, the third approach for bias-adjusting the simple estimate of the average treatment effect again restricts the slope coefficients to be equal in Equations (18.6) and (18.7). To estimate the adjustment coefficients, however, instead of regressing the difference in observed outcomes, $Y_i^{\text{obs}} - Y_{m_i^c}^{\text{obs}}$, on the matching discrepancy D_i, this approach instead estimates the regression function on the pooled sample of size $2 \cdot N_t$ constructed by stacking the treatment and control elements of each of the N_t pairs, that is, by ignoring the matching structure.

More formally, for each unit in this pooled sample of $2 \cdot N_t$ units (two from each matched pair), we record the unit's outcome, $\tilde{Y}_i$, its covariate value, $\tilde{X}_i$, and an indicator for whether it was a treated or a control unit, $\tilde{W}_i$. Note also that, by construction, we have exactly as many treated as control units in this pooled sample.

Given this artificial sample, we regress the outcome variable on a constant, the covariate values, and the treatment status indicator:

$$\tilde{Y}_i = \alpha_p + \tau_p \cdot \tilde{W}_i + \tilde{X}_i \beta_p + \nu_i. \qquad (18.16)$$

Then we estimate the average treatment effect as

$$\hat{\tau}_t^{\text{adj},p} = \frac{1}{N_t} \sum_{i:W_i=1} \left(Y_i(1) - \hat{Y}_i(0) \right)$$

$$= \frac{1}{N_t} \sum_{i:W_i=1} \left(Y_i^{\text{obs}} - Y_{m_i^c}^{\text{obs}} - (X_i - X_{m_i^c})\hat{\beta}^p \right) = \hat{\tau}_t^{\text{match}} - \overline{D}\hat{\beta}_p, \qquad (18.17)$$

which is numerically equivalent to the least squares coefficient $\hat{\tau}_p$ from the regression (18.16). The difference with the adjustments in (18.13) and (18.15) is the least squares estimator $\hat{\beta}_p$.

18.8.4 Bias-Adjustment for the Card-Krueger Data

Let us now see how these three bias-adjustment approaches work in our subsample of twenty observations from the Card and Krueger minimum wage data. Returning to our results from Section 18.6, the top panel of Table 18.4 gives the matched pairs, when we match, without replacement, the five treated (New Jersey) restaurants, using a greedy algorithm and the sum-of-squared differences as our naive distance metric. For these units, Table 18.5 presents some additional information: the covariate values (BK and KFC, with KFC coded as 1, and initial employment) for the treated and control members of each pair (X_i and $X_{m_i^c}$), the matching discrepancy D_i, the outcome variables (Y_i^{obs} and $Y_{m_i^c}^{obs}$), and the associated within-pair simple estimate of the treatment effect, $\hat{\tau}_i^{unadj}$. For example, in the first pair, the treated unit, unit $i = 1$, is a Burger King with an initial employment of 22.5 workers, $X_1 = (0, 22.5)'$, and its control match, unit $m_1^c = 11$, is also a Burger King with initial employment of 20.0, $X_{m_1^c} = (0, 20.0)'$. Hence the matching discrepancy for the first pair is $D_1 = (0, 2.5)'$. For all three bias-adjustment approaches, the adjustment would be zero if the matching were perfect with zero matching discrepancies.

In the first bias-adjustment approach, we regress, for the N_t pairs, the simple difference in matched outcomes, $\hat{\tau}_i^{unadj} = Y_i^{obs} - Y_{m_i^c}^{obs}$, on a constant and the matching discrepancies, $D_{i,1}$ and $D_{i,2}$. Using the five pairs listed in Table 18.5, the estimated regression function (listed in the first column of Table 18.6) is

$$\widehat{Y_i^{obs} - Y_{m_i^c}^{obs}} = -1.30 - 1.20 \cdot D_{i,1} + 1.43 \cdot D_{i,2}.$$

We can use these estimated regression coefficients to adjust the outcomes for the match within each pair, in this case the five controls. Following the approach in Section 18.8.1, our adjusted estimate of the unobserved potential outcomes therefore equals

$$\hat{Y}_i(0) = Y_{m_i^c}(0) + (X_i - X_{m_i^c})\hat{\beta}_d.$$

Applying these coefficients to our data, for the first matched pair we observe the control outcome $Y_{m_1^c}^{obs} = 19.5$ for unit 11 with covariate values $X_{m_1^c,1} = 0$ and $X_{m_1^c,2} = 20.0$. Because the covariate for the treated unit is $X_1 = (0, 22.5)$, the match discrepancy is $D_1 = (0, 2.5)$. Hence we adjust the imputed control outcome for the match, $\hat{Y}_1(0)$, from 19.5 to

$$\hat{Y}_1(0) = Y_{m_1^c} + D_1\hat{\beta}_d = 19.5 - 1.20 \cdot D_{1,1} + 1.43 \cdot D_{1,2}$$

$$= 19.5 - 1.20 \cdot 0 + 1.43 \cdot 2.5 = 23.1.$$

This gives an adjusted control outcome, $\hat{Y}_1(0)$, equal to $Y_{m_1^c}(0) + 3.6 = 19.5 + 3.6 = 23.1$, and an adjusted estimate of the unit-level treatment effect, $\hat{\tau}_1^{adj} = Y_1(1) - \hat{Y}_1(0)$, equal to 16.9. Following this same procedure for all five pairs, we find the adjusted control outcomes listed in Table 18.7. Averaging the corresponding adjusted estimates of the unit-level treatment effects gives a bias-adjusted estimate of the average causal effect for the New Jersey restaurants equal to 0.63 employees.

In the second bias-adjustment method, we estimate the regression function $\mu_c(x)$ separately using the N_t matched control units to get $\hat{\beta}_c$. Using our five pairs, regressing the five observed outcome values $Y_{m_i^c}^{obs}$ on a constant, $X_{m_i^c,1}$ and $X_{m_i^c,2}$, gives the following coefficients (listed in Column 2 of Table 18.6):

$$\hat{Y}_{m_i^c} = 4.21 + 2.65 \cdot X_{m_i^c,1} + 0.62 \cdot X_{m_i^c,2}.$$

Table 18.5. *Matching Discrepancy, Match Order is 1,2,3,4,5, Metric is $x_1^2 + x_2^2$, Matching without Replacement, for the 20 Units from the Card-Krueger New Jersey and Pennsylvania Minimum Wage Data*

i	m_i^c	Y_i^{obs}	$Y_{m_i^c}^{obs}$	$\hat{\tau}_i^{match}$	$X_{i,1}$	$X_{i,2}$	$X_{m_i^c,1}$	$X_{m_i^c,2}$	$D_{i,1}$	$D_{i,2}$
1	11	40.0	19.5	20.5	0	22.5	0	20.0	0	2.5
2	7	12.5	17.0	−4.5	1	14.0	1	13.8	0	0.2
3	15	20.0	22.5	−2.5	0	37.5	0	32.5	0	5.0
4	8	3.5	10.5	−7.0	1	9.0	1	8.5	0	0.5
5	20	5.5	8.0	−2.5	1	8.0	0	8.0	1	0

Table 18.6. *Bias-Adjustment Regression Coefficients for the 20 Units from the Card-Krueger New Jersey and Pennsylvania Minimum Wage Data*

	Difference Regression (Approach #1)	Control Regression (Approach #2)	Pooled Regression (Approach #3)
Regression coefficients			
Intercept	−1.30	4.21	12.01
Treatment indicator	−	−	1.63
Restaurant chain	−1.20	2.65	−7.32
Initial employment	1.43	0.62	0.39

Table 18.7. *First Bias-Adjustment Approach: Difference Regression for the 20 Units from the Card-Krueger New Jersey and Pennsylvania Minimum Wage Data*

i	m_i^c	$Y_i(1)$	$Y_{m_i^c}(0)$	$X_{i,1}$	$X_{i,2}$	$X_{m_i^c,1}$	$X_{m_i^c,2}$	$D_{i,1}$	$D_{i,2}$	$D_i\hat{\beta}_d^T$	$\hat{Y}_i(0)$
1	11	40.0	19.5	0	22.5	0	20.0	0	2.5	3.6	23.1
2	7	12.5	17.0	1	14.0	1	13.8	0	0.2	0.3	17.3
3	15	20.0	22.5	0	37.5	0	32.5	0	5.0	7.1	29.6
4	8	3.5	10.5	1	9.0	1	8.5	0	0.5	0.7	11.2
5	20	5.5	8.0	1	8.0	0	8.0	1	0	−1.2	6.8

$$\hat{\tau}_t^{match} = +0.8 \qquad\qquad \hat{\tau}_t^{adj} = -1.3$$

For the first pair this gives an adjusted control outcome of

$$\hat{Y}_1(0) = Y_{m_1^c}^{obs} + 2.65 \cdot D_{1,1} + 0.62 \cdot D_{1,2} = 19.5 + 2.65 \cdot 0 + 0.62 \cdot 2.5 = 21.1.$$

Following this same procedure for the remaining four pairs (summarized in Table 18.8), and averaging the unit-level results, leads to a bias-adjusted estimate of the average causal effect for the New Jersey restaurants equal to 0.74 employees.

In the third bias-adjustment method, we pool the data (so we have $2 \cdot N_t$ observations), and regress the unit-level outcome $\tilde{Y}_i$ on a constant, the two covariates $\tilde{X}_{i,1}$ and $\tilde{X}_{i,2}$, and

Table 18.8. *Second Bias-Adjustment Approach: Control Regressions for the 20 Units from the Card-Krueger New Jersey and Pennsylvania Minimum Wage Data*

i	m_i^c	$Y_i(1)$	$Y_{m_i^c}(0)$	$X_{i,1}$	$X_{i,2}$	$X_{m_i^c,1}$	$X_{m_i^c,2}$	$D_{i,1}$	$D_{i,2}$	$D_i\hat{\beta}_c^T$	$\hat{Y}_i(0)$
1	11	40.0	19.5	0	22.5	0	20.0	0	2.5	1.5	21.0
2	7	12.5	17.0	1	14.1	1	13.8	0	0.2	0.1	17.1
3	15	20.0	22.5	0	37.5	0	32.5	0	5.0	3.1	25.6
4	8	3.5	10.5	1	9.0	1	8.5	0	0.5	0.3	10.8
5	20	5.5	8.0	1	8.0	0	8.0	1	0	2.7	10.7

$$\hat{\tau}_t^{\text{match}} = +0.8 \qquad \hat{\tau}_t^{\text{adj}} = -0.7$$

Table 18.9. *Third Bias-Adjustment Approach: Pooled Regression for the 20 Units from the Card-Krueger New Jersey and Pennsylvania Minimum Wage Data*

i	m_i^c	$Y_i(1)$	$Y_{m_i^c}(0)$	$X_{i,1}$	$X_{i,2}$	$X_{m_i^c,1}$	$X_{m_i^c,2}$	$D_{i,1}$	$D_{i,2}$	$D_i\hat{\beta}_p^T$	$\hat{Y}_i(0)$
1	11	40.0	19.5	0	22.5	0	20.0	0	2.5	1.0	20.5
2	7	12.5	17.0	1	14.0	1	13.8	0	0.2	0.1	17.1
3	15	20.0	22.5	0	37.5	0	32.5	0	5.0	1.9	24.4
4	8	3.5	10.5	1	9.0	1	8.5	0	0.5	0.2	10.7
5	20	5.5	8.0	1	8.0	0	8.0	1	0	−7.3	0.7

$$\hat{\tau}_t^{\text{match}} = +0.8 \qquad \hat{\tau}_t^{\text{adj}} = +1.6$$

an indicator for the treatment received, $\tilde{W}_i$. The results for this regression using our five pairs (summarized in Column 3 of Table 18.6) are

$$\tilde{Y}_i = 12.01 + 1.63 \cdot \tilde{W}_i - 7.32 \cdot \tilde{X}_{i,1} + 0.39 \cdot \tilde{X}_{i,2}.$$

In this method, as in the first, we can read the bias-adjusted estimate of the average causal effect for the New Jersey restaurants directly from these results, here as the estimated coefficient on the treatment indicator $\tilde{W}_i$, equal to $+1.63$ employees. We can find this same result by using these coefficients to adjust the observed control outcomes. For the first pair the adjustment is now equal to

$$\hat{B}_i = -7.32 \cdot D_{1,1} + 0.39 \cdot D_{1,2} = -7.32 \cdot 0 + 0.39 \cdot 2.5 = 0.98,$$

and the adjusted control outcome is therefore $\hat{Y}_1(0) = Y_{m_1}(0) + 0.98 = 20.48$. Doing the same across all pairs and averaging (Table 18.9), we get a bias-adjusted estimate equal to $+1.63$, as expected.

We conclude this section with some general comments regarding the choice between the three bias-adjustment methods just discussed. There are some theoretical arguments in favor of the second. With sufficient data, one can make the associated regression function more flexible by including higher-order terms, allowing for approximations for $\mu_c(x)$ that become arbitrarily accurate. A comparable regression involving the differenced covariates (the first method) would have to involve differences in higher-order moments of the covariates – rather than higher-order moments of the matching discrepancy – in order to obtain accurate approximations of $\mu_c(X_i) - \mu_c(X_{m_i^c})$.

In practice, however, the choice between the three bias-adjustment approaches is likely to be less important than the decision whether or not to use a bias-adjustment method. In many cases, all three methods are preferable to that based on the simple average of within-pair differences, and, from limited experience, all are likely to be closer to one another than to the unadjusted estimate. In our example with only five matched pairs this is not the case, but as we will see in Section 18.12, when we expand the analysis to the full Card and Krueger data set, this similarity of answers does in fact hold.

18.9   MATCHING WITH REPLACEMENT

In this and the next two sections we study the second set of modifications to the basic matching estimator. This set of modifications includes changes to the matching approach in which there is no longer a single, distinct, match for each treated unit, either because we match and replace control units (this section), we use more than one match (Section 18.10), or we match both treated and control units (Section 18.11).

In this section we consider matching with replacement. Allowing a control unit to be used as a match more than once has both advantages and disadvantages. The first advantage is that it eases the computational burden. Now finding an optimal set of matches is straightforward: for each treated unit we choose its closest match within the entire set of control units. Recall that, for matching without replacement, the choices were either an optimal matching algorithm that was computationally cumbersome in large samples, or a sequential (greedy) matching algorithm. When we match with replacement, there is no such trade-off.

The second advantage of matching with replacement is that matching with replacement may reduce the bias of the matching estimators. Because we no longer restrict the set of matches, and thus allow some matches that were not available with distinct control matches, the discrepancy in pre-treatment covariates across matched pairs is reduced.

A disadvantage of matching with replacement is that the sampling variance of estimators based on matching with replacement is typically larger than the sampling variance of estimators based on matching without replacement. Intuitively, because control units can be used as matches more than once, the resulting estimator is typically based on fewer control units, which increases its sampling variance. A second drawback of matching with replacement is that the sampling variance is more difficult to estimate because using a control more than once creates correlations across pairs that share the same control matches.

Initially we ignore the possibility of ties. Let the first treated unit to be matched be unit $i = 1$. For this unit the optimal match is now m_1^c,

$$m_1^c = \text{argmin}_{j \in \mathbb{I}_c} \|X_1 - X_j\|.$$

Solving the same minimization problem for all treated units, we obtain a set of N_t pairs (i, m_i^c), for $i = 1, \ldots, N_t$. This set does not depend on the ordering of treated units, because the set from which we choose the match does not change. The average treatment effect for the treated is then estimated as

$$\hat{\tau}_t^{\text{repl}} = \frac{1}{N_t} \sum_{i:W_i=1} \left(Y_i^{\text{obs}} - Y_{m_i^c}^{\text{obs}} \right) = \frac{1}{N_t} \sum_{i:W_i=1} \left(Y_i(1) - Y_{m_i^c}(0) \right). \tag{18.18}$$

Now that we are matching with replacement, an important variable is the number of times each control unit is used as a match – let us call this $L(i) = \sum_{j=1}^{N} \mathbf{1}_{i \in \mathcal{M}_j^c}$ for control unit $i \in \mathbb{I}_c$; $L(i) = 0$ for all $i \in \mathbb{I}_t$ and a non-negative integer for all $i \in \mathbb{I}_c$, with $\sum_{i=1}^{N} L(i) = N_t$.[6] (When matching without replacement, $L(i) \in \{0, 1\}$ for all units.)

The simple matching estimator of the sample average treatment effect on the treated can be written as

$$\hat{\tau}_t^{\text{repl}} = \frac{1}{N_t} \sum_{i=1}^{N} \left(W_i \cdot Y_i^{\text{obs}} - (1 - W_i) \cdot L(i) \cdot Y_i^{\text{obs}} \right) \tag{18.19}$$

$$= \frac{1}{N_t} \sum_{i=1}^{N} \left(W_i \cdot Y_i(1) - (1 - W_i) \cdot L(i) \cdot Y_i(0) \right).$$

Notice that here we sum over *all* N units in the sample – hence the notation $Y_i(0)$ rather than $Y_{m_i^c}(0)$ – but continue to divide by N_t, the number of treated units and thus the number of matched pairs. This representation illustrates that the matching estimator is a weighted average of treated and control outcomes within the full sample. For the treated units the weights are all $1/N_t$, and for the control units the weights sum to one, but vary, with the value of the weight reflecting each control units' relative value as a comparison unit for the treated units.

18.10   THE NUMBER OF MATCHES

Although the discussion so far has focused on pairwise matching, where each observation is matched to a single unit, it is also possible to use multiple matches. Especially when the pool of possible control units is large relative to the number of treated units, one may be able to improve the precision of the resulting estimator by using more than one match. However, using multiple matches tends to increase the bias of the resulting estimator by increasing the average covariate discrepancy within pairs. With a sufficiently large number of possible matches, this need not be a problem, but it should be clear that using multiple matches does not come without possible costs.

Although the precision of the matching estimator can be improved by using multiple matches, the improvement is somewhat limited. To see this, consider the case where we match each treated unit to M controls. Let $\mathcal{M}_i^c$ represent the set of matches for unit i, with cardinality $\#\mathcal{M}_i^c = M$. (Before we considered the case with a single match so that the set $\mathcal{M}_i^c$ contained just a single element.) Suppose we have sufficient observations to find M exact matches for each treated unit without using the same control more than once.

---

[6]   Remember that we are still assuming no ties. As we discuss later, once we allow ties, $L(i)$ can take on non-integer values.

Let σ_c^2 and σ_t^2 be the super-population variances of $Y_i(0)$ and $Y_i(1)$ conditional on the covariates used for matching, respectively (implicitly assuming homoskedasticity with respect to the covariates). In that case the simple matching estimator using M matches is equal to

$$\hat{\tau}_t^{\text{match},M} = \frac{1}{N_t} \sum_{i=1}^{N_t} \left(Y_i(1) - \frac{1}{M} \sum_{j \in \mathcal{M}_i^c} Y_j(0) \right),$$

and the sampling variance of this estimator is

$$\mathbb{V}(\hat{\tau}_t^{\text{match},M}) = \frac{1}{N_t} \left(\sigma_t^2 + \frac{\sigma_c^2}{M} \right).$$

If we simplify by assuming that the two variances are equal, $\sigma_c^2 = \sigma_t^2$, the proportional reduction in sampling variance from using M matches rather than just a single match is equal to

$$\frac{\mathbb{V}(\hat{\tau}_t^{\text{match},1}) - \mathbb{V}(\hat{\tau}_t^{\text{match},M})}{\mathbb{V}(\hat{\tau}_t^{\text{match},1})} = \frac{M-1}{2M}.$$

Thus, using two matches reduces the sampling variance by 25% relative to using a single match, and using three reduces it by 33%. Increasing M, the reduction in sampling variance will rise toward 50%, but no higher. Thus, going beyond two or three matches can only lead to small improvements in the sampling precision in this simple setting.

We now describe how to implement the matching estimator using the M nearest matches. Let $m_i^{c,k} \in \mathbb{I}_c$ be the index for the control unit that solves

$$\sum_{j \in \mathbb{I}_c} \mathbf{1}_{\left\{ \|X_i - X_j\| \le \|X_i - X_{m_i^{c,k}}\| \right\}} = k, \tag{18.20}$$

that is, $m_i^{c,k}$ is the index of the control that is the kth closest unit to observation i. The set $\mathcal{M}_i^c$ now includes the closest M matches for unit i:

$$\mathcal{M}_i^c = \{m_i^{c,1}, m_i^{c,2}, \ldots, m_i^{c,M}\}.$$

Finally, defining

$$\widehat{Y_i(0)} = \frac{1}{M} \sum_{j \in \mathcal{M}_i^c} Y_j^{\text{obs}},$$

we can define the matching estimator for the average treatment effect on the treated as

$$\hat{\tau}_t^{\text{match},M} = \frac{1}{N} \sum_{i \in \mathbb{I}_t} \left(Y_i(1) - \widehat{Y_i(0)} \right) = \frac{1}{N} \sum_{i=1}^{N} \left(W_i - \frac{L(i)}{M} \right) \cdot Y_i^{\text{obs}}. \tag{18.21}$$

When there are ties for the Mth closest control match for treated unit i, this will mean that more than M units are at least as close to unit i as is unit $m_i^{c,M}$. If, as before, we use all ties, the number of units matched to unit i can therefore be greater than M. In this case, let M_i be the number of matches for unit i, again letting $\mathcal{M}_i^c$ denote the set of indices of those matches. The estimator is then the same as in Equation (18.21), but with M_i replacing M.

18.11 MATCHING ESTIMATORS FOR THE AVERAGE TREATMENT EFFECT FOR THE CONTROLS AND FOR THE FULL SAMPLE

So far we have focused the discussion on estimating the average effect of the treatment on the subpopulation of treated units. However, especially once we allow for matching with replacement, we can apply the same ideas to estimate the average effect of the treatment for the control units. Combining estimates for the average effect of the treatment for the controls and for the treated, we can also estimate the overall average effect of the treatment. In this section we discuss details of these extensions.

We focus on the bias-adjusted matching estimator for the treated units, based on matching with replacement, with a single match, and the bias adjustment based on the control regression:

$$\hat{\tau}_t^{\text{adj}} = \frac{1}{N_t} \sum_{i:W_i=1} \left(Y_i^{\text{obs}} - Y_{m_i^c}^{\text{obs}} - (X_i - X_{m_i^c})\hat{\beta}_c \right). \tag{18.22}$$

Here the matching set of controls for treated unit i is $\mathcal{M}_i^c = \{m_i^c\}$, with

$$m_i^c = \arg\min_{j:W_j=0} \|X_j - X_i\|,$$

based on, say the Mahalanobis metric and matching with replacement. The adjustment coefficient $\hat{\beta}_c$ is based on the regression of the outcomes for the N_t control matches on the covariates as in (18.15).

Let us first focus on estimating the average effect of the treatment for the controls. The analogous estimator is

$$\hat{\tau}_c^{\text{adj}} = \frac{1}{N_c} \sum_{i:W_i=0} \left(Y_{m_i^t}^{\text{obs}} - Y_i^{\text{obs}} - (X_{m_i^t} - X_i)\hat{\beta}_t \right). \tag{18.23}$$

Here the set of (treated) matches for *control* unit i is $\mathcal{M}_i^t = \{m_i^t\}$, with m_i^t the closest unit with the opposite treatment level:

$$m_i^t = \arg\min_{j:W_j=1} \|X_j - X_i\|,$$

based on, say the Mahalanobis metric and matching with replacement. The adjustment coefficient $\hat{\beta}_t$ is based on the regression of the outcomes for the N_c treated matches on the covariates as in analogy with (18.15).

Next, consider the case where we are interested in using a matching estimator for the average effect of the treatment for the entire sample, rather than only for the subsample of treated units or only the subsample of controls. Here we simply sum the estimates for the average treatment effect for the controls, $\hat{\tau}_c^{adj}$, and the average treatment effect for the treated, $\hat{\tau}_t^{adj}$, weighted by their shares in the sample, N_c/N and N_t/N, respectively, leading to

$$\hat{\tau}^{adj} = \frac{N_c}{N_c + N_t} \cdot \hat{\tau}_c^{adj} + \frac{N_t}{N_c + N_t} \cdot \hat{\tau}_t^{adj}. \qquad (18.24)$$

18.12   MATCHING ESTIMATES OF THE EFFECT OF THE MINIMUM WAGE INCREASE

Now we return to the full Card-Krueger data set with 347 restaurants, 279 in New Jersey and 68 in Pennsylvania. First we compare, for four different matching methods, the normalized average within-match difference in covariates. The second column in Table 18.10 gives the normalized differences in the seven covariates in the full sample, identical to those presented in Column 8 in Table 18.1. We then present, for various matching estimators, the average difference in covariates for the matched samples, normalized by $\sqrt{(s_c^2 + s_t^2)/2}$, where s_c^2 and s_t^2 are calculated on the full sample to facilitate the comparison with the balance in the full sample. Because we are primarily interested in the effect of the minimum wage increase in New Jersey, we initially match only the 279 New Jersey restaurants, not the 68 Pennsylvania restaurants.

The first matching estimator uses a single match, with replacement, using the Mahalanobus metric based on the average of the within-treatment group sample covariance matrices. The third column in Table 18.10 reveals that this greatly reduces the imbalance in the seven covariates. In the full sample one normalized difference is as large as 0.28, and four out of the seven normalized differences exceed 0.10. In the matched sample, all normalized differences are less than 0.10, with the largest equal to 0.07. Next, we use the Euclidean metric, ignoring correlations between the covariates. Third, in an attempt to decrease the sampling variance of the corresponding estimator, we increase the number of matches to three, albeit at the risk of increasing bias. And fourth and last, again with only one match, we use the Mahalanobis metric, but modified as discussed in Section 18.5 to first match exactly on restaurant chain. The results in Columns 4–6 in Table 18.10 show that the choice of matching method itself does not matter much for covariate balance in this example: all four methods lead to greatly improved balance compared to the full sample.

Table 18.11 reports the estimates of the average causal effect of the minimum wage increase on the New Jersey restaurants. To provide a baseline estimate, Table 18.11 first reports simple ordinary least squares estimates from the full sample, first without covariates (the simple difference between average outcomes for treated and controls, $\overline{Y}_t^{obs} - \overline{Y}_c^{obs}$), and second with the six covariates, `initial empl`, `burger king`, `kfc`, `roys`, `initial wage`, and `time until raise` (omitting `wendys`, because the four chain indicators add up to one). Ignoring the covariates gives an estimated treatment effect of -0.22 employees. Using covariates the estimator switches signs, to $+1.35$ employees.

Table 18.10. *Average Normalized Covariate Differences for the Card-Krueger New Jersey and Pennsylvania Minimum Wage Data*

Variable	Full Sample	Matched Samples			
		Euclidean	Euclidean	Mahalanobis	Exact on Chain Euclid on Others
		$M = 1$	$M = 4$	$M = 1$	$M = 1$
Initial employment	−0.28	0.06	0.10	0.06	0.07
Restaurant chain:					
Burger King	−0.02	−0.01	−0.01	−0.01	0.00
KFC	0.20	0.00	0.00	0.00	0.00
Roys	0.00	0.01	0.01	0.01	0.00
Wendys	−0.18	0.00	0.00	0.00	0.00
Starting wage	−0.05	0.07	−0.01	0.06	0.07
Time till first raise	−0.11	−0.01	0.05	−0.01	−0.01

Table 18.11. *Estimated Effect of Minimum Wage Increase on Employment for the Card-Krueger New Jersey and Pennsylvania Minimum Wage Data*

Estimand	Method	M	Metric	Estimate
	OLS, no controls			−0.22
New Jersey	OLS, controls			1.35
New Jersey	Match	1	Mahalanobis	0.89
New Jersey	Match	4	Mahalanobis	1.01
New Jersey	Match	1	Euclidean	0.93
New Jersey	Match	1	Exact on Chain, Mahal. on Others	0.92
Pennsylvania	Match	1	Mahalanobis	0.63
All	Match	1	Mahalanobis	0.84
New Jersey	Bias adj, dif regress	1	Mahalanobis	0.51
New Jersey	Bias adj, control regress	1	Mahalanobis	0.71
New Jersey	Bias adj, pooled regress	1	Mahalanobis	0.79

The next four estimates rely on the four matching methods with replacement for which we gave the covariate balance in Table 18.10 to motivate adjusting for covariate differences. The first matching estimator listed in Table 18.11 is for the average treatment effect for the New Jersey restaurants based on the Mahalanobis metric and a single match. As one can see in Table 18.11, this approach gives an estimated treatment effect equal to +0.89 employees. When we increase the number of matches to four, this gives an estimated treatment effect of +1.01.

Next consider the matching estimator with replacement based on the Euclidean metric and one match; this gives an estimated average effect for the restaurants in New Jersey equal to +0.93 employees. Thus, as we might predict, given comparable covariate distributions in the two matched samples, in this data set, using Mahalanobis versus the Euclidean distance has little effect because the covariates are nearly uncorrelated. Insisting that the matches are exact on the four-valued indicator for restaurant chain before matching the other covariates, the estimate drops slightly to +0.92 employees.

Table 18.12. *Bias-Adjusted Matching Estimators for the Card-Krueger New Jersey and Pennsylvania Minimum Wage Data*

Variable	Regression Coefficients		
	Difference Regression	Control Regression	Pooled Regression
Initial employment	0.50	0.12	0.35
Restaurant chain:			
KFC	−23.27	4.05	2.03
Roys	−	−3.62	−3.03
Wendys	−	−3.23	−2.00
Starting wage	−3.20	7.07	2.13
Time till first raise	−0.01	0.12	0.07
$\hat{\tau}_t^{adj}$	0.51	0.71	0.79

The next two entries in Table 18.11 report matching estimates of the average treatment effect for the controls – the expected effect on employment levels if Pennsylvania were to institute a comparable minimum wage increase – and the average treatment effect overall. Matching using the Mahalanobis metric and a single match gives an average effect for the restaurants in Pennsylvania equal to +0.63 employees and a sample average effect estimator of +0.84. Hence neither estimate varies substantially from our estimate of the average treatment effect for the New Jersey restaurants.

Returning to the original matched sample, based on a single match and the Euclidean metric, we explore the effect of using the bias-adjustment approaches discussed in Section 18.8. The estimated regression coefficients are reported in Table 18.12. When we apply the first approach – regressing the within-pair outcome difference $Y_i^{obs} - Y_{m_i^c}^{obs}$ on the matching discrepancy D_i – this gives a bias-adjusted estimate of the average effect for the New Jersey restaurants equal to +0.51 employees. Using the second approach, estimating the bias-adjustment coefficients by estimating $\mu_c(x)$, we get an estimated treatment effect equal to +0.71 employees. Using the third approach, estimating the bias-adjustment coefficients by estimating a regression using the pooled $2 \cdot N_t$ observations, gives an estimate of +0.79.

Overall, this exercise with a full data set illustrates the possible benefit of using the matching approach – its robustness to minor changes in its implementation. Unlike the two naive least squares estimates, which are very different from one another (even with different signs), all of the matching estimators are relatively close to one another, despite their conceptual differences. This robustness in this one example does not imply that these estimates are correct. But, as seen in this example, their robustness is a possible attraction of using matching methods in observational studies.

18.13   CONCLUSION

In this chapter we discuss matching methods for estimating causal effects. Whereas in Chapter 15 we discussed matching as a method for obtaining samples balanced in terms of covariate distributions, here we focus on the use of matching methods to construct estimators. We discuss matching with and without replacement, as well as cases where the estimand is the effect for the treated units, the control units, or the overall average causal effect. We look at different matching metrics and discuss the differences between them, and the use of linear regression methods on the set of units chosen by matching. Applying these methods to a data set collected by Card and Krueger suggest that these methods lead to robust estimates.

NOTES

There is a large literature on matching in statistics and social sciences, starting with more informal discussions (e.g., Peters and Van Voorhis, 1941, and Cochran, 1965) and continuing to the recent, more rigorous literature, that we view as starting with Cochran (1968), followed by Rubin (1970, 1973a, 1976ab). The literature continues at this moment, and more developments are likely. See Rubin (2006) for a number of influential papers going back to the early 1970s, and the introductions therein for a personal overview. Rosenbaum (1989ab, 1995, 2002, 2009) contain detailed discussions of various aspects of matching methods. For formal results in the econometrics literature see Abadie and Imbens (2006, 2009, 2012), and for an overview of the econometric literature, see Imbens (2004) and Imbens and Wooldridge (2009).

Gu and Rosenbaum (1993) discuss various matching algorithms, including optimal algorithms, as well as greedy algorithms that use sequential matching. They make the distinction between evaluating matching methods in terms of distance between matched units and in terms of balance in distributions, without regard to which units are matched (see also Rosenbaum and Rubin, 1984). Gu and Rosenbaum also suggest ordering the units by the propensity score before matching. Whereas in Chapter 15 we focused on global balance, in this chapter the goal is to estimate treatment effects. Cochran and Rubin (1973), Rubin (1973b, 1979), Quade (1982), Rubin and Thomas (2000), Espindle (2004), Abadie and Imbens (2006, 2009), and Rubin and Stuart (2006) discuss various aspects of matching. Gutman and Rubin (2014) discuss bias removal through the combination of spline regression and matching. Our discussion of the various specific bias-reduction methods in this chapter follows partly the discussions in Rubin (1973b) and Abadie and Imbens (2011). Abadie and Imbens (2006) establish large-sample properties regarding the bias of matching estimators with and without bias reduction. Abadie, Drukker, Herr, and Imbens (2003) describe implementations in STATA.

Most of the statistical literature has focused on matching without replacement, so that matched pairs are distinct and the focus is on average effects for the subpopulation of the treated units. Matching with replacement, which introduces complications when estimating sampling variances due to the common units across matched pairs, is discussed extensively in Abadie and Imbens (2006, 2008, 2009, 2010, 2012). We address sampling variance estimation in Chapter 19.

Other recently developed matching methods include genetic matching (Diamond and Sekhon, 2013), entropy matching (Hainmueller, 2012), and optimal full matching (Hansen and Klopfer, 2006). Heckman, Ichimura, and Todd (1997, 1998) study kernel matching methods where the multiple matches are weighted by their distance to the units being matched.

Matching on the estimated propensity score is discussed in Rosenbaum and Rubin (1983a, 1984). Formal asymptotic properties for such matching methods are derived in Abadie and Imbens (2012). These include the asymptotic variances for matching estimators for the average effect and the average effect for the treated. Influential applications include Dehejia (2005ab), Dehejia and Wahba (1999, 2002), Lechner (2002), and Smith and Todd (2001, 2005).

There are extensive simulation studies of matching methods in the literature. Cochran and Rubin (1973) focus on the average effect of the treatment for the treated, comparing regression estimators, matching estimators, and matching estimators with bias adjustment based on control regressions. Rubin (1973b) studies the properties of matching estimators for the average effect for the treated using the range of regression methods for bias adjustment discussed in the current chapter. Rubin (1979) also focuses on various bias adjustment methods in combination with single-nearest-neighbor matching. Rubin and Thomas (2000) compare covariate and propensity score matching methods, both in combination with regression adjustments. Waernbaum (2010) compares doubly robust estimators and matching estimators. Abadie and Imbens (2009) look at matching estimators with a substantial number of covariates and study the effect of bias adjustments based on linear regression. Frölich (2004ab), Zhao (2004), and Busso, DiNardo, and McCrary (2009) compare matching and weighting estimators. A common finding in these simulations is that the combination of regression adjustment with matching is superior to simply matching.

An alternative matching strategy uses outcome data to form matches based on best predictors of the outcomes given covariates. Such "predictive mean matching" strategies, also used in general missing data settings, are discussed in Rubin (1986b), Heitjan and Little (1991), Hansen (2008), and Frölich (2004).

Software for particular matching methods is available in R, Matlab, and STATA and at various websites for the authors of the articles cited previously. See Becker and Ichino (2002), Abadie, Drukker, Herr, and Imbens (2003), and Sekhon (2004–2013).

Card and Krueger (1994) do not use matching methods in their original analysis of the minimum wage data. Instead they use difference-in-difference methods. Rosenbaum (2002) re-analyzes their data using matching methods. The Card and Krueger data are available at http://www.princeton.edu/.

The employment variables used in this discussion are created as follows: initial employment = `emppt` × 0.5 + `empft`, and final employment = `emppt2` × 0.5 + `empft2`, where `emppt` refers to part-time employees, `empft` to full-time employees, and "2" refers to the post-minimum-wage measures. We use only those observations with complete data for each of these four employment variables, as well as for the other three matching variables.

## CHAPTER 19

A General Method for Estimating Sampling Variances for Standard Estimators for Average Causal Effects

19.1 INTRODUCTION

In Chapters 17 and 18, two general frequentist approaches for estimating causal effects were discussed, with special focus on estimating average causal effects. In order to conduct inference in those settings, it is important to have methods for estimating sampling variances so that we can construct large-sample confidence intervals. In the current chapter we discuss such methods. In doing so, a number of issues arise.

The first issue we raise concerns the choice of estimand. If we are interested in the average effect of the treatment, we need to be explicit about whether we are interested in the average effect in the sample, or in the average effect in the super-population from which the sample is hypothetically randomly drawn. Although this choice is generally immaterial for the estimation of causal effects, the associated sampling variances generally differ, even in large samples, and so will the corresponding estimators for the sampling variances, at least in settings allowing for heterogeneity in the causal effects. Thus, in such settings, the researcher faces a choice regarding the estimand and the estimator for the associated sampling variance.

Second, we face the choice as to whether we should construct estimators for the sampling variance tied to the specific method for estimating the average treatment effects or estimators that apply more generally. In the current chapter we emphasize the second approach, exploiting some of the properties shared by most standard estimators for average causal effects, and develop a general method for estimating sampling variances for such estimators. A key insight is that nearly all the estimators discussed in the previous chapters, as well as most others proposed in the literature, have a common structure. These estimators can be written as the difference between two terms, both weighted averages of observed outcomes. The first term is a weighted average of the observed outcomes for the treated units, and the second term is a weighted average of the observed outcomes for control units. The weight on the observed outcome for unit i depends on the level of the treatment for unit i, the levels of the treatment assignment for the other units, and the values of the set of pre-treatment variables (including the pre-treatment variables for other units). The weight is free, however, of dependence on any missing or observed potential outcomes for any unit. In addition, the weights in the first term (the weighted sum of the treated units) sum up to one, and the weights in the second term (the weighted sum of the control units) sum up to one. As a result, these estimators share the following three desirable properties, which we collectively refer to as *affine consistency*: (*i*), adding a constant c_t to all observed outcomes for treated units increases the estimated average causal effect by c_t; (*ii*), adding a constant c_c to all observed outcomes for control units decreases the estimated average effect by c_c; and (*iii*), changing the scale of the outcome by multiplying all observed outcomes by a constant c_s changes the estimated average effect by a factor c_s. All standard estimators for average causal effects proposed in the literature have this form and differ only in the functional form of the dependence of the weights on the treatment assignments and pre-treatment variables.

The sampling variance of any affinely consistent estimator for average treatment effects can be written as a simple function of the conditional unit-level potential outcome variances given covariates, the covariate values, and the treatment indicators. We discuss a matching-based method for estimating these unit-level conditional variances, using ideas from Chapter 18. We discuss how simple versions of these matching estimators for the unit-level variance may be improved by bias-adjustment methods. We also discuss, for both the blocking and the matching estimators discussed in detail in Chapters 17 and 18, specific estimators for the sampling variance appropriate for the particular estimation methods. Other options for estimating the sampling variances discussed in the current chapter include resampling methods such as the bootstrap, although there is both theoretical and simulation evidence that such methods may not work well for matching estimators.

To discuss the properties of the methods for estimating sampling variances in this chapter, we take a super-population perspective, where the sample of N units is viewed as a random sample from an infinite super-population, with the random sampling and randomization of the assignment vector given covariates together generating a joint distribution on the quadruple of covariates, treatment indicator, and the two potential outcomes. We should also note that the perspective taken here is entirely frequentist. Alternative approaches use multiple imputation to simulate draws of the missing potential outcomes under a Bayesian model on the potential outcomes, but currently there are only a few examples of such approaches in the literature, although they appear promising.

The data set used in this chapter to illustrate the methods is the Imbens-Rubin-Sacerdote lottery data set we previously used in Chapters 14 and 17. We briefly revisit these data in Section 19.2. In Section 19.3 we discuss possible estimands, and the implications the choice of estimand has for the sampling variance of estimators. In Section 19.4 we formulate the common structure of standard estimators for average causal effects. Next, in Section 19.5, we derive the general expression for the sampling variance conditional on covariates and treatment assignments. In Section 19.6 we propose estimators for the unit-level conditional sampling variance, including methods that use regression adjustment to account for inexact matching. In Section 19.7 we develop estimators for the sampling variance for the estimator for the sample average causal effect. In 19.8 we modify the methods for settings where the focus is on the average effect for the subsample of treated units. In Section 19.9 we discuss the problem of estimating the sampling variance when the focus is on estimating the super-population average treatment effect. In Section 19.10 we discuss two alternatives to the matching-based sampling variance estimators: first, one based on covariance adjustment methods, and second, methods based on resampling techniques such as the bootstrap. Section 19.11 concludes.

Table 19.1. *Summary Statistics for the Trimmed Sample, IRS Lottery Data*

	Losers ($N_c = 172$)		Winners ($N_t = 151$)		Nor
Covariate	Mean	(S.D.)	Mean	(S.D.)	Dif
Year Won	6.40	(1.12)	6.32	(1.18)	−0.06
# Tickets	2.40	(1.88)	3.67	(2.95)	0.51
Age	51.5	(13.4)	50.4	(13.1)	−0.08
Male	0.65	(0.48)	0.60	(0.49)	−0.11
Education	14.01	(1.94)	13.03	(2.21)	−0.47
Work Then	0.79	(0.41)	0.80	(0.40)	0.03
Earn Year -6	15.5	(14.0)	13.0	(12.4)	−0.19
Earn Year -5	16.0	(14.4)	13.3	(12.7)	−0.20
Earn Year -4	16.4	(14.9)	13.4	(12.7)	−0.22
Earn Year -3	16.8	(15.6)	14.3	(13.3)	−0.18
Earn Year -2	17.8	(16.4)	14.7	(13.8)	−0.20
Earn Year -1	18.4	(16.6)	15.4	(14.4)	−0.19
Pos Earn Year -6	0.71	(0.46)	0.71	(0.46)	−0.00
Pos Earn Year -5	0.70	(0.46)	0.74	(0.44)	0.10
Pos Earn Year -4	0.71	(0.46)	0.74	(0.44)	0.06
Pos Earn Year -3	0.70	(0.46)	0.72	(0.45)	0.03
Pos Earn Year -2	0.70	(0.46)	0.72	(0.45)	0.05
Pos Earn Year -1	0.72	(0.45)	0.71	(0.46)	−0.01

19.2   THE IMBENS-RUBIN-SACERDOTE LOTTERY DATA

We illustrate the ideas in this chapter using the Imbens-Rubin-Sacerdote lottery data, previously used in Chapters 14 and 17. The specific sample we use in this chapter is trimmed
using the propensity score, following the method discussed in Chapter 16, which leaves
us with a sample of size $N = 323$, of whom $N_c = 172$ are "losers" (people who won
only small, one-time prizes) and $N_t = 151$ are "winners" (people who won big prizes,
paid out in yearly installments over twenty years). Table 19.1 presents summary statistics for the trimmed sample for all eighteen basic pre-treatment variables, including the
averages and standard deviations by treatment status, and the normalized differences
$(\bar{X}_t - \bar{X}_c)/\sqrt{(s_t^2 + s_c^2)/2}$. (Note that these normalized differences are based on sample
variances in the trimmed sample, in contrast to the normalized difference in Table 17.1
in Chapter 17, where the focus was on the change in normalized differences when going
from the full sample to the trimmed sample.)

As before, we are interested in the average effect of winning a big prize in the lottery versus being a loser on subsequent earnings for some set of units to be specified
subsequently. The specific outcome we use is the average of yearly earnings over the
first six years after winning the lottery, measured by averaging social security earnings
in thousands of 1995 dollars. We apply three estimators for average treatment effects to
this sample. First, we implement the blocking estimator described in detail in Chapter 17
with the tuning parameters recommended in that chapter. As reported in Chapter 17, this
leads to five subclasses based on the estimated propensity score and, after least squares

regression in each subclass with the full set of eighteen covariates, a point estimate for the average treatment effect equal to a reduction in annual labor earnings of 5.74 (in thousands of 1995 dollars). Second, we apply a bias-adjusted matching estimator discussed in Chapter 18. We use the Mahalanobis metric based on all eighteen covariates with a single match ($M = 1$), with replacement, followed by bias-adjustment based on all eighteen covariates; this leads to a point estimate of -4.54. Third, we use the same matching estimator with $M = 4$ matches, leading to a point estimate of -5.03.

19.3 ESTIMANDS

First let us discuss the choice of estimand. This discussion builds on the discussion of finite-sample and super-population average treatment effects in the context of randomized experiments in Chapter 6, but in the current context, there are some additional implications of this choice that are often ignored in the empirical literature. Recall the definition of the finite-sample average effect of the treatment, averaged over the N units in the finite sample,

$$
\tau_{\text{fs}} = \frac{1}{N} \sum_{i=1}^{N} \big(Y_i(1) - Y_i(0) \big),
$$

and the super-population average treatment effect,

$$
\tau_{\text{sp}} = \mathbb{E}_{\text{sp}} \big[Y_i(1) - Y_i(0) \big] = \mathbb{E}_{\text{sp}} \big[\tau_{\text{fs}} \big],
$$

where, as before, the subscript "sp" on the expectation operator indicates that the expectation is taken over the distribution induced by random sampling from an (infinite) super-population. In most of this chapter we focus on average effects for the entire sample or population, rather than for the subsample or subpopulation of the treated. Conceptually the extension to the case where the focus is on the average effect for the treated is straightforward, and we discuss this extension in Section 19.8.

The difference between the two estimands, τ_{fs} and τ_{sp}, is *not* important for estimation in a setting where we have a random sample from the population because the random sampling implies $\tau_{\text{sp}} = \mathbb{E}_{\text{sp}}[\tau_{\text{fs}}]$; this in turns implies that an estimator $\hat{\tau}$ that is attractive for estimating the sample average treatment effect is, in this setting, also attractive for estimating the population average effect. Therefore, the researcher need not make a distinction between the estimands for the purpose of point estimation. The difference between the estimands, τ_{fs} and τ_{sp}, is important, however, for inference (i.e., interval estimation): the sampling variance for a generic estimator $\hat{\tau}$ is

$$
\mathbb{V}_W \left(\hat{\tau} \right) = \mathbb{E}_W \left[\left(\hat{\tau} - \tau_{\text{fs}} \right)^2 \right],
$$

(where, as before, the subscript "W" on the expectation or variance operators indicates that expectations are taken only over the randomization distribution induced by the assumed regular assignment mechanism) is, in general, different from

$$\mathbb{V}\left(\hat{\tau}\right) = \mathbb{E}\left[\left(\hat{\tau} - \tau_{sp}\right)^2\right].$$

(Recall that expectations and variances without a subscript "W" or "sp" are taken over both the randomized treatment assignment and over the random sampling from the super-population.) As we will see, the approximate difference is

$$\mathbb{V}(\hat{\tau}) - \mathbb{V}_W(\hat{\tau}) \approx \mathbb{V}_{sp}(\tau(X_i))/N.$$

To illustrate this difference in sampling variances, let us start with a simple example. Suppose we have a single, binary, pre-treatment variable, for example, sex, $X_i \in \{f, m\}$. Let $N(f)$ and $N(m)$ be the number of females (units with $X_i = f$) and males (units with $X_i = m$) respectively in the finite sample. For $x \in \{f, m\}$, let $N_c(x)$, $N_t(x)$, and $N(x)$ denote the number of control, treated, and all units with $X_i = x$, and let $\overline{Y}_c^{obs}(x)$ and $\overline{Y}_t^{obs}(x)$ denote the average observed outcomes for control and treated units with covariate value $X_i = x$:

$$N_c(x) = \sum_{i:X_i=x}(1 - W_i), \quad N_t(x) = \sum_{i:X_i=x} W_i, \quad N(x) = N_c(x) + N_t(x),$$

$$\overline{Y}_c^{obs}(x) = \frac{1}{N_c(x)}\sum_{i:X_i=x}(1 - W_i) \cdot Y_i^{obs}, \quad \text{and} \quad \overline{Y}_t^{obs}(x) = \frac{1}{N_t(x)}\sum_{i:X_i=x} W_i \cdot Y_i^{obs}.$$

Finally, let $\tau_{fs}(x)$ and $\tau_{sp}(x)$ denote the average causal effect for units with $X_i = x$ in the sample and the population respectively, for $x = f, m$:

$$\tau_{fs}(x) = \frac{1}{N(x)}\sum_{i:X_i=x}\left(Y_i(1) - Y_i(0)\right), \quad \text{and} \quad \tau_{sp}(x) = \mathbb{E}_{sp}\left[Y_i(1) - Y_i(0)|X_i = x\right].$$

Suppose that treatment assignment is super-population unconfounded,

$$W_i \perp\!\!\!\perp \left(Y_i(0), Y_i(1)\right) \mid X_i,$$

and suppose there is at least some overlap in the covariate distributions in the sample, so that $N_c(f)$, $N_c(m)$, $N_t(f)$, and $N_t(m)$ are all strictly positive. Under these assumptions, natural estimators for $\tau_{fs}(x)$ and $\tau_{sp}(x)$ are

$$\hat{\tau}^{dif}(x) = \overline{Y}_t^{obs}(x) - \overline{Y}_c^{obs}(x), \quad \text{for} \quad x = f, m. \tag{19.1}$$

A natural estimator for the sample average treatment effect, τ_{fs}, is the weighted average of the estimators for the two subsamples, with the weights equal to the proportions of the two subsamples:

$$\hat{\tau}^{strat} = \frac{N(f)}{N(f) + N(m)} \cdot \hat{\tau}^{dif}(f) + \frac{N(m)}{N(f) + N(m)} \cdot \hat{\tau}^{dif}(m). \tag{19.2}$$

This estimator, $\hat{\tau}$, is also a natural estimator for τ_{sp}, unless we have additional information about the proportions of males and females in the super-population beyond the sample proportions.

Now let us consider the sampling variances of these estimators, as well as estimators for these sampling variances. First we focus on the estimators for the within-subpopulation average treatment effects $\hat{\tau}^{dif}(x)$, for $x \in \{f, m\}$, and then we turn to the estimator for the overall average effect, $\hat{\tau}$. The sampling variance for $\hat{\tau}^{dif}(x)$, given random assignment conditional on the pre-treatment variable, following the discussion in Chapter 6 (see in particular Equation 6.4) is

$$\mathbb{V}_W(\hat{\tau}^{dif}(x)) = \mathbb{E}_W\left[\left(\hat{\tau}^{dif}(x) - \tau_{fs}(x)\right)^2\right] = \frac{S_c^2(x)}{N_c(x)} + \frac{S_t^2(x)}{N_t(x)} - \frac{S_{ct}^2(x)}{N(x)}.$$

The numerators of the first two terms in the variance are

$$S_c^2(x) = \frac{1}{N(x) - 1} \sum_{i:X_i=x} \left(Y_i(0) - \frac{1}{N(x)} \sum_{i':X_{i'}=x} Y_{i'}(0)\right)^2,$$

and

$$S_t^2(x) = \frac{1}{N(x) - 1} \sum_{i:X_i=x} \left(Y_i(1) - \frac{1}{N(x)} \sum_{i':X_{i'}=x} Y_{i'}(1)\right)^2,$$

respectively. Recall, by analogy with the discussion in Chapter 6 on Neyman's repeated sampling perspective, that the numerator in the third term equals the variance of the unit-level treatment effect in the subsample with $X_i = x$:

$$S_{ct}^2(x) = \frac{1}{N(x) - 1} \sum_{i:X_i=x} \left(Y_i(1) - Y_i(0) - \tau_{fs}(x)\right)^2,$$

which vanishes if the treatment effect is constant in the subsample with $X_i = x$. In Chapter 6 we discussed in detail the difficulties with estimating the third term, and the reasons for commonly ignoring this term. As a result, we commonly estimate the (so-called conservative) sampling variance

$$\mathbb{V}_W\left(\hat{\tau}^{dif}(x)\right) = \frac{S_c^2(x)}{N_c(x)} + \frac{S_t^2(x)}{N_t(x)}. \tag{19.3}$$

The two numerators in the expression for the sampling variance in (19.3), $S_c^2(x)$ and $S_t^2(x)$, are unknown, but an unbiased estimator for (19.3) is available (again, see the discussion in Chapter 6). Letting

$$s_c^2(x) = \frac{1}{N_c(x) - 1} \sum_{i:W_i=0,X_i=x} \left(Y_i^{obs} - \overline{Y}_c^{obs}(x)\right)^2,$$

and

$$s_t^2(x) = \frac{1}{N_t(x) - 1} \sum_{i:W_i=1,X_i=x} \left(Y_i^{obs} - \overline{Y}_t^{obs}(x)\right)^2,$$

we have the following, Neyman-type, statistically conservative estimator for the sampling variance of $\hat{\tau}(x)$:

$$\hat{\mathbb{V}}_W(\hat{\tau}^{dif}(x)) = \frac{s_c(x)^2}{N_c(x)} + \frac{s_t(x)^2}{N_t(x)}. \tag{19.4}$$

Now let us turn to the sampling variance of $\hat{\tau}^{dif}(x)$ as an estimator of the super-population average effect $\tau_{sp}(x)$. Using the results from Chapter 6 (see in particular Equation 6.14), we find:

$$\mathbb{V}\left(\hat{\tau}^{dif}(x)\right) = \mathbb{E}\left[\left(\hat{\tau}^{dif}(x) - \tau_{sp}(x)\right)^2\right] = \frac{\sigma_c^2(x)}{N_c(x)} + \frac{\sigma_t^2(x)}{N_t(x)},$$

where $\sigma_c^2(x)$ and $\sigma_t^2(x)$ are the super-population variances of $Y_i(0)$ and $Y_i(1)$ in the sub-population with $X_i = x$, respectively. We do not know $\sigma_c^2(x)$ and $\sigma_t^2(x)$, but unbiased estimators for these variances exist in the form of $s_c^2(x)$ and $s_t^2(x)$, leading to an estimated sampling variance identical to (19.4). Thus, in terms of the estimated sampling variance of $\hat{\tau}^{dif}(x)$, it is immaterial whether we focus on $\hat{\tau}^{dif}(x)$ as an estimator for the finite-sample estimand $\tau_{fs}(x)$, or as an estimator for the super-population estimand $\tau_{sp}(x)$ – in both cases the expression in (19.4) gives a natural estimator for the sampling variance, in the former case generally an upwardly biased estimator, and in the latter case an unbiased estimator.

This situation, however, changes when we focus on the estimator $\hat{\tau}^{strat}$ for the overall average treatment effect. First, the sampling variance of $\hat{\tau}^{strat}$ in (19.2) as an estimator of the sample average effect τ_{fs} is

$$\mathbb{V}_W\left(\hat{\tau}^{strat}\right) = \mathbb{E}_W\left[\left(\hat{\tau}^{strat} - \tau_{fs}\right)^2\right]$$

$$= \left(\frac{N(f)}{N(f) + N(m)}\right)^2 \cdot \left(\frac{S_c^2(f)}{N_c(f)} + \frac{S_t^2(f)}{N_t(f)} - \frac{S_{ct}^2(f)}{N(f)}\right)$$

$$+ \left(\frac{N(m)}{N(f) + N(m)}\right)^2 \cdot \left(\frac{S_c^2(m)}{N_c(m)} + \frac{S_t^2(m)}{N_t(m)} - \frac{S_{ct}^2(m)}{N(m)}\right).$$

The natural (but conservative) estimator for this sampling variance is based on ignoring the $S_{ct}^2(f)$ and $S_{ct}^2(m)$ terms, and replacing $S_t(x)^2$ by $s_t(x)^2$, and $S_c(x)^2$ by $s_c(x)^2$ for $x = f, m$, leading to:

$$\hat{\mathbb{V}}_W(\hat{\tau}^{strat}) = \left(\frac{N(f)}{N(f) + N(m)}\right)^2 \cdot \left(\frac{s_c^2(f)}{N_c(f)} + \frac{s_t^2(f)}{N_t(f)}\right) \tag{19.5}$$

$$+ \left(\frac{N(m)}{N(f) + N(m)}\right)^2 \cdot \left(\frac{s_c^2(m)}{N_c(m)} + \frac{s_t^2(m)}{N_t(m)}\right).$$

Second, consider the sampling variance of $\hat{\tau}^{\text{strat}}$ in (19.2) as an estimator of the population average effect, τ_{sp}:

$$
\begin{aligned}
\mathbb{V}(\hat{\tau}^{\text{strat}}) &= \mathbb{E}\left[\left(\hat{\tau}^{\text{strat}} - \tau_{\text{sp}}\right)^2\right] \\
&= \mathbb{E}_{\text{sp}}\left[\left(\left(\hat{\tau} - \left(\frac{N(f)}{N(f) + N(m)} \cdot \tau_{\text{sp}}(f) + \frac{N(m)}{N(f) + N(m)} \cdot \tau_{\text{sp}}(m)\right)\right)\right.\right. \\
&\qquad\qquad \left.\left. + \left(\left(\frac{N(f)}{N(f) + N(m)} \cdot \tau_{\text{sp}}(f) + \frac{N(m)}{N(f) + N(m)} \cdot \tau_{\text{sp}}(m)\right) - \tau_{\text{sp}}\right)\right)^2\right] \\
&= \mathbb{E}\left[\left(\frac{N(f)}{N(f) + N(m)}\right)^2 \cdot \left(\hat{\tau}^{\text{dif}}(f) - \tau_{\text{sp}}(f)\right)^2 + \left(\frac{N(m)}{N(f) + N(m)}\right)^2\right. \\
&\qquad \left. \cdot \left(\hat{\tau}^{\text{dif}}(m) - \tau_{\text{sp}}(m)\right)^2 + \left(\frac{N(f)}{N(f) + N(m)} - q(f)\right)^2 \cdot \left(\tau_{\text{sp}}(f) - \tau_{\text{sp}}(m)\right)^2\right].
\end{aligned}
$$

A natural estimator for the sampling variance of $\hat{\tau}^{\text{strat}}$ as an estimator of τ_{sp} is

$$
\begin{aligned}
\hat{\mathbb{V}}(\hat{\tau}^{\text{strat}}) &= \left(\frac{N(f)}{N(f) + N(m)}\right)^2 \cdot \left(\frac{s_c^2(f)}{N_c(f)} + \frac{s_t^2(f)}{N_t(f)}\right) + \left(\frac{N(m)}{N(f) + N(m)}\right)^2 \\
&\quad \cdot \left(\frac{s_c^2(m)}{N_c(m)} + \frac{s_t^2(m)}{N_t(m)}\right) + \frac{1}{N} \cdot \frac{N(f) \cdot N(m)}{(N(f) + N(m))^2} \cdot \left(\hat{\tau}^{\text{dif}}(f) - \hat{\tau}^{\text{dif}}(m)\right)^2 \\
&= \hat{\mathbb{V}}_W(\hat{\tau}^{\text{strat}}) + \frac{N(f) \cdot N(m)}{N^3} \cdot \left(\hat{\tau}^{\text{dif}}(f) - \hat{\tau}^{\text{dif}}(m)\right)^2. \qquad (19.6)
\end{aligned}
$$

Because

$$
\mathbb{V}_{\text{sp}}(\tau(X_i)) = \frac{N(f) \cdot N(m)}{N^2} \cdot (\tau(f) - \tau(m))^2,
$$

the difference between $\hat{\mathbb{V}}(\hat{\tau}^{\text{strat}})$ and $\hat{\mathbb{V}}_W(\hat{\tau}^{\text{strat}})$, the final term on the right-hand side of (19.6), can be approximated by

$$
\hat{\mathbb{V}}(\hat{\tau}^{\text{strat}} - \tau_{\text{sp}}) - \hat{\mathbb{V}}_W(\hat{\tau}^{\text{strat}} - \tau_{\text{fs}}) \approx \frac{1}{N} \cdot \mathbb{V}_{\text{sp}}(\tau(X_i)),
$$

the variance, over the super-population, in the treatment effect conditional on the pre-treatment variable. The interpretation of this difference is that if we are interested in the average effect for the super-population, and if the treatment effect varies by the value of the pre-treatment variables (here, if $\tau(f) \neq \tau(m)$), we need to take into account the difference between the distribution of the pre-treatment variable in our sample and its distribution in the population. In the example with the binary covariate, sex, the proportion of women in the sample is $\hat{q}(f) = N(f)/(N(f) + N(m))$, but in the population it is $q(f)$, with the sampling variance of the difference between these two proportions equal to $q(f)q(m)/N$, traditionally estimated as $\hat{q}(f)\hat{q}(m)/N = N(f)N(m)/N^3$. Because the last term in (19.6) is of the same order of magnitude as the other terms, taking it into account will generally matter, even in large samples.

Although the extension from the scalar binary pre-treatment variable to the general case with multiple, and multi-valued, pre-treatment variables is algebraically messy, a similar distinction arises between the sampling variance of an estimator of the sample average effect and the sampling variance of an estimator of the population average effect, with approximately,

$$\mathbb{V}(\hat{\tau}^{\,\text{strat}}) \approx \mathbb{V}_W(\hat{\tau}^{\,\text{strat}}) + \mathbb{V}_{\text{sp}}\Big(\tau(X_i)\Big)/N. \qquad (19.7)$$

In this chapter, we present estimators for the general version of both (19.5), in Section 19.7, and (19.6), in Section 19.9. However, our view is that, in general, one should focus on the sampling variance of an estimator viewed as an estimator of the sample average effect rather than viewed as an estimator of the super-population average effect. Thus we recommend focusing on the generalization of (19.5), rather than taking into account differences between the distribution of the pre-treatment variables in the sample and the analogous distribution in a somewhat vague, hypothetical, and often ill-defined, super-population.

19.4   THE COMMON STRUCTURE OF STANDARD ESTIMATORS FOR AVERAGE TREATMENT EFFECTS

Most estimators for average treatment effects, including those discussed in Chapters 12, 17, and 18, have a common structure, which is that each can be written as a linear combination of observed outcomes, with specific restrictions on the coefficients. Viewed as a property of estimators, we refer to this structure as *affine consistency* of the estimators, defined in Section 19.1. This property has intuitive appeal, and estimators that do not have this property often have particular unattractive features. In this section we explore this structure, and in Sections 19.5–19.7 we exploit it to develop expressions and estimators for their sampling variances.

19.4.1 Weights

Most estimators for average treatment effects that are used in practice can be written as the difference between two terms, the first an average of observed outcomes for treated units and the second an average of observed outcomes for control units:

$$\hat{\tau} = \hat{\tau}(\mathbf{Y}^{\text{obs}}, \mathbf{W}, \mathbf{X}) = \frac{1}{N_{\text{t}}} \sum_{i:W_i=1} \lambda_i \cdot Y_i^{\text{obs}} - \frac{1}{N_{\text{c}}} \sum_{i:W_i=0} \lambda_i \cdot Y_i^{\text{obs}}, \qquad (19.8)$$

with weights λ_i/N_{t} for treated units and weights and λ_i/N_{c} for control units. We refer to the λ_i as the normalizaed weights. For all the estimators we have considered so far, the normalized weights λ_i share a number of properties. First, they can be written as a function of the treatment indicator and pre-treatment variables for unit i, W_i, X_i, and the treatment indicators and covariate values for other units, $\mathbf{W}_{(-i)}$ and $\mathbf{X}_{(-i)}$, where $\mathbf{W}_{(-i)}$ is the $N-1$ vector of treatment indicators omitting the ith indicator W_i, and $\mathbf{X}_{(-i)}$ is the $(N-1) \times K$ dimensional matrix equal to $\mathbf{X}$ with the ith row omitted:

$$\lambda_i = \lambda(W_i, X_i, \mathbf{W}_{(-i)}, \mathbf{X}_{(-i)}),$$

with $\lambda(W_i, X_i, \mathbf{W}_{(-i)}, \mathbf{X}_{(-i)})$ a row exchangeable function in $(\mathbf{W}_{(-i)}, \mathbf{X}_{(-i)})$. The specific form of the weight function $\lambda(W_i, X_i, \mathbf{W}_{(-i)}, \mathbf{X}_{(-i)})$ depends on the estimator. The normalized weights also satisfy two summation restrictions:

$$\sum_{i:W_i=0} \lambda_i = N_{\text{c}}, \quad \text{and} \quad \sum_{i:W_i=1} \lambda_i = N_{\text{t}}, \tag{19.9}$$

so that the average of the normalized weights is equal to one. Expression (19.8), with the restrictions in (19.9) that capture affine consistency, is a natural form for estimators for average treatment effects.

Now let us return to some of the estimators discussed in the previous chapters to illustrate the forms of the weights and to document that these estimators are affinely consistent.

Difference Estimator

First, the simple difference between average outcome for treated and control units, $\hat{\tau}^{\text{dif}} = \overline{Y}_{\text{t}}^{\text{obs}} - \overline{Y}_{\text{c}}^{\text{obs}}$ corresponds to $\lambda_i^{\text{dif}} = 1$, for all i.

Regression Estimator

Second, consider a regression estimator where $\hat{\tau}^{\text{ols}}$ is the least squares estimator in the regression with a scalar covariate X_i (affine consistency also holds in the case with multiple pretreatment variables, but the form of the weights is more complicated algebraically):

$$Y_i^{\text{obs}} = \alpha + \tau \cdot W_i + \beta \cdot X_i + \varepsilon_i.$$

This implies

$$\lambda_i^{\text{ols}} = \begin{cases} -(1/N) \cdot \dfrac{S_X^2(N_{\text{t}}(N-1)/N^2)+(N_{\text{t}}/N)(N_{\text{c}}/N)(\overline{X}_{\text{t}}-\overline{X}_{\text{c}})(X_i-\overline{X})}{S_X^2(N_{\text{c}}N_{\text{t}}(N-1)/N^3)-(N_{\text{t}}/N)^2(N_{\text{c}}/N)^2(\overline{X}_{\text{t}}-X_{\text{c}})^2}, & \text{if } W_i = 0, \\[2ex] (1/N) \cdot \dfrac{S_X^2(N_{\text{c}}(N-1)/N^2)-(N_{\text{t}}/N)(N_{\text{c}}/N)(\overline{X}_{\text{t}}-\overline{X}_{\text{c}})(X_i-\overline{X})}{S_X^2(N_{\text{c}}N_{\text{t}}(N-1)/N^3)-(N_{\text{t}}/N)^2(N_{\text{c}}/N)^2(\overline{X}_{\text{t}}-\overline{X}_{\text{c}})^2}, & \text{if } W_i = 1, \end{cases}$$

for all i, and where $S_X^2 = \sum_{i=1}^N (X_i - \overline{X})^2 / (N-1)$ is the sample variance of X_i. Note that in this case, the weights need not all be non-negative.

Weighting Estimator

Third, consider weighting proportional to the inverse of the true propensity score $e(X_i)$. In that case the estimator is

$$\hat{\tau}^{\text{ht}} = \sum_{i:W_i=1} \frac{Y_i^{\text{obs}}}{e(X_i)} \Big/ \sum_{i':W_{i'}=1} \frac{1}{e(X_{i'})} - \sum_{i:W_i=0} \frac{Y_i^{\text{obs}}}{1 - e(X_i)} \Big/ \sum_{i':W_{i'}=0} \frac{1}{1 - e(X_{i'})},$$

(where the superscript "ht" stands for Horvitz-Thompson) so that

$$
\lambda_i^{\text{ht}} = \begin{cases} \frac{N_c}{1-e(X_i)} \Big/ \sum_{i':W_{i'}=0} \frac{1}{1-e(X_{i'})}, & \text{if } W_i = 0, \\ \frac{N_t}{e(X_i)} \Big/ \sum_{i':W_{i'}=1} \frac{1}{e(X_{i'})}, & \text{if } W_i = 1. \end{cases}
$$

The same argument applies to the case where we use the estimated propensity score to construct the weights, with the difference that the weights are now a more complicated function of all the pre-treatment variables and treatment indicators. In both cases, however, the weights are all positive.

Subclassification Estimator

Fourth, consider the simple, unadjusted, subclassification estimator. Let the number of units in subclass j be equal to $N(j)$, and the number of control and treated units in this subclass be equal to $N_c(j)$ and $N_t(j)$ respectively, and let $B_i(j) \in \{0, 1\}$ be a binary indicator for unit i falling in subclass j. Then

$$
\lambda_i^{\text{strat}} = \begin{cases} \sum_{j=1}^{J} B_i(j) \cdot (N_c/N_c(j)) \cdot (N(j)/N), & \text{if } W_i = 0, \\ \sum_{j=1}^{J} B_i(j) \cdot (N_t/N_t(j)) \cdot (N(j)/N), & \text{if } W_i = 1. \end{cases}
$$

Using regression within the subclasses maintains the affine consistency property, with the weights now a more complicated function of the pre-treatment variables for other units. Because of the regression adjustment, the weights can in that case be negative.

Matching Estimator

Finally, let us consider matching estimators. A simple matching estimator with M matches for each treated and control unit has the form (see Chapter 18 for details)

$$
\hat{\tau}^{\text{match}} = \frac{1}{N} \sum_{i=1}^{N} \left(\hat{Y}_i(1) - \hat{Y}_i(0) \right),
$$

where

$$
\hat{Y}_i(w) = \begin{cases} Y_i^{\text{obs}} & \text{if } W_i = w, \\ \sum_{j \in \mathcal{M}_i^c} Y_j^{\text{obs}}/M & \text{if } W_i = 1, w = 0, \\ \sum_{j \in \mathcal{M}_i^t} Y_j^{\text{obs}}/M & \text{if } W_i = 0, w = 1, \end{cases}
$$

ensuring that $\hat{Y}_i(w)$ is a linear combination of Y_j^{obs} with weights summing to one, and therefore satisfying affine consistency. The affine consistency is maintained if we combine the matching with regression adjustment, but again this can lead the weights to become negative.

19.4.2 Weights for the Lottery Data

To illustrate the weighting representations of the subclassification and matching estimators, we calculate the weights for the regression-adjusted version of these two estimators

Table 19.2. *Summary Statistics for the Normalized Weights for Different Estimators, for the IRS Lottery Data*

Trimmed Sample	Blocking with Regression		Matching with Regression		Weighting	
($N_c = 172, N_t = 151$)	Controls	Treated	Controls	Treated	Controls	Treated
Mean	1.00	1.00	1.00	1.00	1.00	1.00
Median	1.03	0.73	0.53	0.49	0.74	0.72
Standard deviation	1.09	0.87	0.94	0.95	0.87	0.82
Minimum	-1.98	-0.74	-0.11	-0.14	0.55	0.47
Maximum	3.87	3.55	6.62	6.59	9.68	6.45
Full Sample	Blocking with Regression		Matching with Regression		Weighting	
($N_c = 259, N_t = 237$)	Controls	Treated	Controls	Treated	Controls	Treated
Mean	1.00	1.00	1.00	1.00	1.00	1.00
Median	0.79	0.80	0.48	0.52	0.57	0.61
Standard deviation	1.57	1.43	1.47	1.45	2.69	1.34
Minimum	-1.13	-1.88	-1.08	-0.44	0.48	0.50
Maximum	9.42	7.35	14.22	14.18	41.7	13.2

for the lottery data, as well as for the simple weighting estimator. Table 19.2 reports some summary statistics for the normalized weights λ_i, including the mean and median weight, the standard deviation of the weights, and the minimum and maximum value of the weights. Note that the average of the normalized weights is exactly equal to one by affine consistency. We report the summary statistics for the weights for two samples, in the first panel for the trimmed sample, and in the second panel for the full sample, for three estimators: the subclassification estimator with regression adjustment, matching with a single match and regression adjustment, and weighting on the estimated propensity score.

First consider the results for the trimmed sample. For all three estimators, the regression-adjusted subclassification and matching estimators, and the weighting estimator, the standard deviation of the weights is approximately one, in both treatment groups. The largest value of the weights is markedly larger for the matching and the weighting estimators than for the subclassification estimator. For both the subclassification and matching estimators, the weights are negative for some units, which occurs because in both cases we use least squares covariance adjustment, either within subclasses or over the matched pairs. For the simple weighting estimator, the weights are non-negative. In general, it is useful to inspect the weights for any particular estimator. If some of the weights are extreme, the resulting estimator is likely to be sensitive to small changes in the specific implementation. With the lottery data, the relatively large weights for the simple weighting estimator suggest that this estimator may be an unattractive choice in this setting.

Next, consider the weights for the full sample. For all three estimators the weights are now substantially more variable. In particular for the weighting estimator, some units have fairly extreme weights, as large as 41, which occurs because of the bigger difference between covariate distributions for controls and treated in the full sample, and is another way of highlighting the consequences for inference of limited overlap in covariate distributions.

19.5   A GENERAL FORMULA FOR THE CONDITIONAL SAMPLING VARIANCE

Using the notation introduced in Chapter 7, let $\mu_c(x)$ and $\mu_t(x)$ denote the super-population expected values of the potential outcomes $Y_i(0)$ and $Y_i(1)$ in the subpopulation with $X_i = x$ respectively, and let $\sigma_c^2(x)$ and $\sigma_t^2(x)$ denote the super-population variances of $Y_i(0)$ and $Y_i(1)$ in the subpopulation with $X_i = x$, respectively. By super-population unconfoundedness it follows that these expectations and variances satisfy

$$\mu_c(x) = \mathbb{E}_{sp}\left[Y_i(0)\middle|X_i = x\right] = \mathbb{E}_{sp}\left[Y_i^{obs}\middle|W_i = 0, X_i = x\right],$$

$$\mu_t(x) = \mathbb{E}_{sp}\left[Y_i(1)\middle|X_i = x\right] = \mathbb{E}_{sp}\left[Y_i^{obs}\middle|W_i = 1, X_i = x\right],$$

$$\sigma_c^2(x) = \mathbb{V}_{sp}\left(Y_i(0)\middle|X_i = x\right) = \mathbb{V}_{sp}\left(Y_i^{obs}\middle|W_i = 0, X_i = x\right),$$

and

$$\sigma_t^2(x) = \mathbb{V}_{sp}(Y_i(1)|X_i = x) = \mathbb{V}_{sp}(Y_i^{obs}|W_i = 1, X_i = x).$$

Also define the unit-level conditional expectations and variances:

$$\mu_i = \mathbb{E}_{sp}\left[Y_i^{obs}|W_i, X_i\right] = \begin{cases} \mu_c(X_i), & \text{if } W_i = 0, \\ \mu_t(X_i), & \text{if } W_i = 1, \end{cases}$$

$$\sigma_i^2 = \mathbb{V}_{sp}\left(Y_i^{obs}|W_i, X_i\right) = \begin{cases} \sigma_c^2(X_i), & \text{if } W_i = 0, \\ \sigma_t^2(X_i), & \text{if } W_i = 1. \end{cases}$$

Using this notation, we can write a generic affinely consistent estimator $\hat{\tau}$ for the average effect, with the representation in (19.8), as

$$\hat{\tau} = \frac{1}{N_t} \sum_{i:W_i=1} \lambda_i \cdot Y_i^{obs} - \frac{1}{N_c} \sum_{i:W_i=0} \lambda_i \cdot Y_i^{obs} \qquad (19.10)$$

$$= \left(\frac{1}{N_t} \sum_{i:W_i=1} \lambda_i \cdot \mu_i - \frac{1}{N_c} \sum_{i:W_i=0} \lambda_i \cdot \mu_i\right)$$

$$+ \left(\frac{1}{N_t} \sum_{i:W_i=1} \lambda_i \cdot (Y_i^{obs} - \mu_i) - \frac{1}{N_c} \sum_{i:W_i=0} \lambda_i \cdot (Y_i^{obs} - \mu_i)\right).$$

The difference between the first pair of terms on the right-hand side of (19.10), $\sum_{i:W_i=1} \lambda_i \cdot \mu_i/N_t - \sum_{i:W_i=0} \lambda_i \cdot \mu_i/N_c$, and the estimand τ_{fs} equals the conditional bias. With a sufficiently flexible estimator, this term will generally be small. We ignore this term for the purpose of inference for the estimand. The second pair of terms on the right-hand side in (19.10), $\sum_{i:W_i=1} \lambda_i \cdot (Y_i^{obs} - \mu_i)/N_t - \sum_{i:W_i=0} \lambda_i \cdot (Y_i^{obs} - \mu_i)/N_c$, has expectation equal to zero, over the distribution induced by random sampling from the super-population and conditional on $(\mathbf{X}, \mathbf{W})$. Hence, conditional on $(\mathbf{X}, \mathbf{W})$, the sampling variance of $\hat{\tau}$ in (19.8) is equal to the variance of the second term:

$$\mathbb{V}_{\text{sp}}(\hat{\tau}|\mathbf{X}, \mathbf{W}) = \frac{1}{N_{\text{t}}^2} \sum_{i:W_i=1} \lambda_i^2 \cdot \sigma_i^2 + \frac{1}{N_{\text{c}}^2} \sum_{i:W_i=0} \lambda_i^2 \cdot \sigma_i^2. \tag{19.11}$$

Because the weights λ_i are, for a specific estimator, a known function of the covariates and the assignment vector, the only unknown components of the conditional sampling variance of $\hat{\tau}$ given $(\mathbf{W}, \mathbf{X})$ are the conditional unit-level potential outcome variances σ_i^2. Our proposed estimator for the sampling variance substitutes estimators $\hat{\sigma}_i^2$ for σ_i^2, leading to the following generic estimator for the conditional sampling variance:

$$\widehat{\mathbb{V}}_{\text{sp}}(\hat{\tau}|\mathbf{X}, \mathbf{W}) = \frac{1}{N_{\text{t}}^2} \sum_{i:W_i=1} \lambda_i^2 \cdot \hat{\sigma}_i^2 + \frac{1}{N_{\text{c}}^2} \sum_{i:W_i=0} \lambda_i^2 \cdot \hat{\sigma}_i^2. \tag{19.12}$$

The next section discusses specific estimators for σ_i^2.

19.6   A SIMPLE ESTIMATOR FOR THE UNIT-LEVEL CONDITIONAL SAMPLING VARIANCE

In this section we discuss a general approach to estimating σ_i^2 for all units. We first discuss the simplest case, followed by an illustration based on a subset of the lottery data consisting of ten treated units. Then we introduce two extensions, again followed by an illustration, now based on the trimmed lottery sample with $N = 323$ units.

19.6.1   A Single Exact Match

Suppose we wish to estimate the conditional variance, σ_i^2, for a particular unit i, and suppose this unit received the active treatment, so that $W_i = 1$. Suppose there is a second unit, say unit i', with an identical value for the pre-treatment variables, and which also received the active treatment, so that $W_{i'} = W_i = 1$ and $X_{i'} = X_i = x$. Then the expected outcomes for these units, conditional on $W_{i'} = W_i = 1$ and $X_{i'} = X_i = x$, based on the distribution generated by random sampling from the super-population, are equal:

$$\mathbb{E}_{\text{sp}}\left[Y_i^{\text{obs}} - Y_{i'}^{\text{obs}}\,\middle|\, X_i = X_{i'} = x, W_i = W_{i'} = 1\right]$$

$$= \mathbb{E}_{\text{sp}}\left[\left(\mu_i + (Y_i^{\text{obs}} - \mu_i)\right) - \left(\mu_{i'} + (Y_{i'}^{\text{obs}} - \mu_{i'})\right)\,\middle|\, X_i = X_{i'} = x, W_i = W_{i'} = 1\right]$$

$$= \mathbb{E}_{\text{sp}}\left[(Y_i^{\text{obs}} - \mu_i) - \left((Y_{i'}^{\text{obs}} - \mu_{i'})\right)\,\middle|\, X_i = X_{i'} = x, W_i = W_{i'} = 1\right] = 0,$$

exploiting the fact that, because $X_i = X_{i'} = x$ and $W_i = W_{i'} = 1$, it follows that $\mu_i = \mu_{i'} = \mu_{\text{t}}(x)$. Hence, the expected square of the difference in outcomes, conditional on $X_i = X_{i'} = x$ and $W_i = W_{i'} = 1$, is

$$\mathbb{E}_{\mathrm{sp}}\left[\left(Y_i^{\mathrm{obs}} - Y_{i'}^{\mathrm{obs}}\right)^2 \middle| X_i = X_{i'} = x, W_i = W_{i'} = 1\right]$$

$$= \mathbb{E}_{\mathrm{sp}}\left[\left(Y_i^{\mathrm{obs}} - \mu_i\right)^2 + \left(Y_{i'}^{\mathrm{obs}} - \mu_{i'}\right)^2 \middle| X_i = X_{i'} = x, W_i = W_{i'} = 1\right]$$

$$= \mathbb{V}_{\mathrm{sp}}\left(Y_i^{\mathrm{obs}} \middle| X_i = x, W_i = 1\right) + \mathbb{V}_{\mathrm{sp}}\left(Y_{i'}^{\mathrm{obs}} \middle| X_{i'} = x, W_{i'} = 1\right) = 2 \cdot \sigma_{\mathrm{t}}^2(x),$$

by random sampling from the super-population. Thus, we can estimate the conditional variance $\sigma_i^2 = \sigma_{\mathrm{t}}^2(X_i)$ as

$$\hat{\sigma}_i^2 = \left(Y_i^{\mathrm{obs}} - Y_{i'}^{\mathrm{obs}}\right)^2 / 2. \tag{19.13}$$

This estimator for the unit-level sampling variance is unbiased for σ_i^2 conditional on $\mathbf{W}$ and $\mathbf{X}$: $\mathbb{E}_{\mathrm{sp}}[\hat{\sigma}_i^2 | \mathbf{X}, \mathbf{W}] = \sigma_i^2$. However, it is not consistent, meaning that even in large samples, the difference between $\hat{\sigma}_i^2$ and σ_i^2 does not converge to zero, because its sampling variance does not vanish. Nevertheless, despite $\hat{\sigma}_i^2$ in (19.13) being an imprecise estimator of the sampling variance of $Y_i(w)$, we obtain an attractive estimator for the conditional sampling variance of $\hat{\tau}$ by substituting this estimator $\hat{\sigma}_i^2$ into the expression for the sampling variance for $\hat{\tau}$, which averages N such noisy (but unbiased) estimates:

$$\widehat{\mathbb{V}}_{\mathrm{sp}}(\hat{\tau} | \mathbf{X}, \mathbf{W}) = \frac{1}{N_{\mathrm{t}}^2} \sum_{i:W_i=1} \lambda_i^2 \cdot \hat{\sigma}_i^2 + \frac{1}{N_{\mathrm{c}}^2} \sum_{i:W_i=0} \lambda_i^2 \cdot \hat{\sigma}_i^2.$$

Under mild regularity conditions, the difference between this estimator and its target, normalized by the sample size, will converge to zero:

$$N \cdot \left(\widehat{\mathbb{V}}_{\mathrm{sp}}(\hat{\tau} | \mathbf{X}, \mathbf{W}) - \mathbb{V}_{\mathrm{sp}}(\hat{\tau} | \mathbf{X}, \mathbf{W})\right)$$

$$= \frac{N}{N_{\mathrm{t}}^2} \sum_{i:W_i=1} \lambda_i^2 \cdot \left(\hat{\sigma}_i^2 - \sigma_i^2\right) + \frac{N}{N_{\mathrm{c}}^2} \sum_{i:W_i=0} \lambda_i^2 \cdot \left(\hat{\sigma}_i^2 - \sigma_i^2\right) \longrightarrow 0.$$

Even though the differences $\hat{\sigma}_i^2 - \sigma_i^2$ do not vanish for a particular i with an increasing sample size, summing these differences over all units, suitably weighted, leads to an asymptotically attractive estimator for the normalized sampling variance of $\hat{\tau}$.

19.6.2 A Single Approximate Match

In general we may not be able to find for each unit i a matching unit i' with the same treatment level and exactly the same covariate values. Nevertheless, if we look for the most similar unit (in terms of covariate values) in the set of units with the same level of the treatment, we can obtain an approximately unbiased estimator for σ_i^2. Here we use the same ideas as we used in developing matching estimators in Chapter 18. There is one key difference: we now match treated units to treated units and control units to control units. Formally, we match treated unit i to the closest treated unit. Let, as in Chapter 18, $\mathbb{I}_{\mathrm{c}} = 1, \ldots, N_{\mathrm{t}}$ be the set of indices for the control units and $\mathbb{I}_{\mathrm{t}} = 1, \ldots, N_{\mathrm{t}} + N_{\mathrm{c}}$ the set of indices for the treated units. Then, let $\mathcal{M}_i^c$ be the set of control matches for unit i and $\mathcal{M}_i^t$ the set of treated matches for this unit, in both cases excluding unit i itself. In Chapter 18 we focused on control matches for treated units and treated matches for control units. Here the key difference is that we focus on control matches for control units and treated matches for treated units. Initially we will let $\mathcal{M}_i^c$ and $\mathcal{M}_i^t$ be singletons, with its element denoted by m_i^c and m_i^t, respectively. Then

$$m_i^c = \arg \min_{i' \in \mathbb{I}_c, i' \neq i} \|X_{i'} - X_i\|,$$

and

$$m_i^t = \arg \min_{i' \in \mathbb{I}_t, i' \neq i} \|X_{i'} - X_i\|.$$

Also define

$$\ell_i = \left\{ \begin{array}{ll} m_i^t & \text{if } W_i = 1, \\ m_i^c & \text{if } W_i = 0. \end{array} \right. \tag{19.14}$$

Then we estimate σ_i^2 as

$$\hat{\sigma}_i^2 = \left(Y_i^{\text{obs}} - Y_{\ell_i}^{\text{obs}} \right)^2 / 2. \tag{19.15}$$

This estimator for the unit-level conditional potential outcome variance σ_i^2 can be written as

$$\hat{\sigma}_i^2 = \left(\mu_i - \mu_{\ell_i} + (Y_i^{\text{obs}} - \mu_i) - (Y_{\ell_i}^{\text{obs}} - \mu_{\ell_i}) \right)^2 / 2.$$

Taking the expectation of this squared difference, conditional on $(\mathbf{X}, \mathbf{W})$, over the distribution induced by random sampling from the super-population, and subtracting the true variance σ_i^2, gives

$$\mathbb{E}_{\text{sp}} \left[\hat{\sigma}_i^2 \mid \mathbf{X}, \mathbf{W} \right] / 2 - \sigma_i^2 = \left(\mu_i - \mu_{\ell_i} \right)^2 \Big/ 2 + \left(\sigma_{\ell_i}^2 - \sigma_i^2 \right) / 2.$$

There are two reasons why this difference is not equal to zero, that is, why the estimator is biased for σ_i^2. First, because the match is not exact ($X_i \neq X_{\ell_i}$), the two conditional expectations μ_i and μ_{ℓ_i} are not identical, and so the first term generally differs from zero. Second, the two conditional variances are not the same. The second component of the bias can be positive or negative, but will tend to average to zero over all units in large samples. The first component of the bias is always positive, and it will vanish as the sample size increases, at least if we ignore measure-theoretic details. In Section 19.6.4 we discuss methods to reduce this first component of the bias.

Regarding the choice of metric, the same issues arise here that were discussed in Chapter 18. In the illustrations in this chapter we use the Mahalanobis metric.

Table 19.3. *Ten Treated Observations from the IRS Lottery Data*

Unit	Earn Year -1	Outcome	ℓ_i	$\hat{\sigma}_i^2$
1	29.7	3.4	6	27.3^2
2	19.7	6.4	10	2.6^2
3	0.8	0.0	5, 9	0.8^2
4	28.8	25.5	1	15.6^2
5	0.0	0.0	9	1.0^2
6	30.3	42.0	1	27.3^2
7	39.4	25.4	8	12.0^2
8	39.9	42.4	7	12.0^2
9	0.0	1.4	5	1.0^2
10	19.3	10.1	2	2.6^2

19.6.3 An Illustration

Let us illustrate the ideas developed thus far in this chapter with a subset of the lottery data introduced earlier. Table 19.3 presents information on ten treated units (winners) from the Imbens-Rubin-Sacerdote lottery data set. In the table we report the value of only one of the covariates, Earn Year -1 (earnings the year before playing the lottery) and the outcome (the average of six years of earnings after winning the lottery).

We wish to estimate, for each of these ten individuals (all winners), the conditional variance of the outcome, by matching each unit to the closest winner in terms of prior earnings. Consider the first individual. The value of the covariate for this individual is $X_1 = 29.7$ (corresponding to earnings equal to $29,700 in the year prior to winning the lottery), and the value of the outcome is $Y_1^{\text{obs}} = 3.4$. The closest individual, in terms of prior earnings, to this individual is unit $\ell_1 = 6$, with prior earnings equal to $X_{\ell_1} = X_6 = 30.3$, and outcome $Y_{\ell_1}^{\text{obs}} = Y_6^{\text{obs}} = 42.0$. The difference in outcomes is therefore $Y_1^{\text{obs}} - Y_{\ell_1}^{\text{obs}} = 38.6$, leading to an estimate for σ_1^2 equal to $\hat{\sigma}_1^2 = 38.6^2/2 = 27.3^2$. Analogously, the second individual, with $X_2 = 19.7$, is matched to $\ell_2 = 10$, with $X_{\ell_2} = X_{10} = 19.3$. For this pair the difference in outcomes is $Y_2^{\text{obs}} - Y_{\ell_2}^{\text{obs}} = 6.4 - 10.1$, leading to $\hat{\sigma}_2^2 = (6.4 - 10.1)^2/2 = 2.6^2$.

Matching the third individual leads to a minor complication: this individual, with $X_3 = 0.8$, is equally close to individuals 5 and 9, with $X_5 = X_9 = 0.0$. We therefore use both as matches, and estimate the conditional variance for unit 3 as the sample variance for the three units, unit 3 and the two units that are equally close:

$$\hat{\sigma}_3^2 = \frac{1}{2} \cdot \left(\left(Y_3^{\text{obs}} - \overline{Y}_3 \right)^2 + \left(Y_5^{\text{obs}} - \overline{Y}_3 \right)^2 + \left(Y_9^{\text{obs}} - \overline{Y}_3 \right)^2 \right) = 0.8^2,$$

where $\overline{Y}_3 = \left(Y_3^{\text{obs}} + Y_5^{\text{obs}} + Y_9^{\text{obs}} \right)/3 = 0.5$.

Table 19.3 presents the results of this matching exercise for all ten units.

19.6.4 A Bias-Adjusted Variance Estimator

As we discussed before, the bias of the unit-level conditional variance estimator is

$$\mathbb{E}_{\text{sp}}\left[\hat{\sigma}_i^2 \,\middle|\, \mathbf{X}, \mathbf{W}\right] /2 - \sigma_i^2 = (\mu_i - \mu_{\ell_i})^2 /2 + \left(\sigma_{\ell_i}^2 - \sigma_i^2\right)/2.$$

If the number of covariates is large, this expectation may be substantially different from the unit-level conditional variance σ_i^2. This bias has two components. The unit-level conditional variance at the match, $\sigma_{\ell_i}^2$, may be different from that at the ith unit itself, σ_i^2. Unless there is substantial heteroskedasticity, this is unlikely to be a problem, and we ignore it in this discussion. The other, and the more likely source of bias, is the difference in conditional expectations, $\mu_i - \mu_{\ell_i}$. To remove some of this bias, it is useful to apply some of the bias-reduction methods we used for matching estimators in Chapter 18.

To reduce the bias, we approximate the conditional expectation of the potential outcomes as linear and estimate the regression functions

$$\mathbb{E}_{\text{sp}}[Y_i^{\text{obs}}|X_i, W_i = 1] = X_i\beta_{\text{t}}, \quad \text{and} \quad \mathbb{E}_{\text{sp}}[Y_i^{\text{obs}}|X_i, W_i = 0] = X_i\beta_{\text{c}}.$$

Given the two estimated regression functions, we calculate the residuals

$$\hat{\varepsilon}_i = \begin{cases} Y_i^{\text{obs}} - X_i\hat{\beta}_{\text{c}} & \text{if } W_i = 0, \\ Y_i^{\text{obs}} - X_i\hat{\beta}_{\text{t}} & \text{if } W_i = 1. \end{cases}$$

Now we estimate the unit-level conditional variance σ_i^2 using the same match defined in (19.14), and the same estimator as in (19.16), with observed outcome Y_i^{obs} replaced by the residual $\hat{\varepsilon}_i$:

$$\hat{\sigma}_i^{2,\text{adj}} = \left(\hat{\varepsilon}_i - \hat{\varepsilon}_{\ell_i}\right)^2 /2. \tag{19.16}$$

If instead of the estimated residuals $\hat{\varepsilon}_i$, we used the true deviations from the conditional means, $Y_i^{\text{obs}} - \mu_i$, this would eliminate the $(\mu_i - \mu_{\ell_i})^2$ term from the bias of the unit-level conditional variance estimator.

The corresponding bias-adjusted estimator for the sampling variance of the estimator for the average treatment effect is

$$\widetilde{\mathbb{V}}_{\text{sp}}(\hat{\tau}|\mathbf{X}, \mathbf{W}) = \frac{1}{N_{\text{t}}^2}\sum_{i:W_i=1} \lambda_i^2 \cdot \hat{\sigma}_i^{2,\text{adj}} + \frac{1}{N_{\text{c}}^2}\sum_{i:W_i=0} \lambda_i^2 \cdot \hat{\sigma}_i^{2,\text{adj}}. \tag{19.17}$$

19.6.5 Multiple Matches

In the discussion in the previous section, we use only the square of the difference in outcomes between unit i and its closest match to estimate σ_i^2. More generally, we may be able to improve the precision of the estimator for σ_i^2 by using multiple matches or additional model-based adjustments. Specifically, one can for some $M \geq 1$ use the closest M units to unit i in terms of covariate values, so that $\mathcal{M}_i^c$ and $\mathcal{M}_i^t$ are sets with L elements.

Table 19.4. *Unit-Level Standard Deviation Estimates ($\hat{\sigma}_i$) for the IRS Lottery Data*

	Unadjusted			Adjusted		
	$M = 1$	$M = 4$	$M = 10$	$M = 1$	$M = 4$	$M = 10$
Mean	4.9	6.8	7.7	4.8	6.4	7.0
Median	2.5	6.4	8.0	2.6	5.3	6.6
Standard deviation	6.2	5.7	5.1	5.4	4.7	4.0
Min	0.0	0.0	0.0	0.0	0.3	1.1
Max	29.8	21.5	20.0	33.2	21.1	19.0
Proportion equal to zero	0.22	0.16	0.11	0.00	0.00	0.00

Then we can estimate the conditional variance σ_i^2 using all units in these sets. For example, if unit i is a treated unit:

$$\hat{\sigma}_{i,M}^2 = \frac{1}{2 \cdot M} \cdot \sum_{i' \in \mathcal{M}_i^t} \left(Y_{i'}^{obs} - Y_i^{obs} \right)^2, \tag{19.18}$$

and analogously for control units.

What are the trade-offs when choosing the number of matches M? Using more than one match increases the precision in the estimator for σ_i^2, because the estimator is now based on a larger sample. The disadvantage is that, when using more matches, the quality of the typical match decreases. In other words, the difference between the pre-treatment variables for a unit and its typical match, $X_i - X_{i'}$, increases, and thus we introduce an additional upward bias in the estimation of σ_i^2. In general the increase in the bias may be the bigger concern, because the averaging of the $\hat{\sigma}_i^2$ in the variance estimator $\widehat{\mathbb{V}}_{sp}(\hat{\tau}|\mathbf{X}, \mathbf{W})$ suggests that the precision is of less concern. However, if the weights λ_i^2 on the different $\hat{\sigma}_i^2$ vary widely, the precision of $\hat{\sigma}_i^2$ may be more of a concern. In practice we recommend a small number of matches, between one and four.

19.6.6 An Illustration with the Trimmed Lottery Data Set

Here we estimate the unit-level sampling variances on the lottery data for the purpose of estimating the sampling variance of the subclassification estimator $\hat{\tau}^{\text{strata}}$. We consider three values for the number of matches, $M = 1, 2$, and 4. Table 19.4 reports summary statistics for the estimates of the 323 standard deviations σ_i. The median estimate of the standard deviation in the single match case is 2.8. Using a larger value for M leads to a larger average estimate but a smaller standard deviation. Note that there is a substantial fraction of the units for whom the conditional variance σ_i^2 is estimated to be zero. This happens for units with outcome equal to zero for both the unit and its closest matches. To put the values for these conditional variances in perspective, the standard deviation of the outcome in the trimmed sample is $s_Y = 15.5$.

19.7 AN ESTIMATOR FOR THE SAMPLING VARIANCE OF $\hat{\tau}$ CONDITIONAL ON COVARIATES

To estimate the sampling variance of $\hat{\tau}$, the estimator for the average treatment effect, conditional on the covariates, we substitute the unit-level sampling variance estimates using a single match into the expression for the conditional sampling variance given in (19.12):

$$\hat{\mathbb{V}}_{M=1} = \frac{1}{N_t^2} \sum_{i:W_i=1} \lambda_i^2 \cdot \hat{\sigma}_i^2 + \frac{1}{N_c^2} \sum_{i:W_i=0} \lambda_i^2 \cdot \hat{\sigma}_i^2. \tag{19.19}$$

Let us again return to the lottery data. In Table 19.5 we present some of the estimates for the sampling variances. First we estimate the sampling variance with a single match, $\hat{\mathbb{V}}_{M=1}$. For the subclassification estimator, with a single match, the sampling variance is estimated to be $\hat{\mathbb{V}}_{M=1} = 1.53^2$. Using $M = 4$ matches leads to a small decrease in the estimated sampling variance, to $\hat{\mathbb{V}}_{M=4} = 1.47^2$. With $M = 10$ matches, we find $\hat{\mathbb{V}}_{M=10} = 1.52^2$. For the matching estimator, we find estimates ranging from 1.32^2 to 1.42^2.

If we are willing to assume homoskedasticity, so that $\sigma_t^2(x) = \sigma_c^2(x) = \sigma^2$ for all x, one can first average the unit-level variance estimates $\hat{\sigma}_i^2$ to estimate the common variance σ^2,

$$\hat{\sigma}^2 = \frac{1}{N} \sum_{i=1}^{N} \hat{\sigma}_i^2,$$

and then combine this estimator with the weights to estimate the sampling variance of the estimator for the average treatment effect as

$$\hat{\mathbb{V}}^{\text{homosk}} = \hat{\sigma}^2 \cdot \left(\frac{1}{N_t^2} \sum_{i:W_i=1} \lambda_i^2 + \frac{1}{N_c^2} \sum_{i:W_i=0} \lambda_i^2 \right). \tag{19.20}$$

In the lottery data set, $\hat{\mathbb{V}}^{\text{homosk}} = 1.34^2$, for the case with $M = 1$. Assuming homoskedasticity does not change the sampling variance estimates substantially in this example.

19.8 AN ESTIMATOR FOR THE SAMPLING VARIANCE FOR THE ESTIMATOR FOR THE AVERAGE EFFECT FOR THE TREATED

So far we focused on the overall average effect of the treatment in the full sample, $\tau_{\text{fs}} = \frac{1}{N} \sum_{i=1}^{N} (Y_i(1) - Y_i(0))$. In some cases researchers are interested in the average effect of the treatment only for those who actually received the treatment,

$$\tau_{\text{fs,t}} = \frac{1}{N_t} \sum_{i:W_i=1} (Y_i(1) - Y_i(0)).$$

Table 19.5. *Estimated Standard Errors for Average Treatment Effect Estimates for the IRS Lottery Data*

	Blocking plus Regression	Matching plus Regression	
		$(M = 1)$	$(M = 4)$
Point estimate $\longrightarrow$	-5.74	-4.54	-5.03
Method for calculating standard error $\downarrow$			
Matching, heteroskedastic ($M = 1$)	(1.53)	(1.40)	(1.40)
Matching, heteroskedastic ($M = 4$)	(1.47)	(1.32)	(1.32)
Matching, heteroskedastic ($M = 10$)	(1.52)	(1.41)	(1.41)
Matching, homoskedastic ($M = 1$)	(1.36)	(1.34)	(1.34)
Matching, homoskedastic ($M = 4$)	(1.41)	(1.39)	(1.39)
Matching, homoskedastic ($M = 10$)	(1.48)	(1.46)	(1.46)
Analytic	(1.37)	(1.18)	
Bootstrap	(2.09)	(1.43)	

In this section we discuss the modification of the estimator for the sampling variance for settings where the focus is on $\tau_{\text{fs},t}$.

Like its counterpart for the overall average, the generic estimator for $\tau_{\text{fs},t}$ can be written as a weighted average of the observed outcomes,

$$\hat{\tau}_{\text{fs,t}} = \frac{1}{N_\text{t}} \sum_{i:W_i=1} \lambda_i \cdot Y_i^{\text{obs}} - \frac{1}{N_\text{c}} \sum_{i:W_i=0} \lambda_i \cdot Y_i^{\text{obs}}.$$

Again the weights λ_i are functions of the matrix of pre-treatment variables $\mathbf{X}$ and the vector of treatment assignments $\mathbf{W}$, and average to one for the treated units and to one for the control units. The only difference is that the values of the weights are different for estimators of $\tau_{\text{fs},t}$. Typically λ_i is equal to 1 for all treated units in this case.

The conditional variance has the same form as before:

$$\hat{\mathbb{V}}_W \left(\hat{\tau}_{\text{fs,t}} \right) = \hat{\mathbb{V}}_W \left(\hat{\tau}_{\text{fs,t}} \big| \mathbf{X}, \mathbf{W} \right) = \frac{1}{N_\text{t}^2} \sum_{i:W_i=1} \lambda_i^2 \cdot \hat{\sigma}_i^2 + \frac{1}{N_\text{c}^2} \sum_{i:W_i=0} \lambda_i^2 \cdot \hat{\sigma}_i^2.$$

We can use the same estimator for σ_i^2 as in Section 19.6, and substitute that into this expression for the sampling variance to get

$$\hat{\mathbb{V}}_W \left(\hat{\tau}_{\text{fs,t}} \right) = \sum_{i=1}^{N} \lambda_i^2 \cdot \hat{\sigma}_i^2.$$

19.9   AN ESTIMATOR FOR THE SAMPLING VARIANCE FOR THE POPULATION AVERAGE TREATMENT EFFECT

In the previous two sections we focused on estimating $\mathbb{V}_W(\hat{\tau})$ for a generic estimator $\hat{\tau}$. In some cases the researcher may be interested in estimating the sampling variance of $\hat{\tau}$ as an estimator for the population average treatment effect τ_{sp} and therefore wish to estimate $\mathbb{V}(\hat{\tau})$. In this section we develop general methods for doing so.

As noted in Section 19.3, the difference between $\mathbb{V}(\hat{\tau})$ and $\mathbb{V}_W(\hat{\tau})$ is the super-population variance of the average treatment effect conditional on the pre-treatment variable, $\mathbb{V}(\tau(X_i))/N$. Given that we developed, in Section 19.7, an estimator for the finite-sample variance $\mathbb{V}_W(\hat{\tau})$, it now suffices to develop an estimator for the sampling variance of the average effect conditional on the pre-treatment variables, $\mathbb{V}(\tau(X_i))$.

The proposed estimator for this sampling variance is based on a preliminary matching estimator of the type discussed in Chapter 18. For simplicity we focus on a matching estimator with a single match. For each unit we find the closest unit, in terms of pre-treatment variables, with the alternative value for the treatment. For unit i, let the index of this match be denoted by ℓ_i. We estimate the unit-level treatment effect for unit i as

$$\hat{\tau}^{\text{match}} = \hat{Y}_i(1) - \hat{Y}_i(0),$$

where

$$\hat{Y}_i(0) = \begin{cases} Y_i^{\text{obs}} & \text{if } W_i = 0, \\ Y_{\ell_i}^{\text{obs}} & \text{if } W_i = 1, \end{cases} \quad \text{and} \quad \hat{Y}_i(1) = \begin{cases} Y_{\ell_i}^{\text{obs}} & \text{if } W_i = 0, \\ Y_i^{\text{obs}} & \text{if } W_i = 0. \end{cases}$$

We can write

$$\hat{\tau}^{\text{match}} = \tau_i + (2 \cdot W_i - 1) \cdot \left(\mu_i - \mu_{\ell_i}\right) + (2 \cdot W_i - 1) \cdot \left((Y_i^{\text{obs}} - \mu_i) - (Y_{\ell_i}^{\text{obs}} - \mu_{\ell_i})\right).$$

In sufficiently large samples, the second term on the right-hand side of this expression will be small relative to the other terms, and so we will ignore it and write

$$\hat{\tau}^{\text{match}} \approx \tau_i + (2 \cdot W_i - 1) \cdot \left((Y_i^{\text{obs}} - \mu_i) - (Y_{\ell_i}^{\text{obs}} - \mu_{\ell_i})\right).$$

Now suppose we observe τ_i. In that case we could estimate $\mathbb{V}(\tau(X_i))$ as

$$\widehat{\mathbb{V}}(\tau(X_i)) = \frac{1}{N-1} \sum_{i=1}^{N} \left(\tau_i - \frac{1}{N}\sum_{j=1}^{N} \tau_i\right)^2 = \frac{1}{N-1}\sum_{i=1}^{N}(\tau_i - \tau_{\text{fs}})^2.$$

However, we do not observe τ_i, only the estimate $\hat{\tau}_i^{\text{match}}$. Let us therefore examine the average squared difference between $\hat{\tau}_i^{\text{pair}}$ and the average $\tau_{\text{fs}} = \sum_{i=1}^{N} \tau_i/N$:

$$\mathbb{E}\left[\frac{1}{N}\sum_{i=1}^{N}\left(\hat{\tau}_i^{\text{match}} - \tau_{\text{fs}}\right)^2\right] = \mathbb{E}\left[\frac{1}{N}\sum_{i=1}^{N}(\tau_i - \tau_{\text{fs}})^2\right] + \mathbb{E}\left[\frac{1}{N}\sum_{i=1}^{N}\left(\hat{\tau}_i^{\text{match}} - \tau_i\right)^2\right].$$

$$(19.21)$$

First consider the second term. Ignoring the terms involving $\mu_i - \mu_{\ell_i}$, this average squared difference is, in expectation, approximately equal to

$$\mathbb{E}\left[\frac{1}{N}\sum_{i=1}^{N}\left(\hat{\tau}_i^{\text{match}} - \tau_i\right)^2\right] \approx \mathbb{E}\left[\frac{1}{N}\sum_{i=1}^{N}\left(\tau_i + (2 \cdot W_i - 1)\right.\right.$$

$$\left.\left. \cdot \left((Y_i^{\text{obs}} - \mu_i) - (Y_{\ell_i}^{\text{obs}} - \mu_{\ell_i})\right) - \tau_i\right)^2\right]$$

$$= \frac{1}{N}\sum_{i=1}^{N}\mathbb{E}\left[\left((2 \cdot W_i - 1) \cdot \left((Y_i^{\text{obs}} - \mu_i) \cdot -(Y_{\ell_i}^{\text{obs}} - \mu_{\ell_i})\right)\right)^2\right]$$

$$= \frac{1}{N}\sum_{i=1}^{N}\left(\sigma_i^2 + \sigma_{\ell_i}^2\right) \approx \frac{2}{N}\sum_{i=1}^{N}\sigma_i^2.$$

Thus,

$$\mathbb{V}_{\text{sp}}(\tau_i) \approx \mathbb{E}\left[\frac{1}{N}\sum_{i=1}^{N}\left(\hat{\tau}_i^{\text{match}} - \tau_{\text{fs}}\right)^2\right] - \frac{2}{N}\sum_{i=1}^{N}\sigma_i^2,$$

which we can estimate as

$$\hat{\mathbb{V}}_{\text{sp}}(\tau_i) \approx \mathbb{E}\left[\frac{1}{N}\sum_{i=1}^{N}\left(\hat{\tau}_i^{\text{match}} - \hat{\tau}\right)^2\right] - \frac{2}{N}\sum_{i=1}^{N}\hat{\sigma}_i^2.$$

Thus, our proposed estimator for the sampling variance for the estimated population average treatment effect is

$$\hat{\mathbb{V}}_{\text{sp}}(\hat{\tau}) = \hat{\mathbb{V}}_W(\hat{\tau}) + \frac{1}{N} \cdot \hat{\mathbb{V}}_{\text{sp}}(\tau(X_i)) = \sum_{i=1}^{N}\hat{\sigma}_i^2 \cdot \left(\lambda_i^2 - \frac{2}{N^2}\right) + \frac{1}{N^2}\sum_{i=1}^{N}\left(\hat{\tau}_i^{\text{match}} - \hat{\tau}\right)^2.$$

$$(19.22)$$

Let us return to the lottery data again. Using a single match to estimate σ_i^2, we estimate the variance of $\tau(X_i)$ to be

$$\hat{\mathbb{V}}_{\text{sp}}(\tau(X_i)) \approx \mathbb{E}\left[\frac{1}{N}\sum_{i=1}^{N}\left(\hat{\tau}_i^{\text{match}} - \hat{\tau}\right)^2\right] - \frac{2}{N}\sum_{i=1}^{N}\hat{\sigma}_i^2 = 2.9^2.$$

Thus, the estimate of the sampling variance of $\hat{\tau}$ as an estimator of the super-population average treatment effect is

$$\hat{\mathbb{V}}(\hat{\tau}) = \sum_{i=1}^{N}\hat{\sigma}_i^2 \cdot \left(\lambda_i^2 - \frac{2}{N^2}\right) + \mathbb{E}\left[\frac{1}{N^2}\sum_{i=1}^{N}\left(\hat{\tau}_i^{\text{match}} - \hat{\tau}\right)^2\right] = 1.41^2,$$

slightly larger than the sampling variance of the finite-sample average treatment effect (which we estimated to be 1.40^2 in Table 19.5).

19.10   ALTERNATIVE ESTIMATORS FOR THE SAMPLING VARIANCE

In this section we discuss two alternative estimators for the sampling variance of τ. Neither of these methods is, in our view, to be recommended, and we mention them largely to contrast them with the methods discussed so far, and also because versions of these methods have been used, perhaps ill-advisedly so, in practice. The first alternative is based on conventional least squares standard errors. Both of the estimators we recommend use least squares regression to estimate the average effect, not applied to the original sample but in combination with initial adjustment based on subclassification or matching. In Section 19.10.1 we use the regression step to motivate an estimator for the sampling variance. The second alternative is based on resampling. For simplicity we focus on the simplest version of the bootstrap.

19.10.1 Least Squares Sampling Variance Estimators

Least Squares Sampling Variance Estimators for the Subclassification Estimator

Consider the subclassification estimator. First we construct the subclasses. Suppose there are J subclasses, with, as before, $B_i(j)$ the zero-one indicator for the event that unit i belongs to subclass j. We then estimate the average effect in subclass j, denoted by $\tau(j)$, by least squares regression of the outcome Y_i^{obs} on an intercept, the indicator for receipt of the treatment, W_i, and the vector of covariates (or pre-treatment) variables X_i. Let Z_i be the vector $(W_i, 1, X_i)$. Then let the least squares estimator be $\hat{\beta}(j)$, defined by

$$\hat{\beta}(j) = \left(\sum_{i:B_i(j)=1} Z_i^T \cdot Z_i \right)^{-1} \left(\sum_{i:B_i(j)=1} Z_i^T \cdot Y_i^{\text{obs}} \right).$$

The estimator for the average treatment in subclass j is the first element of the vector $\hat{\beta}(j)$, or $\hat{\tau}^{\text{ols}}(j) = \hat{\beta}_1(j)$. The conventional least squares estimator of the sampling variance for $\hat{\tau}^{\text{ols}}(j)$ is the $(1, 1)$ element of

$$\hat{\mathbb{V}}(\hat{\beta}(j)) = \hat{\sigma}_j^2 \cdot \left(\sum_{i:B(j)=1} Z_i^T \cdot Z_i \right)^{-1},$$

where

$$\hat{\sigma}_j^2 = \frac{1}{N - K - 2} \sum_{i:B_i(j)=1} \left(Y_i^{\text{obs}} - Z_i \hat{\beta}(j) \right)^2,$$

and K is the number of elements of the vector of pre-treatment variables X_i. Let $\hat{\mathbb{V}}(\hat{\tau}^{\text{ols}}(j))$ denote this estimate, the $(1, 1)$ element of $\hat{\mathbb{V}}\left(\hat{\beta}(j) \right)$. The estimator for the average effect of the treatment is a weighted average of the within-block estimators:

$$\hat{\tau}^{\text{strat}} = \sum_{j=1}^{J} \frac{N_c(j) + N_t(j)}{N_{ji}} \cdot \hat{\tau}^{\text{ols}}(j).$$

Table 19.6. *Estimates and Estimated Standard Errors by Subclass for the IRS Lottery Data*

Subclass	Estimate	$\widehat{(\text{s. e.})}$	Weight	$\hat{\sigma}^{2,\text{block}}(j)$
1	-8.20	(3.19)	0.25	9.63
2	-6.74	(3.84)	0.12	6.93
3	-2.19	(4.13)	0.13	9.78
4	-7.30	(2.01)	0.25	7.84
5	-3.06	(2.82)	0.25	9.26
Overall	-5.74	(1.37)	1	

The corresponding estimator for the sampling variance of the subclass estimator for the overall average treatment effect is

$$\hat{\mathbb{V}}\left(\hat{\tau}^{\text{strat}}\right) = \sum_{j=1}^{J}\left(\frac{N_{\text{c}}(j) + N_{\text{t}}(j)}{N_j}\right)^2 \cdot \hat{\mathbb{V}}(\hat{\tau}^{\text{ols}}(j)).$$

Let us illustrate this approach with the lottery data. Our algorithm for the subclassification estimator led to five subclasses. The first and last two subclasses each have approximately 25% of the units, and the second and third each have between 12% and 13%. In Table 19.6 we present point estimates and estimated standard errors for each of the five subclasses, and the standard error for the point estimate of the overall average treatment effect. The estimated standard error for the overall estimate is equal to 1.37, somewhat smaller than the matching-based estimated standard errors. The within-subclass estimates of the conditional variances, the $\hat{\sigma}_j^2$, are slightly larger than the matching-based estimated conditional sampling variances.

A Sampling Variance Estimator for the Matching Estimator for Paired Randomization

The simple (i.e., without bias adjustment) matching estimator with M matches has the form

$$\hat{\tau} = \frac{1}{N}\sum_{i=1}^{N}\left(\hat{Y}_i(1) - \hat{Y}_i(0)\right), \tag{19.23}$$

where the, partly imputed, potential outcomes $\hat{Y}_i(w)$ have the form

$$\hat{Y}_i(0) = \begin{cases} Y_i^{\text{obs}} & \text{if } W_i = 0, \\ \frac{1}{M}\sum_{j\in\mathcal{M}_i^c} Y_j^{\text{obs}} & \text{if } W_i = 1, \end{cases} \quad \text{and} \quad \hat{Y}_i(1) = \begin{cases} Y_i^{\text{obs}} & \text{if } W_i = 1, \\ \frac{1}{M}\sum_{j\in\mathcal{M}_i^t} Y_j^{\text{obs}} & \text{if } W_i = 0. \end{cases}$$

Let us first consider the case with a single match, $M = 1$, so that $\mathcal{M}_i^c = \{\ell_i^c\}$ and $\mathcal{M}_i^t = \{\ell_i^t\}$, and with matching without replacement. In that case, all the pairs $(\hat{Y}_i(0), \hat{Y}_i(1))$ correspond to outcomes for distinct units, exactly like a paired randomized experiment. Hence, a natural estimator for the sampling variance is

$$\hat{\mathbb{V}} = \hat{\sigma}^2/N,$$

where $\hat{\sigma}^2$ is the obvious estimator for the sampling variance of the treatment effect, that is,

$$\hat{\sigma}^2 = \frac{1}{N-1} \sum_{i=1}^{N} \left(\hat{Y}_i(1) - \hat{Y}_i(0) - \hat{\tau} \right)^2. \tag{19.24}$$

There are two complications that make estimating the sampling variance more complicated for our matching estimator. First, we match with replacement, which introduces some dependence because the ith pair $(\hat{Y}_i(0), \hat{Y}_i(1))$ may have one or two outcomes in common with the i'^{th} pair $(\hat{Y}_{i'}(0), \hat{Y}_{i'}(1))$. To capture the dependence that results from this overlap, define the $N \times N$ matrix Ω, with

$$\Omega_{ii'} = \begin{cases} 1 & \text{if } i = i', \\ 1 & \text{if } \ell_i = j, \ell_{i'} = i, \\ 1/2 & \text{if } \ell_i = i', \ell_{i'} \neq i, \\ 1/2 & \text{if } \ell_{i'} = i, \ell_i \neq i', \\ 0 & \text{otherwise.} \end{cases}$$

For matching without replacement, Ω would be equal to the identity matrix, and $\hat{\mathbb{V}} = \hat{\sigma}^2 \iota_N' \Omega \iota_N / N^2$. With the modified Ω, we can estimate the sampling variance of $\hat{\tau}$ in (19.23) as

$$\hat{\mathbb{V}} = \frac{\hat{\sigma}^2}{N^2} \cdot \iota_N' \Omega \iota_N, \tag{19.25}$$

where ι_N is the vector of dimension N with all elements equal to unity, and $\hat{\sigma}^2$ is as in Equation (19.24).

The second complication arises from the use of multiple matches. Let M be the number of matches. For any pair of units i and i' let $M_{ii'}$ be the number of shared matches:

$$M_{ii'} = \begin{cases} 0 & \text{if } i = i', \\ 0 & \text{if } W_i \neq W_{i'}, \\ \#\left\{ \mathcal{M}_i^c \cap \mathcal{M}_{i'}^c \right\} & \text{if } W_i = W_j = 1, \\ \#\left\{ \mathcal{M}_i^t \cap \mathcal{M}_{i'}^t \right\} & \text{if } W_i = W_j = 0. \end{cases}$$

Then define Ω as the $N \times N$ with typical element

$$\Omega_{ii'} = \begin{cases} 1 & \text{if } i = i', \\ 2/(M+1) & \text{if } i \neq i', W_i = 0, W_{i'} = 1, i' \in \mathcal{M}_i^t, i \in \mathcal{M}_{i'}^c, \\ 2/(M+1) & \text{if } i \neq i', W_i = 1, W_{i'} = 0, i' \in \mathcal{M}_i^c, i \in \mathcal{M}_{i'}^t, \\ 1/(M+1) & \text{if } i \neq i', W_i = 0, W_{i'} = 1, i' \in \mathcal{M}_i^t, i \notin \mathcal{M}_{i'}^c, \\ 1/(M+1) & \text{if } i \neq i', W_i = 1, W_{i'} = 0, i' \in \mathcal{M}_i^c, i \notin \mathcal{M}_{i'}^t, \\ 1/(M+1) & \text{if } i \neq i', W_i = 0, W_{i'} = 1, i \in \mathcal{M}_{i'}^t, i' \notin \mathcal{M}_i^c, \\ 1/(M+1) & \text{if } i \neq i', W_i = 1, W_{i'} = 0, i \in \mathcal{M}_{i'}^c, i' \notin \mathcal{M}_i^t, \\ M_{i'}/(M(M+1)) & \text{if } i \neq i', W_i = W_{i'}, \end{cases}$$

and we can estimate the sampling variance again as $\hat{\mathbb{V}} = \iota_N' \Omega \iota_N \hat{\sigma}^2 / N^2$.

For the bias-adjusted matching estimator, we first define

$$\hat{X}_i(0) = \begin{cases} X_i & \text{if } W_i = 0, \\ \frac{1}{M} \sum_{j \in \mathcal{M}_i^c} X_j & \text{if } W_i = 1, \end{cases} \quad \text{and} \quad \hat{X}_i(1) = \begin{cases} X_i & \text{if } W_i = 1, \\ \frac{1}{M} \sum_{j \in \mathcal{M}_i^c} X_j & \text{if } W_i = 0. \end{cases}$$

Next we define

$$
\tilde{Y}_i(0) = \begin{cases} \hat{Y}_i(0) & \text{if } W_i = 0, \\ \hat{Y}_i(0) + \left(X_i - \hat{X}_i(0) \right) \hat{\beta}_0 & \text{if } W_i = 1, \end{cases}
$$

and

$$
\tilde{Y}_i(1) = \begin{cases} \hat{Y}_i(1) & \text{if } W_i = 1, \\ \hat{Y}_i(1) + \left(X_i - \hat{X}_i(1) \right) \hat{\beta}_1 & \text{if } W_i = 0. \end{cases}
$$

Then, the bias-adjusted matching estimator is

$$
\hat{\tau}^{\,\mathrm{adj}} = \frac{1}{N} \sum_{i=1}^{N} \left(\tilde{Y}_i(1) - \tilde{Y}_i(0) \right).
$$

We use the sampling variance estimator in (19.25), replacing $\hat{\sigma}^2$ in this expression with

$$
\tilde{\sigma}^2 = \frac{1}{N-1} \left(\tilde{Y}_i(1) - \tilde{Y}_i(0) - \hat{\tau}^{\,\mathrm{adj}} \right)^2.
$$

The estimator for the sampling variance of $\hat{\tau}_{\mathrm{adj}}$ is then

$$
\hat{\mathbb{V}}^{\mathrm{adj}} = \frac{\tilde{\sigma}^2}{N^2} \cdot \iota_N' \Omega \iota_N. \tag{19.26}
$$

For the matching estimator based on the trimmed lottery sample, and a single match, using the variance estimator in (19.26) leads to an estimated sampling variance of

$$
\hat{\mathbb{V}}^{\mathrm{adj}} = 1.18^2.
$$

19.10.2 Bootstrap Sampling Variance Estimators

In this section we discuss resampling methods for estimating the sampling variance of estimators for average treatment effects. Resampling methods have become popular in the empirical literature, partly due to the lack of guidance in the theoretical literature regarding sampling variance estimation, and partly due to its conceptual simplicity and computational ease of implementation. Nevertheless, for two reasons we do not generally recommend the bootstrap here. First of all, there is theoretical evidence against its validity. The intuition for the theoretical results rests on the non-smooth nature of matching estimators. For example, if one matches treated units, adding a replicate of a control unit to a bootstrap sample does not affect the point estimate of the matching estimator. Second, at best it delivers the sampling variance for the estimator with estimand equal to the super-population average treatment effect, rather than the sample average treatment effect, and we are often interested in the sampling variance of estimators for the sample average treatment effect.

Here we implement a simple version of the bootstrap. We bootstrap separately the control and treated subsamples, to create a bootstrap sample of size N, with N_c units in the control group and N_t units in the treatment group. Given this bootstrap sample, we follow exactly the same procedure as applied to the original sample to calculate the bootstrap estimate. For the subclassification estimator, this procedure includes re-estimating the propensity score, choosing the optimal number of subclasses again, and averaging the within-subclass estimates over the blocks. For the matching estimator, this includes re-normalizing the pre-treatment variables, and then matching the treated and control units again. Note that, in the bootstrap sample, there will likely be many ties, even if in the original sample there are no ties. This is one reason for the failure of the bootstrap to deliver valid confidence intervals for matching estimators.

Given the B bootstrap estimates, denoted by $\hat{\tau}_b$, $b = 1, \ldots, B$, we calculate the bootstrap variance as the sampling variance over the bootstrap estimates, $\hat{\mathbb{V}}^{\text{boot}} = \sum_b (\hat{\tau}_b - \overline{\tau}_{\text{boot}})/(B-1)$, where $\overline{\tau}_{\text{boot}} = \sum_b \hat{\tau}_b/B$ is the average over the bootstrap estimates.

There is no formal justification for the bootstrap for either the subclassification or the matching estimator. In fact, it has been shown that using the bootstrap sampling variance estimator can lead to confidence intervals with over, or under, coverage for matching estimators.

19.11 CONCLUSION

In this chapter we discuss an approach to frequentist inference for average treatment effects that applies to many estimators. The approach relies on the characterization of estimators as weighted averages of the observed outcomes, with the weights known functions of the covariates and treatment indicators. Given this characterization, the only unknown component of the sampling variance of the estimator is the unit-level outcome variance conditional on specific covariate values. We propose an estimator for this unit-level variance, and show how it can be used to estimate the sampling variance of estimators for the average treatment effect.

We briefly compare this estimator for the sampling variance to two alternatives, one analytic and one based on resampling.

NOTES

The theoretical discussion in this chapter builds heavily on the papers by Abadie and Imbens (2006, 2008, 2009, 2010). These studies also present simulation evidence for the effectiveness of the matching estimators of sampling variances, at least in certain situations, as well as of evidence of theoretical problems with the bootstrap in the same situations. Simulation evidence demonstrating problems with the bootstrap are also presented in Du (1998). For general bootstrap discussions and alternative resampling strategies, see Efron and Tibshirani (1993), Horowitz (2002), and Politis and Romano (1999).

CHAPTER 20

Inference for General Causal Estimands

20.1   INTRODUCTION

Much of the discussion in the fourth part of the book focused on an average treatment effect as the causal estimand of primary interest. Although this is an important case, many of the analyses extend to other causal estimands in a conceptually straightforward manner. In this chapter we discuss some examples of other estimands, and show how some of the earlier analyses apply with other estimands.

In many cases concerning causal questions, average effects are the most obviously interesting objects. Sometimes the focus is on average effects after taking some transformation of the outcome, possibly involving pre-treatment variables, but this does not lead to any conceptual problems or operational difficulties when applying the analyses from the previous chapters. In other cases, however, the causal estimands are conceptually distinct from average treatment effects. This includes situations where the average effect is just one of the objects of interest, as well as settings where the primary object is not an average effect. For example, policy makers may be interested in the effect of a new program on specific parts of the distribution of outcomes. In a labor market training program, policy makers may be less interested in the effect of the program on relatively high-earning individuals, instead being more concerned about the effect on the left tail of the distribution. In that case, differences between quantiles of the two potential outcome distributions may be more interesting estimands. Alternatively, policy makers may be interested in the effect of a new program on inequality in outcomes, say, through the effect of the treatment on the variance or the inter-quartile range of the distribution of outcomes.

The approach to estimation and inference that is the focus here is model-based imputation, which has a number of conceptual advantages relative to other approaches. The most important one is that once the missing potential outcomes are imputed, any causal estimand of the type we consider can be directly calculated. As a result, under this approach, estimation of and inference for any causal estimand are conceptually straightforward. We can therefore consider a variety of estimands given the same model for the potential outcomes. In contrast, if one uses, say, regression estimates, one would implicitly be using different models for the potential outcomes when focusing on the average

461

effect versus the median effect of the treatment. The main alternative to using model-based imputation is weighting. Weighting approaches also can be used to estimate a variety of estimands, including some of the causal estimands considered in this chapter. As discussed in Chapter 12, a concern with weighting methods, specifically when weights must be estimated, is that the resulting estimators for causal effects can be particularly sensitive to the model for the propensity score. As a result, relatively minor changes in the specification for the propensity score can lead to substantial changes in the estimates of causal effects.

To implement the imputation of the missing potential outcomes, in our preferred approach we first estimate the propensity score. Next we block on the estimated propensity score. Within blocks defined by the estimated propensity score, we build parametric models for the outcome distributions conditional on the covariates, possibly with cross-block restrictions. We then use these models to impute the missing potential outcomes. Note that different models for imputation will generally be used for different outcome variables, an approach that fundamentally differs from the weighting or the pure propensity score approaches in important ways that give the model-based approaches substantial flexibility to obtain reliable causal effect estimates.

The rest of this chapter is organized as follows. In the next section we describe the data used in this chapter, originally collected and analyzed by Lalonde (1986), and previously used in Chapter 14, and we conduct some preliminary analyses on the data based on the previous chapters. In Section 20.3 we introduce some causal estimands that are of interest in the context of this application. In Section 20.4 we discuss the models for the potential outcomes used in this chapter. Next, in Section 20.5 we discuss the implementation of the methods. In Section 20.6 we return to the Lalonde data and report results for the application. Finally, Section 20.7 concludes.

20.2   THE LALONDE NSW OBSERVATIONAL JOB-TRAINING DATA

Here we return to the non-experimental part of the Lalonde data that we previously used in Chapter 14. The treated subsample consists of 185 men, and the control sample consists of 15,992 men. We first estimate the propensity score on the full sample of 16,177 men. As discussed in Chapter 14, there are substantial differences in the covariate distributions between the treated and control subsamples. We then use the trimming described in Chapter 16 to construct subsamples with more overlap. The estimated optimal threshold based on the methods from Chapter 16 is 0.0792. Dropping men with an estimated propensity score below 0.0792 or above $1 - 0.0792 = 0.9208$ leaves us with a subsample consisting of $N_c = 282$ men in the control sample and $N_t = 151$ men who received the job training. Table 20.1 gives summary statistics for this trimmed sample. In the trimmed sample, the overlap is still limited, with the normalized difference for some covariates as large as 0.54. Nevertheless, this is a substantial improvement over the original sample where some normalized differences were in excess of 2.0 (see Table 14.7 in Chapter 14).

Next we re-estimate the propensity score. This time the algorithm from Chapter 13 selects eight linear terms and six second-order terms. The parameter estimates for the propensity score models are reported in Table 20.2. Given this estimate of the propensity score, we construct blocks based on the methods from Chapter 17. The algorithm

Table 20.1. *Summary Statistics for Trimmed Lalonde Non-Experimental Data*

	Controls ($N_c = 282$)		Trainees ($N_c = 151$)		Nor Dif	Log Ratio of STD
	mean	(S.D.)	mean	(S.D.)		
black	0.92	(0.27)	0.95	(0.21)	0.15	−0.27
hispanic	0.06	(0.23)	0.03	(0.18)	−0.12	−0.26
age	25.13	(7.64)	25.70	(7.02)	0.08	−0.08
married	0.26	(0.44)	0.13	(0.34)	−0.32	−0.25
nodegree	0.64	(0.48)	0.74	(0.44)	0.22	−0.09
education	10.54	(3.05)	10.26	(2.05)	−0.11	−0.40
earn '74	2.75	(4.63)	1.67	(4.64)	−0.23	0.00
unempl '74	0.52	(0.50)	0.77	(0.42)	0.54	−0.17
earn '75	1.84	(2.66)	1.01	(1.97)	−0.36	−0.30
unempl '75	0.39	(0.49)	0.66	(0.48)	0.56	−0.03
pscore	0.26	(0.19)	0.51	(0.24)	1.15	0.22
linearized pscore	−1.26	(1.12)	0.07	(1.15)	1.18	0.03

Table 20.2. *Estimated Parameters of Propensity Score for the Trimmed Lalonde Non-Experimental Data*

Variable	Est	(s. e.)	t-Stat
Intercept	−11.65	(0.13)	−92.6
Linear terms			
earn '74	0.15	(0.04)	3.4
unempl '74	−1.76	(1.17)	−1.5
earn '75	0.45	(0.38)	1.2
unempl '75	−0.95	(1.18)	−0.8
married	−3.15	(0.79)	−4.0
black	2.70	(0.55)	4.9
nodegree	1.33	(0.35)	3.8
age	0.55	(0.12)	4.7
Second-order terms			
age × age	−0.01	(0.00)	−5.1
married × nodegree	2.16	(0.86)	2.5
unempl '74 × age	0.12	(0.05)	2.4
earn '74× nodegree	−0.10	(0.05)	−2.0
earn '75 × black	−0.58	(0.38)	−1.5
unempl '74 × unempl '75	1.89	(1.20)	1.6

from that chapter leads to eight blocks. Summary statistics for the blocks are reported in Table 20.3.

Using the blocking estimator discussed in Chapter 17, including within-block regression adjustment, we obtain an estimate for the average effect for the treated equal to 2.33 (in thousands of dollars), with a standard error of 0.92. In this chapter, however, we are

Table 20.3. *Optimal Subclassification for the Trimmed Lalonde Non-Experimental Data*

Subclass	Min P-Score	Max P-Score	# Controls	# Treated	t-Stat
1	0.00	0.17	96	4	0.8
2	0.17	0.18	11	5	−0.1
3	0.18	0.22	46	10	0.9
4	0.22	0.29	37	19	−0.2
5	0.29	0.40	38	19	0.4
6	0.40	0.47	15	19	0.4
7	0.47	0.73	29	38	0.5
8	0.73	1.00	10	38	1.6

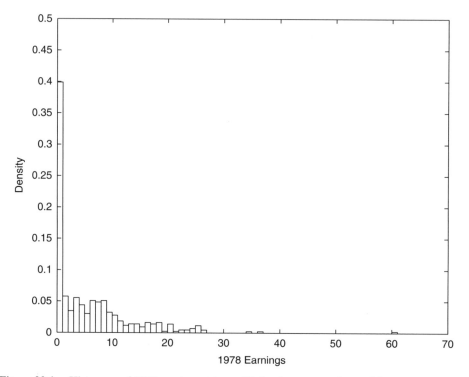

Figure 20.1. Histogram of 1978 earnings, trimmed Lalonde non-experimental data

interested in different causal estimands, and we will therefore build more flexible models for the conditional potential outcome distributions given the covariates and treatment levels. To inform the choice of such models, it is useful to inspect the marginal distributions of the observed outcomes, earnings in 1978 in thousands of dollars, either for the full trimmed sample, or separately by treatment group. Figure 20.1 presents a histogram of the outcome for the trimmed sample with 433 men. Two key features are the large proportion of individuals with zero earnings and the excess skewness and kurtosis of the distribution of earnings conditional on earnings being positive. In the trimmed sample, the proportion of men with zero earnings is 0.29, and the skewness among those positive earnings is 2.0, and the kurtosis is 10.7. It should also be noted that there is an extreme value for the outcome. One individual in the trainee sample had subsequent 1978 earnings over $60,000. The next highest earning individual had yearly earnings less than $40,000. To put this in context, the average earnings for trainees in 1974 and 1975, respectively, are $1,670 and $1,010, with maximum values in the sample in those years equal to $31,000 and $11,500. Given that there are only 151 trainees in our sample, changing the 1978 earnings for this one man from over $60,000 to less than $40,000. would lower the point estimate of the average treatment effect substantially, from $2,327 to $2,170. We will attempt to take these features into account when developing models for the conditional distributions of the potential outcomes.

20.3   CAUSAL ESTIMANDS

At the very beginning of this book, in Chapter 1, we defined causal estimands to be a general function of the potential outcomes, the covariates, and the vector of treatment assignments,

$$\tau = \tau(\mathbf{Y}(0), \mathbf{Y}(1), \mathbf{X}, \mathbf{W}). \tag{20.1}$$

Because of tradition and mathematical tractability we often focused on the finite-sample average effect

$$\tau_{\text{fs}} = \frac{1}{N} \sum_{i=1}^{N} \left(Y_i(1) - Y_i(0) \right),$$

or the super-population average treatment effect

$$\tau_{\text{sp}} = \mathbb{E}_{\text{sp}}[Y_i(1) - Y_i(0)].$$

In this chapter we consider two alternatives. For example, in a job-training program, policy makers may be interested in the effects of the program on the lower tail of the distribution. We can do this in a variety of ways. We may simply look at the average effect on a transformation of the original outcome. For example, we could define as the outcome whether an individual has positive earnings, or earnings exceeded some threshold level, such as some measure of the poverty level. Such transformations do not require any conceptual change in the methods discussed in previous chapters. Here we discuss some causal estimands that cannot be written as average effects on transformations of the original outcomes.

20.3.1 Quantile Treatment Effects

Distributional effects may conveniently be summarized by the difference in quantiles of the empirical distribution of the potential outcomes. For any outcome Y, with observations on N units $Y_1, \ldots, Y_N$, define q_Y^s to be the sth quantile of the empirical distribution of Y_i:

$$q_Y^s = \inf_q \left\{ q \in (-\infty, \infty) \;\middle|\; \frac{1}{N} \sum_{i=1}^N \mathbf{1}_{Y_i \leq q} \geq s \right\}.$$

Then we can define the sth quantile treatment effect as the difference of the sth quantile of the $Y_i(1)$ and $Y_i(0)$ distributions:

$$\tau_{\text{quant}}^s = q_{Y(1)}^s - q_{Y(0)}^s.$$

We can estimate quantile treatment effects at different quantiles. Using the median gives a more robust estimate of a "typical" effect, although it should be kept in mind that the difference in medians by treatment status is generally *not* the median of the unit-level treatment effects. We can also look at differences in lower or higher quantiles to assess the effect of the treatment at the bottom or top of the distribution.

20.3.2 Causal Effects on Dispersion and Inequality

A conceptually very different estimand we consider in this chapter is a measure of inequality of the outcome distributions. A simple measure of this would be the difference in standard deviations in the two potential outcome distributions:

$$\tau_{\text{sd}} = \sqrt{\frac{1}{N-1} \sum_{i=1}^N \left(Y_i(1) - \overline{Y}(1) \right)^2} - \sqrt{\frac{1}{N-1} \sum_{i=1}^N \left(Y_i(0) - \overline{Y}(0) \right)^2}.$$

Such a measure may be sensitive to the presence of outliers, in which case a more robust measure might be the interquartile range:

$$\tau_{\text{iqr}} = \left(q_{Y(1)}^{0.75} - q_{Y(1)}^{0.25} \right) - \left(q_{Y(0)}^{0.75} - q_{Y(0)}^{0.25} \right).$$

Alternatively, a common scale-free measure of inequality widely used in the social sciences is the so-called *Gini coefficient*. The Gini coefficient is often used to measure inequality of wealth. Given the ordered non-negative values $0 \leq Y_1 < Y_2 < \ldots < Y_N$, define the *Lorenz curve* as the piece-wise linear function $L^Y(y) : [0, 1] \mapsto [0, 1]$, going through the $N + 1$ pairs of values $(F_0^Y, L_0^Y), \ldots, (F_N^Y, L_N^Y)$ where $(F_0^Y, L_0^Y) = (0, 0)$, and, for $i = 1, \ldots, N$,

$$F_i^Y = \frac{i}{N}, \quad \text{and} \quad L_i^Y = \frac{\sum_{j=1}^i Y_i}{\sum_{j=1}^N Y_i}.$$

The Lorenz curve $L^Y(y)$ for, say, wealth, at a value $y \in [0, 1]$, measures the share of the total wealth held by the bottom y proportion of the population. If wealth is shared equally, the Lorenz curve is equal to the forty-five-degree line, $L^Y(y) = y$. The Gini coefficient, denoted by G, is a scalar functional of the Lorenz curve, measuring the area between the forty-five-degree line and the Lorenz curve as a share of the area underneath the forty-five-degree line:

$$G_Y = 1 - 2 \int_0^1 L^Y(y)\,dy. \tag{20.2}$$

If all values Z_i are identical, there is no inequality, the Lorenz curve equals to the forty-five-degree line, and the Gini coefficient is zero. The other extreme value is one, which occurs when all values Z_i are zero other than Z_N (and so all wealth is concentrated in the hands of one extremely wealthy individual). There are other measures of inequality available in the literature, but the Gini coefficient is widely used.

The causal estimand we focus on is the difference in Gini coefficients, the causal effect of the program on the Gini coefficient of the outcome distribution:

$$\tau^{\text{gini}} = G_{Y(1)} - G_{Y(0)}.$$

Policy makers may be interested to know whether the program increases inequality in earnings in the population.

20.3.3 Other Estimands

Here we focus primarily on two estimands, τ^s_{quant} for some specific values of s, and τ^{gini}. Many other estimands are possible. It is important, however, to note one common aspect of the two estimands we consider here. Both are functionals of the two marginal distributions of the potential outcomes, rather than functionals of the full joint distribution of the pair of potential outcomes. An example of a functional of the (full) joint distribution that cannot be written as a functional of the two marginal potential outcome distributions, and that is sometimes discussed in the literature, is the sth quantile of the difference in potential outcomes, which is generally different from the difference in the sth quantiles. In contrast to the distinction between the quantile of the difference and the difference in the quantiles, the average of the treatment effects is identical to the difference in the averages of the potential outcomes, because of the linearity of the expectations operator. In principle, the methods discussed in this chapter apply equally to estimands such as the median of the treatment effect, and we can directly apply the methods discussed in this chapter. In practice, though, it can be difficult to draw precise inferences about causal estimands that depend on the dependence structure of the potential outcomes. As discussed in the chapters on model-based inference in randomized experiments (Chapter 8), the data are not directly informative about the conditional dependence structure of the potential outcomes given covariates, and therefore prior information about the dependence structure may have important effects on posterior distributions, even in large samples. A question that sometimes arises is which object is of more interest, the median of the differences or the differences in the medians. In general, that question is difficult to answer without context. However, often policy makers contemplate exposing all units in a population (possibly homogeneous in characteristics) to the treatment versus no units. In that case, their decision should be based solely on the two marginal potential outcome distributions, not on the joint distribution of potential outcomes.

20.4   A MODEL FOR THE CONDITIONAL POTENTIAL OUTCOME DISTRIBUTIONS

The main model we consider is similar to the fourth model in Chapter 8 used for the model-based analysis of the experimental part of the Lalonde data. First we describe the general model and then extend it to the current setting that has substantial differences in the covariate distributions between the two treatment groups.

20.4.1 Single Block – Model I

We separately model the distribution of potential outcomes for each of the two treatment levels. For each treatment group we build a model for two parts of the conditional distribution given the pre-treatment variables X_i (all ten covariates other than the two indicators for ethnicity, `black` and `hispanic`, because there is little variation in ethnicity in the sample). Note that by unconfoundedness, these conditional distributions are free of dependence on W_i, that is, the same for units with $W_i = 0$ and $W_i = 1$. First, consider the probability of a positive value for $Y_i(0)$. A possible model for the event of a positive value for $Y_i(0)$ is a logistic model:

$$\Pr(Y_i(0) > 0 | X_i, \theta) = \frac{\exp(\gamma_{c,0} + X_i \gamma_{c,1})}{1 + \exp(\gamma_{c,0} + X_i \gamma_{c,1})}, \tag{20.3}$$

and, analogously, we model the probability of a treated potential outcome for treated outcome as

$$\Pr(Y_i(1) > 0 | X_i, \theta) = \frac{\exp(\gamma_{t,0} + X_i \gamma_{t,1})}{1 + \exp(\gamma_{t,0} + X_i \gamma_{t,1})}.$$

Second, we build models for the distributions of $Y_i(0)$ and $Y_i(1)$ conditional on a positive value for the potential outcome. Here, taking into account the excess skewness, we assume that the logarithm of the potential outcomes have normal distributions. Thus for the potential control outcome, we assume

$$\ln(Y_i(0)) | Y_i(0) > 0, X_i, \theta \sim \mathcal{N}\left(\beta_{c,0} + X_i \beta_{c,1}, \sigma_c^2\right), \tag{20.4}$$

and for the potential treated outcome,

$$\ln(Y_i(1)) | Y_i(1) > 0, X_i, \theta \sim \mathcal{N}\left((\beta_{t,0} + X_i \beta_{t,1}, \sigma_t^2\right),$$

where $Y_i(0)$ and $Y_i(1)$ are independent conditional on X_i and the parameters. Let $\theta = (\gamma_c, \gamma_t, \beta_c, \beta_t, \sigma_c^2, \sigma_t^2)$ denote the full parameter vector for these two distributions.

For convenience in conveying ideas, we specify a prior distribution for θ that is independent in its components and relatively dispersed, and for the regression parameters γ_c, γ_t, β_c, and β_t, we use normal prior distributions centered at zero with the variance equal to 10^2 times the identity matrix to capture relative ignorance about the components of these parameters. Similarly, for the variance parameters, σ_c^2 and σ_t^2, we use inverse Chi-squared distributions with parameters 1 and 0.01 respectively.

   The implementation of this model using Markov-Chain Monte Carlo methods is similar to that in Chapter 8. In Table 20.4 we report summary statistics for the posterior distributions of the parameters. These are not of intrinsic interest but are useful to ensure that the posterior distribution is reasonable. Next we report in Table 20.5 the results for the causal estimands.

20.4.2 A Model with Multiple Blocks – Model II

The model in the previous subsection is a reasonable one in experimental settings where the covariate distributions are similar for treated and control units. However, in the current setting, even after the trimming, the covariate distributions are substantially different in the two treatment groups. We therefore consider a different model to allow for more flexibility. Specifically, we estimate separate models in each of the eight blocks of the propensity score. In this section we ignore the covariates. Using the methods from Chapter 17, we partition the range of the propensity score into $J = 8$ blocks, that is, intervals of the type $[b_{j-1}, b_j)$, where $b_0 = 0$ and $b_J = 1$, so that $\cup_{j=1}^{J}[b_{j-1}, b_j) = [0, 1)$, where $B_i(j) \in \{0, 1\}$ is an indicator. Let $B_i(j) \in \{0, 1\}$ be an indicator for unit i being in block j, for $j = 1, \ldots, J$:

$$B_i(j) = \begin{cases} 1 & \text{if } b_{j-1} \le \hat{e}(X_i) < b_j, \\ 0 & \text{otherwise.} \end{cases}$$

   Within each block and treatment level, we again specify a model for the event that the outcome is equal to zero, and a model for the outcome conditional on being positive. Specifically, for the control potential outcome $Y_i(0)$ in block j, we specify the model

$$\Pr(Y_i(0) > 0 | B_i(j) = 1, \theta) = \frac{\exp \gamma_c(j)}{1 + \exp \gamma_c(j)}, \tag{20.5}$$

and analogously, we model the probability of a positive outcome for treated outcome in this block as

$$\Pr(Y_i(1) > 0 | B_i(j) = 1, \theta) = \frac{\exp \gamma_t(j)}{1 + \exp \gamma_t(j)}.$$

Next, we build a model for the distribution of $Y_i(0)$ and $Y_i(1)$ in block j conditional on a positive value for the potential outcome. We assume

$$\ln(Y_i(0)) \,|\, Y_i(0) > 0, B_i(j) = 1, \theta \sim \mathcal{N}\left(\beta_c(j), \sigma_c^2\right), \tag{20.6}$$

and for the potential treated outcome,

$$\ln(Y_i(1)) \,|\, Y_i(1) > 0, B_i(j) = 1, \theta \sim \mathcal{N}\left(\beta_t(j), \sigma_t^2\right),$$

where $Y_i(0)$ and $Y_i(1)$ are assumed to be independent conditional on the block and the parameter. Here, for simplicity, we let the conditional variances differ by treatment status but not by block.

Table 20.4. Single Block Model for Trimmed Lalonde Non-Experimental Data

Sample	Controls						Treated					
	γ_c			β_c			γ_t			β_t		
	$q^{0.025}$	med	$q^{0.975}$	$q^{0.025}$	med	$q^{0.975}$	$q^{0.025}$	med	$q^{0.975}$	$q^{0.025}$	med	$q^{0.975}$
Intercept	0.25	**1.39**	3.20	1.02	**1.25**	1.48	0.89	**4.00**	9.05	1.43	**1.66**	1.89
age	−0.25	**−0.09**	0.05	−0.05	**−0.01**	0.03	−0.16	**0.06**	0.39	−0.04	**−0.01**	0.03
married	−2.82	**−0.00**	2.95	−0.38	**0.20**	0.78	−3.34	**4.15**	17.68	−0.52	**0.08**	0.69
nodegree	−2.55	**0.70**	4.02	−0.97	**−0.29**	0.38	−8.66	**−1.64**	4.19	−0.97	**−0.34**	0.30
education	−0.44	**0.04**	0.50	−0.03	**0.08**	0.18	−1.84	**−0.25**	0.95	−0.12	**0.01**	0.14
earn '74	−0.34	**0.06**	0.57	−0.07	**−0.01**	0.05	−0.71	**0.14**	1.45	−0.06	**0.01**	0.08
unempl '74	−3.02	**0.59**	4.85	−1.29	**−0.66**	−0.03	−3.98	**6.35**	23.43	−0.35	**0.47**	1.27
earn '75	−0.29	**0.44**	1.82	−0.01	**0.09**	0.20	−2.93	**−0.12**	2.22	−0.15	**0.03**	0.21
unempl '75	−4.87	**−0.94**	2.30	−0.42	**0.23**	0.88	−28.20	**−6.64**	4.07	−0.81	**−0.10**	0.61
σ				1.30	**1.43**	1.59				0.95	**1.08**	1.24

Table 20.5. *Model-Based Analysis for Various Estimands for Trimmed Lalonde Non-Experimental Data*

Model	τ_{avg}			τ_{med}			$Y_i > 0$			$Y_i > 1$			Gini		
	$q^{.025}$	**med**	$q^{.975}$	$q^{.025}$	**med**	$q^{.975}$	$q^{.025}$	**med**	$q^{.975}$	$q^{.025}$	**med**	$q^{.975}$	$q^{.025}$	**med**	$q^{.975}$
I,fs	1.71	**3.11**	4.00	−0.52	**1.65**	3.35	−0.09	**0.11**	0.31	0.01	**0.18**	0.34	−0.23	**−0.16**	−0.07
I,sp	−0.10	**2.43**	4.82	−0.97	**2.17**	5.28	−0.16	**0.13**	0.40	−0.03	**0.22**	0.45	−0.25	**−0.13**	0.01
II,fs	1.12	**2.61**	4.00	−1.41	**0.85**	2.74	−0.07	**0.07**	0.24	−0.01	**0.12**	0.27	−0.20	**−0.13**	−0.06
II,sp	0.08	**2.25**	4.43	−1.81	**1.11**	4.02	−0.10	**0.12**	0.34	−0.01	**0.20**	0.40	−0.24	**−0.13**	−0.01
III,fs	2.09	**3.19**	4.00	−0.53	**1.48**	3.07	−0.02	**0.11**	0.24	0.07	**0.19**	0.30	−0.22	**−0.16**	−0.10
III,sp	0.38	**2.11**	3.85	−0.89	**1.77**	4.37	−0.08	**0.10**	0.26	0.02	**0.19**	0.34	−0.19	**−0.11**	−0.02

Note: Model I: single block, with covariates; Model II, eight blocks, no covariates; Model III: eight blocks, with covariates; fs, focus on finite sample causal estimand; sp, focus on super-population causal estimand.

Let $\theta = (\gamma_c(j), \gamma_t(j), \beta_c(j), \beta_t(j), j = 1, \ldots, J, \sigma_c^2, \sigma_t^2)$ denote the full parameter vector. Again, we use independent, fairly dispersed prior distributions for all elements of θ.

20.4.3 Multiple Blocks and Covariates – Model III

In our final model we incorporate both the block information and the covariates, and we combine the previous two specifications, using the analogous two-part model. We now specify for the control outcome the probability of a positive outcome as

$$\Pr\left(Y_i(0) > 0 | X_i, B_i(j) = 1, \theta\right) = \frac{\exp\left(\gamma_{c,0}(j) + X_i \gamma_{c,1}\right)}{1 + \exp\left(\gamma_{c,0}(j) + X_i \gamma_{c,1}\right)},$$

with, for positive income,

$$\ln\left(Y_i(0)\right) | Y_i(0) > 0, X_i, B_i(j) = 1, \theta \sim \mathcal{N}\left(\beta_{c,0}(j) + \beta_{c,1} X_i, \sigma_c^2\right),$$

and analogously for the treated outcome $Y_i(1)$. Thus we allow the intercepts in both models to be block-specific but restrict the slope coefficients to be identical. This restriction is partly motivated by the modest sample size. In some of the blocks there are only a few treated or only a few control units, so that it would be impossible to estimate precisely the slope coefficients separately within each block. An alternative would be a hierarchical structure where the parameters in each block are allowed to be different but are linked through a hierarchical structure through their prior distributions.

20.5   IMPLEMENTATION

We use Markov-Chain Monte Carlo methods to obtain draws from the posterior distribution of the parameters. Then we use two methods to obtain draws from the posterior distribution of the causal estimands. The first method follows closely that of Chapter 8. In this approach we draw values of the parameters from the posterior distribution given the observed data. We then use those parameter values in combination with the statistical model to impute the missing potential outcomes. Finally we calculate the estimand as a function of observed and imputed potential outcomes. Doing so repeatedly gives us the draws from the posterior distribution of the causal estimand.

However, this method does not always give credible results, and it is useful to sound a cautionary note. Specifically, in order for this first method to give accurate results, it relies heavily on the statistical model being a good approximation to the underlying distribution with regard to the particular estimand. For example, suppose we are interested in the average treatment effect for the treated, and we use the two-part model described in the previous section, with no covariates and a single block. We estimate the model for the control outcome using the control units. For this subsample, the proportion of zero outcomes is 0.31. Among the 69% control units with positive outcomes, the average and standard deviation of the logarithm of the outcome are 1.39 and 1.49 respectively. This implies, under the two-part model, that the expected value should be approximately $0.69 \cdot \exp\left(1.39 + 1.49^2/2\right) = 8.41$, whereas the actual average is 7.71. Because the model is non-linear, at the fitted values the implied expectation is not necessarily equal to the sample average. In this simple example one could address this by estimating a linear model, but when we look at different estimands, unless the model fits the data well, it will not necessarily give good results for all estimands.

The second method addresses this as follows. We again draw parameter values from the posterior distribution of the parameters given the observed data. Now, however, for all units, we draw values for both potential outcomes. We then calculate the causal estimand as a function of these imputed potential outcomes, instead of combining observed and imputed outcomes. Implicitly this changes the focus from the sample causal estimand to the super-population causal estimand.

20.6 RESULTS FOR THE LALONDE DATA

For the Lalonde data we focus on estimands for the subsample of treated men. There is no interest in extending the labor market training program to the control individuals, only in assessing the benefits, if any, of the training program to those who took part in it. We focus on five estimands. The first is, for comparison purposes with earlier analyses, the average effect of the treatment on the treated:

$$\tau_{\text{fs},t} = \frac{1}{N_t} \sum_{i:W_i=1} \left(Y_i(1) - Y_i(0)\right).$$

The second estimand is, the difference in medians of $Y_i(1)$ and of $Y_i(0)$ for the treated units. First, extending the earlier definitions, we define the quantiles for the treated subsample as

$$q_{Y,t}^s = \inf_q \left\{ q : \frac{1}{N_t} \sum_{i:W_i=1} \mathbf{1}_{Y_i \leq q} \geq s \right\}.$$

Then we define the sth quantile treatment effect for the treated as the difference of the sth quantile of the $Y_i(1)$ and $Y_i(0)$ distributions for the treated:

$$\tau_{\text{quant},t}^s = q_{Y(1),t}^s - q_{Y(0),t}^s.$$

Here we focus on the difference in medians,

$$\tau_t^{\text{med}} = q_{Y(1),t}^{1/2} - q_{Y(0),t}^{1/2}.$$

The next estimand is the causal effect of the treatment on the probability of having positive earnings,

$$\tau_t^{\text{pos}} = \frac{1}{N_t} \sum_{i:W_i=1} \left(\mathbf{1}_{Y_i(1)>0} - \mathbf{1}_{Y_i(0)>0}\right),$$

and the probability of having earnings exceeding 1 ($1,000),

$$\tau_t^{\geq 1} = \frac{1}{N_t} \sum_{i:W_i=1} \left(\mathbf{1}_{\{Y_i(1)>1\}} - \mathbf{1}_{\{Y_i(0)>1\}} \right).$$

Finally, the fifth estimand is the difference in Gini coefficients in the $Y_i(1)$ and $Y_i(0)$ distributions for the treated units. Let $G_{Y(1),t}$ denote the Gini coefficient for the $Y_i(1)$ distribution among the treated. Then the causal estimand we focus on is

$$\tau_t^{\text{gini}} = G_{Y(1),t} - G_{Y(0),t}.$$

We estimate all three models, the one with covariates (Model I), with block indicators (Model II), and with both covariates and block indicators (Model III) on the Lalonde data.

In Table 20.4 we present posterior percentiles for the parameters of the first model. These parameter estimates are not of intrinsic interest and are presented here for completeness. In Table 20.5 we present posterior percentiles for the causal estimands. There are two rows for the first model, one for the finite-sample causal estimand, where only the control outcomes are imputed for all treated units, and one for the super-population causal estimand, where both control and treated outcomes are effectively imputed in the super-population. In general the estimates suggest that there is a positive effect of the treatment, as seen by the posterior medians for the average and median effects. It also suggests that the program may have led to a modest decrease in inequality as measured by the effect on the Gini coefficient.

20.7 CONCLUSION

In this chapter we discuss estimation of and inference for estimands other than average treatment effects. Under our preferred, model-based approach, there are no conceptual difficulties to studying general causal estimands. The approach of imputing the missing potential outcomes is valid in general. The main issue is that, in many cases, it becomes obvious that one has to be more careful in the choice of models. Depending on the choice of estimand, the results may be sensitive to particular modeling choices.

NOTES

Quantile treatment effects, defined as the difference in quantiles, as in the current chapter, have been considered previously by Lehman (1974). Causal effects of treatments on inequality measured through their effect on the Gini coefficient has been considered by Firpo (2003, 2007). In applied work, Bitler, Gelbach, and Hoynes (2006) study distributional effects beyond the average treatment effects and find, in the context of a randomized labor market program, that the effects are bigger at the lower tail of the distribution than in the upper tail.

For an early study of the sensitivity of estimates of causal effects to the choice of Bayesian model, see Rubin (1983); this example is discussed further in Gelman, Carlin, Stern, and Rubin (1995).

The discussion regarding the difference between, on the one hand, the difference between the medians of the potential outcomes by treatment status and, on the other hand, the median of the difference in potential outcomes is an old one. See for recent comments on this Manski (1996), Deaton (2010), and Imbens (2010).

The first general discussion of the imputation approach to inference for general causal estimands beyond average treatment effects is in Rubin (1978). Althauser and Rubin (1970) discuss computational issues.

Dehejia (2005b) and Manski (2013) discuss decision problems in a treatment effect context, where the intermediate focus is often on more complex estimands than simple average treatment effects.

## PART V

Regular Assignment Mechanisms: Supplementary Analyses

## CHAPTER 21

Assessing Unconfoundedness

21.1 INTRODUCTION

The previous three chapters assume a regular assignment mechanism, requiring the assignment mechanism to be individualistic, probabilistic, and unconfounded. In this chapter we maintain the first two conditions, which are often uncontroversial, and focus on the plausibility of the third, most controversial assumption, unconfoundedness. Formally, unconfoundedness requires that the probability of treatment assignment is free of dependence on the potential outcomes. Specifically, the super-population version implies, by Theorem 12.1, first, that the conditional distribution of the outcome under the control treatment, $Y_i(0)$, given receipt of the active treatment and given covariates, is identical to its distribution conditional on receipt of the control treatment and conditional on covariates, and second, that, analogously, the conditional distribution of the outcome under the active treatment, $Y_i(1)$, given receipt of the control treatment and conditional on covariates, is identical to its distribution given receipt of the active treatment and conditional on covariates. Informally, unconfoundedness requires that we have a sufficiently rich set of pre-treatment variables so that adjusting for differences in values for observed pre-treatment variables removes systematic biases from comparisons between treated and control units. This critical assumption is not testable. The issue is that the data are not directly informative about the distribution of the control outcome $Y_i(0)$ for those who received the active treatment (for those with $W_i = 1$, we never observe $Y_i(0)$), nor are they directly informative about the distribution of the active treatment outcome given receipt of the control treatment (for those with $W_i = 0$, we never observe $Y_i(1)$). Thus, the data cannot directly provide evidence on the validity of the unconfoundedness assumption. Nevertheless, here we consider ways to assess the plausibility of this assumption from the data at hand.

The analyses discussed in this chapter are *supporting* or *supplementary analyses* that can, depending on their results, increase or reduce the credibility of the main analyses. These supporting analyses focus on estimating, and doing inference for, "pseudo"-causal estimands with *a priori* known values, under assumptions more restrictive than unconfoundedness. If these analyses suggest that the null hypotheses assessing whether these pseudo-causal effects are equal to their null values are not supported by the data, then the unconfoundedness assumption will be viewed as less plausible than in cases where these null hypotheses are supported by the data. How much the results of these analyses change our assessment of the unconfoundedness assumption depends on specific aspects of the substantive application at hand, in particular on the richness of the set of pre-treatment variables, their number and type.

The results of these assessments of the unconfoundedness assumption may suggest that unconfoundedness is less plausible than we thought beforehand, and thus that important pre-treatment differences between treated and control units may not have been measured. An important point is that finding pseudo-causal effects different from their known values generally will *not* suggest an alternative approach to estimating the causal estimands. Establishing that methods based on adjusting for observed differences between control and treated units may be unlikely to be adequate for drawing credible causal inferences does not imply the existence of credible alternative methods for causal inferences based on alternative assumptions. The implication may, therefore, be that, given the current data, it is simply not possible to estimate credibly and precisely the causal effects of interest and that one may either have to abandon any attempt to do so without either additional information or without richer data, or at least should be explicit about the lack of credibility.

The specific methods discussed in this chapter are divided here into three classes. The first class of methods can be viewed as comprising a *design* approach, not requiring data on the outcome variable. We partition the full set of pre-treatment variables into two parts, the first set consisting of some selected pre-treatment variables, and the second set consisting of the remaining covariates. Typically the first set consists of a single pre-treatment variable, but in principal it can consist of multiple pre-treatment variables. It takes the first set of selected pre-treatment variables and analyzes them as *pseudo-outcomes*, known *a priori* not to be affected by versus treatment assignment. These pre-treatment variables will be viewed in this approach as "proxy" variables for the potential outcomes, variables likely to be statistically associated with the potential outcomes. In principle, such a proxy variable can be any pre-treatment variable. However, the most compelling case arises when the proxy variable is a lagged outcome, that is, a measure of the same substantive quantity as the outcome but measured at a point in time prior to the receipt of treatment. We then assess the null hypothesis that there are no systematic differences in the pseudo-outcome by treatment status, after adjusting for the second set of covariates. We refer to this hypothesis as *pseudo-outcome unconfoundedness*. A finding that one cannot reject the null hypothesis that these pseudo-causal effects are zero (or small) lends credibility to the unconfoundedness assumption based on the full set of covariates. Again, this approach is not a direct assessment of the unconfoundedness assumption. Even if the null hypothesis of no treatment effect on the pseudo-outcome is found to be implausible, one cannot be confident that the hypothesis underlying the planned main analysis, unconfoundedness, is violated. Nevertheless, if two conditions are satisfied, we will argue that the assessment is informative about the credibility of the analysis under unconfoundedness. The first is that the pseudo-outcome used in this assessment is a good proxy for one of the potential outcomes. The second condition is that *subset unconfoundedness* (which requires that the pseudo-outcome is not required for unconfoundedness to hold) is plausible – in other words, that there should be an *a priori* argument that the pseudo-outcome is not essential for removing biases in comparisons between treated and control units given the second set of pre-treatment variables. This is often most compelling if the pseudo-outcome differs from some of the remaining pre-treatment variables only in the time of measurement. In cases where all the pre-treatment variables are qualitatively different, it may be more diffi-cult to argue that any of the pre-treatment variables can be omitted without leading to violations of unconfoundedness. This first approach to assessing unconfoundedness is a *design* approach that does not use data on the outcome variable.

The second approach to assessing the unconfoundedness assumption focuses on pseudo-causal effects for the original outcome. Instead of focusing on causal effects of the actual treatment, this approach analyzes the effects of a pseudo-treatment that is known *a priori* not to have a causal effect. We refer to it as a *semi-design* approach, using only outcome data for the units in the control group. Originally proposed by Rosenbaum (1987), this approach relies on the presence of multiple control groups. Suppose the researcher has available two potential comparison groups consisting of units not exposed to the active treatment. In the main analysis, one may have combined the two comparison groups into a single control group to estimate the treatment effects of interest. However, one can also compare the two comparison groups to each other, arbitrarily viewing one of them as being a "pseudo-treatment" group and the other as a "clean" control group under the stronger version of unconfoundedness that we call *group unconfoundedness*. Because neither group did, in fact, receive any active treatment, there should be no causal effects of this pseudo-treatment. Statistical evidence of the presence of systematic differences between the two control groups after adjusting for pre-treatment differences (non-zero "pseudo-causal effects") implies that unconfoundedness is violated for at least one of the comparison groups. Finding no evidence of a difference between the two groups does not imply the unconfoundedness assumption is valid, because it could be that both com-parison groups exhibit the same bias for comparing the actual treated and control units after adjusting for differences in pre-treatment variables. However, if *a priori* any poten-tial biases in treatment-control differences between the two control groups are judged to be different, the finding that the hypothesis of a zero effect of the pseudo-treatment is consistent with the data makes the analysis under unconfoundedness more plausible. The key for the force of this approach is to have control groups that are systematically differ-ent and, as a result, are likely to exhibit different biases in treatment-control comparisons, if they have any biases at all.

The third class of methods focuses on the robustness of estimates of causal effects to the choice of pre-treatment variables. Here we require outcome data for both the treat-ment and control groups, and so this is not a design-stage analysis. We again partition the full set of pre-treatment variables into two parts, the first set consisting of some selected pre-treatment variables, and the second set consisting of the remaining covari-ates. We then assume subset unconfoundedness, where unconfoundedness is assumed to hold conditional only on the remaining set of pre-treatment variables. Given sub-set unconfoundedness, we estimate the causal effects of the treatment on the actual outcome and compare the results to those based on (full) unconfoundedness to assess the null hypothesis that both unconfoundedness and subset unconfoundedness hold. If we find substantial and statistically significant differences, we would conclude that either (*i*) the first set of selected pre-treatment variables is critical for unconfounded-ness (subset unconfoundedness is violated), or (*ii*) unconfoundedness does not hold. If we *a priori* view that the substantive difference between unconfoundedness and subset unconfoundedness is minor, the implication is that we should be concerned with the unconfoundedness assumption. Clearly this assertion depends on the context and the nature of the variables in the two sets.

In all three approaches, we are interested in assessing the presence and magnitude of causal effects under an unconfoundedness assumption. In practice, we often focus on testing whether an estimated average pseudo-causal effect under unconfoundedness is different from its presumed known value (typically zero). It should be noted, though, that in principle the interest here is in assessing whether there is *any* effect of the (pseudo)-treatment on the (pseudo)-outcome different from its presumed known value, not just a zero average difference. That is, we may therefore wish to go beyond studying average effects and also investigate differences in distributions of outcomes, as well as average outcomes by subpopulations. In doing so, we are interested in both statistically and substantially significant differences between the comparison groups.

The remainder of this chapter is organized as follows. In Section 21.3, we discuss the role of pseudo-outcomes for assessing the unconfoundedness assumption. In the next section, Section 21.4, we discuss how one can exploit the presence of multiple control groups. In Section 21.5 we focus on assessing the robustness of the causal effect estimates to changes in the set of pre-treatment variables. In Section 21.6 we illustrate some of the methods using the Imbens-Rubin-Sacerdote lottery data, which we previously used in Chapters 14, 17, and 19. Section 21.7 concludes.

21.2 SETUP

The setup in this section is largely the same as in the previous chapters. For unit i we postulate the existence of two potential outcomes $Y_i(0)$ and $Y_i(1)$, a treatment indicator $W_i \in \{0, 1\}$, and a vector of covariates or pre-treatment variables X_i. We observe the triple consisting of the vector of covariates X_i, the treatment indicator W_i, and the realized and observed outcome

$$Y_i^{\text{obs}} = Y_i(W_i) = \begin{cases} Y_i(0) & \text{if } W_i = 0, \\ Y_i(1) & \text{if } W_i = 1. \end{cases}$$

We consider the super-population unconfoundedness assumption,

$$W_i \perp\!\!\!\perp Y_i(0), Y_i(1) \mid X_i \quad \text{(unconfoundedness)}, \tag{21.1}$$

where the dependence on the parameter ϕ is suppressed. This assumption is not testable, as discussed in Chapter 12.

21.3 ESTIMATING EFFECTS ON PSEUDO-OUTCOMES

In this section we discuss the first approach to assessing unconfoundedness, where we focus on tests for causal effects on pseudo-outcomes. This is an approach that can be used at the design stage, without access to outcome data. First we introduce some additional notation. We partition the vector of covariates X_i into two parts, the first denoted by X_i^{p} ("p" for pseudo), and the remainder denoted by X_i^{r}, so that the full vector of pre-treatment variables can be written as $X_i = (X_i^{\text{p}}, X_i^{\text{r}})$.

Instead of testing whether the conditional independence assumption in (21.1) holds, which we showed before is impossible to do from the data at hand, we shall test whether the following conditional independence relation, which we label *pseudo-outcome unconfoundedness*, holds:

$$W_i \perp\!\!\!\perp X_i^{\text{p}} \,\Big|\, X_i^{\text{r}} \quad \text{(pseudo-outcome unconfoundedness).} \tag{21.2}$$

The two issues are, first, the interpretation of assumption (21.2) and specifically its connection to full unconfoundedness (21.1), which is of primary interest, and second, the implementation of the assessment.

21.3.1 Interpretation

The first issue concerns the link between the conditional independence relation in (21.2) and unconfoundedness in (21.1). This link is indirect, because unconfoundedness cannot be tested directly. Here we lay out the arguments for the connection. First consider an additional version of unconfoundedness, which we label *subset unconfoundedness*

$$W_i \perp\!\!\!\perp \big(Y_i(0), Y_i(1)\big) \,\Big|\, X_i^{\text{r}} \quad \text{(subset unconfoundedness).} \tag{21.3}$$

Subset unconfoundedness is not testable for the same reasons full unconfoundedness is not testable: we do not observe $Y_i(0)$ if $W_i = 1$, and we do not observe $Y_i(1)$ if $W_i = 0$. Here we explore an alternative approach to assess it. Suppose we have a proxy for either of the potential outcomes, and in particular a proxy or *pseudo-outcome* whose value is observed irrespective of the realized treatment status; one can test independence of that proxy variable and the treatment indicator W_i. We use the selected pre-treatment variable X_i^{p} as such a pseudo-outcome or proxy variable. For example, we view X_i^{p} as a proxy for $Y_i(0)$, and assess (21.3) by testing (21.2), which involves only observed variables.

The most compelling applications of these assessments are settings where the two steps in going from unconfoundedness (21.1) to the testable condition (21.2) are plausible. One such example occurs when X_i contains multiple lagged measures of the outcome, as in the Imbens-Rubin-Sacerdote lottery study that we use to illustrate these methods in this chapter. The pre-treatment variables X_i in that application consist of some time-invariant pre-treatment variables V_i (e.g., age, education), and some lagged outcomes (earnings), $(Y_{i,-1}, \ldots, Y_{i,-T})$. One can implement these ideas using earnings for one of the most recent pre-program years $(Y_{i,-1}, \ldots, Y_{i,-T})$ as the pseudo-outcome X_i^{p}. Under unconfoundedness, $Y_i(0)$ is independent of W_i given $Y_{i,-1}, \ldots, Y_{i,-6}$ and V_i, which would suggest that it is also plausible that $Y_{i,-1}$ is independent of W_i given $Y_{i,-2}, \ldots, Y_{i,-6}$ and V_i. Given those arguments, one can plausibly assess unconfoundedness by testing whether

$$W_i \perp\!\!\!\perp Y_{i,-1} \mid V_i, Y_{i,-2}, \ldots, Y_{i,-6}. \tag{21.4}$$

The claim now is that finding that (21.4) is not supported by the data would lower the credibility of an analysis that relies on unconfoundedness (21.1), relative to a finding that the relation in (21.4) is consistent with the data.

21.3.2 Implementation

Now we turn again to the implementation of this assessment of unconfoundedness. One approach to test the conditional independence assumption in (21.2) is to estimate the average difference in X_i^{p} by treatment status, after adjusting for differences in X_i^{r}. This is exactly the same problem as estimating the average effect of the treatment, using X_i^{p} as the pseudo-outcome and X_i^{r} as the vector of pre-treatment variables. We can do this using any of the methods discussed in the previous chapters, such as blocking or matching, ideally in combination with model-based adjustment.

The main limitation of this approach, testing whether an adjusted average difference is equal to zero, is that it does not test all aspects of the conditional independence restriction. It effectively tests only whether

$$\mathbb{E}\Big[\mathbb{E}\big[X_i^{\mathrm{p}} \mid W_i = 1, X_i^{\mathrm{r}} \big] - \mathbb{E}\big[X_i^{\mathrm{p}} \mid W_i = 0, X_i^{\mathrm{r}} \big] \Big] = 0.$$

Pseudo-outcome unconfoundedness (21.2) implies two additional sets of restrictions. First, of all, it implies that

$$\mathbb{E}\Big[\mathbb{E}\big[g(X_i^{\mathrm{p}}) \mid W_i = 1, X_i^{\mathrm{r}} \big] - \mathbb{E}\big[g(X_i^{\mathrm{p}}) \mid W_i = 0, X_i^{\mathrm{r}} \big] \Big] = 0,$$

for any function $g(\,\cdot\,)$, not just the identity function. We can implement this by comparing average outcomes for different transformations of the pseudo-outcome and testing jointly whether any of the averages effects are zero. For example, for a pseudo-outcome bounded between zero and one, one might test jointly whether the effects of the treatment on $1_{X_i^{\mathrm{p}} \leq 0.2}$, $1_{X_i^{\mathrm{p}} \leq 0.4}$, $1_{X_i^{\mathrm{p}} \leq 0.6}$, and $1_{X_i^{\mathrm{p}} \leq 0.8}$ are all zero. For non-negative outcomes such as earnings, we may wish to test whether the average value of earnings, as well as the fraction of individuals with positive earnings, are equal in treatment and control groups. Of course one has to be careful here doing multiple comparisons, because in that case some contrasts may appear substantial just by chance.

Second, the conditional independence restriction in (21.2) implies that, not only on average, but conditional on $X_i^{\mathrm{r}} = x^{\mathrm{r}}$, for all x^{r},

$$\mathbb{E}\big[X_i^{\mathrm{p}} \mid W_i = 1, X_i^{\mathrm{r}} = x^{\mathrm{r}} \big] - \mathbb{E}\big[X_i^{\mathrm{p}} \mid W_i = 0, X_i^{\mathrm{r}} = x^{\mathrm{r}} \big] = 0.$$

One can therefore also consider tests of the restriction

$$\mathbb{E}\Big[\mathbb{E}\big[g(X_i^{\mathrm{p}}) \mid W_i = 1, X_i^{\mathrm{r}} \big] - \mathbb{E}\big[g(X_i^{\mathrm{p}}) \mid W_i = 0, X_i^{\mathrm{r}} \big] \Big| X_i^{\mathrm{r}} \in \mathbb{X}_j^{\mathrm{r}} \Big] = 0, \tag{21.5}$$

for some partitioning $\{\mathbb{X}_j^r\}_{j=1}^J$ of the support $\mathbb{X}^r$ of the set of remaining covariates X_i^r. That is, rather than testing whether the overall average effect of the treatment on the pseudo-outcome differs from zero, one might wish to test the null hypothesis that the average effect of the treatment on the pseudo-outcome in subpopulations differ from zero.

21.4   ESTIMATING EFFECTS OF PSEUDO-TREATMENTS

We now discuss the second approach to assessing unconfoundedness, which focuses on tests for non-zero causal effects of pseudo-treatments.

21.4.1 Setup

This approach to assess the plausibility of the unconfoundedness assumption relies on the presence of additional control information, specifically, a two-component control group. For this approach, we require outcome data for the control group but not for the treatment group. It could therefore be called a *semi-design stage* method. We change notation in a subtle way. Let G_i be an indicator variable denoting the generalized treatment group that unit i is a member of. This indicator variable takes on three or more values. For ease of exposition we focus on the case with two control groups and thus three values for G_i, $G_i \in \{c_1, c_2, t\}$. Units with $G_i = c_1$ or c_2 receive the control treatment, $W_i = 0$, and units with $G_i = t$ receive the active treatment, $W_i = 1$:

$$W_i = \begin{cases} 0 & \text{if } G_i = c_1, c_2, \\ 1 & \text{if } G_i = t. \end{cases}$$

Unconfoundedness only requires that

$$W_i \perp\!\!\!\perp \left(Y_i(0), Y_i(1) \right) \Big| X_i, \tag{21.6}$$

which is not testable with the data at hand. Instead we focus on testing an implication of the stronger conditional independence relation, which we label *group unconfoundedness*:

$$G_i \perp\!\!\!\perp \left(Y_i(0), Y_i(1) \right) \Big| X_i, \quad \text{(Group Unconfoundedness)} \tag{21.7}$$

This independence condition implies unconfoundedness, but in contrast to unconfoundedness, it has testable restrictions. In particular, we focus on the implication that

$$G_i \perp\!\!\!\perp Y_i(0) \Big| X_i, G_i \in \{c_1, c_2\},$$

which is equivalent to

$$G_i \perp\!\!\!\perp Y_i^{\text{obs}} \Big| X_i, G_i \in \{c_1, c_2\}, \tag{21.8}$$

because $G_i \in \{c_1, c_2\}$ implies that $W_i = 0$, and thus $Y_i^{\text{obs}} = Y_i(W_i) = Y_i(0)$. This conditional independence condition has the same form as (21.2), and we test it in the same fashion. Again we discuss first the link between (21.8) and unconfoundedness, (21.1), and second the implementation of tests of this conditional independence assumption.

21.4.2 Interpretation

Because condition (21.12) is strictly stronger than unconfoundedness, (21.1), the question is whether there are interesting settings where the weaker and untestable condition of unconfoundedness holds but not the stronger condition. To discuss this question, it is useful to consider two alternative unconfoundedness-like conditional independence conditions, both of which are implied by (21.6):

$$W_i \perp\!\!\!\perp \left(Y_i(0), Y_i(1)\right) \,\Big|\, X_i, G_i \in \{c_1, t\}, \tag{21.9}$$

and

$$W_i \perp\!\!\!\perp \left(Y_i(0), Y_i(1)\right) \,\Big|\, X_i, G_i \in \{c_2, t\}. \tag{21.10}$$

If (21.9) holds, then we can estimate causal effects by invoking the unconfoundedness assumption using only the first control group. Similarly, if (21.10) holds, then we can estimate causal effects by invoking the unconfoundedness assumption using only the second control group. The point is that it is difficult to envision a situation where unconfoundedness based on the two comparison groups (21.6) holds, but using only one of the two comparison groups the unconfoundedness condition fails (i.e., neither (21.9) nor (21.10) holds). So, in practice, if unconfoundedness holds, typically also the stronger condition (21.6) would hold, and we have the testable implication (21.8). Again, there is no theorem here, but an implication that when stronger conditional independence assumptions are false, weaker conditional independence assumptions are more likely also to be false.

21.4.3 Implementation

The implementation of the test follows the same pattern as the implementation of the tests of (21.2). We test whether there is a difference in average values of Y_i^{obs} between the two control groups, after adjusting for differences in X_i. That is, we effectively test whether

$$\mathbb{E}\left[\mathbb{E}\left[Y_i^{\mathrm{obs}}\Big| G_i = c_1, X_i\right] - \mathbb{E}\left[Y_i^{\mathrm{obs}}\Big| G_i = c_2, X_i\right]\right] = 0.$$

We can then extend the test by simultaneously testing whether the average value of transformations of the form $g(Y_i^{\mathrm{obs}})$ differs by group, that is, whether

$$\mathbb{E}\left[\mathbb{E}\left[g(Y_i^{\mathrm{obs}})\Big| G_i = c_1, X_i\right] - \mathbb{E}\left[g(Y_i^{\mathrm{obs}})\Big| G_i = c_2, X_i\right]\right] = 0.$$

In addition we can extend the tests by assessing the null hypothesis whether, given a partition $\{\mathbb{X}_j\}_{j=1}^{J}$ of the support $\mathbb{X}$ of X_i,

$$\mathbb{E}\left[\mathbb{E}\left[g(Y_i^{\mathrm{obs}})\Big| W_i = 1, X_i\right] - \mathbb{E}\left[g(Y_i^{\mathrm{obs}})\Big| W_i = 0, X_i\right]\Big| X_i \in \mathbb{X}_j\right] = 0, \tag{21.11}$$

for all subsets $\mathbb{X}_j$, for $j = 1, \ldots, J$.

21.5   ROBUSTNESS TO THE SET OF PRE-TREATMENT VARIABLES

Here we discuss the third approach to assessing unconfoundedness: investigating the sensitivity of the estimates of causal effects to the choice of pre-treatment variables used for adjustments.

21.5.1 Subset Unconfoundedness and Robustness

We use the same notion of partitioning the set of pre-treatment variables into two parts that we introduced in Section 21.3. Again let us consider subset unconfoundedness:

$$ W_i \perp\!\!\!\perp Y_i(0), Y_i(1) \mid X_i^{\mathrm{r}} \quad \text{(subset unconfoundedness).} \tag{21.12} $$

If this subset unconfoundedness condition were to hold, one could use the adjustment methods described in Chapters 17 and 18, using only the subset of covariates X_i^{r}, instead of the full vector of pre-treatment variables X_i to obtain approximately unbiased estimates of treatment effects. Although this is not a formal result, subset unconfoundedness in (21.3) is intuitively a more restrictive condition than the original unconfoundedness condition (21.1). One has to be careful because it is theoretically possible that conditional on a subset of the covariates (e.g., X_i^{r}) subset unconfoundedness (21.3) holds, but at the same time, unconfoundedness (21.1) does not hold conditional on the full set of covariates $(X_i^{\mathrm{r}}, X_i^{\mathrm{p}})$. In practice, however, this situation is rare if all covariates are proper pre-treatment variables. For example, it is difficult to imagine in an evaluation of a labor market program where unconfoundedness would hold given age, last year's earnings, and the level of education, but not hold if one additionally conditions on sex. Generally having subpopulations that are more homogeneous in pre-treatment variables improves the plausibility of unconfoundedness, although, again, theoretically it is possible that the biases are exactly canceled out if one of the pre-treatment variables is omitted from the analysis. This possibility, however, appears to be of little practical interest.

The main concern for the application of this approach is not this remote possibility of canceling biases but the very real possibility that the pseudo-outcome X_i^{p} may be critical to unconfoundedness, and so that (21.1) may hold, but not (21.3). This is likely to be the case if X_i^{p} is qualitatively different from the variables in X_i^{r}. Again this reinforces the idea that this approach is most valuable when X_i contains several variables that differ from each other only in their time of measurement.

On its own, the assumption of subset unconfoundedness is not directly testable for the same reason that unconfoundedness is not testable: it restricts distributions of missing potential outcomes in terms of distributions of observed potential outcomes. However, the combination of the two assumptions, subset unconfoundedness (21.3) and unconfoundedness (21.1), both not testable on their own, does have testable implications. The combination implies that adjusting for differences in the subset of covariates X_i^{r} and adjusting for differences in the full set of covariates X_i should give similar point estimates (but not necessarily precisions). Thus, we can compare point estimates based on adjusting for the full set of covariates and the subset of covariates. If we find that the results are statistically different for the two sets of covariates, it must be that at least one of the two assumptions, (full) unconfoundedness or subset unconfoundedness, does not hold. The fact that, in that case, the presence of X_i^{p} is critical for the adjustment suggests that there may be concerns about unconfoundedness in general. On the other hand, if we find that the point estimates based on the two assumptions are similar, one may be more confident in the underlying unconfoundedness assumption.

One of the leading examples occurs when X_i contains multiple lagged measures of the outcome. For example, in the evaluation of the effect of a labor market program or lottery on annual earnings, one might have observations on earnings for multiple years prior to the program. Consider the Imbens-Rubin-Sacerdote lottery data, where we have six years of annual earnings prior to winning the lottery. Denote these lagged outcomes by $Y_{i,-1}, \ldots, Y_{i,-6}$, where $Y_{i,-1}$ is the most recent and $Y_{i,-6}$ is the most distant (in time) pre-lottery earnings measure, and denote the remaining covariates by V_i, so that $X_i = (Y_{i,-1}, \ldots, Y_{i,-6}, V_i)$. Unconfoundedness corresponds to the independence relation

$$W_i \perp\!\!\!\perp Y_i(0), Y_i(1) \mid V_i, Y_{i,-1}, Y_{i,-2}, \ldots, Y_{i,-6}. \tag{21.13}$$

This assumption is not testable with the data at hand. However, one could implement the foregoing ideas using earnings for the most recent pre-program year $Y_{i,-1}$ as the selected pre-treatment variable X_i^{p}, so that the vector of remaining pre-treatment variables X_i^{r} would still include the five prior years of pre-program earnings, $Y_{i,-2}, \ldots, Y_{i,-6}$, and the additional pre-treatment variables V_i. In that case, one might reasonably argue that, on *a priori* grounds, unconfoundedness is viewed as reasonable given the presence of six years of pre-program earnings (i.e., (21.13) holds), and it is plausible that it would also hold given only five years of pre-program earnings, so that also

$$W_i \perp\!\!\!\perp Y_i(0), Y_i(1) \mid V_i, Y_{i,-2}, \ldots, Y_{i,-6}. \tag{21.14}$$

21.5.2 Implementation

Here we discuss the implementation of this First we focus on a specific testable implication, and then we discuss more general results. Let $\tau_{\mathrm{SP}} = \mathbb{E}[Y_i(1) - Y_i(0)]$ be the super-population average causal effect of the treatment. Under (super-population) unconfoundedness,

$$\tau_{\mathrm{sp}} = \mathbb{E}\left[\mathbb{E}\left[Y_i^{\mathrm{obs}} \mid W_i = 1, X_i\right] - \mathbb{E}\left[Y_i^{\mathrm{obs}} \mid W_i = 0, X_i\right]\right].$$

Under subset unconfoundedness, it is also true that adjusting solely for differences in X_i^{r} removes biases from comparisons between treated and control units:

$$\tau_{\mathrm{sp}} = \mathbb{E}\left[\mathbb{E}\left[Y_i^{\mathrm{obs}} \mid W_i = 1, X_i^{\mathrm{r}}\right] - \mathbb{E}\left[Y_i^{\mathrm{obs}} \mid W_i = 0, X_i^{\mathrm{r}}\right]\right].$$

These two results imply the testable restriction that

$$\mathbb{E}\left[\mathbb{E}\left[Y_i^{\mathrm{obs}} \mid W_i = 1, X_i\right] - \mathbb{E}\left[Y_i^{\mathrm{obs}} \mid W_i = 0, X_i\right]\right]$$
$$= \mathbb{E}\left[\mathbb{E}\left[Y_i^{\mathrm{obs}} \mid W_i = 1, X_i^{\mathrm{r}}\right] - \mathbb{E}\left[Y_i^{\mathrm{obs}} \mid W_i = 0, X_i^{\mathrm{r}}\right]\right].$$

We can implement this assessment by estimating the average effect of the treatment using the full set of covariates and comparing that to the estimate of the average treatment effect based on the subset of covariates.

We compare two quantities that both estimate the average causal effect under the combination of two assumptions, unconfoundedness and subset unconfoundedness. If both assumptions hold, the treatment effects should also be identical for subpopulations and for causal estimands other than the average effect, and for comparisons of both potential outcomes separately. We can capture these additional implications by focusing on comparisons of more general estimands. Let $f_{Y_i^{\text{obs}}|W_i=w,X_i=x}(y|w,x)$ be the conditional distribution in the super-population of Y_i^{obs} conditional on $W_i = w$ and the full set of covariates $X_i = x$, and similarly for $f_{Y_i^{\text{obs}}|W_i=w,X_i^{\text{r}}=x^{\text{r}}}(y|w,x^{\text{r}})$, where we only condition on X_i^{r}. By definition, the distribution conditioning on the subset of the covariates X_i^{r} can be written as

$$f_{Y_i^{\text{obs}}|W_i=w,X_i^{\text{r}}=x^{\text{r}}}(y|w,x^{\text{r}}) = \int f_{Y_i^{\text{obs}}|W_i=w,X_i=x}(y|w,x) \cdot f_{X_i|W_i=w,X_i^{\text{r}}=x^{\text{r}}}(x|w,x^{\text{r}})dx.$$

At its most general level, the implication of the combination of the two assumptions, unconfoundedness and subset unconfoundedness, is that

$$f_{Y_i^{\text{obs}}|W_i=w,X_i^{\text{r}}=x^{\text{r}}}(y|w,x^{\text{r}}) = \int f_{Y_i^{\text{obs}}|W_i=w,X_i=x}(y|w,x) \cdot f_{X_i|X_i^{\text{r}}=x^{\text{r}}}(x|x^{\text{r}})dx.$$

Hence, the hypothesis that is being assessed is whether the conditioning on $W_i = w$ in the conditional distribution of X_i given $X_i^{\text{r}} = x^{\text{r}}$ in this integral matters:

$$\int f_{Y_i^{\text{obs}}|W_i=w,X_i=x}(y|w,x) \cdot \left(f_{X_i|W_i,X_i^{\text{r}}=x^{\text{r}}}(x|w=1,x^{\text{r}}) - f_{X_i|W_i,X_i^{\text{r}}=x^{\text{r}}}(x|w=0,x^{\text{r}})\right)dx = 0. \tag{21.15}$$

Directly comparing the two complete conditional distributions is complicated, so here we focus on a different set of comparisons. Let $\mathbb{X}^{\text{r}}$ be the support of X_i^{r}, and let $\mathbb{X}_1^{\text{r}}, \ldots, \mathbb{X}_J^{\text{r}}$ partition this space. Then consider, for some function $g(\cdot)$ of the outcome, conditional on $X_i^{\text{r}} \in \mathbb{X}_j^{\text{r}}$ the conditional average outcome,

$$\mathbb{E}\left[g(Y_i^{\text{obs}})\middle| W_i = w, X_i^{\text{r}} \in \mathbb{X}_j^{\text{r}}\right]. \tag{21.16}$$

If we maintain both assumptions, unconfoundedness and subset unconfoundedness, we can estimate the expectation in (21.16) in two different ways. First, under unconfoundedness, it is equal to

$$\mathbb{E}\left[\mathbb{E}\left[g(Y_i^{\text{obs}})\middle| W_i = w, X_i\right]\middle| W_i = 0, X_i^{\text{r}} \in \mathbb{X}_j^{\text{r}}\right]. \tag{21.17}$$

Second, under subset unconfoundedness, the expectation in (21.17) is also equal to

$$\mathbb{E}\left[\mathbb{E}\left[g(Y_i^{\text{obs}})\middle| W_i = w, X_i^{\text{r}}\right]\middle| W_i = 1, X_i^{\text{r}} \in \mathbb{X}_j^{\text{r}}\right],$$

leading to the restriction that for all functions $g(\,\cdot\,)$, for all subsets $\mathbb{X}_j^{\mathrm{r}}$ and for both $w = 0, 1$,

$$\mathbb{E}\left[\mathbb{E}\left[g(Y_i^{\mathrm{obs}})\middle|\, W_i = w, X_i\right]\middle|\, W_i = 0, X_i^{\mathrm{r}} \in \mathbb{X}_j^{\mathrm{r}}\right]$$

$$= \mathbb{E}\left[\mathbb{E}\left[g(Y_i^{\mathrm{obs}})\middle|\, W_i = w, X_i^{\mathrm{r}}\right]\middle|\, W_i = 1, X_i^{\mathrm{r}} \in \mathbb{X}_j^{\mathrm{r}}\right].$$

To gain some insight into this approach, let us consider a simple case. In this example we focus on the case with $g(y) = y$, leading to the restriction

$$\mathbb{E}\left[\mathbb{E}\left[Y_i^{\mathrm{obs}}\middle|\, W_i = 1, X_i\right]\middle|\, W_i = 0, X_i^{\mathrm{r}} \in \mathbb{X}_j^{\mathrm{r}}\right]$$

$$- \mathbb{E}\left[\mathbb{E}\left[Y_i^{\mathrm{obs}}\middle|\, W_i = 1, X_i^{\mathrm{r}}\right]\middle|\, W_i = 1, X_i^{\mathrm{r}} \in \mathbb{X}_j^{\mathrm{r}}\right]. \tag{21.18}$$

Moreover, suppose that the conditional expectations are linear, $\mathbb{E}[Y_i^{\mathrm{obs}}|W_i = 1, X_i] = X_i\beta_{\mathrm{t}}$, with $\beta_{\mathrm{t}} = (\beta_{\mathrm{t}}^{\mathrm{p}}, \beta_{\mathrm{t}}^{\mathrm{r}})$ corresponding to X_i^{p} and X_i^{r}, so that $\mathbb{E}[Y_i^{\mathrm{obs}}|W_i = 1, X_i] = X_i^{\mathrm{p}}\beta_{\mathrm{t}}^{\mathrm{p}} + X_i^{\mathrm{r}}\beta_{\mathrm{t}}^{\mathrm{r}}$. Then:

$$\mathbb{E}\left[\mathbb{E}\left[Y_i^{\mathrm{obs}}\middle|\, W_i = 1, X_i\right]\middle|\, W_i = 0, X_i^{\mathrm{r}} \in \mathbb{X}_j^{\mathrm{r}}\right] = \mathbb{E}\left[X_i^{\mathrm{p}}\beta_{\mathrm{t}}^{\mathrm{p}} + X_i^{\mathrm{r}}\beta_{\mathrm{t}}^{\mathrm{r}}\middle|\, W_i = 0, X_i^{\mathrm{r}} \in \mathbb{X}_j^{\mathrm{r}}\right]$$

$$= X_i^{\mathrm{r}}\beta_{\mathrm{t}}^{\mathrm{r}} + \mathbb{E}\left[X_i^{\mathrm{p}}\middle|\, W_i = 0, X_i^{\mathrm{r}} \in \mathbb{X}_j^{\mathrm{r}}\right]\beta_{\mathrm{t}}^{\mathrm{p}},$$

and the difference (21.18) reduces to

$$\left(\mathbb{E}[X_i^{\mathrm{p}}|W_i = 0, X_i^{\mathrm{r}}] - \mathbb{E}[X_i^{\mathrm{p}}|W_i = 1, X_i^{\mathrm{r}}]\right)\beta_{\mathrm{t}}^{\mathrm{p}}. \tag{21.19}$$

In this linear conditional expectation case, the combination of unconfoundedness and subset unconfoundedness implies that the association (i.e., correlation here) of the selected covariates X_i^{p} with the outcome conditional on the remaining covariates is zero on average.

21.6   THE IMBENS-RUBIN-SACERDOTE LOTTERY DATA

In this section we apply the methods developed in this chapter to the lottery data previously analyzed in Chapter 13. We start with the full sample of 496 individuals. In Table 21.1 we present summary statistics for these 496 individuals. We focus on estimates based on the blocking or subclassification methods from Chapter 17, after using the trimming approach from Chapter 16. In that chapter we estimated the average effect of winning a big prize on average earnings for the next six years to be approximately $-\$6,000$ per year, with a standard error of approximately $\$1,000$. In this section we investigate the plausibility of the unconfoundedness assumption for this data set.

21.6.1 Testing for Effects on Pseudo Outcomes

The data are well-suited for using the methods discussed in Section 21.3, because the data set contains information on earnings (whose value after winning the lottery is the outcome of primary interest) for six years prior to winning the lottery prize, making

Table 21.1. *Summary Statistics for Selected Lottery Sample for the IRS Lottery Data*

Variable	Label	All ($N=496$) Mean	(S.D.)	Non-Winners ($N_t=259$) Mean	Winners ($N_c=237$) Mean	[t-Stat]	Nor Dif
Year Won	(X_1)	6.23	(1.18)	6.38	6.06	−3.0	−0.27
Tickets Bought	(X_2)	3.33	(2.86)	2.19	4.57	9.9	0.90
Age	(X_3)	50.22	(13.68)	53.21	46.95	−5.2	−0.47
Male	(X_4)	0.63	(0.48)	0.67	0.58	−2.1	−0.19
Years of Schooling	(X_5)	13.73	(2.20)	14.43	12.97	−7.8	−0.70
Working Then	(X_6)	0.78	(0.41)	0.77	0.80	0.9	0.08
Earnings Year -6	(Y_{-6})	13.84	(13.36)	15.56	11.97	−3.0	−0.27
Earnings Year -5	(Y_{-5})	14.12	(13.76)	15.96	12.12	−3.2	−0.28
Earnings Year -4	(Y_{-4})	14.21	(14.06)	16.20	12.04	−3.4	−0.30
Earnings Year -3	(Y_{-3})	14.80	(14.77)	16.62	12.82	−2.9	−0.26
Earnings Year -2	(Y_{-2})	15.62	(15.27)	17.58	13.48	−3.0	−0.27
Earnings Year -1	(Y_{-1})	16.31	(15.70)	18.00	14.47	−2.5	−0.23
Pos Earnings Year -6	($Y_{-6}>0$)	0.69	(0.46)	0.69	0.70	0.3	0.03
Pos Earnings Year -5	($Y_{-5}>0$)	0.71	(0.45)	0.68	0.74	1.6	0.14
Pos Earnings Year -4	($Y_{-4}>0$)	0.71	(0.45)	0.69	0.73	1.1	0.10
Pos Earnings Year -3	($Y_{-3}>0$)	0.70	(0.46)	0.68	0.73	1.4	0.13
Pos Earnings Year -2	($Y_{-2}>0$)	0.71	(0.46)	0.68	0.74	1.6	0.15
Pos Earnings Year -1	($Y_{-1}>0$)	0.71	(0.45)	0.69	0.74	1.2	0.10

these variables attractive candidates to play the role of pseudo-outcomes. In this first set of assessments, we focus on analyses using either $X_i^p = Y_{i,-1}$ (earnings in the last year before winning) as the selected pre-treatment variable, or $X_i^p = Y_{i,-6}$ (earnings in the sixth year before winning) as the selected pre-treatment variable, and in each case the remaining pre-treatment variables as X_i^r.

The first analysis is design-based and uses $X_i^p = Y_{i,-1}$ as the pseudo-outcome. First, note that the difference in average prior earnings for winners and losers is $14.47 - 18.00 = -3.53$ (in thousands of dollars; see Table 21.1). This raw difference is substantial, relative to the estimated effect of winning the lottery of -5.74, and it is statistically significantly different from zero at conventional significance levels. Because this difference cannot be a causal effect of winning the lottery, it must be due to pre-existing differences between the winners and the losers. The question is whether adjusting for the remaining pre-treatment variables removes this difference.

To implement the analyses discussed in this chapter, recall that in the analysis that led to the point estimate of -5.74, we included automatically in the propensity score the selected covariates X_{i2} (Tickets Bought), X_{i5} (Years of Schooling), X_{i6} (Working Then), and the most recent earnings, $Y_{i,-1}$. To make the analysis with the pseudo-outcome as similar as possible to the main analyses, we always include in the propensity score the covariates X_{i2}, X_{i5}, X_{i6}, and the most recent earnings $Y_{i,-2}$. The blocking estimate based on this setup is

$$\hat{\tau}^{strat} = -0.53 \quad (\widehat{s.e.} = 0.58),$$

This is statistically not significantly different from zero at conventional significance levels, and substantively unimportant. It is also small compared to the effect we find for the actual outcome, that is, -5.74.

Table 21.2. *Estimates of Average Treatment Effect on Pseudo-outcome for the IRS Lottery Data*

Pseudo-Outcome	Remaining Covariates	Selected Covariates	Est	$(\widehat{s.e.})$
Y_{-1}	$X_1, \ldots, X_6, Y_{-6}, \ldots, Y_{-2}, Y_{-6} > 0, \ldots, Y_{-2} > 0$	X_2, X_5, X_6, Y_{-2}	-0.53	(0.58)
$\frac{Y_{-1}+Y_{-2}}{2}$	$X_1, \ldots, X_6, Y_{-6}, \ldots, Y_{-3}, Y_{-6} > 0, \ldots, Y_{-3} > 0$	X_2, X_5, X_6, Y_{-3}	-1.16	(0.71)
$\frac{Y_{-1}+Y_{-2}+Y_{-3}}{3}$	$X_1, \ldots, X_6, Y_{-6}, Y_{-5}, Y_{-4}, Y_{-6} > 0, Y_{-5} > 0, Y_{-4} > 0$	X_2, X_5, X_6, Y_{-4}	-0.39	(0.77)
$\frac{Y_{-1}+\ldots+Y_{-4}}{4}$	$X_1, \ldots, X_6, Y_{-6}, Y_{-5}, Y_{-6} > 0, Y_{-5} > 0$	X_2, X_5, X_6, Y_{-5}	-0.56	(0.89)
$\frac{Y_{-1}+\ldots+Y_{-5}}{5}$	$X_1, \ldots, X_6, Y_{-6}, Y_{-6} > 0$	X_2, X_5, X_6, Y_{-6}	-0.49	(0.87)
$\frac{Y_{-1}+\ldots+Y_{-6}}{6}$	$X_1, \ldots, X_6$	X_2, X_5, X_6	-2.56	(1.55)
Actual outcome Y	$X_1, \ldots, X_6, Y_{-6}, \ldots, Y_{-1}, Y_{-6} > 0, \ldots, Y_{-1} > 0$	X_2, X_5, X_6, Y_{-1}	-5.74	(1.14)

Next, we repeat this for five additional choices of the pseudo-outcome. In each of the five additional analyses we take the average of the J most recent pre-treatment earnings as the pseudo-outcome, and use the remaining pre-treatment earnings as pre-treatment variables. The results for all six tests are in Table 21.2. We find that, as long as there are some pre-treatment earnings left in the set of covariates used to remove biases, the estimates are statistically and substantively insignificant. Only with the average of all pre-treatment earnings used as the pseudo-outcome, so that there are no earnings variables among the remaining pre-treatment variables to be used in the adjustment, do we find a substantially and statistically significant estimate. In that case, the point estimate is -2.56 with an estimated standard error of 1.55. It appears that some measures of pre-treatment earnings are required to remove biases and make unconfoundedness plausible, but we do not appear to need more than one such measure.

Finally, we return to the case with the pseudo-outcome equal to the most recent pre-treatment earnings. Now we look at both the effect on the pseudo-outcome, and on the indicator that the pseudo-outcome is positive. In addition, we do this separately for those with positive and zero earnings in the second year prior to winning the lottery (the most recent pre-treatment year left in the set of pre-treatment variables) in order to assess whether the distribution of the pseudo-outcome differs between treatment groups conditional on covariates, excluding the pseudo-outcomes. The number of individuals with positive earnings in the second year prior to winning the lottery is 351, with 145 individuals with zero earnings in that year. In Table 21.3 we present the four estimates separately, as well as a p-value for the overall test. The p-value of 0.13 suggests that there is relatively little evidence that the distributions of the most recent pre-treatment earnings differ by treatment group conditional on the remaining pre-treatment variables.

Overall the pseudo-outcome assessments suggest that, with the rich set of covariates used, for the selected sample with overlap, the unconfoundedness assumption may be a reasonable assumption, and therefore the estimate of -5.74 for the effect of winning the lottery on subsequent earnings is credible.

Table 21.3. *Estimates of Average Treatment Effect on Transformations of Pseudo-Outcome for Subpopulations for the IRS Lottery Data*

Pseudo-Outcome	Subpopulation	Est	$(\widehat{s.e.})$
$\mathbf{1}_{Y_{-1}=0}$	$Y_{-2} = 0$	-0.05	(0.04)
$\mathbf{1}_{Y_{-1}=0}$	$Y_{-2} > 0$	-0.04	(0.03)
Y_{-1}	$Y_{-2} = 0$	-1.46	(0.92)
Y_{-1}	$Y_{-2} > 0$	-0.59	(0.58)
		statistic	p-value
Combined statistic (chi-squared, df 4)		5.51	(0.24)

21.6.2 Assessing Effects of Pseudo-Treatments

Next we investigate the plausibility of unconfoundedness through the second approach of testing for the presence of effects of pseudo-treatments, the so-called semi-design approach. In the context of the lottery study it would have been most useful to have a comparison group, say of individuals who did not play the lottery at all. Such individuals might have been expected to be quite different from lottery players in terms of earnings levels and growth. Then we would have two possible control groups: first lottery players who did not win a major prize ("losers"), and second non-lottery players. Then we could have compared the outcome distributions for these two possible control groups. In that case finding that the observed covariates were sufficient to remove differences between non-lottery players and losers would have lent substantial support to the results based on unconfoundedness, precisely because non-lottery players and lottery players might *a priori* have been expected to be quite different in terms of their unobserved economic behavior. However, in the lottery sample we do not have a second control group for whom we are confident that there is no causal effect. Nevertheless, we have a subsample that is almost as good as that, and which we will use for that purpose. Specifically, a subset of the treatment group of lottery winners will be used to serve as such a pseudo-control group. We take the subsample of winners whose yearly prizes were relatively small. For this subset we expect, *a priori*, the causal effects of winning to be modest.

First we define what we mean by "small yearly prize winners." In our sample the median yearly prize is 31.8, and the average is 55.2, all in thousands of dollars. The 75th percentile is 63.0 per year. First, we take the subsample of 111 yearly winners who won an annual prize less than or equal to \$30,000. Even if there is some effect of such a prize on subsequent earnings, one would expect it to be modest compared to the effects of a bigger prize.

Thus, for the purpose of this illustrative analysis, we view those who won more than \$30,000 per year as members of the treatment group, and both winners who won a large enough amount to be paid in yearly installments, but less than \$30,000 per year, and losers as part of the control group.

The results for these analyses are in Table 21.4. For the winners of prizes less than \$30,000, we find that the differences from the losers, after adjusting for all observed covariates, are substantially small and statistically insignificant at conventional levels.

Table 21.4. *Estimates of Average Difference in Outcomes for Controls and Small Winners (less than $30,000) for the IRS Lottery Data*

Outcome	Subpopulation	Est	$(\widehat{s.e.})$
Y_i	All	-0.82	(1.37)
$\mathbf{1}_{Y_i=0}$	$Y_{i,-1}=0$	-0.02	(0.05)
$\mathbf{1}_{Y_i=0}$	$Y_{i,-1}>0$	0.07	(0.05)
Y_i	$Y_{i,-1}=0$	-1.18	(1.10)
Y_i	$Y_{i,-1}>0$	-0.16	(0.69)
		statistic	p-value
Combined statistic (chi-squared, dof 4)		1.24	(0.87)

21.6.3 Assessing Robustness

Finally, we carry out the robustness analysis from Section 21.5.2. To make the analysis fully comparable to those in Chapter 17, we start by trimming the sample using $X_i^r = Z_i, Y_{i,-2}, \ldots, Y_{i,-6}$ as the pre-treatment variables to create a common sample with which to compare confoundedness and subset unconfoundedness. Starting with the full sample with 259 control units and 237 treated units, for a total of 496 units, this leads to a trimmed sample with $N_c = 179$ control units and $N_t = 148$ treated units for a total of $N = 327$ units. Given the trimmed sample, we estimate the average treatment effect, first using the full set of covariates (justified by unconfoundedness), and second using the restricted set of covariates (justified by subset unconfoundedness). The estimates, based on subclassification on the estimated propensity score with additional adjustment within the blocks by linear regression, are

$$\hat{\tau}_{sp}^X = -6.94 \ (\widehat{s.e.} = 1.20), \qquad \hat{\tau}_{sp}^{X^r} = -5.92 \ (\widehat{s.e.} = 1.16),$$

for the estimate based on the full and restricted sets of covariates respectively, and based on the selected sample of 327 units. The difference in the estimates is relatively modest, supportive of unconfoundedness.

We also look directly at the differences in adjusted average outcomes,

$$\left(\mathbb{E}[X_i^p | W_i = 0, X_i^r] - \mathbb{E}[X_i^p | W_i = 1, X_i^r]\right) \beta_t^p.$$

Approximating the two conditional expectations by linear functions,

$$\mathbb{E}[X_i^p | W_i = 0, X_i^r] = X_i^r \gamma_c, \qquad \mathbb{E}[X_i^p | W_i = 1, X_i^r] = X_i^r \gamma_t,$$

we find

$$\frac{1}{N} \sum_{i=1}^N \left(\hat{\mathbb{E}}[X_i^p | W_i = 0, X_i^r] - \hat{\mathbb{E}}[X_i^p | W_i = 1, X_i^r]\right) \hat{\beta}_t^p = -0.13 \ \widehat{s.e.} = 0.12$$

and

$$\frac{1}{N} \sum_{i=1}^N \left(\hat{\mathbb{E}}[X_i^p | W_i = 0, X_i^r] - \hat{\mathbb{E}}[X_i^p | W_i = 1, X_i^r]\right) \hat{\beta}_c^p = -0.19 \ \widehat{s.e.} = 0.11,$$

in both cases small relative to the average causal effect estimate of -5.74.

Again the overall conclusion from these supporting analyses is that unconfoundedness appears to be credible for this data set.

21.7 CONCLUSION

In this chapter we discuss how one can assess the critical unconfoundedness assumption. Although this assumption is not testable, there are three broad classes of methods that can, in some settings, be used to assess whether unconfoundedness is plausible. One of the three classes is design based, not requiring the use of outcome data. One is semi-design, only using control outcome data. The third uses treated and control outcome data. All three classes estimate pseudo-causal effects known, or presumably known, to be equal to zero. If one cannot reject the null hypothesis that (all of) these pseudo-causal effects are equal to zero, one may, cautiously, and with caveats, proceed and accept the unconfoundedness assumption. Rejections of the hypotheses of zero effects, however, do not suggest alternatives to the unconfoundedness assumption. Instead such rejections may simply suggest that it may be impossible to obtain credible inferences regarding the causal effects of interest with the data at hand.

NOTES

Rosenbaum (1987) was one of the first to stress formally the benefits of having multiple control groups when assessing unconfoundedness. His ideas have been used in a variety of ways. Sometimes researchers use the multiple control groups to obtain multiple estimates of the effects of interest and compare those. For a leading example, pre-dating Rosenbaum's work, see Lalonde (1986). Lalonde was interested in comparing experimental estimates to non-experimental estimates of a job-training program. For his non-experimental estimates, he uses comparison groups constructed from the Panel Study of Income Dynamics (PSID) and from the Current Population Survey (CPS). Lalonde then compares estimates of the average effect of the treatment, the job-training program, based on the two different comparison groups. This is a somewhat indirect way of comparing the adjusted differences between the two comparison groups that we discuss in the current chapter.

The idea of using estimates of the effect of the treatment on pseudo-outcomes known not to be affected by the intervention has also a long history. Most often this is in the context of settings with lagged outcomes where one analyzes the data as if the intervention has occurred prior to the time it was actually implemented. See, for example, Heckman and Hotz (1989) and Crump, Hotz, Imbens, and Mitnik (2008).

## CHAPTER 22

Sensitivity Analysis and Bounds

22.1 INTRODUCTION

Part IV of this text focused on estimation and inference under regular assignment mechanisms, that is, ones that are individualistic with probabilistic assignment, as well as unconfounded. In Part V we study methods that confront the unconfoundedness assumption. In Chapter 21 we discussed methods to assess the plausibility of this assumption by combining it with additional assumptions. In the current chapter we relax the unconfoundedness assumption without replacing it with additional assumptions, and so do not focus on obtaining point estimates of the causal estimands of interest. Instead we end up with ranges of plausible values for these estimands, with the width of these ranges corresponding to the extent to which we allow the unconfoundedness assumption to be violated.

We consider two approaches that have much in common. The first, developed by Manski in a series of studies (e.g., Manski, 1990, 1996, 2003, 2013), allows for arbitrarily large violations of the unconfoundedness assumption. This *bounds* or *partial identification* approach, as it is called, leads to sharp results, but at the same time will be seen to limit severely the types of inferences about causal effects that can be drawn from observational data. The second approach, following work in this area by Rosenbaum and Rubin (1983) and Rosenbaum (1995), with important antecedents in the work by Cornfield et al. (1959), works from the other extreme in the sense that unconfoundedness is the starting point, and only limited violations from it need to be considered. If we allow for large violations in the Rosenbaum-Rubin approach, it will often lead to essentially the same results as the Manski bounds approach, but with limited violations of the unconfoundedness assumption, the sensitivity approach results in narrower ranges for the estimands than the partial identification approach.

The key to any sensitivity analysis will be how to assess the magnitude of violations from unconfoundedness. The setup in the current chapter assumes that unconfoundedness is satisfied conditional on an additional, unobserved covariate. If, conditional on the other, observed, covariates, this unobserved covariate is independent of the potential outcomes, or if, again conditional on the observed covariates, it is independent of treatment assignment, unconfoundedness holds even without conditioning on this additional covariate. If, however, this additional, unobserved covariate is associated both with the potential outcomes and with the treatment indicator, biases will result from estimates based on the assumption of unconfoundedness. The magnitude of the bias depends on the strength of the associations between the unobserved covariate and the potential outcomes and treatment indicator.

In the Rosenbaum-Rubin sensitivity approach we consider the range of implied treatment effects as a function of the magnitude of the associations between the unobserved covariate and the potential outcomes and treatment indicator. To assess what reasonable magnitudes are for those associations, we compare them to the associations between observed covariates and the potential outcomes and treatment indicators in the current data, or in cases where other more extensive data are available, to those data.

We also consider a second approach to sensitivity analyses developed by Rosenbaum (1995). Here the sensitivity analyses only require the researcher to specify the magnitude of the association between the unobserved components and the treatment assignment, taking a Manski-style attitude to the associations between the hidden covariate and the potential outcomes. Without making assumptions about associations with the potential outcomes we again obtain ranges of average treatment effects consistent with the evidence in the current study.

Throughout this chapter we take a super-population approach where the sample is viewed as a random sample from an infinite population, with the random sampling generating a distribution for the potential outcomes. In Section 22.2 we describe the subset of the lottery data that will be used to illustrate the sensitivity analyses. Next, in Section 22.3, we study the Manski bounds approach. In Section 22.4 we study the Rosenbaum-Rubin sensitivity approach for the case with binary outcomes. Next, in Section 22.5 we discuss Rosenbaum's approach. Section 22.6 concludes.

22.2   THE IMBENS-RUBIN-SACERDOTE LOTTERY DATA

Here we use again the lottery data originally collected by Imbens, Rubin, and Sacerdote (2001) that we used previously in Chapters 14, 17, 19, and 21. In Chapter 14 we assessed the overlap in covariate distributions for the lottery data and found that overlap was substantial, although there were subsets of covariate values with little overlap. In Chapter 17 we used the methods from Chapter 16 to trim the sample, originally consisting of 496 units, which led to the creation of a sample containing information on $N = 323$ individuals, of whom $N_c = 172$ are losers and $N_t = 151$ are winners, which comprise the sample that is the basis for the analyses in this chapter. The outcome that we are studying is the indicator for having positive earnings during the six-year period (essentially being employed full time during each of these six years) following the lottery.

Assuming unconfoundedness, and using the subclassification estimator developed in Chapter 17, the point estimate of the average effect of winning the lottery on the outcome is -0.134, with an estimated standard error equal to 0.049.

22.3   BOUNDS

We start by focusing on a simple case with no covariates. For unit i, there are two potential outcomes, $Y_i(0)$ and $Y_i(1)$. For illustrative purposes, we consider the average effect of the treatment in the super-population,

$$\tau_{sp} = \mathbb{E}_{sp} \left[Y_i(1) - Y_i(0) \right].$$

In this section we restrict the discussion to the case with binary outcomes, $Y_i(0)$ and $Y_i(1) \in \{0, 1\}$ (some period of non-employment during the six years post lottery versus full-time employment during this period) to allow a sharper focus on the key conceptual issues.

We observe for unit i the treatment received, W_i, and the realized outcome, $Y_i^{obs} = Y_i(W_i)$. In the case without covariates, super-population unconfoundedness simply corresponds to independence of the treatment indicator and the potential outcomes:

$$W_i \perp\!\!\!\perp \left(Y_i(0), Y_i(1) \right).$$

Under random assignment we can unbiasedly estimate the average treatment effect as the difference in average observed outcomes by treatment status, which for the lottery data leads to:

$$\hat{\tau}^{dif} = \overline{Y}_t^{obs} - \overline{Y}_c^{obs} = 0.4106 - 0.5349 = -0.1243.$$

Using Neyman's approach (see Chapter 6), it follows that, if assignment were completely random, $\hat{\tau}^{dif}$ would be unbiased for both the finite-sample average treatment effect τ_{fs} and for the super-population average treatment effect τ_{sp}, with associated standard sampling variance estimate $\hat{\mathbb{V}}^{neyman} = 0.055^2$. We also calculate the exact Fisher p-value assuming complete randomization, using the difference in average outcomes for treated and control units as the statistic, leading to a p-value of 0.034.

Now suppose we do not wish to assume unconfoundedness and, moreover, we do not wish to make any alternative assumptions (but we maintain the stability assumption, SUTVA). What can we learn about τ_{sp} in the absence of this assumption? Manski's approach to this problem is as follows. Suppose we observe for all units in the super-population the treatment indicator W_i and the realized outcome Y_i^{obs}, $Y_i^{obs} = 1$ indicating employment every year versus $Y_i^{obs} = 0$ if individual i was unemployed for at least one year during the six-year post-lottery period. We can obtain method-of-moments estimates for, or using the terminology from the econometric literature, in large samples we can *identify*, three quantities. First, in the super-population share of treated units, $p = \mathbb{E}[W_i] = \Pr(W_i = 1)$, which, in this case without covariates, is also the propensity score for each unit. Second, we can similarly estimate the population distribution of $Y_i(0)$ conditional on $W_i = 0$. Because $Y_i(0)$ is binary, this distribution can be summarized by the scalar $\mu_{c,0} = \Pr(Y_i(0) = 1 | W_i = 0) = \mathbb{E}[Y_i(0) | W_i = 0]$. Finally, we can estimate the population distribution of $Y_i(1)$ given $W_i = 1$, $\mu_{t,1} = \Pr(Y_i(1) = 1 | W_i = 1) = \mathbb{E}[Y_i(1) | W_i = 1]$. In addition, define the super-population quantities

$$\mu_{c,1} = \mathbb{E}[Y_i(0) | W_i = 1], \quad \text{and} \quad \mu_{t,0} = \mathbb{E}[Y_i(1) | W_i = 0].$$

Note that if super-population unconfoundedness holds, then $\mu_{c,1}$ and $\mu_{t,0}$ are equal to

$$\mu_{c,1} = \mu_{c,0} = \mathbb{E}[Y_i(0)] = \mathbb{E}[Y_i^{obs} | W_i = 0],$$

and

$$\mu_{t,0} = \mu_{t,1} = \mathbb{E}[Y_i(1)] = \mathbb{E}[Y_i^{obs}|W_i = 1],$$

respectively, so that under super-population unconfoundedness

$$\tau_{sp} = \mu_{t,1} - \mu_{c,0} = \mathbb{E}[Y_i(1)|W_i = 1] - \mathbb{E}[Y_i(0)|W_i = 0]$$
$$= \mathbb{E}[Y_i^{obs}|W_i = 1] - \mathbb{E}[Y_i^{obs}|W_i = 0],$$

and $\hat{\tau}^{dif}$ is unbiased for τ_{sp}. Without the unconfoundedness assumption, however, we cannot infer τ_{sp} from only these three quantities, p, $\mu_{c,0}$, and $\mu_{t,1}$.

In general, without assuming unconfoundedness, we can rewrite τ_{sp} as the difference in the average of the potential outcomes,

$$\tau_{sp} = \mu_t - \mu_c,$$

where

$$\mu_t = \mathbb{E}[Y_i(1)] = p \cdot \mu_{t,1} + (1-p) \cdot \mu_{t,0},$$

and

$$\mu_c = \mathbb{E}[Y_i(0)] = p \cdot \mu_{c,1} + (1-p) \cdot \mu_{c,0}.$$

Without unconfoundedness (and without making any additional assumptions to replace it), the data are not informative about $\mu_{t,0}$ or $\mu_{c,1}$ beyond the obvious fact that, because the outcomes are binary, these quantities must lie inside the interval $[0, 1]$. These *natural bounds* on $\mu_{t,0}$ and $\mu_{c,1}$ imply bounds on μ_t and μ_c:

$$\mu_c \in \left[(1-p) \cdot \mu_{c,0}, (1-p) \cdot \mu_{c,0} + p\right],$$

and

$$\mu_t \in \left[p \cdot \mu_{t,1}, p \cdot \mu_{t,1} + (1-p)\right].$$

These ranges on μ_t and μ_c in turn imply bounds on the estimand, the population average effect τ_{sp}:

$$\tau_{sp} \in \left[p \cdot \mu_{t,1} - p - (1-p) \cdot \mu_{c,0}, p \cdot \mu_{t,1} + (1-p) - (1-p) \cdot \mu_{c,0}\right]. \tag{22.1}$$

These bounds on the average treatment effect are *sharp*, in the sense that any value of τ_{sp} inside these bounds is consistent with the data if we are not assuming unconfoundedness. In other words, we cannot rule out, even in an infinitely large sample, any value inside these bounds. If we wish to obtain sharper inferences for τ_{sp}, we need to make stronger assumptions about the distribution of the potential outcomes, the assignment mechanism, or both. It is useful to see precisely why the bounds are sharp. Consider the upper bound in (22.1), $p \cdot \mu_{t,1} + (1-p) - (1-p) \cdot \mu_{c,0}$. What is the joint distribution of the potential outcomes and the assignment mechanism that would lead to this value for the average treatment effect? In order for τ to be equal to this upper bound, it must be the case that $\mu_{t,0} = 1$ (i.e., all the units who received the control treatment would have positive earnings given the active treatment), and $\mu_{c,1} = 0$ (i.e., all the units receiving the active treatment would have zero earnings had they received the control treatment). Although such a scenario appears extreme, there is nothing in the data that formally rules out this possibility.

In this specific setting, the bounds are arguably not very informative. Note that without any data, we can infer from the fact that the outcomes are binary that the average effect τ_{sp} must lie in the interval $[-1, 1]$, with the width of that interval equal to two. The data, with everyone exposed to treatment or control, but without the unconfoundedness assumption, can narrow this range to Equation (22.1). Inspection of these bounds shows that they are of the form $[-c, 1-c]$ for some $c \in [0, 1]$. Thus, in this case, the bounds always have range one, and always include zero (corresponding to the Fisher null hypothesis of no effect of the treatment for any unit), irrespective of the data. The fact that the bounds must include zero follows immediately from the fact that nothing in our setup so far rules out the possibility that the treatment effect is zero for all units. The fact that the width of this bounding interval is always one follows from the fact that the width of the interval without the data is two, in combination with the fact that exactly half the potential outcomes are missing.

For the IRS lottery data, the fraction treated is $N_t/N = 0.4675$, and the fraction of individuals with positive earnings in the control and treatment groups are $\overline{Y}_c^{obs} = 0.5349$ and $\overline{Y}_t^{obs} = 0.4106$, respectively. Replacing p, $\mu_{t,1}$ and $\mu_{c,0}$ by N_t/N, $\overline{Y}_t^{obs}$, and $\overline{Y}_c^{obs}$, respectively, in Equation (22.1) leads to a lower and upper bound for the super-population average treatment effect, without additional assumptions, equal to:

$$\tau_{sp} \in \left[p \cdot \mu_{t,1} - p - (1-p) \cdot \mu_{c,0}, p \cdot \mu_{t,1} + (1-p) - (1-p) \cdot \mu_{c,0} \right]$$

$$= [-0.56, 0.44].$$

22.4 BINARY OUTCOMES: THE ROSENBAUM-RUBIN SENSITIVITY ANALYSIS

Now let us study the same setting from a different perspective, the sensitivity analysis approach developed by Rosenbaum and Rubin (1983). Rosenbaum and Rubin start with the assumption that super-population unconfoundedness holds given an unobserved scalar covariate. Let us denote this unobserved covariate by U_i. Super-population unconfoundedness given this unobserved covariate, in the absence of observed covariates, requires that

$$W_i \perp\!\!\!\perp (Y_i(0), Y_i(1)) \mid U_i. \tag{22.2}$$

It is convenient, at least initially, to model U_i as binary with

$$q = \Pr(U_i = 1) = 1 - \Pr(U_i = 0).$$

Now let us build parametric models for the relations between the unobserved covariate U_i and both the treatment indicator and both potential outcomes. In principle we would like a model for

$$f(W_i, Y_i(0), Y_i(1)|U_i).$$

By Equation (22.2) W_i is independent of $(Y_i(0), Y_i(1))$ given U_i, so we can write this as

$$f(W_i|U_i) \cdot f(Y_i(0), Y_i(1)|U_i).$$

As discussed in Chapters 6 and 8, the data are not informative about the dependence structure between $Y_i(0)$ and $Y_i(1)$, so here, for simplicity, we model them as independent conditional on U_i. Thus, we need to specify models for $f(W_i|U_i), f(Y_i(0)|U_i)$, and $f(Y_i(1)|U_i)$. We use the following specifications, taking into account the fact that $Y_i(0)$ and $Y_i(1)$ are binary:

$$\Pr(W_i = 1|U_i = u) = \frac{\exp(\gamma_0 + \gamma_1 \cdot u)}{1 + \exp(\gamma_0 + \gamma_1 \cdot u)},$$

$$\Pr(Y_i(1) = 1|U_i = u) = \frac{\exp(\alpha_0 + \alpha_1 \cdot u)}{1 + \exp(\alpha_0 + \alpha_1 \cdot u)},$$

and

$$\Pr(Y_i(0) = 1|U_i = u) = \frac{\exp(\beta_0 + \beta_1 \cdot u)}{1 + \exp(\beta_0 + \beta_1 \cdot u)},$$

where dependence on the parameters is notationally suppressed to avoid clutter.

There are seven scalar components of the parameter $\theta = (q, \gamma_1, \alpha_1, \beta_1, \gamma_0, \alpha_0, \beta_0)$, which we partition into two subvectors. The first, $\theta_s = (q, \gamma_1, \alpha_1, \beta_1)$, comprises the *sensitivity* parameters, which we do not attempt to estimate. Instead we postulate (ranges of) values for them *a priori*. We discuss later how we select the particular values, or rather the range of values for these parameters, but now we discuss how to proceed conditional on postulated values for these parameters. Conditional on values for the sensitivity parameters $(q, \gamma_1, \alpha_1, \beta_1)$, we estimate the remaining parameters, that is the *estimable parameters* $\theta_e = (\gamma_0, \alpha_0, \beta_0)$, from the data and infer the average treatment effect τ_{sp}. The approach has in common with the bounds approach that even in large samples we cannot reject any combination of values for $(q, \gamma_1, \alpha_1, \beta_1)$: the data do not lead to unbiased method-of-moment estimates of these parameters even in infinite samples, or, in the econometric terminology, these parameters are *not identified*.

Let us look at this argument in more detail. The data allow for unbiased estimates of $p = \mathbb{E}[W_i]$, $\mu_{t,1} = \mathbb{E}[Y_i^{\text{obs}}|W_i = 1] = \mathbb{E}[Y_i(1)|W_i = 1]$ and $\mu_{c,0} = \mathbb{E}[Y_i^{\text{obs}}|W_i = 0] = \mathbb{E}[Y_i(0)|W_i = 0]$. Let us take those parameters as known, and ignore for the moment the sampling variation in their estimates. These three estimable quantities relate to the parameters $(q, \gamma_1, \alpha_1, \beta_1)$ and $(\gamma_0, \alpha_0, \beta_0)$ through the three equalities

$$p = q \cdot \frac{\exp(\gamma_0 + \gamma_1)}{1 + \exp(\gamma_0 + \gamma_1)} + (1 - q) \cdot \frac{\exp(\gamma_0)}{1 + \exp(\gamma_0)}, \qquad (22.3)$$

$$\mu_{t,1} = \Pr(U_i = 1 | W_i = 1) \cdot \mathbb{E}[Y_i(1) | W_i = 1, U_i = 1]$$
$$+ (1 - \Pr(U_i = 1 | W_i = 1)) \cdot \mathbb{E}[Y_i(1) | W_i = 1, U_i = 0]$$

$$= \frac{q \cdot \frac{\exp(\gamma_0 + \gamma_1)}{1 + \exp(\gamma_0 + \gamma_1)}}{q \cdot \frac{\exp(\gamma_0 + \gamma_1)}{1 + \exp(\gamma_0 + \gamma_1)} + (1 - q) \cdot \frac{\exp(\gamma_0)}{1 + \exp(\gamma_0)}} \cdot \frac{\exp(\alpha_0 + \alpha_1)}{1 + \exp(\alpha_0 + \alpha_1)} \tag{22.4}$$
$$+ \frac{(1 - q) \cdot \frac{\exp(\gamma_0)}{1 + \exp(\gamma_0)}}{q \cdot \frac{\exp(\gamma_0 + \gamma_1)}{1 + \exp(\gamma_0 + \gamma_1)} + (1 - q) \cdot \frac{\exp(\gamma_0)}{1 + \exp(\gamma_0)}} \cdot \frac{\exp(\alpha_0)}{1 + \exp(\alpha_0)},$$

and

$$\mu_{c,0} = \frac{q \cdot \frac{1}{1 + \exp(\gamma_0 + \gamma_1)}}{q \cdot \frac{1}{1 + \exp(\gamma_0 + \gamma_1)} + (1 - q) \cdot \frac{1}{1 + \exp(\gamma_0)}} \cdot \frac{\exp(\beta_0 + \beta_1)}{1 + \exp(\beta_0 + \beta_1)}$$
$$+ \frac{(1 - q) \cdot \frac{1}{1 + \exp(\gamma_0)}}{q \cdot \frac{1}{1 + \exp(\gamma_0 + \gamma_1)} + (1 - q) \cdot \frac{1}{1 + \exp(\gamma_0)}} \cdot \frac{\exp(\beta_0)}{1 + \exp(\beta_0)}. \tag{22.5}$$

It is straightforward to see that for all values of $(q, \gamma_1, \alpha_1, \beta_1)$, and for all distributions of the observed data (captured by the values for the triple $(p, \mu_{t,1}, \mu_{c,0})$), we can find values for the triple $(\gamma_0, \alpha_0, \beta_0)$ such that all three of these equalities hold. Moreover, these values for the estimable parameters $(\gamma_0, \alpha_0, \beta_0)$ are unique for all values of $\mu_{c,0}, \mu_{t,1}$, and p, and for all values of the sensitivity parameters. For example, If $\gamma_0 \to -\infty$, the right-hand side of the first equality goes to zero, and if $\gamma_0 \to \infty$, the right-hand side goes to one. Because the right-hand side is strictly increasing in γ_0, there must be a unique value such that (22.3) holds for any $p \in (0, 1)$. Let us write these implied values for $(\gamma_0, \alpha_0, \beta_0)$ as functions of the data and $(q, \gamma_1, \alpha_1, \beta_1)$:

$$\gamma_0(q, \gamma_1, \alpha_1, \beta_1 | \text{data}), \qquad \alpha_0(q, \gamma_1, \alpha_1, \beta_1 | \text{data}),$$

and

$$\beta_0(q, \gamma_1, \alpha_1, \beta_1 | \text{data}),$$

where, ignoring sampling variation, the data consist of the triple

$$\text{data} = (p, \mu_{t,1}, \mu_{c,0}).$$

Given the postulated values for $(q, \gamma_1, \alpha_1, \beta_1)$, and given the values for $(\gamma_0, \alpha_0, \beta_0)$ that are implied by the combination of the data and the postulated values for $(q, \gamma_1, \alpha_1, \beta_1)$, there are implied values for $\mu_{t,0}$ and $\mu_{t,1}$. In terms of $\theta = (q, \gamma_1, \alpha_1, \beta_1, \gamma_0, \alpha_0, \beta_0)$, we can write

$$\mu_{t,0}(q, \gamma_1, \alpha_1, \beta_1, \gamma_0, \alpha_0, \beta_0) = \mathbb{E}[Y_i(1) | W_i = 0]$$

$$= \frac{q \cdot \frac{1}{1 + \exp(\gamma_0 + \gamma_1)}}{q \cdot \frac{1}{1 + \exp(\gamma_0 + \gamma_1)} + (1 - q) \cdot \frac{1}{1 + \exp(\gamma_0)}} \cdot \frac{\exp(\alpha_0 + \alpha_1)}{1 + \exp(\alpha_0 + \alpha_1)}$$
$$+ \frac{(1 - q) \cdot \frac{1}{1 + \exp(\gamma_0)}}{q \cdot \frac{1}{1 + \exp(\gamma_0 + \gamma_1)} + (1 - q) \cdot \frac{1}{1 + \exp(\gamma_0)}} \cdot \frac{\exp(\alpha_0)}{1 + \exp(\alpha_0)},$$

and

$$\mu_{c,1}(q, \gamma_1, \alpha_1, \beta_1, \gamma_0, \alpha_0, \beta_0) = \mathbb{E}[Y_i(0)|W_i = 1]$$

$$= \frac{q \cdot \frac{\exp(\gamma_0+\gamma_1)}{1+\exp(\gamma_0+\gamma_1)}}{q \cdot \frac{\exp(\gamma_0+\gamma_1)}{1+\exp(\gamma_0+\gamma_1)} + (1-q) \cdot \frac{\exp(\gamma_0)}{1+\exp(\gamma_0)}} \cdot \frac{\exp(\beta_0 + \beta_1)}{1 + \exp(\beta_0 + \beta_1)}$$

$$+ \frac{(1-q) \cdot \frac{\exp(\gamma_0)}{1+\exp(\gamma_0)}}{q \cdot \frac{\exp(\gamma_0+\gamma_1)}{1+\exp(\gamma_0+\gamma_1)} + (1-q) \cdot \frac{\exp(\gamma_0)}{1+\exp(\gamma_0)}} \cdot \frac{\exp(\beta_0)}{1 + \exp(\beta_0)},$$

where the conditioning on parameters is notationally suppressed. Then, finally, we can write the average treatment effect τ_{sp} as

$$\tau_{sp} = \mu_t - \mu_c = p \cdot (\mu_{t,1} - \mu_{c,1}) + (1-p) \cdot (\mu_{t,0} - \mu_{c,0}).$$

In summary, given the (super-population) data $= (p, \mu_{t,1}, \mu_{c,0})$, there is a function that gives τ_{sp} as a function of $(p, \mu_{t,1}, \mu_{c,0})$ and the sensitivity parameters:

$$\tau_{sp} = \tau(q, \gamma_1, \alpha_1, \beta_1 | \text{data}) = \tau(q, \gamma_1, \alpha_1, \beta_1 | p, \mu_{t,1}, \mu_{c,0}). \tag{22.6}$$

It is this function in which we are interested. Given the data, that is, for fixed values for $(p, \mu_{t,1}, \mu_{c,0})$, we wish to inspect how sensitive the average treatment effect τ_{sp} is to assumptions about the sensitivity parameters $(q, \gamma_1, \alpha_1, \beta_1)$.

There are two special sets of values for the sensitivity parameters. First, if we fix $\gamma_1 = 0$, then we are back assuming unconfoundedness (or a completely randomized experiment in this case without covariates), and

$$\tau_{sp} = \mu_{t,1} - \mu_{c,0}.$$

The same holds if we fix both $\alpha_1 = \beta_1 = 0$. Note that it is not necessary that both $\gamma_1 = 0$, *and* $\alpha_1 = \beta_1 = 0$, for there to be no bias associated with estimates based on unconfoundedness ignoring U_i. It is sufficient if (a) the unobserved covariate does not affect assignment ($\gamma_1 = 0$), or (b) the unobserved covariate is not associated with either potential outcome ($\alpha_1 = \beta_1 = 0$).

Second, suppose we fix $q = p$, and let $\gamma_1 \to \infty$. In that case W_i and U_i become perfectly correlated. If we also let $\alpha_1 \to -\infty$ and $\beta_1 \to -\infty$, then

$$\tau_{sp} \to p \cdot \mu_{t,1} + (1-p) - (1-p) \cdot \mu_{c,0},$$

which equals to the upper limit in the Manski bounds, showing that the setup with unconfoundedness given an unobserved binary covariate is not technically restrictive in this sense. Similarly, if we again fix $q = p$, and let $\gamma_1 \to \infty$, $\alpha_1 \to \infty$, and $\beta_1 \to \infty$, then

$$\tau_{sp} \to p \cdot \mu_{t,1} - p - (1-p) \cdot \mu_{c,0},$$

which is equal to the lower limit in the Manski bounds. This demonstrates that, in this setting, the bounds analysis can be viewed as an extreme version of a sensitivity analysis, or taking the opposite perspective, the sensitivity analysis can be viewed as a generalization of the bounds analysis.

Outside of these special values, the key question concerns the set of reasonable values for the sensitivity parameters $\theta_s = (q, \gamma_1, \alpha_1, \beta_1)$. Given a set of reasonable values Θ_s for θ_s, we calculate a lower and upper bound of the average treatment effect τ_{sp} over that set,

$$\tau_{\text{low}} = \inf_{(q,\gamma_1,\alpha_1,\beta_1)\in\Theta_s} \tau(q, \gamma_1, \alpha_1, \beta_1 | p, \mu_{\text{t},1}, \mu_{\text{c},0}),$$

and

$$\tau_{\text{high}} = \sup_{(q,\gamma_1,\alpha_1,\beta_1)\in\Theta_s} \tau(q, \gamma_1, \alpha_1, \beta_1 | p, \mu_{\text{t},1}, \mu_{\text{c},0}),$$

leading to the range

$$\tau_{\text{sp}} \in \left[\tau_{\text{low}}, \tau_{\text{high}} \right].$$

We generally do not have any substantive judgment regarding q, and one could simply investigate all values for q. Often results are not sensitive to intermediate values for q, and q can be taken to be equal to $\mathbb{E}[W_i] = p$. The remaining parameters are more interesting. The sensitivity parameter γ_1 represents the effect on the log odds ratio of receiving the treatment of a change from $U_i = 0$ to $U_i = 1$. In cases where the researcher has specific variables in mind that could bias the results based on assuming unconfoundedness, this can be a meaningful, interpretable parameter. In specific cases, one could be able to make an informed judgment about reasonable values for this parameter. Note that the Manski bounds on τ_{sp} implicitly allow U_i to be perfectly correlated with the receipt of treatment W_i, corresponding to $\gamma_1 \to \infty$. In settings where researchers have attempted to record all relevant determinants of treatment assignment, such a correlation may be viewed as logically too extreme. On the other hand, it may be difficult to specify a number that would meet with widespread agreement as a bound for α_1, and our preferred strategy is therefore to report the sensitivity of τ_{sp} to changes in these parameters.

One specific strategy, in cases where covariates are available, is to consider the association between the observed covariates and both the treatment assignment and the potential outcomes, assuming unconfoundedness, and use the estimated associations as indicative of ranges of reasonable values for the association between the unobserved covariate and the treatment indicator and the potential outcomes.

We illustrate this strategy with the lottery data where we observe eighteen covariates. For each covariate we estimate two logistic regression models. Denote the kth covariate, after normalizing by its standard deviation, by X_{ik}. We estimate a model for the treatment indicator conditional on the covariate,

$$\Pr(W_i = 1 | X_{ik}) = \frac{\exp\left(\delta_{k0} + \delta_{k1} \cdot X_{ik}\right)}{1 + \exp\left(\delta_{k0} + \delta_{k1} \cdot X_{ik}\right)},$$

and another model for the outcome conditional on the covariate and the treatment indicator,

$$\Pr(Y_i^{\text{obs}} = 1 | W_i, X_{ik}) = \frac{\exp\left(\zeta_{k0} + \zeta_{k1} \cdot X_{ik} + \zeta_{k2} \cdot W_i\right)}{1 + \exp\left(\zeta_{k0} + \zeta_{k1} \cdot X_{ik} + \zeta_{k2} \cdot W_i\right)},$$

again with dependence on parameters notationally suppressed.

Estimating these two models for each covariate X_{ik} leads to eighteen estimates $\hat{\delta}_{k1}$ and $\hat{\zeta}_{k2}$, $k = 1, \ldots, 18$. The largest values, in absolute value, were $|\hat{\delta}_{2,1}| = 0.56$ (the association between the number of tickets bought and winning the lottery) and $|\hat{\zeta}_{18,1}| = 1.61$ (the association between the indicator for positive earnings in the year prior to winning the lottery and post-lottery employment). We use these two values to anchor the sensitivity parameters. The idea is to limit the association between the unobserved binary covariate U_i and the treatment indicator and potential outcomes by assuming that they are bounded by the strongest marginal associations of the observed covariates. If one has made a good-faith effort to collect all relevant covariates, it may be difficult to see how one would miss covariates more important than any of those observed, at least unless there are specific reasons, such as confidentiality concerns. If $q = 1/2$, the standard deviation of U_i is 1/2, so we implement the sensitivity analysis by letting γ_1 range over the interval $[-0.56/(1/2), 0.56/(1/2)] = [-1.12, 1.12]$ and α_1 and β_1 over the interval $[-1.61/(1/2), 1.61/(1/2)] = [-3.22, 3.22]$ (multiplying the maximum of the coefficients by a factor two, equal to the ratio of the standard deviation of the normalized covariates and the maximum standard deviation of U, to take account of the normalization of the covariates). We let q range over the interval $[0, 1]$ because there is no substantive argument to restrict its range. Choosing values for the sensitivity parameters in this range leads to values for the average treatment effect in the interval

$$\tau_{\text{sp}} \in \left[-0.28, 0.05\right] \,\big|\, \{q \in [0, 1], \gamma_1 \in [-1.12, 1.12], \alpha_1 \in [-3.22, 3.22],$$
$$\beta_1 \in [-3.22, 3.22]\}.$$

Substantively this range suggests that the unobserved covariate would have to be fairly strongly associated with both treatment and potential outcomes to change the conclusion in the lottery example that the treatment has a positive and substantial effect on employment. We do not know whether such a covariate exists, but it would have to be somewhat stronger than any of the covariates the researchers managed to collect in terms of its association with the outcome and the treatment indicator. Note that in these calculations we allow γ_1 to be as large as the effect of any observed covariate on the log odds ratio for receiving the treatment, and simultaneously allow α_1 and β_1 to be as large as the effect of any observed covariates on the log odds ratio for the potential outcome. No single covariate in the sample had such strong effects on both simultaneously. In fact, for each covariate separately, the range of values for the average treatment effect τ associated with letting $q \in [0, 1]$, $\gamma_1 \in [-2 \cdot |\hat{\delta}_{k1}|, 2 \cdot |\hat{\delta}_{k1}|]$, and $\alpha_1, \beta_1 \in [-2 \cdot |\hat{\zeta}_{k1}|, 2 \cdot |\hat{\zeta}_{k1}|]$, for some $k = 1, \ldots, 18$, the widest range we find for the average treatment effect is $[-0.18, -0.07]$, with these values corresponding to the 12th covariate, earnings in the year prior to winning the lottery, with $\hat{\delta}_{12,1} = -0.1891$ and $\hat{\zeta}_{12,1} = 1.3257$.

Another approach for assessing the sensitivity that does not directly require us to postulate reasonable values for the sensitivity parameters is to explore the magnitude necessary for $(\gamma_1, \alpha_1, \beta_1)$ in order to change the sign for the estimated average treatment effect found under unconfoundedness. There are trade-offs between the parameters, because with a larger value for γ_1, the required values for α_1 and β_1 will not be quite as large. For example, and this is also useful for a subsequent comparison with the Rosenbaum sensitivity analysis, it is also interesting to look at the range of values for the average treatment effect given that $|\gamma_1| \leq 0.52$, with α_1 and β_1 essentially unrestricted. Then,

$$\tau_{\text{sp}} \in [-0.22, 0.48] \,|\{q \in [0, 1], \gamma_1 \in [-0.52, 0.52], \alpha_1 \in (-\infty, \infty), \beta_1 \in (-\infty, \infty)\},$$

just on the margin where the sign of the average treatment effect switches.

22.5 BINARY OUTCOMES: THE ROSENBAUM SENSITIVITY ANALYSIS FOR P-VALUES

Rosenbaum (1995) is interested in calculating Fisher p-values under the sharp null hypothesis of no treatment effects and wishes to assess how sensitive the conclusions under unconfoundedness are to that assumption. In principle applying these methods requires knowledge of the propensity score. Although we do not know the propensity score in observational studies, under unconfoundedness we can estimate the propensity score for each unit. Let these estimated propensity score values be denoted by $\hat{e}_i$. Given these values, we can use Fischer's exact p-value approach to obtain p-values for the null hypothesis of no effect whatsoever of the treatment. In the IRS lottery data, still without covariates, using the difference in average ranks as the statistic, the p-value is 0.034. Assuming random assignment, we can be very confident that the treatment has some effect on employment.

Now suppose unconfoundedness does not hold. In that case it is no longer the case that the estimated probability of the treatment is $\hat{e}_i$, where $\hat{e}_i$ was estimated under the assumption of unconfoundedness. Let us denote the actual treatment probability by p_i. Rosenbaum then limits the difference between the actual probability p_i and the estimated probability under unconfoundedness $\hat{e}_i$. Specifically, he assumes that the difference in log odds ratios, under the assumption of unconfoundedness, and based on the true assignment probabilities, is bounded by a pre-specified constant Γ:

$$\left| \ln \left(\frac{\hat{e}_i}{1 - \hat{e}_i} \right) - \ln \left(\frac{p_i}{1 - p_i} \right) \right| \leq \Gamma, \tag{22.7}$$

for all $i = 1, \ldots, N$. We can relate this to the analysis in the previous subsection by specifying a model for the treatment assignment as a function of an unobserved binary covariate u:

$$p_i = \Pr(W_i = 1 | U_i = u) = \frac{\exp(\gamma_0 + \gamma_1 \cdot u)}{1 + \exp(\gamma_0 + \gamma_1 \cdot u)}, \tag{22.8}$$

so that the logarithm of the true odds ratio is $\ln(p_i/(1 - p_i)) = \gamma_0 + \gamma_1 \cdot u$. If we approximate the average propensity score, averaged over the distribution of U_i, by the propensity score at the average value of U_i, $q = \mathbb{E}[U_i]$, so that $e_i = \exp(\gamma_0 + \gamma_1 \cdot q)/(1 + \exp(\gamma_0 + \gamma_1 \cdot q))$, the implied logarithm of the odds ratio is $\ln(e_i/(1 - e_i)) = \gamma_0 + \gamma_1 \cdot q$. The difference between the log odds ratio for the average propensity score under unconfoundedness and the log odds ratio for the true treatment probability is then $\ln(e_i/(1 - e_i)) - \ln(p_i/(1 - p_i)) = \gamma_1 \cdot (q - U_i)$. The Rosenbaum restriction implies we should consider all possible values for (q, γ_1) such that

$$q \cdot |\gamma_1| < \Gamma, \quad \text{and} \quad (1 - q) \cdot |\gamma_1| < \Gamma.$$

We can simplify the problem in this context by allowing for all possible values for q in the interval $[0, 1]$, and all possible values for γ_1 such that $|\gamma_1| < \Gamma$, thus requiring the difference in log odds ratios for units with $U_i = 1$ and units with $U_i = 0$ to be restricted to

$$\left| \ln\left(\frac{\Pr(W_i = 1 | U_i = 1)}{1 - \Pr(W_i = 1 | U_i = 1)} \right) - \ln\left(\frac{\Pr(W_i = 1 | U_i = 0)}{1 - \Pr(W_i = 1 | U_i = 0)} \right) \right| = \gamma_1 \leq \Gamma.$$

The question we now address is, given that we restrict γ_1 but place no restrictions on q, what is the evidence in the data against the null hypothesis of no effect whatsoever of the treatment? It is immediately clear that without any restriction on γ_1 there is no evidence against the null hypothesis that there is no effect of the treatment: if we let $\gamma_1 \to \infty$, then W_i and U_i are perfectly correlated.

Let us consider a particular statistic. In this case with a binary outcome, the natural statistic is the difference in means, $T^{\text{dif}} = \overline{Y}_t^{\text{obs}} - \overline{Y}_c^{\text{obs}}$. The value of this statistic for the lottery data is -0.12, with an exact Fisher p-value for the null hypothesis of no effects, calculated under complete random assignment, equal to 0.034. To make the comparison with the Rosenbaum sensitivity analysis easier, it is useful to change the assignment mechanism slightly; from a completely randomized experiment to a Bernoulli experiment with assignment probability 0.47, the p-value changes to 0.026. Now, pick a particular value for (q, γ_1). Given these values, the probability of receiving the treatment for unit i can be either

$$p_{\text{low}} = \frac{\exp(\gamma_0)}{1 + \exp(\gamma_0)}, \quad \text{or} \quad p_{\text{high}} = \frac{\exp(\gamma_0 + \gamma_1)}{1 + \exp(\gamma_0 + \gamma_1)},$$

with the first probability corresponding to the case where unit i has $U_i = 0$, and the second corresponding to the case where unit i has $U_i = 1$. Now suppose we assign each unit a value for U_i, and thus implicitly assign the unit a value for the assignment probability. Denote this assignment probability for unit i by p_i. Given that assignment probability, we can calculate the p-value for any statistic under its randomization distribution. The statistic we focus on is the difference in average outcomes by treatment status, $T^{\text{dif}} = \overline{Y}_t^{\text{obs}} - \overline{Y}_c^{\text{obs}}$. The fact that the assignment probabilities are not all equal does not create any problems when calculating or simulating the p-values.

The question now is what the most extreme (and in particular what the largest) p-value is we can find by assigning the unobserved covariate U_i to each unit for a given value of γ_1, allowing q to range over the interval $[0, 1]$. We can again turn to the associations between covariates and the treatment indicator to find a possibly reasonable value for γ_1. The largest value we found for δ_{k1}, which captures the relationship between an observed covariate (normalized to have unit variance) and the treatment indicator was approximately 0.56. (Recall that this corresponds to the number of lottery tickets bought.) This suggests that limiting γ_1 to be less than or equal to $2 \cdot 0.56 = 1.12$ (where the factor 2 captures the fact that the standard deviation of the binary covariate U is bounded by 1/2) may present a reasonable range of values for γ_1. This changes the p-value from 0.026 to 0.99, suggesting that such an association between the treatment indicator and the unobserved covariate eliminates any evidence of a negative effect of the treatment. Instead, using the δ_k for earnings, 0.19, to bound γ to less than 0.38 in absolute value leads to p-value of 0.27. Finally, using an upper bound on γ_1 equal to 0.52 leads to a p-value equal to 0.50.

22.5.1 The Rosenbaum Sensitivity Analysis for Average Treatment Effects

It is instructive, for the purpose of understanding the similarities and differences between the two approaches to sensitivity analyses, to modify Rosenbaum's approach to derive a range of feasible values for the average treatment effect. Instead of looking at the p-values associated with a pair of values for (q, γ_1), we again look at a range of values for the average treatment effect. Using the derivations from Section 22.4, we look at the range of values for τ if we allow $q \in [0, 1]$, $\gamma_1 \in [-\Gamma, \Gamma]$. In addition, we allow $\alpha_1 \in (-\infty, \infty)$ and $\beta_1 \in (-\infty, \infty)$, which reveals how the Rosenbaum sensitivity approach differs from the Cornfield-Rosenbaum-Rubin method for assessing sensitivity. In the latter we restrict α_1 and β_1, in addition to γ_1, whereas the former approach only restricts γ_1. This modification obviously leads to a wider range of possible values for τ. Restricting γ_1 to be less than 1.12 in absolute value, without restricting α_1 or β_1, leads to a range of possible values for the average effect of the treatment equal to

$$\tau_{sp} \in [-0.62, 0.48] \mid \{q \in [0, 1], \gamma_1 \in [-1.12, 1.12], \alpha_1 \in (-\infty, \infty), \beta_1 \in (-\infty, \infty)\},$$

considerably wider than the values we found before when we also restricted α_1 and β_1.

It is also interesting to restrict γ_1 to be less than 0.52 in absolute value, without restricting α_1 or β_1. This leads to a range of possible values for the average effect of the treatment equal to

$$\tau_{sp} \in [-0.22, 0.48] \mid \{q \in [0, 1], \gamma_1 \in [-0.52, 0.52], \alpha_1 \in (-\infty, \infty), \beta_1 \in (-\infty, \infty)\}.$$

Now the set of estimates (ignoring sampling uncertainty) has zero as its upper limit. This corresponds to the case where the upper bound on the p-values is equal to 0.50.

22.6 CONCLUSION

In this chapter we present methods for assessing the sensitivity of results obtained under unconfoundedness. The unconfoundedness assumption can be controversial, and the analyses discussed here allow the researcher to quantify how much the estimates and p-values rely on the full force of this assumption. Finding that particular results are, or are not, sensitive to this assumption helps evaluate the results of any analysis under unconfoundedness.

However, in our limited experience, the application and value of such sensitivity analyses depend rather critically on the context of the study and general scientific knowledge that the investigators can bring to bear on the problem at hand.

NOTES

The key papers underlying the first sensitivity analysis in this chapter are Cornfield et al. (1959) and Rosenbaum and Rubin (1983a). Rosenbaum and Rubin focus on the case with a binary outcome, where, in their example, the sample is divided into five subclasses or blocks, and use the analysis where the sensitivity parameters are restricted to the same values in each block. They directly limit the values of the sensitivity parameters γ_1, α_1, and β_1 to be less than or equal to three in absolute value. Rosenbaum (1995, 2002) developed the sensitivity analysis that restricts only the assignment probabilities. Imbens (2003) applies the Rosenbaum-Rubin sensitivity analysis and is the original source for the suggestion to anchor the thresholds to values based on the association between treatment and observed covariates and between the outcomes and observed values. For another recent application, see Ichino, Mealli, and Nannichini (2008).

Manski (1990, 1996, 2003, 2013) in a series of papers proposed calculating worst-case bounds of the type discussed in this chapter, with earlier results for special cases in Cochran (1977). Manski, Sandefur, McLanahan, and Powers (1992) present an early application.

## PART VI

Regular Assignment Mechanisms with Noncompliance: Analysis

Instrumental Variables Analysis of Randomized Experiments with One-Sided Noncompliance

23.1 INTRODUCTION

In this chapter we discuss a second approach to analyzing causal effects when unconfoundedness of the treatment of interest is questionable. In Chapter 22 we also relaxed the unconfoundedness assumption, but there we did not make any additional assumptions. The resulting sensitivity and bounds analyses led to a range of estimated values for treatment effects, all of which were consistent with the observed data. Instead, in this chapter we consider alternatives to the standard unconfoundedness assumption that still allow us to obtain essentially unbiased point estimates of some treatment effects of interest, although typically *not* the overall average effect. In the settings we consider, there is, on substantive grounds, reason to believe that units receiving and units not receiving the treatment of interest are systematically different in characteristics associated with the potential outcomes. Such cases may arise if receipt of treatment is partly the result of deliberate choices by units, choices that take into account perceptions or expectations of the causal effects of the treatment based on information that the analyst may not observe. In order to allow for such violations of unconfoundedness, we rely on the presence of additional information and consider alternative assumptions regarding causal effects. More specifically, a key feature of the Instrumental Variables (IV) approach, the topic of the current chapter and the next two, is the presence of a secondary treatment, in the current setting the *assignment* to treatment instead of the *receipt* of treatment, where by "secondary" we do not mean temporally but secondary in terms of scientific interest. This secondary treatment is assumed to be unconfounded. In fact, in the randomized experiment setting of the current chapter, the assignment to treatment is unconfounded by design. This implies we can, using the methods from Part II of the book, unbiasedly estimate causal effects of the *assignment* to treatment. The problem is that these causal effects are not the causal effects of primary interest, which are the effects of the receipt of treatment. Assumptions that allow researchers to link these causal effects are at the core of the instrumental variables approach.

This chapter is the first of three chapters on instrumental variables approaches. For readers unfamiliar with this terminology, instrumental variables methods refer to a set of techniques originating in the econometrics literature, starting in the 1920s with work by Wright (1927), Tinbergen (1930), and later Haavelmo (1943). A central role in these methods is played by a variable, the so-called *instrument* or *instrumental variable*, which is a variable known *a priori* almost certainly to have a causal effect on the treatment of primary interest, W_i. The key characteristic of this instrument, here denoted by Z_i, is the *a priori* assumed absence of a "direct" causal effect of the instrument on the outcome of interest Y_i, with any causal effect of Z_i on Y_i "passing through" a causal effect of the instrument on the treatment W_i, where these terms will become clear shortly. More generally, *principal stratification* refers to settings with *latent unconfoundedness* of the primary treatment, where, conditional on an only partially observed covariate, unconfoundedness holds. In the special case of instrumental variables, this latent unconfoundedness applies with the latent compliance status to assigned secondary treatment, more precisely defined later, playing the role of the partially unobserved covariate.

We start this instrumental variables discussion in the simplest setting of a completely randomized experiment with one-sided noncompliance. By noncompliance we refer to the situation where some units who are assigned to receive a particular treatment level do not comply with their assignment and instead receive an alternative treatment. In this chapter, compliance is assumed to be all or nothing: units cannot receive, or be exposed to, only part of the treatment. By one-sided, we mean that the noncompliance is asymmetric in the sense that only units assigned to receive the active treatment can potentially circumvent their assigned treatment and receive the control treatment. In contrast, all units assigned to receive the control treatment do, in fact, comply with this assignment. This type of noncompliance is common in settings with individual people as the units of analysis, where receipt of the active treatment requires individuals to take, or subject themselves to, a particular action, such as undergoing surgery or entering a job-training program. In such cases, it is often difficult, or even impossible, to compel individuals to undergo the active treatment if assigned to it, even if individuals give consent prior to the randomization. As a result, compliance among those assigned to the active treatment is often imperfect. In contrast, those assigned to receive the control treatment can often effectively be denied access to the active treatment, so the noncompliance is one-sided. In this setting, the assignment to treatment is the instrument Z_i, and the receipt of treatment is the treatment of primary interest W_i.

Many traditional formal statistical analyses of randomized experiments with noncompliance focus on the relation between the random assignment and the outcome of interest, discarding entirely any information about the treatment, in the current setting actually received, that is, ignoring W_i. Such an approach is generally referred to as an *intention-to-treat* (ITT) analysis. In our setting of a completely randomized experiment, ITT analyses are validated by the randomization of the assignment to treatment, without the need for additional assumptions beyond SUTVA. The main drawback of these ITT analyses is that they do not answer questions about causal effects of the receipt of treatment itself, only about causal effects of the assignment to treatment. Two other simple analyses, focusing directly on causal effects of the treatment of interest, but neither of which is generally valid, are sometimes conducted in such settings. First, *per protocol* analyses, where units that are observed not to comply with the treatment assigned are discarded (i.e., units with $Z_i \neq W_i$), and the data for all units who are observed to comply with their assigned treatment (i.e., units with $Z_i = W_i$) are analyzed as if they came from a randomized experiment with full compliance; that is, the analysis is as if W_i were randomized for units who appear to comply, discarding units who are observed to be noncompliers. A second simple alternative is an *as-treated* analysis where data from all units are analyzed as if they had been randomly assigned to the treatment they actually received, ignoring information on assignment Z_i, and simply comparing treated units having $W_i = 1$ with control units having $W_i = 0$, as if W_i were randomized for all units. Both of these naive analyses are generally invalid as we discuss in Section 23.9.

In this chapter we focus on defining causal estimands and on the additional assumptions that allow us to go beyond the global effect of assignment that is the focus of ITT analyses, and estimate "local" average effects for the treatment of interest, that is, averages for subsets of units. Although we briefly mention some traditional econometric, moment-based, estimators for simple cases with no covariates, we leave the main discussion of our preferred model-based estimators and inference to Chapter 25.

In order to obtain alternatives to the assumption of unconfoundedness of the receipt of the treatment, we consider separately the nature of the noncompliance and the causal effects of the assignment to treatment for what we will call *compliers* and *noncompliers*. These groups are defined by their partly unobserved compliance behavior, and thus define *latent strata*. A key insight is that, although unconditionally receipt of treatment is confounded, within these latent strata the receipt of treatment is unconfounded. We then consider assumptions that rule out effects of assignment to the treatment on outcomes for certain groups but allow for general differences between units who comply and those who do not comply with their treatment assignment. Assessment of the plausibility of these assumptions relies heavily on subject-matter knowledge, in addition to the design of the assigned treatment.

In general there are two key assumptions justifying instrumental variables approaches. The first is that, although the receipt of the treatment is generally confounded when noncompliance occurs, the assignment to the treatment *is* unconfounded. As a result of unconfoundedness, we can estimate the effect of the assignment to treatment on both the outcome of interest, and on the receipt of treatment, that is, the two ITT effects. The unconfoundedness of assignment assumption is satisfied by design in the completely randomized experiment setting considered in this chapter, although in other applications of IV methods, this assumption can be controversial. The second key assumption is that the assignment to treatment has no effect on the final outcome of interest for those units whose receipt of treatment is unaffected by the assignment. For instance, for those who do not take the drug even when assigned to take it, the assignment itself is assumed to have no effect on the final outcome. We refer to this assumption as an *exclusion restriction*, because the instrument is excluded from affecting the outcome of interest for noncompliers. This assumption can be justified by design, for example, using double-blind experiments, where neither the unit nor the physician knows which treatment was assigned, thereby supporting the exclusion restriction. The key result in this chapter is that the exclusion assumption, when combined with the unconfoundedness assumption, enables us to estimate causal effects of the assignment to treatment on the principal outcome, Y_i, for the subpopulation of compliers, known as the *local average treatment effect* (LATE) or the *complier average causal effect* (CACE). The estimand, the average effect for compliers, is equal to the ratio of the ITT effect of Z_i on the outcome of interest, Y_i, and the ITT effect of Z_i on the receipt of treatment W_i. In other words, under the exclusion restriction, the ITT effect of assignment on the outcome of interest is due entirely to those units for whom receipt of treatment W_i is always identical to the assignment to treatment Z_i, irrespective of their assignment. In many cases, it may then be reasonable to attribute the causal effect of *assignment* for the compliers to the causal effect of the *receipt* of treatment, the same way researchers often do, typically implicitly, in completely randomized experiments with full compliance.

We must emphasize from the outset that the assumptions underlying the instrumental variables approach, most importantly various forms of exclusion restrictions, are often controversial. When appropriate, these assumptions allow the researcher to make more interesting, and stronger, inferences than those obtained from ITT analyses. However, these assumptions are not always appropriate. Moreover, unlike the unconfoundedness assumption, the validity of the exclusion restriction cannot be guaranteed solely by physical randomization, requiring in addition double blinding. Therefore, like SUTVA, its validity often relies on subject-matter knowledge.

The rest of the chapter is organized as follows. In Section 23.2 we describe the data set that will be used to illustrate the theoretical concepts introduced in this chapter. Next, in Section 23.3 we extend the potential outcomes notation to account for the instrumental variables setup. In the following section, Section 23.4, we analyze intention-to-treat effects. We define compliance behavior in Section 23.5. In Section 23.6 we discuss the instrumental variables estimand. In Section 23.7 we briefly discuss traditional moment-based estimators for the instrumental variables estimand. Then, in Section 23.8 we relate the discussion to traditional, linear-model-based instrumental variables methods. In Section 23.9 we discuss three naive methods for analyzing data from a randomized experiment with one-sided noncompliance. Section 23.10 concludes.

23.2   THE SOMMER-ZEGER VITAMIN A SUPPLEMENT DATA

We illustrate the methods discussed in this chapter using data previously analyzed by Sommer and Zeger (1991). Sommer and Zeger study the effect of vitamin A supplements on infant mortality in Indonesia. The vitamin supplements were randomly assigned to villages, but some of the individuals in villages assigned to the treatment group failed to receive them. None of the individuals assigned to the control group received the supplements, so noncompliance is one-sided. In this study, outcomes are observed for $N = 23{,}682$ infants. The observed outcome of interest, denoted by Y_i^{obs}, is a binary variable, indicating survival of an infant. Receipt of the vitamin supplements, which is considered the treatment of interest, is denoted by $W_i^{obs} \in \{0, 1\}$. In a slight departure from the notation in previous chapters, we add here the superscript "obs" to W_i for reasons that will become apparent later. Assignment to the supplements, the instrument, is denoted by $Z_i \in \{0, 1\}$. This assignment varies only at the village level. We ignore the clustering of the assignment at the village level because we do not have indicators for villages; this will tend to lead us to understate standard errors.

With all three observed variables binary, there are, in principal, eight different possible values for the triple $(Z_i, W_i^{obs}, Y_i^{obs})$. Because of the noncompliance, there may be units with $Z_i \neq W_i^{obs}$, but because $Z_i = 0$ implies $W_i^{obs} = 0$, there are only six values of the triple with positive counts in our sample. Table 23.1 contains the counts of the six observed values for the triple in the data set, with a total sample size of $N = 23{,}682$.

Table 23.1. *Sommer–Zeger Vitamin Supplement Data*

Compliance Type	Assignment Z_i	Vitamin Supplements W_i^{obs}	Survival Y_i^{obs}	Number of Units ($N = 23{,}682$)
co or nc	0	0	0	74
co or nc	0	0	1	11,514
nc	1	0	0	34
nc	1	0	1	2385
co	1	1	0	12
co	1	1	1	9663

23.3 SETUP

First, let us expand the potential outcomes notation to fit the IV setting. Given the divergence between assignment to, and receipt of, treatment, the potential outcomes notation becomes more complex than in previous chapters. We maintain throughout this chapter the SUTVA assumption, that (*i*) there are no versions of the treatments, and (*ii*) there are no causal effects of one unit's treatment assignment on another unit's outcome. We focus on the case where the assignment Z_i takes on two values, $Z_i = 0$ if unit *i* is assigned to the control group, and $Z_i = 1$ if unit *i* is assigned to the treatment group. The treatment of primary interest (in the Sommer-Zeger application, the receipt of the vitamin supplements) is denoted by W_i^{obs}. Formally recognizing the role of this variable as an outcome, possibly affected by the assignment to treatment Z_i, we postulate the existence of two potential outcomes, $W_i(0)$ and $W_i(1)$, describing the treatment that would be received under each of the two values of the assignment Z_i. Thus $W_i(0)$ is the treatment unit *i* would receive if assigned to the control, $Z_i = 0$, and $W_i(1)$ is the treatment unit *i* would receive if assigned to the active treatment, $Z_i = 1$. Both $W_i(0)$ and $W_i(1)$ take values in $\{0, 1\}$. For unit *i*, the realized or observed treatment status, W_i^{obs}, equals

$$W_i^{obs} = W_i(Z_i) = \begin{cases} W_i(0) & \text{if } Z_i = 0, \\ W_i(1) & \text{if } Z_i = 1. \end{cases}$$

In contrast to earlier chapters, we use the superscript "obs" for the treatment here to distinguish the observed value of the primary treatment from the potential primary treatment, which is generally a function of the secondary treatment, Z_i.

For the outcome of interest we take into account that there are, in the noncompliance setting, two "treatments," assignment to treatment Z_i and receipt of treatment W_i. Each takes on two values, so to be general we postulate four potential outcomes, $Y_i(z, w)$, describing the outcome observed if unit *i* were assigned treatment *z* and actually received treatment *w*. For each unit, only two of these four potential outcomes can possibly be observed, $Y_i(0, W_i(0))$ and $Y_i(1, W_i(1))$. The remaining two, $Y_i(0, 1 - W_i(0))$ and $Y_i(1, 1 - W_i(1))$, cannot be observed irrespective of the assignment, and so we refer to the last two as *a priori counterfactuals*. The observed outcome for unit *i* in our sample, denoted by Y_i^{obs}, is

$$Y_i^{\text{obs}} = Y_i(Z_i, W_i^{\text{obs}}) = Y_i(Z_i, W_i(Z_i)) = \begin{cases} Y_i(0,0), & \text{if } Z_i = 0, W_i^{\text{obs}} = 0, \\ Y_i(1,0), & \text{if } Z_i = 1, W_i^{\text{obs}} = 0, \\ Y_i(1,1), & \text{if } Z_i = 1, W_i^{\text{obs}} = 1. \end{cases}$$

Note that because the noncompliance is one-sided, there are no units for whom we observe $Y_i(0, 1)$. As usual, we think of the population of interest as the N units for which we observe: the instrument Z_i, the treatment received W_i^{obs}, and the outcome Y_i^{obs}.

In this chapter we consider both (*a*) averages over observations by treatment received, and (*b*) averages by treatment assigned. It is therefore useful to have formal notation for these. For notational clarity, the subscripts 0 and 1 denote treatment assignment levels, and the subscripts c and t denote the level of receipt of treatment. Define the subsample sizes by treatment assignment:

$$N_0 = \sum_{i=1}^{N} (1 - Z_i), \quad N_1 = \sum_{i=1}^{N} Z_i,$$

sample sizes by treatment received:

$$N_c = \sum_{i=1}^{N} (1 - W_i^{\text{obs}}), \quad \text{and} \quad N_t = \sum_{i=1}^{N} W_i^{\text{obs}}$$

and sample sizes by both treatment assignment and receipt:

$$N_{0c} = \sum_{i=1}^{N} (1 - Z_i) \cdot (1 - W_i^{\text{obs}}), \quad N_{0t} = \sum_{i=1}^{N} (1 - Z_i) \cdot W_i^{\text{obs}},$$

$$N_{1c} = \sum_{i=1}^{N} Z_i \cdot (1 - W_i^{\text{obs}}), \quad \text{and} \quad N_{1t} = \sum_{i=1}^{N} Z_i \cdot W_i^{\text{obs}}.$$

Analogously, define the average outcomes and average treatment received by assignment subsample:

$$\overline{Y}_0^{\text{obs}} = \frac{1}{N_0} \sum_{i=1}^{N} (1 - Z_i) \cdot Y_i^{\text{obs}}, \quad \overline{Y}_1^{\text{obs}} = \frac{1}{N_1} \sum_{i=1}^{N} Z_i \cdot Y_i^{\text{obs}},$$

$$\overline{W}_0^{\text{obs}} = \frac{1}{N_0} \sum_{i=1}^{N} (1 - Z_i) \cdot W_i^{\text{obs}}, \quad \overline{W}_1^{\text{obs}} = \frac{1}{N_1} \sum_{i=1}^{N} Z_i \cdot W_i^{\text{obs}},$$

average outcomes by treatment received:

$$\overline{Y}_c^{\text{obs}} = \frac{1}{N_c} \sum_{i=1}^{N} (1 - W_i^{\text{obs}}) \cdot Y_i^{\text{obs}}, \quad \text{and} \quad \overline{Y}_t^{\text{obs}} = \frac{1}{N_t} \sum_{i=1}^{N} W_i^{\text{obs}} \cdot Y_i^{\text{obs}};$$

and, finally, average outcomes by both treatment assignment and treatment receipt:

$$\overline{Y}_{0c}^{\text{obs}} = \frac{1}{N_{0c}} \sum_{i=1}^{N} (1 - Z_i) \cdot (1 - W_i^{\text{obs}}) \cdot Y_i^{\text{obs}}, \quad \overline{Y}_{0t}^{\text{obs}} = \frac{1}{N_{0t}} \sum_{i=1}^{N} (1 - Z_i) \cdot W_i^{\text{obs}} \cdot Y_i^{\text{obs}},$$

$$\overline{Y}_{1c}^{\text{obs}} = \frac{1}{N_{1c}} \sum_{i=1}^{N} Z_i \cdot (1 - W_i^{\text{obs}}) \cdot Y_i^{\text{obs}}, \quad \text{and} \quad \overline{Y}_{1t}^{\text{obs}} = \frac{1}{N_{1t}} \sum_{i=1}^{N} Z_i \cdot W_i^{\text{obs}} \cdot Y_i^{\text{obs}}.$$

Some of the N_{zw} may be zero (in fact, N_{0t} is zero in the current chapter with one-sided compliance), and the corresponding $\overline{Y}_{zw}$ would not be defined in that case.

23.4 INTENTION-TO-TREAT EFFECTS

The first step in our discussion of the IV approach is to study intention-to-treat (ITT) estimands. As we mentioned in Section 23.1, ITT analyses entirely avoid the problem of noncompliance by focusing only on the relationship between the random assignment of Z_i and the outcome, because inference for such effects relies solely on the randomization of the assignment. In contrast to many conventional ITT analyses, we consider two versions of such analyses: analyzing, as "outcomes," both the receipt of treatment (receipt of vitamin A supplements) and the final outcome (survival).

23.4.1 ITT Estimands

Let us first consider the intention-to-treat effect on the receipt of treatment. The unit-level effect of the assignment on the receipt of treatment is

$$\text{ITT}_{\text{W},i} = W_i(1) - W_i(0).$$

The ITT effect on the receipt of treatment is the average of this over all units:

$$\text{ITT}_{\text{W}} = \frac{1}{N} \sum_{i=1}^{N} \text{ITT}_{\text{W},i} = \frac{1}{N} \sum_{i=1}^{N} \left(W_i(1) - W_i(0) \right). \tag{23.1}$$

Because noncompliance is one-sided, $W_i(0) = 0$ for all i, and the expression in Equation (23.1) simplifies to

$$\text{ITT}_{\text{W}} = \frac{1}{N} \sum_{i=1}^{N} W_i(1).$$

Next, let us consider the outcome of primary interest, Y_i. The unit-level intention-to-treat effect is equal to the difference in unit-level outcomes Y_i by assignment status Z_i:

$$\text{ITT}_{\text{Y},i} = Y_i(1, W_i(1)) - Y_i(0, W_i(0)),$$

for $i = 1, \ldots, N$. The average ITT effect on Y is therefore

$$\text{ITT}_Y = \frac{1}{N} \sum_{i=1}^{N} \text{ITT}_{Y,i} = \frac{1}{N} \sum_{i=1}^{N} \left(Y_i(1, W_i(1)) - Y_i(0, W_i(0)) \right).$$

The key assumption for identifying the ITT effects in the simple setting of the Sommer-Zeger data set is that the assignment is random. (More generally, we could allow for unconfounded treatment assignment.) Here we formulate that in terms of the extended potential outcome notation, by assuming the distribution of Z_i is free from dependence on all potential outcomes, including the two potential treatments $W_i(z)$ and four potential outcomes $Y_i(z, w)$:

Assumption 23.1 (Random Assignment of Z_i)

$$\Pr(Z_i = 1 \mid W_i(0), W_i(1), Y_i(0, 0), Y_i(0, 1), Y_i(1, 0), Y_i(1, 1)) = \Pr(Z_i = 1).$$

From a super-population perspective, the assumption, is in the Dawid conditional-independence notation,

$$Z_i \perp\!\!\!\perp W_i(0), W_i(1), Y_i(0, 0), Y_i(0, 1), Y_i(1, 0), Y_i(1, 1).$$

23.4.2 Estimating the ITT Effect for the Receipt of Treatment

Given Assumption 23.1, we can estimate ITT_W following Neyman's approach, outlined in Chapter 6. Complete randomization of the assignment implies that an unbiased estimator for the average causal effect ITT_W exists in the form of the average difference in treatment status by assignment status:

$$\widehat{\text{ITT}_W} = \overline{W}_1^{\text{obs}} - \overline{W}_0^{\text{obs}} = \overline{W}_1^{\text{obs}},$$

where we use the fact that $W_i(0) = 0$ for all units. Following the derivation presented in Chapter 6, the general form of the (conservative) estimator for the finite-sample sampling variance of $\widehat{\text{ITT}_W}$, under the randomization distribution, is

$$\widehat{\mathbb{V}}(\widehat{\text{ITT}_W}) = \frac{s_{W,0}^2}{N_0} + \frac{s_{W,1}^2}{N_1},$$

where $s_{W,0}^2$ and $s_{W,1}^2$ are the sample variances of $W_i(z)$ within each assignment arm. Because $W_i(0) = 0$, it follows that

$$s_{W,0}^2 = \frac{1}{N_0 - 1} \sum_{i:Z_i=0} \left(W_i^{\text{obs}} - \overline{W}_0^{\text{obs}} \right)^2 = 0,$$

and we are concerned only with

$$s_{W,1}^2 = \frac{1}{N_1 - 1} \sum_{i:Z_i=1} \left(W_i^{\text{obs}} - \overline{W}_1^{\text{obs}} \right)^2 = \frac{N_1}{N_1 - 1} \cdot \overline{W}_1^{\text{obs}} \cdot (1 - \overline{W}_1^{\text{obs}}).$$

Hence the estimator for the sampling variance of $\widehat{\text{ITT}_\text{W}}$ reduces to

$$\widehat{\mathbb{V}}(\widehat{\text{ITT}_\text{W}}) = \frac{1}{N_1 - 1} \cdot \overline{W}_1^{\text{obs}} \cdot (1 - \overline{W}_1^{\text{obs}}).$$

Recall, from the discussion of randomized experiments in Chapter 6, that this is also a valid estimator for the sampling variance of $\widehat{\text{ITT}_\text{W}}$ when it is viewed as an estimator of the super-population average treatment effect. Using a normal approximation to the sampling distribution, we can construct a randomization-distribution-based, large-sample, 95% confidence interval for ITT_W as

$$\text{CI}^{0.95}(\text{ITT}_\text{W}) = \left(\widehat{\text{ITT}_\text{W}} - 1.96 \cdot \sqrt{\widehat{\mathbb{V}}(\widehat{\text{ITT}_\text{W}})}, \widehat{\text{ITT}_\text{W}} + 1.96 \cdot \sqrt{\widehat{\mathbb{V}}(\widehat{\text{ITT}_\text{W}})} \right).$$

Let us illustrate this using the Sommer-Zeger vitamin A data. For these data we find

$$\overline{W}_1^{\text{obs}} = 0.8000, \quad \text{and} \quad s_{W,1}^2 = 0.4000^2.$$

Given that $N_1 = 12,094$ individuals were assigned to receive the vitamin supplements, it follows that

$$\widehat{\text{ITT}_\text{W}} = 0.8000, \quad \text{and} \quad \widehat{\mathbb{V}}\left(\widehat{\text{ITT}_\text{W}}\right) = 0.0036^2,$$

leading to a 95% large-sample confidence interval for ITT_W equal to

$$\text{CI}^{0.95}(\text{ITT}_\text{W}) = \left(0.7929, 0.8071 \right).$$

Thus, we obtain for the Sommer-Zeger data a precise estimate of the ITT effect of assignment to treatment on the receipt of treatment (with the caveat that we ignore the clustered randomization).

23.4.3 Estimating the ITT Effect for the Outcome of Interest

Next let us consider the outcome of primary interest, Y_i. Because the assignment Z_i is unconfounded, we can unbiasedly estimate the conventional intention-to-treat estimand, ITT_Y. Using the analysis for randomized experiments from Chapter 6, an unbiased estimator for this effect can be obtained by differencing the average outcomes for those assigned to the treatment and those assigned to the control:

$$\widehat{\text{ITT}_\text{Y}} = \overline{Y}_1^{\text{obs}} - \overline{Y}_0^{\text{obs}},$$

where $\overline{Y}_1^{\text{obs}}$ and $\overline{Y}_0^{\text{obs}}$ are as defined in Section 23.3. The sampling variance for this estimator can also be estimated using the methods from Chapter 6:

$$\widehat{\mathbb{V}}(\widehat{\text{ITT}_\text{Y}}) = \frac{s_{Y,1}^2}{N_1} + \frac{s_{Y,0}^2}{N_0},$$

where

$$s_{Y,0}^2 = \frac{1}{N_0 - 1} \sum_{i:Z_i=0} \left(Y_i^{obs} - \overline{Y}_0^{obs} \right)^2, \quad \text{and} \quad s_{Y,1}^2 = \frac{1}{N_1 - 1} \sum_{i:Z_i=1} \left(Y_i^{obs} - \overline{Y}_1^{obs} \right)^2.$$

Let us return again to the vitamin A supplement data. Using the survival indicator Y_i^{obs} as the outcome, we find:

$$\overline{Y}_0^{obs} = 0.9956, \quad \overline{Y}_1^{obs} = 0.9962, \quad s_{Y,0}^2 = 0.0797^2, \quad \text{and} \quad s_{Y,1}^2 = 0.0616^2.$$

Given that $N_1 = 12,094$ individuals were assigned to receive the supplements, and $N_0 = 11,588$ were assigned to receive no supplements, it follows that

$$\widehat{ITT}_Y = 0.0026, \quad \text{and} \quad \widehat{\mathbb{V}}\left(\widehat{ITT}_Y \right) = 0.0009^2,$$

leading to a large-sample 95% confidence interval for ITT_Y:

$$CI^{0.95}(ITT_Y) = \left(0.0008, 0.0044 \right).$$

We conclude that the estimated ITT effect of assignment to supplements on survival is positive and statistically different from zero at conventional significance levels. If all we were interested in is these ITT effects, we could stop here. In many cases, however, there is also interest in the causal effect of *taking* the supplements as opposed to the causal effect of *being assigned to take* them. Part of the motivation is that one may believe that the causal effect of actually taking the treatment has more external validity, that is, is more likely to generalize to other settings and populations, than the causal effect of being assigned to take them. The argument for this is that the ITT effect combines partly the biological effect of taking the supplements, and the psychological effect of assignment to take the supplements on actually taking them. When this is true, the causal effect of taking the supplements may be more relevant than the causal effect of assigning individuals to take the supplements for policy makers who are considering making them available in other parts of the country or on a wider scale, with more or less encouragement to them than in the current experiment. This point is particularly compelling when the reasons for the noncompliance are idiosyncratic to the setting in which the experiment was conducted, so that in different settings, compliance may be substantially different.

23.5   COMPLIANCE STATUS

A crucial role in the analyses discussed in this chapter is played by the compliance behavior of the units. Here we continue our analysis of the IV approach with a detailed discussion of this behavior, captured by the pair of potential outcomes $(W_i(0), W_i(1))$. A key feature of our approach is that we view the compliance behavior in this study when assigned not to take ($W_i(0)$) and when assigned to take ($W_i(1)$) as reflecting partially observed characteristics of each unit.

Table 23.2. *Possible Compliance Status by Observed Assignment and Receipt of Treatment for the Sommer-Zeger Vitamin Supplement Data*

		Assignment Z_i	
		0	1
Receipt of treatment W_i^{obs}	0	nc or co	nc
	1	–	co

Note: One-sided noncompliance rules out the $Z_i = 0$ $W_i^{\text{obs}} = 1$ cell.

23.5.1 Compliers and Noncompliers

Let us return to the two potential outcomes for the treatment received, $W_i(0)$ and $W_i(1)$. By the assumption that noncompliance is one-sided, it follows that all units assigned to the control in fact receive the control, thus $W_i(0) = 0$ for all i. In contrast, $W_i(1)$, the treatment unit i would receive if assigned to the active treatment, can equal either 0 or 1. Units with $W_i(1) = 1$ will be observed to comply with their assignment, irrespective of what that assignment is, whereas those with $W_i(1) = 0$ will be observed not to comply if assigned to $Z_i = 1$. We therefore label the former group *compliers* and the latter group *noncompliers*. In a randomized experiment with full compliance, $W_i(z)$ would be equal to z for all units, and as a result, all units would be compliers. Note that this definition of compliance status is based solely on a unit's behavior given assignment to the active treatment in this experiment. Because all units assigned to the control can be prevented from receiving the active treatment, all units will be observed to comply when assigned $Z_i = 0$. Thus we can only distinguish, by observation, compliers from noncompliers in the subgroup assigned to the treatment. For the purposes of our discussion, compliance status will be denoted by a group indicator $G_i \in \{\text{co}, \text{nc}\}$, with $G_i = $ co for compliers and $G_i = $ nc for noncompliers:

$$
G_i = \begin{cases} \text{co} & \text{if } W_i(1) = 1, \\ \text{nc} & \text{if } W_i(1) = 0. \end{cases}
$$

Table 23.2 illustrates the compliance status and its relation to the observed assignment Z_i and the observed receipt of the treatment W_i^{obs}. The "–" entry, corresponding to $Z_i = 0$ and $W_i^{\text{obs}} = 1$, indicates that by the fact that noncompliance is one-sided, there are no units with $Z_i = 0$ and $W_i^{\text{obs}} = 1$.

When we consider two-sided noncompliance in the next chapter, we generalize these ideas to allow for the possibility that some of those assigned to the control group in fact can receive the active treatment, and thus allow $W_i(0)$ to differ from zero.

Let N_{co} and N_{nc} denote the number of units of each type in the sample:

$$
N_{\text{co}} = \sum_{i=1}^{N} \mathbf{1}_{G_i = \text{co}}, \quad \text{and} \quad N_{\text{nc}} = \sum_{i=1}^{N} \mathbf{1}_{G_i = \text{nc}} = N - N_{\text{co}},
$$

and let π_{co} and π_{nc} denote the sample fractions of compliers and noncompliers:

$$
\pi_{\text{co}} = \frac{N_{\text{co}}}{N}, \quad \text{and} \quad \pi_{\text{nt}} = \frac{N_{\text{nc}}}{N} = 1 - \pi_{\text{co}}.
$$

In the potential outcomes notation, it becomes clear that the compliance status in this experiment is a latent characteristic of an individual unit. It is a characteristic in the sense that compliance status is not affected by outside manipulation (specifically, it is not affected by the assignment to treatment Z_i); it is latent because we cannot observe its value for all units: that is, for those units assigned to the control group, we do not observe their compliance status. In contrast, for units assigned to receive the active treatment, we do observe whether they are compliers or noncompliers (although this will change when we allow for two-sided noncompliance in the next chapter). Hence the three key features of this latent compliance status are: (*i*) it is a function of the two (secondary) potential outcomes, which describe the receipt of treatment for different values of the assignment Z_i; (*ii*) the value of the characteristic is not affected by the assignment to treatment, although which value is observed is affected by the assignment; and (*iii*) it cannot always be entirely inferred from the observed values for assignment and treatment, Z_i and W_i^{obs}. This last feature is illustrated in Table 23.2 by the fact that the ($Z_i = 0$, $W_i^{\text{obs}} = 0$) cell contains a mixture of compliers and noncompliers.

23.5.2 The ITT Effect on the Treatment Received by Compliance Status

First let us consider the population ITT effect on the secondary outcome, treatment received, separately by compliance status. For noncompliers, $W_i(z) = 0$ for $z = 0, 1$. Hence

$$\text{ITT}_{\text{W,nc}} = \frac{1}{N_{\text{nc}}} \sum_{i:G_i=\text{nc}} \left(W_i(1) - W_i(0) \right) = \frac{1}{N_{\text{nc}}} \sum_{i:G_i=\text{nc}} W_i(1) = 0.$$

For compliers, $W_i(z) = z$ for $z = 0, 1$. Hence

$$\text{ITT}_{\text{W,co}} = \frac{1}{N_{\text{co}}} \sum_{i:G_i=\text{co}} \left(W_i(1) - W_i(0) \right) = \frac{1}{N_{\text{co}}} \sum_{i:G_i=\text{co}} W_i(1) = 1.$$

The overall ITT effect on treatment received is a weighted average of the within-compliance subpopulation ITT effects:

$$\text{ITT}_{\text{W}} = \pi_{\text{nc}} \cdot \text{ITT}_{\text{W,nc}} + \pi_{\text{co}} \cdot \text{ITT}_{\text{W,co}} = \pi_{\text{co}},$$

and $\pi_{\text{nc}} = 1 - \text{ITT}_{\text{W}}$. In words, the ITT effect on treatment received is equal to the population fraction of compliers. Note that this does not rely on any assumptions. It simply follows from the definition of compliance behavior and the existence of the potential outcomes.

23.5.3 The ITT Effect on the Primary Outcome by Compliance Status

The next step is to decompose the intention-to-treat effect for the primary outcome, ITT_{Y}, into a weighted average of the intention-to-treat effects by compliance status. Define

$$\text{ITT}_{\text{Y,co}} = \frac{1}{N_{\text{co}}} \sum_{i:G_i=\text{co}} \left(Y_i(1, W_i(1)) - Y_i(0, W_i(0)) \right),$$

and

$$\mathrm{ITT}_{Y,\mathrm{nc}} = \frac{1}{N_{\mathrm{nc}}} \sum_{i:G_i=\mathrm{nc}} \left(Y_i(1, W_i(1)) - Y_i(0, W_i(0)) \right),$$

so that we can write

$$\mathrm{ITT}_Y = \mathrm{ITT}_{Y,\mathrm{co}} \cdot \pi_{\mathrm{co}} + \mathrm{ITT}_{Y,\mathrm{nc}} \cdot \pi_{\mathrm{nc}} \tag{23.2}$$

$$= \mathrm{ITT}_{Y,\mathrm{co}} \cdot \mathrm{ITT}_W + \mathrm{ITT}_{Y,\mathrm{nc}} \cdot (1 - \mathrm{ITT}_W).$$

Let us consider directly the ITT_Y effects by compliance type. The average ITT effect for noncompliers is

$$\mathrm{ITT}_{Y,\mathrm{nc}} = \frac{1}{N_{\mathrm{nc}}} \sum_{i:G_i=\mathrm{nc}} \left(Y_i(1, 0) - Y_i(0, 0) \right).$$

Note, however, that this ITT effect for noncompliers is not informative about the effect of the primary treatment: it compares two potential outcomes for a group of units, all of which always receive the control treatment.

For compliers the ITT effect is generally more interesting for the causal effects of the receipt of treatment. The average ITT_Y effect for compliers is

$$\mathrm{ITT}_{Y,\mathrm{co}} = \frac{1}{N_{\mathrm{co}}} \sum_{i:G_i=\mathrm{co}} \left(Y_i(1, 1) - Y_i(0, 0) \right).$$

This ITT effect is at least potentially informative about the effect of the primary treatment, because it is based on a comparison of potential outcomes when receiving the active treatment and when not receiving the active treatment for the subpopulation of compliers.

The two ITT effects on Y by complier status, $\mathrm{ITT}_{Y,\mathrm{co}}$ and $\mathrm{ITT}_{Y,\mathrm{nc}}$, cannot be estimated directly from the observable data, because we cannot infer the latent compliance status for units assigned to the control group. Nevertheless, because receipt of treatment, W_i^{obs}, is unconfounded conditional on compliance status given randomization of the assignment, we can still disentangle the ITT effects by compliance type under an additional assumption: the exclusion restriction.

It is important here that the receipt of treatment is unconfounded within subpopulations defined by compliance status. This follows from Assumption 23.1, that Z_i is randomly assigned, in combination with the fact that W_i^{obs} is a deterministic function of Z_i given compliance status.

Lemma 23.1 (Super-Population Unconfoundedness of Receipt of Treatment Given Compliance Status)
Suppose Assumption 23.1 holds. Then, for $g \in \{\mathrm{co}, \mathrm{nc}\}$,

$$\Pr\left(W_i^{\mathrm{obs}} = 1 \,\middle|\, Y_i(0,0), Y_i(0,1), Y_i(1,0), Y_i(1,1), G_i = g \right) = \Pr\left(W_i^{\mathrm{obs}} = 1 \,\middle|\, G_i = g \right),$$

or

$$W_i^{\text{obs}} \perp\!\!\!\perp Y_i(0,0), Y_i(0,1), Y_i(1,0), Y_i(1,1) \mid G_i.$$

To see this, consider the two compliance types separately. First, for noncompliers ($G_i = \text{nc}$), we always have $W_i^{\text{obs}} = 0$, so unconfoundedness holds trivially. For compliers ($G_i = \text{co}$), $W_i^{\text{obs}} = Z_i$, and thus Lemma 23.1 holds by Assumption 23.1, random assignment of Z_i. The problem is that we cannot directly exploit the latent unconfoundedness result in Lemma 23.1 (latent, because it only holds given a partially unobserved covariate), because compliance type is only partially observed. We therefore rely on indirect methods for exploiting this latent unconfoundedness property.

23.6 INSTRUMENTAL VARIABLES

In this section we discuss the key assumption underlying the method of instrumental variables, and present the main result of this chapter that, under that key assumption, we can estimate the average ITT effect for compliers, $\text{ITT}_{Y,\text{co}}$. We discuss the interpretation of this ITT effect and how it may be related to the causal effect of the receipt of treatment. We then discuss two approaches to inference for this average effect.

23.6.1 Exclusion Restriction for Noncompliers

First we discuss the key assumption that underlies, in some form or another, all instrumental variables analyses.

Assumption 23.2 (Exclusion Restriction for Noncompliers) *For all noncompliers, that is, all units with $G_i = \text{nc}$,*

$$Y_i(0,0) = Y_i(1,0).$$

This assumption, the exclusion restriction, rules out, for noncompliers, an effect of the assignment, the instrument Z_i, on the outcome of interest Y_i. It states that changing the assignment has no causal effect on the outcome, for those units for whom the level of the primary treatment W_i does not change with the change in assignment.

This exclusion restriction is the key assumption underlying the instrumental variables approach. Unlike the latent unconfoundedness assumption, however, it is not implied by the randomization of the assigned treatment. Instead, it is a substantive assumption that need not be appropriate in all randomized experiments with noncompliance, although it can be made plausible by design features such as double-blinding.

A slightly weaker version of the exclusion restriction for noncompliers requires the exclusion restriction to hold in distribution for the super-population:

Assumption 23.3 (Stochastic Exclusion Restriction for Noncompliers)

$$Z_i \perp\!\!\!\perp Y_i(Z_i, W_i(Z_i)),$$

for all noncompliers, that is, all units with $G_i = \text{nc}$.

This assumption implies that the super-population distribution of $Y_i(0, 1)$ is the same as that of $Y_i(1, 0)$ for noncompliers with $W_i(0) = W_i(1) = 0$. One advantage of this assumption is that there is a natural way to relax it in the presence of pre-treatment variables by requiring the independence to hold only conditional on the pre-treatment variables.

23.6.2 Exclusion Restriction for Compliers

Because of the central role of the exclusion restriction, some general comments about the applicability of this assumption are in order. Before doing so, let us also formulate a second exclusion restriction, this time for compliers.

Assumption 23.4 (Exclusion Restrictions for Compliers) *For all units with $G_i =$ co, that is, all compliers,*

$$Y_i(0, w) = Y_i(1, w)$$

for both levels of the treatment w.

This is an assumption of a very different nature from the exclusion restriction for non-compliers. It restricts, for compliers, $Y_i(0, 0)$ to be equal to $Y_i(1, 0)$, and restricts $Y_i(0, 1)$ to be equal to $Y_i(1, 1)$. But for compliers, we observe either $Y_i(0, 0)$ or $Y_i(1, 1)$, and never observe $Y_i(0, 1)$ or $Y_i(1, 0)$, and so these restrictions have no empirical consequences, either in the current form or in a stochastic version, unlike the exclusion restriction for noncompliers. In a sense, this restriction is essentially an *attribution* of the ITT effect for compliers to the causal effect of the receipt of treatment, rather than to its assignment. It is primarily about the interpretation of this ITT effect, not about issues concerning estimating it from the data.

Note that the exclusion restriction for compliers is routinely made, often implicitly, in randomized experiments with full compliance (in that case all units are compliers). For instance, when analyzing and interpreting the results from double-blind randomized drug trials with full compliance, one often implicitly assumes that the estimated effect is due to the *receipt* of the drug, not to the *assignment* to receive the drug. Thus, the assumption is implicitly made that similar unit-level treatment effects will occur if the assignment mechanism is changed from randomized assignment to either voluntary assignment or full adoption. Specifically, suppose a drug company estimates the efficacy of a new drug in a randomized trial. Implicitly the assumption is that, had, at the start of the trial, all individuals been told that they would receive the new active drug and that no one would receive the control treatment, the typical outcome would have been approximately the same as the typical outcome observed in the subsample actually assigned to the treatment. Moreover, after the drug is approved, physicians will presumably prescribe the new drug without using randomization. Again the presumption is that their patients will respond to the prescribed treatment in the same way that similar subjects in the randomized trial responded to assignment to the possibly unknown, blinded, treatment.

Yet the fact that this assumption is often implicitly made does not mean that this exclusion restriction is innocuous. There are many examples of studies where assignment did make an important difference, separate from receipt of the active treatment. Concerns about potential complications from such direct effects of assignment motivate the use of placebos, and blinding or double blinding, in clinical trials with human subjects. If individuals do not know their values of assignment, it is difficult to see how the assignments could affect their outcomes, except through the biological effect of the treatment received. But, again, receipt of a known, approved drug is not necessarily the same as receipt of a blinded drug being evaluated in the experiment.

23.6.3 Discussion of the Exclusion Restrictions

In some settings where noncompliance is an issue, however, placebos and (double-) blinding are often infeasible. If the treatment is an invasive procedure or requires active participation on the part of the individual, the researcher typically cannot hide the nature of the treatment. Even in randomized eligibility designs, where access (eligibility) to the treatment is randomized, the exclusion restriction may be violated. Individuals assigned to the active treatment may refuse to accept it but, in response to the notification of eligibility, may take actions they would not have taken otherwise. For example, consider the evaluation of a smoking cessation program. Suppose the program is offered to a random sample of smokers. Some may be unwilling to go through the program if it takes a large amount of time or effort. Yet in response to the assignment such individuals may still change their lifestyles, including their smoking habits in ways that affect their subsequent health outcomes. In that case, health outcomes would differ by assignment for such individuals, even though they are noncompliers who do not participate in the program irrespective of their assignment. Examples such as these illustrate that the exclusion restriction requires careful consideration of the various paths through which assignment may affect outcomes.

One should note, however, that the exclusion restrictions, Assumptions 23.2 and 23.4, do *not* in any way restrict compliance behavior itself. For example, it allows for the possibility that individuals know their outcomes under both treatments and deliberately choose to comply when assigned to the active treatment only if it will benefit them. Specifically, suppose that all those with $Y_i(1, 1) > Y_i(1, 0)$ (those whose health status would improve with the receipt of the treatment) choose to comply, and all those with $Y_i(1, 1) \leq Y_i(1, 0)$ choose not to. Such behavior would imply that the receipt of treatment W_i^{obs} is confounded, and it is often exactly this type of systematic noncompliance behavior that motivates researchers to consider instrumental variable analyses. Such behavior is not, however, inconsistent with the exclusion restriction and thus will be compatible with the analyses developed here.

Let us consider the exclusion restriction for noncompliers for the Sommer-Zeger vitamin A supplement data. This restriction requires that, for those individuals who would not receive the supplements even if assigned to take them, the potential outcomes are unaffected by assignment. This assumption seems fairly plausible. If some mothers living in villages assigned to the treatment did not receive the supplements because of administrative mishaps, or through lack of interest, it is quite likely that the infants of such mothers would not have had different outcomes had their village been assigned to the control group, except if there are fewer contiguous infant diseases in the villages that were assigned the vitamin supplements. Nevertheless, this is a key assumption for the validity of the IV approach, and even in this example it is not necessarily satisfied. Violations of this assumption could arise if the reason these women did not receive the supplements was related to other health improvement measures taken in some villages but not in others. For example, suppose that noncompliance was high in some villages because the administrators in those villages, if assigned to receive the supplements, diverted the program funding toward other health care improvements that would have been otherwise unaffordable. In that case, outcomes for noncomplying mothers would differ by assignment, even though none took the supplements, violating the exclusion restriction. Such a story may seem fairly implausible in this case, but such stories are important to consider. We will return to discuss such violations in other examples in subsequent chapters.

23.6.4   Local Average Treatment Effects

In this section we discuss the most important result in this chapter. Consider the average ITT effect in the population, decomposed by compliance status:

$$ITT_Y = ITT_{Y,co} \cdot ITT_W + ITT_{Y,nc} \cdot (1 - ITT_W), \qquad (23.3)$$

using the fact that the one-sided nature of the noncompliance implies that $ITT_W = \pi_{co}$. The exclusion restriction for noncompliers implies that for noncompliers $Y_i(0,0) = Y_i(1,0)$, and thus,

$$ITT_{Y,nc} = 0.$$

Hence, the second term on the right-hand side of (23.3) is zero, reducing the global ITT on the outcome to the product of two ITT effects, the "local" ITT effect on the outcome for the compliers, and the global ITT effect on the receipt of treatment:

$$ITT_Y = ITT_{Y,co} \cdot ITT_W. \qquad (23.4)$$

We now rearrange Equation (23.4) to give our formal result:

Theorem 23.1 (Local Average Treatment Effect)
Suppose that Assumption 23.2 holds. Then

$$\tau_{late} = ITT_{Y,co} = \frac{ITT_Y}{ITT_W}.$$

In other words, under the exclusion restriction for noncompliers, the ratio of the ITT effect on the outcome to the ITT effect on the treatment is equal to the ITT effect on the outcome for compliers, or what is called the Local Average Treatment Effect (LATE), or, synonymously, the Complier Average Causal Effect (CACE).

   If we are also willing to assume the second exclusion restriction, the exclusion restriction for compliers given in Assumption 23.4, we can interpret this local average treatment effect as the average causal effect of the receipt of treatment for compliers. Thus, given both exclusion restrictions and the randomization assumption, we can learn about the effect of the primary treatment for the subpopulation of compliers, because we can unbiasedly estimate both the numerator and the denominator of τ_{late}.

To give a different interpretation for the result in Theorem 23.1, suppose for a moment that we could observe compliance status for all units. By Lemma 23.1, receipt of treatment is unconfounded given compliance status G_i, and so we could then analyze the data separately for noncompliers and compliers. Within these subpopulations, we can compare outcome by treatment status. For noncompliers, there would be no information in the data regarding the effect of the primary treatment on the outcome, because no noncomplier ever receives the active treatment. The data from noncompliers would therefore be discarded because of the absence of units who received the active treatment. For compliers, receipt of treatment is identical to assignment, and for this subpopulation we can therefore consistently estimate effects of the receipt of the treatment on the outcome, because, by the second exclusion restriction, it equals the intention-to-treat effect of the assignment on the final outcome. The only, but crucial, missing piece in this argument is that we do not observe the compliance status for all units. However, given the exclusion restriction, we can disentangle the potential outcome distributions for compliers and noncompliers from the mixture of noncompliers and compliers in the subpopulation assigned to the control treatment, through, for example, an imputation-based approach such as that outlined in Chapter 25 or the moment-based approach introduced here.

23.7 MOMENT-BASED INSTRUMENTAL VARIABLES ESTIMATORS

Summarizing the discussion so far, the overall ITT effect consists of two parts, the ITT effect for compliers and the ITT effect for noncompliers, weighted by their population proportions. The exclusion restriction for noncompliers implies that the ITT effect for noncompliers is zero. Hence, under the exclusion restriction for noncompliers, the ratio of the overall ITT effect, to the population proportion of compliers, is equal to the ITT effect for compliers.

In Section 23.4 we discussed how to estimate and conduct inference for ITT_W and ITT_Y. Given those two unbiased estimators, a simple moment-based instrumental variables (iv) estimator for τ_{late} is the ratio of estimated ITT effects,

$$\hat{\tau}^{\text{iv}} = \frac{\widehat{\text{ITT}_Y}}{\widehat{\text{ITT}_W}}.$$

This simple estimator has some drawbacks, and in Chapter 25 we discuss model-based methods that have more attractive statistical properties, especially in small samples. One of the reasons is that it does not necessarily satisfy all the restrictions implied by Assumptions 23.1 and 23.2. We will discuss these restrictions in more detail in Chapter 25, but as a simple example, suppose that $\text{ITT}_W = 0$. In that case there are no compliers, and by the exclusion restriction for noncompliers, it must be the case that $\text{ITT}_Y = 0$. More generally, the restrictions imply that the joint distribution of the data is consistent with the subpopulation of $(Z_i = 0, W_i^{\text{obs}} = 0)$ being a mixture of compliers and noncompliers, and the outcome distribution for noncompliers being the same as that for units with $(Z_i = 1, W_i^{\text{obs}} = 0)$.

The sampling variance calculations for the two ITT effects separately followed from the Neyman approach discussed in Chapter 6. Here we discuss the extension to the sampling variance for $\hat{\tau}^{\mathrm{iv}}$. Here we take explicitly a super-population perspective. That is, we view our sample as a random sample from a large population. In that large population, there is an average ITT effect for compliers, $\mathrm{ITT}_{Y,co} = \mathbb{E}[Y_i(1, W_i(1)) - Y_i(0, W_i(0))|G_i] = co$. We consider the sampling variance of $\hat{\tau}^{\mathrm{iv}} - \mathrm{ITT}_{Y,co}$. To calculate the sampling variance of the IV estimator $\hat{\tau}^{\mathrm{iv}}$ requires estimation of the sampling covariance between $\widehat{\mathrm{ITT}_W}$ and $\widehat{\mathrm{ITT}_Y}$. With that covariance, we can use the delta method to estimate the large-sample sampling variance of the ratio of ITT effects. The result is that in large samples, $\hat{\tau}^{\mathrm{iv}}$, as an estimator of the super-population ITT effect for compliers, will be approximately normally distributed with sampling variance

$$\mathbb{V}_{\mathrm{sp}}(\hat{\tau}^{\mathrm{iv}}) = \frac{1}{\mathrm{ITT}_W^2} \cdot \mathbb{V}(\widehat{\mathrm{ITT}_Y}) + \frac{\mathrm{ITT}_Y^2}{\mathrm{ITT}_W^4} \cdot \mathbb{V}(\widehat{\mathrm{ITT}_W}) \tag{23.5}$$

$$- 2 \cdot \frac{\mathrm{ITT}_Y}{\mathrm{ITT}_W^3} \cdot \mathbb{C}(\widehat{\mathrm{ITT}_Y}, \widehat{\mathrm{ITT}_W}),$$

where $\mathbb{C}(\cdot, \cdot)$ denotes the covariance of two random variables. A simple estimator for the sampling variance can be based on substituting estimates for the components of this sampling variance. Using this to construct confidence intervals raises some issues, such as if the denominator of (23.5), ITT_W, is close to zero, normality is likely to be a poor approximation to the sampling distribution of the estimator.

Returning to the vitamin A supplement data, using our earlier estimates for ITT_Y, $\mathbb{V}(\widehat{\mathrm{ITT}_Y})$, ITT_W, and $\mathbb{V}(\widehat{\mathrm{ITT}_W})$, in combination with the estimate for the covariance of $\widehat{\mathrm{ITT}_Y}$ and $\widehat{\mathrm{ITT}_W}$, $\hat{\mathbb{C}}(\widehat{\mathrm{ITT}_Y}, \widehat{\mathrm{ITT}_W}) = -0.00000017$ (corresponding to a correlation between $\widehat{\mathrm{ITT}_Y}$ and $\widehat{\mathrm{ITT}_W}$ equal to -0.0502), we find that the method-of-moments IV estimate for the effect of taking vitamin A supplements on survival is

$$\hat{\tau}^{\mathrm{iv}} = \frac{\widehat{\mathrm{ITT}_Y}}{\widehat{\mathrm{ITT}_W}} = 0.0032, \quad \text{and} \quad \mathbb{V}(\hat{\tau}^{\mathrm{iv}}) = 0.0012^2,$$

leading to a 95% large-sample confidence interval for $\mathrm{ITT}_{Y,co}$ (or τ_{late}) equal to

$$\mathrm{CI}^{0.95}(\mathrm{ITT}_{Y,co}) = \left(0.0010, 0.0055\right).$$

Because the ITT effect on the receipt of treatment is precisely estimated, and far from zero, the 95% confidence interval is likely to be valid (in the statistically conservative sense), with the qualification that we ignored the clustering of the experiment by village.

If in addition to the exclusion restriction for noncompliers, we are willing to assume the exclusion restriction for compliers, this estimated ITT effect for compliers can be interpreted as equal to the estimated average effect of the primary treatment on the primary outcome for compliers.

23.8   LINEAR MODELS AND INSTRUMENTAL VARIABLES

Even for readers familiar with traditional discussions of instrumental variables in econometric textbooks, the discussion thus far may look unfamiliar. In this section we discuss the link between the approach advocated in this book and conventional econometric instrumental variables analyses. Readers not familiar with the textbook econometrics approach may wish to skip this section.

The traditional use of instrumental variables in the economics literature relies heavily on linear parametric specifications, even though some of these are not critical. It also takes a super-population perspective, where the sample at hand is assumed to be a random sample from an infinitely large population, and the estimands are population average causal effects. We maintain here both exclusion restrictions, for noncompliers and compliers. As a result we can drop the dependence of the potential outcome $Y_i(z, w)$ on z and write, without ambiguity, $Y_i(w)$, as a function of the receipt of treatment alone. In order to see the connection with our framework, it is useful to assume initially a constant treatment effect: $Y_i(1) - Y_i(0) = \tau$ for all i. We relax this assumption later. Define $\alpha = \mathbb{E}_{sp}[Y_i(0)]$ to be the super-population average outcome given the control treatment, so that we can write

$$\mathbb{E}_{sp}[Y_i(w)] = \alpha + \tau \cdot w,$$

for $w = \{0, 1\}$. We define the residual $\varepsilon_i = Y_i(0) - \alpha$ to be the unit-level deviation of the control outcome from its population mean, so that we can further write

$$Y_i(w) = \alpha + \tau \cdot w + \varepsilon_i. \tag{23.6}$$

Equation (23.6) is what is known in the econometric literature as a *structural* or *behavioral* equation: it relates treatments to outcomes in a causal way. For a given unit i (and thus, for a fixed value ε_i), $Y_i(w)$ is the outcome we would observe if we fixed (*set* in Pearl's (2000) terminology) $W_i = w$.

Equation (23.6) is *not*, however, a conventional regression function. Note that it is not written in terms of observed quantities. Substituting observed values for the treatment and outcome we can instead write

$$Y_i^{obs} = Y_i(W_i^{obs}) = Y_i(0) + W_i^{obs} \cdot (Y_i(1) - Y_i(0)) = \alpha + \tau \cdot W_i^{obs} + \varepsilon_i. \tag{23.7}$$

Yet, as written, Equation (23.7) remains a behavioral equation, not a conditional expectation: in general it is *not* true that $\mathbb{E}[Y_i^{obs}|W_i^{obs} = w] = \alpha + \tau \cdot W_i^{obs}$. The coefficient τ for the treatment indicator W_i^{obs} represents the *causal* effect of the treatment on the outcome; it is not equal to the ratio of the super-population covariance of Y_i^{obs} and W_i^{obs}, to the variance of W_i^{obs}.

The key factor distinguishing Equation (23.7) from a standard regression function is that the regressor, the receipt of treatment W_i^{obs}, is possibly correlated with $Y_i(0)$, and thus with the residual ε_i. To see this, let us first calculate the conditional mean of ε_i given W_i^{obs} in the super-population. Here let π_g be the share in the super-population of compliance type $G_i = g$. Remember that ε_i is defined as the difference between the observed and expected control outcome: $\varepsilon_i = Y_i(0) - \alpha = Y_i(0) - \mathbb{E}_{sp}[Y_i(0)]$. Given $W_i^{obs} = 1$ we have:

$$\mathbb{E}_{sp}[\varepsilon_i|W_i^{obs} = 1] = \mathbb{E}_{sp}[\varepsilon_i|G_i = co]$$

$$= \mathbb{E}_{sp}[Y_i(0)|G_i = co] - \mathbb{E}_{sp}[Y_i(0)]$$

$$= \mathbb{E}_{\mathrm{sp}}[Y_i(0)|G_i = \mathrm{co}] - \left(\mathbb{E}_{\mathrm{sp}}[Y_i(0)|G_i = \mathrm{co}] \cdot \pi_{\mathrm{co}} + \mathbb{E}_{\mathrm{sp}}[Y_i(0)|G_i = \mathrm{nc}] \cdot \pi_{\mathrm{nc}}\right)$$

$$= \pi_{\mathrm{nc}} \cdot \left(\mathbb{E}_{\mathrm{sp}}[Y_i(0)|G_i = \mathrm{co}] - \mathbb{E}_{\mathrm{sp}}[Y_i(0)|G_i = \mathrm{nc}]\right)$$

$$= \pi_{\mathrm{nc}} \cdot \Delta_{\mathrm{co,nc}},$$

where $\Delta_{\mathrm{co,nc}}$ is defined as the difference in average control outcome for compliers and noncompliers, $\Delta_{\mathrm{co,nc}} = \mathbb{E}_{\mathrm{sp}}[Y_i(0)|G_i = \mathrm{co}] - \mathbb{E}_{\mathrm{sp}}[Y_i(0)|G_i = \mathrm{nc}]$. To calculate $\mathbb{E}_{\mathrm{sp}}[\varepsilon_i|W_i^{\mathrm{obs}} = 0]$, first decompose $\mathbb{E}_{\mathrm{sp}}[\varepsilon_i] = 0$:

$$0 = \mathbb{E}_{\mathrm{sp}}[\varepsilon_i] = \mathbb{E}_{\mathrm{sp}}[\varepsilon_i|W_i^{\mathrm{obs}} = 1] \cdot \mathrm{Pr}(W_i^{\mathrm{obs}} = 1) + \mathbb{E}_{\mathrm{sp}}[\varepsilon_i|W_i^{\mathrm{obs}} = 0] \cdot \mathrm{Pr}(W_i^{\mathrm{obs}} = 0).$$

Given that the probability $\mathrm{Pr}(W_i^{\mathrm{obs}} = 1)$ is equal to $p_Z \cdot \pi_{\mathrm{co}}$, and thus $\mathrm{Pr}(W_i^{\mathrm{obs}} = 0) = (1 - p_Z \cdot \pi_{\mathrm{co}})$, it follows that

$$\mathbb{E}_{\mathrm{sp}}[\varepsilon_i|W_i^{\mathrm{obs}} = 0] = -\frac{\pi_{\mathrm{nc}} \cdot p_Z \cdot \pi_{\mathrm{co}}}{1 - p_Z \cdot \pi_{\mathrm{co}}} \cdot \Delta_{\mathrm{co,nc}}.$$

These expectations $\mathbb{E}_{\mathrm{sp}}[\varepsilon_i|W_i^{\mathrm{obs}} = w]$ are typically not zero. In econometric terminology, the explanatory variable W_i^{obs} is *endogenous*, and least squares methods do not lead to consistent estimation of τ.

Although the receipt of treatment, W_i^{obs}, is *not* independent of ε_i, the assignment to treatment, or the instrument Z_i *is* independent of ε_i. This follows from the random assignment assumption and the definition of ε_i in terms of the potential outcomes. This independence of Z_i and ε_i can be exploited through what is known in econometrics as Two-Stage-Least-Squares (TSLS) estimation. First, this independence implies that the conditional expectation of ε_i given Z_i is zero. This in turn implies that the conditional expectation of Y_i^{obs} given Z_i equals,

$$\mathbb{E}_{\mathrm{sp}}[Y_i^{\mathrm{obs}}|Z_i] = \alpha + \tau \cdot \mathbb{E}_{\mathrm{sp}}[W_i^{\mathrm{obs}}|Z_i] + \mathbb{E}_{\mathrm{sp}}[\varepsilon_i|Z_i] = \alpha + \tau \cdot \mathbb{E}_{\mathrm{sp}}[W_i^{\mathrm{obs}}|Z_i].$$

This conditional expectation of Y_i^{obs} given Z_i is linear in $\mathbb{E}_{\mathrm{sp}}[W_i^{\mathrm{obs}}|Z_i]$, with coefficient equal to the treatment effect of interest τ. We can therefore write

$$Y_i^{\mathrm{obs}} = \alpha + \tau \cdot \left(\mathbb{E}_{\mathrm{sp}}[W_i^{\mathrm{obs}}|Z_i] + \left(W_i^{\mathrm{obs}} - \mathbb{E}_{\mathrm{sp}}[W_i^{\mathrm{obs}}|Z_i]\right)\right) + \varepsilon_i$$

$$= \alpha + \tau \cdot \mathbb{E}_{\mathrm{sp}}[W_i^{\mathrm{obs}}|Z_i] + \eta_i, \tag{23.8}$$

where the composite residual is $\eta_i = \tau \cdot (W_i^{\mathrm{obs}} - \mathbb{E}_{\mathrm{sp}}[W_i^{\mathrm{obs}}|Z_i]) + \varepsilon_i$. By random assignment (Assumption 23.1), both ε_i and this unit-level difference $W_i^{\mathrm{obs}} - \mathbb{E}_{\mathrm{sp}}[W_i^{\mathrm{obs}}|Z_i]$ are uncorrelated with Z_i. Thus, the composite residual η_i is uncorrelated with Z_i. This in turn implies that least squares regression of Y_i^{obs} on the conditional expectation $\mathbb{E}_{\mathrm{sp}}[W_i^{\mathrm{obs}}|Z_i]$ will lead to an unbiased estimate of τ, the treatment effect of interest.

Unfortunately this linear regression is infeasible because we do not know the conditional expectation $\mathbb{E}_{\mathrm{sp}}[W_i^{\mathrm{obs}}|Z_i]$. However, we can estimate this conditional expectation. First let us write out the expected value of W_i^{obs} given Z_i as a function of Z_i – for those familiar with IV, the first-stage equation:

$$\mathbb{E}_{\mathrm{sp}}[W_i^{\mathrm{obs}}|Z_i] = \pi_0 + \pi_1 \cdot Z_i,$$

where $\pi_0 = \mathbb{E}_{\mathrm{sp}}[W_i^{\mathrm{obs}}|Z_i = 0]$ and $\pi_1 = \mathbb{E}_{\mathrm{sp}}[W_i^{\mathrm{obs}}|Z_i = 1] - \mathbb{E}_{\mathrm{sp}}[W_i^{\mathrm{obs}}|Z_i = 0]$. Given one-sided noncompliance, $\pi_0 = 0$ ($Z_i = 0$ implies $W_i^{\mathrm{obs}} = 0$), and π_1 equals $\mathbb{E}_{\mathrm{sp}}[W_i^{\mathrm{obs}}|Z_i = 1]$, which is equal to the super-population proportion of compliers, π_{co}. Hence $\mathbb{E}_{\mathrm{sp}}[W_i^{\mathrm{obs}}|Z_i] = \pi_1 \cdot Z_i = \pi_{\mathrm{co}} \cdot Z_i$.

Using this expression we can rewrite Equation (23.8):

$$Y_i^{\mathrm{obs}} = \alpha + \gamma \cdot Z_i + \eta_i, \quad \text{where } \gamma = \tau \cdot \pi_{\mathrm{co}}. \tag{23.9}$$

Equation (23.9) is known as a *reduced form* in econometric terminology. Here the regression function does represent a conditional expectation, and as a result, its parameters can be consistently estimated by ordinary least squares. The least squares estimator, equal to the ratio of the covariance of Z_i and Y_i^{obs}, and the variance of Z_i will give an unbiased estimator of the composite coefficient $\gamma = \tau \cdot \pi_{\mathrm{co}}$. With Z_i binary, this estimator will be equal to the difference in average outcomes by assignment, $\hat{\gamma} = \widehat{\mathrm{ITT}_Y} = \overline{Y}_1^{\mathrm{obs}} - \overline{Y}_0^{\mathrm{obs}}$. Similarly, given the unconfoundedness of Z_i, regressing W_i^{obs} on Z_i will give an unbiased estimate of π_{co}. The estimator, with Z_i binary, equals $\hat{\pi}_{\mathrm{co}} = \widehat{\mathrm{ITT}_W} = \overline{W}_1^{\mathrm{obs}} - \overline{W}_0^{\mathrm{obs}}$.

Dividing the least squares estimator $\hat{\gamma} = \widehat{\mathrm{ITT}_Y}$, by the estimator $\hat{\pi}_{\mathrm{co}} = \widehat{\mathrm{ITT}_W}$, gives the instrumental variables estimator $\hat{\tau}^{\mathrm{iv}} = \widehat{\mathrm{ITT}_Y}/\widehat{\mathrm{ITT}_W}$ given earlier. For noncompliers, $W_i(z) = 0$ for $z = 0, 1$. Hence, given a binary assignment and treatment, using the linear parametric specification leads to an estimator identical to the moment-based estimator based on the potential outcomes approach. This estimator is also identical to that based on regressing Y_i on $\hat{\pi}_{\mathrm{co}} \cdot Z_i$. The mechanical two-stage procedure of first regressing the receipt of treatment on the instrument to get an estimate of $\mathbb{E}_{\mathrm{sp}}[W_i^{\mathrm{obs}}|Z_i]$, followed by regressing the outcome of interest on this predicted value of the receipt of treatment, is what led to the econometric terminology of TSLS, and the IV estimator is therefore also known as the TSLS estimator.

As just noted, we assumed in this derivation that the treatment effect is constant. Yet we did not make this same assumption in our potential outcomes discussion of the instrumental variables approach. As it turns out, this assumption is not necessary in either approach. Without it, we end up estimating the average treatment effect for compliers. More precisely, the numerical equivalence of the linear-equation IV estimand to the ratio of ITT effects does not rely on the assumption of a constant treatment effect. To see this, let τ_{late} be the average treatment effect for compliers, or the local average treatment effect, $\tau_{\mathrm{late}} = \mathbb{E}_{\mathrm{sp}}[Y_i(1) - Y_i(0)|G_i = \mathrm{co}]$, and let ν_i be the unit-level difference between τ_i and τ, $\nu_i = Y_i(1) - Y_i(0) - \tau$. Again let $\alpha = \mathbb{E}_{\mathrm{sp}}[Y_i(0)]$, and $\varepsilon_i = Y_i(0) - \alpha$. As before

$$Y_i^{\mathrm{obs}} = Y_i(0) + W_i^{\mathrm{obs}} \cdot (Y_i(1) - Y_i(0)),$$

which, given the definitions provided here, can be rewritten as

$$Y_i^{\mathrm{obs}} = \alpha + W_i^{\mathrm{obs}} \cdot \tau_{\mathrm{late}} + \varepsilon_i + W_i^{\mathrm{obs}} \cdot \nu_i. \tag{23.10}$$

We now have a new composite disturbance term, $\varepsilon_i + W_i^{\mathrm{obs}} \cdot \nu_i$, which again is potentially correlated with W_i^{obs}. Thus an ordinary least squares regression of Y_i^{obs} on W_i^{obs} will not provide an unbiased estimate of τ.

However, just as ε_i is uncorrelated with Z_i, the second component of this new error term, $W_i^{\mathrm{obs}} \cdot \nu_i$, is also uncorrelated with Z_i. To see this, consider this expectation separately for $Z_i = 0$ and 1. Because $Z_i = 0$ implies $W_i^{\text{obs}} = 0$, it follows that $\mathbb{E}_{\text{sp}}[W_i^{\text{obs}} \cdot v_i | Z_i = 0] = 0$. To calculate the expectation given $Z_i = 1$, begin by expanding the expectation for both possible values of W_i^{obs}:

$$\mathbb{E}_{\text{sp}}[W_i \cdot v_i | Z_i = 1] = \mathbb{E}_{\text{sp}}[0 \cdot v_i | Z_i = 1, W_i = 0] \cdot \Pr(W_i = 0 | Z_i = 1)$$

$$+ \mathbb{E}_{\text{sp}}[1 \cdot v_i | Z_i = 1, W_i = 1] \cdot \Pr(W_i = 1 | Z_i = 1)$$

$$= \mathbb{E}_{\text{sp}}[v_i | Z_i = 1, G_i = \text{co}] \cdot \pi_{\text{co}} = \mathbb{E}_{\text{sp}}[Y_i(1) - Y_i(0) - \tau | Z_i = 1, G_i = \text{co}]$$

$$= \mathbb{E}_{\text{sp}}[Y_i(1) - Y_i(0) - \tau | G_i = \text{co}] \cdot \pi_{\text{co}} = 0,$$

by the definition of τ as the average treatment effect for compliers. Hence, looking at Equation (23.10), given that Z_i is uncorrelated with both elements of the error term, we can use the same argument as used earlier to motivate the moment estimator $\hat{\tau}^{\text{iv}}$.

23.9   NAIVE ANALYSES: "AS-TREATED," "PER PROTOCOL," AND UNCONFOUNDEDNESS

To put the simple instrumental variables analysis that is the main topic of this chapter in perspective, we conclude this chapter by discussing three other analyses, two of which are occasionally used in randomized experiments with noncompliance, and one of which serves to provide some perspective. (Note that we have already discussed one such alternative, the intention-to-treat analysis.) Like the IV approach, but unlike the ITT approach, these two additional analyses focus on the receipt of treatment, not merely on the causal effect of the assignment to treatment. Four analyses, IV, ITT, As-Treated, and Per Protocol, are identical when observed compliance is perfect, but they generally differ from one another when compliance is less than perfect. As will be seen here, however, in the presence of noncompliance, there is no compelling justification for these two other approaches. We present them merely to provide a better understanding of the competing intention-to-treat and instrumental variables methods.

23.9.1   As-Treated Analyses

The first of these two analyses is the "as-treated" approach. In this approach, the causal effect of the receipt of treatment is estimated as the difference in average outcomes by treatment received, W_i^{obs}:

$$\hat{\tau}_{\text{at}} = \overline{Y}_{\text{t}}^{\text{obs}} - \overline{Y}_{\text{c}}^{\text{obs}}. \tag{23.11}$$

This approach would be justified, in the sense that it would give an unbiased estimate of the average treatment effect, if receipt of treatment W_i^{obs} were unconfounded. In general, however, it will not estimate a causal estimand. Here we explore the properties of this estimator. It will be convenient to take a super-population perspective, where we take the expectation over the randomization as well as over the distribution generated by random sampling from a large population.

The expectation of this estimator in the super-population is

$$\tau_{at} = \mathbb{E}_{sp}\left[Y_i^{obs}\,|\,W_i^{obs} = 1\right] - \mathbb{E}_{sp}\left[Y_i^{obs}\,|\,W_i^{obs} = 0\right].$$

Let us look at this difference in expectations under the two instrumental variables assumptions, random assignment and the exclusion restriction on noncompliers. Note that in our one-sided noncompliance case, units receiving the treatment must have $Z_i = 1$ and be compliers. Hence $\mathbb{E}_{sp}[Y_i^{obs}|W_i^{obs} = 1] = \mathbb{E}_{sp}[Y_i(1)|G_i = \text{co}]$. The second half of Equation (23.11) shows that units not receiving the treatment are a mixture of those assigned to the control and those assigned to the treatment who did not comply:

$$\mathbb{E}_{sp}[Y_i^{obs}|W_i^{obs} = 0] = \mathbb{E}_{sp}[Y_i^{obs}|W_i^{obs} = 0, Z_i = 0] \cdot \text{Pr}_{sp}(Z_i = 0|W_i^{obs} = 0)$$

$$+ \mathbb{E}_{sp}[Y_i^{obs}|W_i^{obs} = 0, Z_i = 1] \cdot \text{Pr}(Z_i = 1|W_i^{obs} = 0). \quad (23.12)$$

With $p_Z = \text{Pr}_{sp}(Z_i = 1)$, Bayes rule implies that, the probability that $Z_i = 1$ among those who do not take the treatment is equal to

$$\text{Pr}_{sp}(Z_i = 1|W_i^{obs} = 0) = \frac{\pi_{nc} \cdot p_Z}{\pi_{nc} \cdot p_Z + 1 \cdot (1 - p_Z)}.$$

In the two expectations on the right-hand side of Equation (23.12), the second is simply the expected outcome for noncompliers under the control treatment. The first expectation in Equation (23.12) is a mixture of the expected value given the control treatment, for both compliers and noncompliers:

$$\mathbb{E}_{sp}[Y_i^{obs}|Z_i = 0, W_i^{obs} = 0] = \mathbb{E}_{sp}[Y_i(0)|G_i = \text{co}] \cdot \pi_{co} + \mathbb{E}_{sp}[Y_i(0)|G_i = \text{nc}] \cdot \pi_{nc}.$$

Combining all of the above, we can rewrite the expectation of the as-treated estimator as

$$\tau_{at} = \text{ITT}_{Y,co} + \Delta_{co,nc} \cdot \frac{\pi_{nc}}{p_Z \cdot \pi_{nc} + 1 - p_Z},$$

where, as before, $\Delta_{co,nc}$ is the expected difference in control outcomes for compliers and noncompliers:

$$\Delta_{co,nc} = \mathbb{E}_{sp}[Y_i(0)|G_i = \text{co}] - \mathbb{E}_{sp}[Y_i(0)|G_i = \text{nc}].$$

Unless compliance is perfect and there are no noncompliers ($\pi_{nc} = 0$), or the average control outcome is the same for compliers and noncompliers ($\Delta_{co,nc} = 0$, as implied by unconfoundedness of the treatment W_i), the expected value of $\hat{\tau}_{at}$ differs from the complier average causal effect.

This bias is easy to interpret: τ_{at} compares the average observed outcome given the active treatment to the average observed outcome given the control treatment. The first term is the average outcome given the active treatment for compliers, but the second term is an average of expected control outcome for compliers and noncompliers. If, as estimated in our example, noncompliers have lower average outcomes *without* the active treatment than compliers *without* the active treatment, this lowers the average outcome in the as-treated "control" group. Hence, the as-treated approach will overestimate the average treatment effect for compliers.

Let us illustrate this using the vitamin supplement data. In this sample the estimate of the average outcomes, with and without the supplements, are

$$\overline{Y}_c^{obs} = \frac{11,514 + 2,385}{11,514 + 2,385 + 74 + 34} = 0.9923,$$

and

$$\overline{Y}_t^{obs} = \frac{9,663}{9,663 + 12} = 0.9988.$$

Hence the as-treated estimate is

$$\hat{\tau}_{at} = 0.9988 - 0.9923 = 0.0065 \quad (\widehat{s.e.}\ 0.0008).$$

This estimator differs substantially from the IV estimate of 0.0033 calculated earlier. The reason can be seen by considering the estimates of the average outcomes of those assigned to the control for compliers and noncompliers separately. For noncompliers we estimated $\widehat{\mathbb{E}}_{sp}[Y_i(0)|G_i = \text{nc}] = 0.9859$, whereas for compliers we estimated $\widehat{\mathbb{E}}_{sp}[Y_i(0)|G_i = \text{co}] = 0.9955$, considerably higher. If the exclusion restriction holds, and hence our estimates of $\mathbb{E}_{sp}[Y_i(0)|G_i = \text{co}]$ and $\mathbb{E}_{sp}[Y_i(0)|G_i = \text{nc}]$ are unbiased, the fact that the average outcome under the control treatment is higher for compliers than for noncompliers will lead the as-treated estimator to overestimate the complier average causal treatment effect.

23.9.2 Per Protocol Analyses

Now let us look at a second alternative to ITT and IV analyses, the per protocol analysis, in which only those units who are observed to comply with their assigned status are compared. In this analysis we therefore discard all observed noncompliers assigned to the treatment. Given the observable data, however, we cannot discard noncompliers assigned to the control. By one-sided noncompliance, these individuals automatically take the control; we would only be able to observe their compliance status if we instead saw them assigned to the treatment. If we could, in fact, discard *all* noncompliers, we would be left with only compliers, and then comparing their average outcomes by treatment status would estimate the average effect of receipt of treatment for compliers.

The per protocol analysis, however, discards only those noncompliers who do not comply with their *observed* treatment assignment and *not* those noncompliers who were assigned to the control group. The result is that the per protocol estimator, $\hat{\tau}_{pp}$, compares units receiving the treatment, that is, the compliers assigned to the treatment, to all units assigned to the control, with the latter a mixture of both compliers and noncompliers:

$$\hat{\tau}_{pp} = \overline{Y}_t^{obs} - \overline{Y}_0^{obs} = \frac{1}{N_t} \sum_{i=1}^{N} W_i^{obs} \cdot Y_i^{obs} - \frac{1}{N_0} \sum_{i=1}^{N} (1 - Z_i) \cdot Y_i^{obs},$$

which is biased for τ_{co}. Its expectation is:

$$\tau_{pp} = \mathbb{E}[Y_i^{obs}|W_i^{obs} = 1, Z_i = 1] - \mathbb{E}[Y_i^{obs}|W_i^{obs} = 0, Z_i = 0]$$

$$= \mathbb{E}_{sp}[Y_i(1)|G_i = \text{co}] - \mathbb{E}_{sp}[Y_i(0)]. \tag{23.13}$$

The last term in this expression is equal to $\mathbb{E}[Y_i(0)|G_i = \text{co}] \cdot \pi_{\text{co}} - \mathbb{E}[Y_i(0)|G_i = \text{nc}] \cdot \pi_{\text{nc}}$; hence we can rewrite τ_{pp} as

$$\tau_{\text{pp}} = \mathbb{E}[Y_i(1) - Y_i(0)|G_i = \text{co}] \cdot \pi_{\text{co}} + (\mathbb{E}[Y_i(0)|G_i = \text{co}] - \mathbb{E}[Y_i(0)|G_i = \text{nc}]) \cdot \pi_{\text{nc}}$$

$$= \text{ITT}_{Y,\text{co}} + \pi_{\text{nc}} \cdot \Delta_{\text{co,nc}}.$$

Again, unless either π_{nc} or $\Delta_{\text{co,nc}}$ (or both) are equal to zero, $\hat{\tau}_{\text{pp}}$ will not give an unbiased estimate of the average effect of the treatment on compliers, even under the exclusion restriction for noncompliers.

To illustrate this, we again use the Sommer-Zeger data to estimate τ_{pp}. Given these data, the first term of the estimand, $\mathbb{E}_{\text{sp}}[Y_i(1)|G_i = \text{co}] = \mathbb{E}_{\text{sp}}[Y_i^{\text{obs}}|W_i^{\text{obs}} = 1]$, is estimated as 0.9988 ($\widehat{\text{s.e.}}$ 0.0004), and the second, $\mathbb{E}_{\text{sp}}[Y_i(0)] = \mathbb{E}_{\text{sp}}[Y_i^{\text{obs}}|Z_i = 0]$, as $11,514/(11,514 + 74) = 0.9936$ ($\widehat{\text{s.e.}}$ 0.0007). Thus the per protocol estimate,

$$\hat{\tau}_{\text{pp}} = 0.9988 - 0.9936 = 0.0051 \qquad (\widehat{\text{s.e.}}\ 0.0008),$$

is again much larger than our estimate of the local average treatment effect, $\hat{\tau}_{\text{late}} = 0.0033$.

23.9.3 Analyses under Conditional Unconfoundedness Given the Instrument

A final analysis we wish to discuss briefly assumes unconfoundedness, like the "as-treated" analysis, but only conditional on the instrument. That is, it focuses on comparisons of units receiving and not receiving the treatment within subpopulations receiving the same level of assignment. Implicitly it treats the instrument as a covariate or pre-treatment variable that needs to be controlled for. In the current setting, with one-sided noncompliance among the subpopulation of units assigned to the control group, there are no units receiving the treatment, so we can do this only for the units assigned to the treatment.

The conditional unconfoundedness (cu) statistic focuses, for units assigned to the treatment, on the difference in average outcomes by receipt of treatment:

$$\hat{\tau}_{\text{cu}} = \overline{Y}_{1t} - \overline{Y}_{1c}.$$

This approach would be justified if, conditional on the assignment, receipt of treatment is random. Of course, the concern is that the very fact that these units, although assigned to the same level of the treatment, receive different levels of the treatment reflects systematic differences between these units. Let us look at the interpretation of this estimand under the instrumental variables assumptions. Given the definition of the compliance types, $\hat{\tau}_{\text{cu}}$ estimates

$$\tau_{\text{cu}} = \mathbb{E}_{\text{sp}}[Y_i(1)|G_i = \text{co}] - \mathbb{E}_{\text{sp}}[Y_i(0)|G_i = \text{nc}].$$

It is fundamentally comparing different subpopulations of units, under different treatment levels. More interesting, from a perspective of understanding the differences between the units, is to estimate the average outcomes for compliers and noncompliers under the control treatment:

$$\Delta_{co,nc} = \mathbb{E}_{sp}[Y_i(0)|G_i = co] - \mathbb{E}_{sp}[Y_i(0)|G_i = nc],$$

because this compares the same potential outcomes for different subpopulations.

For the Sommer-Zeger data, we find

$$\hat{\tau}_{cu} = 0.9988 - 0.9859 = 0.0128 \qquad (\widehat{s.e.}\ 0.0024).$$

Survival rates for compliers assigned to the control treatment are substantially higher than for noncompliers assigned the active treatment, despite the fact that neither group took any active treatment.

23.10   CONCLUSION

The discussion in this chapter describes the instrumental variables approach to estimation of causal effects in randomized experiments with one-sided noncompliance, in settings where unconfoundedness of the receipt of treatment of interest is viewed as untenable. The approach exposited here relies on two key assumptions, which together replace the assumption of unconfoundedness of the receipt of treatment. The two assumptions are: unconfoundedness of the assignment to the active treatment (the instrument), rather than the receipt of treatment; and an exclusion restriction that rules out an effect of assignment on the outcome of interest for noncompliers. The first of these assumptions is implied by design in the randomized experiment setting. The second assumption relies more heavily on subject-matter knowledge, although it can be made more plausible by design measures such as double-blinding. Under those two assumptions, we can estimate the average effect of the treatment on a subset of the population, the so-called compliers, who comply with the treatment assignment irrespective of what that assignment is.

NOTES

Instrumental variables analyses have a long tradition in econometrics. The first cases of such analyses include S. Wright (1921, 1923), P. Wright (1928), Tinbergen (1930), and Haavelmo (1943). See Stock and Tregbi (2003) for a fascinating historical perspective. In these early analyses, as in most of the subsequent econometric discussions, models were typically specified in terms of linear equations. There was a clear sense, however, of what these equations meant: by assumption they describe behavioral or causal relationships between variables, not correlations, and thus they do not necessarily (although they may do so accidentally) describe conditional expectations.

Early on these models were characterized by constant treatment effects and tight parametric and distributional assumptions. More recently researchers have tried to relax these models by allowing for heterogeneity in the treatment effects and flexible functional forms. Heckman (1990) showed that conditions required for identification of the population average treatment effect in these models were very strong: essentially they required that the instruments changed the probability of receiving the treatment from zero to one so that for an identifiable subset of the population there was a randomized experiment.

For discussions on intention-to-treat effects, see Fisher, Dixon, Herson, Frankowski, Hearron, and Peace (1990) and Meier (1991).

Starting with the work by Imbens and Angrist (1994), Angrist, Imbens, and Rubin (1996), and Imbens and Rubin (1997ab), explicit connections were made between the Rubin Causal Model, or the potential outcomes perspective, and instrumental variables. Imbens and Angrist referred to the average effect for compliers as the *Local Average Treatment Effect*. Imbens and Rubin referred to it as the *Complier Average Causal Effect*. Sheiner and Rubin (1995) discuss the links to ITT effects. Other recent theoretical work in econometrics using the potential outcome framework in instrumental variables settings includes Abadie, Angrist, and Imbens (2002), Abadie (2002, 2003), and Chernozhukov and Hansen (2005). Rosenbaum (1996) and Imbens and Rosenbaum (2005) discuss randomization inference in instrumental variables settings. Athey and Stern (1998) discuss settings in which the exclusion restriction arises naturally from substantive assumptions in economics. Interesting applications in economics include Angrist and Krueger (1999) and Angrist and Lavy (1999).

Traditionally in statistics such structural equation methods, and specifically instrumental variables, were largely ignored. Noncompliance was viewed as a nuisance and largely ignored by focusing on intention-to-treat effects. More recently, this has changed. Independently of the econometric work, researchers analyzed issues in design of clinical trials with noncompliance. Zelen in a series of papers proposed designing experiments to avoid problems with subject consent by randomizing individuals to treatments before seeking consent of those assigned to the active treatment. Such designs were originally called *Randomized Consent Designs* and have sometimes been referred to as *Zelen's Design* (e.g., Zelen, 1979, 1990; Baker, 2000; Torgerson and Roland, 1998). Also related are *Randomized Encouragement Designs* where individuals are randomly assigned to receive encouragement or incentives to take part in an active treatment (Powers and Swinton, 1984; Holland, 1988). Bloom (1984) also studied the one-sided noncompliance case, allowing for heterogeneity in the causal effects. Robins (1986) analyzed models more closely related to the econometric tradition with noncompliance. Cuzick, Edwards, and Segnan (1997) independently derived the relation between the ratio of ITT effects and the average effect for compliers. Rubin (1998) studies Fisher-style p-value calculations in these settings.

The data used in this chapter were previously analyzed by Sommer and Zeger (1991) and Imbens and Rubin (1997b). They come from an experiment conducted in Indonesia in the early 1980s. For more detail on the specific issues in this evaluation, see the Sommer and Zeger paper and Sommer, Tarwotjo, Djunaedi, West, Loeden, Tilden, and Mele (1986).

Mealli and Rubin (2002ab) discuss extensions to missing data. Lui (2011) focuses on the case with binary outcome data. McNamee (2009) compares per protocol, intention-to-treat, and instrumental variables approaches.

APPENDIX

We first approximate the super-population joint sampling distribution of the two ITT estimators by a normal distribution centered around the ITT_Y and ITT_W:

$$
\begin{pmatrix} \widehat{\text{ITT}_Y} \\ \widehat{\text{ITT}_W} \end{pmatrix} \approx \mathcal{N} \left(\begin{pmatrix} \text{ITT}_Y \\ \text{ITT}_W \end{pmatrix}, \begin{pmatrix} \widehat{\mathbb{V}}(\widehat{\text{ITT}_Y}) & \widehat{\mathbb{C}}(\widehat{\text{ITT}_Y}, \widehat{\text{ITT}_W}) \\ \widehat{\mathbb{C}}(\widehat{\text{ITT}_Y}, \widehat{\text{ITT}_W}) & \widehat{\text{Var}}(\widehat{\text{ITT}_W}) \end{pmatrix} \right).
$$

We have already seen in Chapter 6 how to estimate $\mathbb{V}(\widehat{\text{ITT}_Y})$ and $\mathbb{V}(\widehat{\text{ITT}_W})$; thus the only remaining element is the covariance of $\widehat{\text{ITT}_Y}$ and $\widehat{\text{ITT}_W}$. To estimate this covariance, first note that the covariance between $\overline{Y}_1^{\text{obs}}$ and $\overline{W}_0^{\text{obs}}$ and the covariance between $\overline{Y}_0^{\text{obs}}$ and $\overline{W}_1^{\text{obs}}$ are both zero, because these averages are estimated on different subsamples. In addition, $\overline{W}_0^{\text{obs}} = 0$. Hence the covariance between $\widehat{\text{ITT}_Y} = \overline{Y}_1^{\text{obs}} - \overline{Y}_0^{\text{obs}}$ and $\widehat{\text{ITT}_W} = \overline{W}_1^{\text{obs}} - \overline{W}_0^{\text{obs}}$ is equal to the covariance between $\overline{Y}_1^{\text{obs}}$ and $\overline{W}_1^{\text{obs}}$. $\mathbb{C}\left(\overline{Y}_1^{\text{obs}}, \overline{W}_1^{\text{obs}}\right)$ is just the covariance between two sample averages:

$$
\mathbb{C}(\widehat{\text{ITT}_Y}, \widehat{\text{ITT}_W}) = \mathbb{C}\left(\overline{Y}_1^{\text{obs}}, \overline{W}_1^{\text{obs}}\right)
$$

$$
= \frac{1}{N_1 \cdot (N_1 - 1)} \sum_{i: Z_i = 1} \left(Y_i^{\text{obs}} - \overline{Y}_1^{\text{obs}}\right) \cdot \left(W_i^{\text{obs}} - \overline{W}_1^{\text{obs}}\right).
$$

Given this quantity, we can estimate the sampling variance of $\hat{\tau}^{\text{iv}}$ by substituting our estimates for ITT_Y, ITT_W, $\mathbb{V}(\widehat{\text{ITT}_Y})$, $\mathbb{V}(\widehat{\text{ITT}_W})$, and $\mathbb{C}(\widehat{\text{ITT}_Y}, \widehat{\text{ITT}_W})$ into Equation (23.5).

In the Sommer-Zeger example,

$$
\mathbb{C}(\widehat{\text{ITT}_Y}, \widehat{\text{ITT}_W}) = -0.00000017,
$$

corresponding to a correlation between $\widehat{\text{ITT}_Y}$ and $\widehat{\text{ITT}_W}$, equal to -0.0502.

## CHAPTER 24

Instrumental Variables Analysis of Randomized Experiments with Two-Sided Noncompliance

24.1 INTRODUCTION

In this chapter we extend the instrumental variables analyses discussed in Chapter 23 to allow for two-sided noncompliance in a randomized experiment. In the discussion on one-sided noncompliance, only those units assigned to the active treatment could choose whether or not to comply with their assignment. Now we allow for the possibility that some of the units assigned to the control group do in fact receive the active treatment. In terms of the notation introduced in Chapter 23, we allow the value of the potential receipt of treatment given assignment to the control group, $W_i(0)$, to be 1. This generalization implies that there are now possibly four different compliance types, defined by the pair of values of potential treatment responses, $(W_i(0), W_i(1))$, instead of two as in the one-sided compliance case. As in Chapter 23, these compliance types play a key role in our analysis.

Critical again in our analysis are assumptions about the absence of effects of assignment on the primary outcome for subgroups for which the assignment has no effect on the receipt of treatment. These are assumptions that we referred to as *exclusion restrictions* in the previous chapter. A new type of assumption in this chapter is what we refer to as *monotonicity*. This assumption rules out the presence of units who always, in this experiment, that is, under both values of the assignment, do the opposite of their assignment; such units are characterized by $W_i(z) = 1 - z$ for $z = 0, 1$, that is, $W_i(0) = 1$ and $W_i(1) = 0$. Units with such compliance behavior are sometimes referred to as *defiers*. The monotonicity assumption, which rules out the presence of these defiers, implies that $W_i(z)$ is weakly monotone in z for all units and is also referred to as the *no-defier* assumption. In many applications this assumption is a plausible one, but in some cases it can be controversial. In the previous chapter it was satisfied by construction because no one assigned to the control group could receive the active treatment. In the two-sided noncompliance setting, monotonicity is a substantive assumption that need not always be satisfied. Given monotonicity and exclusion restrictions, we can identify causal effects of the receipt of treatment for the subpopulation of compliers, as we discuss in this chapter.

This chapter is organized as follows. In the next section, Section 24.2, we discuss the data used in this chapter. These data are from a seminal study by Angrist (1990) that spawned a resurgence of interest in instrumental variables analyses in economics.

Building on work by Hearst, Newman, and Hulley (1986), Angrist (1990) is interested in estimating the causal effect of serving in the military during the Vietnam War on earnings. To address possible concerns with unobserved differences between veterans and non-veterans, he used the random assignment to draft priority status as an instrument. In Section 24.3 we discuss compliance status in the two-sided noncompliance setting. In Section 24.4 we look at the intention-to-treat effects. Next, in Section 24.5 we study the critical assumptions for instrumental variables analyses. We discuss the arguments for and against validity of the key assumptions in the Angrist application and illustrate what can be learned using the instrumental variables perspective. In Section 24.6 we take a detour and look at more traditional econometric analyses and see how they relate to our approach. Section 24.7 concludes.

24.2   THE ANGRIST DRAFT LOTTERY DATA

Angrist (1990) is concerned with the possibility that veterans and non-veterans are systematically different in unobserved ways, even after adjusting for differences in observed covariates, and that these unobserved differences may correspond to systematic differences in their earnings. For example, to serve in the military, drafted individuals need to pass medical tests and to have achieved minimum education levels. These variables are known to be associated with differences in earnings, and might imply that veterans would have had higher earnings than non-veterans, had they not served in the military. On the other hand, individuals with attractive civilian labor market prospects may have been less likely to volunteer for military service, which could imply that the civilian earnings of veterans, had they not served in the military, would have been lower than those of non-veterans. As a result of these unobserved differences, simple comparisons of earnings between veterans and non-veterans are arguably not credible estimates of causal effects of serving in the military. Adjusting for covariates that are associated with both civilian labor market prospects, as well as the decision to enroll in the military, may improve such comparisons but ultimately may not be sufficient to remove all biases. Thus, a strategy based on unconfoundedness of military service is unlikely to be satisfactory in the absence of detailed background information beyond what is available.

Angrist exploits the implementation of the draft during the Vietnam War. During this conflict all men of a certain age were required to register for the draft. However, the military did not need all men in these cohorts, and for birth cohorts 1950–1953 established a policy to determine draft priority that would make all men within a birth year cohort *a priori* equally likely to be drafted. Ultimately draft priority was assigned based on a random ordering of birth dates within birth year cohorts. Thus, for birth year 1950, a random ordering of the 365 days was constructed. Eventually, although this was not known in advance, all men born in 1950 with birth dates corresponding to draft lottery numbers less than or equal to 195 were drafted, and those with birth dates corresponding to draft lottery numbers larger than 195 were not. For the birth cohorts from 1951 and 1952, these thresholds were 125, and 95, respectively. (No one born in 1953 was drafted although all men in this birth year were required to register for the draft and draft priority numbers were assigned.)

Table 24.1. *Summary Statistics for the Angrist Draft Lottery Data*

	Non-Veterans ($N_c = 6{,}675$)				Veterans ($N_t = 2{,}030$)			
	Min	Max	Mean	(S.D.)	Min	Max	Mean	(S.D.)
Draft eligible	0	1	0.24	(0.43)	0	1	0.40	(0.49)
Yearly earnings (in \$1,000's)	0	62.8	11.8	(11.5)	0	50.7	11.7	(11.8)
Earnings positive	0	1	0.88	(0.32)	0	1	0.91	(0.29)
Year of birth	50	52	51.1	(0.8)	50	52	50.9	(0.8)

Let Z_i be a binary indicator for being draft eligible, meaning that the individual had a draft lottery number less than or equal to the threshold for their birth year. Angrist uses this binary indicator as an instrument for serving in the military (described subsequently as "veteran status"). Observed veteran status for individual i is denoted by W_i^{obs}. We focus on civilian earnings in thousands of dollars in 1978 as the outcome of interest, with the realized and observed value for the ith person in our sample denoted by Y_i^{obs}.

Table 24.1 presents some summary statistics for the three "birth-year" cohorts (1950–1952) used in our analyses. We see that veterans have approximately the same average earnings as non-veterans (11.8 for non-veterans, and 11.7 for veterans, in thousands of dollars per year) but are slightly more likely to be employed (91% versus 88%). However, the concern is that these simple comparisons of veterans and non-veterans, yielding a point estimate of -0.2 ($\widehat{s.e.}$ 0.2), for annual earnings, and 0.03 ($\widehat{s.e.}$ 0.01) for employment, are not credible estimates of causal effects of veteran status because of the anticipated systematic observable and unobservable differences between veterans and non-veterans just discussed.

24.3   COMPLIANCE STATUS

As in Chapter 23, we postulate the existence of a pair of compliance potential responses to assignment, $W_i(z)$, for $z = 0, 1$. The first, $W_i(0)$, describes for unit i the treatment response to being assigned to the control group. If unit i would receive the treatment (serving in the military in the draft-lottery application) when assigned to the control group, then $W_i(0) = 1$, otherwise $W_i(0) = 0$, and similarly for $W_i(1)$. Compliance status refers to a unit's response to the assignment, for both values of the assignment whether that status is observed or unobserved. Formally, it is a function of the pair of potential responses $(W_i(0), W_i(1))$. Because both $W_i(0)$ and $W_i(1)$ are binary indicators, there are four possible values for the pair of potential responses to treatment assignment. Let us consider the four groups in turn. We continue to refer to those who always comply with their assignment in the context of this study, units with $W_i(z) = z$ for $z = 0, 1$, and thus $(W_i(0) = 0, W_i(1) = 1)$, as *compliers*. All others units are *noncompliers*, but they can be of different noncomplier types.

We distinguish three distinct types of noncompliers. Those who never (in the context of these drafts) take the treatment, irrespective of their assignment

$(W_i(0) = 0, W_i(1) = 0)$, will be referred to as *nevertakers*. Those who would, in this study, always take the treatment, irrespective of their assignment $(W_i(0) = 1, W_i(1) = 1)$, will be referred to as *alwaystakers*. Finally, those who, in the context of this study, irrespective of the value of their assignment, would do the opposite of their assignment, that is, units with $(W_i(0) = 1, W_i(1) = 0)$, will be referred to as *defiers*. We denote the compliance type by G_i, taking values in {nt, at, co, df}:

$$
G_i = g(W_i(0), W_i(1)) = \begin{cases} \text{nt} & \text{if } W_i(0) = 0, W_i(1) = 0, \\ \text{co} & \text{if } W_i(0) = 0, W_i(1) = 1, \\ \text{df} & \text{if } W_i(0) = 1, W_i(1) = 0, \\ \text{at} & \text{if } W_i(0) = 1, W_i(1) = 1. \end{cases}
$$

Here the function $g(\,\cdot\,)$ emphasizes the fact that compliance status is a deterministic function of the two potential outcomes, $W_i(0)$ and $W_i(1)$. Let $\pi_g = \Pr(G_i = g)$, for $g \in \{\text{nt, at, co, df}\}$ denote the shares of the four compliance types in the super-population.

The compliance type of a unit is not directly observable. We observe the realized treatment status

$$
W_i^{\text{obs}} = W_i(Z_i) = \begin{cases} W_i(0) & \text{if } Z_i = 0, \\ W_i(1) & \text{if } Z_i = 1, \end{cases}
$$

but not the value of $W_i^{\text{mis}} = W_i(1 - Z_i)$. In this regard, the two-sided noncompliance case analyzed in this chapter is more complicated than the one-sided case. In the one-sided noncompliance case, we could infer the compliance type for at least some units; specifically, we could infer for all units with $Z_i = 1$ what compliance type they were. For units with $(Z_i = 1, W_i^{\text{obs}} = 0)$ we could infer that they must be noncompliers with $(W_i(0), W_i(1)) = (0, 0)$, and for units with $(Z_i = 1, W_i^{\text{obs}} = 1)$ we could infer that they must be compliers with $(W_i(0), W_i(1)) = (0, 1)$. However, for units with $Z_i = 0$, we could not infer what type they were. Here we cannot tell the compliance status of any particular unit without additional assumptions. For unit i we observe Z_i and $W_i^{\text{obs}} = W_i(Z_i)$, but we do not know what that unit would have done had it received the alternative assignment, $1 - Z_i$. Because noncompliance is two-sided, for all values of Z_i, the unobserved $W_i^{\text{mis}} = W_i(1 - Z_i)$ can take either the value 0 or 1.

As a result, there will always be two compliance types that are consistent with the observed behavior of a specific unit. For example, if we observe unit i assigned to the control group and taking the treatment, we can infer that unit i is *not* a complier or nevertaker, but we cannot infer whether unit i is a defier or an alwaystaker. For a unit assigned to the control group and not taking the treatment, we can infer that such a unit is not an alwaystaker or a defier, but the observed behavior is consistent with that unit being a complier or a nevertaker. If unit i is assigned to the treatment group and takes the treatment, we can only infer that this unit is an alwaystaker or a complier. Finally if unit i is assigned to the treatment group and does not receive the treatment, we can only infer that unit i is a nevertaker or a defier. Tables 24.2 and 24.3 summarize this discussion by describing the compliance status and the extent to which we can learn about compliance status from the data on assignment and receipt of treatment.

Table 24.2. *Compliance Status in the Case with Two-Sided Noncompliance, for the Angrist Draft Lottery Data*

		$W_i(1)$	
		0	1
$W_i(0)$	0	nt	co
	1	df	at

Table 24.3. *Possible Compliance Status by Observed Assignment and Observed Receipt of Treatment in the Case with Two-Sided Noncompliance, for the Angrist Draft Lottery Data*

		Z_i	
		0	1
W_i^{obs}	0	nt/co	nt/df
	1	at/df	at/co

We use the compliance status as a *latent pre-treatment variable* or *latent characteristic*. It is a pre-treatment variable or characteristic because it is not affected by either the assigned treatment or the received treatment. It is latent because it is not fully observed.

24.4 INTENTION-TO-TREAT EFFECTS

Let us briefly look at the Intention-To-Treat (ITT) effects in this setting. This analysis is largely unchanged from that in the previous chapter on one-sided noncompliance.

First consider the ITT effect on the treatment received. The unit-level effect of treatment assigned on treatment received is equal to 1 for compliers, 0 for both nevertakers and alwaystakers, and -1 for defiers, so that the super-population average intention-to-treat effect on the receipt of treatment is

$$\text{ITT}_W = \mathbb{E}_{sp}\left[W_i(1) - W_i(0)\right] = \pi_{co} - \pi_{df},$$

the difference in population fractions of compliers and defiers. Here the expectations are taken over the distribution induced by random sampling from the super-population. The ITT effect on the primary outcome is, as in the previous chapter,

$$\text{ITT}_Y = \mathbb{E}_{sp}\left[Y_i(1, W_i(1)) - Y_i(0, W_i(0))\right].$$

As before, we assume that assignment is super-population unconfounded and completely randomized.

Assumption 24.1 (Super-Population Random Assignment)

$$Z_i \perp\!\!\!\perp \left(W_i(0), W_i(1), Y_i(0,0), Y_i(0,1), Y_i(1,0), Y_i(1,1) \right).$$

We can relax this assumption by requiring it to hold only within homogeneous subpopulations defined by fully observed pre-treatment variables, thus combining an analysis based on unconfoundedness with an instrumental variables analysis. However, in the draft lottery example, the physical randomization of the draft lottery ensures that Assumption 24.1 holds by design. In other applications, this assumption may be substantive, rather than satisfied by design, and as a result more controversial. This assumption validates two intention-to-treat analyses, one with the receipt of treatment as the outcome, and one with the primary outcome, for example, earnings in the Angrist example.

Given a random sample and random assignment, we can estimate the average causal effect of assignment on W_i in the super-population as

$$\widehat{\mathrm{ITT}_{\mathrm{W}}} = \overline{W}_1^{\mathrm{obs}} - \overline{W}_0^{\mathrm{obs}},$$

with the (Neyman) sampling variance estimated as

$$\widehat{\mathbb{V}}(\widehat{\mathrm{ITT}_{\mathrm{W}}}) = \frac{s_{W,0}^2}{N_0} + \frac{s_{W,1}^2}{N_1},$$

Here, for $z = 0, 1$, $N_z = \sum_{i=1}^{N} \mathbf{1}_{Z_i=z}$, $\overline{W}_z^{\mathrm{obs}} = \sum_{i:Z_i=z} W_i^{\mathrm{obs}}/N_z$, and $S_{W,z}^2 = \sum_{i:W_i^{\mathrm{obs}}=z} (W_i^{\mathrm{obs}} - \overline{W}_z^{\mathrm{obs}})^2/(N_z - 1) = \overline{W}_z(1 - \overline{W}_z)/(N_z - 1)$.

Let us illustrate these ideas using the Angrist draft lottery data. Of the $N = 8{,}705$ men in our sample, $N_0 = 6{,}293$ had a draft lottery number exceeding the threshold (and so were not draft eligible), and $N_1 = 2{,}412$ had a draft lottery number less than or equal to the threshold for their birth year. Thus we find:

$$\widehat{\mathrm{ITT}_{\mathrm{W}}} = \overline{W}_1^{\mathrm{obs}} - \overline{W}_0^{\mathrm{obs}} = 0.3387 - 0.1928 = 0.1460,$$

with the sampling variance for the super-population average treatment effect estimated as

$$\widehat{\mathbb{V}}(\widehat{\mathrm{ITT}_{\mathrm{W}}}) = \frac{s_{W,0}^2}{N_0} + \frac{s_{W,1}^2}{N_1} = 0.0108^2,$$

leading to a large-sample 95% confidence interval for $\mathrm{ITT}_{\mathrm{W}}$ equal to

$$\mathrm{CI}^{0.95}(\mathrm{ITT}_{\mathrm{W}}) = (0.1247, 0.1672).$$

Thus, unsurprisingly, we find that being draft eligible (having a low draft lottery number) leads to a substantially, and at conventional levels statistically significant, higher probability of subsequently serving in the military.

Next, let us consider estimation of the super-population ITT effect on the primary outcome. As in the case for the ITT effect on the treatment received, this analysis is identical to that in Chapter 23. We estimate ITT_Y as the difference in average outcomes by assignment status,

$$\widehat{ITT_Y} = \overline{Y}_1^{obs} - \overline{Y}_0^{obs}.$$

The sampling variance for this estimator of the ITT effect is, using Neyman's approach, estimated as

$$\widehat{\mathbb{V}}(\widehat{ITT_Y}) = \frac{s_{Y,1}^2}{N_1} + \frac{s_{Y,0}^2}{N_0}.$$

Let us return to the Angrist draft lottery data. Here we find

$$\widehat{ITT_Y} = \overline{Y}_1^{obs} - \overline{Y}_0^{obs} = 11.634 - 11.847 = -0.2129,$$

a drop in annual earnings of \$212.90, and,

$$\widehat{\mathbb{V}}(\widehat{ITT_Y}) = \frac{s_{Y,1}^2}{N_1} + \frac{s_{Y,0}^2}{N_0} = 0.1980^2,$$

and thus we have the 95% large-sample confidence interval

$$CI^{0.95}(ITT_Y^{earn}) = (-0.6010, 0.1752).$$

We may also wish to look at the effect of draft eligibility on employment (measured as having positive annual earnings). Here we find a point estimate of -0.005, with a 95% large-sample confidence interval equal to

$$CI^{0.95}(ITT_Y^{emp}) = (-0.018, 0.011).$$

In a traditional ITT analysis, we are essentially done. One might not even estimate the ITT effect on the treatment received, because this estimate has little relevance for the causal effects of interest, those on the outcome. However, this ITT analysis does not really answer the question of interest: What is the causal effect on earnings of actually serving in the military? Instead, it informs us about the effect of changing the draft priority on earnings. If, in a future conflict, there were again to be a military draft, it would likely be implemented in a very different way. The effect of the lottery number on earnings is therefore of limited interest. Of considerably more interest is the effect of actually serving on future earnings, as this may be of use in predicting the effect, or cost, of military service in subsequent drafts.

24.5 INSTRUMENTAL VARIABLES

In this section we discuss the main results of this chapter, which extend the analyses from the previous chapter to allow for two-sided noncompliance. We consider the assumptions underlying instrumental variables and use those to draw additional inferences regarding the relation between the outcome of interest and the treatment of primary interest beyond what can be learned from the ITT analyses. Much of this analysis is about extending the ITT analyses by obtaining separate ITT effects by compliance status:

$$\text{ITT}_{W,g} = \mathbb{E}_{\text{SP}}\left[Y_i(1, W_i(1)) - Y_i(0, W_i(0))|G_i = g\right],$$

for $g \in \{\text{nt}, \text{at}, \text{co}, \text{df}\}$. The challenge is that this decomposition is not immediately feasible because compliance status is only partly observed. However, if we were to observe compliance status directly, one could simply estimate the ITT effects separately by compliance status. In that case, the ITT effects for nevertakers and alwaystakers would obviously not be informative about the causal effect of the receipt of treatment, because there is no variation in the receipt of treatment for these two subgroups of units. In contrast, for defiers and compliers there is variation in the receipt of treatment. In fact, for compliers and defiers, receipt of treatment and assignment to treatment are perfectly (positively for compliers and negatively for defiers) correlated, and the strategy will be to *attribute* the causal effect of the assignment to treatment to the effect of the receipt of treatment, W_i^{obs}.

24.5.1 Exclusion Restrictions

The first set of assumptions we consider are exclusion restrictions. As in the previous chapter, we consider multiple versions of these restrictions. All versions capture the notion that there is no effect of the assignment on the outcome, in the absence of an effect of the assignment of treatment on the treatment received, the treatment of primary interest. The first set of exclusion restrictions rules out dependence of the potential outcomes on the assignment:

Assumption 24.2 (Exclusion Restriction for Nevertakers) *For all units i with $G_i = \text{nt}$,*

$$Y_i(0,0) = Y_i(1,0).$$

This assumption requires that changing z for nevertakers does not change the value of the realized outcome.

We can make a similar assumption for alwaystakers:

Assumption 24.3 (Exclusion Restriction for Alwaystakers) *For all units i with $G_i = \text{at}$,*

$$Y_i(0,1) = Y_i(1,1).$$

We also state exclusion restrictions for compliers and defiers:

Assumption 24.4 (Exclusion Restriction for Compliers) *For units with $G_i = \text{co}$,*

$$Y_i(0,w) = Y_i(1,w),$$

for both levels of the treatment w.

Assumption 24.5 (Exclusion Restriction for Defiers) *For units with $G_i = \text{df}$,*

$$Y_i(0,w) = Y_i(1,w),$$

for both levels of the treatment w.

A key feature of these exclusion restrictions is that they are, at their core, substantive assumptions, requiring judgment regarding subject-matter knowledge. It is rarely satisfied by design outside of settings with double-blinding. In settings where units are individuals who are aware of their assignment and treatment, one needs to consider the incentives and restrictions faced by units assigned and not assigned to receive the treatment, and argue on the basis of such considerations whether each of the exclusion restrictions is plausible. In many cases they need not be satisfied for all groups, but in some classes of applications, they may be useful approximations to the underlying process. At some level this is not so different from the type of assumptions we have considered before. In particular, the stable-unit-treatment-value assumption required that there was no interference between units. This required substantive judgments about the possibility of interference: applying fertilizer in area A may well affect crops in area B if there is some possibility of leaching, but this is less plausible if the areas are sufficiently separated. The differences between the exclusion restrictions and SUTVA is a matter of degree: often the subject-matter knowledge required to assess the plausibility of exclusion restrictions is more subtle than that required to evaluate SUTVA, especially for some subgroups such as compliers.

Let us consider the exclusion restriction for alwaystakers and nevertakers in the draft lottery application. Consider first the subpopulation of nevertakers. These are men who would not serve in the military, irrespective of whether they had a high or a low lottery number. One can think of different types of men in this subpopulation of nevertakers. Some may have had medical exemptions for the draft. For such men it would appear reasonable that the lottery number had no effect on their subsequent lives. Especially if these men already knew, prior to the allocation of their draft lottery number, that they would not be required to serve in the military, there is no reason to expect that any decisions these men made would be affected by the lottery number they were assigned. On the other hand, there may also be individuals whose educational or professional career choices allowed them exemptions from military service. For some of these individuals, these choices would have been made irrespective of the value of the draft lottery number assigned to them. Again, for such individuals the exclusion restriction appears plausible. For other individuals, however, it may be the case that a low lottery number allowed them to change their plans so that they would not have to serve in the military. For example, men intent on avoiding military service may have decided to enter graduate school or to move to Canada to avoid the draft. However, these men would need to do so only if they were assigned a low lottery number, because with a high lottery number they would not get drafted anyway. For such men, even though the lottery number did not affect their veteran status, it could have affected their outcomes, and thus the exclusion restriction could be violated. This example illustrates that in many cases there are reasons to doubt the exclusion restriction, and an assessment as to whether it provides a sufficiently accurate description of the underlying processes is important for the credibility of any subsequent analyses based on the assumption.

For compliers, the exclusion restriction is again one of attribution. It implies that the causal effect of assignment to the treatment for these units can be attributed to the causal effect of the receipt of treatment. For defiers, the substantive content of the exclusion restriction is the same as for the compliers. However, in practice it is less important because we often are willing to make the monotonicity (no-defier) assumption that implies that the proportion of defiers in the population is zero.

We can weaken the exclusion restriction for alwaystakers and nevertakers by requiring the equality to hold in distribution in the super-population:

Assumption 24.6 (Stochastic Exclusion Restriction for Nevertakers)

$$Z_i \perp\!\!\!\perp Y_i(Z_i, W_i(Z_i)) \mid G_i = \text{nt}.$$

Assumption 24.7 (Stochastic Exclusion Restriction for Alwaystakers)

$$Z_i \perp\!\!\!\perp Y_i(Z_i, W_i(Z_i)) \mid G_i = \text{at}.$$

These versions of the assumption require that there is no difference between the distribution of outcomes for nevertakers or alwaystakers with given assignment to control or treatment group. It weakens the non-stochastic versions of the assumption; rather than requiring the effect to be identically zero for all units, they only require the difference to be zero in a distributional sense, similar to the difference between the Fisher and Neyman null hypotheses of no effect of the treatment in a randomized experiment. An important advantage of the stochastic versions of the exclusion restrictions are that covariates are easily incorporated: in that case we need the independence in Assumptions 24.6 and 24.7 to hold only conditional on covariates.

24.5.2 The Monotonicity Assumption

The next assumption is special to the two-sided noncompliance setting. We rule out the presence of defiers or, in other words, restrict the sign of the effect of the assignment on the treatment:

Assumption 24.8 (Monotonicity/No Defiers)
There are no defiers: $W_i(1) \geq W_i(0)$.

In the one-sided noncompliance case analyzed in Chapter 23, this assumption was automatically satisfied because $W_i(0) = 0$ for all units, ruling out the presence of both defiers and alwaystakers. In that case monotonicity was essentially verifiable. Here it is a substantive assumption, that is not directly testable (beyond the implication that ITT_W is non-negative: if we find that our estimate of ITT_W is negative and statistically significant at conventional levels, we may want to reconsider the entire model!). Given monotonicity, Table 24.3 simplifies to Table 24.4. Now we can infer, at least for units with $W_i^{\text{obs}} \neq Z_i$, which compliance type they are: for units with $Z_i = 0$, $W_i^{\text{obs}} = 1$, we observe $W_i(0) = 1$, and we can, because of the monotonicity assumption, infer the value of $W_i(1) = 1$, so such units are alwaystakers. Similarly, for units with $Z_i = 1$, $W_i^{\text{obs}} = 0$ we observe $W_i(1) = 0$, and thus can, because of monotonicity, infer the value of $W_i(0) = 0$, and therefore such units are nevertakers. For units whose realized treatment is identical to the assigned treatment, we cannot infer what type they are: if $W_i^{\text{obs}} = Z_i = 0$, unit i could be a nevertaker or complier, and observing $W_i^{\text{obs}} = Z_i = 1$ is consistent with unit i being an alwaystaker or complier.

Table 24.4. *Compliance Status by Observed Assignment and Observed Receipt of Treatment with the Monotonicity Assumption in the Case with Two-Sided Noncompliance, for the Angrist Draft Lottery Data*

		Z_i	
		0	1
W_i^{obs}	0	nt/co	nt
	1	at	at/co

In the draft lottery example, monotonicity appears to be a reasonable assumption. Having a low draft lottery number imposes restrictions on individuals' behaviors: it requires individuals to prepare, if fit for military service, to serve in the military, where having a high lottery number would not require them to do so. The monotonicity assumption asserts that, in response to these restrictions, individuals are more likely to serve in the military, and that no one responds to this restriction by serving only if they are not required to do so. It is of course possible that there are some individuals who would be willing to volunteer if they are not drafted but would resist the draft if assigned a low lottery number. It seems likely that, in actual fact, this is a small fraction of the population, and we will ignore this possibility here, and so accept monotonicity. In Section 24.5.5 we return to a discussion of the implications of violations of this assumption. Similarly, in a randomized experiment, it is often plausible that there are no individuals who would take the treatment if assigned to the control group and not take the treatment if assigned to the treatment. It seems reasonable to view the assignment to the treatment as increasing the incentives for the individual to take the treatment. These incentives need not be strong enough to induce everybody to take the treatment, but in many situations (e.g., drug trials) these incentives would rarely be perverse in the sense that individuals would do the opposite of their assignment. In many applications the instrument has this interpretation of increasing the incentives to participate in or to be exposed to a treatment, and in such cases the monotonicity assumption is often plausible, but this conclusion is not automatic.

Let us return to the ITT effect on the treatment received and investigate the implications of monotonicity for this ITT effect. The effect of the assignment on the receipt of treatment by compliance status, in the super-population, can be written as

$$\text{ITT}_W = \mathbb{E}_{sp}\left[W_i(1) - W_i(0)\right]$$

$$= \sum_{g \in \{co,nt,at,df\}} \mathbb{E}_{sp}\left[W_i(1) - W_i(0)| G_i = g\right] \cdot \Pr_{sp}(G_i = g)$$

$$= \mathbb{E}_{sp}\left[W_i(1) - W_i(0)| G_i = co\right] \cdot \Pr_{sp}(G_i = co)$$

$$+ \mathbb{E}_{sp}\left[W_i(1) - W_i(0)| G_i = nt\right] \cdot \Pr_{sp}(G_i = nt)$$

$$+ \mathbb{E}_{sp}\left[W_i(1) - W_i(0)| G_i = df\right] \cdot \Pr_{sp}(G_i = df)$$

$$+ \mathbb{E}_{sp} [W_i(1) - W_i(0) | G_i = \text{at}] \cdot \text{Pr}_{sp} (G_i = \text{at})$$

$$= \text{Pr}(G_i = \text{co}) - \text{Pr}(G_i = \text{df}) = \pi_{co} - \pi_{df},$$

the difference in proportions of compliers and defiers. By the monotonicity or no-defiers assumption, this is equal to the proportion of compliers π_{co}. Thus, under two-sided noncompliance, as long as there are no defiers, the ITT effect on the treatment received still equals the proportion of compliers, just as we found in the one-sided noncompliance case.

24.5.3 Local Average Treatment Effects under Two-Sided Noncompliance

Now consider the intention-to-treat effect, the average effect of assignment on the outcome. Again we decompose this super-population ITT effect into four local effects by the four compliance types:

$$\text{ITT}_Y = \mathbb{E}_{sp}[Y(1, D(1)) - Y(0, D(0))]$$

$$= \sum_{g \in \{co, nt, at, df\}} \mathbb{E}_{sp} [Y_i(1, W_i(1)) - Y_i(0, W_i(0)) | G_i = g] \cdot \text{Pr}_{sp}(G_i = g)$$

$$= \mathbb{E}_{sp} [Y_i(1, W_i(1)) - Y_i(0, W_i(0)) | G_i = \text{co}] \cdot \text{Pr}_{sp}(G_i = \text{co})$$

$$+ \mathbb{E}_{sp} [Y_i(1, W_i(1)) - Y_i(0, W_i(0)) | G_i = \text{nt}] \cdot \text{Pr}_{sp}(G_i = \text{nt})$$

$$+ \mathbb{E}_{sp} [Y_i(1, W_i(1)) - Y_i(0, W_i(0)) | G_i = \text{at}] \cdot \text{Pr}_{sp}(G_i = \text{at})$$

$$+ \mathbb{E}_{sp} [Y_i(1, W_i(1)) - Y_i(0, W_i(0)) | G_i = \text{df}] \cdot \text{Pr}_{sp}(G_i = \text{df}).$$

Under either the deterministic (Assumptions 24.2 and 24.3) or the stochastic (Assumptions 24.6 and 24.7) version of the exclusion restrictions, the super-population average ITT effect for nevertakers and alwaystaker is zero, and hence the ITT effect on the primary outcome is equal to

$$\text{ITT}_Y = \mathbb{E}_{sp} [Y_i(1, 1) - Y_i(0, 0) | G_i = \text{co}] \cdot \pi_{co}$$

$$- \mathbb{E}_{sp} [Y_i(0, 1) - Y_i(1, 0) | G_i = \text{df}] \cdot \pi_{df}.$$

Maintaining the monotonicity assumption implies the proportion of defiers is zero, and so this expression further simplifies to

$$\text{ITT}_Y = \mathbb{E}_{sp} [Y_i(1, 1) - Y_i(0, 0) | G_i = \text{co}] \cdot \pi_{co},$$

or, dropping the Z argument in the potential outcomes because under the exclusion restriction it is redundant,

$$\text{ITT}_Y = \mathbb{E}_{sp} [Y_i(1) - Y_i(0) | G_i = \text{co}] \cdot \pi_{co}.$$

In other words, under the exclusion restrictions and the monotonicity assumption, the ITT effect on the primary outcome can be attributed entirely to the compliers. The non-compliers either have a zero effect (this holds for nevertakers and alwaystakers by the exclusion restrictions), or they are absent from the population (this holds for defiers by the monotonicity assumption).

Now consider the ratio of average effects of assignment:

Theorem 24.1 (Local Average Treatment Effect)
Suppose that Assumptions 24.1–24.3 (or 24.1, 24.6, 24.7) and 24.8 hold. Then

$$\tau_{\text{late}} = \frac{\text{ITT}_Y}{\text{ITT}_W} = \mathbb{E}_{\text{SP}}\left[Y_i(1) - Y_i(0) \middle| G_i = \text{co} \right].$$

This local average treatment effect is also referred to as the *complier average causal effect*.

Note that by assuming monotonicity, we extend the main result from the one-sided noncompliance case.

Let us return to the draft lottery application. Previously we estimated the two ITT effects:

$$\widehat{\text{ITT}_W} = 0.1460 \;\; (\widehat{\text{s. e.}}\; 0.0108), \quad \text{and} \quad \widehat{\text{ITT}_Y} = -0.21 \;\; (\widehat{\text{s. e.}}\; 0.20).$$

The analysis in this section implies that, under the stated assumptions, the ratio of the two estimated intention-to-treat effects can be interpreted as a simple method-of-moments estimator of the average effect of serving in the military for compliers:

$$\hat{\tau}^{\text{iv}} = \frac{\widehat{\text{ITT}_Y}}{\widehat{\text{ITT}_W}} = -\frac{0.21}{0.1460} = -1.46 \quad (\widehat{\text{s. e.}}\; 1.36),$$

with the estimated standard error based on the same type of calculation as in the previous chapter and the appendix thereof.

24.5.4 Inspecting Outcome Distributions for Compliers and Noncompliers

We cannot estimate the effect of the treatment for the subpopulations of alwaystakers or nevertakers, because each group appears in only one of the two treatment arms. Nevertheless, we can compare their potential outcome distributions given the treatment they are exposed to and compare them to the potential outcome distributions given the same treatment for compliers. The latter relies on the insight that the data are not just informative about the average of $Y_i(1) - Y_i(0)$ for compliers, they are also informative about the entire potential outcome distributions for compliers. This result follows from the mixture structure of the distribution of observed data. Comparing, say, the distribution of $Y_i(0)$ for nevertakers and compliers is useful to assess the plausibility of generalizing the local average treatment effect for compliers to other subpopulations, something about which these data are not directly informative.

Consider the distribution of observed outcomes for units assigned to the control group, who receive the control treatment. By the definition of the compliance types, this subpopulation consists of compliers and nevertakers, with shares proportional to their population shares. Thus, the distribution of the observed outcome in this subpopulation has a mixture structure

$$f(Y_i^{\text{obs}}|W_i^{\text{obs}} = 0, Z_i = 0) = \frac{\pi_{\text{nt}}}{\pi_{\text{nt}} + \pi_{\text{co}}} \cdot f(Y_i(0)|G_i = \text{nt}) + \frac{\pi_{\text{co}}}{\pi_{\text{nt}} + \pi_{\text{co}}} \cdot f(Y_i(0)|G_i = \text{co}).$$

Note that these distributions are induced by the random sampling from the super-population. For this result we use the fact that if $G_i = \text{nt}$, then $Y_i^{\text{obs}} = Y_i(0)$, and if $G_i = \text{co}$ and $Z_i = 0$, then $Y_i^{\text{obs}} = Y_i(0)$. Moreover, units with $W_i^{\text{obs}} = 0$ and $Z_i = 1$ must be nevertakers, and thus the distribution of observed outcomes in this subpopulation estimates

$$f(Y_i^{\text{obs}}|W_i^{\text{obs}} = 0, Z_i = 1) = f(Y_i(0)|G_i = \text{nt}, Z_i = 1),$$

where, by random assignment of the instrument Z_i we can drop the conditioning on Z_i, and this distribution is therefore equal to $f(Y_i(0)|G_i = \text{nt})$. We can disentangle these mixtures to obtain the distribution of $Y_i(0)$ for compliers:

$$f(Y_i(0)|G_i = \text{co}) = \frac{\pi_{\text{nt}} + \pi_{\text{co}}}{\pi_{\text{co}}} \cdot f(Y_i^{\text{obs}}|W_i^{\text{obs}} = 0, Z_i = 0)$$

$$- \frac{\pi_{\text{nt}}}{\pi_{\text{co}}} \cdot f(Y_i^{\text{obs}}|W_i^{\text{obs}} = 0, Z_i = 1).$$

By a similar argument we can obtain the distribution of $Y_i(1)$ for compliers:

$$f(Y_i(1)|G_i = \text{co}) = \frac{\pi_{\text{at}} + \pi_{\text{co}}}{\pi_{\text{co}}} \cdot f(Y_i^{\text{obs}}|W_i^{\text{obs}} = 1, Z_i = 1) - \frac{\pi_{\text{at}}}{\pi_{\text{co}}}$$

$$\cdot f(Y_i^{\text{obs}}|W_i^{\text{obs}} = 1, Z_i = 0).$$

Thus, the data are indirectly informative about four distributions, $f(Y_i(0)|G_i = \text{co})$, $f(Y_i(1)|G_i = \text{co}), f(Y_i(0)|G_i = \text{nt})$, and $f(Y_i(1)|G_i = \text{at})$.

Estimating the average annual earnings for compliers with and without military service in this manner, using method-of-moments estimators, leads to

$$\widehat{\mathbb{E}}[Y_i(0)|G_i = \text{co}] = 13.22, \qquad \widehat{\mathbb{E}}[Y_i(1)|G_i = \text{co}] = 11.77.$$

For nevertakers and alwaystakers we estimate

$$\widehat{\mathbb{E}}[Y_i(0)|G_i = \text{nt}] = 11.60, \qquad \widehat{\mathbb{E}}[Y_i(1)|G_i = \text{at}] = 11.65.$$

Thus, earnings for compliers who do not serve appear to be substantially higher than earnings for nevertakers, but compliers who serve in the military appear to have earnings comparable to those of alwaystakers.

24.5.5   Relaxing the Monotonicity Condition

Suppose we do not assume monotonicity. In that case the ITT effect of assignment on treatment received is the difference in population proportions of compliers and defiers:

$$\text{ITT}_{\text{W}} = \pi_{\text{co}} - \pi_{\text{df}}.$$

The ITT effect on the primary outcome is

$$\text{ITT}_Y = \mathbb{E}_{\text{sp}} \left[Y_i(1) - Y_i(0) \middle| G_i = \text{co} \right] \cdot \pi_{\text{co}} - \mathbb{E}_{\text{sp}} \left[Y_i(1) - Y_i(0) \middle| G_i = \text{df} \right] \cdot \pi_{\text{df}}.$$

Thus, the ratio of average effects of the assignment on outcome and treatment is equal to

$$\mathbb{E}_{\text{sp}}[Y_i(1) - Y_i(0)|G_i = \text{co}] \cdot \frac{\pi_{\text{co}}}{\pi_{\text{co}} - \pi_{\text{df}}} - \mathbb{E}_{\text{sp}}[Y_i(1) - Y_i(0)|G_i = \text{df}] \cdot \frac{\pi_{\text{df}}}{\pi_{\text{co}} - \pi_{\text{df}}}.$$

Without the monotonicity assumption, the ratio is equal to a weighted average of the ITT effects for compliers and defiers. Although the weights add up to one, the weight on the average effect of the treatment for defiers is always negative, which implies that the weighted average can be outside the range spanned by the average effects for compliers and defiers. As a result, modest violations of the monotonicity assumption are therefore not critical to the interpretation of instrumental variables estimates, but in settings with substantial heterogeneity of causal effects, substantial violations of the monotonicity assumption may lead to instrumental variables estimates that are not representative of causal effects of the treatment of primary interest.

24.6   TRADITIONAL ECONOMETRIC METHODS FOR INSTRUMENTAL VARIABLES

As in Chapter 23, we will compare the methods developed so far to the traditional equation-based approach originally developed in the econometrics literature. Again, the goal is primarily to link the two approaches and illustrate the benefits of the framework presented in this chapter. It will be seen that the two approaches lead to the same estimands and estimators in this simple case without covariates, although they get there in different ways, with the traditional approach appearing to rely on restrictive and unnecessary linearity assumptions. We first go through the mechanics of the traditional econometrics approach and then discuss the traditional formulation of the critical assumptions.

Traditional econometric analyses start with a linear relation between the outcome and the primary treatment. Here we derive that relation in terms of population parameters. Let $\tau_{\text{late}} = \mathbb{E}_{\text{SP}}[Y_i(1) - Y_i(0)|G_i = \text{co}]$ be the average treatment effect for compliers. Also define

$$\alpha = \pi_{\text{nt}} \cdot \mathbb{E}_{\text{sp}}[Y_i(0)|G_i = \text{nt}] + \pi_{\text{co}} \cdot \mathbb{E}_{\text{sp}}[Y_i(0)|G_i = \text{co}]$$
$$+ \pi_{\text{at}} \cdot \mathbb{E}_{\text{sp}}[Y_i(1)|G_i = \text{at}] - \pi_{\text{at}} \cdot \tau.$$

Finally, define the residual

$$\varepsilon_i = Y_i(0) - \alpha + W_i^{\text{obs}} \cdot (Y_i(1) - Y_i(0) - \tau_{\text{late}}).$$

Now we can write the observed outcome as a function of the residual and the treatment received:

$$Y_i^{\text{obs}} = \alpha + W_i^{\text{obs}} \cdot \tau_{\text{late}} + \varepsilon_i. \tag{24.1}$$

This is the key equation, and in fact the starting point, of traditional econometric analyses. Equation (24.1) is viewed as describing a causal or *structural* relationship between the treatment W_i^{obs} and the outcome Y_i^{obs}. Typically τ_{late} is interpreted as the (constant across units) causal effect of the receipt of treatment on the outcome. However, this relationship cannot be estimated by standard regression methods. The problem is that the residual ε_i is potentially correlated with the regressor W_i^{obs}. Units with large unobserved values of the residual may be more or less likely to receive the treatment. Therefore, least squares methods will not work. The critical assumption in the traditional econometric approach is that

$$\mathbb{E}_{\text{sp}}[\varepsilon_i | Z_i = z] \quad \text{does not depend on } z.$$

We will first show that, using the potential outcomes framework, by construction, the residual is uncorrelated with the instrument. Consider the expectation of the residual given $Z_i = z$, first given $Z_i = 0$. We decompose it out by compliance status, taking into account the absence of defiers:

$$
\begin{aligned}
\mathbb{E}_{\text{sp}}[\varepsilon_i | Z_i = 0] &= \mathbb{E}_{\text{sp}}\left[Y_i^{\text{obs}} - \alpha - W_i^{\text{obs}} \cdot \tau^{\text{iv}} \,\middle|\, Z_i = z \right] \\
&= \pi_{\text{nt}} \cdot \mathbb{E}_{\text{sp}}[Y_i(0) - \alpha + W_i^{\text{obs}} \cdot (Y_i(1) - Y_i(0) - \tau^{\text{iv}}) | G_i = \text{nt}, Z_i = 0] \\
&\quad + \pi_{\text{at}} \cdot \mathbb{E}_{\text{sp}}[Y_i(0) - \alpha + W_i^{\text{obs}} \cdot (Y_i(1) - Y_i(0) - \tau^{\text{iv}}) | G_i = \text{at}, Z_i = 0] \\
&\quad + \pi_{\text{co}} \cdot \mathbb{E}_{\text{sp}}[Y_i(0) - \alpha + W_i^{\text{obs}} \cdot (Y_i(1) - Y_i(0) - \tau^{\text{iv}}) | G_i = \text{co}, Z_i = 0] \\
&= \pi_{\text{nt}} \cdot (\mathbb{E}_{\text{sp}}[Y_i(0) | G_i = \text{nt}] - \alpha) + \pi_{\text{at}} \cdot (\mathbb{E}_{\text{sp}}[Y_i(1) | G_i = \text{at}] - \alpha) - \pi_{\text{at}} \cdot \tau^{\text{iv}} \\
&\quad + \pi_{\text{co}} \cdot \mathbb{E}_{\text{sp}}([Y_i(0) | G_i = \text{co}] - \alpha) \\
&= \pi_{\text{nt}} \cdot \mathbb{E}_{\text{sp}}[Y_i(0) | G_i = \text{nt}] + \pi_{\text{at}} \cdot \mathbb{E}_{\text{sp}}[Y_i(1) | G_i = \text{at}] \\
&\quad + \pi_{\text{co}} \cdot \mathbb{E}_{\text{sp}}[Y_i(0) | G_i = \text{co}] - \pi_{\text{at}} \cdot \tau^{\text{iv}} - \alpha \\
&= 0.
\end{aligned}
$$

Similarly,

$$
\begin{aligned}
\mathbb{E}_{\text{sp}}[\varepsilon_i | Z_i = 1] &= \pi_{\text{nt}} \cdot \mathbb{E}_{\text{sp}}[Y_i(0) - \alpha + W_i^{\text{obs}} \cdot (Y_i(1) - Y_i(0) - \tau) | G_i = \text{nt}, Z_i = 1] \\
&\quad + \pi_{\text{at}} \cdot \mathbb{E}_{\text{sp}}[Y_i(0) - \alpha + W_i^{\text{obs}} \cdot (Y_i(1) - Y_i(0) - \tau) | G_i = \text{at}, Z_i = 1] \\
&\quad + \pi_{\text{co}} \cdot \mathbb{E}_{\text{sp}}[Y_i(0) - \alpha + W_i^{\text{obs}} \cdot (Y_i(1) - Y_i(0) - \tau_{\text{late}}) | G_i = \text{co}, Z_i = 1] \\
&= \pi_{\text{nt}} \cdot (\mathbb{E}_{\text{sp}}[Y_i(0) | G_i = \text{nt}] - \alpha) + \pi_{\text{at}} \cdot (\mathbb{E}_{\text{sp}}[Y_i(1) | G_i = \text{at}] - \alpha) - \pi_{\text{at}} \cdot \tau \\
&\quad + \pi_{\text{co}} \cdot \mathbb{E}_{\text{sp}}[Y_i(0) | G_i = \text{co}] - \alpha) + \pi_{\text{co}} \cdot \mathbb{E}_{\text{sp}}[Y_i(1) - Y_i(0) - \tau | G_i = \text{co}] = 0.
\end{aligned}
$$

Thus, $\mathbb{E}_{\text{sp}}[\varepsilon_i | Z_i = z] = 0$ for $z = 0, 1$, and ε_i is uncorrelated with Z_i.

Exploiting the zero correlation between Z_i and ε_i, we can use the same approach as in the one-sided noncompliance case. Consider the conditional expectation of the outcome of interest given the instrument:

$$\mathbb{E}_{\text{sp}}[Y_i^{\text{obs}} | Z_i] = \alpha + \tau_{\text{late}} \cdot \mathbb{E}_{\text{sp}}[W_i^{\text{obs}} | Z_i].$$

Hence we can write a new regression function with a different explanatory variable but the same coefficients as (24.1):

$$Y_i^{\text{obs}} = \alpha + \mathbb{E}_{\text{sp}}[W_i^{\text{obs}}|Z_i] \cdot \tau^{\text{iv}} + \eta_i, \tag{24.2}$$

where the new residual is a composite of two residuals:

$$\eta_i = \varepsilon_i + (W_i^{\text{obs}} - \mathbb{E}_{\text{sp}}[W_i^{\text{obs}}|Z_i]) \cdot \tau^{\text{iv}}.$$

Because $(W_i^{\text{obs}} - \mathbb{E}_{\text{sp}}[W_i^{\text{obs}}|Z_i])$ is by definition uncorrelated with Z_i, it follows that the composite disturbance term $\eta_i = \varepsilon_i + (W_i^{\text{obs}} - \mathbb{E}_{\text{sp}}[W_i^{\text{obs}}|Z_i]) \cdot \tau$ is uncorrelated with Z_i. Moreover, this composite residual is also uncorrelated with functions of Z_i, such as $\mathbb{E}_{\text{sp}}[W_i^{\text{obs}}|Z_i]$. If we observed $\mathbb{E}_{\text{sp}}[W_i^{\text{obs}}|Z_i]$, we could therefore estimate the regression function (24.2) by least squares. We do not observe $\mathbb{E}_{\text{sp}}[W_i^{\text{obs}}|Z_i]$, so this is not feasible, but we can follow the same two-stage least squares (TSLS) procedure as in the previous chapter. First regress, using ordinary least squares, the indicator for receipt of treatment W_i^{obs} on the instrument Z_i to get an estimate for $\mathbb{E}_{\text{sp}}[W_i^{\text{obs}}|Z_i]$. Let $\widehat{\mathbb{E}_{\text{sp}}[W_i^{\text{obs}}|Z_i]}$ be the predicted value from this estimated regression function. Second, regress the outcome of interest using ordinary least squares on the predicted value of the treatment indicator:

$$Y_i^{\text{obs}} = \alpha + \widehat{\mathbb{E}_{\text{sp}}[W_i^{\text{obs}}|Z_i]} \cdot \tau^{\text{iv}} + \eta_i.$$

The coefficient on $\widehat{\mathbb{E}_{\text{sp}}[W_i^{\text{obs}}|Z_i]}$ is the TSLS estimator for the average treatment effect for compliers. In this case with no additional covariates, this TSLS estimate is numerically identical to the ratio of ITT effects. This is easy to see here:

$$\widehat{\mathbb{E}_{\text{sp}}[W_i^{\text{obs}}|Z_i]} = \overline{W}_1 \cdot Z_i + \overline{W}_0 \cdot (1 - Z_i) = \overline{W}_0 + Z_i \cdot \left(\overline{W}_1 - \overline{W}_0 \right).$$

Hence the regression coefficient on $\widehat{\mathbb{E}_{\text{sp}}[W_i^{\text{obs}}|Z_i]}$ is simply the regression coefficient in a regression on Z_i (which itself is the ITT effect on Y_i), divided by $\left(\overline{W}_1 - \overline{W}_0 \right)$.

Now let us return to the formulations of the critical assumptions in the traditional econometric approach. The starting point is equation (24.1). The key assumption in many econometric analyses is that

$$\mathbb{E}_{\text{sp}}\left[\varepsilon_i | Z_i = z\right] = 0,$$

for all z. This assumption captures implicitly the exclusion restriction by excluding Z_i from the structural function (24.1). It also captures the independence assumption by requiring the residual to be uncorrelated with the instrument. It is therefore a mix of substantive and design-related assumptions, making it difficult to assess its plausibility. Perhaps most clearly this is shown by the role of the randomization of the instrument. Clearly, randomization of the instrument makes an instrumental variables strategy more plausible. However, it does not imply that the instrument is uncorrelated with the residual η_i. The separation of the critical assumptions into some that are design-based and implied by randomization, and some that are substantive and unrelated to the randomization, clarifies the benefits of randomization and of the substantive assumptions.

24.7   CONCLUSION

In this chapter we extend the discussion of instrumental variables methods from the setting of randomized experiments with one-sided noncompliance to the setting with two-sided noncompliance. We introduce an additional assumption, the monotonicity or no-defier assumption. We also introduce types of noncompliance. With stronger forms of the exclusion restrictions, distinct for each type of noncomplier, we show that one can again estimate, using the method-of-moments, the causal effect of the treatment for the subpopulation of compliers.

NOTES

The traditional econometric approach to instrumental variables can be found in many textbooks. See, for example, Wooldridge (2002), Angrist and Pischke (2008), and Greene (2011). Imbens and Angrist (1994) and Angrist, Imbens, and Rubin (1996) developed the link to the potential outcomes framework. Björklund, and Moffitt (1987) use a more model-based approach.

Frumento, Mealli, Pacini, and Rubin (2012) consider various versions of exclusion restrictions in the context of the evaluation of a labor market program with random assignment, noncompliance, and missing data.

## CHAPTER 25

Model-Based Analysis in Instrumental Variable Settings: Randomized Experiments with Two-Sided Noncompliance

25.1 INTRODUCTION

In this chapter we develop a multiple-imputation, or model-based alternative, to the Neyman-style moment-based analyses for super-population average treatment effects introduced in Chapters 23 and 24. The model-based approach discussed in this chapter has a number of advantages over the moment-based approach, both conceptual and practical. First, it offers a principled way to incorporate the restrictions on the joint distribution of the observed variables that arise from the various exclusion restrictions and the monotonicity assumption. Second, it allows for a straightforward and flexible way to incorporate covariates. In the current chapter we allow for continuous covariates, or at least covariates taking on too many values for analyses to be feasibly conducted separately on subpopulations homogeneous in the covariates' values. Therefore, we focus on a model-based approach, similar to that in Chapter 8 used in completely randomized experiments. As in Chapter 8, we start by building statistical models for the potential outcomes. A distinct feature of the approach in this chapter is that we also build a statistical model for the compliance behavior. We use these models to simulate the missing potential outcomes *and* the missing compliance behaviors, and use those in turn to draw inferences regarding causal effects of the primary treatment for the subset of units who would always comply with their assignment.

The remainder of this chapter is organized as follows. In the next section we introduce the data used to illustrate the concepts and methods discussed in this chapter. These data come from a randomized experiment designed to evaluate the effect of an influenza vaccine on hospitalization rates. Rather than randomly giving or withholding the flu vaccine itself (the latter was considered unethical), encouragement to vaccinate was randomized, making this what Holland (1988) called a *randomized encouragement design*, closely related to Zelen's (1979, 1990) *randomized consent design*. In Section 25.2 we also carry out some preliminary analyses of the type discussed in the previous chapters, including simple intention-to-treat analyses. In Section 25.3 we discuss the implications of the presence of covariates and formulate critical assumptions to account for them. Next, in Section 25.4, we introduce the model-based imputation approach to the Instrumental Variables (IV) setting. In Section 25.5 we discuss simulation methods to obtain draws from the posterior distribution of the causal estimands. In the following section, Section

25.6, we return to the flu-vaccination example and develop a model for that application as well as discuss a scientifically motivated analysis that can easily be conducted from the model-based perspective. In Section 25.7 we discuss the results for the flu-shot data. Section 25.8 concludes.

25.2 THE MCDONALD-HIU-TIERNEY INFLUENZA VACCINATION DATA

To illustrate these methods, we re-analyze a subset of the data set on influenza vaccinations previously analyzed by McDonald, Hiu, and Tierney (1992) and Hirano, Imbens, Rubin, and Zhou (2000). In the original study, a population of physicians was selected. A random subset of these physicians was sent a letter encouraging them to vaccinate patients deemed at risk for influenza. The remaining physicians were not sent such a letter. In a conventional moment-based analysis, the sending of the letter would play the role of the instrument. The treatment of primary interest is each patient's actual receipt or not of the influenza vaccine, not the randomly assigned encouragement in the form of the letter sent to each patient's physician. In this discussion, the units are patients; in particular, we focus on the subset of female patients. The outcome we focus on in this discussion is whether or not the patient was hospitalized for influenza-related reasons.

For each unit (patient) we observe whether the patient's physician was sent the letter encouraging vaccination for at-risk patients, denoted by Z_i, equal to one if a letter was sent to the physician, and zero otherwise. For the ith patient we also observe whether a flu shot was received, denoted by the binary indicator W_i^{obs}; whether the patient was hospitalized for flu-related illnesses, Y_i^{obs}; and a set of pre-treatment variables, X_i. Note that the design of the experiment involved randomization of physicians rather than patients. Some physicians in our sample have multiple patients, which may lead to correlated outcomes between patients. Although we do not have information on the clustering of patients by doctor, we do have some covariate information on patients. We therefore assume exchangeability of patients conditional on these covariates. To the extent that outcomes and compliance behavior are associated with missing cluster (physician) indicators, even after conditioning on the covariates, our analysis may lead to incorrect posterior inferences, typically the underestimation of posterior uncertainty. This situation has this feature in common with the example in Chapter 23 on randomized experiments with one-sided noncompliance.

There are 1,931 female patients in our sample, and Table 25.1 presents averages by treatment and assignment group for outcomes and covariates. Table 25.2 presents the number of individuals in each of the eight subsamples defined by the binary assignment, binary treatment received and binary outcome, as well as averages for four basic covariates (i.e., pre-treatment variables), age (age measured in years), copd (a binary indicator for chronic obstructive pulmonary disease), and heart (an indicator for prior heart problems), possible latent compliance status.

Before discussing the model-based analyses with covariates that are the main topic of this chapter, let us apply the methods introduced in Chapters 23 and 24 to these data. A standard intention-to-treat (ITT) analysis suggests, not surprisingly, a relatively strong effect of sending the letter encouraging vaccination on the receipt of the influenza

Table 25.1. *Summary Statistics for Women by Assigned Treatment, Received Treatment: Covariates and Outcome for Influenza Vaccination Data*

	Mean	STD	Means No Letter $Z_i = 0$	Means Letter $Z_i^{obs} = 1$	t-Stat dif	Means No Flu Shot $W_i^{obs} = 0$	Means Flu Shot $W_i^{obs} = 1$	t-Stat dif
letter (Z_i)	0.53	(0.50)	0	1	–	0.49	0.63	[7.8]
flu shot (W_i^{obs})	0.24	(0.43)	0.18	0.29	[7.7]	0	1	–
hosp (Y_i^{obs})	0.08	(0.27)	0.09	0.06	[−3.2]	0.08	0.07	[−0.4]
age	65.4	(12.8)	65.2	65.6	[1.1]	64.9	67.1	[4.9]
copd	0.20	(0.40)	0.21	0.20	[−1.3]	0.20	0.23	[2.4]
heart	0.56	(0.50)	0.56	0.57	[0.6]	0.55	0.60	[2.4]

Table 25.2. *Summary Statistics for Women by Assigned Treatment, Received Treatment and Outcome, and Possible Latent Compliance Status for Influenza Vaccination Data*

Type under Monotonicity and Exclusion Restr.	Assign. (Letter) Z_i	Receipt of Flu Shot W_i^{obs}	Hosp. Y_i^{obs}	# of Units 1,931	Means age	copd	heart
Complier or nevertaker	0	0	0	685	64.7	0.18	0.524
Complier or nevertaker	0	0	1	64	62.9	0.33	0.77
Alwaystaker	0	1	0	148	67.8	0.28	0.60
Alwaystaker	0	1	1	20	68.9	0.30	0.70
Nevertaker	1	0	0	672	65.4	0.19	0.55
Nevertaker	1	0	1	51	62.0	0.29	0.69
Complier or alwaystaker	1	1	0	277	66.6	0.20	0.57
Complier or alwaystaker	1	1	1	14	67.3	0.21	0.79

vaccination. Patients whose physicians were not sent a letter were vaccinated at a rate of 18%, whereas those patients whose physicians were sent the letter were vaccinated at a rate of 29%, equivalent to roughly a 50% increase in the proportion of female patients vaccinated. The difference is a method-of-moments estimate for the ITT effect on the treatment received, $\text{ITT}_W = \mathbb{E}[W_i(1) - W_i(0)]$:

$$\widehat{\text{ITT}_W} = \overline{W}_1^{obs} - \overline{W}_0^{obs} = 0.104 \ (\widehat{\text{s.e.}} \ 0.019).$$

Here, as in the previous two chapters, subscripts 0 and 1 refer to levels of the assignment Z_i, and subscripts c and t refer to levels of the treatment received W_i^{obs}. The estimated effect is substantial and statistically significant at conventional levels. Clearly the sending of the letter was effective in encouraging actual vaccination, although it is also clear from Table 25.2 that many patients whose physicians were sent the letter did not receive a flu shot (71%), and many patients whose physicians were not sent the letter nevertheless received a flu shot (18%).

Next, let us consider the ITT effect on the outcome of interest, the hospitalization rate: 6.4% of patients whose physicians received the letter were hospitalized for flu-related reasons, whereas 9.2% of patients whose physicians did not receive the letter were hospitalized for flu-related reasons, which suggests a substantial effect of assignment on hospitalization rates:

$$\widehat{\text{ITT}}_Y = \overline{Y}_1^{\text{obs}} - \overline{Y}_0^{\text{obs}} = -0.028 \ (\widehat{\text{s. e.}} \ 0.012),$$

approximately a 50% decrease. The estimated effect is substantial in terms of percentage reduction, and statistically significant at the 5% level.

Now, let us look at some IV analyses ignoring covariates, maintaining both the exclusion restrictions for all compliance types and the monotonicity assumption, largely following the discussion from Chapter 24. Define, as in the previous chapter, the four compliance groups as

$$G_i = \begin{cases} \text{nt} & \text{if } W_i(0) = 0, \ W_i(1) = 0, \\ \text{co} & \text{if } W_i(0) = 0, \ W_i(1) = 1, \\ \text{df} & \text{if } W_i(0) = 1, \ W_i(1) = 0, \\ \text{at} & \text{if } W_i(0) = 1, \ W_i(1) = 1. \end{cases}$$

First, let us interpret the ITT effect on the receipt of treatment, under the monotonicity assumption. Without the monotonicity assumption, this ITT effect is equal to the difference in proportions of compliers and defiers. The monotonicity assumption rules out the presence of defiers, so in that case this ITT effect is equal to the proportion of compliers, and the share of compliers in the super-population is estimated by method-of-moments to be

$$\hat{\pi}_{\text{co}} = \widehat{\text{ITT}}_W = 0.104 \ (\widehat{\text{s. e.}} \ 0.019).$$

The population proportion of those receiving the vaccination, despite their physician not being sent the encouragement letter, equals the population proportion of alwaystakers. The sample proportion of those receiving the vaccination, even though their physician was not sent the letter equals 0.183, so that the population share of alwaystakers is estimated, by a simple method-of-moments procedure, to be

$$\hat{\pi}_{\text{at}} = \frac{N_{0t}}{N_{0t} + N_{0c}} = 0.183 \ (\widehat{\text{s. e.}} \ 0.013).$$

The sample proportion of individuals not vaccinated among those whose physicians were sent the letter is a simple method-of-moments estimator of the proportion of nevertakers in the population, also equal to one minus the proportions of compliers and alwaystakers. With our data the resulting estimate equals

$$\hat{\pi}_{\text{nt}} = 1 - \hat{\pi}_{\text{co}} - \hat{\pi}_{\text{at}} = \frac{N_{1c}}{N_{1c} + N_{1t}} = 0.713 \ (\widehat{\text{s. e.}} \ 0.014).$$

Next, let us consider the primary outcome, hospitalization for flu-related reasons, by treatment assigned and treatment received. Table 25.3 gives average outcomes and their associated standard errors for the four groups defined by treatment assigned and treatment received. From this table we can obtain method-of-moments estimates of average potential outcomes for nevertakers and alwaystakers. Patients who did not receive the

Table 25.3. *Average Outcomes and Estimated Standard Errors by Treatment Assigned and Treatment Received, for Influenza Vaccination Data*

		Assignment of Treatment Z_i			
		0		1	
Receipt of treatment W_i^{obs}	c	$\overline{Y}_{0c}^{obs} = 0.085$	(0.010)	$\overline{Y}_{1c}^{obs} = 0.071$	(0.010)
	t	$\overline{Y}_{0t}^{obs} = 0.112$	(0.025)	$\overline{Y}_{1t}^{obs} = 0.048$	(0.013)

vaccine, despite their physician having been sent the encouragement letter, must be nevertakers given that, under monotonicity, defiers do not exist. Hence we can estimate the super-population average outcome for nevertakers as

$$\hat{\mathbb{E}}[Y_i(0)|G_i = \text{nt}] = \overline{Y}_{1c}^{obs} = 0.071 \ (\widehat{\text{s. e.}} \ 0.010).$$

Similarly, patients who got vaccinated, even though their physicians were not sent the letter, must be alwaystakers, and thus

$$\hat{\mathbb{E}}[Y_i(1)|G_i = \text{at}] = \overline{Y}_{0t}^{obs} = 0.119 \ (\widehat{\text{s. e.}} \ 0.025).$$

Those assigned to the control group (i.e., those patients whose physicians were not sent a letter) who did not receive the flu shot can be one of two types: their observed behavior is consistent with being a complier or a nevertaker. The expected proportion of each in this subgroup is the same as their relative proportions in the population. Hence, within the subgroup of those assigned to the control group who did not receive the flu shot, the (*ex ante*) proportion of compliers is $\pi_{co}/(\pi_{co} + \pi_{nt})$. The population share of compliers is estimated to be $\hat{\pi}_{co} = 0.104$. The share of nevertakers is estimated to be $\hat{\pi}_{nt} = 0.713$. Hence the share of compliers among those not receiving the flu shot is estimated as $\hat{\pi}_{co}/(\hat{\pi}_{co} + \hat{\pi}_{nt}) = 0.127$. The average outcome in the control group who were not assigned the treatment is estimated as $\overline{Y}_{0c}^{obs} = 0.085$. This reflects a mixture of compliers, with an estimated share equal to 0.127, and nevertakers, with an estimated share of 0.713. In terms of super-population quantities,

$$\mathbb{E}\left[Y_i^{obs} \middle| Z_i = 0, W_i^{obs} = 0\right]$$
$$= \frac{\pi_{co}}{\pi_{co} + \pi_{nt}} \cdot \mathbb{E}[Y_i(0)|G_i = \text{co}] + \frac{\pi_{nt}}{\pi_{co} + \pi_{nt}} \cdot \mathbb{E}[Y_i(0)|G_i = \text{nt}].$$

Because we estimated the average outcome for nevertakers to be $\hat{\mathbb{E}}[Y_i(0)|G_i = \text{nt}] = 0.071$, we can estimate the average control potential outcome for compliers as

$$\hat{\mathbb{E}}[Y_i(0)|G_i = \text{co}] = \frac{\hat{\mathbb{E}}\left[Y_i^{obs} \middle| Z_i = 0, W_i^{obs} = 0\right] - \hat{\mathbb{E}}[Y_i(0)|G_i = n] \cdot \hat{\pi}_{nt}/(\hat{\pi}_{co} + \hat{\pi}_{nt})}{\hat{\pi}_{co}/(\hat{\pi}_{co} + \hat{\pi}_{nt})}$$

$$= \frac{\overline{Y}_{0c}^{obs} - \overline{Y}_{1c}^{obs} \cdot \hat{\pi}_{nt}/(\hat{\pi}_{co} + \hat{\pi}_{nt})}{\hat{\pi}_{co}/(\hat{\pi}_{co} + \hat{\pi}_{nt})} = 0.188 \ (\widehat{\text{s. e.}} \ 0.092).$$

Similarly, those assigned to the control group who did receive the vaccination must be alwaystakers, again because, by monotonicity, there are no defiers. Hence we can estimate the average treatment potential outcome for compliers as

$$\hat{\mathbb{E}}[Y_i(1)|G_i = \text{co}] = \frac{\hat{\mathbb{E}}\left[Y_i^{\text{obs}}\,|\,Z_i = 1, W_i^{\text{obs}} = 1\right] - \hat{\mathbb{E}}[Y_i(1)|G_i = \text{at}] \cdot \hat{\pi}_{\text{at}}/(\hat{\pi}_{\text{co}} + \hat{\pi}_{\text{at}})}{\hat{\pi}_{\text{co}}/(\hat{\pi}_{\text{co}} + \hat{\pi}_{\text{at}})}$$

$$= \frac{\overline{Y}_{1t}^{\text{obs}} - \overline{Y}_{0t}^{\text{obs}} \cdot \hat{\pi}_{\text{at}}/(\hat{\pi}_{\text{co}} + \hat{\pi}_{\text{at}})}{\hat{\pi}_{\text{co}}/(\hat{\pi}_{\text{co}} + \hat{\pi}_{\text{at}})} = -0.077 \ (\widehat{\text{s. e.}}\ 0.054).$$

Thus, the method-of-moment-based IV estimate of the local average treatment effect (or average causal effect) for women in the flu-shot data set is

$$\hat{\tau}^{\text{iv}} = \hat{\mathbb{E}}[Y_i(1)|G_i = \text{co}] - \hat{\mathbb{E}}[Y_i(0)|G_i = \text{co}] = -0.265 \quad (\widehat{\text{s. e.}}\ 0.110).$$

We could repeat this analysis for subpopulations defined by covariates, but that would not work well in settings with covariates taking on many values. That is a key reason why, in this chapter, we pursue a model-based strategy to incorporate the covariates, similar to that in Chapter 8 on randomized experiments.

Note that the moment-based estimate of $\mathbb{E}[Y_i(1)|G_i = \text{co}]$ is negative, -0.077. Because this is the probability of an event, $\mathbb{E}[Y_i(1)|G_i = \text{co}] = \Pr(Y_i(1) = 1|G_i = \text{co})$, the true value of $\mathbb{E}[Y_i(1)|G_i = \text{co}]$ is obviously bounded by zero and one. The moment-based IV estimate does not impose these restrictions, and as a result it does not make efficient use of the data. The model-based strategy discussed in the current chapter provides a natural way to incorporate these restrictions efficiently, and this is an important benefit of the model-based strategy. Note also that one would not necessarily have realized that implicitly the moment-based IV estimate of -0.265 is based on a negative estimate of $\mathbb{E}[Y_i(1)|G_i = \text{co}]$ without performing the additional calculations in the previous paragraph, that is, additional to calculating mechanically the moment-based instrumental variables estimate.

Let us return briefly to the ITT analyses. As an alternative to the Neyman-style analyses for ITT$_Y$ and ITT$_W$, we could apply the methods discussed in the model-based chapter for randomized experiments, Chapter 8. In the specific context of the flu-shot study, both the primary outcome (hospitalization for flu-related reasons) and the secondary outcome (receipt of influenza vaccination) are binary. Hence natural models for these potential outcomes are binary regression (e.g., logistic) models. We consider these models with the additional assumption that the $Z_i = 0$ and $Z_i = 1$ potential outcomes conditional on G_i are independent.

First, consider ITT$_W$, where we continue for the moment to ignore the presence of covariates. Recall that there are two inputs into a model-based analysis: first, the joint (i.i.d.) distribution for the potential outcomes given a parameter, and second, a prior distribution on that parameter. We begin by specifying

$$\Pr(W_i(z) = 1|\theta^W) = \frac{\exp(\theta_z^W)}{1 + \exp(\theta_z^W)},$$

for $z = 0, 1$, with $\theta^W = (\theta_0^W, \theta_1^W)$, and where we assume independence of $W_i(0)$ and $W_i(1)$ given θ^W, and the θ_z^W are functions of a global parameter θ. In this simple case, we could have specified a binomial model, $W_i(z) \sim \mathcal{B}(N, p_z^W)$, for $z = 0, 1$, but we use the logistic specification to facilitate the generalization to models with covariates. Given this model, the super-population average ITT effect is

$$\text{ITT}_W = \frac{\exp(\theta_1^W)}{1 + \exp(\theta_1^W)} - \frac{\exp(\theta_0^W)}{1 + \exp(\theta_0^W)}.$$

The second input is the prior distribution on θ. In the current setting, with a fairly large data set and the focus on the ITT effects, the choice of prior distribution is largely immaterial, but again to facilitate the comparison with the models discussed later in this chapter, we use a prior distribution specifically designed for logistic regression models. The prior distribution can be interpreted as introducing N_{prior} artificial observations, divided equally over $(z, w) \in \{(0, 0), (0, 1), (1, 0), (1, 1)\}$, so that the prior distribution is

$$p(\theta_0^W, \theta_1^W) \propto \left(\frac{\exp(\theta_0^W)}{1 + \exp(\theta_0^W)} \right)^{N_{\text{prior}}/4} \left(\frac{1}{1 + \exp(\theta_0^W)} \right)^{N_{\text{prior}}/4}$$

$$\times \left(\frac{\exp(\theta_1^W)}{1 + \exp(\theta_1^W)} \right)^{N_{\text{prior}}/4} \left(\frac{1}{1 + \exp(\theta_1^W)} \right)^{N_{\text{prior}}/4}.$$

In the calculations, we use $N_{\text{prior}} = 4$.

Using simulation methods to obtain draws from the posterior distribution, we find that the mean and variance of the posterior distribution for ITT_W are

$$\mathbb{E}\left[\text{ITT}_W | \mathbf{Y}^{\text{obs}}, \mathbf{W} \right] = 0.104, \quad \mathbb{V}\left(\text{ITT}_W | \mathbf{Y}^{\text{obs}}, \mathbf{W} \right) = 0.0191^2.$$

These numbers are very similar to the point estimate and standard error based on the Neyman-style analysis, which is not surprising given the size of the data, the simple model being used, and the choice of relatively diffuse prior distributions.

To conduct the model-based analysis for ITT_Y, we reinterpret the current setup slightly. We specify the following model for the two potential outcomes:

$$\Pr(Y_i(z, W_i(z)) = 1 | \theta^Y) = \frac{\exp(\theta_z^Y)}{1 + \exp(\theta_z^Y)},$$

for $z = 0, 1$, with $\theta^Y = (\theta_0^Y, \theta_1^Y)$ independent of θ^W, and where, as stated earlier, $Y_i(0, W_i(0))$ is independent of $Y_i(1, W_i(1))$ conditional on θ^Y, so that the super-population average ITT effect is

$$\text{ITT}_Y = \frac{\exp(\theta_1^Y)}{1 + \exp(\theta_1^Y)} - \frac{\exp(\theta_0^Y)}{1 + \exp(\theta_0^Y)}.$$

Using the same prior distribution with four artificial observations for (θ_0^Y, θ_1^Y) as we used for (θ_0^W, θ_1^W), we find for the posterior mean and variance for the intention-to-treat effect ITT_Y,

$$\mathbb{E}\left[\text{ITT}_Y \mid \mathbf{Y}^{\text{obs}}, \mathbf{W}\right] = -0.028, \qquad \mathbb{V}\left(\text{ITT}_Y \mid \mathbf{Y}^{\text{obs}}, \mathbf{W}\right) = 0.012^2.$$

Again, not surprisingly, these numbers are very similar to those based on the Neyman moment-based analysis.

25.3   COVARIATES

Now we consider settings where, in addition to observing the outcome, the treatment, and the instrument, we observe for each unit a vector of covariates (i.e., pre-treatment variables), denoted by X_i. As in the earlier chapters on analyses under unconfoundedness, the key requirement for these covariates is that they are not affected by the treatment. In the current setting that requirement extends to being unaffected by the instrument or the primary treatment. The pre-treatment variables may include permanent characteristics of the units, or pre-treatment outcomes whose values were determined prior to the determination of the value of the instrument and the treatment. In the presence of covariates we can relax the critical assumptions underlying the analyses discussed in the previous chapter.

The generalization of the random assignment assumption is standard, and mirrors the unconfoundedness assumption for regular assignment mechanisms. It requires that conditional on the pre-treatment variables the assignment is effectively random:

Assumption 25.1 (Super-Population Unconfoundedness of the Instrument)

$$Z_i \perp\!\!\!\perp \left(W_i(0), W_i(1), Y_i(0,0), Y_i(0,1), Y_i(1,0), Y_i(1,1)\right) \mid X_i.$$

Because the compliance type G_i is a one-to-one function of $(W_i(0), W_i(1))$, we can also write this assumption as

$$Z_i \perp\!\!\!\perp \left(G_i, Y_i(0,0), Y_i(0,1), Y_i(1,0), Y_i(1,1)\right) \mid X_i.$$

In the flu-shot application, this assumption is satisfied by design both with, and without, the pre-treatment variables, because the instrument, Z_i, is randomly assigned independent of the values of the pre-treatment variables. In observational studies, unconfoundedness of the instrument is often a substantive assumption, and it may represent an important relaxation of the stronger assumption that Z_i is independent of all potential outcomes without any conditioning. For example, a class of applications of instrumental variables methods in public health settings uses distance from a patient's residence to the nearest medical facility (with particular capabilities or expertise) as an instrument for the use of those capabilities (e.g., McClellan and Newhouse, 1994). To be specific, one might be interested in the effect of the presence of advanced neo-natal urgent care facilities in a hospital on outcomes for prematurely born infants. Typically, the distance to hospitals with such facilities is a strong predictor of the use of those facilities. However, families do not choose location of their residence randomly, and full independence of the distance from their residence to such hospitals may not be plausibly viewed as random. It may be more plausible to view the distance as essentially random given other characteristics of the location such as median housing cost, population density, and distance to nearest medical facility of any kind.

The presence of covariates does not affect the deterministic version of the exclusion restriction substantially. We continue to make the assumption that for nevertakers and alwaystakers the change in instrument does not affect the outcome.

Assumption 25.2 (Exclusion Restriction for Nevertakers)

$$Y_i(0, W_i(0)) = Y_i(1, W_i(1)),$$

for all nevertakers, that is units i with $G_i = $ nt.

Assumption 25.3 (Exclusion Restrictions for Alwaystakers)

$$Y_i(0, W_i(0)) = Y_i(1, W_i(1)),$$

for all alwaystakers, that is units i with $G_i = $ at.

The generalization of the stochastic exclusion restriction is more subtle. This is stated as a conditional independence assumption and therefore can be weakened to hold only conditional on covariates:

Assumption 25.4 (Stochastic Exclusion Restrictions for Nevertakers)

$$Z_i \perp\!\!\!\perp Y_i(Z_i, W_i(Z_i)) \mid X_i, G_i = \text{nt}.$$

Assumption 25.5 (Stochastic Exclusion Restrictions for Alwaystakers)

$$Z_i \perp\!\!\!\perp Y_i(Z_i, W_i(Z_i)) \mid X_i, G_i = \text{at}.$$

In practice these may not be substantial weakenings of the exclusion restrictions relative to the deterministic versions of the restrictions.

25.4 MODEL-BASED INSTRUMENTAL VARIABLES ANALYSES FOR RANDOMIZED EXPERIMENTS WITH TWO-SIDED NONCOMPLIANCE

Now we turn to a model-based strategy for estimating treatment effects in randomized experiments with two-sided noncompliance in settings with covariates. We maintain the exclusion restrictions for nevertakers and alwaystakers, Assumptions 25.2 and 25.3.

We develop the likelihood function using a missing data approach, similar to that used in the model-based chapter on completely randomized experiments, Chapter 8. The key difference is that in the current setting there are, for each unit, two missing potential outcomes. The first missing potential outcome is the primary outcome corresponding to the treatment not received, and the second one is for the secondary outcome, the treatment that would be received under the alternative assignment.

25.4.1 Notation

As before, let $\mathbf{W}(0)$ and $\mathbf{W}(1)$ be the N-vectors of secondary potential outcomes with i^{th} element equal to $W_i(0)$ and $W_i(1)$, indicating the primary treatment received under assignment to $Z_i = 0$ and $Z_i = 1$ respectively, and let $\mathbf{W} = (\mathbf{W}(0), \mathbf{W}(1))$.

For the primary outcomes the notation we use is more subtle. Because we maintain the exclusion restriction for nevertakers and alwaystakers, we drop the z (assignment) argument of the potential outcomes $Y_i(z, w)$ and write, without ambiguity, $Y_i(w)$, indexed only by the level of the primary treatment of interest. For compliers, $Y_i(0) = Y_i(0, W_i(0)) = Y_i(0, 0)$ and $Y_i(1) = Y_i(1, W_i(1)) = Y_i(1, 1)$. For nevertakers, $Y_i(0) = Y_i(0, W_i(0)) = Y_i(1, W_i(1))$, but $Y_i(1)$ is not defined, and to be mathematically precise we use the notation $Y_i(1) = \star$ to capture this for nevertakers. For alwaystakers, $Y_i(1) = Y_i(0, W_i(0)) = Y_i(1, W_i(1))$, but $Y_i(0)$ is not defined, and we use $Y_i(0) = \star$ for them. Given this notation, let $\mathbf{Y}(0)$ denote the N-vector of primary potential outcomes under receipt of the control treatment, with the ith element equal to $Y_i(0)$, and $\mathbf{Y}(1)$ the N-vector of potential outcomes under receipt of the active treatment, with i^{th} element equal to $Y_i(1)$. Let $\mathbf{Y} = (\mathbf{Y}(0), \mathbf{Y}(1))$ be the corresponding $N \times 2$ matrix formed by combining $\mathbf{Y}(0)$ and $\mathbf{Y}(1)$.

We focus on causal estimands that can be written as functions of $(\mathbf{Y}, \mathbf{W}, \mathbf{X}, \mathbf{Z})$, although in practice interesting causal estimands rarely depend on $\mathbf{Z}$ or on $\mathbf{Y}$ values equal to $\star$. Let $G_i \in \{\text{nt}, \text{df}, \text{co}, \text{at}\}$ denote the compliance type of unit i, and let $\mathbf{G}$ denote the N-vector with i^{th} element equal to G_i. Then $\mathbf{G}$ is a one-to-one function of $\mathbf{W}$, so we can also write the causal estimands as functions of $(\mathbf{Y}, \mathbf{G}, \mathbf{X}, \mathbf{Z})$. This class of estimands includes, for example, the average effect for compliers,

$$\tau_{\text{late}} = \frac{1}{N_{\text{co}}} \sum_{i:G_i=\text{co}} \left(Y_i(1) - Y_i(0) \right),$$

where, for $g \in \{\text{nt}, \text{df}, \text{co}, \text{at}\}$, N_g is the number of units of type $G_i = g$: $N_g = \sum \mathbf{1}_{G_i=g}$. The class of estimands also includes other functions of the potential outcomes for compliers, or even functions of outcomes for the different types of noncompliers, as we see later.

Recall the definitions of the missing and observed outcomes from Chapter 3. Here we define missing and observed values for the treatment received in similar fashion:

$$W_i^{\text{mis}} = W_i(1 - Z_i) = \begin{cases} W_i(0) & \text{if } Z_i = 1, \\ W_i(1) & \text{if } Z_i = 0, \end{cases}$$

and

$$W_i^{\text{obs}} = W_i(Z_i) = \begin{cases} W_i(0) & \text{if } Z_i = 0, \\ W_i(1) & \text{if } Z_i = 1, \end{cases}$$

where $\mathbf{W}^{\text{mis}}$ and $\mathbf{W}^{\text{obs}}$ are the N-vectors with i^{th} elements equal to W_i^{mis} and W_i^{obs} respectively. For compliers we define

$$Y_i^{\text{mis}} = Y_i(1 - W_i^{\text{obs}}) = \begin{cases} Y_i(0) & \text{if } W_i^{\text{obs}} = 1, \\ Y_i(1) & \text{if } W_i^{\text{obs}} = 0, \end{cases}$$

and

$$Y_i^{\text{obs}} = Y_i(W_i^{\text{obs}}) = \left\{ \begin{array}{ll} Y_i(0) & \text{if } W_i^{\text{obs}} = 0, \\ Y_i(1) & \text{if } W_i^{\text{obs}} = 1. \end{array} \right.$$

For nevertakers

$$Y_i^{\text{obs}} = Y_i(0), \quad \text{and} \quad Y_i^{\text{mis}} = \star,$$

and for alwaystakers

$$Y_i^{\text{obs}} = Y_i(1), \quad \text{and} \quad Y_i^{\text{mis}} = \star.$$

Finally, let $\mathbf{Y}^{\text{mis}}$ and $\mathbf{Y}^{\text{obs}}$ be the N-vectors with i^{th} elements equal to Y_i^{mis} and Y_i^{obs} respectively.

Any causal estimand of the form $\tau(\mathbf{Y}, \mathbf{G}, \mathbf{X}, \mathbf{Z})$ can be written in terms of observed and missing variables as $\tau(\mathbf{Y}^{\text{obs}}, \mathbf{Y}^{\text{mis}}, \mathbf{W}^{\text{obs}}, \mathbf{W}^{\text{mis}}, \mathbf{X}, \mathbf{Z})$. The estimand is unknown because $(\mathbf{Y}^{\text{mis}}, \mathbf{W}^{\text{mis}})$ are not observed. In order to derive the posterior distribution of $\tau(\mathbf{Y}, \mathbf{G}, \mathbf{X}, \mathbf{Z})$, we therefore need to derive the predictive distribution of the missing data $(\mathbf{Y}^{\text{mis}}, \mathbf{W}^{\text{mis}})$ given the observed data $(\mathbf{Y}^{\text{obs}}, \mathbf{W}^{\text{obs}}, \mathbf{X}, \mathbf{Z})$, that is, the posterior predictive distribution of the missing data.

25.4.2 The Inputs into a Model-Based Approach

We do not directly specify the posterior predictive distribution of the missing data, $f(\mathbf{Y}^{\text{mis}}, \mathbf{W}^{\text{mis}}|\mathbf{Y}^{\text{obs}}, \mathbf{W}^{\text{obs}}, \mathbf{X}, \mathbf{Z})$. As in the randomized experiment setting, such a task would generally be difficult, because this predictive distribution combines features of the assignment mechanism with those of the distribution of the potential outcomes. Instead we start with three inputs into the analyses. The first two together specify the joint distribution of all four potential outcomes, $(Y_i(0), Y_i(1), W_i(0), W_i(1))$ given covariates and parameters, or equivalently, because G_i is a one-to-one function of $(W_i(0), W_i(1))$, the joint distribution of the two primary potential outcomes and the compliance type, $(Y_i(0), Y_i(1), G_i)$, given covariates and parameters. We factor this joint distribution into two parts. First, a model for the primary potential outcomes given covariates, compliance status, and its parameters, denoted

$$f(Y_i(0), Y_i(1)|G_i, X_i; \theta),$$

which implies that the joint distribution

$$f(\mathbf{Y}|\mathbf{G}, \mathbf{X}; \theta) = \prod_{i=1}^{N} f(Y_i(0), Y_i(1)|G_i, X_i; \theta),$$

has independence of the units conditional on the unknown parameter θ. As in the simpler situation of Chapter 8, such an i.i.d specification follows from the unit exchangeability and an appeal to de Finetti's theorem. Often we specify distributions for each of the compliance types and for potential outcomes given compliance types with parameters *a priori* independent. For example, with continuous outcomes, we could use Gaussian models. For compliers,

$$Y_i(0)|G_i = \text{co}, X_i; \theta \sim \mathcal{N}(X_i\beta_{\text{co,c}}, \sigma^2_{\text{co,c}}),$$

and, independently,

$$Y_i(1)|G_i = \text{co}, X_i; \theta \sim \mathcal{N}(X_i\beta_{\text{co,t}}, \sigma^2_{\text{co,t}});$$

and for nevertakers,

$$Y_i(0)|G_i = \text{nt}, X_i; \theta \sim \mathcal{N}(X_i\beta_{\text{nt}}, \sigma^2_{\text{nt}});$$

and finally, for alwaystakers,

$$Y_i(1)|G_i = \text{at}, X_i; \theta \sim \mathcal{N}(X_i\beta_{\text{at}}, \sigma^2_{\text{at}}),$$

in combination with *a priori* independence of the parameters. For notational convenience we include an intercept in X_i. Alternatively we may wish to impose some restrictions, for example, assuming the slope coefficients, or conditional variances, for different complier types are equal. Note that we do not model $Y_i(1)$ for nevertakers or $Y_i(0)$ for alwaystakers because these are *a priori* counterfactual and values cannot be observed for them in the current setting.

The second input is a model for compliance status given covariates and the parameters:

$$f(\mathbf{G}|\mathbf{X}; \theta) = \prod_{i=1}^{N} f(G_i|X_i, \theta).$$

Here one natural model is a multinomial logit model, where again we include the intercept in X_i,

$$\Pr(G_i = \text{co}|X_i, \theta) = \frac{1}{1 + \exp(X_i\gamma_{\text{nt}}) + \exp(X_i\gamma_{\text{at}})},$$

$$\Pr(G_i = \text{nt}|X_i, \theta) = \frac{\exp(X_i\gamma_{\text{nt}})}{1 + \exp(X_i\gamma_{\text{nt}}) + \exp(X_i\gamma_{\text{at}})},$$

and

$$\Pr(G_i = \text{at}|X_i, \theta) = \frac{\exp(X_i\gamma_{\text{at}})}{1 + \exp(X_i\gamma_{\text{nt}}) + \exp(X_i\gamma_{\text{at}})}.$$

If we generalize this specification to include functions of the basic covariates, this general specification is essentially without loss of generality.

An alternative to the multinomial logistic model would be a sequence of two binary response models. The first binary response model specifies the probability of being a complier *versus* a noncomplier (nevertaker or alwaystaker):

$$\Pr(G_i = \text{co}|X_i, \theta) = \frac{\exp(X_i\alpha_{\text{co}})}{1 + \exp(X_i\alpha_{\text{co}})}.$$

The second model specifies the probability of being a nevertakers conditional on being a noncomplier:

$$\Pr(G_i = \text{nt}|X_i, G_i \in \{\text{nt}, \text{at}\}, \theta) = \frac{\exp(X_i\alpha_{\text{nt}})}{1 + \exp(X_i\alpha_{\text{nt}})}.$$

The third input is the prior distribution for the unknown parameter $\theta = (\beta_{\text{co,c}}, \beta_{\text{co,t}}, \beta_{\text{nt}}, \beta_{\text{at}}, \gamma_{\text{nt}}, \gamma_{\text{at}})$:

$$p(\theta) = p(\beta_{\text{co,c}}, \beta_{\text{co,t}}, \beta_{\text{nt}}, \beta_{\text{at}}, \gamma_{\text{nt}}, \gamma_{\text{at}}).$$

We will generally attempt to specify prior distributions that are not dogmatic and instead are relatively diffuse. Because of the delicate mixture nature of the model, it will be important to specify proper prior distributions to stabilize estimation. We will specify prior distributions that correspond to the introduction of a few artificial units for whom we observe the covariate values, the compliance types, and different values of the outcome. We specify the prior distribution in such a way that these artificial units carry little weight relative to the observed data. Their presence, in combination with the fact that for these artificial units we observe the compliance types that we may not in reality observe for any units in our sample, ensures that the posterior distribution is always proper and well behaved in a sense that is clear in particular cases, though vague in general.

Now we follow the same four steps described in Chapter 8 to derive the posterior predictive distribution of the estimands (i.e., given the data). Here we do this for general specifications of the potential outcome distributions. Next, we discuss simulation-based methods for approximating these posterior distributions. Later we provide more detail in the specific context of the flu-shot application.

25.4.3 Derivation of $f(\mathbf{Y}^{\text{mis}}, \mathbf{W}^{\text{mis}}|\mathbf{Y}^{\text{obs}}, \mathbf{W}^{\text{obs}}, \mathbf{X}, \mathbf{Z}, \theta)$

The first step is to derive the conditional distribution of the missing data, $(\mathbf{Y}^{\text{mis}}, \mathbf{W}^{\text{mis}})$, given the observed data, $(\mathbf{Y}^{\text{obs}}, \mathbf{W}^{\text{obs}}, \mathbf{X}, \mathbf{Z})$, *and* the parameter θ, from the specifications just described.

Given the specifications of $f(\mathbf{Y}|\mathbf{G}, \mathbf{X}; \theta)$ and $f(\mathbf{G}|\mathbf{X}; \theta)$, we can infer the joint distribution

$$f(\mathbf{Y}(0), \mathbf{Y}(1), \mathbf{W}(0), \mathbf{W}(1))|\mathbf{X}, \theta),$$

which, because of the unconfoundedness assumption is equal to the conditional distribution

$$f(\mathbf{Y}(0), \mathbf{Y}(1), \mathbf{W}(0), \mathbf{W}(1))|\mathbf{X}, \mathbf{Z}, \theta). \qquad (25.1)$$

Next we use the fact (a special case of which was also exploited in Chapter 8) that there is a one-to-one relation between $(\mathbf{Y}(0), \mathbf{Y}(1), \mathbf{W}(0), \mathbf{W}(1), \mathbf{Z})$ and $(\mathbf{Y}^{\text{mis}}, \mathbf{Y}^{\text{obs}}, \mathbf{W}^{\text{mis}}, \mathbf{W}^{\text{obs}}, \mathbf{Z})$, and therefore we can write

$$(\mathbf{Y}(0), \mathbf{Y}(1), \mathbf{W}(0), \mathbf{W}(1)) = g(\mathbf{Y}^{\text{mis}}, \mathbf{Y}^{\text{obs}}, \mathbf{W}^{\text{mis}}, \mathbf{W}^{\text{obs}}, \mathbf{Z}). \qquad (25.2)$$

We can use this, in combination with (25.1), to derive

$$f(\mathbf{Y}^{\text{mis}}, \mathbf{Y}^{\text{obs}}, \mathbf{W}^{\text{mis}}, \mathbf{W}^{\text{obs}} | \mathbf{X}, \mathbf{Z}, \theta). \qquad (25.3)$$

Then, using Bayes' Rule, we can infer the conditional distribution of the missing potential outcomes given the observed values and θ:

$$\begin{aligned} &f(\mathbf{Y}^{\text{mis}}, \mathbf{W}^{\text{mis}} | \mathbf{Y}^{\text{obs}}, \mathbf{W}^{\text{obs}}, \mathbf{X}, \mathbf{Z}, \theta) \\ &= \frac{f(\mathbf{Y}^{\text{mis}}, \mathbf{Y}^{\text{obs}}, \mathbf{W}^{\text{mis}}, \mathbf{W}^{\text{obs}} | \mathbf{X}, \mathbf{Z}, \theta)}{\int \int f(\mathbf{Y}^{\text{mis}}, \mathbf{Y}^{\text{obs}}, \mathbf{W}^{\text{mis}}, \mathbf{W}^{\text{obs}} | \mathbf{X}, \mathbf{Z}, \theta) d\mathbf{Y}^{\text{mis}} d\mathbf{W}^{\text{mis}}}, \end{aligned}$$

which, because of the conditioning on θ, is the product of N factors.

25.4.4 Derivation of the Posterior Distribution $p(\theta | \mathbf{Y}^{\text{obs}}, \mathbf{W}^{\text{obs}}, \mathbf{X}, \mathbf{Z})$

In the previous subsection, when deriving the conditional distribution of missing potential outcomes given observed data, we derived the joint distribution of missing and observed outcomes given covariates, instruments, and parameter,

$$f(\mathbf{Y}^{\text{mis}}, \mathbf{Y}^{\text{obs}}, \mathbf{W}^{\text{mis}}, \mathbf{W}^{\text{obs}} | \mathbf{X}, \mathbf{Z}, \theta).$$

Integrating out the missing data leads to the joint distribution of observed data given the parameter θ, which, when regarded as a function of the unknown vector parameter θ, given observed data, is proportional to the likelihood function of θ:

$$\begin{aligned} \mathcal{L}(\theta | \mathbf{Y}^{\text{obs}}, \mathbf{W}^{\text{obs}}, \mathbf{X}, \mathbf{Z}) &= f(\mathbf{Y}^{\text{obs}}, \mathbf{W}^{\text{obs}} | \mathbf{X}, \mathbf{Z}, \theta) \\ &= \int \int f(\mathbf{Y}^{\text{mis}}, \mathbf{Y}^{\text{obs}}, \mathbf{W}^{\text{mis}}, \mathbf{W}^{\text{obs}} | \mathbf{X}, \mathbf{Z}, \theta) d\mathbf{Y}^{\text{mis}} d\mathbf{W}^{\text{mis}}. \end{aligned}$$

To obtain analytically the posterior distribution of θ, we multiply this likelihood function of θ by the prior distribution for θ, $p(\theta)$, and find the normalizing constant to make the product integrate to one:

$$p(\theta | \mathbf{Y}^{\text{obs}}, \mathbf{W}^{\text{obs}}, \mathbf{X}, \mathbf{Z}) = \frac{p(\theta) \cdot f(\mathbf{Y}^{\text{obs}}, \mathbf{W}^{\text{obs}} | \mathbf{X}, \mathbf{Z}, \theta)}{f(\mathbf{Y}^{\text{obs}}, \mathbf{W}^{\text{obs}} | \mathbf{X}, \mathbf{Z})},$$

where the denominator,

$$f(\mathbf{Y}^{\text{obs}}, \mathbf{W}^{\text{obs}} | \mathbf{X}, \mathbf{Z}) = \int p(\theta) \cdot f(\mathbf{Y}^{\text{obs}}, \mathbf{W}^{\text{obs}} | \mathbf{X}, \mathbf{Z}, \theta) d\theta,$$

is the marginal distribution of the observed outcomes given θ, integrated over the vector parameter θ.

25.4.5 Derivation of the Posterior Distribution of Missing Potential Outcomes $f(\mathbf{Y}^{\text{mis}}, \mathbf{W}^{\text{mis}} | \mathbf{Y}^{\text{obs}}, \mathbf{W}^{\text{obs}}, \mathbf{X}, \mathbf{Z})$

The third step outlined in Chapter 8 applies to the current situation without any modification. Combining the conditional posterior distribution of the missing potential outcomes, given the parameter θ, $f(\mathbf{Y}^{\text{mis}}, \mathbf{W}^{\text{mis}} | \mathbf{Y}^{\text{obs}}, \mathbf{W}^{\text{obs}}, \mathbf{X}, \mathbf{Z}, \theta)$, with the posterior distribution of θ, $p(\theta | \mathbf{Y}^{\text{obs}}, \mathbf{W}^{\text{obs}}, \mathbf{X}, \mathbf{Z})$, and integrating over θ, we obtain the posterior distribution of the missing potential outcomes:

$$f(\mathbf{Y}^{\text{mis}}, \mathbf{W}^{\text{mis}} | \mathbf{Y}^{\text{obs}}, \mathbf{W}^{\text{obs}}, \mathbf{X}, \mathbf{Z})$$

$$= \int_{\theta} f(\mathbf{Y}^{\text{mis}}, \mathbf{W}^{\text{mis}} | \mathbf{Y}^{\text{obs}}, \mathbf{W}^{\text{obs}}, \mathbf{X}, \mathbf{Z}, \theta) \cdot p(\theta | \mathbf{Y}^{\text{obs}}, \mathbf{W}^{\text{obs}}, \mathbf{X}, \mathbf{Z}) \mathrm{d}\theta.$$

25.4.6 Derivation of the Posterior Distribution of Estimands

The final step is again analogous to that in Chapter 8. We have the distribution of the missing potential outcomes given observed potential outcomes, covariates and instruments, $f(\mathbf{Y}^{\text{mis}}, \mathbf{W}^{\text{mis}} | \mathbf{Y}^{\text{obs}}, \mathbf{W}^{\text{obs}}, \mathbf{X}, \mathbf{Z})$. Because of (25.2) we can rewrite any estimand that is a function of $(\mathbf{Y}(0), \mathbf{Y}(1), \mathbf{W}(0), \mathbf{W}(1), \mathbf{X}, \mathbf{Z})$, as a function of missing and observed potential outcomes and covariates and instruments, $(\mathbf{Y}^{\text{mis}}, \mathbf{Y}^{\text{obs}}, \mathbf{W}^{\text{mis}}, \mathbf{W}^{\text{obs}}, \mathbf{X}, \mathbf{Z})$. We combine these two results to infer the posterior distribution of τ given the observed data, $(\mathbf{Y}^{\text{obs}}, \mathbf{W}^{\text{obs}}, \mathbf{X}, \mathbf{Z})$.

25.5 SIMULATION METHODS FOR OBTAINING DRAWS FROM THE POSTERIOR DISTRIBUTION OF THE ESTIMAND GIVEN THE DATA

In many cases the steps outlined in the previous section are often difficult, and in fact essentially impossible, to implement analytically. In practice we therefore often use simulation and, specifically, data augmentation methods to obtain approximations to the posterior distribution of causal estimands. Here we describe the general outline for these methods in instrumental variables settings, meaning settings where we accept randomization of the instrument, no defiers, and the exclusion restrictions on the primary outcome for the nevertakers and the alwaystakers.

The conditional joint distribution of the matrix of primary outcomes $\mathbf{Y}$ is the product of the N conditional distributions given θ, and assuming conditional independence between the potential outcomes under $W_i = 0$ and $W_i = 1$, this can be written as:

$$f(\mathbf{Y}|\mathbf{G}, \mathbf{X}; \theta) = \prod_{i=1}^{N} f(Y_i(0)|G_i, X_i, \theta) \cdot f(Y_i(1)|G_i, X_i, \theta)$$

$$= \prod_{i:G_i=\text{co}} f(Y_i(0)|G_i = \text{co}, X_i, \theta) \cdot f(Y_i(1)|G_i = \text{co}, X_i, \theta)$$

$$\times \prod_{i:G_i=\text{nt}} f(Y_i(0)|G_i = \text{nt}, X_i, \theta)$$

$$\times \prod_{i:G_i=\text{at}} f(Y_i(1)|G_i = \text{at}, X_i, \theta),$$

because the distributions of $Y_i(0)$ for alwaystakers and $Y_i(1)$ for nevertakers are degenerate. Moreover, as above, assume that $f(Y_i(w)|G_i = g, X_i, Z_i, \theta)$, depends only on a subset of the parameter vector, β_{gw}, and assume for notational simplicity that these distributions have the same functional form for all pairs (g, w), so that we can write $f(y|x; \beta_{gw})$ without ambiguity. Moreover, let us specify the compliance type probabilities as

$$f(\mathbf{G}|\mathbf{X}; \theta) = \prod_{i=1}^{N} p(G_i|X_i, \gamma),$$

depending only on a subvector of the full parameter vector, γ, so that $\theta = (\beta_{co,c}, \beta_{co,t}, \beta_{nt}, \beta_{at}, \gamma)$. (For nevertakers and alwaystakers the distribution of $Y_i(0)$ is identical to that of $Y_i(1)$ by the two exclusion restrictions, so we do not index β_{nt} and β_{at} by the treatment received.)

Now, let us consider the actual observed data likelihood function for θ. There are four possible patterns of missing and observed data corresponding to the four possible values for (Z_i, W_i^{obs}): $(0,0)$, $(0, 1)$, $(1, 0)$, and $(1, 1)$. Partition the set of N units in the sample into the subsets of units exhibiting each pattern of missing and observed data, and denote these subsets by $S(z, w)$, for $z, w \in \{0, 1\}$, with $S(z, w) \subset \{1, \dots, N\}$, and $\cup_{z,w} S(z, w) = \{1, 2, \dots, N\}$, where the sets $S(z, w)$ are disjoint, $S(z, w) \cap S(z', w') = \emptyset$, unless $z = z'$ and $w = w'$.

First consider the set $S(0, 1)$. Under the monotonicity assumption we can infer that units with this pattern of observed compliance behavior are alwaystakers, and we observe $Y_i(1)$ for these units. Hence the likelihood contribution from the ith such unit is proportional to:

$$\mathcal{L}_{(0,1),i} = p(G_i = \text{at}|X_i, Z_i, \gamma) \cdot f(Y_i(1)|G_i = \text{at}, X_i, Z_i, \beta_{at}), \quad i \in S(0, 1). \tag{25.4}$$

Next, for units in the set $S(1, 0)$, we can infer that they are nevertakers. Hence the likelihood contribution from the ith such unit is proportional to:

$$\mathcal{L}_{(1,0),i} = p(G_i = \text{nt}|X_i, Z_i, \gamma) \cdot f(Y_i(0)|G_i = \text{nt}, X_i, Z_i, \beta_{nt}), \quad i \in S(1, 0). \tag{25.5}$$

For units in the two remaining sets, we cannot unambiguously infer the compliance type of such units. Consider first $S(0, 0)$. Receiving the control treatment after being assigned to the control treatment is consistent with being either a nevertaker or a complier. The likelihood contribution for units in this set is therefore a mixture of two outcome distributions. First, the outcome distribution for nevertakers under the control treatment, and second, the outcome distribution for compliers under the control treatment. Using this argument, we can write the likelihood contribution for the ith unit in the set $S(0, 0)$ as proportional to:

$$\mathcal{L}_{(0,0),i} = p(G_i = \text{nt}|X_i, Z_i, \gamma) \cdot f(Y_i(0)|G_i = \text{nt}, X_i, Z_i, \beta_{nt}) \tag{25.6}$$

$$+ p(G_i = \text{co}|X_i, Z_i, \gamma) \cdot f(Y_i(0)|G_i = \text{co}, X_i, Z_i, \beta_{co,c}), \quad i \in S(0, 0).$$

The set $S(1, 1)$ is also a mixture of two types, in this case compliers and alwaystakers. Hence, we can write the likelihood contribution for the ith unit in the subset $S(1, 1)$ as proportional to:

$$\mathcal{L}_{(1,1),i} = p(G_i = \text{at}|X_i, Z_i, \gamma) \cdot f(Y_i(1)|G_i = \text{at}, X_i, Z_i, \beta_{\text{at}}) \tag{25.7}$$

$$+ p(G_i = \text{co}|X_i, Z_i, \gamma) \cdot f(Y_i(1)|G_i = \text{co}, X_i, Z_i, \beta_{\text{co,t}}), \quad i \in \mathcal{S}(1, 1).$$

Combining (25.4)–(25.7), we can write the likelihood function in terms of the observed data as

$$\mathcal{L}_{\text{obs}}(\theta|\mathbf{Z}^{\text{obs}}, \mathbf{W}^{\text{obs}}, \mathbf{Y}^{\text{obs}}, \mathbf{X}^{\text{obs}})$$

$$= \prod_{i \in \mathcal{S}(0,1)} p(G_i = \text{at}|X_i, Z_i, \gamma) \cdot f(Y_i(1)|G_i = a, X_i, Z_i, \beta_{\text{at}})$$

$$\times \prod_{i \in \mathcal{S}(1,0)} p(G_i = \text{nt}|X_i, Z_i, \gamma) \cdot f(Y_i(0)|G_i = \text{nt}, X_i, Z_i, \beta_{\text{nt}})$$

$$\times \prod_{i \in \mathcal{S}(0,0)} \left[p(G_i = \text{nt}|X_i, Z_i, \gamma) \cdot f(Y_i(0)|G_i = \text{nt}, X_i, Z_i, \beta_{\text{nt}}) \right.$$

$$\left. + p(G_i = \text{co}|X_i, Z_i, \gamma) \cdot f(Y_i(0)|G_i = \text{co}, X_i, Z_i, \beta_{\text{co,c}}) \right]$$

$$\times \prod_{i \in \mathcal{S}(1,1)} \left[p(G_i = \text{at}|X_i, Z_i, \gamma) \cdot f(Y_i(1)|G_i = \text{at}, X_i, Z_i, \beta_{\text{at}}) \right.$$

$$\left. + p(G_i = \text{co}|X_i, Z_i, \gamma) \cdot f(Y_i(1)|G_i = \text{co}, X_i, Z_i, \beta_{\text{co,t}}) \right].$$

This likelihood function has a very specific mixture structure. Had we observed the full compliance types of the units, the resulting complete-data likelihood function would factor into five components, each component depending on a single (*a priori* independent) subvector of the full parameter vector:

$$\mathcal{L}_{\text{comp}}(\theta|\mathbf{G}, \mathbf{Z}, \mathbf{W}^{\text{obs}}, \mathbf{Y}^{\text{obs}}, \mathbf{X})$$

$$= \prod_{i:G_i=\text{nt}} f(Y_i(0)|G_i = \text{nt}, X_i, Z_i, \beta_{\text{nt}})$$

$$\times \prod_{i:G_i=\text{at}} f(Y_i(1)|G_i = \text{at}, X_i, Z_i, \beta_{\text{at}}) \prod_{i:G_i=\text{co},Z_i=0} f(Y_i(0)|G_i = \text{co}, X_i, Z_i, \beta_{\text{co,c}})$$

$$\times \prod_{i:G_i=\text{co},Z_i=1} f(Y_i(1)|G_i = \text{at}, X_i, Z_i, \beta_{\text{co,t}}) \prod_{i:G_i=\text{co}} p(G_i = \text{co}|X_i, Z_i, \beta)$$

$$\times \prod_{i:G_i=\text{at}} p(G_i = \text{at}|X_i, Z_i, \beta) \prod_{i:G_i=\text{nt}} p(G_i = \text{nt}|X_i, Z_i, \beta).$$

With conventional models for the distribution of the potential outcomes and the compliance types, and with conventional prior distributions, analyzing posterior distributions given this complete-data likelihood function would be straightforward. Specifically, it would generally be simple to estimate the parameters by maximum likelihood or Bayesian methods. The key complication is that we do not fully observe the compliance types, leading to the mixture observed-data likelihood function. To exploit the simplicity of the complete-data likelihood function, it is useful to use missing data methods, either the Expectation Maximization (EM) algorithm to find the maximum likelihood estimates (i.e., the posterior mode given a flat prior distribution), or Data Augmentation (DA) methods to obtain draws from the posterior distribution.

The key step in these methods is to impute, either stochastically in a DA algorithm or by expectation in the EM algorithm, the missing compliance type, conditional on both current draws or estimates of the parameters, and the observed data. For units with $i \in (\mathcal{S}(0,1) \cup \mathcal{S}(1,0))$, we know the compliance type, either alwaystaker or nevertaker. For units with $i \in (\mathcal{S}(0,0) \cup \mathcal{S}(1,1))$ we do not know the compliance type. Specifically, for units with $i \in \mathcal{S}(0,0)$ we can only infer that they are either nevertakers or compliers. The probability of such a unit being a complier given observed data and parameters is

$$\Pr(G_i = \text{co} | Y_i^{\text{obs}} = y, W_i^{\text{obs}} = 0, Z_i = 0, X_i = x, \theta) \tag{25.8}$$

$$= \frac{p(\text{co}|x, \beta) \cdot f(y|x, \beta_{\text{co,c}})}{p(G_i = \text{co}|x, \beta) \cdot f(y|x, \beta_{\text{co,c}}) + p(G_i = \text{nt}|x, \beta) \cdot f(y|x, \beta_{\text{nt}})}.$$

Similarly,

$$\Pr(G_i = \text{co} | Y_i^{\text{obs}} = y, W_i^{\text{obs}} = 1, Z_i = 1, X_i = x, \theta) \tag{25.9}$$

$$= \frac{p(\text{co}|x, \beta) \cdot f(y|x, \beta_{\text{co,t}})}{p(G_i = \text{co}|x, \beta) \cdot f(y|x, \beta_{\text{co,t}}) + p(G_i = \text{at}|x, \beta) \cdot f(y|x, \beta_{\text{at}})}.$$

It is straightforward to obtain draws from this distribution, or to calculate the numerical conditional probabilities.

To be specific, suppose we wish to obtain draws from the posterior distribution of the average effect of the treatment on the outcome for compliers, the local average treatment effect (or the complier average causal effect),

$$\tau_{\text{late}} = \frac{1}{N_{\text{co}}} \sum_{i:G_i=\text{co}} \left(Y_i(1) - Y_i(0) \right),$$

where $N_{\text{co}} = \sum_{i=1}^{N} \mathbf{1}_{G_i=\text{co}}$ is the number of compliers. To estimate τ_{late} we obtain such draws using the Data Augmentation algorithm. Starting with initial values of the parameter θ, we simulate the compliance type, using (25.10) and (25.9). Given the complete (compliance) data $\mathbf{G}$, and given the parameters, we draw from the posterior predictive distribution of the missing potential outcomes for compliers. That is, for compliers with $Z_i = 0$, we impute $Y_i(1)$, and for compliers with $Z_i = 1$, we impute $Y_i(0)$. With these imputations we can calculate the value of τ_{late}. In the fourth step we update the parameters given $(\mathbf{Y}^{\text{obs}}, \mathbf{G}, \mathbf{Z}, \mathbf{X})$. Then we return to the imputation of the compliance types given the updated parameters.

25.6   MODELS FOR THE INFLUENZA VACCINATION DATA

In this section we illustrate the methods discussed in the previous section using the flu-shot data set with information on 1,931 women. We focus on estimating the average effect of the flu shot on flu-related hospitalizations for compliers. We exploit the presence of the four covariates, age (age minus 65, in tens of years), copd (heart disease), and heart dis (indicator for prior heart conditions), in order to improve precision and to obtain subpopulation causal effects. Because the instrument, receipt of the letter, was randomly assigned irrespective of covariate values, the posterior distribution for the complier average treatment effect is likely to be relatively robust to modeling choices regarding the conditional distributions given the covariates. In settings where the instrument is correlated with the covariates, such assumptions are likely to be more important.

25.6.1 A Model for Outcomes Given Compliance Type

First, we specify a model for the conditional distribution of the potential outcomes given compliance type, covariates, and parameters. We use a logistic regression model, although other binary regression models are certainly possible. For compliers

$$\Pr(Y_i(0) = y | X_i = x, G_i = \text{co}, \theta) = \frac{\exp(y \cdot x\beta_{\text{co,c}})}{1 + \exp(x\beta_{\text{co,c}})},$$

$$\Pr(Y_i(1) = y | X_i = x, G_i = \text{co}, \theta) = \frac{\exp(y \cdot x\beta_{\text{co,t}})}{1 + \exp(x\beta_{\text{co,t}})},$$

for nevertakers,

$$\Pr(Y_i(0) = y | X_i = x, G_i = \text{nt}, \theta) = \frac{\exp(y \cdot x\beta_{\text{nt}})}{1 + \exp(x\beta_{\text{nt}})},$$

and, finally, for alwaystakers,

$$\Pr(Y_i(1) = y | X_i = x, G_i = \text{at}, \theta) = \frac{\exp(y \cdot x\beta_{\text{at}})}{1 + \exp(x\beta_{\text{at}})}.$$

Given this model, the super-population average effect of the treatment on the outcome for compliers with $X_i = x$, is equal to

$$\mathbb{E}\left[Y_i(1) - Y_i(0) | G_i = \text{co}, X_i = x\right] = \frac{\exp(x\beta_{\text{co,t}})}{1 + \exp(x\beta_{\text{co,t}})} - \frac{\exp(x\beta_{\text{co,c}})}{1 + \exp(x\beta_{\text{co,c}})}.$$

However, this super-population average treatment effect is not what we want to estimate. Instead we are interested in the average causal effect for compliers in the sample,

$$\tau_{\text{late}} = \frac{1}{N_{\text{co}}} \sum_{i:G_i=\text{co}} \left(Y_i(1) - Y_i(0)\right).$$

25.6.2 A Model for Compliance Type

The compliance type is a three-valued indicator. We model this through a trinomial logit model:

$$
\Pr(G_i = g | X_i = x) =
\begin{cases}
\dfrac{1}{1 + \exp(x\gamma_{at}) + \exp(x\gamma_{nt})}, & \text{if } g = co, \\[2ex]
\dfrac{\exp(x\gamma_{nt})}{1 + \exp(x\gamma_{at}) + \exp(x\gamma_{nt})}, & \text{if } g = nt, \\[2ex]
\dfrac{\exp(x\gamma_{at})}{1 + \exp(x\gamma_{at}) + \exp(x\gamma_{nt})}, & \text{if } g = at,
\end{cases}
$$

where, again, x includes a constant term. An alternative, discussed in Section 25.4.2, is to model first the probability of being a complier versus a noncomplier, and then the probability of being a nevertaker versus alwaystaker conditional on being a noncomplier.

25.6.3 The Prior Distribution

The prior distribution is based on adding artificial observations. These artificial observations are somewhat special because we assume that for them we observe not only the values of the instruments, the treatment, the outcome, and the covariates, but also their compliance type (which is observed for only some but not all units in the sample even under monotonicity and both exclusion restrictions). Each of the artificial observations is of the type (y_a, z_a, g_a, x_a), where $y_a \in \{0, 1\}$, $z_a \in \{0, 1\}$, $g_a \in \{co, nt, at\}$, and x_a takes on the observed values of the covariates in the sample. Because there are potentially $N = 1,931$ different values of the vector of covariates in the actual sample, there are $2 \times 2 \times 3 \times N$ different values for the quadruple (y_a, z_a, g_a, x_a). For example, there are $4 \times N$ artificial observations that are nevertakers, for each value of x_a, two with $y_a = 0$ and two with $y_a = 1$. Fixing the total weight for these $12 \times N$ artificial observations at N_a, the prior distribution for $\theta = (\beta_{co,c}, \beta_{co,t}, \beta_{nt}, \beta_{at}, \gamma_{nt}, \gamma_{at})$ takes the form

$$
p(\theta) = \left[\prod_{i=1}^{N} \left(\frac{1}{1 + \exp(X_i\beta_{at})} \right)^2 \left(\frac{\exp(X_i\beta_{at})}{1 + \exp(X_i\beta_{at})} \right)^2 \right.
$$

$$
\times \prod_{i=1}^{N} \left(\frac{1}{1 + \exp(X_i\beta_{nt})} \right)^2 \left(\frac{\exp(X_i\beta_{nt})}{1 + \exp(X_i\beta_{nt})} \right)^2
$$

$$
\times \prod_{i=1}^{N} \frac{1}{1 + \exp(X_i\beta_{co,c})} \cdot \frac{\exp(X_i\beta_{co,c})}{1 + \exp(X_i\beta_{co,c})}
$$

$$
\times \prod_{i=1}^{N} \frac{1}{1 + \exp(X_i\beta_{co,t})} \cdot \frac{\exp(X_i\beta_{co,t})}{1 + \exp(X_i\beta_{co,t})}
$$

$$
\times \prod_{i=1}^{N} \frac{1}{1 + \exp(X_i\gamma_{at}) + \exp(X_i\gamma_{nt})}
$$

$$\times \prod_{i=1}^{N} \frac{\exp(X_i \gamma_{\mathrm{at}})}{1 + \exp(X_i \gamma_{\mathrm{at}}) + \exp(X_i \gamma_{\mathrm{nt}})}$$

$$\times \prod_{i=1}^{N} \frac{\exp(X_i \gamma_{\mathrm{nt}})}{1 + \exp(X_i \gamma_{\mathrm{at}}) + \exp(X_i \gamma_{\mathrm{nt}})} \Bigg]^{N_a/(12 \times N)}.$$

By keeping the value of N_a small but positive (with the limiting prior distribution flat, i.e., constant, as $N_a \to 0$), we limit the total influence of the prior distribution, while ensuring that the resulting posterior distribution is proper. In the implementation here, we fix N_a at 30, which is small relative to the actual sample size, which is nearly 100 times as large.

25.6.4 Implementation

Given the specification for the outcome distributions and the specification for the model of compliance type, the full parameter vector is $\theta = (\gamma_{\mathrm{nt}}, \gamma_{\mathrm{at}}, \beta_{\mathrm{nt}}, \beta_{\mathrm{at}}, \beta_{\mathrm{co},c}, \beta_{\mathrm{co},t})$. The likelihood function is

$$\mathcal{L}_{\mathrm{obs}}(\theta | \mathbf{Z}^{\mathrm{obs}}, \mathbf{W}^{\mathrm{obs}}, \mathbf{Y}^{\mathrm{obs}}, \mathbf{X}^{\mathrm{obs}})$$

$$= \prod_{i \in \mathcal{S}(0,0)} \Bigg[\frac{\exp(X_i \gamma_{\mathrm{nt}})}{1 + \exp(X_i \gamma_{\mathrm{at}}) + \exp(X_i \gamma_{\mathrm{nt}})} \cdot \frac{\exp(Y_i^{\mathrm{obs}} \cdot X_i \beta_{\mathrm{nt}})}{1 + \exp(X_i \beta_{\mathrm{nt}})}$$

$$+ \frac{1}{1 + \exp(X_i \gamma_{\mathrm{at}}) + \exp(X_i \gamma_{\mathrm{nt}})} \cdot \frac{\exp(Y_i^{\mathrm{obs}} \cdot X_i \beta_{\mathrm{co},c})}{1 + \exp(X_i \beta_{\mathrm{co},c})} \Bigg]$$

$$\times \prod_{i \in \mathcal{S}(0,1)} \frac{\exp(X_i \gamma_{\mathrm{at}})}{1 + \exp(X_i \gamma_{\mathrm{at}}) + \exp(X_i \gamma_n)} \cdot \frac{\exp(Y_i^{\mathrm{obs}} \cdot X_i \beta_{\mathrm{at}})}{1 + \exp(X_i \beta_{\mathrm{at}})}$$

$$\times \prod_{i \in \mathcal{S}(1,0)} \frac{\exp(X_i \gamma_{\mathrm{nt}})}{1 + \exp(X_i \gamma_{\mathrm{at}}) + \exp(X_i \gamma_{\mathrm{nt}})} \cdot \frac{\exp(Y_i^{\mathrm{obs}} \cdot X_i \beta_{\mathrm{nt}})}{1 + \exp(X_i \beta_{\mathrm{nt}})}$$

$$\times \prod_{i \in \mathcal{S}(1,1)} \Bigg[\frac{\exp(X_i \gamma_{\mathrm{at}})}{1 + \exp(X_i \gamma_{\mathrm{at}}) + \exp(X_i \gamma_{\mathrm{nt}})} \cdot \frac{\exp(Y_i^{\mathrm{obs}} \cdot X_i \beta_{\mathrm{at}})}{1 + \exp(X_i \beta_{\mathrm{at}})}$$

$$+ \frac{1}{1 + \exp(X_i \gamma_{\mathrm{at}}) + \exp(X_i \gamma_{\mathrm{nt}})} \cdot \frac{\exp(Y_i^{\mathrm{obs}} \cdot X_i \beta_{\mathrm{co},t})}{1 + \exp(X_i \beta_{\mathrm{co},t})} \Bigg].$$

Let us consider implementing our methods, using this likelihood function and the specified prior distribution, on the flu data. To be specific, we will focus on obtaining draws from the posterior distribution for

$$\tau_{\mathrm{late}} = \frac{1}{N_{\mathrm{co}}} \sum_{i:G_i = \mathrm{co}} \Big(Y_i(1) - Y_i(0) \Big).$$

We condition on the number of compliers in the sample being positive, so draws from the vector of compliance types with no compliers are discarded.

Let $\theta_{(0)}$ denote the starting values for the parameter vector θ. Given these values we impute the compliance type G_i for units i with $i \in \mathcal{S}(0,0)$ and $i \in \mathcal{S}(1,1)$ (for units with $i \in \mathcal{S}(0,1)$ and $i \in \mathcal{S}(1,0)$ we can directly infer the compliance type). This imputation involves drawing from a binomial distribution. Specifically, consider a unit i with $i \in \mathcal{S}(0,0)$, that is, a unit with $Z_i = 0$ and $W_i^{\text{obs}} = 0$. Suppose this unit has a realized outcome $Y_i^{\text{obs}} = y$. We cannot infer with certainty whether this unit is a complier or a nevertaker, although we can be sure this unit is not an alwaystaker. The probability of this unit being a complier is

$$\Pr(G_i = \text{co}|Y_i^{\text{obs}} = y, W_i^{\text{obs}} = 0, Z_i = 0, X_i = x, \theta) \tag{25.10}$$

$$= \frac{p(\text{co}|x,\beta) \cdot f(y|x,\beta_{\text{co,c}})}{p(\text{co}|x,\beta) \cdot f(y|x,\beta_{\text{co,c}}) + p(\text{nt}|x,\beta) \cdot f(y|x,\beta_{\text{nt}})}$$

$$= \left(\frac{\frac{1}{1+\exp(x\gamma_{\text{at}})+\exp(x\gamma_{\text{nt}})} \cdot \frac{\exp(x\beta_{\text{co,c}})}{1+\exp(x\beta_{\text{co,c}})}}{\frac{1}{1+\exp(x\gamma_{\text{at}})+\exp(x\gamma_{\text{nt}})} \cdot \frac{\exp(x\beta_{\text{co,c}})}{1+\exp(x\beta_{\text{co,c}})} + \frac{\exp(\gamma_{\text{nt}})}{1+\exp(x\gamma_{\text{at}})+\exp(x\gamma_{\text{nt}})} \cdot \frac{\exp(x\beta_{\text{nt}})}{1+\exp(x\beta_{\text{nt}})}} \right)^y$$

$$\times \left(\frac{\frac{1}{1+\exp(x\gamma_{\text{at}})+\exp(x\gamma_{\text{nt}})} \cdot \frac{1}{1+\exp(x\beta_{\text{co,c}})}}{\frac{1}{1+\exp(x\gamma_{\text{at}})+\exp(x\gamma_{\text{nt}})} \cdot \frac{1}{1+\exp(x\beta_{\text{co,c}})} + \frac{\exp(x\gamma_{\text{nt}})}{1+\exp(x\gamma_{\text{at}})+\exp(x\gamma_{\text{nt}})} \cdot \frac{1}{1+\exp(x\beta_{\text{nt}})}} \right)^{1-y}.$$

Similarly, we impute the compliance type for units with $Z_i = 1$ and $W_i^{\text{obs}} = 1$, who may be compliers or nevertakers.

In the second step we impute the missing potential outcomes for all units. For compliers with $Z_i = 0$, this means imputing $Y_i(1)$ from a binomial distribution with mean $(\exp(X_i\beta_{\text{co,c}})/(1 + \exp(X_i\beta_{\text{co,c}}))$, and for compliers with $Z_i = 1$, this means imputing $Y_i(0)$ from a binomial distribution with mean $\exp(X_i\beta_{\text{co,t}})/(1 + \exp(X_i\beta_{\text{co,t}}))$, using the appropriate subvectors of the current parameter value $\theta_{(0)}$. Given these values, we can calculate the average treatment effect for compliers, $\tau_{\text{late},(0)}$, where the second subscript indexes the iteration. We also impute the missing potential outcomes for nevertakers and alwaystakers in order to update the parameter vectors in the third step.

In the third step we update the parameter vector, from $\theta_{(k)}$ to $\theta_{(k+1)}$. For each of the subvectors $(\gamma_{\text{nt}}, \gamma_{\text{at}})$, β_{nt}, β_{at}, $\beta_{\text{co,c}}$, and $\beta_{\text{co,t}}$, we use the corresponding factor of the complete-data likelihood function with a Metropolis-Hastings step (see, e.g., Gelman, Carlin, Stern and Rubin, 1995). In each of these five cases, this is a straightforward step, with the likelihood in four cases corresponding to a binary logit one, and in one case corresponding to a trinomial logit model.

25.7 RESULTS FOR THE INFLUENZA VACCINATION DATA

Now let us apply this approach to the flu-shot data. We implement five versions. First, in the first column of Table 25.4, we report maximum likelihood estimates (posterior modes with a flat prior distribution) assuming no covariates were observed. In the randomized experiment settings we considered in Chapter 8, maximum likelihood estimates for the parameters of the statistical model were generally close to the posterior means, and asymptotic standard errors (estimated using the information matrix) were close to

Table 25.4. *Model-Based Estimates of Local Average Treatment and Intention-to-Treat Effects: Posterior Quantiles, for Influenza Vaccination Data*

	MLE (flat prior)	Post Mode	No Cov $q^{.025}$	med	$q^{.975}$	Parallel $q^{.025}$	med	$q^{.975}$	Unrestricted $q^{.025}$	med	$q^{.975}$
τ_{late}	−0.11	−0.10	−0.32	**−0.15**	−0.02	−0.32	**−0.14**	−0.01	−0.48	**−0.16**	0.13
ITT_W	0.12	0.13	0.09	**0.12**	0.15	−0.03	**−0.02**	−0.00	0.05	**0.10**	0.13
ITT_Y	−0.01	−0.01	−0.04	**−0.02**	−0.00	0.09	**0.12**	0.15	−0.04	**−0.02**	0.01
$\mathbb{E}[Y_i(0)\|G_i=\text{co}]$	0.11	0.15	0.06	**0.19**	0.35	0.69	**0.71**	0.72	0.01	**0.22**	0.53
$\mathbb{E}[Y_i(1)\|G_i=\text{co}]$	0.00	0.06	0.00	**0.03**	0.09	0.16	**0.18**	0.19	0.00	**0.04**	0.31
$\mathbb{E}[Y_i(0)\|G_i=\text{nt}]$	0.08	0.08	0.06	**0.07**	0.08	0.06	**0.18**	0.35	0.06	**0.07**	0.08
$\mathbb{E}[Y_i(1)\|G_i=\text{at}]$	0.11	0.11	0.08	**0.09**	0.10	0.00	**0.04**	0.09	0.06	**0.09**	0.10
$\beta_{\text{co},c,\text{intercept}}$	−2.07	−1.71	−2.64	**−1.42**	−0.53	−3.32	**−2.06**	−1.09	−7.83	**−2.88**	0.17
$\beta_{\text{co},c,\text{age}}$						−0.25	**−0.11**	0.02	−0.04	**0.02**	0.10
$\beta_{\text{co},c,\text{copd}}$						0.08	**0.46**	0.83	−5.06	**1.29**	7.66
$\beta_{\text{co},c,\text{heart}}$						0.42	**0.79**	1.16	−2.35	**1.88**	6.00
$\beta_{\text{co},t,\text{intercept}}$	$-\infty$	−2.82	−4.71	**−3.13**	−2.09	−5.30	**−3.70**	−2.66	−14.05	**−4.18**	0.61
$\beta_{\text{co},t,\text{age}}$						−0.25	**−0.11**	0.02	−0.02	**0.06**	0.20
$\beta_{\text{co},t,\text{copd}}$						0.08	**0.46**	0.83	−6.56	**0.37**	7.32
$\beta_{\text{co},t,\text{heart}}$						0.42	**0.79**	1.16	−4.75	**0.58**	5.34
$\beta_{\text{nt},\text{intercept}}$	−2.41	−2.42	−2.82	**−2.55**	−2.31	−3.56	**−3.16**	−2.79	−3.72	**−3.23**	−2.81
$\beta_{\text{nt},\text{age}}$						−0.25	**−0.11**	0.02	0.02	**0.04**	0.06
$\beta_{\text{nt},\text{copd}}$						0.08	**0.46**	0.83	0.13	**0.75**	1.28
$\beta_{\text{nt},\text{heart}}$						0.42	**0.79**	1.16	0.18	**0.71**	1.23
$\beta_{\text{at},\text{intercept}}$	−2.08	−2.13	−2.58	**−2.18**	−1.82	−3.36	**−2.84**	−2.37	−3.67	**−2.78**	−2.09
$\beta_{\text{at},\text{age}}$						−0.25	**−0.11**	0.02	0.02	**0.04**	0.06
$\beta_{\text{at},\text{copd}}$						0.08	**0.46**	0.83	−0.91	**0.07**	0.95
$\beta_{\text{at},\text{heart}}$						0.42	**0.79**	1.16	−0.25	**0.58**	1.47
$\gamma_{\text{nt},\text{intercept}}$	1.78	1.66	1.47	**1.74**	2.06	1.36	**1.80**	2.39	1.40	**1.91**	3.03
$\gamma_{\text{nt},\text{age}}$						−0.19	**0.02**	0.22	−0.22	**0.04**	0.26
$\gamma_{\text{nt},\text{copd}}$						−0.48	**0.28**	1.50	−0.30	**0.69**	2.36
$\gamma_{\text{nt},\text{heart dis}}$						−0.80	**−0.14**	0.46	−1.18	**−0.12**	0.64
$\gamma_{\text{at},\text{intercept}}$	0.48	0.36	0.05	**0.36**	0.74	−0.35	**0.21**	0.91	−0.28	**0.35**	1.63
$\gamma_{\text{at},\text{age}}$						−0.01	**0.25**	0.50	−0.05	**0.25**	0.53
$\gamma_{\text{at},\text{copd}}$						−0.12	**0.79**	2.10	0.06	**1.21**	2.94
$\gamma_{\text{at},\text{heart}}$						−0.79	**−0.02**	0.73	−1.16	**0.01**	0.88

posterior standard deviations. This is sometimes the case in instrumental variables settings, but there can be substantial differences because of the restrictions on the joint distributions of the observed variables implied by the instrumental variables assumptions. Moreover, in such cases, the curvature of the logarithm of the likelihood function need not provide a good approximation to either the posterior standard deviation or the repeated sampling standard error of the estimates. For the flu-shot data, one of the parameter estimates is zero, which is on the boundary of the parameter space, which is not surprising considering that the simple moment-based estimate of $\mathbb{E}[Y_i(1)|G_i = \text{at}]$ is negative. In the second column of Table 25.4 we report posterior modes given the prior distribution we use, again assuming no covariates were observed.

Next, we report summary statistics for posterior distributions for three models. First, we report summary statistics for the posterior distribution for the model assuming no covariates were observed. We report posterior medians and posterior 0.025 and 0.975 quantiles, in Columns 3–5 of Table 25.4. Here the posterior medians are fairly close

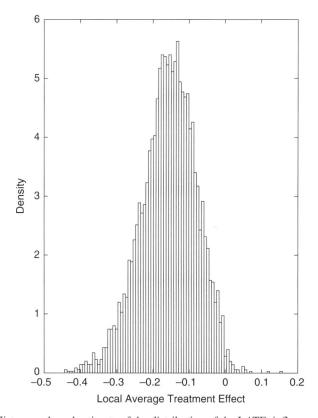

Figure 25.1.   Histogram-based estimate of the distribution of the LATE, influenza vaccination data

to the maximum likelihood estimates for β_{nt} and β_{at}. The posterior medians for the parameters for the outcome distributions for the compliers are somewhat different from the maximum likelihood estimates, which is not surprising considering that one of the maximum likelihood estimates is on the boundary of the parameter space. A normal approximation to the posterior distribution for the local average treatment effect is not particularly accurate. The skewness of the posterior distribution is -0.79, and the kurtosis is 3.29. Interestingly, the posterior probability that τ_{late} is exactly equal to zero is 0.08, which corresponds to the probability that, among the compliers, the fractions of treated and control units that are hospitalized are exactly equal. Figure 25.1 presents a histogram estimate of the marginal posterior distribution of τ_{late} under this model.

The second model includes the three covariates, age, copd, and heart dis. We assume the slope coefficients in the models for the four potential hospitalization outcomes are identical for the groups, nevertakers, alwaystakers, and compliers with and without flu shots. The posterior 0.025 and 0.975 quantiles and the median for this model are reported in Columns 6–8 of Table 25.4. Finally, we relax the model and allow the slope coefficients to differ between the four potential hospitalization outcomes. The results for this model are reported in the Columns 9–11 of Table 25.4.

Note that in all cases the posterior medians are well above the moment-based estimates of the local average treatment effect. The reason for this result is that the moment-based estimate is based implicitly on estimating the probability of being hospitalized

Table 25.5. *Estimated Average Covariate Values by Compliance Type, for Influenza Vaccination Data*

Sample	Models for Potential Outcomes given Covariates		
	No Cov	Parallel	Unrestricted
Compliers			
age	65.5	64.9	64.9
copd	0.21	0.16	0.12
heart	0.57	0.58	0.57
Nevertakers			
age	70.7	70.8	70.7
copd	0.19	0.20	0.20
heart	0.55	0.55	0.55
Alwaystakers			
age	75.0	75.0	75.0
copd	0.24	0.26	0.27
heart	0.60	0.60	0.60

for compliers given the flu shot to be negative. The model-based estimates restrict this to be non-negative, leading to a higher value for the posterior medians than the method-of-moments estimates.

In Table 25.5 we report estimated values for average covariate values for the three covariates, age, copd, and heart, within each of the three complier types to assess how different the three compliance types are. We see that nevertakers are older, less likely to have heart complications compared to compliers. Alwaystakers are even older, and more likely to have heart complications compared to compliers. Such results may be of substantive relevance.

25.8   CONCLUSION

In this chapter we discuss a model-based approach to estimating causal effects in instrumental variables settings. A key conceptual advantage of a model-based approach over a moment-based one is that with a model-based approach, it is straightforward to incorporate the restrictions implied by the instrumental variables assumptions, or to relax them. We discuss in detail the complications in analysis relative to the method-of-moments-based approach in settings with randomized experiments and unconfoundedness.

NOTES

The model-based approach to instrumental variables discussed in the current chapter was developed in Imbens and Rubin (1997b) and Hirano, Imbens, Rubin, and Zhou (2000). Hirano, Imbens, Rubin, and Zhou (2000) also consider alternative models where the exclusion restrictions are relaxed. Rubin and Zell (2010) uses a similar model. Rubin,

Wang, Yin, and Zell (2010) uses the two binary models, one for being a complier or not, and one for being a nevertaker conditional on not being a complier, to model the three-valued compliance status indicator.

An alternative Bayesian posterior predictive check approach to assessing Fishers sharp null hypothesis in in the presence of noncompliance is proposed in Rubin (1998). Although that is not developed in this text, it may have interesting applications.

Distance to facilities that provide particular treatments are a widely used class of instruments. In health care settings examples include McClellan and Newhouse (1994) and Baiocchi, Small, Lorch, and Rosenbaum (2010). In the economics literature examples include the use of distance to college in Card (1995) and Kane and Rouse (1995).

For the Data Augmentation algorithm, see Tanner and Wong (1987), Tanner (1996), and Gelman, Carlin, Stern, and Rubin (1995).

## PART VII

Conclusion

## CHAPTER 26

Conclusions and Extensions

In this text we exposited the potential outcomes approach to causality, also known as the Rubin Causal Model, and hope to have convinced the reader of its usefulness. In this final chapter we briefly summarize this approach and discuss other topics in causal inference where this approach may be useful. Many of these are areas of ongoing research, and we hope to discuss them in more detail in a second volume.

The starting point of our approach is the notion of potential outcomes. For each unit in a population, and for each level of a treatment, there is a potential outcome. Comparisons of these potential outcomes define the causal effects; we view these as well-defined irrespective of the assignment mechanism, and thus irrespective of what we actually are able to observe. We often place restrictions on these potential outcomes. Most important in the current text is the stability assumption, or SUTVA, that rules out differences between potential outcomes corresponding to different levels of the treatment for units other than the unit under consideration, and rules out unrepresented levels of treatments

We can observe at most one of the potential outcomes for each unit. Causal inference is therefore intrinsically a missing data problem. Given the potential outcomes, there is a key role in our approach for the assignment mechanism, which defines which potential outcomes are observed and which are missing. The current text is largely organized by different types of assignment mechanisms. The simplest is that of a classical randomized experiment where the researcher knows the assignment mechanism entirely. Such assignment mechanisms are discussed in Part II of the text. Then, in the main part of the text, Parts III and IV, we discuss regular assignment mechanisms where we know part but not all of the assignment mechanism. We discuss the importance of the design stage of a study for causal effects where the outcome data are not yet used. At this stage a researcher can carry out preliminary analyses that make the final analyses that do involve the outcome data more credible and robust.

In Part V we examine the unconfoundedness assumption, which implies that units with the same values of the pre-treatment variables but different treatment levels are comparable in terms of potential outcome distributions. First we assess its plausibility, and then we discuss the sensitivity of conclusions based on its possible violations.

In Part VI we discuss some particular, non-regular, assignment mechanisms involving noncompliance with assigned treatments, in particular, instrumental variables settings.

There are many areas of causal inference that we do not discuss in the current text, and which we intend to discuss in a second volume. A partial list of such methods where we feel the potential outcome framework can clarify assumptions and methods includes settings where SUTVA is violated because there are network or peer effects. It also includes generalizations of instrumental variables settings to principal stratification where there are latent strata such that unconfoundedness holds generally only within the strata. We will also discuss treatments that take on more than two values, including both finite unordered discrete cases and continuous dose-response cases. We also plan to discuss dynamic, sequential, treatment settings. Another currently active area of research is regression discontinuity designs, both sharp and fuzzy, where the overlap assumption regarding covariate distributions is not necessarily satisfied, but the extrapolation is limited. A set of methods popular in economics is referred to as difference-in-differences. A related set of methods includes the use of artificial control groups. In epidemiological settings case-control studies are popular, which we intend to discuss from our perspective. Causal methods are now also used in duration settings, which we also intend to address.

NOTES

Many of the topics mentioned in this chapter are the subject of active research. General texts on evaluation methods in economics, with a special focus on regression methods, include Angrist and Krueger (2000) and Angrist and Pischke (2008). A more general social science text is Shadish, Campbell, and Cook (2002). See also Gelman and Hill (2006). Papers on difference-in-differences methods include Abadie (2005), Athey and Imbens (2006), and Blundell, Gosling, Ichimura, and Meghir (2007). For regression discontinuity designs, see Thistlewaite and Campbell (1960), Goldberger (1991), Black (1999), Van Der Klaauw (2002), Imbens and Lemieux (2008), Hahn, Todd, and Van Der Klaauw (2000), Porter (2003), Imbens and Kalyanaraman (2012), Lee and Lemieux (2010), and Lee (2008). Artificial control groups were introduced by Abadie, Diamond, and Hainmueller (2010). For discussions in duration models, see Abbring and Van Den Berg (2003) and Ham and Lalonde (1996). For the notion of the generalized propensity score and multi-valued treatments, see Imbens (2000), Hirano and Imbens (2004), Yang, Imbens, Cui, Faries, and Kadziola (2014), and Imai and Van Dyk (2004). Principal stratification was introduced in Frangakis and Rubin (2002), and a recent application is Frumento, Mealli, Pacini, and Rubin (2012).

References

Abadie, A. (2002), "Bootstrap Tests for Distributional Treatment Effects in Instrumental Variable Models," *Journal of the American Statistical Association*, Vol. 97(457): 284–292.

Abadie, A. (2003), "Semiparametric Instrumental Variable Estimation of Treatment Response Models," *Journal of Econometrics*, Vol. 113(2): 231–263.

Abadie, A. (2005): "Semiparametric Difference-in-Differences Estimators," *Review of Economic Studies*, Vol. 72(1): 1–19.

Abadie, A., J. Angrist, and G. Imbens, (2002), "Instrumental Variables Estimation of Quantile Treatment Effects," *Econometrica*, Vol. 70(1): 91–117.

Abadie, A., A. Diamond, and J. Hainmueller, (2010), "Control Methods for Comparative Case Studies: Estimating the Effect of California's Tobacco Control Program," *Journal of the American Statistical Association*, Vol. 105(490): 493–505.

Abadie, A., D. Drukker, H. Herr, and G. Imbens, (2003), "Implementing Matching Estimators for Average Treatment Effects in STATA," *The STATA Journal*, Vol. 4(3): 290–311.

Abadie, A., and G. Imbens, (2006), "Large Sample Properties of Matching Estimators for Average Treatment Effects," *Econometrica*, Vol. 74(1): 235–267.

Abadie, A., and G. Imbens, (2008), "On the Failure of the Bootstrap for Matching Estimators," *Econometrica*, Vol. 76(6): 1537–1557.

Abadie, A., and G. Imbens, (2009), "Bias-Corrected Matching Estimators for Average Treatment Effects," *Journal of Business and Economic Statistics*, Vol. 29(1): 1–11.

Abadie, A., and G. Imbens, (2010), "Estimation of the Conditional Variance in Paired Experiments," *Annales d'Economie et de Statistique*, Vol. 91: 175–187.

Abadie, A., and G. Imbens, (2011), "Bias-Corrected Matching Estimators for Average Treatment Effects," *Journal of Business and Economic Statistics*, Vol. 29(1): 1–11.

Abadie, A., and G. Imbens, (2012), "Matching on the Estimated Propensity Score," National Bureau of Economic Research Working paper 15301.

Abbring, J., and G. van den Berg, (2003), "The Nonparametric Identification of Treatment Effects in Duration Models," *Econometrica*, Vol. 71(5): 1491–1517.

Althauser, R., and D. Rubin, (1970), "The Computerized Construction of a Matched Sample," *The American Journal of Sociology*, Vol. 76(2): 325–346.

Altman, D. (1991), *Practical Statistics for Medical Research*, Chapman and Hall/CRC.

Angrist, J. (1990), "Lifetime Earnings and the Vietnam Era Draft Lottery: Evidence from Social Security Administrative Records," *American Economic Review*, Vol. 80: 313–335.

Angrist, J. (1998), "Estimating the Labor Market Impact of Voluntary Military Service Using Social Security Data on Military Applicants," *Econometrica*, Vol. 66(2): 249–288.

Angrist, J. D., and J. Hahn, (2004) "When to Control for Covariates? Panel-Asymptotic Results for Estimates of Treatment Effects," *Review of Economics and Statistics*, Vol. 86(1): 58–72.

Angrist, J., G. Imbens, and D. Rubin, (1996), "Identification of Causal Effects Using Instrumental Variables," *Journal of the American Statistical Association*, Vol. 91: 444–472.

Angrist, J., and A. Krueger, (1999), "Does Compulsory Schooling Affect Schooling and Earnings," *Quarterly Journal of Economics*, Vol. CVI(4): 979–1014.

Angrist, J. D., and A. B. Krueger, (2000), "Empirical Strategies in Labor Economics," in A. Ashenfelter and D. Card, eds. *Handbook of Labor Economics*, vol. 3. Elsevier Science.

Angrist, J. D., and G. M. Kuersteiner, (2011), "Causal Effects of Monetary Shocks: Semiparametric Conditional Independence Tests with a Multinomial Propensity Score," *Review of Economics and Statistics*, Vol. 93(3): 725–747.

Angrist, J., and V. Lavy, (1999), "Using Maimonides' Rule to Estimate the Effect of Class Size on Scholastic Achievement," *Quarterly Journal of Economics*, Vol. CXIV: 1243.

Angrist, J., and S. Pischke, (2008), *Mostly Harmless Econometrics: An Empiricists' Companion*, Princeton University Press.

Anscombe, F. J. (1948), "The Validity of Comparative Experiments," *Journal of the Royal Statistical Society, Series A*, Vol. 61: 181–211.

Ashenfelter, O. (1978), "Estimating the Effect of Training Programs on Earnings," *Review of Economics and Statistics*, Vol. 60: 47–57.

Ashenfelter, O., and D. Card, (1985), "Using the Longitudinal Structure of Earnings to Estimate the Effect of Training Programs," *Review of Economics and Statistics*, Vol. 67: 648–660.

Athey, S., and G. Imbens, (2006), "Identification and Inference in Nonlinear Difference-In-Differences Models," *Econometrica*, Vol. 74(2): 431–497.

Athey, S., and G. Imbens, (2014), "Supervised Learning Methods for Causal Effects" Unpublished Manuscript.

Athey, S., and S. Stern, (1998), "An Empirical Framework for Testing Theories About Complementarity in Organizational Design," NBER working paper 6600.

Austin, P. (2008), "A Critical Appraisal of Propensity-Score Matching in the Medical Literature Between 1996 and 2003," *Statistics in Medicine*, Vol. 27: 2037–2049.

Baiocchi, M., D. Small, S. Lorch, and P. Rosenbaum, (2010), "Building a Stronger Instrument in an Observational Study of Perinatal Care for Premature Infants," *The Journal of the American Statistical Association*, Vol. 105(492): 1285–1296.

Baker, S. (2000), "Analyzing a Randomized Cancer Prevention Trial with a Missing Binary Outcome, an Auxiliary Variable, and All-or-None Compliance," *The Journal of the American Statistical Association*, Vol. 95(449): 43–50.

Ball, S., G. Bogatz, D. Rubin, and A. Beaton, (1973), "Reading with Television: An Evaluation of The Electric Company. A Report to the Children's Television Workshop," Vol 1 and 2, Educational Testing Service, Princeton NJ.

Barnard, J., J. Du, J. Hill, and D. Rubin, (1998), "A Broader Template for Analyzing Broken Randomized Experiments," *Sociological Methods & Research*, Vol. 27: 285–317.

Barnow, B. S., G. G. Cain, and A. S. Goldberger, (1980), "Issues in the Analysis of Selectivity Bias," in E. Stromsdofer and G. Farkas, eds. *Evaluation Studies*, vol. 5, Sage.

Becker, S., and A. Ichino, (2002), "Estimation of Average Treatment Effects Based on Propensity Scores," *The Stata Journal*, Vol. 2(4): 358–377.

Beebee, H., C. Hitchcock, and P. Menzies, (2009), *The Oxford Handbook of Causation*, Oxford University Press.

Belloni, A., V. Chernozhukov, and C. Hansen, (2014), "Inference on Treatment Effects After Selection Amongst High-Dimensional Controls," *The Review of Economic Studies*, Vol. 81(2): 608–650.

Bertrand, M., and S. Mullainathan, (2004), "Are Emily and Greg More Employable than Lakisha and Jamal? A Field Experiment on Labor Market Discrimination," *American Economic Review*, Vol. 94(4): 991–1013.

Bitler, M., J. Gelbach, and H. Hoynes, (2002), "What Mean Impacts Miss: Distributional Effects of Welfare Reform Experiments," *American Economic Review*, Vol. 96(4): 988-1012.

Björklund, A., and R. Moffitt, (1987), "The Estimation of Wage Gains and Welfare Gains in Self–Selection Models," *Review of Economics and Statistics*, Vol. LXIX: 42–49.

Black, S. (1999), "Do Better Schools Matter? Parental Valuation of Elementary Education," *Quarterly Journal of Economics*, Vol. CXIV: 577.

Bloom, H. (1984), "Accounting for No-Shows in Experimental Evaluation Designs," *Evaluation Review*, Vol. 8: 225–246.

Blundell, R., and M. Costa-Dias, (2000), "Evaluation Methods for Non-Experimental Data," *Fiscal Studies*, Vol. 21(4): 427–468.

Blundell, R., and M. Costa-Dias, (2002), "Alternative Approaches to Evaluation in Empirical Microeconomics," *Portuguese Economic Journal*, Vol. 1(1): 91–115.

Blundell, R., A. Gosling, H. Ichimura, and C. Meghir, (2007), "Changes in the Distribution of Male and Female Wages Accounting for the Employment Composition," *Econometrica*, Vol. 75(2): 323–363.

Box, G., S. Hunter, and W. Hunter, (2005), *Statistics for Experimenters: Design, Innovation and Discovery*, Wiley.

Box, G., and G. Tiao, (1973), *Bayesian Inference in Statistical Analysis*, Addison Wesley.

Breiman, L., and P. Spector, (1992), "Submodel Selection and Evaluation in Regression: The x-Random Case," *International Statistical Review*, Vol. 60: 291–319.

Brillinger, D. R., L. V. Jones, and J. W. Tukey, (1978), "Report of the statistical task force for the weather modification advisory board." *The Management of Western Resources, Vol. II: The Role of Statistics on Weather Resources Management.* Stock No. 003-018-00091-1, Government Printing Office, Washington, DC.

Brooks, S., A. Gelman, G. Jones, and X.-L. Meng, (2011), *Handbook of Markov Chain Monte Carlo*, Chapman and Hall.

Bühlman, P., and S. van der Geer, (2011), *Statistics for High-Dimensional Data: Methods, Theory and Applications*, Springer Verlag.

Busso, M., J. DiNardo, and J. McCrary, (2009), "New Evidence on the Finite Sample Properties of Propensity Score Matching and Reweighting Estimators," Unpublished Working Paper.

Caliendo, M. (2006), *Microeconometric Evaluation of Labour Market Policies*, Springer Verlag.

Card, D. (1995), "Using Geographic Variation in College Proximity to Estimate the Return to Schooling," in Christofides, E. K. Grant, and R. Swidinsky, ed. *Aspects of Labor Market Behaviour: Essays in Honour of John Vanderkamp*, University of Toronto Press.

Card, D., and A. Krueger, (1994), "Minimum Wages and Employment: A Case Study of the Fast-food Industry in New Jersey and Pennsylvania," *American Economic Review*, Vol. 84(4): 772–784.

Card, D., and D. Sullivan, (1988), "Measuring the Effect of Subsidized Training Programs on Movements In and Out of Employment," *Econometrica*, Vol. 56(3), 497–530.

Chernozhukov, V., and C. Hansen, (2005), "An IV Model of Quantile Treatment Effects," *Econometrica*, Vol. 73(1): 245–261.

Chetty, R., J. Friedman, N. Hilger, E. Saez, D. Schanzenbach, and D. Yagan, (2011), "How Does Your Kindergarten Classroom Affect Your Earnings? Evidence from Project STAR," *Quarterly Journal of Economics*, Vol. 126(4): 1593–1660.

Clogg, C., D. Rubin, N. Schenker, B. Schultz, and L. Weidman, (1991), "Multiple Imputation of Industry and Occupation Codes in Census Public-Use Samples Using Bayesian Logistic Regression,", *Journal of the American Statistical Association*, Vol. 86(413): 68–78.

Cochran, W. G. (1965), "The Planning of Observational Studies of Human Populations," *Journal of the Royal Statistical Society, Series A (General)*, Vol. 128(2): 234–266.

Cochran, W. G. (1968) "The Effectiveness of Adjustment by Subclassification in Removing Bias in Observational Studies," *Biometrics*, Vol. 24: 295–314.

Cochran, W. G. (1977), *Sampling Techniques*, Wiley.

Cochran, W. G., and G. Cox, (1957), *Experimental Design*, Wiley Classics Library.

Cochran, W. G., and D. Rubin, (1973), "Controlling Bias in Observational Studies: A Review," *Sankhya*, Vol. 35: 417–46.

Cook, T. (2008), "'Waiting for Life to Arrive': A History of the Regression-Discontinuity Design in Psychology, Statistics, and Economics," *Journal of Econometrics*, Vol. 142(2): 636–654.

Cook, T., and D. DeMets, (2008), *Introduction to Statistical Methods for Clinical Trials*, Chapman and Hall/CRC.

Cornfield et al. (1959), "Smoking and Lung Cancer: Recent Evidence and a Discussion of Some Questions," *Journal of the National Cancer Institute*, Vol. 22: 173–203.

Cox, D. (1956), "A Note on Weighted Randomization," *The Annals of Mathematical Statistics*, Vol. 27(4): 1144-1151.

Cox, D. (1958), *Planning of Experiments*, Wiley Classics Library.

Cox, D. (1992), "Causality: Some Statistical Aspects," *Journal of the Royal Statistical Society, Series A*, Vol. 155: 291–301.

Cox, D., and P. McCullagh, (1982), "Some Aspects of Covariance," (with discussion). *Biometrics*, Vol. 38: 541–561.

Cox, D., and N. Reid, (2000), *The Theory of the Design of Experiments*, Chapman and Hall/CRC.

Crump, R., V. J. Hotz, G. Imbens, and O. Mitnik, (2008), "Nonparametric Tests for Treatment Effect Heterogeneity," *Review of Economics and Statistics*, Vol. 90(3): 389–405.

Crump, R., V. J. Hotz, G. Imbens, and O. Mitnik, (2009), "Dealing with Limited Overlap in Estimation of Average Treatment Effects," *Biometrika*, Vol. 96: 187–99.

Cuzick, J., R. Edwards, and N. Segnan, (1997), "Adjusting for Non-Compliance and Contamination in Randomized Clinical Trials," *Statistics in Medicine,* Vol. 16: 1017–1039.

Darwin, C. (1876), *The Effects of Cross- and Self-Fertilisation in the Vegetable Kingdom*, John Murry.

Davies, O. (1954), *The Design and Analysis of Industrial Experiments*, Oliver and Boyd.

Dawid, P. (1979), "Conditional Independence in Statistical Theory," *Journal of the Royal Statistical Society, Series B*, Vol. 41(1): 1–31.

Dawid, P. (2000), "Causal Inference Without Counterfactuals," *Journal of the American Statistical Association*, Vol. 95(450): 407–424.

Deaton, A. (2010), "Instruments, Randomization, and Learning about Development," *Journal of Economic Literature*, Vol. 48(2): 424–455.

Dehejia, R. (2002), "Was There a Riverside Miracle? A Hierarchical Framework for Evaluating Programs with Grouped Data," *Journal of Business and Economic Statistics*, Vol. 21(1): 1–11.

Dehejia, R. (2005a), "Practical Propensity Score Matching: A Reply to Smith and Todd," *Journal of Econometrics*, Vol. 125: 355–364.

Dehejia, R. (2005b), "Program Evaluation as a Decision Problem," *Journal of Econometrics*, Vol. 125: 141–173.

Dehejia, R., and S. Wahba, (1999), "Causal Effects in Nonexperimental Studies: Reevaluating the Evaluation of Training Programs," *Journal of the American Statistical Association*, Vol. 94: 1053–1062.

Dehejia, R., and S. Wahba, (2002), "Propensity Score-Matching Methods for Nonexperimental Causal Studies," *Review of Economics and Statistics*, Vol. 84(1): 151–161.

Diaconis, P. (1976), "Finite Forms of de Finetti's Theorem on Exchangeability," Technical Report 84, Department of Statistics, Stanford University.

Diamond, A., and J. Sekhon, (2013), "Genetic Matching for Estimating Causal Effects: A General Multivariate Matching Method for Achieving Balance in Observational Studies," *Review of Economics and Statistics*, Vol. 95(3): 932–945.

Diehr, P., D. Martin, T. Koepsell, and A. Cheadle, (1995), "Breaking the Matches in a Paired t-Test for Community Interventions When the Number of Pairs is Small," *Statistics in Medicine*, Vol. 14: 1491–1504.

Donner, A. (1987), "Statistical Methodology for Paired Cluster Designs," *American Journal of Epidemiology*, Vol. 126(5), 972–979.

Du, J. (1998) "Valid Inferences After Propensity Score Subclassification Using Maximum Number of Subclasses as Building Blocks," PhD Thesis, Department of Statistics, Harvard University.

Duflo, E., R. Hanna, and S. Ryan, (2012), "Incentives Work: Getting Teachers to Come to School," *American Economic Review*, Vol. 102(4): 1241–1278.

Efron, B., and D. Feldman, (1992), "Compliance as an Explanatory Variable in Clinical Trials," *Journal of the American Statistical Association*, Vol. 86(413): 9–17.

Efron, B., and R. Tibshirani, (1993), *An Introduction to the Bootstrap*, Chapman and Hall.

Engle, R., D. Hendry, and J.-F. Richard, (1974), "Exogeneity," *Econometrica*, Vol. 51(2): 277–304.

Espindle, L. (2004), "Improving Confidence Coverage for the Estimate of the Treatment Effect in a Subclassification Setting," Undergraduate Thesis, Department of Statistics, Harvard University.

de Finetti, B. (1964), "Foresight: Its Logical Laws, Its Subjective Sources," in Kyburg and Smokler, eds. *Studies in Subjective Probability*, Wiley.

de Finetti, B. (1992), *Theory of Probability: A Critical Introductory Treatment*, Vol. 1 & 2, Wiley Series in Probability & Mathematical Statistics.

Feller, W. (1965), *An Introduction to Probability and its Applications*, Vol. 1, John Wiley and Sons, New York City.

Firpo, S. (2003), "Efficient Semiparametric Estimation of Quantile Treatment Effects," PhD Thesis, Chapter 2, Department of Economics, University of California, Berkeley.

Firpo, S. (2007), "Efficient Semiparametric Estimation of Quantile Treatment Effects,"*Econometrica*, Vol. 75(1): 259-276.

Fisher, L., D. Dixon, J. Herson, R. Frankowski, M. Hearron, and K. Peace, (1990), "Intention to Treat in Clinical Trials," in Peace, ed. *Statistical Issues in Drug Research and Development*, Marcel Dekker, Inc.

Fisher, R. A. (1918), "The Causes of Human Variability," *Eugenics Review*, Vol. 10: 213–220.

Fisher, R. A. (1925), *Statistical Methods for Research Workers*, 1st ed, Oliver and Boyd.

Fisher, R. A. (1935), *Design of Experiments*, Oliver and Boyd.

Fisher, R., and W. MacKenzie, (1923), "Studies in Crop Vacation. II. The Manurial Response of Different Potato Varieties," *Journal of Agricultural Science,* Vol. 13: 311–320.

Fisher, L. et al. (1990), "Intention-to-Treat in Clinical Trials," in Peace ed., *Statistical Issues in Drug Research and Development*, Marcel Dekker.

Fraker, T., and R. Maynard, (1987), "The Adequacy of Comparison Group Designs for Evaluations of Employment-Related Programs," *Journal of Human Resources*, Vol. 22(2): 194–227.

Frangakis, C., and D. Rubin, (2002), "Principal Stratification," *Biometrics*, Vol. 58(1): 21–29.

Freedman, D. A. (2006), "Statistical Models for Causation: What Inferential Leverage Do They Provide," *Evaluation Review* , Vol. 30(6): 691–713.

Freedman, D. A. (2008a), "On Regression Adjustments to Experimental Data, " *Advances in Applied Mathematics* , Vol. 30(6): 180–193.

Freedman, D. A. (2008b), "On Regression Adjustments in Experiments with Several Treatments," *Annals of Applied Statistics*, Vol. 2: 176–196.

Freedman, D. A. (2009), in D. Collier, J. S. Sekhon, and P. B. Stark, eds. *Statistical Models and Causal Inference: A Dialogue with the Social Sciences*, Cambridge University Press.

Freedman, D. A., R. Pisani, and R. Purves, (1978), *Statistics*, Norton.

Friedlander, D., and J. Gueron, (1995), "Are High-Cost Services More Effective Than Low-Cost Services," in C. Manski and I. Garfinkel, eds. *Evaluating Welfare and Training Programs*, Harvard University Press, pp. 143–198.

Friedlander, D., and P. Robins, (1995), "Evaluating Program Evaluations: New Evidence on Commonly Used Nonexperimental Methods," *American Economic Review*, Vol. 85: 923–937.

Frölich, M. (2000), "Treatment Evaluation: Matching versus Local Polynomial Regression," Discussion paper 2000-17, Department of Economics, University of St. Gallen.

Frölich, M. (2004a), "Finite-Sample Properties of Propensity-Score Matching and Weighting Estimators," *The Review of Economics and Statistics*, Vol. 86(1): 77–90.

Frölich, M. (2004b), "A Note on the Role of the Propensity Score for Estimating Average Treatment Effects," *Econometric Reviews*, Vol. 23(2): 167–174.

Frumento, P., F. Mealli, B. Pacini, and D. Rubin, (2012), "Evaluating the Effect of Training on Wages in the Presence of Noncompliance, Nonemployment, and Missing Outcome Data," *Journal of the American Statistical Association*, No. 498: 450–466.

Gail, M. H., S. Mark, R. Carroll, S. Green, and D. Pee, (1996), "On Design Considerations and Randomization-based Inference for Community Intervention Trials," *Statistics in Medicine*, Vol. 15: 1069–1092.

Gail, M. H., W. Tian, and S. Piantadosi, (1988), "Tests for No Treatment Effect in Randomized Clinical Trials," *Biometrika*, Vol. 75(3): 57–64.

Gail, M. H., S. Wieand, and S. Piantadosi, (1984), "Biased Estimates of Treatment Effect in Randomized Experiments with Nonlinear Regressions and Omitted Covariates," *Biometrika*, Vol. 71(3): 431–444.

Gelman, A., J. Carlin, H. Stern, and D. Rubin, (1995), *Bayesian Data Analysis*, Chapman and Hall.

Gelman, A., and J. Hill, (2006), *Data Analysis Using Regression and Multilevel/Hierarchical Models*, Cambridge University Press.

Gill, R., and J. Robins, (2001), "Causal Inference for Complex Longitudinal Data: The Continuous Case," *Annals of Statistics*, Vol. 29(6): 1785–1811.

Goldberger, A. (1991), *A Course in Econometrics*, Harvard University Press.

Graham, B. (2008), "Identifying Social Interactions through Conditional Variance Restrictions," *Econometrica*, Vol. 76(3): 643–660.

Granger, C. (1969), "Investigating Causal Relations by Econometric Models and Cross-spectral Methods," *Econometrica*, Vol. 37(3): 424–438.

Greene, W. (2011), *Econometric Analysis*, 7th Edition, Prentice Hall.

Gu, X., and P. Rosenbaum, (1993), "Comparison of Multivariate Matching Methods: Structures, Distances and Algorithms," *Journal of Computational and Graphical Statistics*, Vol. 2: 405–420.

Guo, S., and M. Fraser, (2010), *Propensity Score Analysis*, Sage Publications.

Gutman, R., and D. Rubin, (2014), "Robust Estimation of Causal Effects of Binary Treatments in Unconfounded Studies with Dichotomous Outcomes," *Statistics in Medicine*, forthcoming.

Haavelmo, T. (1943), "The Statistical Implications of a System of Simultaneous Equations," *Econometrica*, Vol. 11(1):1–12.

Haavelmo, T. (1944), "The Probability Approach in Econometrics," *Econometrica*, Vol. 11.

Hahn, J. (1998), "On the Role of the Propensity Score in Efficient Semiparametric Estimation of Average Treatment Effects," *Econometrica*, Vol. 66(2): 315–331.

Hahn, J., P. Todd, and W. VanderKlaauw, (2000), "Identification and Estimation of Treatment Effects with a Regression-Discontinuity Design," *Econometrica*, Vol. 69(1): 201–209.

Hainmueller, J. (2012), "Entropy Balancing for Causal Effects: A Multivariate Reweighting Method to Produce Balanced Samples in Observational Studies," *Political Analysis*, Vol. 20: 25–46.

Ham, J., and R. Lalonde, (1996), "The Effect of Sample Selection and Initial Conditions in Duration Models: Evidence from Experimental Data on Training," *Econometrica*, Vol. 64: 1.

Hansen, B. (2007), "Optmatch: Flexible, Optimal Matching for Observational Studies," *R News*, Vol. 7(2): 18–24.

Hansen, B. (2008), "The Prognostic Analogue of the Propensity Score," *Biometrika*, Vol. 95(2): 481–488.

Hansen, B., and S. Klopfer, (2006), "Optimal Full Matching and Related Designs via Network Flows," *Journal of Computational and Graphical Statistics*, Vol. 15(3): 609–627.

Hartigan, J. (1983), *Bayes Theory*, Springer Verlag.

Hartshorne, C., and P. Weiss (Eds.), (1931), *Collected Papers of Charles Sanders Peirce* (Vol. 1), Harvard University Press.

Hearst, N., Newman, T., and S. Hulley, (1986), "Delayed Effects of the Military Draft on Mortality: A Randomized Natural Experiment," *New England Journal of Medicine*, Vol. 314 (March 6): 620–624.

Heckman, J., and J. Hotz, (1989), "Alternative Methods for Evaluating the Impact of Training Programs" (with discussion), *Journal of the American Statistical Association*, Vol. 84(804): 862–874.

Heckman, J., H. Ichimura, and P. Todd, (1997), "Matching as an Econometric Evaluation Estimator: Evidence from Evaluating a Job Training Program," *Review of Economic Studies*, Vol. 64: 605–654.

Heckman, J., H. Ichimura, and P. Todd, (1998), "Matching as an Econometric Evaluation Estimator," *Review of Economic Studies*, Vol. 65: 261–294.

Heckman, J., H. Ichimura, J. Smith, and P. Todd, (1998), "Characterizing Selection Bias Using Experimental Data," *Econometrica*, Vol. 66: 1017–1098.

Heckman, J., R. Lalonde, and J. Smith, (2000), "The Economics and Econometrics of Active Labor Markets Programs," in Ashenfelter and Card, eds. *Handbook of Labor Economics*, vol. 3, Elsevier Science.

Heckman, J., and R. Robb, (1984), "Alternative Methods for Evaluating the Impact of Interventions," in Heckman and Singer, eds., *Longitudinal Analysis of Labor Market Data*, Cambridge University Press.

Heckman, J., and E. Vytlacil, (2007a), "Econometric Evaluation of Social Programs, Part I: Causal Models, Structural Models and Econometric Policy Evaluation," in J. Heckman and E. Leamer, eds. *Handbook of Econometrics*, vol. 6B, Chapter 70, 4779-4874, Elsevier Science.

Heckman, J., and E. Vytlacil, (2007b), "Econometric Evaluation of Social Programs, Part II: Using the Marginal Treatment Effect to Organize Alternative Econometric Estimators to Evaluate Social Programs, and to Forecast their Effects in New Environments," in J. Heckman and E. Leamer, eds. Handbook of Econometrics, vol. 6B, Chapter 71, 4875-5143, Elsevier Science.

Heitjan, D., and R. Little, (1991), "Multiple Imputation for the Fatal Accident Reporting System," *Applied Statistics*, Vol. 40: 13–29.

Hendry, D., and M. Morgan, (1995), *The Foundations of Econometric Analysis*, Cambridge University Press.

Hewitt, E., and L. Savage, (1955), "Symmetric Measures on Cartesian Products,"*Transactions of the American Mathematical Society*, Vol. 80: 470-501.

Hinkelmann, K., and O. Kempthorne, (2005), *Design and Analysis of Experiments*, Vol. 2, Advance Experimental Design, Wiley.

Hinkelmann, K., and O. Kempthorne, (2008), *Design and Analysis of Experiments*, Vol. 1, Introduction to Experimental Design, Wiley.

Hirano, K., and G. Imbens, (2001), "Estimation of Causal Effects Using Propensity Score Weighting: An Application of Data on Right Heart Catheterization," *Health Services and Outcomes Research Methodology*, Vol. 2: 259–278.

Hirano, K., and G. Imbens, (2004), "The Propensity Score with Continuous Treatments," in Gelman and Meng, eds. *Applied Bayesian Modelling and Causal Inference from Missing Data Perspectives*, Wiley.

Hirano, K., G. Imbens, and G. Ridder, (2003), "Efficient Estimation of Average Treatment Effects Using the Estimated Propensity Score," *Econometrica*, Vol. 71(4): 1161–1189.

Hirano, K., G. Imbens, D. Rubin, and A. Zhou, (2000), "Estimating the Effect of Flu Shots in a Randomized Encouragement Design," *Biostatistics*, Vol. 1(1): 69–88.

Ho, D., and K. Imai, (2006), "Randomization Inference with Natural Experiments: An Analysis of Ballot Effects in the 2003 California Recall Election," *Journal of the American Statistical Association*, Vol. 101(476): 888–900.

Ho, D., K. Imai, G. King, and E. Stuart, (2007), "Matching as Nonparametric Preprocessing for Reducing Model Dependence in Parametric Causal Inference," *Political Analysis*, Vol. 81: 945–970.

Hodges, J. L., and Lehmann, E., (1970), *Basic Concepts of Probability and Statistics*, 2nd ed., Holden-Day.

Holland, P. (1986), "Statistics and Causal Inference" (with discussion), *Journal of the American Statistical Association*, Vol. 81: 945–970.

Holland, P. (1988), "Causal Inference, Path Analysis, and Recursive Structural Equations Models," (with discussion), *Sociological Methodology*, Vol. 18: 449–484.

Holland, P., and D. Rubin, (1982), "Introduction: Research on Test Equating Sponsored by Educational Testing Service, 1978–1980," in *Test Equating*, Academic Press, Inc. pp. 1–6.

Holland, P., and D. Rubin, (1983), "On Lord's Paradox," in Wainer and Messick, eds. *Principles of Modern Psychological Measurement: A Festschrift for Frederick Lord*, Erlbaum, pp. 3–25.

Hood, W., and T. Koopmans, (1953), *Studies in Econometric Method*, Wiley, New York.

Horowitz, J. (2002), "The Bootstrap," in Heckman and Leamer, eds. *Handbook of Econometrics*, Vol. 5, Elsevier.

Horvitz, D., and D. Thompson, (1952), "A Generalization of Sampling Without Replacement from a Finite Universe," *Journal of the American Statistical Association*, Vol. 47: 663–685.

Hotz, V. J., G. Imbens, and J. Klerman, (2001), "The Long-Term Gains from GAIN: A Re-Analysis of the Impacts of the California GAIN Program," *Journal of Labor Economics,* Vol. 24(3): 521–566.

Hotz, J., G. Imbens, and J. Mortimer, (2005), "Predicting the Efficacy of Future Training Programs Using Past Experiences," *Journal of Econometrics*, Vol. 125: 241–270.

Huber, M., M. Lechner, and C. Wunsch, (2012), "The Performance of Estimators Based on the Propensity Score," *Journal of Econometrics*, Vol. 175(1): 1–21.

Imai, K. (2008), Variance Identification and Efficiency Analysis in Randomized Experiments under the Matched-Pair Design." *Statistics in Medicine*, Vol. 27(24) (October): 4857–4873.

Imai, K., and D. van Dyk, (2004), "Causal Inference with General Treatment Regimes: Generalizing the Propensity Score," *Journal of the American Statistical Association*, Vol. 99: 854–866.

Imai, K., G. King, and E. A. Stuart, (2008), "Misunderstandings among Experimentalists and Observationalists about Causal Inference," *Journal of the Royal Statistical Society, Series A (Statistics in Society)*, Vol. 171(2): 481-502.

Imbens, G. (2000), "The Role of the Propensity Score in Estimating Dose-Response Functions," *Biometrika*, Vol. 87(3): 706–710.

Imbens, G. (2003), "Sensitivity to Exogeneity Assumptions in Program Evaluation," *American Economic Review*, Papers and Proceedings.

Imbens, G. (2004), "Nonparametric Estimation of Average Treatment Effects Under Exogeneity: A Review," *Review of Economics and Statistics*, Vol. 86(1): 1–29.

Imbens, G. (2010), "Better LATE Than Nothing: Some Comments on Deaton (2009) and Heckman and Urzua (2009)," *Journal of Economic Literature*, Vol. 48(2): 399–423.

Imbens, G. (2011), "On the Finite Sample Benefits of Stratification, Blocking and Pairing in Randomized Experiments," Unpublished Manuscript.

Imbens, G. (2014), "Instrumental Variables: An Econometrician's Perspective," *Statistical Science*, Vol. 29(3): 375–379.

Imbens, G., (2015), "Matching Methods in Practice: Three Examples," forthcoming, *Journal of Human Resources.*

Imbens, G., and J. Angrist, (1994), "Identification and Estimation of Local Average Treatment Effects," *Econometrica*, Vol. 61(2): 467–476.

Imbens, G., and K. Kalyanaraman, (2012), "Optimal Bandwidth Choice for the Regression Discontinuity Estimator Review of Economic Studies," *Review of Economic Studies*, Vol. 79(3): 933–959.

Imbens, G., and T. Lemieux, (2008), "Regression Discontinuity Designs: A Guide to Practice," *Journal of Econometrics*, Vol. 142(2): 615–635.

Imbens, G., and P. Rosenbaum, (2005), "Randomization Inference with an Instrumental Variable," *Journal of the Royal Statistical Society, Series A*, Vol. 168(1): 109–126.

Imbens, G., and D. Rubin, (1997a), "Estimating Outcome Distributions for Compliers in Instrumental Variable Models," *Review of Economic Studies*, Vol. 64(3): 555–574.

Imbens, G., and D. Rubin, (1997b), "Bayesian Inference for Causal Effects in Randomized Experiments with Noncompliance," *Annals of Statistics*, Vol. 25(1): 305–327.

Imbens, G., and J. Wooldridge, (2009), "Recent Developments in the Econometrics of Program Evaluation," *Journal of Economic Literature*, Vol. 47(1): 1–81.

Jin, H., and D. B. Rubin, (2008), "Principal Stratification for Causal Inference with Extended Partial Compliance: Application to Efron-Feldman Data," *Journal of the American Statistical Association*, Vol. 103: 101–111.

Kane, T., and C. Rouse, (1995), "Labor-Market Returns to Two- and Four- Year College," *American Economic Review*, Vol. 85(3): 600–614.

Kang, J., and J. Schafer, (2007), "Demystifying Double Robustness: A Comparison of Alternative Strategies for Estimating a Population Mean from Incomplete Data," *Statistical Science*, Vol. 22(4): 523–539.

Kempthorne, O. (1952), *The Design and Analysis of Experiments*, Robert Krieger Publishing Company.

Kempthorne, O. (1955), "The Randomization Theory of Experimental Evidence," *Journal of the American Statistical Association*, Vol. 50927(1): 946–967.

Ketel, N., E. Leuven, H. Oosterbeek, and B. VanderKlaauw, (2013), "The Returns to Medical School in a Regulated Labor Market: Evidence from Admission Lotteries," Unpublished Manuscript.

Koch, G., C. Tangen, J. W Jung, and I. Amara, (1998), "Issues for Covariance Analysis of Dichotomous and Ordered Categorical Data from Randomized Clinical Trials and Non-Parametric Strategies for Addressing Them," *Statistics in Medicine*, Vol. 17: 1863–1892.

Koopmans, T., (1950), *Statistical Inference in Dynamic Economic Models*, Wiley, New York.

Krueger, A. (1999), "Experimental Estimates of Education Production Functions Experimental Estimates of Education Production Functions," *The Quarterly Journal of Economics*, Vol. 114(2): 497-532.

Lalonde, R. J., (1986), "Evaluating the Econometric Evaluations of Training Programs with Experimental Data," *American Economic Review*, Vol. 76: 604–620.

Lancaster, T. (2004), *An Introduction to Modern Bayesian Econometrics*, Blackwell Publishing.

Leamer, E. (1988), "Discussion on Marini, Singer, Glymour, Scheines, Spirtes, and Holland," *Sociological Methodology*, Vol. 18: 485-493.

Lechner, M. (1999), "Earnings and Employment Effects of Continuous Off-the-job Training in East Germany After Unification," *Journal of Business and Economic Statistics*, Vol. 17(1): 74–90.

Lechner, M. (2001), "Identification and Estimation of Causal Effects of Multiple Treatments under the Conditional Independence Assumption," in Lechner and Pfeiffer, eds. *Econometric Evaluations of Active Labor Market Policies in Europe*, Heidelberg.

Lechner, M. (2002), "Program Heterogeneity and Propensity Score Matching: An Application to the Evaluation of Active Labor Market Policies," *Review of Economics and Statistics*, Vol. 84(2): 205–220.

Lechner, M. (2008), "A Note on the Common Support Problem in Applied Evaluation Studies," *Annales d'Economie et de Statistique*, Vol. 91-92: 217–234.

Lee, D. (2008), "Randomized Experiments from Non-random Selection in U.S. House Elections," *Journal of Econometrics*, Vol. 142(2): 675–697.

Lee, D., and T. Lemieux, (2010), "Regression Discontinuity Designs in Economics," *Journal of Economic Literature*, Vol. 48(2): 281–355.

Lee, M.-J. (2005), *Micro-Econometrics for Policy, Program, and Treatment Effects*, Oxford University Press.

Lehman, E. (1974), *Nonparametrics: Statistical Methods Based on Ranks*, Holden-Day.

Lesaffre, E., and S. Senn, (2003), "A Note on Non-Parametric ANCOVA for Covariate Adjustment in Randomized Clinical Trials," *Statistics in Medicine*, Vol. 22: 3583–3596.

Lin, W. (2012), "Agnostic Notes on Regression Adjustments to Experimental Data: Reexamining Freedman's Critique," *Annals of Applied Statistics*.

Lindley, D. V., and N. R. Novick, (1981), "The Role of Exchangeability in Inference," *Annals of Statistics*, Vol. 9: 45–58.

Little, R., and D. Rubin, (2002), *Statistical Analysis with Missing Data*, Wiley.

Lord, F. (1967), "A Paradox in the Interpretation of Group Comparisons," *Psychological Bulletin*, Vol. 68: 304–305.

Lui, K.-J. (2011), *Binary Data Analysis of Randomized Clinical Trials with Noncompliance*, Wiley, Statistics in Practice.

Lynn, H., and C. McCulloch, (1992), "When Does It Pay to Break the Matches for Analysis of a Matched-pair Design," *Biometrics*, Vol. 48: 397–409.

McCarthy, M. D. (1939), "On the Application of the z-Test to Randomized Blocks," *Annals of Mathematical Statistics*, Vol. 10: 337.

McClellan, M., and J. P. Newhouse, (1994), "Does More Intensive Treatment of Acute Myocardial Infarction in the Elderly Reduce Mortality," *Journal of the American Medical Association*, Vol. 272(11): 859–866.

McDonald, C., S. Hiu, and W. Tierney, (1992), "Effects of Computer Reminders for Influenza Vaccination on Morbidity During Influenza Epidemics," *MD Computing*, Vol. 9: 304–312.

McNamee, R. (2009), "Intention to Treat, Per Protocol, as Treated and Instrumental Variable Estimators Given Non-Compliance and Effect Heterogeneity," *Statistics in Medicine*, Vol. 28: 2639–2652.

Mann, H. B., and D. R. Whitney, (1947), "On a Test of Whether One of Two Random Variables Is Stochastically Larger Than the Other," *Annals of Mathematical Statistics*, Vol. 18(1): 50–60.

Manski, C. (1990), "Nonparametric Bounds on Treatment Effects," *American Economic Review Papers and Proceedings*, Vol. 80: 319–323.

Manski, C. (1996), "Learning about Treatment Effects from Experiments with Random Assignment of Treatments," *The Journal of Human Resources*, Vol. 31(4): 709–773.

Manski, C. (2003), *Partial Identification of Probability Distributions*, Springer-Verlag.

Manski, C. (2013), *Public Policy in an Uncertain World*, Harvard University Press.

Manski, C., G. Sandefur, S. McLanahan, and D. Powers, (1992), "Alternative Estimates of the Effect of Family Structure During Adolescence on High School," *Journal of the American Statistical Association*, Vol. 87(417): 25–37.

Marini, M., and B. Singer, (1988), "Causality in the Social Sciences," *Sociological Methodology*, Vol. 18: 347–409.

Mealli, F., and D. Rubin, (2002a), "Assumptions When Analyzing Randomized Experiments with Noncompliance and Missing Outcomes," *Health Services Outcome Research Methodology*, Vol. 3: 225–232.

Mealli, F., and D. Rubin, (2002b), "Discussion of Estimation of Intervention Effects with Noncompliance: Alternative Model Specification by Booil Jo," *Journal of Educational and Behavioral Statistics*, Vol. 27: 411–415.

Meier, P. (1991), "Compliance as an Explanatory Variable in Clinical Trials: Comment," *Journal of the American Statistical Association*, Vol. 86(413):19–22.

Miguel, E., C. Camerer, K. Casey, J. Cohen, K. M. Esterling, A. Gerber, R. Glennerster, D. P. Green, M. Humphreys, G. Imbens, D. Laitin, T. Madon, L. Nelson, B. A. Nosek, M. Petersen, R. Sedlmayr, J. P. Simmons, U. Simonsohn, and M. Van der Laan, (1991), "Promoting Transparency in Social Science Research," *Science*, Vol. 343(6166): 30–31.

Mill, J. S. (1973), *A system of logic* , In Collected Works of John Stuart Mill, University of Toronto Press.

Miratrix, L., J. Sekhon, and B. Yu, (2013), "Adjusting Treatment Effect Estimates by Post-Stratification in Randomized Experiments," *Journal of the Royal Statistical Society*, Series B, 75, 369–396.

Morgan, K., and D. Rubin, (2012), "Rerandomization to Improve Covariate Balance in Experiments," *Annals of Statistics*, Vol. 40(2): 1263–1282.

Morgan, S. (2013), *Handbook of Causal Analysis for Social Research*, Springer.

Morgan, S., and C. Winship, (2007), *Counterfactuals and Causal Inference*, Cambridge University Press.

Morris, C., and J. Hill, (2000), "The Health Insurance Experiment: Design Using the Finite Selection Model," *Public Policy and Statistics: Case Studies from RAND 2953*. Springer, New York.

Morton, R., and K. Williams, (2010), *Experimental Political Science and the Study of Causality*, Cambridge University Press.

Mosteller, F. (1995), "The Tennessee Study of Class Size in the Early School Grades," *The Future of Children: Critical Issues for Children and Youths*, V(1995): 113–127.

Murnane, R., and J. Willett, (2011), *Methods Matter: Improving Causal Inference in Educational and Social Science Research*, Oxford University Press.

Murphy, D., and L. Cluff, (1990), "SUPPORT: Study to understand prognoses and preferences for outcomes and risks of treatments: study design," *Journal of Clinical Epidemiology*, Vol. 43: 1S–123S.

Neyman, J. (1923, 1990), "On the Application of Probability Theory to Agricultural Experiments. Essay on Principles. Section 9," translated in *Statistical Science*, (with discussion), Vol. 5(4): 465–480, 1990.

Neyman, J. (1934), "On the Two Different Aspects of the Representative Method: The Method of Stratified Sampling and the Method of Purposive Selection," *Journal of the Royal Statistical Society*, Vol. 97(4): 558-625.

Neyman, J. with the cooperation of K. Iwaskiewicz and St. Kolodziejczyk, (1935), "Statistical Problems in Agricultural Experimentation" (with discussion), Supplement, *Journal of the Royal Statistical Society, Series B*, Vol. 2: 107–180.

Pattanayak, C., D. Rubin, and E. Zell, (2011), "Propensity Score Methods for Creating Covariate Balance in Observational Studies," *Review of Experimental Cardiology*, Vol. 64(10): 897-903.

Paul, I., J. Beiler, A. McMonagle, M. Shaffer, L. Duda, and C. Berlin, (2007), "Effect of Honey, Dextromerhorphan, and No Treatment on Nocturnal Cough and Sleep Quality for Coughing Childre and Their Parents," *Archives of Pediatric and Adolescent Medicine*, Vol. 161(12): 1140–1146.

Pearl, J. (1995), "Causal Diagrams for Empirical Research," *Biometrika*, Vol. 82: 669–688.

Pearl, J. (2000, 2009), *Causality: Models, Reasoning and Inference*, Cambridge University Press.

Peirce, C., and J. Jastrow, (1885), "On Small Differences in Sensation," *Memoirs of the National Academy of Sciences*, Vol. 3: 73–83.

Peters, C., and W. van Vorhis, (1941), *Statistical Procedures and Their Mathematical Bases*, McGraw-Hill.

Politis, D., and J. Romano, (1999), *Subsampling*, Springer Verlag.

Porter, J. (2003), "Estimation in the Regression Discontinuity Model," Unpublished Manuscript, Harvard University.

Powers, D., and S. Swinton, (1984), "Effects of Self-Study for Coachable Test Item Types," *Journal of Educational Measurement*, Vol. 76: 266–278.

Quade, D. (1982), "Nonparametric Analysis of Covariance by Matching," *Biometrics*, Vol. 38: 597–611.

Reid, C. (1998), *Neyman from Life*, Springer.

Reinisch, J., S. Sanders, E. Mortensen, and D. Rubin, (1995), "In Utero Exposure to Phenobarbital and Intelligence Deficits in Adult Men," *The Journal of the American Medical Association*, Vol. 274(19): 1518-1525.

Robert, C. (1994), *The Bayesian Choice*, Springer Verlag.

Robert, C., and G. Casella, (2004), *Monte Carlo Statistical Methods*, Springer Verlag.

Robins, J. (1986), "A New Approach to Causal Inference in Mortality Studies with Sustained Exposure Periods - Application to Control of the Healthy Worker Survivor Effect," *Mathematical Modelling*, Vol. 7: 1393–1512.

Robins, J., and Y. Ritov, (1997), "Towards a Curse of Dimensionality Appropriate (CODA) Asymptotic Theory for Semi-parametric Models," *Statistics in Medicine*, Vol. 16: 285–319.

Robins, J. M., and A. Rotnitzky, (1995), "Semiparametric Efficiency in Multivariate Regression Models with Missing Data," *Journal of the American Statistical Association*, Vol. 90: 122–129.

Robins, J. M., Rotnitzky, A., and Zhao, L.-P., (1995), "Analysis of Semiparametric Regression Models for Repeated Outcomes in the Presence of Missing Data," *Journal of the American Statistical Association*, Vol. 90: 106–121.

Romer, C. D., and D. H. Romer, (2004), "A New Measure of Monetary Shocks: Derivation and Implications," *The American Economic Review*, Vol. 94(4): 1055–1084.

Rosenbaum, P. (1984a), "Conditional Permutation Tests and the Propensity Score in Observational Studies," *Journal of the American Statistical Association*, Vol. 79: 565–574.

Rosenbaum, P. (1984b), "The Consequences of Adjustment for a Concomitant Variable That Has Been Affected by the Treatment," *Journal of the Royal Statistical Society, Series A*, Vol. 147: 656–666.

Rosenbaum, P. (1987), "The Role of a Second Control Group in an Observational Study," *Statistical Science*, (with discussion), Vol. 2(3), 292–316.

Rosenbaum, P. (1988), "Permutation Tests for Matched Pairs," *Applied Statistics*, Vol. 37: 401–411.

Rosenbaum, P. (1989a), "Optimal Matching in Observational Studies," *Journal of the American Statistical Association*, 84, 1024–1032.

Rosenbaum, P. (1989b), "On Permutation Tests for Hidden Biases in Observational Studies: An Application of Holley's Inequality to the Savage Lattice," *Annals of Statistics*, Vol. 17: 643–653.

Rosenbaum, P. (1995, 2002), *Observational Studies*, Springer Verlag.

Rosenbaum, P. (2009), *Design of Observational Studies*, Springer Verlag.

Rosenbaum, P. (2002), "Covariance Adjustment in Randomized Experiments and Observational Studies," *Statistical Science*, Vol. 17(3): 286–304.

Rosenbaum, P., and D. Rubin, (1983a), "The Central Role of the Propensity Score in Observational Studies for Causal Effects," *Biometrika*, Vol. 70: 41–55.

Rosenbaum, P., and D. Rubin, (1983b), "Assessing the Sensitivity to an Unobserved Binary Covariate in an Observational Study with Binary Outcome," *Journal of the Royal Statistical Society, Series B*, Vol. 45: 212–218.

Rosenbaum, P., and D. Rubin, (1984), "Reducing the Bias in Observational Studies Using Sub-classification on the Propensity Score," *Journal of the American Statistical Association*, Vol. 79: 516–524.

Rosenbaum, P., and D. Rubin, (1985), "Constructing a Control Group Using Multivariate Matched Sampling Methods that Incorporate the Propensity Score," *American Statistician*, Vol. 39: 33–38.

Rubin, D. B. (1973a), "Matching to Remove Bias in Observational Studies," *Biometrics*, Vol. 29: 159–183.

Rubin, D. B. (1973b), "The Use of Matched Sampling and Regression Adjustments to Remove Bias in Observational Studies," *Biometrics*, Vol. 29: 185–203.

Rubin, D. B. (1974), "Estimating Causal Effects of Treatments in Randomized and Non-randomized Studies," *Journal of Educational Psychology*, Vol. 66: 688–701.

Rubin, D. B. (1975), "Bayesian Inference for Causality: The Importance of Randomization," *Proceedings of the Social Statistics Section of the American Statistical Association*, 233–239.

Rubin, D. B. (1976a), "Multivariate Matching Methods That Are Equal Percent Bias Reducing, I: Some Examples," *Biometrics*, Vol. 32(1): 109–120.

Rubin, D. B. (1976b), "Multivariate Matching Methods That Are Equal Percent Bias Reducing, II: Maximums on Bias Reduction for Fixed Sample Sizes," *Biometrics*, Vol. 32(1): 121–132.

Rubin, D. B. (1976c), "Inference and Missing Data," *Biometrika*, (with discussion and reply), Vol. 63(3): 581–592.

Rubin, D. B. (1977), "Assignment to Treatment Group on the Basis of a Covariate," *Journal of Educational Statistics*, Vol. 2(1): 1–26.

Rubin, D. B. (1978), "Bayesian Inference for Causal Effects: The Role of Randomization," *Annals of Statistics*, Vol. 6: 34–58.

Rubin, D. B. (1979), "Using Multivariate Matched Sampling and Regression Adjustment to Control Bias in Observational Studies," *Journal of the American Statistical Association*, Vol. 74: 318–328.

Rubin, D. B. (1980a), "Discussion of "Randomization Analysis of Experimental Data in the Fisher Randomization Test" by Basu," *The Journal of the American Statistical Association*, Vol. 75(371): 591–593.

Rubin, D. B. (1980b), "Bias Reduction Using Mahalanobis' Metric Matching," *Biometrics*, Vol. 36(2): 293–298.

Rubin, D. B. (1986a), "Statistics and Causal Inference: Comment: Which Ifs Have Causal Answers," *Journal of the American Statistical Association*, Vol. 81(396): 961–962.

Rubin, D. B. (1986b), "Statistical Matching Using File Concatenation with Adjusted Weights and Multiple Imputations," *Journal of Business and Economic Statistics*, Vol. 4(1): 87–94.

Rubin, D. B. (1990a), "Formal Modes of Statistical Inference for Causal Effects," *Journal of Statistical Planning and Inference*, Vol. 25: 279–292.

Rubin, D. B. (1990b), "Comment on Neyman (1923) and Causal Inference in Experiments and Observational Studies," *Statistical Science*, Vol. 5(4): 472–480.

Rubin, D. B. (1998), "More Powerful Randomization-Based p-Values in Double-Blind Trials with Non-Compliance," *Statistics in Medicine*, Vol. 17: 371–385.

Rubin, D. B. (2001), "Using Propensity Scores to Help Design Observational Studies: Application to the Tobacco Litigation," *Health Services & Outcomes Research Methodology*, Vol. 2: 169–188.

Rubin, D. B. (2004), "Causal Inference Using Potential Outcomes: Design, Modeling, Decisions," Fisher Lecture, *The Journal of the American Statistical Association*, Vol. 100(469): 322–331.

Rubin, D. B. (2006), *Matched Sampling for Causal Effects*, Cambridge University Press.

Rubin, D. B. (2007), "The Design versus the Analysis of Observational Studies for Causal Effects: Parallels with the Design of Randomized Trials," *Statistics in Medicine*, Vol. 26(1): 20–30.

Rubin, D. B. (2008), "The Design and Analysis of Gold Standard Randomized Experiments. Comment on 'Can Nonrandomized Experiments Yield Accurate Answers? A Randomized Experiment Comparing Random to Nonrandom Assignment' by Shadish, Clark, and Steiner," *Journal of the American Statistical Association*, Vol. 103: 1350–1353.

Rubin, D. B. (2010), "Reflections Stimulated by the Comments of Shadish (2010) and West and Thoemmes (2010)," *Psychological Methods*, Vol. 15(1): 38–46.

Rubin, D. B. (2012), "Analyses That Inform Policy Decisions," *Biometrics*, Vol. 68: 671–775.

Rubin, D., and E. Stuart, (2006), "Affinely Invariant Matching Methods with Discriminant Mixtures of Ellipsoidally Symmetric Distributions," *Annals of Statistics*, Vol. 34(4): 1814–1826.

Rubin, D. B., and N. Thomas, (1992a), "Affinely Invariant Matching Methods with Ellipsoidal Distributions," *Annals of Statistics*, Vol. 20(2): 1079–1093.

Rubin, D. B., and N. Thomas, (1992b), "Characterizing the Effect of Matching Using Linear Propensity Score Methods with Normal Distributions," *Biometrika*, Vol. 79(4): 797–809.

Rubin, D. B., and N. Thomas, (1996), "Matching Using Estimated Propensity Scores: Relating Theory to Practice," *Biometrics* 52: 249-264.

Rubin, D. B., and N. Thomas, (2000), "Combining Propensity Score Matching with Additional Adjustment for Prognostic Covariates," *Journal of the American Statistical Association*, Vol. 95(450): 573–585.

Rubin, D. B., X. Wang, L. Yin, and E. Zell, (2010), "Bayesian Causal Inference: Approaches to Estimating the Effect of Treating Hospital Type on Cancer Survival in Sweden Using Principal Stratification," in A. O'Hagen and M. West, eds. *The Handbook of Applied Bayesian Analysis*, Chapter 24, pp. 679–706.

Rubin, D. B., and E. Zell, (2010), "Dealing with Noncompliance and Missing Outcomes in a Randomized Trial using Bayesian Technology: Prevention of Perinatal Sepsis Clinical Trial, Soweto, South Africa," *Statistical Methodology*, Vol. 7(3): 338–350.

Sabbaghi, A., and D. Rubin, (2014), "Comments on the Neyman-Fisher Controversy and Its Consequences," *Statistical Science*, Vol. 29(2): 267–284.

Samii, C., and P. Aronow, (2012), "Equivalencies Between Design-Based and Regression-Based Variance Estimators for Randomized Experiments," *Statistics and Probability Letters*, Vol. 82: 365–370.

Sekhon, J. (2004–2013), "Matching: Multivariate and Propensity Score Matching with Balance Optimization," http://sekhon.berkeley.edu/matching, http://cran.r-project.org/package=Matching.

Senn, S. (1994), "Testing for Baseline Balance in Clinical Trials," *Statistics in Medicine*, Vol. 13: 1715–1726.

Shadish, W., T. Campbell, and T. Cook, (2002), *Experimental and Quasi-experimental Designs for Generalized Causal Inference*, Houghton and Mifflin.

Sheiner, L., and D. Rubin, (1995), "Intention-to-treat Analysis and the Goals of Clinical Trial," *Clinical Pharmacology and Therapeutics*, Vol. 57: 6–15.

Shipley, M., P. Smith, and M. Dramaix, (1989), "Calculation of Power for Matched Pair Studies when Randomization is by Group," *International Journal of Epidemiology*, Vol. 18(2): 457–461.

Shu, Y., G. Imbens, Z. Cui, D. F., and Z. Kadziola, (2013), "Propensity Score Matching and Subclassification with Multivalued Treatments," Unpublished Manuscript.

Sianesi, B. (2001), "Psmatch: Propensity Score Matching in STATA," University College London, and Institute for Fiscal Studies.

Sims, C. (1972), "Money, Income and Causality," *American Economic Review*, Vol. 62(4): 540-552.

Smith, J. A., and P. E. Todd, (2001), "Reconciling Conflicting Evidence on the Performance of Propensity-Score Matching Methods," *American Economic Review*, Papers and Proceedings, Vol. 91: 112–118.

Smith, J. A., and P. E. Todd, (2005), "Does Matching Address LaLonde's Critique of Nonexperimental Estimators," *Journal of Econometrics*, Vol. 125(12): 305–353.

Snedecor, G., and W. Cochran, (1967, 1989), *Statistical Methods*, Iowa State University Press.

Sommer, A., I. Tarwotjo, E. Djunaedi, K. West, A. Loeden, R. Tilden, and L. Mele, (1986), "Impact of Vitamin A Supplementation on Child Mortality: A Randomized Controlled Community Trial," *Lancet*, Vol. 1: 1169–1173.

Sommer, A., and S. Zeger, (1991), "On Estimating Efficacy from Clinical Trials," *Statistics in Medicine*, Vol. 10: 45–52.

Stigler, S. (1986), *American Contributions to Mathematical Statistics in the Nineteenth Century*, Arno Press.

Stock, J., and F. Trebbi, (2003),"Who Invented Instrumental Variable Regression?" *Journal of Economic Perspectives*, Vol. 17: 177–194.

"Student" (1923), "On Testing Varieties of Cereals," *Biometrika*, Vol. 15: 271–293.

Tanner, M. (1996), *Tools for Statistical Inference: Methods for the Exploration of Posterior Distributions and Likelihood Functions*, Springer Verlag.

Tanner, M., and W. Wong, (1987),"The Calculation of Posterior Distributions by Data Augmentation," *Journal of the American Statistical Association*, Vol. 82(398): 528–540.

Thistlewaite, D., and D. Campbell, (1960), "Regression-Discontinuity Analysis: An Alternative to the Ex-Post Facto Experiment," *Journal of Educational Psychology*, Vol. 51: 309–317.

Tibshirani, R. (1996), "Regression Shrinkage and Selection via the Lasso," *Journal of the Royal Statistical Society, Series B* (Methodological), Vol. 58(1): 267–288.

Tinbergen, J. (1930), "Bestimmung und Deutung von Angebotskurven: in Beispiel," *Zeitschrift für Nationalökonomie*, 669–679.

Torgerson, D., and M. Roland, (1998), "Understanding Controlled Trials: What Is Zelen's Design?" *BMJ*, Vol. 316: 606.

Van Der Klaauw, W. (2002), "A Regression–discontinuity Evaluation of the Effect of Financial Aid Offers on College Enrollment," *International Economic Review*, Vol. 43(4): 1249–1287.

Van Der Laan, M., and J. Robins, (2003), *Unified Methods for Censored Longitudinal Data and Causality*, Springer Verlag.

Van Der Vaart, A., (1998), *Asymptotic Statistics*, Cambridge University Press, Cambridge.

Victora, C., J.-P. Habicht, and J. Bryce, (2004), "Evidence-Based Public Health: Moving Beyond Randomized Trials," *American Journal of Public Health*, Vol. 94(3): 400–405.

Waernbaum, I. (2010), "Model Misspecification and Robustness in Causal Inference: Comparing Matching with Doubly Robust Estimation," *Statistics in Medicine*, Vol. 31(15): 1572–1581.

Welch, B. (1937), "On the z Test in Randomized Blocks and Latin Squares," *Biometrika*, Vol. 29: 21–52.

Wilcoxon, F. (1945), "Individual Comparisons by Ranking Methods," *Biometrics Bulletin*, Vol. 1(6): 80–83.

Wooldridge, J. (2002), *Econometric Analysis of Cross Section and Panel Data*, 2nd edition, MIT Press.

Wright, P. (1928), *The Tariff on Animal and Vegetable Oils*, Macmillan.

Wright, S. (1921), "Correlation and Causation," *Journal of Agricultural Research*, Vol. 20: 257-285.

Wright, S. (1923), "The Theory of Path Coefficients: A Reply to Niles' Criticism," *Genetics*, Vol. 8: 239–255.

Wu, J., and Hamada, M. (2009), *Experiments, Planning, Analysis and Optimization*, Wiley Series in Probability and Statistics.

Yang, S., G. Imbens, Z. Cui, D. Faries, and Z. Kadziola, (2014) "Propensity Score Matching and Subclassification with Multi-level Treatments," unpublished manuscript.

Yule, G. N. (1897), "On the Theory of Correlation," *Journal of the Royal Statistical Society*, 812–854.

Zelen, M. (1979), "A New Design for Randomized Clinical Trials," *New England Journal of Medicine*, Vol. 300: 1242–1245.

Zelen, M. (1990), "Randomized Consent Designs for Clinical Trials: An Update," *Statistics in Medicine*, Vol. 9: 645–656.

Zhao, Z. (2004), "Using Matching to Estimate Treatment Effects: Data Requirements, Matching Metrics and an Application," *Review of Economics and Statistics*, Vol. 86(1): 91–107.

Zhang, J., D. Rubin, and F. Mealli, (2009), "Likelihood-Based Analysis of Causal Effects of Job-Training Programs Using Principal Stratification," *Journal of the American Statistical Association*, Vol. 104(485): 166–176.

Author Index

Subject Index

active treatments: assignment mechanisms and, 33–38, 41; basics of, 4, 8, 11, 13, 16–17, 19; classical randomized experiments and, 47–50, 52; Fisher exact p-values and, 59, 62, 64; instrumental variables analysis and, 514, 517, 523–525, 528, 530, 536, 539–540, 542, 569; labor market and, 246; model-based analysis and, 169–170, 569; Neyman's repeated sampling approach and, 83, 87, 105; pairwise randomized experiments and, 220–221, 225, 227; propensity score and, 282, 307; regression analysis and, 131; sampling variances and, 446; sensitivity analysis and, 500; stratified randomized experiments and, 187, 190, 211; unconfoundedness and, 266, 479, 481, 485

affine consistency, 434, 441–444

AIDS, 12

Aid to Families with Dependent Children (AFDC), 240

alwaystakers: instrumental variables analysis and, 545–546, 549–555, 562t, 563–565, 568–572, 574–579, 581, 583; model-based analysis and, 562t, 563–565, 568–572, 574–579, 581, 583

Analysis of Variance (ANOVA), 298

assignment mechanisms, xviii, 34, 589; *a priori* approach and, 33, 41; assumptions and, 16, 32–34, 36, 39, 42–43; attributes and, 33; balance and, 32, 42; basics of, 3, 7, 13–17, 20–21; before Neyman, 24; causal effects and, 31, 36, 42; causal estimands and, 34; classical randomized experiments and, 47–54; classification of, 31–44; completely randomized experiments and, 41; conditional independence assumption and, 43; control units and, 35, 41–42; covariates and, 20, 31–39, 42; design stage and, 32; drug treatment and, 34, 40; exclusion restrictions and, 42; Fisher exact p-values and, 58; general causal estimands and, 479–495; ignorable assignment and, 39; individualistic assignment and, 31, 37–39, 43, 259, 261–262, 316; instrumental variables analysis and, 43, 527, 567, 570; irregular, 20, 42–43; missingness and, 43; model-based analysis and, 141–143, 151–153, 156, 177, 567, 570; notation and, 32–34, 39; observation and, 31–32, 41–43; observed outcomes and, 33, 36,

41; overlap and, 314–316; pairwise randomized experiments and, 41, 221, 223, 232; populations and, 13, 33–35, 39–41; potential outcomes and, 23–24, 27–40; pre-treatment variables and, 32–33, 42; propensity score and, 35–40, 377, 380; randomized experiments and, 31–32, 35–36, 40–43; regression analysis and, 43; regular, 20, 32, 41 (*see also* regular assignment mechanisms); restrictions on, 37–39; samples and, 31–32, 39–40; sampling variances and, 437; sensitivity analysis and, 496, 499, 507; stratified randomized experiments and, 41–42, 191–192, 201–202; strongly ignorable treatment assignment and, 39, 43, 280; subpopulations and, 20, 35; super-populations and, 39–40; treatments and, 31–43; unbalanced, 32; unconfoundedness and, 31–32, 32, 38–43

assignment probabilities, 20; classical randomized experiments and, 52–53; classification and, 31–32, 34–37, 39, 41, 43; matching and, 402; propensity score and, 282; sensitivity analysis and, 506–507, 509; unconfoundedness and, 257–259, 273

assumptions, 6–8, 589–590; *a priori* approach and, 20 (*see also* a priori approach); assignment mechanisms and, 16, 32–34, 36, 39, 42–43; classical randomized experiments and, 53, 55; Fisher exact p-values and, 58, 62, 67–68, 74; general causal estimands and, 468–469; instrumental variables analysis and, 513–517, 520, 523–539, 542–584; labor market and, 240, 249t, 250; Lord's paradox and, 16–18, 22, 28; matching and, 337, 345–347, 401–402, 404, 405, 418, 425n6, 426; model-based analysis and, 142, 144, 148–151, 153, 155–157, 160, 163, 165–172, 175–176, 181, 560–584; Neyman's repeated sampling approach and, 84, 92, 94, 96, 98, 100–101, 104 overlap and, 309, 314, 332; pairwise randomized experiments and, 226 path diagrams and, 22; potential outcomes and, 25–26; propensity score and, 282, 284, 377–378, 383, 397–398; regression analysis and, 113, 115–116, 118–122, 126, 128, 130, 133–134; sampling variances and, 437, 452; sensitivity analysis and, 496–500, 503–506, 509; stratified